1st International Conference on Green and Sustainable Computing (ICoGeS 2017)

Journal of Physics: Conference Series Volume 1019

Kuching, Malaysia
25 – 27 November 2017

ISBN: 978-1-5108-6587-7
ISSN: 1742-6588

Printed from e-media with permission by:

Curran Associates, Inc.
57 Morehouse Lane
Red Hook, NY 12571

proceedings.com

Some format issues inherent in the e-media version may also appear in this print version.

Printed by Curran Associates, Inc. (2018)

For permission requests, please contact the Institute of Physics
at the address below.

Institute of Physics
Dirac House, Temple Back
Bristol BS1 6BE UK

Phone: 44 1 17 929 7481
Fax: 44 1 17 920 0979

techtracking@iop.org

Additional copies of this publication are available from:

Curran Associates, Inc.
57 Morehouse Lane
Red Hook, NY 12571 USA
Phone: 845-758-0400
Fax: 845-758-2633
Email: curran@proceedings.com
Web: www.proceedings.com

Table of contents

Volume 1019

1st International Conference on Green and Sustainable Computing (ICoGeS) 2017

25–27 November 2017, Kuching, Sarawak, Malaysia

Accepted papers received: 1 May 2018
Published online: 27 June 2018

Preface

Papers

IOP Conf. Series: Journal of Physics: Conf. Series **1019** (2018) 011001 doi:10.1088/1742-6596/1019/1/011001

Preface/Introduction

The quest for knowledge has become part of national main agenda. In line with the country's aspiration to enhance research and development in the country, UniMAP has been proactively involved and initiating programs and events, which would encourage high impact research encompassing various fields of expertise.

The Centre of Excellence Geopolymer and Green Technology (CEGeoGTech) has been playing a significant role in this quest by taking up this challenge. Thus, one of the initiatives grabbed by the Green Advanced Computing and Technology (GREAT), CEGeoGTech is in organizing this conference itself, which provides a fine platform for knowledge sharing and research ideas exchange in the field of Big Data, Cloud Computing, Green and Sustainable Computing, Technology Science and Management in Sports.

This is indeed an opportune moment to strengthen the bond among researchers and professionals from the academics, the industries and the public sectors in sharing new ideas, and in collaborating future research activities. The importance of interaction and networking among peers are the inspiration of this joint conference. I would like to congratulate the organizing committee and all parties involved for their sheer dedication and tireless effort in making this joint conference a great success. I hope that all participants will have a productive and pleasant conference.

IOP Conf. Series: Journal of Physics: Conf. Series **1019** (2018) 011002 doi:10.1088/1742-6596/1019/1/011002

Peer review statement

All papers published in this volume of *Journal of Physics: Conference Series* have been peer reviewed through processes administered by the proceedings Editors. Reviews were conducted by expert referees to the professional and scientific standards expected of a proceedings journal published by IOP Publishing.

IOP Conf. Series: Journal of Physics: Conf. Series **1019** (2018) 012001 doi:10.1088/1742-6596/1019/1/012001

Supporting Sustainability and Healthy Learning Environment through Smart Green Management System (SGMS)

S Abdullah[1a], H Hamdan[1b], S Anuar[2c]

[1]Faculty of Computer Science and Information Technology,
University Putra Malaysia, 43400 UPM Serdang, Selangor, Malaysia.

[2]Universiti Kuala Lumpur Malaysia France Institute,
Section 14, Jalan Teras Jernang, 43650 Bandar Baru Bangi, Selangor, Malaysia.

[a]salfarina@upm.edu.my, [b]hazlina@upm.edu.my, [c]sazly@unikl.edu.my

Abstract. Today, most universities worldwide have initiated the green campus campaign. In line with the campaign intention to sustain and create healthy learning environment, we developed a system, which aims to facilitate the management of recycle program as part of green campus initiative. Smart Green Management System (SGMS) is a mobile crowdsourcing-based system with a capability to help fast-managing and coordinating recycling effort between the green university council and green city council. This system allows access from any mobile devices at anytime and anywhere. Centralized participatory crowdsourcing method is the key approach in developing this system. In this approach, the crowd is tasked with finding and collecting information into a common location and format. It is ideal for information gathering, organization, and reporting problems, which are demonstrated by SGMS.

1. Introduction

A green campus is a concept that emphasizes the resource conservation, efficient usage of energy and enhanced quality of environment in the higher learning institution. It simply educating for sustainability and creating healthy learning environment. Our work is very much inspired by the premise that sustainability educates rational use of natural resources in order to preserve the planet for future generations

Today, most universities worldwide have initiated the green campus campaign. Throughout this campaign, various programs have been conducted to promote sustainability and healthy environment. The sky is the limit for this campaign. It means that it caters from litters and waste, to energy, to water conservation and include a day without vehicles, plastic-free day, free public transportation day, earth hour, cleanup day, plant a tree day, recycle and many more.

In line with the campaign intention, we developed a system, which aims to facilitate the management of recycle program as part of green campus initiative. We name this system Smart Green Management System (SGMS) which reflects its capability to help fast-managing and coordinating recycling effort between the green university council and green city council. This system is a web-based system which allows access from any mobile devices at anytime and anywhere and our target users will be the university community especially students residing in the campus. For our study, we choose University Putra Malaysia (UPM) which comprises of 17 residential colleges with estimated capacity of 17 thousand students. With this huge number of students, we believe our effort and intention is rightly emphasized when a clean and healthy learning environment is mandatory.

The remainder of this paper is organized as follows. Section 2 summarizes some previous works related to our intention. Section 3 explains the SGMS. This is followed by section 4 which discusses briefly our highlight of study. Finally, section 5 presents our conclusion.

2. Recycling Management and Technologies

Recycling is one way in reducing the environmental deterioration by managing the waste properly. Although the awareness campaign of recycling repeatedly carried out and consumers' awareness is increase, participations in recycling are still low. Recycling can be define as a process of collecting and processing waste turning into new products that could benefit the community and environment [1].

In Malaysia, national recycling campaign was launched in 1993. However, only 15% of recycling rate was achieved in 2016 [2]. This is far away from the aim for the 2020 national recycling rate set by the Malaysian Government which is 22 % [3]. The 3R program awareness: reuse, reduce and recycle, is still very low among Malaysian. Changing of lifestyle and attitudes are important to achieve the objective.

Respect environmental issues, awareness and fundamental knowledge of recycling as an important stimulus to encourage household in recycle campaign. The emphasis on recycling as a sustainable waste management strategy has taken a shift paradigm towards a change in its implementation. In addition, the efficient recycling of waste is indispensable and still be regarded as a major challenge for today's society. Although the awareness among household is increase, it is often a question on the management of insufficient collection infrastructures and poor collection efficiencies of the waste.

In Malaysia, current approach that usually implements for recyclables collection is presented of recyclables kiosk or establishing collection centers by local authorities, NGOs and private organizations. Another approach is by the private collectors of recyclable that come to the house premises to collect the goods. Mohamad et al. [3] discuss on the role of religious community in Malaysia in the recycling practice and activities that are well-managed and active. One example of successfully waste recycling program based on religious community with the proper management system is at Surau Al-Husna. Although there are required technical improvement in their recycling programs such as a needed of database management for recording the recycling rate and a better storage policy of recyclable items must be implemented, it could serve as a role model for sustainable recycling programs at the nationwide scale in Malaysia [4].

Convenient collection process, together with establishing cost-effective ways to communicate for users are paramount importance for an efficient recycling management. As the information and communication technologies grows, sustainable solutions to the waste and recycling management through the use of real time technologies and IoT (Internet of Things) connectivity also gained significant attention over the past decade. This can be seen with the large number of mobile applications that have been developed for ease of use in waste management. The Recycle Track Systems (RTS) is one example of innovating waste and recycling management that implements real time technologies, IoT connectivity, and accountability that increases efficiency and insight into waste and recycling service pickups. RTS was employed for the customer in New York City by taking picture of the items that needed to be remove and the application will be allocated the approval with price estimation and time schedule for the pickup. Not only for waste removal services but also for contraction and demolition added on for the dumpster order [5]. Many other applications of waste management convenient for mobile user such as iRecycle [6], Recycle App [7] and etc.

Although a lots of application available, data taken from the customer will be value added to be analyzed to improve the management of the system. Exploiting the ICT and Web 2.0 social software is a mechanism that connects people to raise a valuable data to create new forms of distant search for solutions and decision making for environmental application. In order to ensure the convenient collection process together with establishing cost-effective ways, the idea of crowdsourcing can be implemented in the management of the recycling waste. The digital technologies will be the platform to collect and organise contributions from a crowd such as data, idea, opinions and therefore help in obtain innovative solutions. According to Howe (2006) [8], crowdsourcing can be defined as an open call participation of contributor to produce new content, products or data in the development of creative solutions using the crowdsourcing tools. Crowdsourcing now has been subject of numerous experiments in transforming sustainable cities environment [9], [10]. Therefore, our target is to embed the crowdsourcing-based solution into the recycling waste management.

3. The Smart Green Management System (SGMS)

The main objective of this system is to facilitate the management of recycling program as part of the green campus initiative. We are also aiming at educating the university community so that they are aware of the importance of recycling program and encourage them to play their role as much as possible in supporting the program. Basically, SGMS offers the following features: 1) recycling education, 2) collecting inputs from the users 3) arranging for collection and transportation of recycling items, and finally 4) generating reports periodically and on demand.

SGMS involves three entities: 1) the green university council, 2) the green city council and 3) the university community. The green university council is a dedicated committee (appointed by the university management) whose responsibility is towards preserving clean and healthy campus environment as well as instilling the right attitude of recycling among the university community. The green city council on the other hand represents the authorities responsible for the wellbeing of the people through safe and healthy environment (MPKJ for instance). The university community consists of all UPM Serdang students and staffs. This group is the key entity of the system which relies upon their inputs and cooperation to work effectively as intended.

The following model namely SGMS model depicts the approach chosen into the design of this system.

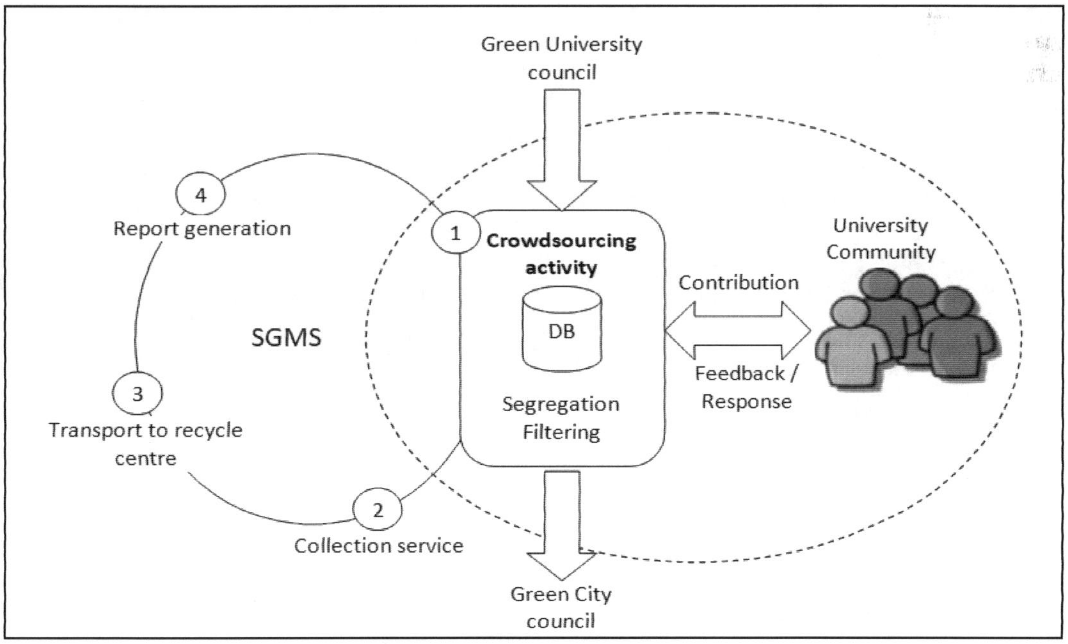

Figure 1. SGMS model.

The SGMS is designed to utilize the idea and advantage of crowdsourcing approach. Our premise is that it allows collection of input from large pool of people, fast, immediate action can be taken, cost effective and smarter solutions can be achieved. Adding to these, crowdsourcing also encourage proactive and caring attitude towards community and surrounding environment. It builds a sense of responsibility and awareness among the community about the importance of smart and healthy living.

SGMS can be accessed anywhere, anytime if the users are connected to the internet by using personal computers, or mobile devices such as tablets or smart phones. This promotes the use of ubiquitous computing services. Moreover, it is easier for the users to submit report using mobile devices in real time and on specific location.

SGMS consists of information concerning recycling campaign including its objectives, activities, benefits, types of recycling items, list of nearby recycle centre, and current market price for the users to view. They can simply take picture of any items of their belongings or waste found somewhere that they think has potential for recycling and submit it with ample description as a report through their mobile devices. They will be informed of their report status by the admin whether it has been received and appropriate action that will follow. This will include the arrangement for collection or meet up.

Basically there are four stages in SGMS. Firstly, based on the input (information), the SGMS will first segregate the input according to the category of information contributed. The categories include report (image with description), complaint, suggestion and general comments. The segregation of information also relies upon the location and time of the item being reported through geo-location data services in the mobile devices. If the input is a report, it will be filtered more distinctively in terms of types of recycling items for example; paper, glass, plastics, aluminum/tin/steel can, or any other indirectly specified items. Next, the system will arrange for collection service according to the availability of the staff in charge. After collection has been made, it will be sent and sold to recycle centre nearby. However, if the input is other than report type, appropriate response will be sent accordingly to the users. The system can generate and produce multiple reports out of the recycling program upon request made by the green university council and green city council.

4. Discussion

Nowadays, crowdsourcing has become the hype in knowledge acquisition and diffusion. It is a powerful method that promotes speedy information gathering and at the same time yields better results compared to other similar methods. Increasingly, crowdsourcing has also been actively used for reaching solutions to complex problems. Based on these premises, as well as to demonstrate our social responsibility towards the surrounding community, this paper aims to put fourth our effort through SGMS.

Mobile devices particularly smart phones are in widespread, everyday use and are always connected. Therefore, they offer a great platform for extending existing web-based crowdsourcing applications to a larger contributing crowd, making contribution easier and omnipresent [11]. Crowdsourcing applications on SGMS can be classified as an extension of web-based application. The crowdsourcing approach used in SGMS is known as participatory crowdsourcing where user generates their own data. It is developed in a centralized manner that ship the data generated and collected from the users to a server. In our study, public participation is our primary motivation. We believe in the power and influence of user participation in supporting our effort through SGMS. The notion of using the community for idea generation through implementation rather than relying on the expertise environmentalists ensures a higher degree of community engagement.

5. Conclusion

The awareness on the recycling program in Malaysia is required. One way is to start the recycle camping at the university campus. The proposed of the Smart Green Management System is to track the recycling rate and create an online database for recycling management systems. In addition, the SGMS allows utilising resources in efficiency and manageable way. Furthermore, the proposed system can monitor all the transaction and generate a report on the performance of the colleges that active in doing the recycle program.

To achieve a sustainable environment, the waste recycling program needs to be directed to proper end-of-life processes from the recycling process to the disposal process. Respect environmental issues, changing of lifestyle, and fundamental knowledge of recycling as important aspects potentially sustaining recycling programs. We hope SGMS which reflects its capability to help fast-managing and coordinating recycling will scaling up recycling practices in university campus.

References

[1] United States Environmental Protection Agency, USEPA 2016, *Recycling basics*. Retrieved from http://www2.epa.gov/recycle/recycling-basics.

[2] The Malay Mail Online 2016, *Minister: Only 15pc of waste is recycled in Malaysia*. Retrieved from http://www.themalaymailonline.com/malaysia/article/minister-only-15pc-of-malaysians-recycle.

[3] Mohamad Z F, Idris N, Baharuddin A, Muhammad A and Sulaiman N M N 2012. The role of religious community in recycling: empirical insights from Malaysia. Resource, Conservation and Recycling, 58 (2012), pp. 143-151.

[4] Tiew K G, Ahmad Basri N E, Watanabe K, et al. 2015. Assessment of the sustainability level of community waste recycling program in Malaysia. *Mater Cycles Waste Management 17*:598–605.

[5] Recycle Track System, RTS 2016, *Recycle Track Systems Launches On-Demand Waste Pick-Up App*. Retrieved from https://www.rts.com/news/date/2016-05-11/recycle_track_systems_ launches_ ondemand_ waste_pickup_app.

[6] Earth911 2016. iRecycle: Find local recycling fast. Retrieved from http://earth911.com/irecycle/.

[7] Appdifferenziata 2016. Recycle App – Differentiated. Retrieved from http://www.appdifferenziata.com/en/.

[8] Howe J 2006. "Crowdsourcing: A Definition". Crowdsourcing: Tracking the rise of the amateur, Retrieved from http://crowdsourcing.typepad.com/cs/2006/06/crowdsourcing_a.html.

[9] Certoma C, Corsini F and Rizzi F 2015. Crowdsourcing urban sustainability. Data, people and technologies in participatory governance. *Futures*, vol. 74, pp. 93–106.

[10] Tatiana D F and Mathieu S 2016. Could "urban crowdsourcing" transform how cities are made? IDDRI, Issue Brief, Issue Brief, n°6, 2016-05.

[11] Chatzimilioudis G, Konstantinidis A, Laoudias C and Zeinalipour-Yazti D 2012. Crowdsourcing with Smartphones. IEEE Internet Comput., 16 (5) (2012), pp. 36-44.

IOP Conf. Series: Journal of Physics: Conf. Series **1019** (2018) 012002 doi:10.1088/1742-6596/1019/1/012002

Critical Review and Simulation of Mid-range Wireless Power Transfer for Electronic Device

Shun Yao Tan and Hui Jing Lee

Institute of Power Engineering (IPE), Universiti Tenaga Nasional, Jalan Ikram-UNITEN, 43000 Kajang, Selangor, Malaysia.

Email: LHJing@uniten.edu.my

Abstract. Functionality of wireless power transfer is yet to be utilized for the purpose of sustaining a green environment. This paper presents a critical review on the functionality of mid-range wireless power transfer using magnetic resonance coupling. Simulation of a 4-coils magnetic resonance for mid-range wireless power transfer is performed using CST microwave studio. The radius of source and load coil are 3 cm, while the radius of the transmitter and receiver coil are 5 cm. Distance between the source coil and load coil is set at 4 cm and resonance magnetic frequency is observed. At the load coil, electric field is achieved at 2140 V/m and surface current is recorded at 52.21 A. The change in impedance along the transmittance are plotted and discussed.

1. Introduction

Wireless Power Transfer (WPT) is defined as power transmission with the absence of any transition medium i.e. transmission grid, cable and wire. It has been widely studied and developed since the past decade [1-3]. WPT can be achieved with several methods such as inductive coupling, magnetic resonant coupling, radio frequency and microwave radiation. Studies on inductive coupling for short-range WPT are scanty; radio frequency or microwave radiation for long-range WPT is questionable in terms of long run health and environment impact, hence the focus is now on optimization of magnetic resonant coupling in the mid-range WPT [4-5].

Magnetic resonant coupling is based on evanescent waves between two resonant couplers which generate and transfer electrical energy through oscillation or variation in magnetic field [6]. The primary coil generates magnetic field, which induces voltage or current produced in the secondary coil with the same resonance frequency and strongly coupled together. When two coils operate at the same frequency, high efficiency of energy can be transferred with a small leakage or loss to the surrounding. Magnetic resonance coupling is environment friendly, with minimum hazardous exposure and line-of-sight transfer requirement. Previous research shows that a maximum efficiency of 97.6% can be achieved for a travelling distance of 0.3 cm [7-8].

Magnetic resonant coupling can transfer at a longer distance as compare to inductive coupling and is able to achieve a higher transmission efficiency as compare to far-range WPT [9-10]. Magnetic resonant coupling is transferred in few Megahertz which have a high quality factor (Q factor). When the charging distance increases, the high quality factor helps to mitigate the sharp decrease in coupling coefficient, and thus charging efficiency. In 2007, MIT proposed Witricity which is able to light up a 60 W light bulb with 40 % efficiency at a 2 meter of transmission distance [11]. At 1 meter of transmittance distance, the power efficiency is achieved at 90 %. However, the limitation of this set up

IOP Conf. Series: Journal of Physics: Conf. Series **1019** (2018) 012002 doi:10.1088/1742-6596/1019/1/012002

is that experimental work is too bulky as the magnetic resonant coupling require a capacitive coil to operate. This means that a proper tuned impedance is required to achieve a high efficiency wireless power transfer [12]. There are two methods in mid-range WPT namely the 2-coils resonance coupled WPT and 4-coils resonance coupled WPT [13-14]. The later method is preferable as it performs impedance matching and hence achieve a higher power transmission efficiency.

WPT technology grows vigorously in the past few years. As the research field in WPT is expanding, such technology would eliminate the usage of high transmission lines and cables, charging wires and cables. This is prudent to sustain a more environmental friendly mother nature. The main challenge in mid-range WPT is to achieve a high power transmission efficiency at an ideal distance between the transmitter and receiver coil. Hence, this paper presents the preliminary simulation result on 4-coils magnetic resonance coupling WPT. The finding presented could be made practical for wireless charging for electronic devices in the near future.

Section 2 discusses about the principle and numerical formula for magnetic coupled resonator applicable in the mid-range wireless power transfer. Functionality of 4-coils magnetic coupled resonator WPT circuitry is discussed. Section 3 presents the simulation result of 4-coils magnetic resonance using CST microwave studio. Magnetic resonance coupling is achieved at 14.175 GHz, electric and magnetic field as well as surface current distribution are presented and discussed.

2. Magnetic Resonance Formulation for Mid-range Wireless Power Transfer

Figure 1 shows the equivalent circuit of 4-coils magnetic coupled resonator WPT system which is made up of the source coil, transmitter (Tx) coil, receiver (Rx) coil and the load coil. The 4 coils comprise of circuit elements such as inductors (*L*), resistors (*R*) and capacitors (*C*). k_{12}, k_{23} and k_{34} are the coupling coefficient which is fundamental to connect the resonant circuit. The coupling effect will affect the amount of energy transferred. High coupling is desired to achieve a high rate of energy transfer [15].

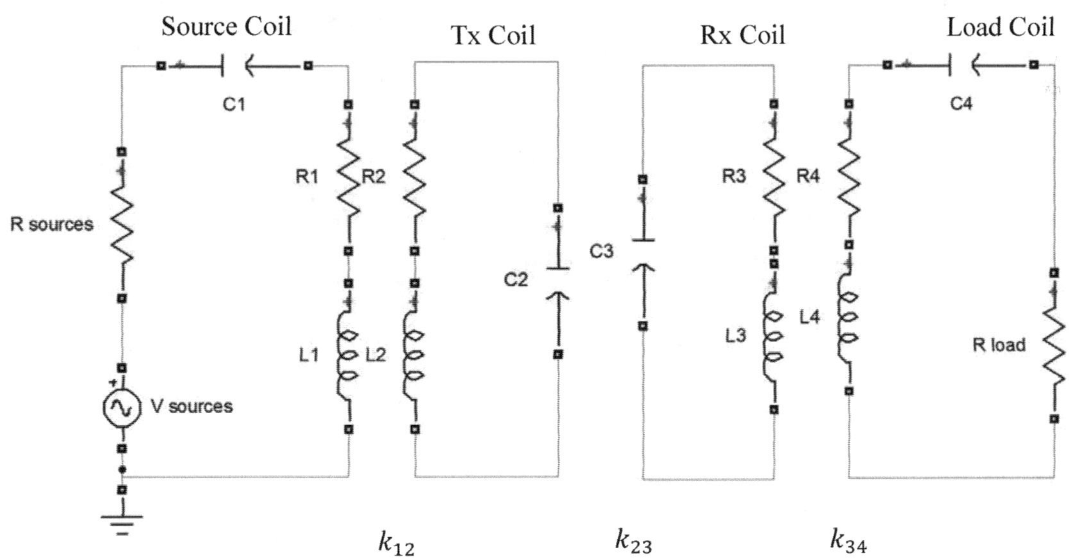

Figure 1. Equivalent circuit of 4-coils magnetic coupled resonator WPT

The current flow in figure 1 can be expressed using the matrix below [16]:

$$\begin{bmatrix} I_1 \\ I_2 \\ I_3 \\ I_3 \end{bmatrix} = \begin{bmatrix} Z_{11} & Z_{12} & Z_{13} & Z_{14} \\ Z_{21} & Z_{22} & Z_{23} & Z_{24} \\ Z_{31} & Z_{32} & Z_{33} & Z_{34} \\ Z_{41} & Z_{42} & Z_{43} & Z_{44} \end{bmatrix}^{-1} \begin{bmatrix} V_S \\ 0 \\ 0 \\ 0 \end{bmatrix} \quad (1)$$

where Z_{11}, Z_{22}, Z_{33}, Z_{44} are the impedance value at source coil, transmitter coil, receiver coil, and load coil, respectively:

$$Z_{ij} = j\omega M_{ij} = j\omega\sqrt{L_i L_j}\ (i, j = 1-4, i \neq j) \quad (2)$$

To calculate the flow of current in both the source and load loop, equation (1) has to be solved with the impedance parameters. To simplify, current at the source loop and load loop are expressed as equation 3 and 4:

$$I_1 = \frac{Z_{22}Z_{33}Z_{44} + (\omega M_{23})^2 Z_{44} + (\omega M_{34})^2 Z_{22}}{Z_{11}Z_{22}Z_{33}Z_{44} + (\omega M_{23})^2 Z_{11}Z_{44} + (\omega M_{34})^2 Z_{11}Z_{22} + (\omega M_{12})^2(\omega M_{34})^2} V_S \quad (3)$$

$$I_4 = \frac{j(\omega M_{12})(\omega M_{23})(\omega M_{34})}{Z_{11}Z_{22}Z_{33}Z_{44} + (\omega M_{23})^2 Z_{11}Z_{44} + (\omega M_{34})^2 Z_{11}Z_{22} + (\omega M_{12})^2(\omega M_{34})^2} V_S \quad (4)$$

In such case, voltage transfer is presented as:

$$|S_{12}| \cong \frac{2\, k_{12}k_{23}k_{34}\, Q_2 Q_3 \sqrt{Q_1 Q_4}}{(1 + k_{12}^2 Q_1 Q_2)(1 + k_{34}^2 Q_3 Q_4) + 1 + k_{23}^2 Q_2 Q_3} \quad (5)$$

The efficiency of transmittance rate which is expressed as ratio of output power versus output power:

$$\eta = \frac{P_{out}}{P_{in}} = \frac{I_4^2 R_L}{I_1^2 \left(\frac{R_S}{4}\right)} = |S_{12}|^2 \quad (6)$$

The efficiency of transmittance rate explains the amount of energy transferred from the transmitter to the receiver. Maximum $|S_{12}|$ is achieved when reasonable amount of k_{23}^* is substituted, where k_{23}^* is the maximum range achievable by the transmitter to transfer power efficiently to transfer power towards receiver. Hence, $|S_{12}|_{max}$ is expressed as:

$$|S_{12}|_{max} = \frac{k_{12}k_{34}Q_1 Q_4 R_L}{k_{23}^* \sqrt{L_1\omega_1 L_4 \omega_4}} \quad (7)$$

Equation 7 shows that a lower k_{23}^* value is desired to achieve a high $|S_{12}|_{max}$.

3. Results and Discussion

Simulation work on 4-coils mid-range magnetic resonance WPT has been performed using CST microwave studio. Figure 2 shows the 4-coils alignment in CST microwave studio, both the source coil and load coil are single turn with 3 cm radius, whereas the transmitter coil and receiver coil are pre-assigned at ten turns with 5 cm radius. A discrete port is inserted at both the source coil and receiver coil to transmit and receive electric and magnetic field. The distance between the transmitter coil and receiver coil is set at 2 cm, while the transmitter coil and receiver coil are set 1 cm apart from the source and load coil. The scattering parameters (S-parameters) is shown in figure 3. Simulation result shows the magnetic resonance occur at 14.17 MHz, and the S-parameter transferred from source to coil is 0.72371. Resonance magnetic coupling is achieved when energy is being transmitted from the transmitter coil to receiver coil at the same frequency. This will generate an oscillating current in the coil system. The energy will deplete with respect to time in a high resonant coil. Therefore, an optimum energy is transferred from the transmitter coil to the receiver coil.

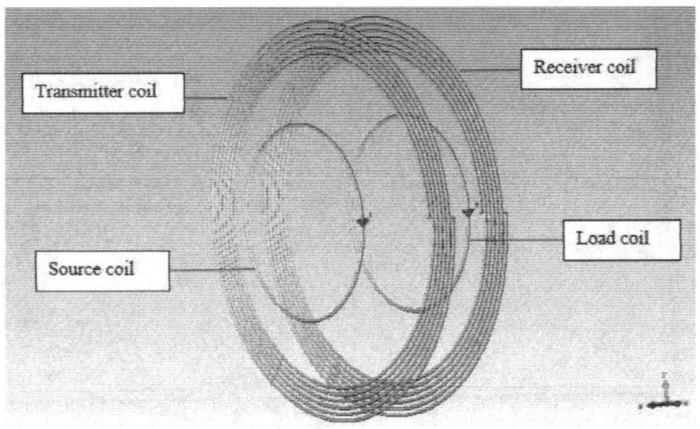

Figure 2. 4-coils design using CST

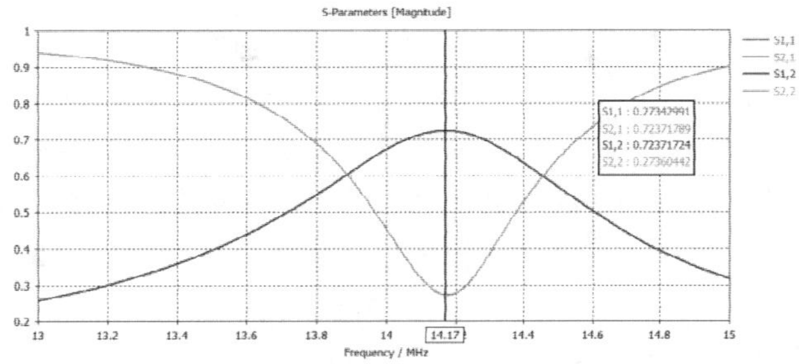

Figure 3. S-parameters

Figure 4 and 5 show the electric field (E) distribution at the resonance frequency (f = 14.17 MHz) and the change in electric field strength at each coil. It is observed that the E-field strength is recorded at 2424.50 V/m at the source coil and 2140 V/m at the load coil. This shows that an optimum electric field transmission is achieved with minimal information loss to surrounding.

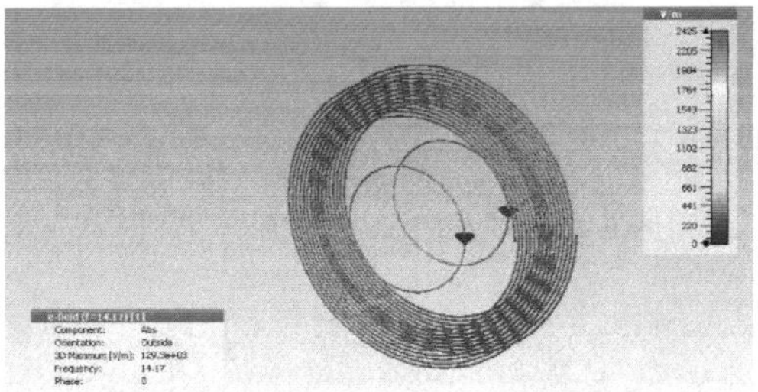

Figure 4. E-field distribution

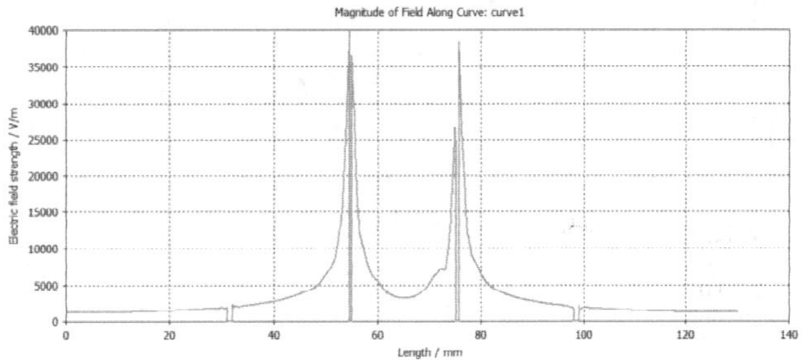

Figure 5. E-field strength versus distance

Figure 6 and 7 show the electric field (H) distribution at the resonance frequency (f =14.17 MHz) and the change in magnetic field strength at all the four coils. The optimum current and voltage transmission will hence make this magnetic resonance coupling method a promising method for wireless power transfer to charge electronic device. The magnetic field at the source coil and load coil are record at 54.35 A/m and 38.71 A/m, respectively.

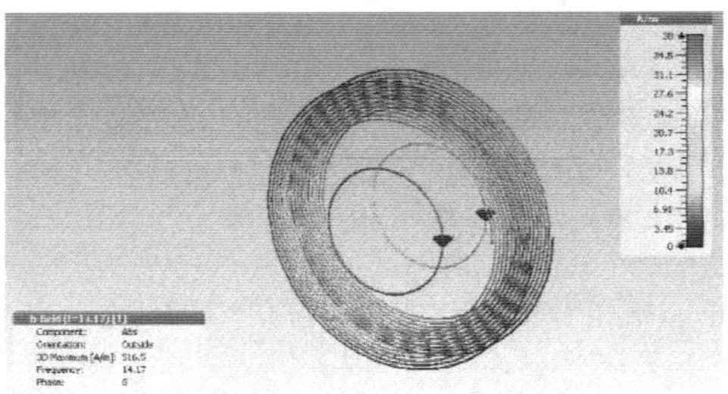

Figure 6. H-field distribution

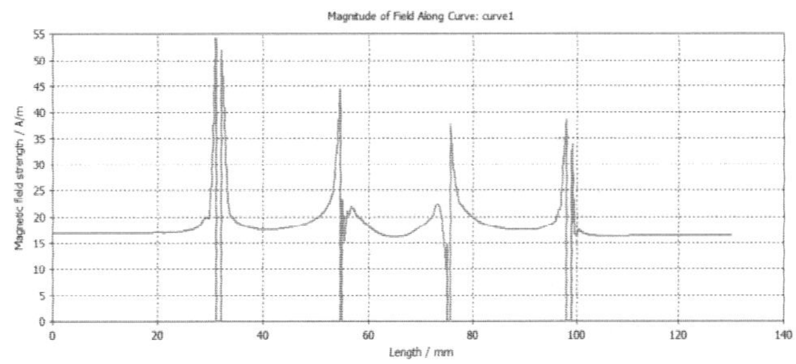

Figure 7. H-field strength versus distance

Figure 8 plots the surface current distribution versus distance. Assuming an ideal wireless transmission with no loss of signal, a surface current of 52.21 A is recorded at the load coil. Figure 9 shows the impedance value, Z at resonance frequency. It is crucial to adapt the ideal impedance parameter in order to identify the suitable inductance (*L*), capacitance (*C*) and resistance (*R*) values in the 4-coils system.

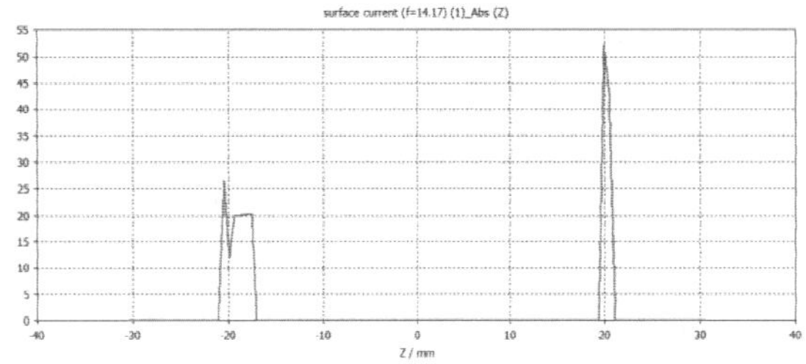

Figure 8. Surface current

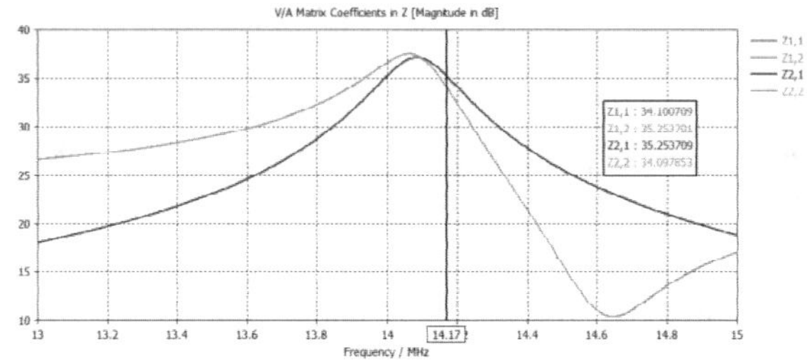

Figure 9. Z-parameters

4. Conclusion and Future Recommendation

This research work is undertaken to perform critical review and analysis on mid-range wireless power transfer. Preliminary simulation result for magnetic resonance coupling wireless power transfer using CST microwave studio is presented and discussed in this research paper. The electric and magnetic field transmission were successfully simulated. Result shows that at 14.17 MHz, magnetic resonance is achieved and the load coil is receiving both electric field and magnetic field from the source coil. Mid
-range wireless power transfer using magnetic resonance coupling would be the new key technology for researchers to investigate and employ into in the next decade. An achievable optimize distance for mid-range wireless transfer with a smaller coils size is yet to be proposed, it would be much convenient if a high efficiency charging process to occur over a distance which is applicable for electronic device such as mobile phone, electronic tabs and laptops. An ideal wireless charging would be an optimum power transmission from single transmitter to multiple receivers at a desired power transmission efficiency.

Acknowledgement
We thank Yap June Xiong from CST microwave studio for the advice and Universiti Tenaga Nasional Internal Grant, RJO10289176 for the support.

References

[1] G. A. Covic and J. T. Boys, "Inductive Power Transfer," *Proceedings of the IEEE*, vol. 101, no. 6, pp. 1276-1289, June 2013.
[2] S. L. Ho, J. Wang, W. N. Fu, and M. Sun, "A Comparative Study Between Novel Witricity and Traditional Inductive Magnetic Coupling in Wireless Charging," IEEE.
[3] H. H. Wu, A. Gilchrist, K. D. Sealy, and D. Bronson, "A High Efficiency 5 kW Inductive Charger for EVs Using Dual Side Control," IEEE Transactions on Industrial Informatics, vol. 8, no. 3, pp. 585-595, Aug. 2012.
[4] B. H. Choi, E. S. Lee, J. H. Kim, and C. T. Rim, "7m-off-long-distance extremely loosely coupled inductive power transfer systems using dipole coils," in 2014 IEEE Energy Conversion Congress and Exposition, ECCE 2014, 2014.
[5] I. Yasar, L. Shi, K. H. Bai, X. Rong, Y. Liu, and X. Wang, "Mobile phone mid-range wireless charger development via coupled magnetic resonance," in 2016 IEEE Transportation Electrification Conference and Expo, ITEC 2016, 2016.

[6] A. Karalis, J. Joannopoulos, and M. Soljacic, "Efficient Wireless Nonradiative Mid-range Energy Transfer," Ann. Phys., vol. 323, no. 1, pp.34-48, 2008.

[7] X. Li, C. Tsui, and W. Ki, "A 13.56 MHz Wireless Power Transfer System With Reconfigurable Resonant Regulating Rectifier and Wireless Power Control for Implantable Medical Devices," IEEE Journal of Solid- State Circuits, vol. 50, no. 4, pp. 978-989, April 2015.

[8] A. Kurs, A. Karalis, R. Moffatt, J. D. Joannopoulos, P. Fisher, and M. Soljacic, "Wireless Power Transfer via Strongly Coupled Magnetic Resonances," Science, vol. 317, no. 5834, pp. 83-86, June 2007.

[9] X. Li, C. Tsui, and W. Ki, "A 13.56 MHz Wireless Power Transfer System With Reconfigurable Resonant Regulating Rectifier and Wireless Power Control for Implantable Medical Devices," IEEE Journal of Solid- State Circuits, vol. 50, no. 4, pp. 978-989, April 2015.

[10] A. Karalis, J. Joannopoulos, and M. Soljacic, "Efficient Wireless Nonradiative Mid-range Energy Transfer," Ann. Phys., vol. 323, no. 1, pp.34-48, 2008.

[11] A. Kurs, A. Karalis, R. Moffatt, J. D. Joannopoulos, P. Fisher, and M. Soljacic, "Wireless Power Transfer via Strongly Coupled Magnetic Resonances," Science, vol. 317, no. 5834, pp. 83-86, June 2007.

[12] A. Kurs, R. Moffatt, and M. Soljacic, "Simultaneous Mid-range Power Transfer to Multiple Devices," Appl. Phys. Lett., vol. 96, pp. 044102-1044102-3, Jan. 2010.

[13] T. Imura T, Y. Hori. Maximizing air gap and efficiency of magnetic resonant coupling for wireless power transfer using equivalent circuit and Neumann formula. IEEE Trans Ind Electron 2011;58(10):4746–52.

[14] S. Cheon, Y. H. Kim, S. Y. Kang, M. L. Lee, J. M. Lee, T. Zyung. Circuit-model-based analysis of a wireless energy transfer system via coupled magnetic resonances. IEEE Trans Ind Electron 2011;58(7):2906–14.

[15] S. Das Barman, A. W. Reza, N. Kumar, and T. Ibne, "Two-side Impedance Matching for Maximum Wireless Power Transmission Two-side Impedance Matching for Maximum Wireless Power Transmission," vol. 2063, no. January, 2016

[16] T. P. Duong and J. W. Lee, "A dynamically adaptable impedance-matching system for midrange wireless power transfer with misalignment," Energies, vol. 8, no. 8, pp. 7593–7617, 2015.

IOP Conf. Series: Journal of Physics: Conf. Series **1019** (2018) 012003 doi:10.1088/1742-6596/1019/1/012003

Measuring System Success In New Context By Adapting DM 2003 Framework With The External Factor Management Support

Ibrahim Eskandar Ibrahim Fadhel[1], Syed Zulkarnain Syed Idrus[2], Amani Ali Elmetwaly Ali Ibrahim[3], Mazni Binti Omar[4], Suhaib Khalaf Albzeirat[5] and Fawzia Baheshwan[6]

[1]National Institute of Science (NIAS), Mukalla - AlShihr & School of Human Development and Techno-communication (iKOM), Universiti Malaysia Perlis, Perlis, Malaysia

[2]School of Human Development and Techno-communication (iKOM) & GREAT Research Group - CEGeoGTech, Universiti Malaysia Perlis, Perlis, Malaysia

[3]School of Human Development and Techno-communication (iKOM), Universiti Malaysia Perlis, Perlis, Malaysia

[4]College of Science and Arts, School of Computing Universiti Utara Malaysia, Kedah, Malaysia

[5]Queen Rania Center for Educational Technology and Information, Hashemite, Kingdom of Jordan

[6]Department of Information Systems, Hadhramout University, Mukalla

ibrahim.eskandar@yahoo.com

Abstract. Rate of failure of software's and information systems are high among the world and there is a dire need for engineering a systems quality framework works a success measure tool in the context of the study domain. Success measure of systems and developing a framework helps in solving the problem of higher rate of systems failure. Success measure is also, an important part of systems development to improve systems performance. However, studies showed a lack of research done on systems success in world universities Arab region universities and Yemeni universities and limited adequate frameworks. Systems especially the web-based faces high rate of failure in world and Arab region, it was observed that many funded system projects failed, it was also, noted that in Yemen systems were completed and implemented without evaluation and success measuring. The Yemen higher education management information system (YHEMIS), is a large-scale application developed and implemented without evaluation. Hence, this research measures the YHEMIS application by investigating into the factors that influence user satisfaction and system use which will further show the benefit of the system. The research findings showed that information quality, system

IOP Conf. Series: Journal of Physics: Conf. Series **1019** (2018) 012003 doi:10.1088/1742-6596/1019/1/012003

quality and management support influenced the use and users' satisfaction and played a vital role in the success of YHEMIS. The findings showed that students' satisfaction have the strongest effect on the perceived net benefits YHEMIS brought for the students. This study provides the first empirical data on the evaluation of web-based systems success conducted on YHEMIS in Yemen. Based on the results, stakeholders can get the feedback to improve the systems and provide input to develop other software projects. The research findings can provide the support to the management of Hadhramout University to other online system projects. Hence, YHEMIS can be said to be successful and users (students) satisfaction can be considered as the indicator of systems success and benefit.

1. Introduction

Organization undergo changes due to external or internal environmental pressures that, in turn, will make new demand on the IS which has been unavoidable. The field of systems is rich with literature in evaluation that effectively measure systems success and there are many theories that can help such as DeLone and McLean and ISO 25010. Yemen is one of the poorest countries, based on the report of Human Development Index (HDI) and report of United Nation. Yemen is also counted among the poorest countries in the Arab world and it is ranked 154. Being placed medium among human development countries, negatively affect all aspect of life in Yemen, especially the field of education and scientific research [1,2].

There has been rapid development in higher education in Yemen in the last years [3]. Unfortunately, this expansion comes with a variety of problems, such as haphazard and risking waste of investments, duplication of efforts, and other levels of inefficiencies lack of resources and technology infrastructure, fail of systems and higher numbers of dissatisfying systems users etc. [1,2,3,4,5].

In fact, administrators and stakeholders don't pay much attention to the value of systems success evaluation especially after the systems implementation. Therefore, per researches, there is lack of studies related to IS success evaluation conducted in the Arab world especially Yemeni context, right now, many institutional organization in Hadhramout governorate such as Hadhramout university started the implementation of ISs and are making efforts for the success of these systems [2]. Supporting the deployment of IS/IT in education organization in Yemen needs more academic research [1]. There is a funded project (Introduce IT based Services to the Yemeni Public Universities) for supporting the universities in Yemen with information systems. Such a project is organized by the Yemeni Center for Information Technology, which is related to The Ministry of Higher Education and is purposed for enhancing the level of Yemen post studies through utilizing the different services of the Information and Communication Technology (ICT) [2].

One of these systems is Yemen Higher Education Management Information System (YHEMIS) that is implemented in 2013. This system is compulsory for students in five different universities of the country to be registered. This web systems project has inspired the researcher to conduct an exploratory research on the evaluation the YHEMIS success and examine some of these important factors for information systems success in Yemen [2].

The students' perspective in Hadhramout University will encourage further studies in this field in Yemen and Arab region. Indeed, DeLone and McLean strongly call for systems success measure and this study aims to be in pursuit of this call [6].

2. Problem Statement

The evolution of software's (SWs) and information systems (ISs) is rooted in many interdisciplinary [1]. Filed of software engineering and information systems becomes more complex, in general facing a challenge in defining and measuring success. Measuring the success of systems based on the context is a prime importance [6].

Organization need a well-defined engineered framework and tools to assist in the assessment of complex web-based systems quality requirements [7,8,9,10]. Information and systems engineering methodology is an engineering approach that aims at providing a framework for planning, evaluating, developing, and implementing applications within an organization. Its primary goal is to enable an organization to improve the administration of its systems. It is seen as a framework outlining a variety of techniques that are used to assess systems effectively [11].

Systems success in Arab countries not well known for the system success is perceived so targeting this gap is the right track [5]. With the high rate of systems failure in world, Arab and Yemen as large and with the large numbers of dissatisfying system users, knowing the level of users' satisfaction on the university web based systems is a recurrent challenge. Omitting the success measure of the systems based on the satisfaction of the users, loyalty, systems benefit and based on the other factors such as quality factors, IT anxiety, management support and etc... is today dilemma and one of the reasons behind the low level of success of the systems [2]. Data and information, which are saved in the IS, could be used to improve decision-making. The failure of information system remains a problem for the organization and thus, there is a need to identify this lack of success. The factor that leads to ISs success is failure identification for the maintenance of the systems. The assessment of Information systems success is necessary as a chief condition to make the IS success better in the future [1]. Indeed, systems evaluation is important for organization despite the concept of success measure and performing systems evaluation has not been subjected to extensive research [12,13].

Providing competitive edge to the business is the role of ISs and IS success evaluation is the recently subject of much debate [14]. Organizations consider the failure of ISs a major concern and to target this failure, information system assessment is a main step and condition to make the rate of success increase. The initiatives of the future of the information systems, opinions of users are deemed as success determinant for any Information system enterprise [15]. The Yemen Higher Education Management Information System (YHEMIS) project is developed during the development process program to support the universities with ICT [2]. Evaluating the success in ISs is a critical issue and it is a key concern by the executives of organization throughout the world to manage carefully and successfully [12]. Many universities, now a day's, deploy and improve their ISs to optimize and maintain the academic performance and to meet the changes occurred in different systems [16].

There is a need for ISs success measure based on student's perspective, because there is a lack of research related to the field ISs success measure. It is urgent to know whether (YHEMIS) is a success or not based on student's perspective. It is necessary to find out whether YHEMIS benefited the students or not. The result can help us in the future to send it to the stakeholders in the organization [2]. The main problem is that the large-scale application Yemen Higher Education Management System was developed and implemented without evaluation. This IS is categorized as large and important system applied in the public Yemeni organizations (Universities). YHEMIS, has not been evaluated before and it is not known either it is success or failure based on the student's perspective. Obviously, it is very important to do an IS evaluation based on student's perspective for the YHEMIS [2].

3. Systems Success Measure in General

Knowing and measuring the level of systems success evaluation identified in IS area as critical issue, many studies such as [15], have been performed to search and explore this value issue. However, proper group of variables that may be used to identify the users understanding of information systems success as a main determinant for the success as ISs are chiefly running to serve the end user. Plethora of IS used in organization such as student information system, human resource management system, e-commerce and many others, so doing IS success evaluation which categorized as complex and illusive is necessary [17]. Based on CHAOS report that eight thousand three hundred eighty, information system applications under development, information system projects deemed as successful in the case of complete on time and budget with satisfaction of all specified function, "16.2%" of projects fell in this category. Information system project deemed as partial failure in case of completed the project but with over cost and time and/or features and functions that already specified are lack, "52.7%" of all studied projects fell into this category [15]. Software's and information systems deemed as full failure in the case of at the some point the project cancelled or abandoned, "31.1%" of all studied projects fell into this category. Project success is little worse than the year 2006 but better than year 1994. He also confirmed that CHAOS report is still to be an important measure for the IT industry in general and IS [18]. Systems evaluation is so important because it is assessing the benefits and values to the organization and it is providing feedback to the stakeholders and peoples in responsible position in the organization [19].

4. Related Work

An empirical and theoretical test on the model proposed by DeLone and McLean and the model by Seddon. One characteristic common to the two models is that they both relate to IS success, although the model by Seddon does not consider the use of IS as a process but rather as a behaviour. Whereas, the initial model by DeLone and McLean views it as a process that leads to organizational or individual impact. The model by Seddon concentrates on the fundamental areas of the taxonomic category's interrelationship [20]. The work by [20] stated that the major thing that makes these two models differ exists in the IS use placement and definition. However, Seddon states that usage precedes benefits and impact, although it is not responsible for the cause. The author further considered the use of IS to be a resultant behaviour reflecting the expected net benefits gained from adopting the information system. Additional information regarding the empirical test by [20], conducted is further explained later. In addition, different researches who have adopted these models is presented next [20].

DeLone and McLean adopted their upgraded model to compose the success figures of e-commerce mentioned in previously existing studies and exhibited how the model can be adopted by considering two example cases. These two cases described usability as a vital measurement for system quality prompting expanded times for access in websites (use) and purchases repeated (user satisfaction). To add to their work, the authors further recommended that e-commerce studies ought to incorporate net benefits measures (such as, market valuation, incremental deals) and not be limited to the collection of surrogate measures, for example, hits on website (use). Then again, to comprehend the results of net benefits, they contended that the users experience quality and the usage by customers of and being satisfied with, the system ought to be measured [21].

An empirical test in quasi-voluntary use of IS with application in student information system (SIS). The SIS gives online access to a database of the academic and personal data of students. The adoption of SIS was not compulsory. The discoveries support DeLone and McLean's perception that IS success models must be deliberately determined in each area. They additionally propose that future study ought to inspect how "performance of IS success models in distinctive areas, including settings that differ from strictly voluntary to strictly involuntary use and prescribe refinements as fitting" [20].

(System quality, information quality, use, user satisfaction, net benefits, service quality and intention to use) [22], tried the IS success model by utilizing field investigation of a compulsory data system. The test was led with Oulu City Council. The council was working on the selection of a new data system and attempting to perform its acknowledgement organizationally. Information gathered by adopting questionnaire approach, which were offered to new data system's basic users. The questionnaire administered considered standard measures. System quality was measured by six factors: Language, system flexibility, time for response, system integration, recovery of error and access convenience [22]. Information quality was likewise measured by six factors: Precision, completeness, accuracy, currency, reliability and output format. The outcomes demonstrated that apparent system quality and information quality perceived were noteworthy indicators of user fulfilment or satisfaction with the system, however they did not matter to the use of the system. User fulfilment was a solid indicator of individual effect [23]. A contribution to IS success research provided form [23], by developing and empirically testing of a procedure arranged framework of IS success that considered the framework of DeLone and McLean.

They analysed the impact of low-level immaterial IS and Information Technology (IT) advantages on high-level monetary measures. They likewise presented IS quality plan as a predecessor to the model's variables inputted. The outcomes of the test supported a procedure arranged perspective of the advantages from the IS and demonstrated how the impacts of IS along a way can prompt better performance organization-ally, for their situation, lower general expenses. Standard measures are presented in this study's appendix. Their study used seven-point Likert scale secured by "Strongly Disagree" to "Strongly Agree" or scale from "Not Much" to "Extensively" [23]. A knowledge management systems (KMS) success framework is stated and experimentally evaluated by [24]. Construct with respect to an investigation of current approach of knowledge management and in addition the DeLone and McLean's framework where five subordinate variables were used (quality of the system, perceived KMS benefits, quality of knowledge or information, use of system and satisfaction of user) in assessing KMS success. The meanings and success measures are introduced [24].

5. Hypothesis and Framework

To examine this framework these hypotheses are suggested:

- H1, Information Quality significantly affect use of (YHEMIS)
- H2, System Quality significantly affect use of (YHEMIS)
- H3, Services Quality significantly affect use of (YHEMIS)
- H4, Management Support significantly affect use of (YHEMIS)
- H5, Information Quality significantly affect students' satisfaction of (YHEMIS)
- H6, System Quality significantly affect students' satisfaction of (YHEMIS)
- H7, Services Quality significantly affect students' satisfaction of (YHEMIS)
- H8, Management Support significantly affect students' satisfaction of (YHEMIS)
- H9, Use significantly affect students' satisfaction of (YHEMIS)
- H10, Students' Satisfaction significantly affect on use of (YHEMIS)
- H11, Use significantly affect Net benefits of (YHEMIS)
- H12, Students' Satisfaction significantly affect Net benefits of (YHEMIS)

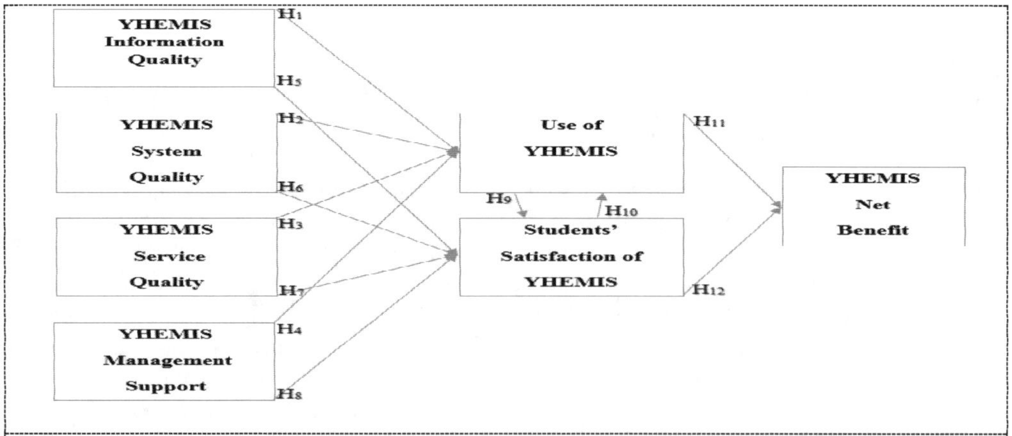

Figure 1. The Framework.

6. Methodology

This study used the quantitative approach where a cross sectional study of the YHEMIS will be evaluated. The quantitative approach enable data to be collected from different users to provide more information. The evaluation of an information system like YHEMIS involve various users (students), thus of data collection must be done depending on the number of respondents and type of data to be collected. Quantitative is suitable because in this study, data collected be in a form of survey using an instrument that adapted from other studies. This study collected opinions of the students, through questionnaires. The YHEMIS is used by most of the level one students. In this study, stratified random sampling is used. Data analysis using descriptive and inferential statistics were used. The aim of this study was to evaluate (YHEMIS) net benefits based on students' perspective. The study based on survey method due to the huge number of students who use the system, to collect the data for analysis. Thus, proper data collection sampling and analysis done. In this study, Statistical Package for the Social Science (SPSS) tool was used to convert the multiple-regression results and population data to be presents in graph, pie chart, table and to provide descriptive statistics. This gives a clear picture of the result. The Partial Least Squares (PLS) tool is used to manage and convert the correlation results to be presents in good manner.

7. Data Collection

Data collected through, questionnaire. Random sampling method was adapted to collect the samples where two hundred seventy-eight respondents were used for the data analysis. The questionnaire adapted from other researches gives the ability to compare the result and provides reliability [25]. An instrument that used in this study is a questionnaire. The researcher communicated with several professors to validate the instrument. After validation process has been done a certified translator translated the instrument to Arabic language by hand. The reliability of the questionnaire tested in a pilot study.

8. Pilot Study

A total of 33 students participated in the pilot study. The test of reliability consistency is Cronbach's coefficient alpha which is used for multipoint scale items [26,27]. The higher amount of coefficient indicates the better measures. Ideally, the Cronbach's coefficient alpha should be greater than 0.70 [28]. Table I indicates the results of reliability for pilot study.

Table 1. Pilot Study Result.

Construct	*No. Items*	Cronbach's alpha
Information Quality	8	0.961
System Quality	8	0.865
Services Quality	6	0.735
Management Support	4	0.966
Use	6	0.725
Student Satisfaction	4	0.963
Net Benefit	3	0.953

9. Population

Unit of analysis for this research is the individual end users, students of level one who use the system. The undergraduate degrees in Yemen consist of four Levels. End users who used the YHEMIS will be selected to answer the questioner.

10. Sampling

The unit of analysis in this study involves the main users of YHEMIS of Hadhramout university that are the students who are in level one. Researcher selected random students to answer the questionnaire that's used the YHEMIS to meet the research aims. Determining the sample size is important to estimate the characteristics of the population [29]. The sample size is larger than thirty and less than five hundred is proper and adequate for analysis, preferably 10 times or more than the number of the variables in the study [30,31].

IOP Conf. Series: Journal of Physics: Conf. Series **1019** (2018) 012003 doi:10.1088/1742-6596/1019/1/012003

11. Data Analysis and Findings

A total of 278 questionnaires were distributed to the students in Hadhramout University in Mukalla, Yemen. Out of 278 distributed questionnaires, 266 were filled and returned by the respondents showing a response rate of 96 percent. In data cleaning process, it was found that five questionnaires were not usable for analysis. Therefore, the number of usable questionnaires are 261 showing 94 percent response rate available for data analysis. Moreover, appropriate sample size was obtained for the study as the rules of thumb that a sample size between 30 and 500 is appropriate for data analysis (preferably 10 times or more than that of number of variables involved in the study) [30,31].

Of the 261 respondents in this research 179 or 68.6% were male and 82 or 31.4% were female. Hadhramout University in Yemen-Mukalla has many males as compared to females who are students in the university.

In terms of age, 51 respondents or (19.5%) of the total less than 18 years, whereas 173 respondents (66.3%) were within the age of 18-20 years and 37 respondents (14.2%) were within the age of greater than 20 years.

In terms of colleges, 67 respondents or (25.7%) from Administrative Science college, Whereas 90 respondents (34.5%) from Science college and 104 respondents (39.8%) from Education college.

Table 2. Cross Tabulation for the Sample.

Colleges	A	B	C	Total
Population Size	256	327	395	978
	26%	34%	40%	100%
Sample Size	72	95	111	278

Table 3. Respondent's Frequency of Gender.

Gender	Frequency	%
Male	179	68.6%
Female	82	31.4
Total	261	100%

Table 4. Respondent's Frequency of Age.

Age	Frequency	%
Less than 18	51	19.5
18-20	173	66.3
More than 20	37	14.2
Total	261	100%

Table 5. Respondent's Frequency of colleges.

Gender	Frequency	%
Administrative Science (A)	67	25.7
Science (B)	90	34.5
Education (C)	104	39.8
Total	261	100

12. Hypothesis Testing

To test the hypotheses suggested for this research, a Pearson correlation analysis was conducted to see the connection between the variables. If the value of correlation is equal to 1.0, it indicates that there is perfect positive or negative relationship. If it is equal to zero it means that there is no relationship [32]. The coefficient of correlation shows the direction and strength of linear association of variables in the study and the significance level of all coefficients is also given [30]. Positive sign with coefficient of correlation indicted that both variables move in same direction while negative sign designates the opposite direction of two variables [33].

Table 6. Reliability Analysis of the Variables.

Variables	No. items	Cronbach's alpha
YHEMIS Information Quality	8	0.965
YHEMIS System Quality	8	0.956
YHEMIS Services Quality	6	0.817
Management Support	4	0.957
Use	6	0.987
User Satisfaction	4	0.702
YHEMIS Net Benefit	3	0.761

Table 7. Hypothesis Summery.

H	B	T	P	Status
H1	0.312	3.380	0.001	Support
H2	0.313	3.849	0.000	Support
H3	0.051	1.157	0.248	Not Support
H4	0.259	3.067	0.002	Support
H5	0.133	2.037	0.042	Support
H6	0.309	4.690	0.000	Support
H7	0.029	0.620	0.535	Not Support
H8	0.227	4.687	0.000	Support
H9	0.311	5.338	0.000	Support
H10	0.421	5.122	0.000	Support
H11	0.052	0.732	0.465	Not Support
H12	0.732	11.503	0.000	Support

The correlation of the variables of YHEMIS. It can be noted that there is positive correlation between the YHEMIS variables and all are accepted. For this study, simple regressions were carried out to predict net benefit based on use and students' satisfaction considering information quality, system quality, service quality and management support, The PLS procedure was applied to estimate the dependent variable of the research framework. PLS algorithm used two times to mutual influence between use and user satisfaction we estimated two frameworks: framework 1 testing the whole framework with relation use to students' satisfaction, framework 2 analysing the whole framework with reversibly students' satisfaction to use [22,34]. Additionally, we used a bootstrapping procedure and generated 500 bootstrap samples to test the significance of the path estimates [35,36].

13. Discussion and Implications

The idea of this study is to determine what factors influence use and students' satisfaction of YHEMIS and to know whether students perceived benefit or not. After measuring the six dimensions and net benefit allowed researcher to understand the relationship between the six dimensions and net benefit. The results of this study are positively related to students' perceived net benefit. With the recognition that information quality, system quality and management support suggest positive significant affect on use and students' satisfaction of YHEMIS while service quality has not significantly affect the use and students' satisfaction of YHEMIS. Use and Students' satisfaction of YHEMIS have a high positive significant effect on each other. On the other hand, students' satisfactions have a high positive significant affect toward net benefit while use has not significantly affect the net benefit.

Hypothesis (H3 service quality significant affect the YHEMIS use, H7 service quality significant affect the students' satisfaction and H11 use significant affect net benefit) not supported and rejected due to being insignificant. The other hypothesis (H1, H2, H4, H5, H6, H8, H9, H10 and H12) are accepted due to being significant. Further, squared multiple correlations R2 that represent the explanatory power of the structured framework should be greater than 0.33 [35]. H9 and H10 comprise a mutual influence between use and user satisfaction we estimated two frameworks: framework 1 testing H9 from use to user satisfaction, framework 2 analysing H10 reversibly [22,34].

Regarding to our framework the R2 for use (0.639), students' satisfaction (0.767) and R2 net benefit (0.598) are high and same value in both frameworks 1 and 2. In summary nine out of twelve hypotheses were supported. This framework accounted for 60% of the variance in perceived net benefit, with students' satisfaction exerting a stronger direct effect than use on perceived net benefit. 69% of the variance in use was explained by information quality, system quality, service quality, management support and students' satisfaction, while 77% of the variance in students' satisfaction was explained by information quality, system quality, service quality, management support and use. The direct and total effect of students' satisfaction on perceived net benefit was 0.732, while the direct and total effect of use on perceived net benefit was 0.053. That means students' satisfaction exhibited stronger direct and total effects on perceived net benefit than use. Among information quality, system quality, service quality and management support the direct and total effect of system quality was the strongest effect on use 0.313 and on the students' satisfaction 0.310. Per the proposed framework, students' satisfaction is a closer measure of YHEMIS success than the other success measures.

Perceived net benefit should develop if the formation of perceived quality, system use, management support and user satisfaction is appropriately managed. Thus, management attention might more fruitfully focus on the development of these psychological and behavioral processes. To increase students-perceived net benefit, YHEMIS administrators need to develop it more with a good service quality, information quality, system quality and provide more management support which, in turn, will influence student's system usage behaviour and satisfaction evaluation and the corresponding perceived net benefit. In this framework, system students' satisfaction was found to have the strongest direct and total effect on perceived net benefit, indicating the importance of system satisfaction in promoting students-perceived net benefit.

14. Future Work

The higher numbers of dissatisfying systems users and fail of software's, information systems, mobile application and web based systems in the World, Arab countries and in Yemen as large clear the way and make it necessary to conduct a comprehensive research and developing a framework that can help the organization and the world. The success measure of web applications based on the satisfaction of the users is rather a recent and frequently neglected issue. This study which is the first study on the evaluation of information system website success conducted for Yemen universities has opened the path for other Yemeni researchers to conduct future researches on evaluation of information success system generally. Researchers call for further investigation for the framework in different context. The studies in Arab in general are lack and in the Yemeni context is almost non-existent. Researchers open the way for the Yemeni researchers to conduct studies in the information system websites success evaluation in Yemeni context. Researchers recommended investigating the factor management support in other frameworks.

Researchers recommended investigating the relationship between user's satisfaction with loyalty and Benefit (success) and investigate the significant affect between the factors that related to the framework ISO 20510 and their effect on users' satisfaction. Also, investigate the significant affect between the quality factors that related to the DM and their effect on users' satisfaction in the domain of universities web-based systems is highly recommended. Yemeni universities web-based systems remain without success evaluation and strongly need for framework that can help and solve the problem. Henceforth, the web portal of Arab countries universities and Yemeni universities in an overall manner, especially of the Hadhramout universities require a robust assessment to outline how the customers are experiencing them and what is their overall level of satisfaction with these websites, what is the Net Impact of these systems towards the students and it is highly recommended to investigate the relationship between students' satisfaction and loyalty of the students towards the university web site. Sadly, there have been no effective studies in this prospect. Moreover, inconsistent results pertaining to the influence of quality components like service, information systems with satisfaction demands for further empirical attention.

Likewise, the literature also suggests that software engineering frameworks are also inconsistent in connection to satisfaction. On parallel, there lies a dearth of empirical results pertaining to ISO 25010 framework and how it can help in evaluating and measuring success of a system. Thus, the influence of student satisfaction on the loyalty towards university info-share systems is unclear. In simple, there is lack of evidence pertaining to the impact of students` satisfaction on their loyalty with the respective university web portals. Likewise, there are inconsistent results pertaining to how the information system assessment can be an important prospect for enterprises to sustain and remain competitive. Investigation this link is becoming tougher and challenging due to heavy intervention of internet and advanced technologies. Since internet comes with the facility of uninterrupted and seamless communication, the assessment of the effectiveness of information systems has become a critical task as now it requires a lot of features and components to be considered simultaneously. The assessment of effectiveness of web-based is essential to remain competitive and offer greater service features to predict student satisfaction.

This importance coupled with complexity of assessment of effective information systems has raised several factors to consider. Precisely, it demands for a more comprehensive framework to help assist corporations as well as educational institutions to understand and outline the effective design and development of robust information systems for profound services to end users. At present, information systems have transformed into web-based information systems (WIS) due to advancements in internet and technology. Therein, it provides a holistic one stop platform to cater for all information need pertaining to the organization and its offerings. Henceforth, it essential for businesses to have strong hold WIS that could enable them to reach up to the expectations of the end users and help enterprise achieve its planned goals. Nevertheless, the assessment of information systems especially of the ones that are web enabled is essential for making sure that the end user requirements are fulfilled. Such assessment will also be very healthy in helping resolve critical issues pertaining to the existing information system and/or website through highlighting its strengths and areas that it requires further improvement.

15. Conclusion

This study planned to fill the gap of lack of studies and empirical data of IS success evaluation and to open the way to do more studies in other universities in Arab region. Furthermore, aimed to enhance our understanding of how IS success evaluation is an important for an organization and it is an integral part of IS's investment. Research in information systems success in public universities is relatively few if compared to study done in the private organizations. Last the findings of this study confirmed that students of Hadhramout University are satisfied with Yemen Higher Education Management Information System. In general, the system is success and benefited the Hadhramout University students. The factors information quality, system quality and the external factor management support were playing a vital role in the success of YHEMIS whereas service quality not.

References

[1] I. E. I. Fadhel, "An evaluation of information system success based on students' perspective: The case of Hadramount University." Universiti Utara Malaysia, 2015.

[2] F. Baheshwan, "Evaluation OF Information Systems In Yemen Based On User's' Perspective?," Hadramout University, Mukalla, 2016.

[3] H. Rakels, "The future of ICT in Yemeni higher Education," Yemen, 2013.

[4] AbdulMonsif, "Status of computer and administrative academic research in Yemen," Hadramout University, Mukalla, 2015.

[5] D. A. Khred, "systems development in our country," 2017.

[6] W. H. DeLone and E. R. McLean, "Information Systems Success Measurement," *Found. Trends® Inf. Syst.*, vol. 2, no. 1, pp. 1–116, 2016.

[7] A. I. ELdesouky, H. Arafat, and H. Ramzey, "Toward complex academic Web-Sites Quality evaluation method (QEM) framework: quality requirements phase definition and specification," *Comput. Syst. Eng. Dep.*, 2008.

[8] N. Forsgren, A. Durcikova, P. F. Clay, and X. Wang, "The Integrated user satisfaction model: Assessing information quality and system quality as second-order constructs in system administration," *Commun. Assoc. Inf. Syst.*, vol. 38, no. 1, p. 39, 2016.

[9] E. W. Mwangi, "Information quality assessment framework: case of the national safety net program single registry system." Strathmore University, 2016.

[10] V. Teeroovengadum, T. J. Kamalanabhan, and A. K. Seebaluck, "Measuring service quality in higher education: Development of a hierarchical model (HESQUAL)," *Qual. Assur. Educ.*, vol. 24, no. 2, pp. 244–258, 2016.

[11] P. Isaias and T. Issa, *Information System Development Life Cycle Models. In High Level Models and Methodologies for Information Systems*. Springer, 2015.

[12] P. E. D. Irani Zahir, "The propagation of technology management taxonomies for evaluating investments in information systems," *J. Manag. Inf. Syst.*, vol. 17, no. 3, pp. 161–177, 2000.

[13] N. Urbach and B. Müller, "The updated DeLone and McLean model of information systems success," in *Information systems theory*, Springer, 2012, pp. 1–18.

[14] A. N. H. Zaied, "An integrated success model for evaluating information system in public sectors," *J. Emerg. Trends Comput. Inf. Sci.*, vol. 3, no. 6, pp. 814–825, 2012.

[15] R. Moh'd Al-adaileh, "An evaluation of information systems success: A user perspective-The case of Jordan telecom group," *Eur. J. Sci. Res.*, vol. 37, no. 2, pp. 226–239, 2009.

[16] F. Dadmand, "Evaluating information system success in university: an empirical test of the delone-mclean model," *A J. Multidiscip. Res.*, vol. 3, no. 2, pp. 38–50, 2014.

[17] S. Petter, W. DeLone, and E. McLean, "Measuring information systems success: models, dimensions, measures, and interrelationships," *Eur. J. Inf. Syst.*, vol. 17, no. 3, pp. 236–263, 2008.

[18] J. Dominguez, "The curious case of the chaos report 2009," *Proj. Smart*, pp. 1–16, 2009.

[19] Z. Irani and P. E. D. Love, "Evaluating information systems : public and private sector," 2008.
.

[20] A. Rai, S. S. Lang, and R. B. Welker, "Assessing the validity of IS success models: An empirical test and theoretical analysis," *Inf. Syst. Res.*, vol. 13, no. 1, pp. 50–69, 2002.

[21] W. H. Delone and E. R. Mclean, "Measuring e-commerce success: Applying the DeLone & McLean information systems success model," *Int. J. Electron. Commer.*, vol. 9, no. 1, pp. 31–47, 2004.

[22] J. Iivari, "An empirical test of the DeLone-McLean model of information system success," *ACM SIGMIS Database*, vol. 36, no. 2, pp. 8–27, 2005.

[23] T. A. Byrd, E. H. Thrasher, T. Lang, and N. W. Davidson, "A process-oriented perspective of IS success: Examining the impact of IS on operational cost," *Omega*, vol. 34, no. 5, pp. 448–460, 2006.

[24] J.-H. Wu and Y.-M. Wang, "Measuring KMS success: A respecification of the DeLone and

McLean's model," *Inf. Manag.*, vol. 43, no. 6, pp. 728–739, 2006.
[25] M. N. K. Saunders, M. Saunders, P. Lewis, and A. Thornhill, *Research methods for business students, 5/e*. Pearson Education India, 2011.
[26] A. Davarpanah and N. Mohamed, "Human Resource Information Systems (HRIS) success factors in a public higher education institution context," in *Research and Innovation in Information Systems (ICRIIS), 2013 International Conference on*, 2013, pp. 79–84.
[27] L. J. Cronbach, "Response sets and test validity," *Educ. Psychol. Meas.*, vol. 6, no. 4, pp. 475–494, 1946.
[28] J. Pallant, *SPSS survival manual*. McGraw-Hill Education (UK), 2013.
[29] J. F. Hair, W. C. Black, B. J. Babin, R. E. Anderson, and R. L. Tatham, *Multivariate data analysis*, vol. 5, no. 3. Prentice hall Upper Saddle River, NJ, 2010.
[30] U. Sekaran and R. J. Bougie, *Research methods for business: A skill building approach*. John Wiley & Sons, 2016.
[31] J. T. Roscoe, *Fundamental research statistics for the behavioral sciences [by] John T. Roscoe.* Holt, Rinehart and Winston, New York, NY., 2016.
[32] C. Augustie, "An analysis on factors influencing international online shopping: Indonesian perspective." Universiti Utara Malaysia, 2014.
[33] U. Sekaran, *Research methods for business: A skill building approach*. John Wiley & Sons, 2003.
[34] S. Dernbecher, "Having the Mind in the Cloud: Organizational Mindfulness and the Successful Use of Desktop as a Service," in *System Sciences (HICSS), 2014 47th Hawaii International Conference on*, 2014, pp. 2137–2146.
[35] W. W. Chin, "The partial least squares approach to structural equation modeling," *Mod. methods Bus. Res.*, vol. 295, no. 2, pp. 295–336, 1998.
[36] M. Tenenhaus, V. E. Vinzi, Y.-M. Chatelin, and C. Lauro, "PLS path modeling," *Comput. Stat. Data Anal.*, vol. 48, no. 1, pp. 159–205, 2005.

IOP Conf. Series: Journal of Physics: Conf. Series **1019** (2018) 012004 doi:10.1088/1742-6596/1019/1/012004

Comparative Study and Simulation of Mid-range Wireless Power Transfer at Various Transmission Distance

Shun Yao Tan[1], June Xiong Yap[1], and Hui Jing Lee[1]

[1]Institute of Power Engineering (IPE), Universiti Tenaga Nasional, Jalan Ikram-UNITEN, 43000 Kajang, Selangor, Malaysia.

Email: LHJing@uniten.edu.my

Abstract. Wireless power transfer would be the niche of research in next decade which is significant to reduce the cable and charges consumption for electronic devices. This paper presents the simulation result of electric and magnetic wave behaviour in mid-range wireless power transfer and various transmission distance in the range of 8 – 16 cm. Simulation result using CST microwave studio shows that scattering parameter in the power transmission decreases from 0.6509 down to 0.5137 with the increase in distance. The transmitted electric field at the receiver decreases from 1615 V/m to 1086 V/m whereas the magnetic field decreases from 32.63 A/m down to 24.85 A/m. Results show that 4-coils magnetic resonance coupling is an ideal approach for mid-range wireless power transfer.

1. Introduction

Fundamental of wireless power transfer (WPT) was firstly introduced back in the 1800s [1]. Today, WPT is gaining strong attention due to its functionality in transmitting power and energy over distance using air as the transmission medium [2]. Such approach is to sustain a greener environment which is free from chargers, cables and transmission wire. The operational WPT occurs in range of millimetres (short range WPT), range of centimetres (mid-range WPT) and range of meters to kilometres (long range WPT). There are numerous research of short range WPT since the past decade, and the operation of long range WPT using radio-frequency or microwave has been ambiguous towards human health impact [3, 4].

In such scenario, mid-range WPT has recently drawn researchers' attention to achieve a high transmission efficiency at a desired transmission distance [5, 6]. Mid-range WPT received great achievement in 2014 when Korea Advanced Institute of Science and Technology (KAIST) produced a 209 W of power to charge 40 smartphones in the 20 kHz of operating frequency [7]. The major drawback of the finding is that the set-up is bulky and immobile. *Song et al.* has recently discovered a dielectric resonator to enhance the performance of mid-range WPT [8]. Tang *et al.* has presented an improved mid-range WPT using segmented coil transformer which made applicable in implantable heart pumps [9]. The improvement in the stability of mid-range WPT was justified by *Lee et al.* using the implementation of relay resonator [10].

Several methods have been proposed to achieve mid-range wireless power transfer namely: 2-coils, 3-coils and 4-coils magnetic resonance coupling [11, 12]. Previous research justified that 4-coils magnetic resonance coupling would be the promising method solely because it functions based on impedance matching which hence optimized the power and energy level received by the load [13]. In

2017, Yang has recently proved that 4-coils magnetic resonance is able to extend the power transmission distance as compared to the 2-coils system [14]. This is because 4-coils system will have a stronger magnetic coupling between the source and load coils.

This paper performs simulation work to observe the coupling between the 4-coils system as well as to study the change in coupling and power transmittance achievable at the load coil when the distance between the source and load coil varies. Section 2 presents the ideal circuitry of transmitter unit and receiver unit for mid-range wireless power transfer which adopts the 4-coils magnetic resonance coupling method and some fundamental equations for the magnetic coupled resonator. The simulation result of wireless power transfer is presented in section 3 using CST microwave studio at various transmission distance in the range of 8-16 cm. Comparative study of the scattering (*S*) parameter, impedance (*Z*) parameter, electric (*E*) and magnetic (*H*) field behavior are presented and discussed.

2. Mid-range Wireless Power Transfer System Design

Figure 1 shows the ideal circuitry structure for a mid-range wireless power transfer. In the transmitter unit, the rectifier will first convert the input AC wave to DC signal. Then the DC signal will be converted to a high AC output via the oscillator. Power amplifier is recommended to amplify the signal before power transmission at the coils. The 4-coils magnetic resonance coupling system comprises of the source coil, transmitter (Tx) coil, receiver (Rx) coil, and the load coil [15]. Resonance coupling will occur between source and transmitter, transmitter and receiver, as well as receiver and coil. Power transmittance and transfer in energy is achieved when the coils are operating at the same frequency. Then, at the receiver unit, the rectifier will convert the AC output from the load coil to a DC output. Finally, the DC output will power up the load connected at the end of receiver unit. It is crucial to ensure that appropriate current and voltage value is received at the load to power up electronic devices such as mobile phone, tablet, laptop etc.

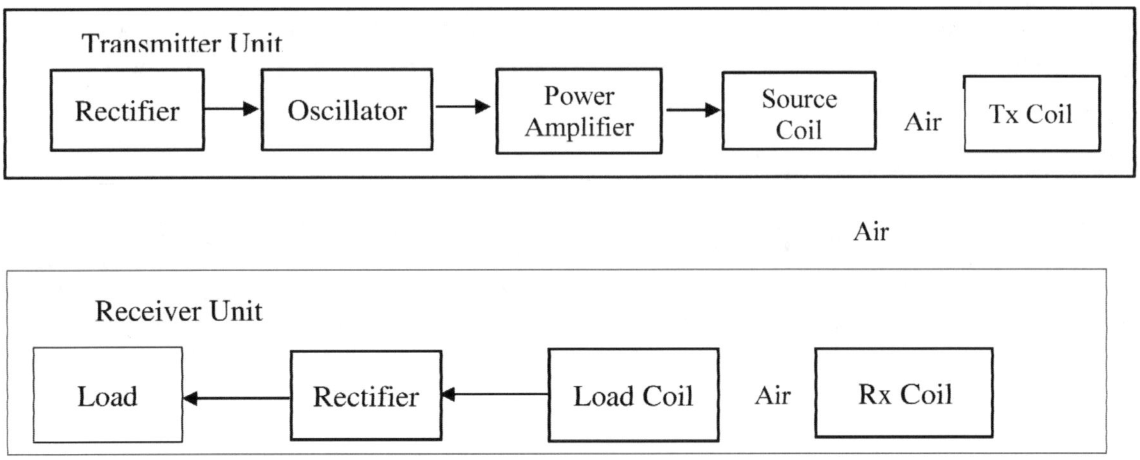

Figure 1. Transmitter and receiver units for mid-range wireless power transfer

The coil inductance, *L* is defined using Wheeler's equation [16]:

$$L = \frac{r^2 N^2}{2.54 \times (8r + 11b)} \quad (1)$$

where *r* is the radius of coil, *N* is the number of turns and *b* is the depth of winding.

A high quality factor (Q) is desired in wireless power transfer to ensure a high power efficiency. The Q-factor of 4-coils system is expressed as [17]:

$$Q = \frac{1}{R}\sqrt{\frac{L}{C}} \tag{2}$$

in which R is resistance, L is the inductance and C is the capacitance value.

The efficiency of transmittance rate which explains the rate of power transfer received from the load coil from the source coil reads as [18]:

$$\eta = \frac{P_{out}}{P_{in}} = \frac{I_4^2 R_L}{I_1^2 \left(\frac{R_S}{4}\right)} = |S_{12}|^2 \tag{3}$$

where I_1 and I_4 are current flow at source coil and load coil, respectively. In addition, R_S and R_L represent the resistance and source and load coils.

Maximum power transferred from the source coil to the load coil, $|S_{12}|_{\max}$ is expressed as [15, 18]:

$$|S_{12}|_{\max} = \frac{k_{12}k_{34}Q_1Q_4R_L}{k_{23}^{*}\sqrt{L_1\omega_1L_4\omega_4}} \tag{4}$$

where Q_1 and Q_2 are quality factors at the source and load coils, ω_1 and ω_4 are the operating frequency at source and load coil, k_{12} is the magnetic resonance coupling resonance between the source and transmitter coil, k_{23} is the coupling between transmitter and receiver coil and k_{34} is coupling between the receiver and load coil. From equation 4, a strong coupling of k_{12} and k_{34} together with a low coupling of k_{23}would be ideal case to achieve maximum energy transfer.

3. Results and Discussion

This research was undertaken to perform simulation of 4-coils magnetic resonance coupling which is applicable in the mid-range wireless power transfer. Figure 2 shows the alignment of 4-coils structure which comprises of the source coil, transmitter coil, receiver coil and the load coil. A discrete port is placed at the source coil and load coil to transmit and receive signal, respectively.

Table 1 lists the variation of distance, d between source and transmitter, transmitter and receiver, as well as receiver and load coil. This variation is significant to perform comparative study of wireless power transfer at various transmitting distance.

Simulation is performed using CST microwave studio and results are presented in figure 3, 4, and 5. The electric field and magnetic field behaviours are observed when the coils are operating at a resonance frequency. For all the simulation circuitry presented in figure 3-5, the radius of source and load coils are 30 mm, whereas transmitter and receiver coils are ten turns with 50 mm of radius.

Table 1. 3 sets of separation distance between source coil and load coil

Fig	d between source and transmitter coil (cm)	d between transmitter and receiver coil (cm)	d between receiver and load coil (cm)	d between source and load coil (cm)
3	2	4	2	8
4	3	6	3	12
5	4	8	4	16

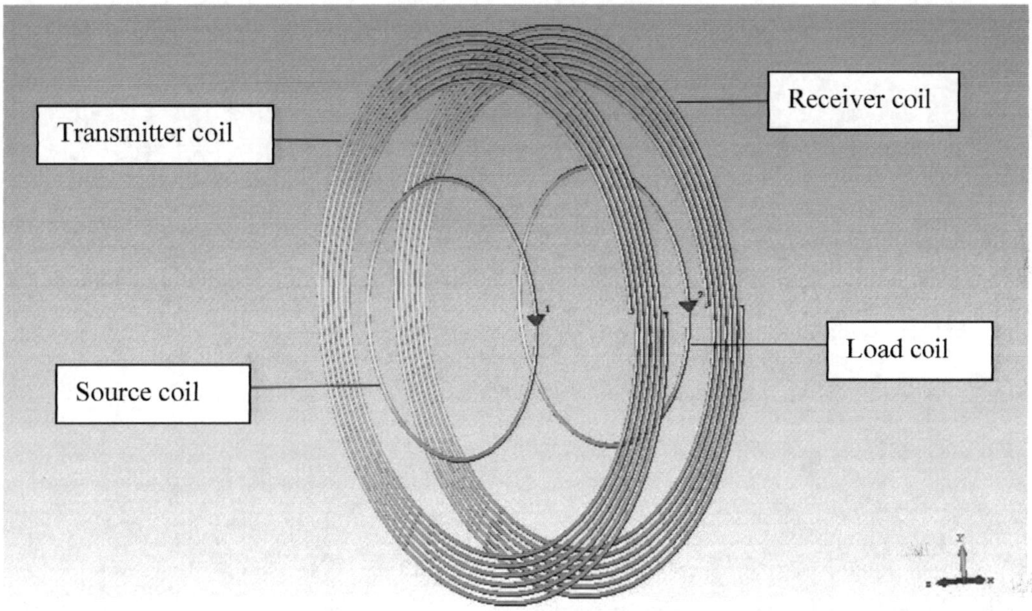

Figure 2. 4-coils system for magnetic resonance

Table 2 lists the performance parameters of 4-coils system design in figure 2 at three various distance between the source coil and load coil at 8, 12, and 16 cm of transmittance distance. The resonance frequency is identified at 17.151 MHz, 18.551 MHz and 21.485 MHz. The resonance frequency is identified at the maximum S-parameter value where the strongest coupling between the coils occurs.

Simulation result shows that the S-parameter for power transmitted from the source coil to load coil (S_{12}) as presented in equation 4 decreases from 0.6509 to 0.5137 when the distance of transmission increases from 8 cm to 16 cm. Such result reflects that the transmittance efficiency decreases from 42.36 % down to 26.38 %. In addition, table 2 presents the impedance parameters and the variation of electric field and magnetic field at the source and load coil.

Table 2. S-parameters, Z-parameters, E-field and H-field distribution at various distance between source coil and load coil

Fig	S-parameters		Z-parameters		E-field (V/m)		H-field (A/m)	
	S_{11}	S_{12}	Z_{11}	Z_{12}	Source	Load	Source	Load
3	0.3457	0.6509	34.67	42.62	1990.9	1615	54.94	32.63
4	0.4228	0.5745	27.19	32.99	1559.5	1546	55.30	31.56
5	0.4782	0.5137	28.55	28.69	1263.8	1086	53.62	24.85

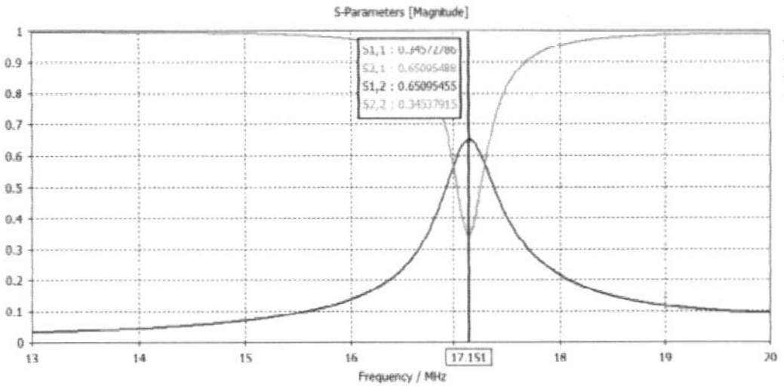

a) S-parameters

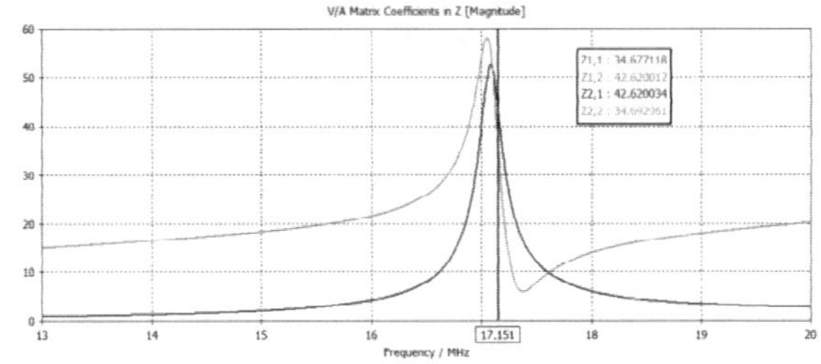

b) Z-parameters

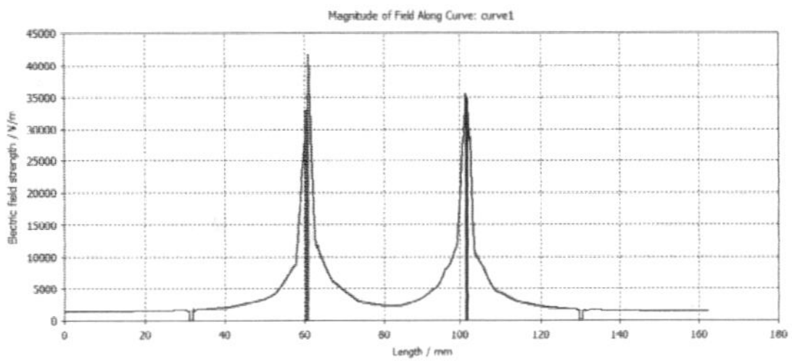

c) E-field distribution

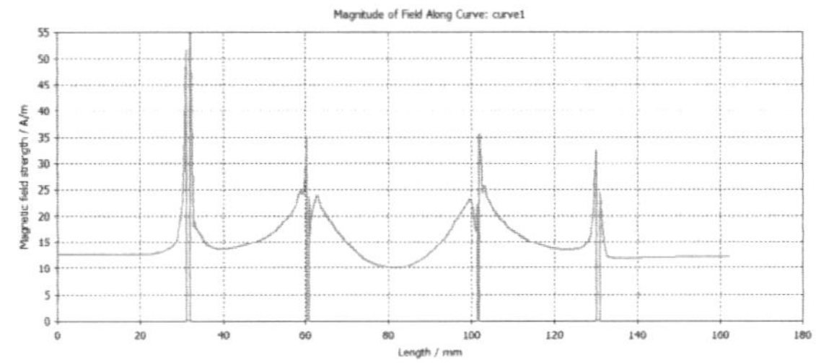

d) H-field distribution

Figure 3. Output parameters for 8 cm distance between source coil and load coil

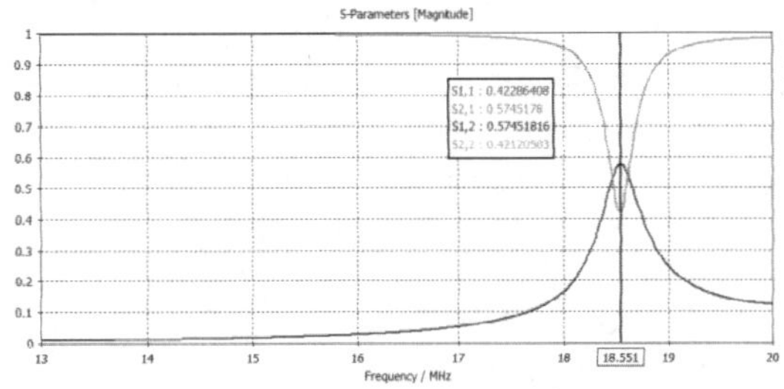

a) S-parameters

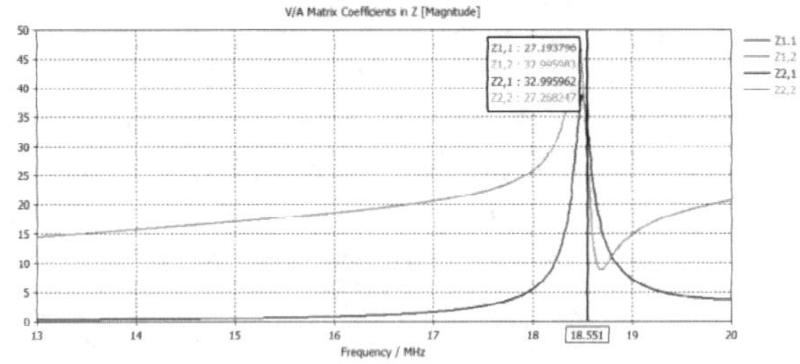

b) Z-parameters

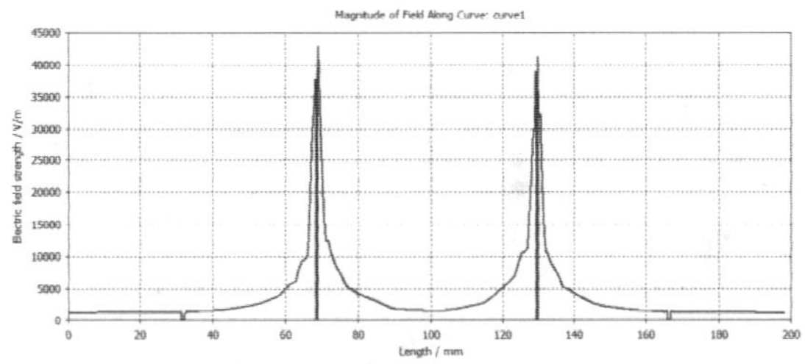

c) E-field distribution

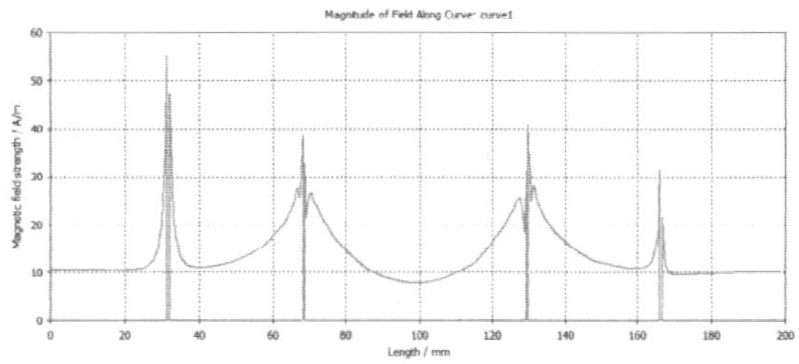

d) H-field distribution

Figure 4. Output parameters for 12 cm distance between source coil and load coil

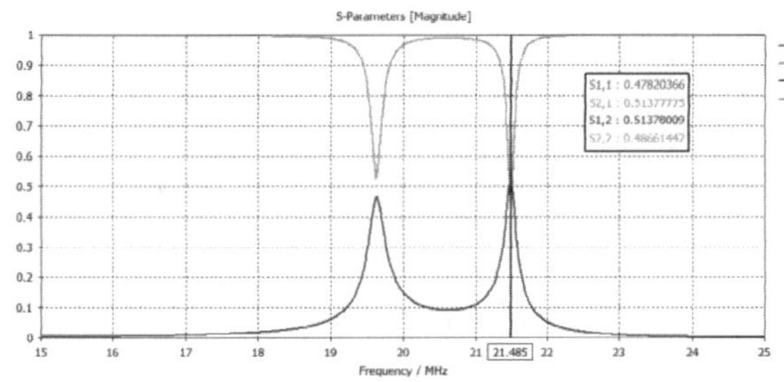

a) S-parameters

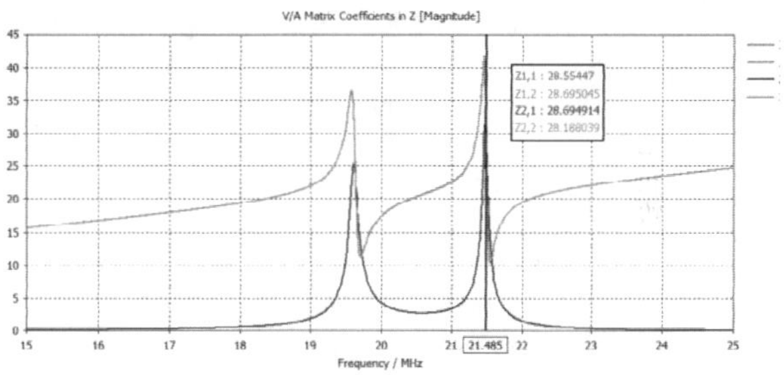

b) Z-parameters

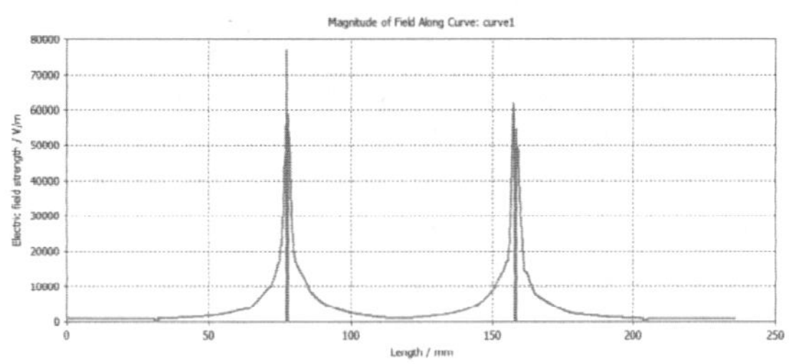

c) E-field distribution

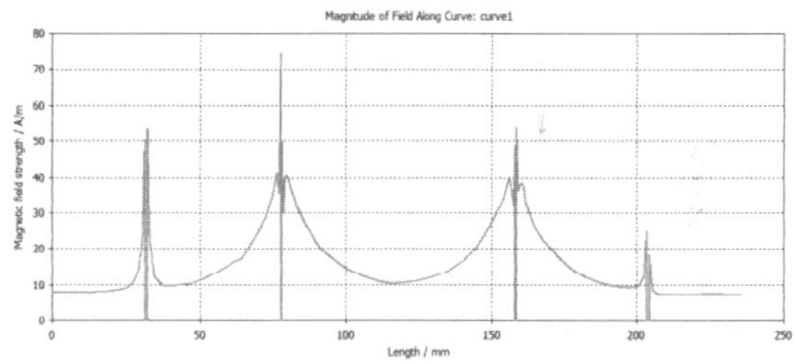

d) H-field distribution

Figure 5. Output parameters for 16 cm distance between source coil and load coil

4. Conclusion and Future Recommendation

This paper performs a comparative study on mid-range wireless power transfer at various transmittance distance using magnetic resonance coupling. Simulation results shows that the resonance frequency increases with respect to the transmission distance. The ratio of output power and input power drops from 0.6509 down to 0.5137 when the transmission distance increases from 8 cm to 16 cm. In addition, the impedance parameters, electric and magnetic field behavior are compared. Hence, result shows that magnetic resonance coupling is an ideal mechanism in mid-range wireless power transfer. Power transmittance could be achieved at a higher transmittance distance, given that the coils resonate in the same frequency. Other design parameters such as the ratio of distance between source coil and transmitter coil with respect to transmitter coil and receiver coil as well as number of turns or radius of the coil could be further investigated to achieve an optimum performance result in mid-range wireless power transfer.

Acknowledgement
We thank Universiti Tenaga Nasional Internal Grant, RJO10289176 for the support.

References

[1] Miller, J. M., Scudiere, M. B., McKeever, J. W., & White, C. (2011). Wireless power transfer. In *Oak ridge National Laboratery's Power Electronics Symposium*.

[2] Imura, T., Okabe, H., & Hori, Y. (2009, September). Basic experimental study on helical antennas of wireless power transfer for electric vehicles by using magnetic resonant couplings. In *Vehicle Power and Propulsion Conference, 2009. VPPC'09. IEEE* (pp. 936-940). IEEE.

[3] Hui, S. Y. R., Zhong, W., & Lee, C. K. (2014). A critical review of recent progress in mid-range wireless power transfer. *IEEE Transactions on Power Electronics*, *29*(9), 4500-4511.

[4] Lyu, Y. L., Meng, F. Y., Yang, G. H., Che, B. J., Wu, Q., Sun, L., & Li, J. L. W. (2015). A method of using nonidentical resonant coils for frequency splitting elimination in wireless power transfer. *IEEE Transactions on Power Electronics*, *30*(11), 6097-6107.

[5] Kurs, A., Karalis, A., Moffatt, R., Joannopoulos, J. D., Fisher, P., & Soljačić, M. (2007). Wireless power transfer via strongly coupled magnetic resonances. *science*, *317*(5834), 83-86.

[6] Zhong, W. X., & Hui, S. Y. R. (2015). Maximum energy efficiency tracking for wireless power transfer systems. *IEEE Transactions on Power Electronics*, *30*(7), 4025-4034.

[7] Yoon, U. (2017). Electrification of Other Transportation Systems. In *The On-line Electric Vehicle* (pp. 261-268). Springer International Publishing.

[8] Song, M., Belov, P. A., & Kapitanova, P. V. (2017, May). Dielectric resonators for mid-range wireless power transfer application. In *Wireless Power Transfer Conference (WPTC), 2017 IEEE* (pp. 1-3). IEEE.

[9] Tang, S. C., Lun, T. L. T., Guo, Z., Kwok, K. W., & McDannold, N. J. (2017). Intermediate Range Wireless Power Transfer with Segmented Coil Transmitters for Implantable Heart Pumps. *IEEE Transactions on Power Electronics*, *32*(5), 3844-3857.

[10] Lee, J., Lee, K., & Cho, D. H. (2017). Stability Improvement of Transmission Efficiency Based on a Relay Resonator in a Wireless Power Transfer System. *IEEE Transactions on Power Electronics*, *32*(5), 3297-3300.

[11] Zhong, W. X., Zhang, C., Liu, X., & Hui, S. R. (2015). A methodology for making a three-coil wireless power transfer system more energy efficient than a two-coil counterpart for extended transfer distance. *IEEE Transactions on Power Electronics*, *30*(2), 933-942.

[12] Huang, R., & Zhang, B. (2015). Frequency, impedance characteristics and HF converters of two-coil and four-coil wireless power transfer. *IEEE Journal of Emerging and selected topics in Power Electronics*, *3*(1), 177-183.

[13] Kiani, M., & Ghovanloo, M. (2012). The circuit theory behind coupled-mode magnetic resonance-based wireless power transmission. *IEEE Transactions on Circuits and Systems I: Regular Papers*, *59*(9), 2065-2074.

[14] Yang, D., Won, S., & Hong, H. (2017, May). Design of Range Adaptive Wireless Power Transfer System Using Non-coaxial Coils. In *IOP Conference Series: Materials Science and Engineering* (Vol. 199, No. 1, p. 012008). IOP Publishing.

[15] Duong, T. P., & Lee, J. W. (2015). A dynamically adaptable impedance-matching system for midrange wireless power transfer with misalignment. *Energies*, *8*(8), 7593-7617.

[16] Wheeler, H. A. (1958). The spherical coil as an inductor, shield, or antenna. *Proceedings of the IRE*, *46*(9), 1595-1602.

[17] Barman, S. D., Reza, A. W., Kumar, N., & Anowar, T. I. (2016). Two-side Impedance Matching for Maximum Wireless Power Transmission. *IETE Journal of Research*, *62*(4), 532-539.

[18] Cross, V. R., Hester, R. K., & Waugh, J. S. (1976). Single coil probe with transmission-line tuning for nuclear magnetic double resonance. *Review of Scientific Instruments*, *47*(12), 1486-1488.

IOP Conf. Series: Journal of Physics: Conf. Series **1019** (2018) 012005 doi:10.1088/1742-6596/1019/1/012005

Headship Rate Projections for Housing Demand in Johor, Malaysia

A R Alias[1], N Y Zainun[1], I A Rahman[1], A Suratkon[1], N Sulaiman[2], F E Mohamed Ghazali[3], N Shamsuddin[4]

[1] Sustainable Construction Management, Jamilus Research Centre,
Faculty of Civil and Environmental Engineering,
Universiti Tun Hussein Onn Malaysia, 86400 Parit Raja, Batu Pahat, Johor, Malaysia.
[2]KANZU Research: Resilient Built Environment,
Universiti Tun Hussein Onn Malaysia, 86400 Parit Raja, Batu Pahat, Johor, Malaysia.
[3]School of Civil Engineering,
Universiti Sains Malaysia, 14300 Nibong Tebal, Penang, Malaysia.
[4]SIRIM Berhad, Section 2, 40700 Shah Alam, Selangor, Malaysia.

Corresponding authors: nryasmin@uthm.edu.my

Abstract. Household formation trend is closely linked with the numbers of population. Meanwhile, the formation of households is largely influenced by the composition of the adult population that are measured by headship rate. This paper attempts to project headship rate which acts as a main predictor in determine potential housing demands. Population and Housing Census data from year 2000 and 2010 has been used in this projection headship rate for Johor state. Modified two-point exponential method was applied to determine the headship rate. All the data was categorized into five years interval for the ages between 15 to 80+ years old. Results indicate that the projected headship rates for age 15-19 and 20-24 are significantly decreased. However for the age's groups 40-44, 50-54, 55-59 75-79 and 80+ years old, the projected headship rate increased until year 2020. This demonstrates, the decreasing ability of young people under 25 years to become homeownership as compared to people from the age groups of 40 years and above. The findings of this study could be used as a basis for the government to identify numbers of potential housing demand in Johor.

1. Introduction

In 2010, 7,345,900 housing units were recorded in Malaysia with a 3.2% average growth rose rate compared to year 2000 [1]. According to [2] & [3], high population will increase potential for housing demand. Therefore, many researchers use the population as an indicator to predict the number of housing needs or demands for the long term [4, 5, 6]. In Malaysia, the projection of housing is made for immediate, medium and long term based on population growth and household size [4]. However, based on the report issued by Valuation and Property Services Department [7] & [8], there are a lot of numbers overhang and unsold properties recorded around the country. Under that report, Johor is identify retained the highest number of overhang in the country.

Pozdena [7] highlights the assessment of housing demand or needs is not just about the number of population growth, the most importance thing that need to know is related to individual (residents).

Thus, to understand the behavior of housing demand, the process of household formation, number of households and the composition of the household is necessary to be known before determining housing demand [7].

According to [8], if population increase, the households formation is expected to be increased, but in different rates. This happen because household formation was influenced by the composition of the adult population, rather than population growth only consider the numbers of people. Headship rate method commonly used to determine the composition of the adult population. Generally, the definition of headship rate can be determine as the ratio of heads to the size of the adult population outside of group quarters [9]. In Malaysia, head of household is based on who give high contributes in the economy of a household, and it is basically related to men [10]. In this paper, it should be highlighted that the headship rate multiplied by the corresponding population in the number of households by age group are key variables for the analysis of the potential demand which reflect to homeownership.

2. Population, household formation and headship rate

Refer to [11] and the Government of Uganda reports [12], the number of housing units required are usually measured through population growth rates. National Economic and Social Council [13], in its report states, the change in household number due to population growth and migration rates affecting the number of household formation and housing demand. While [14] found the growth of population and the characteristics of household formation will increasing of housing demand.

Reference [15], explains the things that must be considered before projecting housing demand is population growth. Clara [16], highlights the number of housing in an area affected by the number of household formation, which leads to the growth of housing demand. Therefore, extensive research were made by researchers to see the relationship between housing demand and household formation. The findings shows that there is an association between population, household formation and age levels in determining housing demand [17],[18] & [19].

The level of household composition among young adults living families (parents) had an impact on home ownership and the number of new housing construction [20]. Results from the study shows, the percentage of young adults aged 18 to 34 years old living with family (parents) increased from 27% to 31% (approximately 22 million people) from 1990 to 2006. Those age group traditionally been the biggest contributor in the purchase of the first house [20]. Because of that, the rate of household formation and housing demand a bit low compare to other years.

The prototype of headship rate has begun to be used in the United States since 1938 to project the number of households with the assumption that the headship rate is constant throughout the year [21]. Until 1993, the headship rate projection was used by institutions and researchers at least 15 countries [22]. In recent decades, many researchers come out with household modelling and household projecting methods such as dynamic macro-demographic models, micro-simulation and probabilistic household forecast [23]. Despite this, however, statistical agencies widely use the simplest method by headship rates and its extensions. The headship rate method were most often used by researchers and institutions in determining household formation because it is not too complex in term of understanding and an analysis [24]. In addition, this method have advantages over other methods, because it reflects with expected changes in the population age structure and sex [25]. Due to the changing of time, this method is suitable to use for a long term projection.

3. The conceptual framework

3.1 Headship rate method

The headship rate method is based on the assumption that the number of households is equivalent to the number of housing demand. Basically, the determination of headship rates can be made in a various ways including age-populations, age-sex, age-ethnic and age-marital status [26]. According to [17] & [18], in order to find the headship rate, only one census data is needed, where the number of households is classified according to the age of the household head. In this paper, headship rate is

IOP Conf. Series: Journal of Physics: Conf. Series **1019** (2018) 012005 doi:10.1088/1742-6596/1019/1/012005

determine using age-specific headship rates decomposed by household size suggested by [23] using equation (1).

$$HSt = \frac{Ht}{Pt} \tag{1}$$

Where,
HSt = headship rate
Ht = number of head in specific group
Pt = number of population in specific group

A total of 2% population census data in Johor were used in determining the headship rate for year 2000 and 2010 respectively (obtained from Department of Statistics, Malaysia). For the first step, manual calculation was used to calculate exactly number of head of household in each age groups for specific data census. Then, both data census (2000 & 2010) were used to make projection headship rate.

3.2 Projection headship rate

The constant headship rate method was widely used in various government agencies and institutions in 1950s to determine household and family projections using one census data [22]. While Statistics Research Agency [27], has been using the modified two-point exponential method in determining projection headship rate in United Kingdom. Through that method, projections headship rate are determined through two years of census data (current and base year). In this paper, modified two-point exponential method has been implemented because availability of data and it seen more accurate because based on current changing of data. The general equation for the two-point exponent modification method is shown in equation (2).

$$y_i = k + ab^{x_i} \tag{2}$$

Where,
i = the year of the projections
y_i = headship rate in year i
k = 1 if the current household head rate is ≥ base year
0 if the current household head rate is ≤ base year
a = headship rate of base year - k
b = (headship rate year i – k)/(headship rate base year – k)
x_i = (i –base year)/(current year-base year)

3.3 Classification of data

Based on data released by the Department of Statistics, Malaysia through population quick info, age distributions are listed by groups (0 to 80+ years) with 5 years intervals [28]. Therefore, same approach have used in this paper. For initial age groups set up, some modification has been made in line with the studies conducted by [25], [27], [28], [29] that using 15 year old. The age selection was made to counter overall level of headship rate for all backgrounds including the younger generation (15-24), working age (25-65) and the elderly (65 years and above) [31].

3.4 Period of projection

The Population and Housing Census in Malaysia is conducted once in every 10 years [32]. Therefore in projecting headship rate, the data was set until 2020 because the next census is expected to be carried out in 2020. Once the new census data is ready, the process will be repeated with 2020 as a current year and 2010 as a base year.

4. Results and discussions

4.1 Headship Rate

Figure 1 shows the pattern of headship rate for each group with difference census data (2000 and 2010) in Johor.

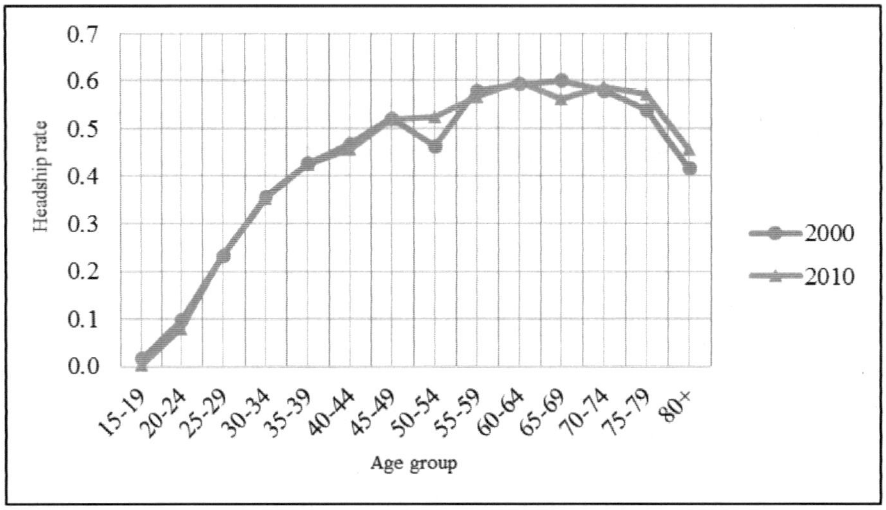

Figure 1. Headship rate for census data 2000 and 2010

Referring to Figure 1, the age group 15-19 years old recorded the lowest headship rate for both census data with an average of 0.011 (1.1%). Based on that figure, it can be noted that the probability of a person between the ages of 15-19 to be homeownership is only 1% in that particular group age. This happen because most people in that particular group age are still in school and do not have stable job and income to buy a house. For age groups 20-24 until 40-44, the pattern of headship rate start increasing without any significant difference between year 2000 and 2010. Means that, probability to become homeownership for those groups are remaining same. The percentage of average headship rate for group 45-49 until 75-79 is very high, which is more than 0.5 (50%) recorded until it reach almost 0.6 (60%) for age group 65-69. Furthermore, the pattern of headship rate declining until it reached 0.41 (41%) for age groups over 80 years.

4.2 Projection headship rate

Using the data calculation from figure 1, the projection of headship rate for 5 years were determined up to year 2020 as shown in Figure 2. Example calculation for headship rate group 15-19 years old for year 2018 using equation (2) is shown below:

Data from headship rate analysis (Figure 1),
Headship rate year 2000 (base line) = 0.016
Headship rate year 2010 (current year) = 0.002

$$y_{2018} = k + ab^{x_i}$$

Where,
$k =$ 0 (Headship rate decreased over the period)
$a =$ 0.016 – 0 = 0.016

b = (0.002-0)/(0.016-0) = 0.125
x_i = (2018-2000)/(2010-2000) = 1.8

Therefore,
y_{2018} = 0+0.016*0.1251.8= 0.000379 ~ 0.0004

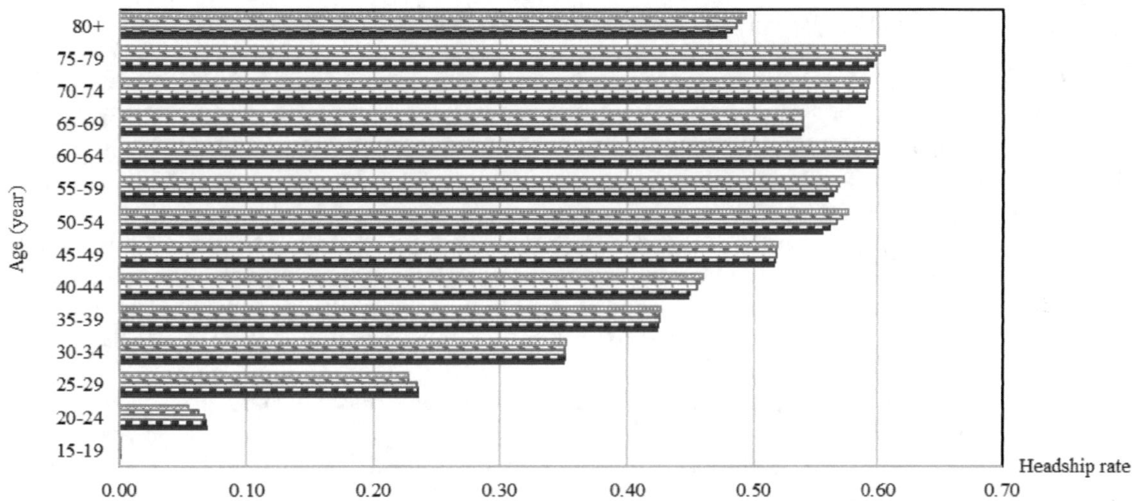

	15-19	20-24	25-29	30-34	35-39	40-44	45-49	50-54	55-59	60-64	65-69	70-74	75-79	80+
2020	0.00	0.05	0.23	0.35	0.43	0.46	0.52	0.58	0.57	0.60	0.54	0.59	0.61	0.49
2019	0.00	0.06	0.23	0.35	0.43	0.46	0.52	0.57	0.57	0.60	0.54	0.59	0.60	0.49
2018	0.00	0.07	0.23	0.35	0.43	0.46	0.52	0.57	0.57	0.60	0.54	0.59	0.60	0.49
2017	0.00	0.07	0.24	0.35	0.42	0.45	0.52	0.56	0.56	0.60	0.54	0.59	0.60	0.48
2016	0.00	0.07	0.24	0.35	0.42	0.45	0.52	0.56	0.56	0.60	0.54	0.59	0.59	0.48

Figure 2. Projecting headship rate

Referring to Figure 2, the range of increments and decrements of headship rate from actual data (Figure 1) is from 0 to 0.61. For age group 15-19 and 20-24 are found decrease dramatically throughout the years 2016-2020. Meanwhile, for age group 30-34 years old, headship rate are estimated almost same for all projection years. However, for age group 40-44, 50-54, 55-59 75-79 and 80+ years old and above, dramatic increment can be seen based on projection years. Therefore, it is expected percentage of homeownership is dominated by these groups.

5. Conclusion

Based on projection headship rate, it can be seen that younger group especially for 20-34 years old struggle to own a house. Projection headship rate in age group 40-44 is higher, means more residents can only afford a house when becoming 40 years old and above. This increase the occupancy of squatter houses despite the main objective of the National Housing Policy (DRN) is to provide affordable housing to young married couple and those who aged 30 years and below. Therefore, findings of this research is hoped could be used as a basis for the government to plan for reasonable amount of housing units to accommodate the demand of different group of age.

6. Recommendation

This paper would like to recommend that; (1) Headship rate – It is better to have specific age reflect to actual housing demand instead of age group; (2) Age range – Some of age is not give effect to headship rate can be neglected because it not give any significant effect.

Acknowledgement

This research is fully supported by UTHM and MOSTI. The authors fully acknowledged Ministry of Higher Education (MOHE) and Ministry of Science, Technology and Innovation for the approved Special fund under VOT A046 for which makes this important research viable and effective.

7. References

[1] Kementerian Perumahan dan Kerajaan Tempatan, "KPKT Selected Statistics Until 31 March 2015," Malaysia, 2015.

[2] A. Karantonis, "Population growth and housing affordability in modern city - Sydney a case study." pp. 1–16, 2007.

[3] and J. N. Zeng, Yi, Lan Li, Zhenglian Wang, Helin Huang, "Effects of Changes in Household Structure on Future Housing Demand in Hebei Province , China," no. 2, pp. 85–111, 2013.

[4] N. Sharipah, M. Zulkifli, A. Halim, and H. Abd Rahim, "A Realistic Model to Determine Housing Need in Malaysia: Study of Johor Bahru, Kuala Lumpur and Penang," Malaysia, 2012.

[5] Jabatan Penilaian dan Perkhidmatan Harta, "Laporan Status Pasaran Harta Tanah Q4 2015," Putrajaya, 2015.

[6] Jabatan Penilaian dan Perkhidmatan Harta, "Laporan Status Pasaran Harta Tanah Q4 2014," Putrajaya, 2014.

[7] R. J. Pozdena, *The Modern Economics of Housing: A Guide to Theory and Policy for Finance and Real Estate Professionals*. Greenwood Publishing Group., 1988.

[8] E. S. Belsky, R. B. Drew, and D. Mccue, "Projecting the Underlying Demand for New Housing Units : Inferences from the Past , Assumptions about the Future," no. November, 2007.

[9] A. Paciorek, "The Long and the Short of Household Formation," *Real Estate Econ.*, vol. 44, no. 1, pp. 7–40, 2016.

[10] Lembaga Penduduk dan Pembangunan Keluarga Negara, *Laporan Penemuan Utama Kajian Penduduk dan Keluarga Malaysia Kelima (KPKM-5) 2014*. 2016.

[11] N. Ahmad, "A Joint Model of Tenure Choice and Demand for Housing in the City of Karachi," *Urban Stud.*, vol. 31, no. 10, pp. 1691–1706, 1994.

[12] Government of Uganda, "State Of Uganda Population Report 2007: Planned Urbanization for Uganda's Growing Population," 2007.

[13] National Economic and Social Council, *Housing in Ireland: Performance and Policy*, no. 112 November. Ireland: National Economic & Social, 2004.

[14] Canada Mortgage and Housing Corporation, "Canadian Housing Observer 2003," 2003.

[15] W. S. Thompson, "Population Growth and Housing Demand," *Ann. Am. Acad. Pol. Soc. Sci.*, vol. 190, pp. 131–137, 2013.

[16] H. M. Clara, "Population and Housing : A Two-Sided Relationship," *Demogr. Res.*, vol. 15, no. 13, pp. 401–412, Nov. 2006.

[17] T. M. Crone and L. O. Mills, "Forecasting trends in the housing stock using age-specific demographic projections," *J. Hous. Res.*, vol. 2, no. 1, pp. 1–20, 1991.

[18] J. Ermisch, "The demand for housing in Britain and population ageing: microeconometric evidence," *Economica*, pp. 383–404, 1996.

[19] T. Dunne, "Household formation and the great recession," 2012.

[20] B. Ritholtz, "Why Are Young Adults Living with Their Parents and When Will They Move Out," 2014. [Online]. Available: http://ritholtz.com/barry-ritholtz-curriculum-vitae-2/. [Accessed: 10-May-2016].

[21] P. C. Glick, "American Families," in *Census Monograph Series*, New York: Wiley, 1957, p. 164.

[22] United Nations, "Headship Rate Method," in *Population Models, Projections and Estimates*, S. D. C. Donald J. Bogue, Eduardo E. Arriaga, Douglas L. Anderton, George W. Rumsey. Chicago, Illinois, Ed. Department of Economic and Social Affairs, 1993.

[23] J. Leiwen and B. C. O. Neill, "Toward a New Model for Probabilistic Household Forecasts," vol. 72, no. 1, pp. 51–64, 2014.

[24] A. Mason and R. Racelis, "A Comparison of Four Methods for Projecting Households," *Int. J. Forecast.*, vol. 8, pp. 509–527, 1992.

[25] M. V. George, S. K. Smith, D. A. Swanson, and J. Tayman, "Population projections," in *The Methods and Materials of Demography*, Elsevier Academic Press, 2004.

[26] United Nation, *Methods of Projecting Households and Families Methods of Projecting*. 1973.

[27] Statistics Research Agency, "Household Projections Across the United Kingdom Technical Report," 2011.

[28] Department of Statistic, "Population Quick Info," 2015. [Online]. Available: http://pqi.stats.gov.my/searchBI.php.

[29] J. Leiwen and B. C. O. Neill, "Household Projections for Rural and Urban Areas of Major Regions of the World," 2009.

[30] L. Goodman, R. Pendall, and J. Zhu, *Headship and Homeownership What Does the Future Hold?*, no. June. Urban Institute, 2015.

[31] United Nations, "Conceptual Issues and Methods for Preparing Demographic Projections," in *Projection Methods for Integrating Population Variables into Development Planning*, 1989, pp. 121–196.

[32] Department of Statistic Malaysia, "Population and Demography," 2017. [Online]. Available: https://www.dosm.gov.my/v1/index.php?r=column/cone&menu_id=ZHJlbWFBSTVEcHY1ait6akR3WmtVUT09. [Accessed: 18-Jul-2017].

IOP Conf. Series: Journal of Physics: Conf. Series **1019** (2018) 012006 doi:10.1088/1742-6596/1019/1/012006

A review of data collection approaches in linear wireless sensor networks (LWSNs)

Yousif Khalid Yousif[1], R Badlishah[2], N. Yaakob[1], A Amir[1]

[1]School of Computer and Communication Engineering (SCCE) Universiti Malaysia Perlis (UniMAP), 02600 Arau, Perlis, Malaysia
[2]Deputy Vice Chancellor (Research and Innovation), Universiti Sultan Zainal Abidin (UniSZA) Kampus Gong Badak, 21300 Kuala Nerus, Terengganu Darul Iman, MALAYSIA.

Yousef_khalid88@yahoo.com,{badli, naimahyaakob, amizaamir}@unimap.edu.my

Abstract. Using wireless sensor nodes in monitoring systems have attracted significant attention over the past few years. Wireless sensor networks (WSNs) are key technology that extensively applied in many fields, such as transportation, health-care and environment monitoring. Different applications of WSN require different structure of nodes deployment form. For example, pipelines, border and Highway monitoring system, these applications have produced a special class of WSN in which nodes deployed in linear structure to form a linear wireless sensor network (LWSN) that extended to long distance area. In this paper, Applications and classification of LWSN are presented. Also, related works of data collection approaches in LWSN are provided, which classified into two categories named multi-hop data collection approach and mobile data collection approach (MDC). In addition, cons and pros in the existing approaches and future trends are also discussed

1. Introduction

In recent years, evolution and advance in micro electro-mechanical systems (MEMS) has enabled the development of low cost, low-power and small size micro sensors [1]. A wireless sensor network can be described as a composition of tiny sensor nodes that communicate wirelessly with each other to sense some specific actions in an interested field. Each sensor node consists of main components[2]: processing capability, memory, RF transceiver, power source and in some cases equipped with GPS. WSNs have deployed in different fields in order to provide online insights into the conditions, demands and performance of the observed infrastructure. WSN can be deployed in different structure of network topology according to applications used in. In some critical applications, WSN can form a linear or semi linear network topology which is called Linear Wireless sensor network (LWSN). Linear Wireless sensor Networks (LWSN) is one of special classes of network topology for wireless sensor networks [3]. In addition, LWSNs have unique features such as long distance and narrow network compared to other network topology. LWSNs have been used in different monitoring applications such as border, river, railway, power line cable, road conditions and etc[4]which will be explained in details in next sections. In LWSN, sensor nodes are distributed in linear structure with

narrow region along the network with distance reaches to hundreds of kilometres [5]. Therefore, it's not possible to communicate with the sink in single hop form. Moreover, traffic generated by far sensors from the sink will have to use other nodes to transfer their generated traffic. Therefore data traffic will be delivered within significant delay[6]. Proposed schemes for balancing energy in WSNS are not suitable for LWSN as nodes in LWSN have unique characteristic [6]. Nodes near the sink deplete their energy quickly as they deal with heavy traffic compared with other nodes in the network[3]. Load balancing between sensor nodes in LWSN is challenge For example network structure of LWSN is not like mesh topology as only few nodes can communicate with sink node.
Due to restricted constraints WSNs such as energy, memory capacity and processor in general, it is very important to reduce energy consumption and provide reliable data collection in LWSN. Therefore different works have been proposed to optimize the data collection in LWSN by using different techniques. However, there are many challenges and issues need to be taken into consideration while developing data collection scheme for LWSN. Therefore, this paper is presented to give an insight of LWNS. In addition, different approaches that aimed to provide energy efficient and reliable data collection in LWSN are surveyed. Moreover, we find out that the related works can be categorized into two categories named multi-hop approach and mobile data collection (MDC) approach, each approach used in data collocation in LWSN has its own cons and pros which are also discussed. Finally, some future trends also provided in order to improve the existing schemes.

2. Classification of LWSN

In this section presents a classifications LWSN which can be classified into different categories according to topological point of view as follows [5]

I. **Thin LWSN:** it describes a LWSN structure when nodes are lined up in one dimensional geographic area. Also, Sensor nodes can be distributed as straight line or curved line depending of network topology.
II. **Thick LWSN:** it means that sensor nodes are distributed into two or three dimensional geographic area.
III. **Very thick:** Sensor nodes are distributed randomly and lined up in between two lines.

3. Applications of LWSN

Some of the applications of WSN where nodes deployed in linear structure include.

3.1. Highway Road Monitoring

Using WSN for Highway Road monitoring with less human efforts is very important in our daily life. Many sensors are lined up along a road for detecting traffic jam, crossing the limited speed, and the continued monitoring of the road. A WSN can work during different weather condition such as fog or presence of dust, rain and even at night[7]. Using sensor nodes in road monitoring can lead to collect real information on road situation which is very necessary. By analyzing the collected information, traffic jam and car accidents can be avoided and reduced [8-9]. Another use of road monitoring is that to give an alarm to driver when he/she crosses the speed limit or any accident happened along the road [6].

3.2. Monitoring of AC power lines

Collecting and monitoring the environmental information of the long-distance transmission electrical lines by the WSN technique is one of the important application. In monitoring system of power line, information which is collected by sensor node would be valuable for the utility company to diagnosis the problems that can happen caused by a certain failure in the equipment or overusing and loading of AC power lines. By monitoring a power line conditions, such as voltage, temperature, humidity, and others conditions, provider companies can able to anticipate any equipment failures and unexpected problems. This result in creating a better maintenance schedules and improving company productivity and early predication of any equipment failure, therefore maintenance cost is reduced [6][10].

3.3. *Border Monitoring*

Border monitoring [11]-[12] is one of the main concerns of every country in order to ensure of security and protection from any illegal entrance or intruders. The long border has made a border monitoring is very difficult matter and costly, Therefore, the advancement of technology has released WSN which can play an important role in such application. WSN monitoring system in border surveillance, applications, focuses on collecting information by using different types of sensors, such as seismic, camera, thermal camera, and motion detectors. Thus, sensor nodes can form a line control and notify if any intruders cross the predefined specific area. Finally, sensor nodes process this information and send to the central station which is responsible to take an appropriate defense action.

3.4. *Oil and gas Pipeline Monitoring*

In most petroleum producer countries there are wide network of pipelines carrying oil and gas. These networks involve many parts from pumps, control station, pipelines and other equipment. As oil and gas have key role in world's economy, managing and controlling the transportation mainly depends upon huge pipeline. Pipeline infrastructures need to be monitored frequently to know the pressure of the oil and gas, temperature, the flow rate, quantity flown and other necessary measurements. The knowledge of these measurements is very essential to avoid failures or to know places of failures and other such as leakages. The place of measurement should be along the pipelines and involve remote area that cannot protect or monitored always. Thus the best solution to such problem is the use wireless sensor network along the pipelines transportation.

3.5. *Railroad/subway monitoring*

Railroad and subway [6] are one of most important infrastructures of each county which serves to support the transportation service. Security and safety of this infrastructure is the most concern in many countries. Therefore, many researchers have investigated the use of WSN in railway and subway infrastructure to monitor the health condition. For example, sensors can monitor dynamic strain caused by the passing trains as well as provide early detection of critical and dangerous cracks.

3.6. *Some of Structural Health Monitoring*

Another area of interest where sensor nodes can be used in Structural Health of tunnel[13], road [8] and bridge [14] in order to provides a costly and timely inspection compared to the typical structural safety assessment which depends on manual and visual inspection. Moreover, damage can't be seen when it located inside the structural elements, thus using sensor nodes in SHM systems is very important to check the structural integrity and condition in an automatic, efficient manner and remotely

3.7. *River Monitoring*

In recent years, environmental monitoring has become one of important concerns. To provide a real time, efficient, and economic monitoring, hence WSN has been used to for a river application monitoring to provide some important measurements regarding to river application, instead of using the traditional way of measurements which can be time consuming and costly. For example, sensor node can measure the level of the water and river pollution to avoid any further risks and potential hazards of natural disasters[15][16].

IOP Conf. Series: Journal of Physics: Conf. Series **1019** (2018) 012006 doi:10.1088/1742-6596/1019/1/012006

4. Related works

For data collection in LWSN, there are different approaches have been done by number of researchers to tackle the obstacles in data collection for this type of network structure. These approaches can be categorized into multi-hop and mobile data collection approaches. For data collection in LWSN based Multi hop approach, authors in [17] proposed a hierarchical network architecture where three types of nodes are introduced and defined with different functionalities named basic sensor nodes (BANs), data relay nodes (DRNs) and data dissemination nodes (DDNs). Each kind of nodes perform different task according to its function. BANs are the basic node which is responsible for performing the sensing function and transmit their sensory data to closest DRNs. Where, DRNs perform data gathering for the data generated from BANs. The distance between these nodes is specified by the communication range of the networking MAC protocol used. Subsequently, DRNs nodes transmit their data to DDNs which is in charge of transmitting the collected data to the Network control centre. Figure 1 shows the hierarchical relationship between the different types of nodes in the networks. The technology which can be used to discharges the data from DDNs to the network control centre can vary such as Satellite cellular technology.

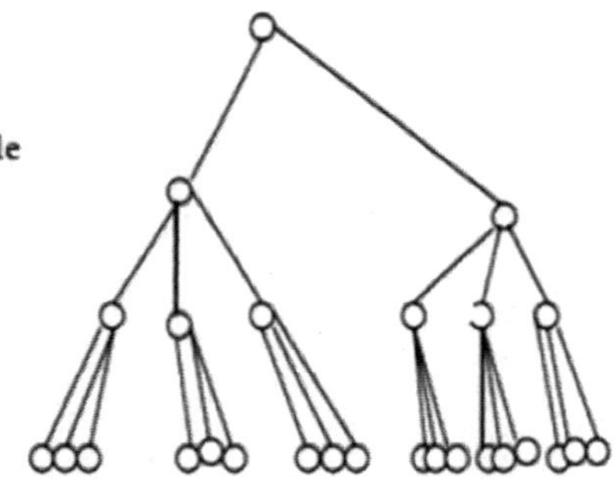

Figure 1. A hierarchical representation of the linear structure sensor network

In providing energy efficiency and avoid hotspot problem in LWSN, the authors in[18] proposed node placement scheme for LWSN. In this scheme, network is divided into multiple distribution points which can be represented as clusters and there is one cluster head (CH) in each. The optimal spacing between each distribution points calculated and it shows that the distance between the distribution points is getting closer as distance to the BS getting close. In addition, optimal density of nodes in each distribution point is found and it shows that the differences between the numbers of nodes in each distribution point is very smaller relatively to the distance from BS. To reduce the communication distance in LWSN, a new liner cluster handling (LCH) technique towards energy efficiency in LWSN with multiple static sink is proposed [19]. Authors assumed that a single static sink in LWSN reduces the lifetime and throughput of the network. Therefore, three statistic sinks and clustering method are used to provide energy efficiency. Also, network topology is divided into four equal regions and equal number of nodes. Sinks are distributed into two at the corner and one at the centre of the network. Moreover, CH is formed in each region to ultimately reduce the energy consumption and the communication distance. CH is selected randomly. Subsequently, CH performs data aggregation for the data received from member nodes and then transmits data to the nearest sink to it. In [20], the authors proposed a Reliable and Energy-efficient Chain-cluster based routing protocol (REC). This protocol aimed to provide a multi-hop reliability and energy efficient. The operation of routing protocol is divided into three phases where in the first phase the network is divided into multiple clusters and this is done by base station (BS). Subsequently, in the second BS selects the node with the maximum link reliability and energy is selected as cluster head. Proper cluster size is also taken into consideration in REC to provide a load balancing between the clusters. Moreover, to ensure a reliable data transmission between CHs and sensor nodes, BS associates the nodes to CHs according to link reliability between them. After the end of the cluster head selection and node association phase, the chain cluster formation is constructed between the CHs. The CH which has highest energy and can communicate directly with the base station is selected as Chain head. Also, in case of any CH cannot communication with next CH, relay node is introduced to works intermediate node between the two CHs. However, results of this protocol shows that REC outperforms a number of approaches such as LEACH and PEGASIS protocol

While in MDC based LWSN , researchers in [5]proposed a mobile data collection (MDC) based LWSN where mobile node visits every sensor nodes in the network and perform data collection from sensor nodes and then carry the data collected to the sink. Mobile node can be easily deployed and utilized for data collection in LWSN as the network takes a liner thus the route the need to followed by Mobile node is already known while others topologies such as mesh mobile node needs to communicate with senor nodes to specify the followed route. Also, another work for using mobile data collection is presented in [3],[21] where authors proposed to use Unmanned Aerial Vehicles (UAVs) for data collection in LWSN. In their works, different nodes types are referred as sensor nodes (SNs), relay nodes (RNs), UAVs, and sinks as shown in figure 2.

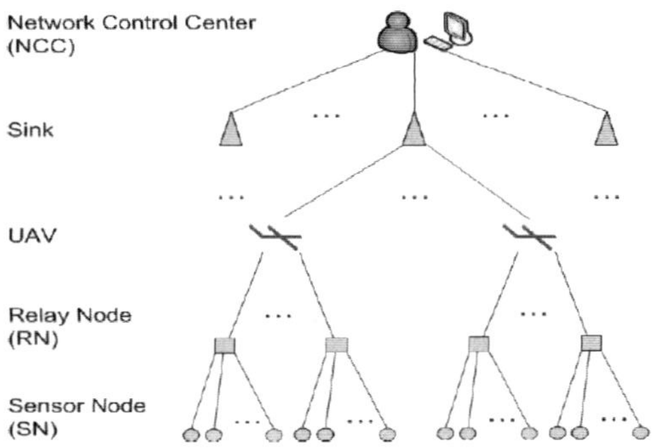

Figure 2. The various nodes at the different levels

Each type has specific function and role. The first type is sensor nodes (SNs), theses nodes represent the ordinary sensor nodes in the network. SNs transmit the sensed data to the nearest RNs. Communication between RNs and sensor nodes is done either directly or in multi-hop form, where RNs nodes act as cluster heads that collect data from the ordinary sensor nodes. Subsequently, RNs communicate with UAV for data discharging when it comes in its communication range. UAV carries the data collected from all RNs to the. Sink node which in charge to transmit the collected data to the network control centre. However, different speeds for UAV are utilized to experiment the end-to-end delay, buffer overflow of RNs.

5. Discussion and future trends

Using sensor nodes in critical infrastructure protection applications such as pipeline for gas and oil requires a high reliability and a long life working for sensor node. As reliable data collection in LWSN plays an important role in success of the event detection application which requires collecting all the data without loss. For example, an event of moving any object in border monitoring system must be reported to the network control centre so that an action can be taken.

Data collection LWSNs is categorized into multi-hop approach and MDC approach in previous section. Each approach has its own cons and pros. The table below summaries the pros and cons in term of energy efficiency, delay and packet dropped.

IOP Conf. Series: Journal of Physics: Conf. Series **1019** (2018) 012006 doi:10.1088/1742-6596/1019/1/012006

Table 1. Comparison between Multi-hop and MDC

Approach	energy consumption	end-to-end delay	packets drop
multi-hop	Medium	Low	Medium
MDC	Low	High	High

From Table 1, it can be seen that using multi-hop approach could lead to deplete the energy of sensor nodes quickly in LWSN as network topology could be extended to hundreds of kilometres and data packets need to be routed all long way in order reach the base station while using MDC could prolong the network lifetime and reduce the energy consumption. However, LWSN based MDC can result in significant end to end delay as also the UAV needs to collect data either from each sensor nodes individually or from RNs, as routing the data packets in multi-hop form can lead to reduce the end-to-end delay and packets drop. In MDC, sensor nodes require to keep the data packets in their buffer till UAV comes within their communication range and this can results in buffer overflow due to the limited memory capacity of sensor nodes To enhance data collection in LWSN and solve the above issues, there are some future trends for both approaches, which are listed as follows.

- For multi-hop approach in LWSN, it is very good to reduce the distance of network by using multiple sink nodes so that the hotspot problem will be avoided and this can also reduce the number of hops which leads to reduce the energy consumption. For example in [19] multi-sink technique is used, but the CHs selection mechanism is performed randomly which can reduce the network lifetime, therefore using multi-sink with [22] where as CHs selected according to different parameter can provide better network performance. Also, sensor nodes may suddenly suffer from energy depletion or hardware damages and this result in disturb the connectivity of network especially in multi-hop approach where there is no alternative route. Therefore, a fault tolerance scheme in this work should be considered.

- MDC approach offers a better performance in term of energy consumption, but high end-to-end delay and number of packets dropped. Therefore dividing the network into multiple areas and deploying one UAV for each area can enhance the data collection in LWSN. Also, it is very necessary to take into consideration the speed of UAV as a parameter while dividing the network into multiple areas. In addition, packets dropping caused by buffer overflow reduces the reliability of monitoring system thus using a buffer management algorithm like in [23] could solve this problem.

- To provide energy efficiency and avoid hotspot problem with minimum end-to-end delay data collection in LWSN, hybrid approach can be used such as[24]. In this hybrid approach, data collection is done using both mobile data collection and multi-hop approach which results in better performance from using each approach individually.

6. Conclusion

The liner structure form of sensor nodes in applications like pipelines and bridge monitoring system and the long distance topology pose some challenges in data collection from sensor nodes. Recent works for data collection in LWSN were based on either Multi-hop approach or MDC. The Multi-hop approach, showed a good results in term of end-to-end delay and packets delivery ratio, but it is very costly in term of energy consumption while minimum energy consumption and significant end-to-end delay and packets dropped when using MDC. Therefore, some future trends are suggested such as using multiple sink nodes, deploying multiple UAVs and hydride approach aimed to improve the data collection in LWSN.

References

[1] I. F. Akyildiz, W. Su, Y. Sankarasubramaniam, and E. Cayirci, "Wireless sensor networks: a survey," *Computer Networks*, vol. 38, no. 4. pp. 393–422, 2002.

[2] H. Karl and A. Willig, *Protocols and Architectures for Wireless Sensor Networks*. Chichester, UK: John Wiley & Sons, Ltd, 2005.

[3] I. Jawhar, N. Mohamed, J. Al-Jaroodi, and S. Zhang, "A Framework for Using Unmanned Aerial Vehicles for Data Collection in Linear Wireless Sensor Networks," *J. Intell. Robot. Syst.*, vol. 74, no. 1–2, pp. 437–453, 2013.

[4] C. M. Imran, M. Aldukhail, N. Almezeini, and M. Alnuem, "Potential Applications of Linear Wireless Sensor Networks : A Survey," vol. 4, no. 6, pp. 183–200, 2016.

[5] I. Jawhar, M. Ammar, S. Zhang, J. Wu, and N. Mohamed, "Ferry-based linear wireless sensor networks," *GLOBECOM - IEEE Glob. Telecommun. Conf.*, pp. 304–309, 2013.

[6] I. Jawhar, N. Mohamed, and D. P. Agrawal, "Linear wireless sensor networks: Classification and applications," *J. Netw. Comput. Appl.*, vol. 34, no. 5, pp. 1671–1682, 2011.

[7] M. Collotta, G. Pau, V. M. Salerno, and G. Scatá, "Wireless Sensor Networks to Improve Road Monitoring," 2012.

[8] M. Collotta, G. Pau, V. M. Salerno, and S. G., "A novel road monitoring approach using Wireless Sensor Networks," *Proc. - 2012 6th Int. Conf. Complex, Intelligent, Softw. Intensive Syst. CISIS 2012*, pp. 376–381, 2012.

[9] M. A. Kafi, Y. Challal, D. Djenouri, M. Doudou, A. Bouabdallah, and N. Badache, "N/A - A Study of Wireless Sensor Networks for Urban Traffic Monitoring: Applications and Architectures," *Procedia Comput. Sci.*, vol. 19, pp. 617–626, 2013.

[10] J. Yu and X. Zhang, "A Cross-Layer Wireless Sensor Network Energy-Efficient Communication Protocol for Real-Time Monitoring of the Long-Distance Electric Transmission Lines," 2015.

[11] E. Felemban, "Advanced Border Intrusion Detection and Surveillance Using Wireless Sensor Network Technology," *Int'l J. Commun. Netw. Syst. Sci.*, vol. 06, no. 05, pp. 251–259, 2013.

[12] G. S. Sankar and S. Angadi, "Wireless Sensor Network for Border Monitoring 1, 1," pp. 65–68, 2013.

[13] B. He and G. Li, "PUAR : Performance and Usage Aware Routing Algorithm for Long and Linear Wireless Sensor Networks," vol. 2014, 2014.

[14] A. Yin, B. Wang, Z. Liu, and X. Hu, "Design and Implementation of the Structure Health Monitoring System for Bridge Based on Wireless Sensor Network," *Lect. Notes Comput. Sci.*, vol. 5553, pp. 915–922, 2009.

[15] T. R. Wagh and A. P. Rao, "River Monitoring an Application for Wireless Sensor Network Platform," vol. 4, no. 11, pp. 511–515, 2014.

[16] L. C. Degrossi, G. G. Amaral, E. S. M. De Vasconcelos, J. P. De Albuquerque, and J. Ueyama, "Using Wireless Sensor Networks in the Sensor Web for Flood Monitoring in Brazil," *Proc. 10th Int. ISCRAM Conf. – Baden-Baden, Ger. May 2013*, no. May, pp. 458–462, 2013.

[17] I. Jawhar, N. Mohamed, K. Shuaib, and N. Kesserwan, "Monitoring linear infrastructures using wireless sensor networks," *IFIP Int. Fed. Inf. Process.*, vol. 264, pp. 185–196, 2008.

[18] L. Hong and S. Xu, "Energy-efficient node placement in linear wireless sensor networks," *2010 Int. Conf. Meas. Technol. Mechatronics Autom. ICMTMA 2010*, vol. 2, pp. 104–107, 2010.

[19] M. Sajid, K. Khan, U. Qasim, Z. a. Khan, S. Tariq, and N. Javaid, "A New Linear Cluster Handling (LCH) Technique Toward's Energy Efficiency in Linear WSNs," *2015 IEEE 29th Int. Conf. Adv. Inf. Netw. Appl.*, vol. 7, pp. 389–393, 2015.

[20] Z. Taghikhaki, N. Meratnia, and P. J. M. Havinga, "A reliable and energy-efficient chain-cluster based routing protocol for Wireless Sensor Networks," *2013 IEEE Eighth Int. Conf. Intell. Sensors, Sens. Networks Inf. Process.*, pp. 248–253, 2013.

[21] I. Jawhar, N. Mohamed, and J. Al-jaroodi, "Data Communication in Linear Wireless Sensor Networks Using Unmanned Aerial Vehicles," no. 31, 2014.

[22] Z. Taghikhaki, N. Meratnia, and P. J. M. Havinga, "A reliable and energy-efficient chain-cluster based routing protocol for Wireless Sensor Networks," in *Proceedings of the 2013 IEEE 8th International Conference on Intelligent Sensors, Sensor Networks and Information Processing: Sensing the Future, ISSNIP 2013*, 2013, vol. 1, pp. 248–253.

[23] P. Söderman, K.-J. Grinnemo, M. Hidell, and P. Sjödin, "Mind the SmartGap: A Buffer Management Algorithm for Delay Tolerant Wireless," *Proc. Eur. Conf. Embed. Wirel. Syst. Networks*, pp. 104–119, 2015.

[24] B. Nazir and H. Hasbullah, "Energy efficient and QoS aware routing protocol for Clustered Wireless Sensor Network," *Comput. Electr. Eng.*, vol. 39, no. 8, pp. 2425–2441, 2013.

IOP Conf. Series: Journal of Physics: Conf. Series **1019** (2018) 012007 doi:10.1088/1742-6596/1019/1/012007

An Energy Efficient and Load Balancing Clustering Scheme for Wireless Sensor Network (WSN) Based on Distributed Approach

Yousif KhalidYousif[1], R Badlishah[2], N. Yaakob[1], A Amir[1]

[1]School of Computer and Communication Engineering (SCCE)Universiti Malaysia Perlis (UniMAP), 02600 Arau, Perlis, Malaysia
Yousef_khalid88@yahoo.com,{badli, naimahyaakob, amizaamir}@unimap.edu.my
[2]Deputy Vice Chancellor (Research and Innovation), Universiti Sultan Zainal Abidin (UniSZA) Kampus Gong Badak, 21300 Kuala Nerus, Terengganu Darul Iman, MALAYSIA.

Abstract. One of the most critical problem and challenging in Wireless Sensor Networks (WSNs) is to reduce energy consumption in order to prolong network lifetime of WSNs. Clustering technique is one of the techniques which have been used to provide energy efficiency.. However Most of clustering schemes select the cluster head either randomly without considered important parameters or based on centralized approach by utilizing the base station which can affect the network scalability. In addition, single hop communication is used by CHs to forward their sensed data to the CH which lead to increases energy consumption of CHs in large scale network. Therefore in this paper clustering scheme is proposed based on distributed approach, different parameters are consider for cluster head selection and also for multi-hop communication. Results shows that the proposed scheme have better performance in term of energy consumption and number of alive sensor nodes

1. Introduction

The design and advance in sensor technology such as low power CMOS technology, microprocessors and low power radio frequency (RF) have made it feasible to develop cheap tiny sensors with wireless network. These Sensors will be utilized for monitoring different environments for example, battle fields, tracking objects and construction distortion detection. One of the most critical issues in WSN is power source as powered by limited small batteries that can't be replaced or recharged. So having power efficient plan for delivering packets to BS is so important due to maximizing life time of network. One of the techniques used to reduce energy consumption and results in prolong network lifetime of WSN is to partition the network into clusters [1]. Clustering technique has a lot of advantages in comparison with flat routing protocols in WSNs, since it adds more scalability, less load, reduce energy consumption. Scalability since only CHs nodes are responsible for data dissemination thus the size of routing table is reduced at the individual sensor node. Also, sensor node can generate redundant data therefore in Clustering algorithm CH node aggregation the received data from its member's nodes using data aggregation method, which in turn help to reduce redundant data and thus reduce the size of the data packets thus energy is saved [2, 3]. In clustering schemes, there are mainly two major phases. In first phase, some nodes are appointed as CHs and formulate different cluster while second phase is usually related to data gathering and forwarding. Within each cluster, all member nodes transfer their sensory information towards their respective CH, which perform further processing (i.e. aggregation, compression, and scheduling and resource allocation) and forward their member data towards sink node. As CHs is responsible for several functions thus consume more energy compared to ordinary nodes[4]. Therefore, CH selection plays a vital role and has great impact on network lifetime [5]. However, Most of cluster mechanism selects the CH randomly without considered important parameters where this can lead to select the deficient node as CH. Others clustering mechanisms are based on centralized approach by utilizing the base station which can affect the network scalability as every time nodes require sending its parameters such as energy, number of neighbours to the BS which in turn leads to increases network overhead. Therefore, this paper presents an energy efficient and load balancing clustering scheme based on distributed approach. To optimize the cluster heads selection two parameters are consider for cluster head selection. In addition, multi-hop communication between CHs is employed to reduce the energy consumption in large scale network.

2. Related works

Low Energy Adaptive Clustering Hierarchy (LEACH), which was. proposed in [6] and it is well-known as LEACH protocol. LEACH was aimed to provide a solution to resolve the energy consumption in WSN and prolong the network lifetime. LEACH is a probabilistic based on both distributed and single-hop approach. It structures the network into clusters based on the strength of receiving signals. In LEACH, nodes are either ordinary SNs or CHs. Every SN sends their sensing data to its CH. In addition, CHs node work as gateways to the BS, Initially a node in LEACH protocol, generates a random number between 0-1 to decides which node to be a CH and this is done by computed a threshold value T (n), . Generally, LEACH provides a good model for energy consumption. To reduce energy consumption in [7] proposed Power efficient gathering in sensor information systems (PEGASIS) was also proposed to form a chain of sensor nodes starting from the farthest node to closest node towards base station. Each node sends and receives data from its neighbours and takes a turn being a leader for transmission towards the base station or destination. However, such a mechanism can produce high transmission delay with an increase in chain or network size. In [8] proposed LEACH-C each node sends its residual energy level and location information to the sink node at the beginning of each cluster formation,. In turns sink node elects CHs and builds clusters based on simulated annealing algorithm. Once the CHs and associated nodes are determined, the sink broadcasts CH IDs and Time Division Multiple Access (TDMA) schedules for each node. All nodes look for their IDs to be matched as the CH ID, if not matched then the nodes follow the Time Division Multiple Access (TDMA) schedule to broadcast their data. The data transmission phase of LEACHC is identical to that of LEACH. Using this scheme in clustering the network can form more uniform sized clusters, but it can suffer from scalability problem and high overheads. To improve LEACH protocol [9] proposed DBCH scheme where the distance from the base station and the energy of the node is considered while selecting the CHs. Based on the specified parameters for selecting CHs, the node which is nearer to the base station and has more residual energy will be elected as the cluster head. The proposed work shows better performance than LEACH in term of energy consumption which it. However, single hop communication is used by CHs to forward their sensed data to the CH which lead to increases energy consumption of CHs in large scale network

3. Network model

The network model used in this scheme is based on some assumptions which are listed below.

- N of nodes are distributed uniformly in a square area of M x M.
- Base station or sink is located at one certain location out of the network filed.

- Base station and all the nodes are stationary.
- Same initial energy are equipped to all the sensor nodes.
- Unique ID is assigned for each node.
- Nodes able to estimate its distance from its neighbors and base according to the RSSI.

The same energy model which is given in [6] is used where the energy required for transmitting and receiving 'l' bits over a distances 'd' is calculated using the following equations

$$E_{TX}(k,d) = \begin{cases} k * E_{elec} + k * E_{fs} * d^2, & d < d_0 \\ k * E_{elec} + k * E_{mp} * d^4, & d \geq d_0 \end{cases} \quad (1)$$

$$d0 = \sqrt{\frac{E_{fs}}{E_{mp}}} \quad (2)$$

Where E_{elec} is energy consumed to transmit or receive a bit, E_{fs} is transmitter amplifier energy for free-space, E_{mp} is transmitter amplifier energy for multi-path . While receiving 'l' bits, radio consumes energy is calculated as follow

$$E_{rx}(K) = k * E_{elec} \quad (3)$$

4. Proposed scheme

This Clustering scheme configures clusters in every round same as others schemes. The process is divided into two phases such as cluster formation phase and data transmission phase.

4.1 Cluster formation phase

Most of cluster mechanism select the cluster head either randomly without considered important parameters or based on centralized approach by utilizing the base station which can affect the network scalability. Subsequently, distributed cluster head mechanism is utilized that able to reduce the network overheads where nodes only require exchanging its local information with their neighbors in order to calculate their weights. In this scheme, nodes compute their score using weighted parameters, which incorporate the residual energy of nodes and the number of their neighbors. In addition, equation 4showsthat the node with highest candidate weight, it is suitable candidate to be elected as CH.

$$Candidate\ weight = E\ resd + No.of\ nighbours \quad (4)$$

In this scheme, nodes keep track of their residual energy levelsas nodes with sufficient energy resources will be considered for CH selection. Secondly, each node counts the number of its neighbors and the node with highest number of neighbors represents the optimal position inside the cluster.Once, the residual energy and number of neighbors are determined. These values are summed up in weighted form to estimate candidate weight for CH selection.

IOP Conf. Series: Journal of Physics: Conf. Series **1019** (2018) 012007 doi:10.1088/1742-6596/1019/1/012007

4.2 Data transmission phase

In order to reduce energy consumption in large scale network multi-hop is employed between CHs nodes. As CHs which located faraway from BS drain out their energy quickly where energy consumption lineally increases as transmission distance increases therefore, multi hop is used for between CHs node which can reduce the energy consumption for data transmission. However, in this scheme, CH selects one of the CHs as relay node according to several parameters that should be considered in the selection of next relay node. To optimize data forwarding, each CH sends a route message to its upper CHs. This message contains the residual energy and number of member nodes of its cluster. By compromising these parameters with distance from the CHs, relay cost is calculated that can reduce energy consumption in data transmission and provide load balancing in WSN. In the relay cost, CH si would choose CH sj as relay node, if its remaining energy is the largest value and have least cluster member nodes and nearest to it. The relay cost is calculated as shown in equation (5) below

$$Relay\ Cost = \frac{distCHij + No.of membernodes}{Eresd.} \quad (5)$$

5. Simulation results and discussion

A simulation of the proposed approach was performed, and two parameters used to evaluate the performance of the proposed work, including number of alive sensor nodes and energy consumed. Also, the performance of proposed work is compared with LEACH. The parameters used for the simulation are given in Table 1.

Table 1.parameters

Parameter	Value
Sensor field	100 x100
BS location	(150, 50)
Number of nodes	100
Initial energy of nodes	2 J
E_{elec}	50 nJ/bit
E_{fs}	10 pJ/(bit m2)
E_{amp}	0.0013 pJ/(bit m4)

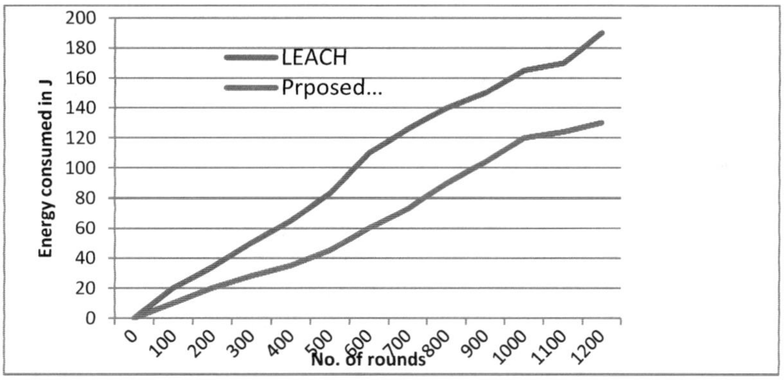

Figure 1. Energy consumed vs number of rounds

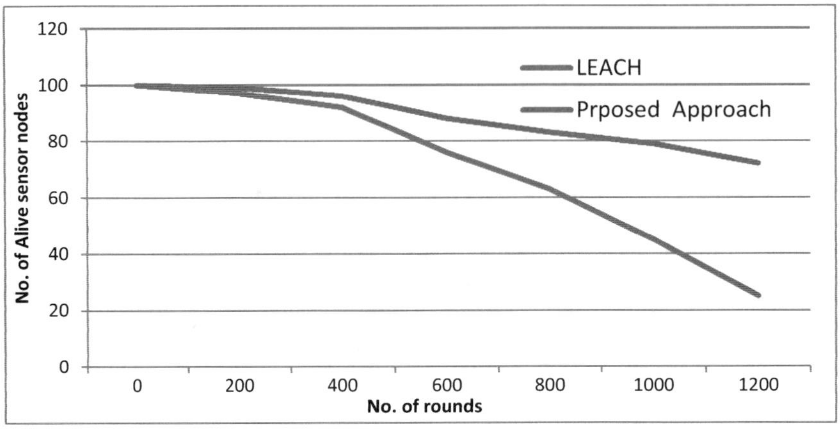

Figure 2: Number of alive sensor nodes VS number of rounds

Fig 1 and Fig 2 show that the proposed work outperform LEACH in term of energy consumption and number of alive sensor nodes. The reason behind this is better selection and positioning of the cluster heads. Proposed work is simple and it consider different parameter for cluster head selection and the relay node through multi-hop approach such as remaining energy, number of neighbour nodes and distance between CHs and number of member nodes in the cluster which in turn enhances the network lifetime.

6. Conclusion

In this paper, we proposed an energy efficient and load balancing clustering scheme based on distributed approach both residual energy and number of neighbours are considered in order to optimize the process of cluster head election and the process of datatransmission of network with lesser overheads. This makes the node with more residual energy and has high degree of neighbour nodes in the network has greater potential to be selected as cluster heads. In the data transmission phase, multi-hop data transmission is employed between CHs to avoid the direct communication between CHs and the base station. The results of simulation indicatethat the proposed scheme reduces the energy consumption, prolongs the network lifetime,and has a better performance than the original LEACH protocol.

References

[1] M. Saadat, R. Saadat, and G. Mirjalily, "Improving threshold assignment for cluster head selection in hierarchical wireless sensor networks," in 2010 5th International Symposium on Telecommunications, IST 2010, 2010, pp. 409–414.

[2] S. Naeimi, H. Ghafghazi, C. O. Chow, and H. Ishii, "A survey on the taxonomy of cluster-based routing protocols for homogeneous wireless sensor networks," Sensors (Switzerland), vol. 12, no. 6, pp. 7350–7409, 2012.

[3] A. . Muthu Krishnan and P. . Ganesh Kumar, "An Effective Clustering Approach with Data Aggregation Using Multiple Mobile Sinks for Heterogeneous WSN," Wirel. Pers. Commun., vol. 90, no. 2, pp. 423–434, 2016.

[4] M. M. Afsar and M. H. Tayarani-N, "Clustering in sensor networks: A literature survey," J. Netw. Comput. Appl., vol. 46, pp. 198–226, 2014.

[5] R. P. Mahapatra and R. K. Yadav, "Descendant of LEACH Based Routing Protocols in Wireless Sensor Networks," Procedia Comput. Sci., vol. 57, pp. 1005–1014, 2015.

[6] W. R. Heinzelman, A. Chandrakasan, and H. Balakrishnan, "Energy-efficient communication protocol for wireless microsensor networks," in Proceedings of the 33rd Annual Hawaii International Conference on System Sciences, 2000, pp. 3005–3014.

[7] S. Lindsey and C. S. Raghavendra, "PEGASIS: Power-efficient gathering in sensor information systems," in IEEE Aerospace Conference Proceedings, 2002, vol. 3, pp. 1125–1130.

[8] W. B. Heinzelman, A. P. Chandrakasan, and H. Balakrishnan, "An application-specific protocol architecture for wireless microsensor networks," IEEE Trans. Wirel. Commun., vol. 1, no. 4, pp. 660–670, 2002.

[9] R. Sharma, N. Mishra, and S. Srivastava, "A proposed Energy Efficient Distance Based Cluster Head (DBCH) Algorithm: An Improvement over LEACH," Procedia Comput. Sci., vol. 57, pp. 807–814, 2015.

IOP Conf. Series: Journal of Physics: Conf. Series **1019** (2018) 012008 doi:10.1088/1742-6596/1019/1/012008

Advanced Encryption Standard with Galois Counter Mode using Field Programmable Gate Array.

Nabihah Ahmad[1*], Lim Mei Wei[1], M. Hairol Jabbar[1]

[1] Faculty of Electrical and Electronic Engineering,
Universiti Tun Hussein Onn Malaysia,
86400 Parit Raja, Batu Pahat, Johor, Malaysia.

*Email: nabihah@uthm.edu.my

Abstract. Nowadays, the protection of transferring data is important to prevent the data hack easily. Advanced Encryption Standard with Galois Counter Mode (AES-GCM) plays an important role to provide high assurance of authenticity and data confidentiality in electronics, computers and other communication applications. This paper presents the implementation of AES-GCM by using Field Programmable Gate Array (FPGA) and AES-GCM designs in parallel- pipelined to achieve high performance in term of throughput and latency. The implementation of AES-GCM in FPGA by using 128-bit of input data block, Initialization vector (IV) and Additional Authenticated Data (AAD) to provide a high speed of authenticated encryption/ decryption. The key length of AES-GCM is 256-bit to provide the high security system and the operation of key expand designed in parallel to optimize operation time of AES-GCM. The proposed architecture is designed in Verilog hardware description language (HDL) and implemented using DE1-SoC with Cyclone V device. A parallel-pipelined of AES-GCM is introduced and it is operated at 10 MHz, achieved throughput of 16.84 Gbps, utilized of 11,196 slices. AES-GCM is carried out with the key-length of 256-bit is suitable to perform at high speed of electronic applications in term of security.

1. Introduction

Advanced Encryption Standard with Galois Counter Mode (AES-GCM) is introduced by United States of America National Institute for Standard and Technology (NIST). AES-GCM is suitable to employ in communication or electronic applications [3]. AES-GCM is a block cipher mode of operation that provides high speed of authenticated encryption and data integrity. Todays, the level of privacy protection is insufficient and make the data is been hacked easily. The FPGA is suitable to implement for AES-GCM by ensuring the confidentiality and integrity of the bit-stream[4].

AES-GCM have two main functions are block cipher encryption and multiplication over the field $GF(2^{128})$[5]. The authenticated encryption operation takes Initialization Vector (IV), Additional Authenticated Data (AAD),secret key and plaintext as an input in128-bit and gives a 128-bit ciphertext and authentication tag,T. The AES-GCM algorithm encrypts or decrypts with 128-bit, 192-bit or 256-bit of cipher key. The number of rounds executed transformations of AES depends on the length of cipher key [6][7][8]. Thus, the number of rounds executed is increased, the time taken to generate output is longer and will affect performance of AES-GCM.

Research of previous works shows the performance of AES-GCM in [17] is the better than other works. The method to design the AES-GCM was right-to-left exponentiation of pipelined GF multipliers. A group of parallel generators are created multiple blocks of input. This makes the more

IOP Conf. Series: Journal of Physics: Conf. Series **1019** (2018) 012008 doi:10.1088/1742-6596/1019/1/012008

efficient of the architecture which achieved the throughput 327.7 Gbps. The design pipelined S-box of AES is the efficient method to achieve the high performance of AES. The multiple of rounds transformation to execute, AES will consume more time as the performance of AES depends the key size in the design.is required. The performance of previous work [5], the throughput 42.7 Gbps, utilize with 2815 slices. The pipelined of S-box is needed to consider when designing the architecture of AES-GCM.

In this paper, the performance of AES-GCM is analyzed when the implementation of AES-GCM encryption using DE1-SoC with Cyclone V device. The performance of AES-GCM is introduced in term of throughput and latency. Verilog HDL is used to describe the behavior of AES-GCM by using Quartus II. The parallel-pipelined of AES-GCM is introduced to achieve the objectives of this paper.

2. Architecture of AES-GCM

The AES architecture is designed in parallel due to higher performance of AES-GCM encryption and decryption. The parallel architecture of key expansion is designed as greater key size is used. Figure 3 shows the key size is 256-bits will cause the execution of 14 rounds substitution and permutations needed. The key size depends on the desired security level. The standard number of transformation rounds of AES-256 is fourteen rounds. As shown in Table 1, AES encrypts a 128-bit of plaintext or decrypts 128-bit of ciphertext by repeatedly applying the same round transformation a number of times depending on the key size.

Table 1. Key-Block-Round Combinations.

	Key Length (32-bit word)	Block Size (32-bit word)	Number of Round
AES-128	4	4	10
AES-192	6	4	12
AES-256	8	4	14

When the key size implemented is larger, the encryption or decryption will take longer time. The main function of AES encryption is according the key schedule. The proposed key size of this project is 256-bit. Thus, the proposed design of AES is in parallel that can optimize the time taken to execute the key expand and encryption of AES. For the AES algorithm, the length of the input block and the output block are 128-bit. The key length designed is 256-bit, which reflects the number of rounds to be performed is 14 rounds. The more rounds to be executed will make the more consuming in time. Therefore, the parallel data paths are designed for improving the performance of AES in term of speed to execute. Figure 1 shown the architecture designed in 256-bit key.

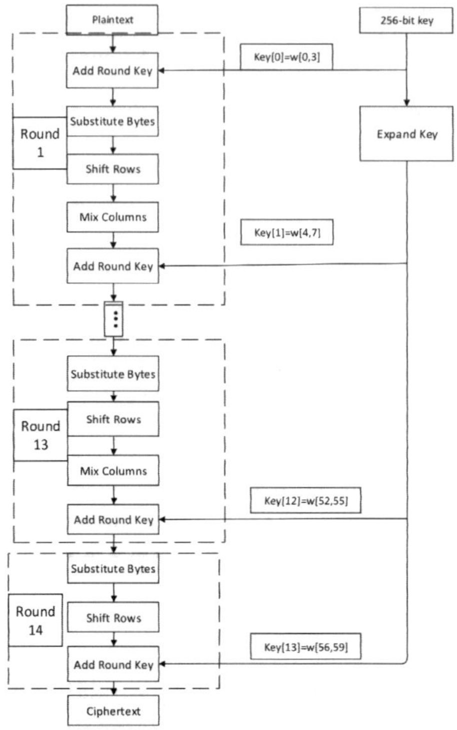

Figure 1. AES encryption with 256-bit of key length.

In Mix Columns, each column of the State is in four term polynomial. By reducing XOR gates in the critical paths, the architecture of Mix Colum is to be more efficient. The standard polynomial equations of the first column of Mix Column as shown Equation 1,

$$s'_{0,c} = \{[02] \cdot s_{0,c}\} \oplus \{[03] \cdot s_{1,c}\} \oplus s_{2,c} \oplus s_{3,c} \quad (1)$$

After the expansion of the equation, the equation shows as Equation 2

$$s'_{0,c} = [02]\, s_{0,c} \oplus [02]\, s_{1,c} \oplus s_{1,c} \oplus s_{2,c} \oplus s_{3,c} \quad (2)$$

The function of multiplication of [02] in hexadecimal is represented by the XTime as shown in Figure 2. After the expansion of the equation, the critical path of the Mix Column transformation is four with the XOR gate. An efficient of Mix Column implementation architecture can be derived by minimizing the number of XOR gates. The equation can be simplified and rewritten as Equation 3

$$s'_{0,c} = [02]\{ s_{0,c} \oplus s_{1,c} \} \oplus \{s_{1,c} \oplus s_{2,c}\} \oplus s_{3,c} \quad (3)$$

As shown in Figure 3, the critical path of the Mix Column is reduced, the three of critical paths XORed to generate the output of Mix Column. Therefore, one XTime is required with the three XOR gates.

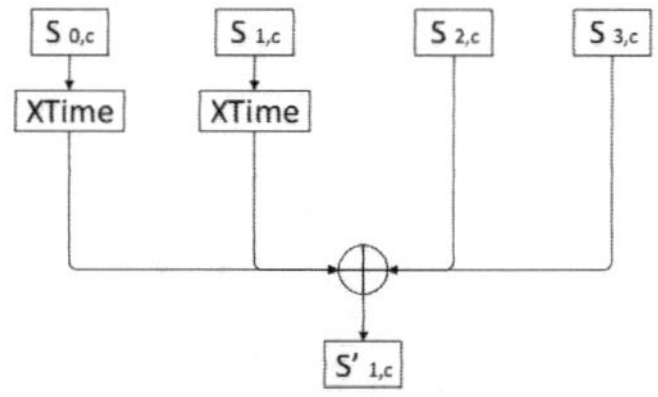

Figure 2. MixColumn transformation.

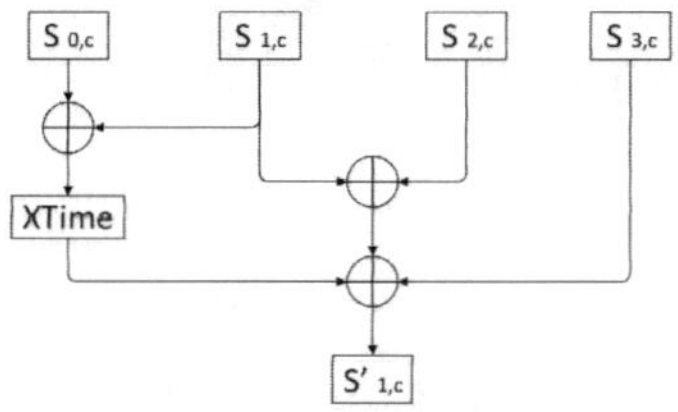

Figure 3. The efficient of Mix Column transformation

AES-GCM has two main components, AES engine and GHASH function as shown in Figure 4. In the hash subkey for GHASH function, 128-bit of "zero"block as an input is encrypted and the hash subkey will store in the register. The 96-bit of IV is appended with $0^{31}||1$ and generated Nonce in counter 1 at the same time the 32-bit incrementing function is applied to form the next counter block. The Nonce acts as an input of AES encryption will generate the intermediate hash value Yi. When it received additional authentication data , AAD and it is appended with ciphertext, GHASH function is generated the authentication tag. The same function of authenticated encryption and authenticated decryption. The output of the authenticated decryption is authentication tag generated need to match the encryption tag. If not matched, the decryption function is failed.

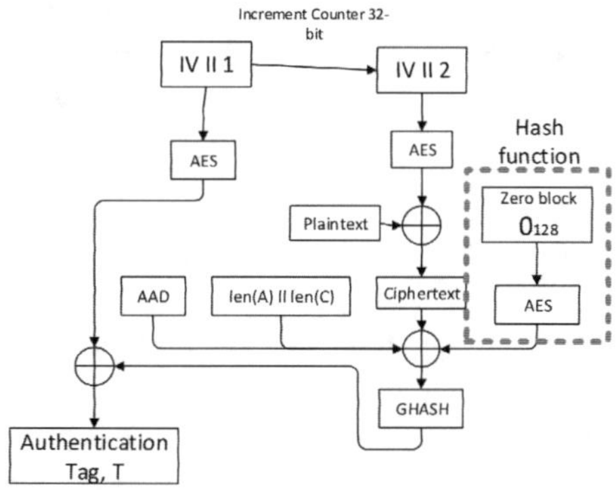

Figure 4. AES-GCM encryption/decryption.

Verilog HDL is used as the hardware description language as the flexibility to exchange among environments. The code is pure Verilog HDL that easy to implement on Cyclone V DE1-SoC board.

2.1. FPGA Implementation of AES-GCM Architecture

The AES-GCM design is programmed in Cyclone V 5CSEMAS5 FPGA on DE1-SoC board. Pins are selected in the pin plan. After programming AES-GCM in DE1-SoC board, the real functionality is done. As shown in Figure 5, DE1-SoC will encrypt and decrypt the data and it will interface with the memory addresses to store or access the data. The output data will display on computer screen. This project is suitable for the electronic applications which requires higher security, such as the Personal Cloud Storage device, Virtual Private Network (VPN) and others.

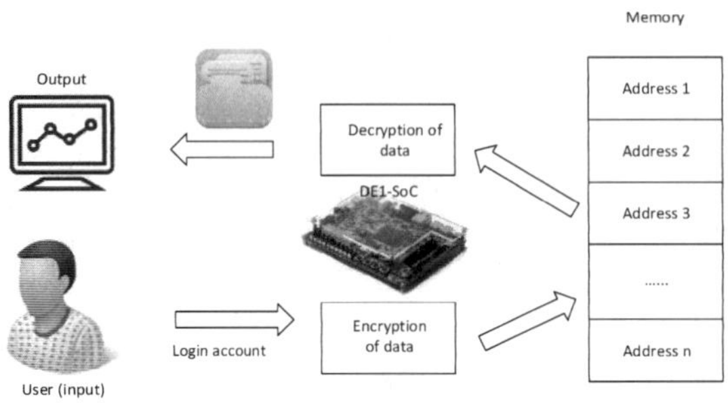

Figure 5. Overall architecture of AES-GCM.

3. Results and discussion

The proposed design is implement with Verilog HDL and it is simulated and synthesized using Quartus II. And ported to Cyclone V DE1-SoC device. The pipelined architecture of GCM operation is proposed.

The performance of AES-GCM is analysed in terms of throughput and latency. Latency is defined as the total time needed to complete the AES-GCM encryption and decryption process of the 128-bit of an input data block. The accurate value is calculated from the time a block of data enters encryption or decryption, until data leaves the process. The latency is calculated as shown in Equation 4:

Latency = Cycles per Encrypted Block(C) x Time period (T) (4)

where C is the clock per byte to encrypt a block of data, T is the reciprocal of the frequency clock, 1/ f_clk.
Latency depends on the key size and the operation mode of AES-GCM. In the parallel-pipelined of AES-GCM designed, asynchronous reset is applied in the circuit. After the reset is 0, the next stage is executed. In the proposed design, there are 10 stages to be executed. The result of latency obtained is 7.6 ns.

The throughput indicated the speed of the AES-GCM encryption and decryption process. It also defined as 128-bit of input can be encrypted and decrypted in a unit of time. Thus, the throughput of AES-GCM is calculated in bit per second (bps) as shown in Equation 5:

Throughput = 128 bits/ latency (5)

= 128-bit/ (7.6×10^{-9})
= 16.84 Gbps

Table 2 shows the performance of proposed work, the throughput and the utilization of the proposed work compared with the previous works.The throughput of proposed design achieved 16.84 Gbps, utilize 11,196 slices. The throughput of proposed design compares with previous works, the proposed work is 94.93% slower than the targeted previous work. The architecture of previous work is a parallel GHASH function while the proposed work is pipelined. In the GHASH function, seven clock cycles are taken for one stage. Therefore the time consumed is more on the GHASH function. The utilization of the proposed work is lesser than the previous work, 93.17% and the architecture of the proposed work will consume less power than the previous work. The pipelined design of GHASH in AES-GCM is consumed less area of FPGA compared with the previous work.

Table 2. Performance of AES-GCM proposed.

Design	Device	Throughput (Gbps)	Frequency (MHz)	Slices
Proposed work	Altera Cyclone V	16.84	10	11,196
[17]	Xilinx Virtex Ultrascale	327.7	320	163,850
[12]	Xilinx Virtex 5	102.4	242	12,161,
[3]	Xilinx Virtex 5	82	641	2,472,
[5]	Synopsys DesignVision	0.118	233	146
[15]	Synopsys DesignVision	42.7	333	2,815
[20]	Xilinx Virtex 4	10.49	82	13,368
[16]	Altera Stratix	5.61	43.86	12,827

4. Conclusion

The proposed design was successfully completed with the implementation of authenticated encryption and decryption of AES-GCM. The proposed design is 16.84 Gbps, utilized with 11,196 slices. Compared with the previous work, the proposed design is slower than 94.93%, utilize is less than previous work 93.17%. The key length of previous work is 128-bit, but the key length proposed is 256-bit. This implementation can be to introduce in confidential corporate documents, government documents, and personal cloud storage device or person information protection which emphasis on the data security. The proposed design of AES-GCM is aimed for the high speed of the encryption and decryption.

Acknowledgement
This work was financial supported by FRGS Grant Vot No. 1538.

References

[1] A. M. Deshpande, M. S. Deshpande and D. N. Kayatanavar 2009 FPGA implementation of AES encryption and decryption *2009 International Conference on Control, Automation, Communication and Energy Conservation* pp. 1-6.

[2] M. Kosug, M. Yasuda and A. Satoh 2015 FPGA implementation of authenticated encryption algorithm Minalpher *2015 IEEE 4th Global Conference on Consumer Electronics (GCCE)* pp. 572-576

[3] M. Pitchaiah, P. Daniel, and Praveen 2012 Implementation of Advanced Encryption Standard Algorithm *Int. J. Sci. Eng. Res.* vol 3 no. 3 pp. 1–6,

[4] V. Arun, K. Vanisree, and D. L. Reddy 2014 Implementation of AES-GCM encryption algorithm for high performance and low power architecture Using FPGA *International Journal of Research and Applications* vol. 1 no. 3 pp. 120–131

[5] D. A. McGrew and J. Viega 2004 The Security and Performance of the Galois/Counter Mode (GCM) of *Operation Proceedings of the 5th International Conference on Cryptology in India*Secur pp. 343–55

[6] K. M. Abdellatif, R. Chotin-Avot, and H. Mehrez 2013 Efficient AES-GCM for VPNs using FPGAs *2013 IEEE 56th International Midwest Symposium on Circuits and Systems (MWSCAS)* pp. 1411–1414

[7] A. Jose 2016 A High Throughput Low Power AES-GCM for FPGAs *International Journal of Advanced Research in Electrical, Electronics and Instrumentation Engineering* vol. 5, no. 4, pp. 183–187

[8] L. Henzen and W. Fichtner 2010 FPGA parallel-pipelined AES-GCM core for 100G ethernet applications *FPGA Parallel-Pipelined AES-GCM Core 100G Ethernet Appl. ESSCIRC 2010 - 36th Eur. Solid State Circuits Conf.* vol. 1 no. 2 pp. 202–205

[9] T. Chen, W. Huo, and Z. Liu 2010 Design and efficient FPGA implementation of Ghash core for AES-GCM *Des. Effic. FPGA Implement. Ghash Core AES-GCM* pp. 1–4

[10] H. Qin, T. Sasao, and Y. Iguchi 2006 A design of AES encryption circuit with 128-bit keys using look-up table ring on FPGA *IEICE Trans. Inf. Syst.* vol. E89–D no. 3 pp. 1139–1147

[11] M. Abdellatif, R. Chotin-Avot, and H. Mehrez 2014 Authenticated encryption on FPGAs from the static part to the reconfigurable part *Microprocess. Microsyst.* vol. 38, no. 6, pp. 526–538

[12] Navabi, Zainalabedin 1999 Verilog digital system design *McGraw-Hill*

[13] Maxfield, Clive 2011 FPGAs: instant access *Newnes*

[14] Nemati, Hamid R. 2010 Applied Cryptography for Cyber Security and Defense: Information Encryption and Cyphering *IGI Global*

[15] B. Sunar, E. Savas, and Ç. K. Koç 2003 Constructing Composite Field Representations for Efficient Conversion *Constr. Compos. F. Represent. Effic. Convers.* vol. 52 no. 11 pp. 1391–1398

[16] M. Mozaffari-Kermani and A. Reyhani-Masoleh 2010 Efficient and high-performance parallel hardware architectures for the AES-GCM *IEEE Trans. Comput.* vol. 61 no. 8 pp. 1165–1178

[17] B. Buhrow, K. Fritz, B. Gilbert, and E. Daniel 2015 A Highly Parallel AES-GCM Core for Authenticated Encryption of 400 Gb / s Network Protocols *2015 International Conference on ReConFigurable Computing and FPGAs (ReConFig)* pp. 1–6

[18] S. M. Wadi and N. Zainal 2013 Rapid Encryption Method based on AES Algorithm for Grey Scale HD Image Encryption *Procedia Technology* vol. 11 no. 12 pp. 51–56

[19] J. Daemen, V. Rijmen, and K. U. Leuven 1999 AES Proposal : The Rijndael Block Cipher pp. 45

[20] N. Fips 2001 197: Announcing the advanced encryption standard (AES) Adv. ENCRYPTION Stand *(AES) Technology Lab. Natl. Inst. Stand.* vol. 2009 pp. 8–12

IOP Conf. Series: Journal of Physics: Conf. Series **1019** (2018) 012009 doi:10.1088/1742-6596/1019/1/012009

Histomorphometric variance of haversian canal in cortical bone of Malaysian ethnic groups

Hadi Abdullah[1], Muhammad Mahadi Abdul Jamil[1], Faridah Mohd Nor[2]

[1]Faculty of Electrical and Electronic Engineering Universiti Tun Hussein Onn Malaysia (UTHM), Batu Pahat, Malaysia.
[2]Department of Pathology, Faculty of Medicine Universiti Kebangsaan Malaysia (UKM), Medical Centre, Kuala Lumpur, Malaysia

mahadi@uthm.edu.my

Abstract. This research presents a comparison of haversian canal based microstructural parameters between Malaysian citizens. The ages ranges from 22-58 which was divided into four groups of each decade. In each decade comparison was done on haversian canal based parameters between the two Malaysian ethnic groups Chinese-Indian. Parameters in long bones femur, tibia, fibula, humerus, ulna and radius were measured under microscope by obtaining the bone thin section of 30 microns from the mid-shaft. Eight microscopic images were taken from each specimen having an observation region of 1.228mm^2 for each image. In the third and sixth decade, haversian canals showed relatively larger total covered area in Indian bone specimens. Haversian canal radius was found to be greater in Chinese bone specimen in the fourth decade. The fifth decade showed no differences. Further studies are required to validate the outcomes, compare other ethnic groups and development of automated microstructures identification and measurement tools.

1. Introduction

Bone histomorphological research over the past few decades have developed various methods to extract information from skeletal remain[1, 2]. Morphological analysis involves study of the main structures of skeleton. Skull, jaw, long bones and pelvis are the main key bones for this analysis[3]. Histological analysis on the other hand is not bound to the main skeletal structures. Researches have been carried out to use only fragments of long bone and analyse them to estimate age, sex, race, height and weight[4-7]. Histological analysis of cortical bone for this information started only few decades ago and research conducted on a particular race of humans didn't show very promising results when applied on others[8-10]. Race and region of human population changes the morphological structures of bone which also effects the microstructures in the cortical bone. This led for stand-alone studies for human bones belonging to various races and region which can be helpful in forensic investigations, archaeological exploration and contribute for future comparisons.

To understand cortical bone microstructures, one must have the knowledge of basic bone structure. Human bones consist of two main types cortical bone and trabecular bone. Cortical bone is the outer hard layer which gives a firm structure to the bones and trabecular bone is the spongy inner part. Cortical bone is build up on basic structures called the osteon systems. The osteon system comprises of blood vessels in the centre called haversian canal and layers of lamella surrounding these canals. Human bones undergo continuous restructuring in the body which is known as bone remodelling. In remodelling process, old bone structure is eaten by cells called osteoclast. Osteoclast leave behind mesh structures which are filled with new bone by the cells known as osteoblast. This process of bone remodelling changes the microstructures in cortical bones[11-13]. These structural changes are the key observation in bone histological analysis for age and sex estimation.

Selection of microstructures and region of microscopic images varies in previous researches (table 1). Most of the microstructures are either observed or derived measurements of osteons and haversian canals. In this study, microstructural parameters related to haversian canal were compared between the two ethnic groups (Chinese-Indian) citizens of Kuala Lumpur, Malaysia. The haversian canal parameters used in this study are mean area (Hcm), total area covered (Hct), mean radius (Hcr), mean perimeter (Hcp) and number (Hcn).

Table 1. Imaging location, bone type and area selected by researchers.

Reference	Region/area	Bone	Imaging location	Purpose
Kerley, 1965[8]	*USA* whites	Femur, tibia, fibula	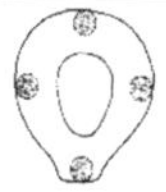	Age regression
IJ Singh, *DL* Gunberg, 1970[14]	Portland	Femur, tibia		Age regression
DD Thompson, 1979, 1981[10, 15]	*USA* whites and Eskimos from St. Lawrence, North Alaska and Canada	Femur		population comparison
David *B* Burr, 1990[16]	Native American and *USA* whites	Femur		Age regression
Ericksen, 1991[9]	*USA* whites	Femur		Age regression
DM Mulhern, 1997[17]	Medieval Nubian population from Kulubnarti, Subanese Nubia	Femur	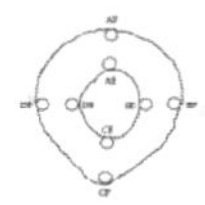	population comparison
KL Bell, 2001[18]	Australia	Femur	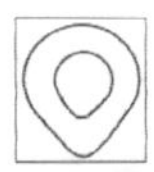	population comparison
C David *L* Thomas, 2005[19]	Australia	Femur		population comparison
F Nor, 2009[20]	Kuala Lumpur, Malaysia	Femur, tibia, ulna, radius, Humerus, fibula		Age regression
HM Britz, 2009[21]	Australia	Femur	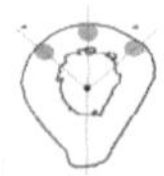	Population comparison
C Hernandez, 2012[22]	*USA* whites	Femur	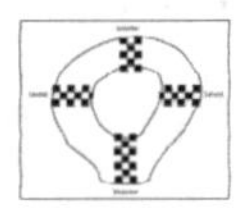	Age regression
This study	Kuala Lumpur, Malaysia	Femur, tibia, ulna, radius, Humerus, fibula	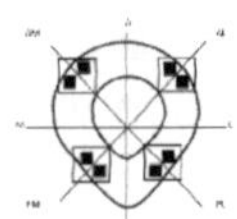	Population comparison

IOP Conf. Series: Journal of Physics: Conf. Series **1019** (2018) 012009 doi:10.1088/1742-6596/1019/1/012009

2. Materials and Methods

2.1. Sample Population

Samples of 33 adults were collocated from Universiti Kebangsaan Malaysia Medical Centre (UKMMC), Kuala Lumpur, Malaysia. The bone samples were collected from individuals having no bone disease in previous medical records. The population of Malaysia is divided into three main ethnic groups Malay, Chinese (from south china) and Indian (from south India)[23]. Among the collected samples 21 were Chinese while 12 were Indian. With respect to sex, 24 samples were males and 9 were females and age of samples ranges from 22-58 The samples are not homogeneously numbered due to the ethnic and religious reservations and availability of cases in forensic medical centre UKMMC. The bone types selected for this work were long bones comprising of humerus, radius, ulna, femur, tibia and fibula. According to previous research conducted on Malaysian samples, haversian canal increases with respect to age. This complies in division of samples into decades for better statistical observation. Figure 1 shows samples orientation with respect to decades and types of bones used.

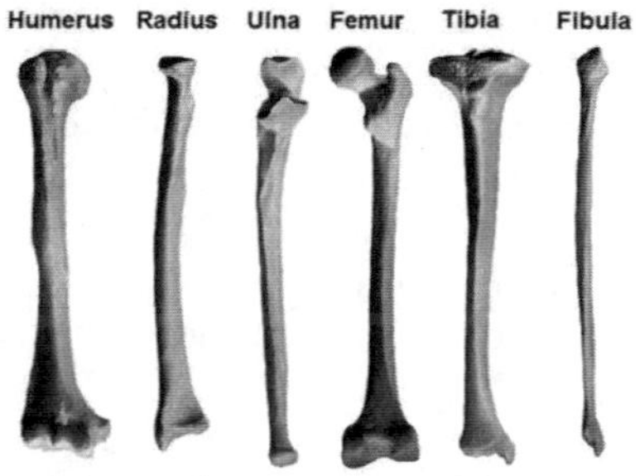

Age	Chinese	Indian
20-30	5	4
30-40	6	3
40-50	5	3
50-60	5	2
Total	21	12

Figure 1. Long bones selected for sampling.

2.2. Samples imaging

The bone is cut from mid shaft at 3cm length. To clean the sample and remove fat, it is placed in diethyl ether solution for 24 hours. Epoxy resin is then mixed with bone along with hardener. To remove bubbles from epoxy bone is place in vacuum chamber half hour. Face cut of 30 micros was done using microtome after gluing it on a glass slide[20].

Eight images of each sample are taken as observed region. Figure 2 (a) shows the selected region. These regions were selected based on previous studies on Malaysian samples which observed more microstructure differences at these locations. Nikon eclipse Ts100 microscope was used with Dino Eye camera to obtain microscopic images. Dino capture 2.0 software was used to calibrate obtained images at 4X magnification. The dimensions of observed region are 1.28mm x 0.96mm which gives us a total region area of 1.228mm^2. Figure 2 (b and c) shows the microscope used and obtained microscopic image. The measurements of haversian can were done manually using measurement tools in Cellprofiler image analysis software.

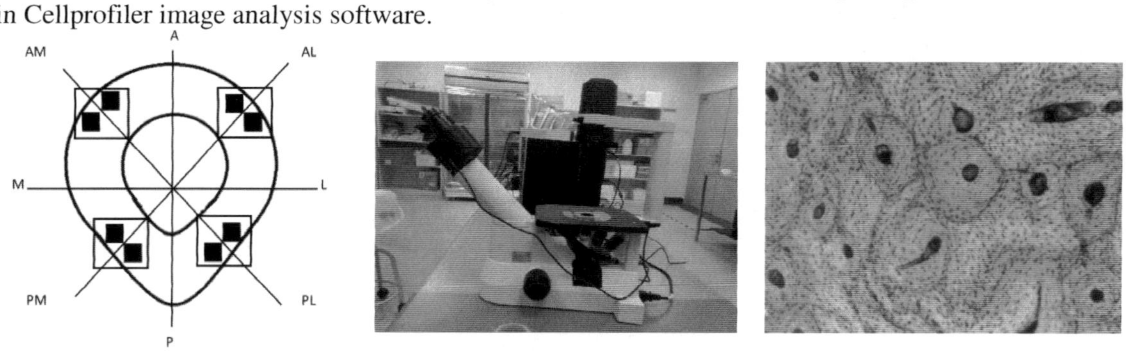

(a) (b) (c)

Figure 2. (a) Location of images from samples. (b) Nikon eclipse Ts100 used to take microscopic images. (c) Obtained image with Observation area of 1.228mm^2.

2.3. Inclusion and exclusion criteria

The criteria for inclusion of haversian canal was slightly different from researches which were focused mainly on osteon systems[4, 5, 21]. Since this research focuses on haversian canal not osteon systems so the shape and intactness of osteon was not taken into consideration. The haversian canal present in observation area with more than fifty percent of its area were considered as full haversian canals. In case where haversian canals are connected to Volksmann's canals, only haverisan canals was measured based on manual observations.

2.4. Statistical tests

The bone samples were divided into four independent age groups of decades from 20 – 60. In each age group five haversian canal based parameters were compared between two sample categories (Chinese-Indian). For comparison of parameters independent samples t-test was used.

3. Results

Measurement of each specimen was done on eight different locations (figure 2 (a)). Mean values of the haversian canal based microstructural parameters measured at these locations are given in table 2. Cumulative fraction distribution was performed to check the contribution of individual haversian canals in total area covered by haversian canals. Figure 3 shows the Hcm cumulative fraction distribution of two groups for ages 20-30. T-test showed significant differences between the two races for Hcm, Hca, Hcr and hcp. It was found that haversian canals in Indians samples were relatively larger than that of Chinese samples ($p < 0.01$), while the number of haversian canals showed no difference (table 3).

Table 2. Descriptive statistics of measured haversian canal parameters in eight selected regions of bone thin sections.

Age groups	Races	Hcm (μm^2)	Hca (mm^2)	Hcr (μm)	Hcp (μm)	Hcn
20-30	Chinese	1336.9	0.16	5.8	149.1	128
	Indian	1833.7	0.21	6.8	173.9	117
30-40	Chinese	3816.3	0.53	9.9	257.4	142
	Indian	3362.8	0.49	8.6	232.4	152
40-50	Chinese	5096.2	0.79	11.7	300.7	150
	Indian	4718.8	0.69	11	290.5	139
50-60	Chinese	8277.5	1.29	14.9	380.3	156
	Indian	9905.7	1.52	16.5	398.4	157

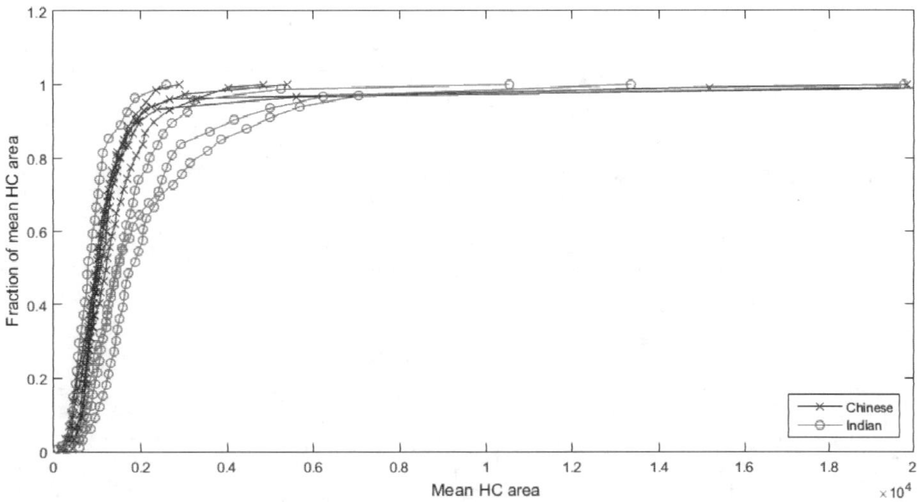

Figure 3. Cumulative fraction distribution graph demonstrating contribution of haversian canals in the total area coved for two races belonging to ages (20-30).

Table 3. Independent sample t-test of microstructural measurements between Malaysian (Chinese-Indian) bone thin sections.

Parameter	Hcm	Hca	Hcr	Hcp	Hcn
t	-3.477	-2.963	-4.261	-4.377	0.982
Sig.	0.00088	0.00419	0.00006	0.00004	0.32955
sd	700.75	10858.1	1.034	26.99	3.47

In the decade group of 30-40 independent t-test showed no significance with respect to Hcm, Hct, Hcp and Hcn. Haversian canal radius on the other hand was observed to be larger in Chinese samples ($p<0.05$). Table 4 shows the independent sample t-test. When compared to the previous decade haversian canal showed less significance. Figure 4 shows the cumulative fraction distribution of haversian canal for this decade group (30-40).

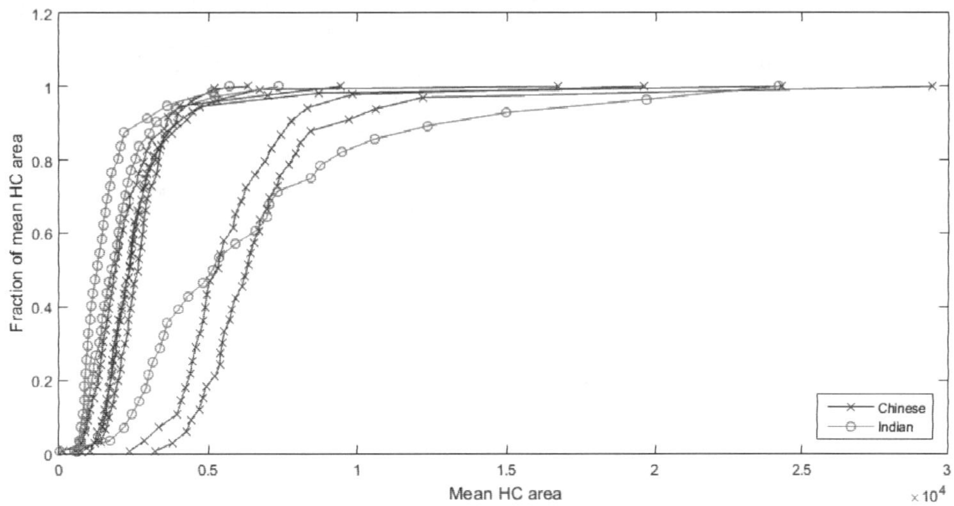

Figure 4. Cumulative fraction distribution graph demonstrating contribution of haversian canals in the total area coved for two races belonging to ages (30-40).

Table 4. Independent sample t-test of microstructural measurements between Malaysian (Chinese-Indian) bone thin sections.

Parameter	Hcm	Hca	Hcr	Hcp	Hcn
T	1.147	0.4638	2.142	1.6458	-1.1452
Sig.	0.2552	0.6442	0.0357	0.1042	0.2559
Sd	2128.65	41321.66	2.61	70.72	3.49

There was no significance seen among any parameters in the age group of 40-50 when t-test was applied. Table 5 shows the Independent sample t-test for this age group and figure 5 shows the cumulative fraction distribution of haversian canal.

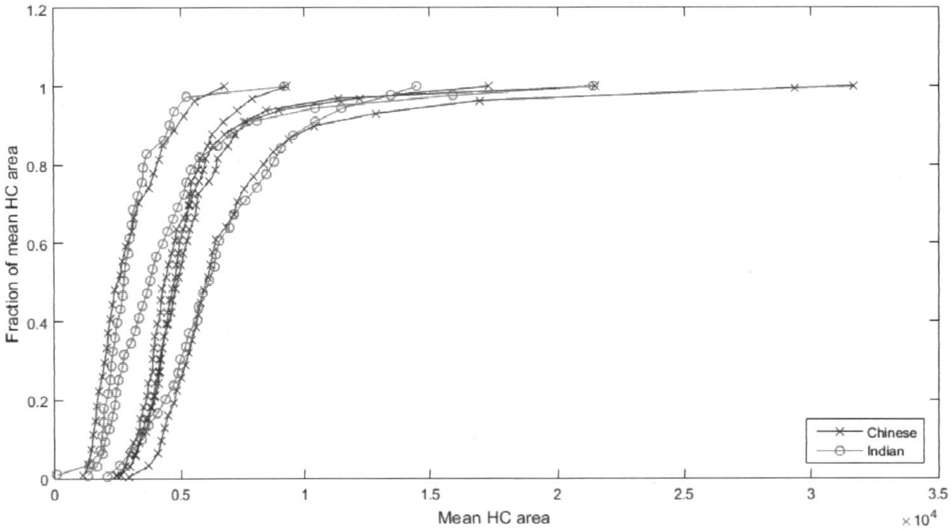

Figure 5. Cumulative fraction distribution graph demonstrating contribution of haversian canals in the total area coved for two races belonging to ages (40-50).

Table 5. Independent sample t-test of microstructural measurements between Malaysian (Chinese-Indian) bone thin sections.

Parameter	Hcm	Hca	Hcr	Hcp	Hcn
T	1.016	0.889	1.502	0.859	1.154
Sig.	0.313	0.377	0.138	0.393	0.252
Sd	1737.966	55657.917	1.99	49.16	5.589

In the fifth decade haversian canals grew larger in size and their number also increased from the previous decades. There was no significant difference found Hcm, Hcn and Hcp, while Hca and Hcr were found to be greater in Indian samples than in Chinese samples ($p<0.05$) (table 6). Figure 6 shows Cumulative fraction distribution of haversian canal area for the fifth decade.

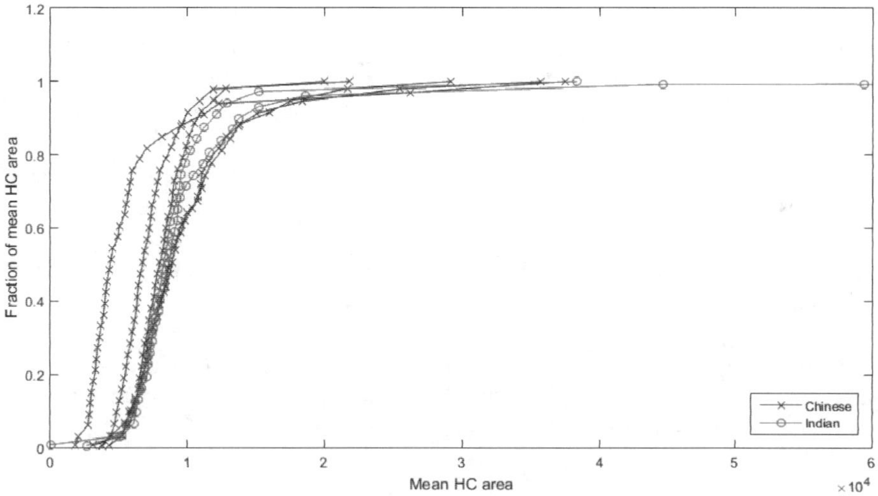

Figure 6. Cumulative fraction distribution graph demonstrating contribution of haversian canals in the total area coved for two races belonging to ages (30-40).

Table 6. Independent sample t-test of microstructural measurements between Malaysian (Chinese-Indian) bone thin sections.

Parameter	Hcm	Hca	Hcr	Hcp	Hcn
T	-1.99	-2.01	-2.77	-1.14	0.16
Sig.	0.0516	0.0492	0.0075	0.2559	0.8723
Sd	2758.3	48755.97	1.928	50.181	4.188

4. Discussion

The haversian canal parameters selected for this research increased in area with each observed decade (table 2). This is in agreement with previous work done[20, 24]. However, their difference with respect to race changes over decades. The third decade showed most differences. In Indian samples haversian canal were relatively larger in size and covered more observed region. The number of haversian canals increased with the progressing decade but showed no difference with respect to races. In the fourth decade haversian canal area in both races showed no differences. Mean radius on the other hand was found to be greater in Chinese samples. The fifth decade showed monotony in all observed parameters. Mean haversian canal radius was found to be greater in Indian samples in the sixth decade. All the samples showed relative measurements except for one Indian sample aged 53. Haversian canals in observation section of AM (figure 2(a)) showed relatively larger haversian canals from any other samples in that decade. Although mean radius was the parameter which showed differences in the third, fourth and sixth decade, the differences did not show continuous significance over the decades.

Cortical bone porosity is generally referred to the network of vascular canals commonly consisting of haversian canals and Volkmann's canals. Haversian canal is the key component to analyse cortical porosity. In this study haversian canal was measured irrespective of the presence of an osteon system or multiple layer of lamella (also referred as non-haversian canals)[25]. This gives us relation of the cortical bone porosity in which percentage of total area covered by haversian in eight sections can be considered as bone porosity.

5. Conclusion

This research presented a comparison of five microstructures in two Malaysian ethnic groups Chinese and Indian. The result showed maximum differences in third and sixth decades where haversian canal total area covered and mean radius were relatively greater in Indian specimens. Although the results were generally encouraging, it is required to take this work as the founding steps and further continue research to justify and expand the work done, which could include comparison of another ethnic group (Malay). A comparison of age graded sex comparison of the Malaysian samples can also be achieved. Computer aided systems are widely designed for medical imaging applications. A system can be designed to extract microstructures parameters using image processing and analyse the given specimens. This could reduce the time consumed in manual measurements and increase the accuracy.

Acknowledgments

The authors wish to express their gratitude to Office for Research, Innovation, Commercialization and Consultancy Management (ORICC), University Tun Hussein Onn Malaysia for providing postgraduate research Grant vote number (U280) to support this research.

References

[1] S. A. Asala, "Sex determination from the head of the femur of South African whites and blacks," *Forensic Sci Int,* vol. 117, pp. 15-22, Mar 01 2001.
[2] N. Márquez-Grant, "An overview of age estimation in forensic anthropology: perspectives and practical considerations," *Annals of Human Biology,* vol. 42, pp. 308-322, 2015/07/04 2015.
[3] F. Geneser, *Textbook of Histology. Copenhagen: Munksgaard.*, 1986.
[4] I. Khan, F. Mohd Nor, and M. M. Abdul Jamil, "A Survey of Human Age Estimation Techniques from Bone Microstructures," in *International Conference for Innovation in Biomedical Engineering and Life Sciences : ICIBEL2015, 6-8 December 2015, Putrajaya, Malaysia*, F. Ibrahim, J. Usman, M. S. Mohktar, and M. Y. Ahmad, Eds., ed Singapore: Springer Singapore, 2016, pp. 203-207.
[5] H. Abdullah, F. Nor, and M. M. Abdul Jamil, "Human Bone Histomorphological Pattern Differences Between Genders: A Review," vol. 56, 2015.
[6] I. Khan, M. M. A. Jamil, T. N. T. Ibrahim, and F. M. Nor, "Analysis of age-related changes in Haversian canal using image processing techniques," in *2016 6th IEEE International Conference on Control System, Computing and Engineering (ICCSCE)*, 2016, pp. 169-172.
[7] I. Khan, M. M. A. Jamil, T. N. T. Ibrahim, and F. M. Nor, "Automated human age estimation at death via bone microstructures," in *2016 6th IEEE International Conference on Control System, Computing and Engineering (ICCSCE)*, 2016, pp. 580-583.
[8] E. R. Kerley, "The microscopic determination of age in human bone," *American Journal of Physical Anthropology,* vol. 23, pp. 149-163, 1965.
[9] M. F. Ericksen, "Histologic estimation of age at death using the anterior cortex of the femur," *Am J Phys Anthropol,* vol. 84, pp. 171-9, Feb 1991.
[10] D. D. Thompson and M. Gunness-Hey, "Bone mineral-osteon analysis of Yupik-inupiaq skeletons," *American Journal of Physical Anthropology,* vol. 55, pp. 1-7, 1981.
[11] L. J. a. J. Carneiro, *Basic Histology: Text and Atlas*, 2005.
[12] Gary, "Gray's Anatomy of the Human Body," in *[*http://www.bartleby.com/107/illus77.h, o. system, Ed., ed, 1918.
[13] D. G. Steele, *The Anatomy and Biology of the Human Skeleton*: Texas A&M University Press, 1988.
[14] I. J. Singh and D. L. Gunberg, "Estimation of age at death in human males from quantitative histology of bone fragments," *Am J Phys Anthropol,* vol. 33, pp. 373-81, Nov 1970.
[15] D. D. Thompson, "The core technique in the determination of age at death of skeletons," *J Forensic Sci,* vol. 24, pp. 902-15, Oct 1979.
[16] D. B. Burr, C. B. Ruff, and D. D. Thompson, "Patterns of skeletal histologic change through time: comparison of an archaic native American population with modern populations," *Anat Rec,* vol. 226, pp. 307-13, Mar 1990.
[17] D. M. Mulhern and D. P. Van Gerven, "Patterns of femoral bone remodeling dynamics in a Medieval Nubian population," *Am J Phys Anthropol,* vol. 104, pp. 133-46, Sep 1997.
[18] K. L. Bell, N. Loveridge, J. Reeve, C. D. Thomas, S. A. Feik, and J. G. Clement, "Super-osteons (remodeling clusters) in the cortex of the femoral shaft: influence of age and gender," *Anat Rec,* vol. 264, pp. 378-86, Dec 1 2001.

[19] C. D. L. Thomas, S. A. Feik, and J. G. Clement, "Regional variation of intracortical porosity in the midshaft of the human femur: age and sex differences," *Journal of Anatomy,* vol. 206, pp. 115-125, 01/05/accepted 2005.

[20] F. M. Nor, R. F. Pastor, and H. Schutkowski, "Age at death estimation from bone histology in Malaysian males," *Med Sci Law,* vol. 54, pp. 203-8, Oct 2014.

[21] H. M. Britz, C. D. Thomas, J. G. Clement, and D. M. Cooper, "The relation of femoral osteon geometry to age, sex, height and weight," *Bone,* vol. 45, pp. 77-83, Jul 2009.

[22] M.-T. J. Cosgriff-Hernandez, "Histomorphometric Estimation of Age at Death Using the Femoral Cortex: A Modification of Established Methods," The Ohio State University, 2012.

[23] S. Swee-Hock, *The Population of Malaysia (Second Edition)*: Institute of Southeast Asian Studies, 2015.

[24] B. K. Samson C, "A new method of estimating age at death from fragmentary and weathered bone. In Bodington A, Garland AN, Janaway RC, editors. Death Decay and Reconstruction Approaches to Archaeology and Forensic Science," *Manchester: Manchester University Press,* pp. 101-108, 1987.

[25] D. M. Cooper, C. E. Kawalilak, K. Harrison, B. D. Johnston, and J. D. Johnston, "Cortical Bone Porosity: What Is It, Why Is It Important, and How Can We Detect It?," *Curr Osteoporos Rep,* vol. 14, pp. 187-98, Oct 2016.

IOP Conf. Series: Journal of Physics: Conf. Series **1019** (2018) 012010 doi:10.1088/1742-6596/1019/1/012010

Bone histology: A key for human sex determination after death

Hadi Abdullah[1], Muhammad Mahadi Abdul Jamil[1], Radzi Ambar[1], Faridah Mohd Nor[2]

[1]Faculty of Electrical and Electronic Engineering, University Tun Hussein Onn Malaysia, Batu Pahat,(86400) Johor, Malaysia.

[2]Medical centre, University Kebangsaan Malaysia, Bandar Tun Razak, (56000) Batu 9 Cheras, Wilayah Persekutuan Kuala Lumpur, Malaysia

Hadi.uthm@yahoo.com

Abstract. Age at death and sex are the key information required at a crime scene investigation or archaeological exploration. A number of morphological techniques are present to estimate age and sex from human bones. These techniques showed good result when major parts of human bones were present. In cases, where only fractions of bone are present, these techniques become irrelevant. Bone histological techniques can be applied on bone fragments to estimate age and sex. Age estimation using these techniques has been the key focus of research over the last few decades. At the early stages of these techniques, some researchers reported no relation of sex with histological changes. This made sex a by-product of age estimation or was left unconcerned. Bone histology of human changes with region, race and age. This paper presents the history of histological sex determination. Comparison of different researches was given with regards to region of research, selection of bone and microstructural parameter selection. A discussion concludes this paper with overall research done, and what future work can be done to improve histological sex determination.

1. Introduction

Information of age at death and sex play a vital role in crime scene investigation, as well as archaeological exploration. Human bones are the strongest tissue in the body, which takes a longer time to decay. Studies have been carried out to analyze human bones to gather information of the person it belonged to. These studies can be divided into two parts, morphological and histological techniques. Morphological techniques are useful when major parts of bones are present. Skull, pelvis and long bones play vital role in morphological analysis [1]. Figure 1 shows major components of human skeleton, which provide good results in using these techniques. Bone morphology fails when no major part of bones are present or only bone fragments are available. In such cases, histological techniques can be used for microscopic analysis of fragments. Kerley in 1965 was the first to give a relational regression equation of bone microstructures, which can be used to calculate age at death [2]. His work opened doors to new research. Thompson in 1981 continued his work to the next level and developed age-graded contrasts in bone cells for both genders[3].

Since then, different researches have been conducted on histological analysis of bone with contradicting outcomes[4-6]. This is mainly because humans belonging to different races and regions, which have different properties of bone structures. These papers are discussed in detail with regards to the history of sex determination from bone histology.

To understand bone histological analysis, one must know the basic structure of bone. Bones are tissues, which provide structural stability to the body, and provide protection to the internal fragile organs. Bones are mainly made of proteins and minerals, making them a reservoir for the body. They have space for bone marrow, and provide perfect environment and protection for bone marrow to provide blood cells to the body.

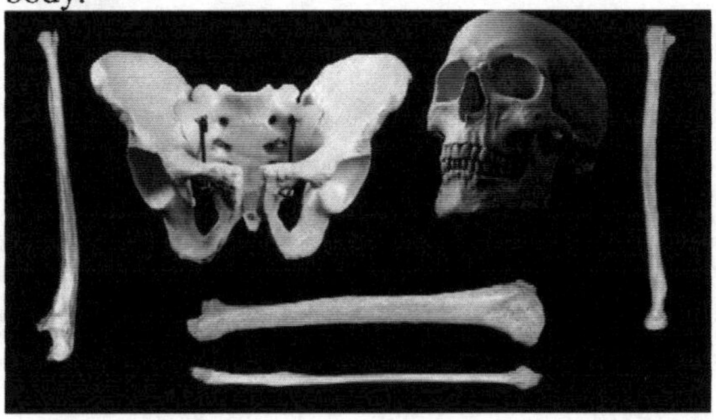

Figure 1. Human skull, pelvis and long bones.

They consist of living cells, which repair and reconstruct bones according to the body requirement [7]. These cells are osteoblasts, osteoclasts and osteocytes. Osteoblasts are bone-making cells. They make bone structures by secreting bone matrix. These mononuclear cells are small in size compared to osteoclasts, and they stick together through short, slender processes[8]. Osteoclasts, on the other hand are multinucleated cells. These cells perform enzymatic digestion of bone, which is also known as bone resorption. Osteocyte forms the chief cellular component of mature bone. During bone formation, osteocyte is formed from resting osteoblast, which is incorporated into a small space called lacuna in the matrix. Osteocytes have a number of canaliculi, which anastomose with canaliculi from adjacent cells [8] for the exchange of tissue fluids and metabolites between osteocytes and blood capillaries Bone remodeling is a process, in which osteoclast and osteoblast take part to resorb old bone and make new bone. This process is carried out at multiple locations in human skeleton [8]. Remodeling process produces a haversian canal system consisting of concentric lamellae. This system is known as an osteon system (Figure 2). Bone is continuously under the process of remodeling, which changes the microscopic structures. These structures are analyzed in many histological techniques mainly for age estimation and sex determination.

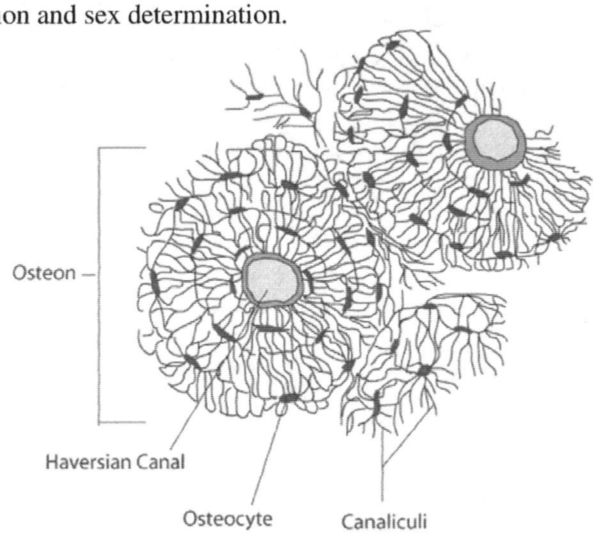

Figure 2. bone cell at microscopic level [9].

2. Bone microstructural parameters

Most common problem encountered in study of sex determination is the differences in selection of microstructural parameters and their classifications by researchers. The outcome of a study to estimate sex and age is strongly dependent on the selection of bone microstructural parameters. During analysis, bone microstructural parameters can be categorized as observed and derived parameters. This section indicated some microstructural parameters, which were generally used by researchers.

2.1. Observed microstructural parameters

In this study observed parameters are referred to the parameters which are directly analysed from the bone microscopic image. Frequently used observed bone microstructural parameters in previous studies are fragmented osteon (FO), secondary osteon (SO), haversian canal count (HCC), osteon area (OA), osteon diameter (OD), haversian canal area (HCA) and haversian canal diameter (HCD).

2.2. Derived microstructural parameters

Derived parameters are normally estimated from observed parameters. The selection of choosing derived parameters or observed parameters in order to estimate sex depends on the researchers work. The following sections will identify the microstructural parameters of both categories analysed by researchers in the literature. Frequently used derived parameters are Fragmented Osteon Density (FOD), Secondary Osteon Density (SOD), Secondary Osteon Region (SOR), Fragmented Osteon Region (FOR), Haversian canal Region (HCR).

3. Sex determination

This section discussed in detail the history of sex determination. Kerley was the first researcher, who analysed bone microstructures, and gave age estimation regression equation. This opened doors from age estimation to sex determination, height and weight estimation[2].

Kerley developed a regression equation for age estimation by analysing modelling and remodelling stages of bone. Four fields were selected for analysis, namely, anterior, posterior, medial and lateral. A 100-power wide field in each of the four anatomic areas were analysed for the number of osteons and its fragments, amount of lamellar bone and the number of non-Haversian canals. Each microscopic field in the image was 1.25 mm in diameter. The inspection area enclosed by a 10-power objective lens combined with 10-power wide-field oculars. After Kerley's work on age estimation, the focus diverted to estimate human age in different races. Since many researchers reported that gender of the human do not affect the histology of bone, therefore researchers were more interested in human age estimation leaving a big space in the field of sex determination. Singh and Gunberg developed age estimation method in 1970 from male samples which was also tested on female samples producing satisfying results. However when this method was experimented on additional amount of female specimen by Samson and Branigan, the accuracy of the estimation decreased[10]. In 1979 a sex determination method was developed by *DD* Thompson which reported that males groups have smaller average haversian canals area than females in similar group ($p< 0.05$)[11]. However in the fifth and eight decade, haversian canal area in males increased by 133% and in females by 52%. Male were reported to have larger haversian canal area in fifth and eighth decade. Female samples showed greater total area of haversian canal ($p< 0.05$). Application of the same regression equations to different populations can be done only when sufficient numbers of skeletons of known age at death from each different population are analysed. Using a small core of bone instead of a complete cross-sectional area minimizes the physical damage to a skeleton, and helps ensure access to skeletons, where the question of population-specific, age-related changes in osteon turnover may be addressed directly. The validity of these regressions was based on the New England white population [11]. In 1981, Thompson analyzed the femoral bone cores of adult Eskimo skeletons from Southampton Island (n = 69), St Lawrence Island (n = 53), Baffin Island (n = 44), Kodiak Island (n = 921) and compared the results with the specimen of U.S. whites (n= 144). Males of Southampton Island Eskimo indicated average haversian canal area (HCA) around 0.084mm^2 / mm^2 and in females average HCA was 0.105mm^2/ mm^2 ($p< 0.05$) [3]. Nine parameters namely bone mineral index, cortical bone density, cortical thickness, haversian canal and primary osteon area, Secondary osteon lamellae area, secondary osteon lamellae plus haversian canal area, secondary osteon number, haversian canal perimeter length and secondary osteon perimeter length were analyzed in this study.

Age at death estimation in Eskimo skeletons of US whites was done using regression equations which showed that ages obtained from this method exceeded from those obtained by morphological methods. It was specifically true for Eskimos morphologically aged less than 35 years, where it may be assumed that morphological ages are the most accurately determined. Age estimates of white samples were calculated accurately with mean absolute difference of 4.76 years between known and estimated age. The absolute difference between histological and morphological age for Southampton Island skeletons was 15.41 years for females and 14.33 years for males. This degree of error related with the histological method among Eskimos is beyond the error, probably due to variances in the rates of change for morphological indicators between the two populations. Differences of similar degree were reported for histologically and morphologically estimated ages in each of the other Eskimo groups.

David B Burr in 1990 compared the study of skeletal remains of an archaic Native American population (Pecos Indian) with the modern American population. The purpose of his study was to compare changes in an archaic Native American population with those in the modern white populations. His study compared age-related patterns of histologic change in an archaic human population with the modern US and British populations. Secondly, he assessed sexual dimorphisms in histologic structure, and examined whether micro- and macro structural adaptations to aging were related. He reported that males in the Pecos population had smaller osteons than females ($p< 0.2$)[12]. Osteonal mean wall thickness was reported to be larger in females. Due to the smaller size of osteons in males, they number, and osteon population density was greater compared to females ($p< 0.06$). This indicates the occurrence of high rate of bone remodelling in males.

MF Erickson in 1991 also studies sex-specified age estimation and divided sex into age-graded groups by decades. This study was conducted in Washington DC [13]. The research comprised 154 female samples with age ranges from 14-97 years and 174 male samples ranging from 16-97 years. Five types of parameters were observed, secondary osteon, type II osteon, fragments, resorption space and non-haversian canal.

Erickson analyzed type II osteons and reported that it had a positive relation with age. However these type II osteons remains constant at a specific stage. Non-haversian canals were reported to have a decrement in both sexes. Furthermore it was reported that in each decade osteon gradually increased in males, while in females, this increase occurred only till sixth decade after which it remained constant. It was reported that in fifth decades of female age, the osteon fragments/mm^2 increased compared to osteons. But, in males, this ratio did not exceed the number of osteons prior to seventh decade. Erickson also analyzed that in eight decade of females, fragmental bone area increased more than osteonal bone and the area of fragmental bone never increased more than the osteonal bone in males.

DM Mulhern (1997) examined remodeling in femurs of Medieval Nubian Population of 1250–1450 AD. Mulhern observed considerable variations in remodeling patterns of both genders. Eight areas in the sample were analyzed under a microscope. Ten microstructural parameters were used in this research which showed that the number of intact osteons was higher in males compare to females, whereas the area of osteons were larger in females compare to males. He did not find any significant statistics in Haversian canal dimensions between sexes. Intact osteons number showed significant correlation with sex ($p< 0.0001$). Average intact osteons number in females was (6.73/mm^2) and (9.74/mm^2) in males [14]. Average fragmented osteons number in females was 4.68/mm^2 and 2.59/mm^2 in males, which showed a significant difference. Osteon size also varied with sex (table 1). The mean osteonal cross-sectional diameter and area are different between both sexes ($p< 0.05$).

KL Bell in 2001 analyzed femoral mid-shaft cross-section of 66 samples and divided them according to their age and sex in groups of two decades each (21-40, 41-60, 61-80, > 80). He reported that the existence of considerable percentage of new young osteon meant that cortical remodeling in female was higher. Females had 34% higher remodeling than males ($p = 0.034$). When combined, both sex showed no significant difference in parameter of osteon density and between ages[15] (figure 3).

Table 1: Microstructural parameter differences in both sexess.

Parameter	Female	Male
Intact osteon	6.73 ± 0.31	9.74 ± 0.39
Fragmented osteon	4.68 ± 0.27	2.59 ± 0.14
Haversian canal area	0.0021 ± 0.0001	0.0022 ± 0.0002
Osteon area	0.040 ± 0.001	0.036 ± 0.002
Average osteon count	12.03 ± 0.47	12.79 ± 0.43

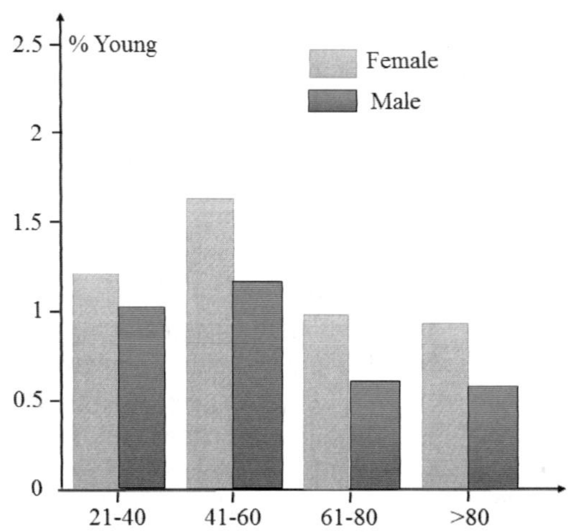

Figure 3. Evaluation of young osteon based on age groups [18].

C David *L* Thomas in 2005 studied sex and age differences in 168 samples (73 males and 75 females) of modern Australian population by analyzing porosity in femur samples. Microradiographs of 100 μm sections were analyzed in which pores in bones were extracted with the help image processing method. First, the samples were categorized in three age groups as young, middle and old (20–44 years, 45-64 years and 65+ years) however later he studied the relation between medullar area with sub periosteal area which produced better outcomes. Cortex of the bone was distributed in three radial rings and eight circumferential segments. The porosity in each section was analyzed and it was reported that with the increasing age the porosity in the inner cortex and porosity in the posterior and anterolateral region increases. The porosity patterns in males developed steadily, however in females the porosity patterns were much different in initial age compare to old age. The porosity patterns were regular with continous bone loss alongside a neutral axis of the cortex, [16].

Faridah Mohd Nor in 2009 analyzed microstructural parameters for age and sex determination in 64 Malaysian samples [17]. And reported that there is no significant difference in number of osteon in both sexes. This was in agreement with the study of Pfeiffer in 1996 and 1998. Nor Faridah also reported that females have less but larger osteons and have larger haversian canals compare to males ($p< 0.05$). These results agreed with Mulhern(1997) and Thompson(1981).

HM Britz in 2009 worked on finding out the relation between weight of the human body and geometry of osteons in human bone. The initial hypothesis of his work stated that weight of human body would be in negative relation with the size of femoral osteon area *(On.Ar)* and femoral osteon diameter (*On.Dm*) if the age, sex and height were kept constant. HM Britz also attempted to find out the link of (osteon circularity; *On.Cr*) with weight. The samples included 45 male and 43 female with age ranging from 17 to 97. Osteons numbers (*n*= 12,690) were counted in the microradiographs of femur samples. Univariate analysis of covariance was used (*n*= 87; 1 outlier) with log-transformed age, weight and height as covariates and sex as a fixed factor [18].

It was reported that weight had a negative correlation with *On.Dm* and *On.Ar* (p= 0.004 and p = 0.006, respectively). There was significant relation between age and osteons, as well as its circularity. Relation of increasing age with *On.Dm* and *On.Ar* was negative, however it was positive with *On.Cr*. The *On.Dm* and *On.Ar* were reported to show significant difference between the both genders (p= 0.019 and p= 0.021), with males having larger osteons. Sex had no significance relation with *On.Cr* (p= 0.449). Height as well, did not show any significant relation with geometric parameters. Considering partial eta-squared values, age had the highest realtion with (*On.Cr*: 30%), (*On.Ar*: 28%), (*On.Dm*: 18%, weight had the second highest relation with (*On.Dm*: 10%) and (*On.Ar*: 9%) of the variance in geometry. Sex had the lowest relation with (*On.Dm*: 7%) and (*On.Ar*: 6%). While other researchers studied relation of osteon size with age/sex, the author of this study believed that these results were the first to describe a link between weight and osteons.

C Hernandez (2012) noted a significant differences in remodeling of femoral cortex in both genders. It was analyzed that with age, fragmented osteon density varies in male more than female by 50%. These variations occurred in medial and anterior region and total cross-section. However the density of fragmented osteon in females was 50% greater than males in the regions of lateral and posterior and total cross-section. In anatomical region of males, 50% or greater intact osteon density was reported. Females had 50% or more intact osteon changes in posterior and complete cross-section. However this study was mainly focused on age estimation which did not present more detail about sex differences [19].

4. Sex determination

In this section, a compilation of research work done on sex determination using bone histology, histomorphology and porosity were compiled and discussed. The main differences in research work were based on the following factors.

- Area and race of human from which samples were selected.
- Selection of part of bone for analysis.
- Selection of microscopic cross-sectional area for analysis.
- Selection of microstructural parameters.

Table 2 shows researchers from different parts of the world with selection of bones from different parts of human bones and selection of microscopic cross-sectional region for analyses and parameters selected by researchers.

Table 2. Field, bone type and area selected by researchers.

Researcher	Region/area	Bone	Field location	Microstructural parameters
Kerley, 1965	*USA* whites	Femur, tibia, fibula		Number of osteons, Number of fragments, Amount of lamellar bone, Number of non-Haversian canals
IJ Singh, *DL* Gunberg, 1970	Portland	Femur, tibia		Number of osteons, Average number of lamella per osteon, Average haversian canal diameter
DD Thompson, 1979, 1981	*USA* whites and Eskimos from St. Lawrence, North Alaska and Canada	Femur		Cortical thickness, Bone mineral index, Cortical bone density, Secondary osteon lamellae area, Haversian canal and primary osteon area, Secondary osteon numbe, Secondary osteon lamellae plus Haversian canal area, Secondary osteon perimeter, Haversian canal perimenter.
David *B* Burr, 1990	Native American and *USA* whites	Femur		Secondary osteon area. Haverisna canal area. Sample area. Osteon wall thickness. Osteon population density. Secondary osteonal bone. Porosity. Percent osteonal refilling. Cortical area. Total area. Cross-sectional inertia. Polar moment of inertia.

Ericksen, 1991	*USA* whites	Femur	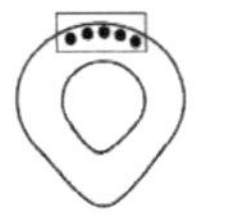	Secondary osteon. Type II osteon. Fragments. Resorption space. Non-Haversian canal.
DM Mulhern, 1997	Medieval Nubian population from Kulubnarti, Subanese Nubia	Femur	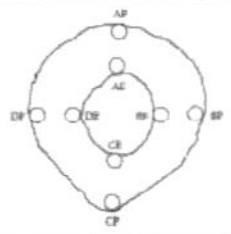	Intact Osteon. Fragmentary Osteon. Haversian Canal Area. Osteon Area. Mean Osteonal area. Cross Sectional Diameter. Osteon Population. Accumulated Osteon. Net Osteonal Remodeling Percent Osteonal Refilling.
KL Bell, 2001	Australia	Femur		Femoral porosity. Young osteon. Remodeling cluster
C David *L* Thomas, 2005	Australia	Femur		Intra cortical porosity.
F Nor, 2009	Kuala lumpur, Malaysia	Femur, tibia, ulna, radius, Humerus, fibula	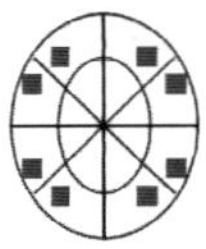	Cortical thickness. Medullary cavity diameter. Osteon count. Osteon diameter. Osteon area. Osteon perimeter. Haversian canal diameter. Haversian canal area. Haversian canal perimeter. Haversian lamellae count.
HM Britz, 2009	Australia	Femur	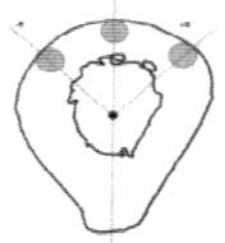	Osteon Area. Osteon Diameter. Osteon Circularity
C Hernandez, 2012	*USA* whites	Femur	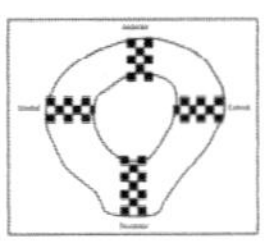	Intact secondary osteons. Fragmentary Secondary Osteons. Treatment of Type II Osteons. Cortical Area Evaluated. Intt. Secondary Osteon Density. Fragments Density. Osteon population. Total subperiosteal Area. Endosteal Area. Cortical Area. Relative Cortical Area.

5. Conclusion

The literature of bone histology does not give strong evidence of changes in bone microstructures based on genders. Several microscopic parameters of different race and regions were analysed by different researchers for sex determination (Table 2). However, the implementation of these methods can be questioned while applying it to all races. The characteristics of bone microstructures varies with race and region. So far, little research has been carried out in the field of human bone histology to determine sex differences and there is little proof of bone microstructural parameters difference between the two genders. No developed method can be applied to identify sex differences in all races, however studies are being conducted to observe sex differences in specific regions and races. Further research is required in this filed to be more precise. A group of clearly defined selected parameters should be studied to determine sex. Use of modern technology could produce better results as computers have become good enough to analyse and extract minor details in medical images. Microscopic images of bone cells are still being inspected manually by experts which makes it a tedious and labour-intensive process. Furthermore, manual inspection of medical images often yields biased results that reliant on the skills of the inspector. Analysis of medical images using modern technology will not only be robust and accurate but will save time and cost. It will also allow objective and reproducible results and could reduce human biased errors. Computer vision systems can assist in storage of large data base of bone images and in transfer of the extracted information from various images. Nevertheless, creating practical automatic systems for bone microstructure image analysis and their explanation has been a major challenge.

References

[1] H. Cho, S. D. Stout, and T. A. Bishop, "Cortical bone remodeling rates in a sample of African American and European American descent groups from the American Midwest: Comparisons of age and sex in ribs," *American Journal of Physical Anthropology,* vol. 130, pp. 214-226, 2006.

[2] E. R. Kerley, "The microscopic determination of age in human bone," *American Journal of Physical Anthropology,* vol. 23, pp. 149-163, 1965.

[3] D. D. Thompson and M. Gunness-Hey, "Bone mineral-osteon analysis of Yupik-inupiaq skeletons," *American Journal of Physical Anthropology,* vol. 55, pp. 1-7, 1981.

[4] I. Khan, F. Mohd Nor, and M. M. Abdul Jamil, "A Survey of Human Age Estimation Techniques from Bone Microstructures," in *International Conference for Innovation in Biomedical Engineering and Life Sciences : ICIBEL2015, 6-8 December 2015, Putrajaya, Malaysia*, F. Ibrahim, J. Usman, M. S. Mohktar, and M. Y. Ahmad, Eds., ed Singapore: Springer Singapore, 2016, pp. 203-207.

[5] I. Khan, M. M. A. Jamil, T. N. T. Ibrahim, and F. M. Nor, "Automated human age estimation at death via bone microstructures," in *2016 6th IEEE International Conference on Control System, Computing and Engineering (ICCSCE)*, 2016, pp. 580-583.

[6] I. Khan, M. M. A. Jamil, T. N. T. Ibrahim, and F. M. Nor, "Analysis of age-related changes in Haversian canal using image processing techniques," in *2016 6th IEEE International Conference on Control System, Computing and Engineering (ICCSCE)*, 2016, pp. 169-172.

[7] D. G. Steele, *The Anatomy and Biology of the Human Skeleton*: Texas A&M University Press, 1988.

[8] L. J. a. J. Carneiro, *Basic Histology: Text and Atlas*, 14 ed.: McGraw-Hill Medical, 2005.

[9] Gary, "Gray's Anatomy of the Human Body," in *[http://www.bartleby.com/107/illus77.h*, o. system, Ed., ed, 1918.

[10] B. K. Samson C, "A new method of estimating age at death from fragmentary and weathered bone. In Bodington A, Garland AN, Janaway RC, editors. Death Decay and Reconstruction Approaches to Archaeology and Forensic Science," *Manchester: Manchester University Press,* pp. 101-108, 1987.

[11] D. D. Thompson, "The core technique in the determination of age at death of skeletons," *J Forensic Sci,* vol. 24, pp. 902-15, Oct 1979.

[12] D. B. Burr, C. B. Ruff, and D. D. Thompson, "Patterns of skeletal histologic change through time: comparison of an archaic native American population with modern populations," *Anat Rec,* vol. 226, pp. 307-13, Mar 1990.

[13] M. F. Ericksen, "Histologic estimation of age at death using the anterior cortex of the femur," *Am J Phys Anthropol,* vol. 84, pp. 171-9, Feb 1991.

[14] D. M. Mulhern and D. P. Van Gerven, "Patterns of femoral bone remodeling dynamics in a Medieval Nubian population," *Am J Phys Anthropol,* vol. 104, pp. 133-46, Sep 1997.

[15] K. L. Bell, N. Loveridge, J. Reeve, C. D. Thomas, S. A. Feik, and J. G. Clement, "Super-osteons (remodeling clusters) in the cortex of the femoral shaft: influence of age and gender," *Anat Rec,* vol. 264, pp. 378-86, Dec 1 2001.

[16] C. D. L. Thomas, S. A. Feik, and J. G. Clement, "Regional variation of intracortical porosity in the midshaft of the human femur: age and sex differences," *Journal of Anatomy,* vol. 206, pp. 115-125, 01/05/accepted 2005.

[17] F. M. Nor, R. F. Pastor, and H. Schutkowski, "Age at death estimation from bone histology in Malaysian males," *Med Sci Law,* vol. 54, pp. 203-8, Oct 2014.

[18] H. M. Britz, C. D. Thomas, J. G. Clement, and D. M. Cooper, "The relation of femoral osteon geometry to age, sex, height and weight," *Bone,* vol. 45, pp. 77-83, Jul 2009.

[19] M.-T. J. Cosgriff-Hernandez, "Histomorphometric Estimation of Age at Death Using the Femoral Cortex: A Modification of Established Methods," The Ohio State University, 2012.

Age regression for Malaysian males using cortical bone Histomorphometry

Ijaz Khan[1], Muhammad Mahadi Bin Abdul Jamil[1], Faridah Mohd Nor[2]

[1]Faculty of Electrical and Electronic Engineering Universiti Tun Hussein Onn Malaysia (UTHM), Batu Pahat , Malaysia.
[2]Department of Pathology, Faculty of Medicine Universiti Kebangsaan Malaysia (UKM), Medical Centre, Kuala lumpur , Malaysia

mahadi@uthm.com

Abstract. Bone microstructures have the capability to record the variations in its structure that occur during the life span. These variations can be analysed to estimate human age at death. This study analysed 30 Malaysian male samples in order to create a regression equation for age estimation. Bone samples were collected from human long bones i.e humerus, radius, ulna, tibia, fibula and femur. Two morphological and five histological variables were selected for analysis. Cortical thickness and medullar cavity diameter did not show any significant relation with increasing age however haversian canal parameters and osteon count showed significant correlation with age. Regression equation with lowest SEE 7.2 was based on Haversian canal area, haversian canal radius and osteon count. The regression equation estimated age at death within 10 years of range for 85% samples. The mean value of the known ages was 39.8 years while the mean value of the estimated ages was 37.03 years. Increasing number of samples including female samples for further research might produce promising results.

1. Introduction

Estimation of human age at death from skeletal remains is one of the main physical aspect in reconstruction demographic profiles. Macroscopic and microscopic analysis of unknown bones might help in estimation of age at death as human skeleton is characterised by both morphological and histological age related changes[1]. Morphological skeletal characteristics are commonly used to estimate age at death in younger individuals [2] while histological features are used in attempt to predict age of older individual [3]. The accuracy of the age estimation techniques is dependent on the diagnosis of several age-related changes in skeleton that occur during the life span. These morphological and histological variation in skeleton are controlled by bone modelling and remodelling. Bone become more porous with increasing age as remodelling results in increases bone resorption and less bone formation. These skeletal changes and growth in sub adults are more controlled and occur over a shorter time span then the breakdown and degeneration of skeleton in adults which means that there is little variation in skeletons of sub adults compare to older adults. This makes age at death estimation of sub adults more accurate while adults age estimation techniques yields estimate with imprecise large ranges. After 50 years of age degeneration patterns in skeleton become unclear to predict their chronological age with fair accuracy.

Different factors such as intrinsic and extrinsic genetic factors, environmental and biomechanical factors alter the histology and morphology of the skeleton [4]. Occupation and different physical activity could also lead to distinct bone mass. Factors such as gender, hormones [5], disease, diet [6], life history [7], ethnicity, nutritional stress [8] and length of daylight [9] influence the rate at which a person's skeleton breaks down.

Several methods based on morphology and histology of bones has been developed to estimate human age at death. Initial macroscopic techniques were qualitative [10],[11],[12] and were dependent of different degenerative stages of skeleton. In latter attempts microscopic histological methods were developed which were applicable to fragments of bones as well. The accuracy of these methods varies with the availability and nature of the samples and age [13],[14].

Jowsey analysed a high amount of bone formation and resorption in young individuals leading to a high rate of bone turnover. In young adults rate of bone turnover is lower while in older adults rate of bone resorption is high [15]. After 70 years of age there is little evidence of increase in bone formation and up to 25% of their endosteal bone surface may be occupied by resorption. Kerley's histomorphometric techniques that used a complete bone cross section is cited to be the most accurate of all age estimation techniques [15] which reported that numbers of osteons and osteon fragments increases with age, while a gradual decrease occur in percentage of circumferential lamella and number of non-haversian canals. However, the definitions of these variables were uncertain as it was difficult to differentiate between complete osteons and osteons fragments. This uncertainty led to increase the observational error and effected the reliability of results [15]. In later techniques, the average of Kerley's regression equations estimates was calculated and more reliable age estimates were derived [16]. The original method was corrected by increasing microscopic filed size from 1.25 mm to 1.62 mm and a correction factor was introduced which should be multiplied with count of osteon, fragmented osteon and non-haversian canal [17].

The existence of difference in patterns of bone microstructures has been an important concern in histological age estimation techniques. Bones with different size of osteons and haversian canals exhibits different osteon population densities and will reach to asymptotes at different stages of age [18]. Bone microstructures were identified to have difference with age [19],[20],[21],[22] and activity and varies with population and ancestry [18], [23],[24],[25]. Some researchers reported the difference in bone microstructures especially osteons with gender [26] while other found no such difference [24]. It is vital in classifying the reliability of age estimation techniques to consider the variation of bone microstructures with population. As these variables exhibit difference at same age in two different population due to under or over ageing. Hence age estimation method developed on the samples of one population might not be reliable for other population. The purpose of this study is to explore the variation in bone microstructures of Malaysian males and develop a regression equation for age estimation methods.

2. Materials and methods

The samples of this study were collected from the mortuaries of Forensic unit, Hospital University Kebangsaan Malaysia and Hospital Kuala Lumpur. Total of 30 men samples were analysed. Medical history including sex, race, ethnicity and age at death of the samples were known. Samples were handled under the Criminal Procedure Code s.331 subsection 2 (CPC s. 331 (2), which stated that: The parts of the human body may be taken and given to the institute for research. Death of most of the samples were reported to occur due to road accidents. The previous medical record reported that all samples were free from pathological conditions. The age range was between 22 to 64 years with mean value of 39.8 years. Only long bones of skeleton were considered in this study, namely humerus, radius, ulna, fibula, tibia and femur. All medical credentials to use bone samples for research analysis were fulfilled.

2.1. Sample preparation and microstructures extraction

Sample preparation technique was based on method described in [4]. A 3 cm length of fragment was cut from the mid shift of long bone and was de-fatted in diethyl ether solution. The fragment was then immersed in mixture of epoxy resin and was placed in vacuum chamber to clear air bubbles and to facilitate resin in fragment of bone. Further sectioning of bone was done with the help of microtome (*LEICA SP* 1600, Germany). The resulting fragment of bone was glued to a glass slide and was exposed to ultra violet light for 30 seconds. The bone sample was then sectioned at 30 microns. The images of the resultant sample were taken with the help of a camera mounted on a transmitted light microscope (Nikon Eclipse E600) at 10x magnification. Figure 1 and Figure 2 show the resultant sample and Microscope used for image acquisition.

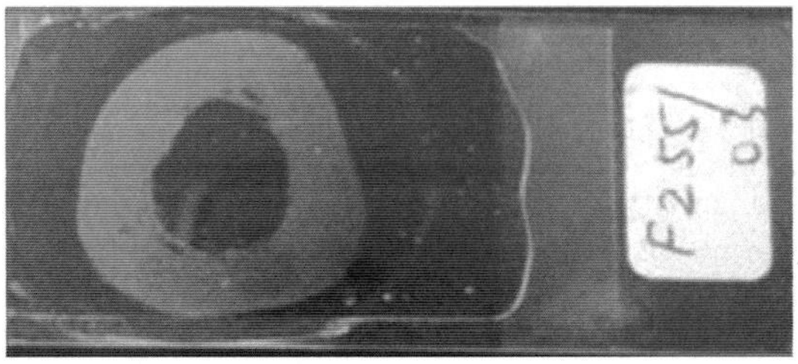

Figure 1. Sample of long bone (radius) used in analysis.

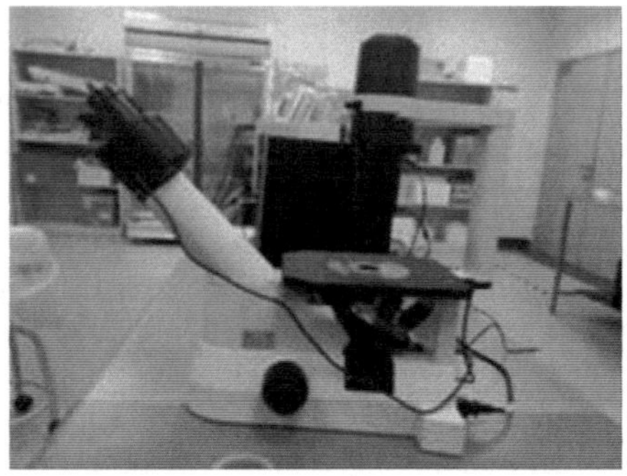

Figure 2. Microscope Nikon Eclipse E600

Total of eight images were taken of each sample on both sides of anteromedial, anterolateral, posteromedial and posterolateral shown in figure 3. The location of image acquisition on both sides of line was separated by a square grid of 10x10 µm (100μm^2).

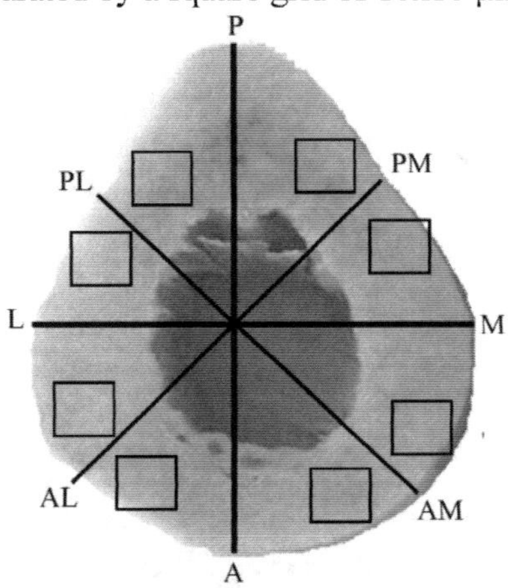

Figure 3. Filed of image acquisition anteromedial (AM), anterolateral (AL), posteromedial (PM), posterolateral (PL).P (posterior), A(anterior),M(Median), L (Lateral).

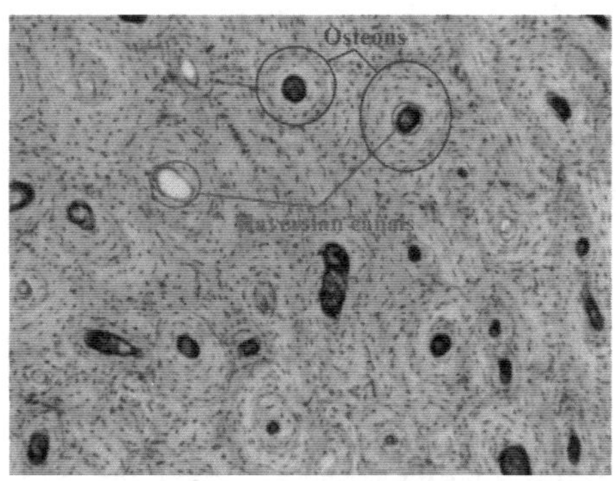

Figure 4. Microscopic image of femur on PM location. Blue circles in image represent osteons and red circles represent haversian canals (HC).

Two morphological variables i.e. cortical thickness (CT) and medullar cavity diameter (MCD) while five histological variables i.e. osteons count (OC), haversian canal number (HCN), haversian canal radius (HCR), haversian canal area (HCA) and haversian canal perimeter (HCP) were taken into consideration. Cortical thickness (CT) and medullar cavity diameter (MCD) were measured using digital Vernier calliper. Osteons having half or more than half of haversian canals present in the visual filed were considered in counting. Fragments of osteons present on the periphery of the filed were considered as well. Haversian canal parameters were extracted and measured using Cell profiler. Haversian canal area (HCA) was calculated by summing up the area of all haversian canals in eight images of one sample.

Haversian canal area (HCA) = $\sum_{8}^{i=1} sum\ of\ HCA\ in\ image_i$

The correlation of morphological and histological variables with age were analysed using Pearson's correlation (R) and regression equations with different combinations of bone microstructures were developed in SPSS software.

3. Results and discussion

In statistical analysis cortical thickness (CT) in human long bones did not show any significant correlation with age ($P > 0.05$, $R=0.005$). Osteon count showed significant correlation with age at ($P<0.03$) and ($R=0.518$). The haversian canal parameters (HCN), (HCR), (HCA) and (HCP) showed significant correlation with increasing age. Haversian canal area (HCA) showed the most significant positive correlation with age with (P<0.01) and (R=0.526). Haversian canal radius (HCR) and haversian canal perimeter (HCP) also showed positive correlation with age. Table 1 shows the correlation and significance values of all parameters with age as dependent variable.

Table 1. Correlation between morphological and histological microstructures with age

No	Bone microstructures	Pearsons's correlation (R	Significance (two tailed)	Total samples (N)
1	Cortical thickness	-.005	.980	30
2	Medullar cavity diameter	-.073	.700	30
3	Osteon count	.518	.001	30
4	Haversian canal area	.526	.001	30
5	Haversian canal radius	.511	.001	30
6	Haversian canal perimeter	.483	.002	30
7	Haversian canal number	.491	.002	30

The correlation of haversian canal parameters shows that HCA increases with age which means that the number of young osteons are present as they have larger haversian canals. The plot diagram of HCA and OC correlation with age is given in figure 5 and figure 6.

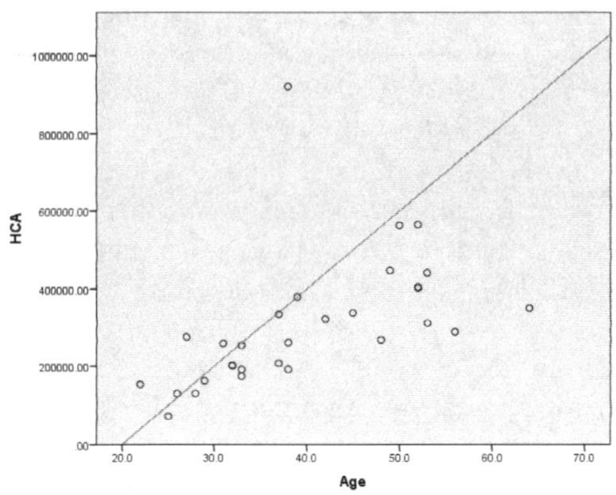

Figure 5. Correlation of HCA with age in μm^2.

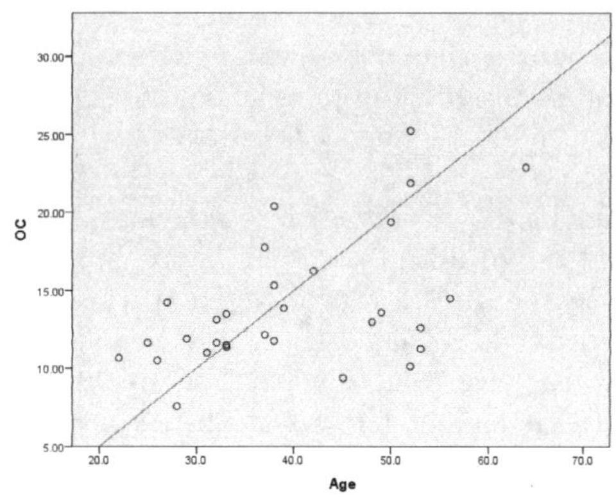

Figure 6. Correlation of OC with age.

Medullar cavity diameter (MCD) shows very low negative correlation with age which suggests that MCD decreases with a very slow rate. Regression equations based on different combination of microstructures were generated.

Table 2. Regression equation for human age estimation in Malaysian males

No	Regression equation	R^2	SEE
1	$Y=16.06+0.5(CT)+1.14(MCD)+5.1x10^{-5}(HCA)-0.007(HCN)-0.21(HCP)+0.44(OC)+3.3(HCR)$	0.6	8.1
2	$Y=16.02+2.41x10^{-5}(HCA)+0.02(HCN)-0.083(HCP)+0.64(OC)+2.47(HCR)$	0.6	7.4
3	$Y=19.83+2.84\ x10^{-5}(HCA)+0.07(HCN)-0.12(HCP)+2.5(HCR)$	0.6	7.6
4	$Y= 4.83+1.48\ x10^{-5}(HCA)+0.026(HCN)+2.2(HCR)+0.67(OC)$	0.5	7.3
5	$Y= 6.66+1.54\ x10^{-5}(HCA)+ 2.33(HCR)+0.74(OC)$	0.5	7.2

The SEE of regression equations shown in table 2 varies between 7.2 and 8.1. As haversian canal area osteon count shows the highest correlation with age in these samples, the equation with the lowest SEE is based on osteon count, haversian canal area and haversian canal radius. Equation 5 in table 2 was tested for cross validation. The estimated age based on this equation varied between 10.4 years for 85% of samples. The estimated age range was within 15 years for samples having age range from 50 to 64. The mean value of the known ages was 39.8 years while the mean value of the estimated ages was 37.03 years.

The results of this analysis is in agreement with [4],[2] studies, which were carried on Asians samples. However, the regression equation of this study was based on long bones of male samples only. Including of female samples and increasing the total numbers of sample might help in achieving more accuracy. Further research on behaviour of different microstructures with increasing age would be needed to generate regression equation for Malaysian females and to achieve better accuracy in existing methods.

4. Conclusion

This study analysed 30 Malaysian male samples to developed age estimation regression equation for Malaysian males. Age related changes were studied in morphological and histological microstructures of long bones. The histological microstructures such as osteon count, haversian canal area and haversian canal radius in long bones showed significant correlation with age. The estimated age for individuals up to 50 years was estimated within 9 years, however the estimated range increased when the chronological age of samples exceeded 50 years. Morphological factors such as cortical thickness and medullar cavity diameter did not show any significant correlation with increasing age. Combination of histological and morphological microstructures were used to generate several regression equations however equation with lowest SEE (7.2) was based on osteon count, haversian canal area and haversian canal radius. Study of different parameters of osteons such as osteon density, osteon area, and osteon population might also help in developing better regression equations. Including samples of females might lead to high rate of accuracy. The results of this study are in agreement with the literature which is based on Asian samples. Further research is needed to reduce the difference between chronological age and estimated age.

Acknowledgments

The authors wish to express their gratitude to Office for Research, Innovation, Commercialization and Consultancy Management (ORICC), University Tun Hussein Onn Malaysia for providing postgraduate research Grant vote number (U279) to support this research.

5. References

[1] C. D. L. Thomas, M. S. Stein, S. A. Feik, J. D. Wark, and J. G. Clement, "Determination of age at death using combined morphology and histology of the femur," *Journal of Anatomy,* vol. 196, pp. 463-471, 2000.
[2] M. Yoshino, K. Imaizumi, S. Miyasaka, and S. Seta, "Histological estimation of age at death using microradiographs of humeral compact bone," *Forensic Sci Int,* vol. 64, pp. 191-8, Feb 1994.
[3] D. Thompson, "Microscopic Determination of Age at Death in an Autopsy Series," 1981.
[4] F. M. Nor, R. F. Pastor, and H. Schutkowski, "Age at death estimation from bone histology in Malaysian males," *Med Sci Law,* vol. 54, pp. 203-8, Oct 2014.
[5] P. Zioupos, A. Williams, G. Christodoulou, and R. Giles, "Determining 'age at death' for forensic purposes using human bone by a laboratory-based biomechanical analytical method," *J Mech Behav Biomed Mater,* vol. 33, pp. 109-23, May 2014.
[6] D. D. Thompson and M. Gunness-Hey, "Bone mineral-osteon analysis of Yupik-inupiaq skeletons," *American Journal of Physical Anthropology,* vol. 55, pp. 1-7, 1981.
[7] D. D. Thompson, "The core technique in the determination of age at death of skeletons," *J Forensic Sci,* vol. 24, pp. 902-15, Oct 1979.
[8] G. SM, "The earlier gain and later loss of cortical bone in nutritional perspective.," *Springfield: Charles C.Thomas, 1970.*
[9] O. D. a. S.-E. F. Richman EA, "Differences in intra-cortical bone remodeling in three aboriginal American populations: possible dietary factors.," *Calcif Tissue Int* pp. 209–214, 1979.
[10] W. M. Krogman, "Skeletal changes in young American males," *American Journal of Orthodontics,* vol. 44, pp. 715-716, 1958.
[11] T. W. Todd, "Age Changes in the Pubic Symphysis: VII. The Anthropoid Strain in Human Pubic Symphyses of the Third Decade," *Journal of Anatomy,* vol. 57, pp. 274-294.12, 1923.
[12] S. T. Brooks, "Skeletal age at death: the reliability of cranial and pubic age indicators," *Am J Phys Anthropol,* vol. 13, pp. 567-97, Dec 1955.
[13] E. Kerley, "Forensic Anthropology and Crimes Involving Children," 1976.
[14] S. S. Katzenberg MA, "Ubelaker DH. Methodological considerations in the forensic applications of human skeletal biology. In: eds. New York,NY: Wiley-Liss " *Biological anthropology of the human skeleton.,* pp. 41–67, 2000.
[15] J. J., "Age changes in human bone.," *Clinical Orthopaedics,* vol. 17, pp. 210-218, 1960.
[16] J. Ahlqvist and O. Damsten, "A modification of Kerley's method for the microscopic determination of age in human bone," *J Forensic Sci,* vol. 14, pp. 205-12, Apr 1969.
[17] E. R. Kerley and D. H. Ubelaker, "Revisions in the microscopic method of estimating age at death in human cortical bone," *Am J Phys Anthropol,* vol. 49, pp. 545-6, Nov 1978.
[18] S. Pfeiffer, C. Crowder, L. Harrington, and M. Brown, "Secondary osteon and Haversian canal dimensions as behavioral indicators," *Am J Phys Anthropol,* vol. 131, pp. 460-8, Dec 2006.
[19] I. Khan, M. M. A. Jamil, T. N. T. Ibrahim, and F. M. Nor, "Analysis of age-related changes in Haversian canal using image processing techniques," in *Proceedings - 6th IEEE International Conference on Control System, Computing and Engineering, ICCSCE 2016*, 2017, pp. 169-172.
[20] I. Khan, M. M. A. Jamil, T. N. T. Ibrahim, and F. M. Nor, "Automated human age estimation at death via bone microstructures," in *Proceedings - 6th IEEE International Conference on Control System, Computing and Engineering, ICCSCE 2016*, 2017, pp. 580-583.

[21] I. Khan, F. Mohd Nor, and M. M. Abdul Jamil, "A survey of human age estimation techniques from bone microstructures," in *IFMBE Proceedings*, 2016, pp. 203-207.

[22] H. Abdullah, F. Mohd Nor, and M. M. Abdul Jamil, "Human bone histomorphological pattern differences between genders: A review," in *IFMBE Proceedings*, 2016, pp. 183-187.

[23] H. M. Frost, "Human Haversian system measurements," *Henry Ford Hosp Med Bull,* vol. 9, pp. 145-7, Mar 1961.

[24] H. Cho, S. D. Stout, R. W. Madsen, and M. A. Streeter, "Population-specific histological age-estimating method: a model for known African-American and European-American skeletal remains," *J Forensic Sci,* vol. 47, pp. 12-8, Jan 2002.

[25] Y. Watanabe, M. Konishi, M. Shimada, H. Ohara, and S. Iwamoto, "Estimation of age from the femur of Japanese cadavers," *Forensic Sci Int,* vol. 98, pp. 55-65, Nov 30 1998.

[26] D. M. Mulhern and D. P. Van Gerven, "Patterns of femoral bone remodeling dynamics in a Medieval Nubian population," *Am J Phys Anthropol,* vol. 104, pp. 133-46, Sep 1997.

IOP Conf. Series: Journal of Physics: Conf. Series **1019** (2018) 012012 doi:10.1088/1742-6596/1019/1/012012

Correlation between Electrical Field Strength and Pulse Width Analysis on Cell Viability

Suhassni Ganeson[1], Muhammad Mahadi Abdul Jamil[1], Hassan Buhari Mamman[1], Nur Adilah Abd Rahman[1]

[1]Biomedical Modelling and Simulation (BIOMEMS) Research Group, Faculty of Electric and Electronic Engineering, Universiti Tun Hussein Onn Malaysia, 86400 Batu Pahat, Johor, Malaysia.

mahadi@uthm.edu.my

Abstract. Electric field pulses can make living cell membrane electroporated either reversibly or irreversibly. This study determined the effects of electric field pulses on HeLa cancer cell viability. First set of experiment- Pulses electric fields (control, 200V/cm, 400V/cm, 600V/cm, 800V/cm, and 1000V/cm) with fixed pulse width (200μs) were performed on HeLa cells. Second set of experiment - Pulses width of (control,100μs, 500μs,1ms,5ms,10ms) with constant electric field strength of (600V/cm) were performed on HeLa cells. Number of pulses fixed to 1 for both sets of experiments. The quantitative determinations of cell viability measurement were performed by the aid of haemocytometer and trypan blue staining. The data shows that HeLa cell viability decreases as the electric field strength and pulse width. This is because electric pulses alter the cell's transmembrane potential, causing disruption of the lipid bilayer, and increase the cell membrane permeability. Efficient electroporation outcome only will be gained by fine tuning the appropriate electric field parameter according to the application needed.

1. Introduction

Electroporation, a microbiology technique being favourably applied in several areas like medicine, biotechnology, biology, food processing and such due to the versatility [1]. It works with any cell, with fewer steps, where it becomes a uniquely beneficial method. In medicine field, electroporation is used for electrochemotherapy. It is a new technological approach treatment; where cytotoxicity of anti-cancer drugs is increased by means of electroporation to the tumorous cells [2]. While in biotechnology, electroporation system used for sterilizing water. Electric field is effective in killing bacteria and pathogens without the chemicals use. Electroporation creates impermanent pores in microorganism cell membranes which decrease dosages of chlorine in water disinfection [3-4].Apart from that, electroporation technique plays a major role in food processing. It has shown a promising outcome in food preservation, where pulsed electric fields used for microbial inactivation with least effect on food quality [5].No matter what application it is, for a successful outcome, pulse parameter is the important key in electroporation-based application. Meaning that, efficient electroporation is achieved only after careful fine-tuning on the parameters which includes pulse amplitude, pulse duration, and number of pulse. However, each application differs in the setting of pulse parameter according to the cell size, cell type, orientation of cells and so on [6]. However, in the present review an attempt has made to investigate the effect of electric field strength on HeLa cancer cell by measuring cell viability percentage.

IOP Conf. Series: Journal of Physics: Conf. Series **1019** (2018) 012012 doi:10.1088/1742-6596/1019/1/012012

2. Methodology

2.1. Cell culture

In this study, HeLa cell were used as an experiment sample. HeLa cell were cultured in RPMI1640 medium supplemented with 10% of fetal bovine serum (FBS) and 1% of antibiotic (penicillin and strepcomycin) from Gibco USA. Cells were kept in incubator at 37°C containing 5% of CO2 in order to grow and maintain its optimal temperature and humidity. HeLa cell will undergo sub-culture process every 3 to 5 days after which it reaches confluency of 80-90%.For sub-culture process, old medium will be aspirated and discard from culture flask. Then cells were washed thrice with 2ml of phosphate buffered saline (PBS) to remove the serum of media, so that trypsin will be able to detach the cell from flask surface. Cells were digested with 2ml of trypsin and incubate for about 5 to 10 minutes. Double volume of complete growth medium added into flask to stop the effect of the trypsin and then centrifuged at 1000 rpm for 5 minutes. Supernatant will be removed and cell pellet will be re-suspended in complete growth of medium for a uniform.

2.2. Electroporation

ECM 830 square wave electroporation system is an appropriate for drug, protein, gene delivery .ECM 830 generators devised for in vitro and in vivo applications. The ECM 830 comes with combination of a wide array of BTX specialty electrodes and accessories to intensify molecular and drug delivery experiments. ECM 830 electroporator were used to generate electric pulses needed to electroporate the cells. This device comes with a wide range of voltages from 5 to 3,000 V and pulse durations from 10µsec to 10sec. It also available with digital display of actual voltage and pulse length delivered.
After detaching process, eight hundred microliter (800µl) of cells suspension, then poured into a 4mm electrode gap cuvette (BTX Harvard Apparatus). A single pulse electric field with an intensity of 200V/cm, 400V/cm, 600V/cm, 800V/cm, 1000V/cm with a constant pulse duration of 200µs and another set of experiment with constant field strength of 600V/cm with a varied pulse duration of 100µs, 500µs, 1ms, 5ms, 10ms were utilized in electroporating the cells. Then, cell suspension from each cuvette, were then seeded into wells of six-well culture flasks containing 2ml of complete growth medium. Similarly, 800µl cells from the same initial flask without electroporation were also seeded in another well containing 2ml of medium. Latterly cells will be incubated at 37oC and 5% of CO2. Viability of cell was observed after 48 hours of incubation.

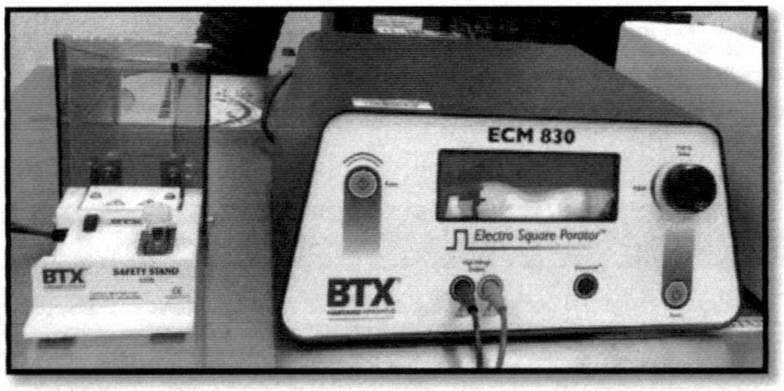

Figure 1. ECM 830 Electroporator

2.3. Measurement of cell viability

Cell viability measured after 48 hours of incubation with the aid of haemocytometer. Haemocytometer is a device where we used to determine the concentration of cells in a given sample. It is a special type of microscope slide consisting of two chambers, which is divided into nine (1.0mm x 1.0mm) large squares which are separated from one another by triple lines.
After cell detachment, 100 µL of cell suspension will be mixed gently with 100 µL of trypan blue. Next by using a pipette, 100 µL of Trypan Blue-treated cell suspension slowly filled on both chambers underneath the coverslip, allowing the cell suspension to be drawn out by capillary action. Next the device will be placed on the microscope stage and the cell suspension is counted. In order to get the

cell viability, add together the live and dead cell to obtain a total cell count. Then, divide the live cell count by the total cell count to calculate the percentage viability.
100µs, 500µs, 1ms, 5ms, 10ms were utilized in electroporating the cells. Then, cell suspension from each cuvette, were then seeded into wells of six-well culture flasks containing 2ml of complete growth medium. Similarly, 800µl cells from the same initial flask without electroporation were also seeded in another well containing 2ml of medium. Latterly cells will be incubated at 37°C and 5% of CO_2. Viability of cell was observed after 48 hours of incubation.

3. Experimental result

3.1. HeLa cell viability

Electroporation is a technique which is used extensively to enhance the permeability of living cell. The proficiency of electroporation depends on the electric field parameters. Cell membrane permeability increases when electric field applied and subsequently allows molecules, drugs, ions to be introduced into cell for an effectual treatment.
Percentage of cell viability was measured. Viability measurements used to estimate the life or dead of cancerous cells. The results obtained in this study, as shown in Table 1 and Figure 2 quantitatively revealed the dependence of cell viability on electrical parameters. Control, 200V/cm, 400V/cm, 600 V/cm, 800V/cm, 1000V/cm electric field strength with a constant pulse width of 200µs showed 100%, 98.2%, 96.5%, 96%,93.3%,89.2% of cell viability respectively. Based on the result obtained, cell survival decreases with an increasing of field strength from 200V/cm to 1000V/cm with fixed pulse width of 200µs and single pulse.

Table 1. Hela cell viability percentage with varied field strength and constant pulse width (200µs).

Electroporation parameter	Cell Viability (%)
Control, 0V/cm	100
200V/cm, 200µs	98.2
400V/cm, 200µs	96.5
600V/cm, 200µs	96
800V/cm, 200µs	93.3
1000V/cm, 200µs	89.2

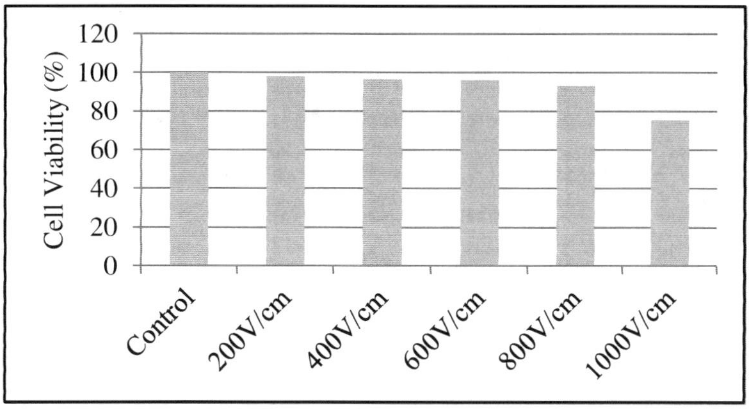

Figure 2. HeLa cell viability percentage with different field strength and fixed pulse width

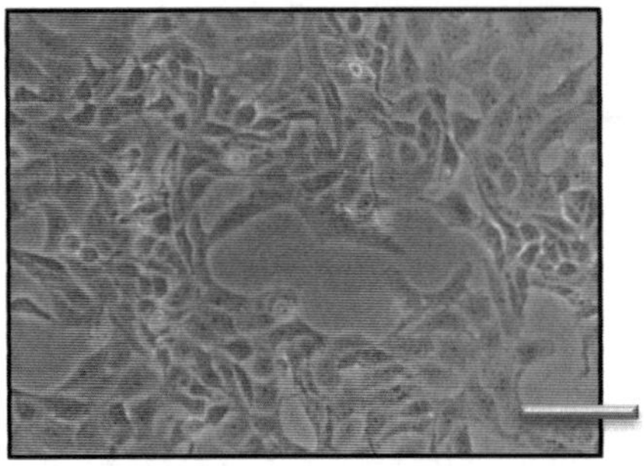

Control, 0V/cm

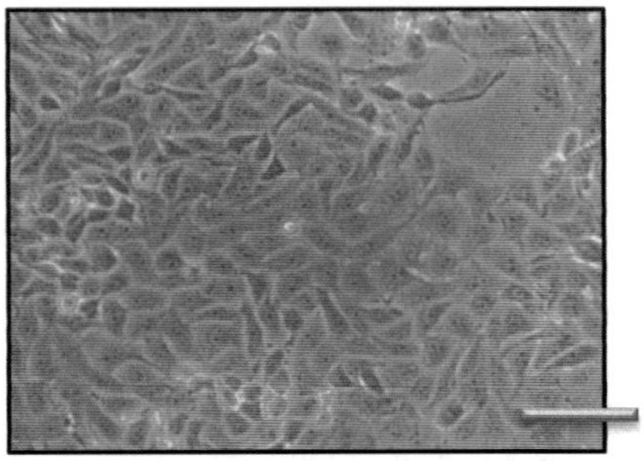

200V/cm, 200μs

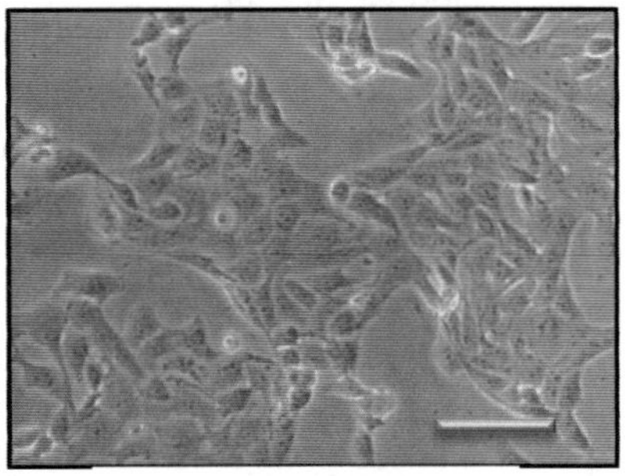

400V/cm, 200μs

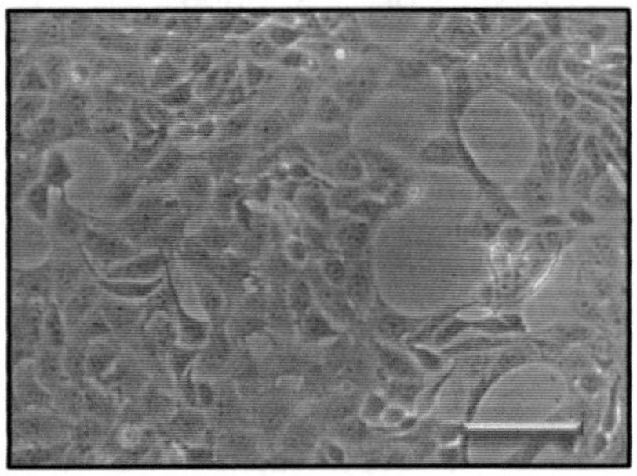

600V/cm, 200μs

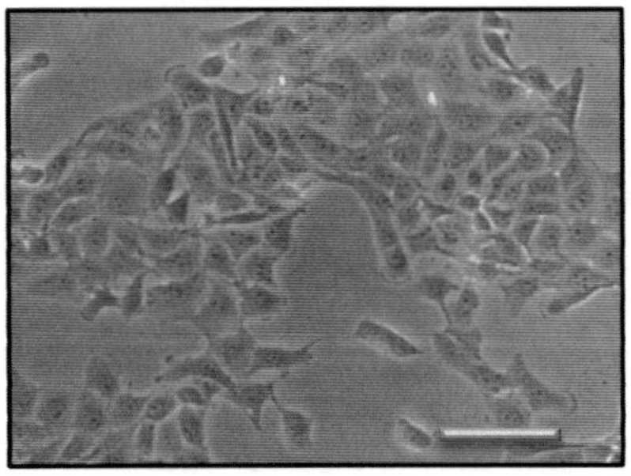

800V/cm, 200μs

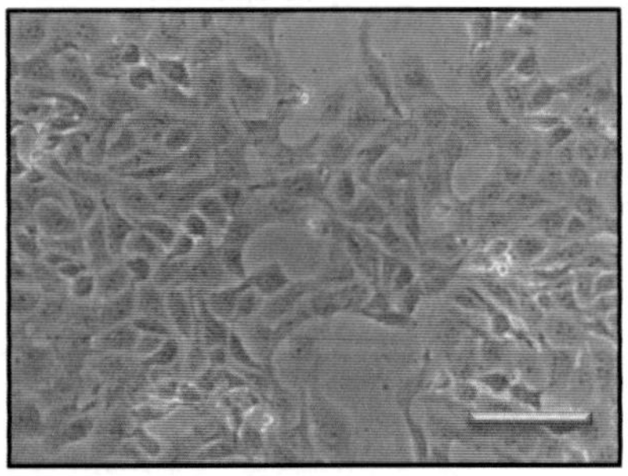

1000V/cm, 200μs

Figure 3. Image of HeLa cell Electroporated with different field strength parameter (Scale bar -50μm).

Another set of experiment carried out, where pulse duration varied (control, 100μs, and 500μs, 1ms, 5ms, and 10ms) and field strength fixed at 600V/cm with single pulse. Result obtained as in Table 2 and Figure 2 showed 95.3%, 97.2%, 95.4%, 86.1%, 75.2%, and 53.3% of cell viability percentage. Cell viability found to be diminishing as the pulse width increases.

Table 2. Hela cell viability percentage with varied pulse width and fixed electric field strength (600V/cm).

Electroporation parameter	Cell Viability (%)
Control, 0V/cm	100
100μs, 600V/cm	97.2

500μs, 600V/cm	95.4
1ms, 600V/cm	86.1
5ms, 600V/cm	75.2
10ms, 600V/cm	53.3

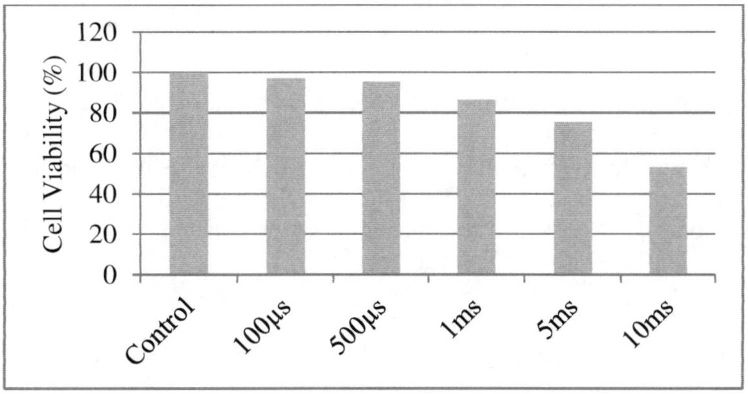

Figure 4. HeLa cell viability percentage with different pulse duration and fixed electric field strength of 600V/cm

In a way to discover the effects of electroporation with different field strength and pulse duration on tumour cell, viability of cell was measured. Experiment results showed that cell survival is significantly affected when treated with pulse electric field. Both set of experiments demonstrates a decreasing number of percentages in HeLa cell viability. Higher the voltage and pulse width, lower the cell viability becomes. Electric pulses alter the cell's transmembrane potential results in membrane permeation. This happens when induced transmembrane potential exceeds the breakdown potential lipid bilayer. Next pores are formed in membranes and maintained by the electrical pulse field. Cell membrane pores created when the energy stored in the membrane capacitor exceed the energy required to keep the membrane impaired against pore expansion. Electric field parameter with high percentage viability stimulates the cells in absorbing more nutrients from the media through the formation of pores. While the higher field strength may lead to cell apoptosis. In these phenomena, mitochondria the power house of biological cells play a potent role. An early signal for the programmed cell death is a decrease of mitochondrial transmembrane potential, which leads to opening of pores in the inner mitochondrial transmembrane. Decrease in mitochondrial transmembrane potential causes the release of an enzyme called cytochrome c from mitochondria into cytosol. This enzyme will activate pro-apoptotic caspases [7]. Qualitative observation of the media 1640 at higher pulse amplitude, showed a change of colour from red to orange. This could be the change of pH of the medium from neutral to acidic due to high number of cell death.

4. Conclusion

In this paper attention was given to the electric field parameter on cell viability. The result of this paper demonstrates that cell viability slowly decreases when the electrical field strength and pulse duration increases. Thus, the result showing that cell viability is influenced by electrical field parameters (field strength and pulse width). Efficient electroporation outcome only will be gained by fine tuning the appropriate electric field parameter according to the application needed. The result of this study can be used for further investigation in order to identify the optimal electroporation condition according to the application needed.

5. Acknowledgement
The author would like to convey her appreciation to supervisor Associate Professor Dr. Muhammad Mahadi Bin Abdul Jamil for the guidance and support. Author wish to acknowledge the financial support of the Ministry of Education Malaysia through MyMasters (MyBrain15) and Contract Grant (U551) ORICC, Universiti Tun Hussein Onn Malaysia.

6. References
[1] L. Rems, D.Miklavcic 2016 Electroporation of cells in complex materials and tissue (Slovenia:Journal of Applied Physics)
[2] M.Cemazar, J.Ambrozic Avgustin, D. Pavlin, G.Sersa, A.Poli, A.Krhac levacic, N.Tesic, U.Lampreht Tratar, M.Rak 2016 Efficacy and safety of electrochemotheraphy combined with peritumoral IL-12 gene electrotransfer of canine mast cell tumours (Slovenia: Veterinary and Comparative Oncology Article)
[3] Lyndon B. Johnson 2005 Electroporation System for sterilizing water (Texas: Biomedical Article)
[4] Schlager, Kenneth J 2001 An electroporation drinking water disinfection system (Bioelectromagnetics Inc.)
[5] Samo Mahnic, Eugene Vorobiev, Damijan Miklavcic 2014 Electroporation in food processing and biorefinery (New York: Springer)
[6] Theodora Tryfona, Mrk T. Bustard 2008 Impact of pulsed electric field on Corynebacterium glutamicum cell membrane permeabilization (UK: Journal of Bioscience and Bioengineering)
[7] Yi Peng, Xiaohong Weng 2008 7th Asian-Pacific Conference on Medical and Biological Engineering: APCMBE (China: IFMBE Proceeding)

Food Code Breaker (FCB) – Developing the Multi-Lingual Food Code Translation Android Application Using Multi–Options Code Reader System

Sharmini Abdullah [1,5], Mohamed Elshaikh [2], Muhammad Bazli Mahmood [1], Norshahrizan Nordin [3,5], Syed Zulkarnain Syed Idrus [4,5], Rozilawati Mahadi[1,5], Umie Naiemah Saraih[3]

[1]Centre of International Language (CIL), Universiti Malaysia Perlis (UniMAP), Malaysia.
[2] School of Computer and Communication Engineering (SCCE),Universiti Malaysia Perlis (UniMAP), Malaysia.
[3] School of Business Innovation and Technopreneurship (PPiPT), Universiti Malaysia Perlis (UniMAP), Malaysia.
[4] School of Human Development and Techno-communication (iKOM) & GREAT Research Group – CEGeoGTech, Universiti Malaysia Perlis (UniMAP), Malaysia.
[5]Centre for Excellence Geopolymer & Green Technology, Universiti Malaysia Perlis (UniMAP), Malaysia.

sharmini@unimap.edu.my,elshaikh@unimap.edu.my,bazlimahmood@unimap.edu.my norshahrizan@unimap.edu.my, syzul@unimap.edu.my, rozilawati@unimap.edu.my, ummi@unimap.edu.my

Abstract.This paper researches a food code and food ingredient translator mobile application based on Android called Food Code Breaker (FCB). FCB functions to decipher food based chemicals or food additives which are represented by unique barcode numbers called E-numbers. FCB incorporates an easy to use and multi-option code reader food code translation platform with an Optical Character Recognition (OCR) system integrated within the application. The aim of this research is to be able to provide an affordable, accessibl and easy to use system that helps consumers to identify and verify the halal, vegan and allergy status of the ingredients listed in food wrappings or packaging via its multi-option code reader and OCR technology. The outcome of this research is an android mobile application that reads food ingredients in either textual or image formats and is able to provide a customizable languages translation feature to help consumers with foreign languages translation. FCB will be particularly useful for travellers with dietary restrictions,whether religious, ethical, or allergy related.

Keywords. Hand-held multimedia device, Text Translation, Text to Speech, Android Application, Optical Character Recognition (OCR) system, E-Codes, Food Code Translator, Multi –Lingual

1. Introduction

Travellers visiting countries which do not speak the same language often face problems in translating the ingredients of the food they buy. Any random food wrappings or packaging be it local or imported would often list their food ingredients either in its native language, foreign language or in E - codes. Therefore, would often have trouble verifying the halal, vegan and allergy status of these E-codes or ingredients listed in a foreign language. The lack of available translation can be more than just an inconvenience, especially if travellers had allergies to specific food ingredients which could cause dangerous after-effects. These travellers or visitors are unable to use online translation services or pocket translators to translate the food ingredients mainly because they are unable to write the foreign text they see on to these hand-held devices. This particular paper proposes an android application called Food Code Breaker (FCB) via mobile technology to solve the aforementioned problem by enabling users to take a quick scan of the QR code from the food packaging or wrappings. Travellers can scroll through the various options and have all the food ingredients displayed in their own language, with several options available including English, Malay, Chinese, Japanese, Thai, and Dutch. FCB provides an easy to use and multi-option code reader food code translation platform. FCB reads the food ingredients in either textual or image formats and is able to provide a customizable languages translation feature by which visitors to the country can quickly scan and translate the ingredients of the food they buy. FCB will be particularly useful for travellers with dietary restrictions, whether religious, ethical, or allergy related.

2. Related Work

Available studies relating to mobile application
translation systems mostly use OCR technology to recognize images, text translation and text to speech mainly to overcome language barriers during travels to
foreign countries. Titus, A. et al (2015) discussed an Android Application ATMA. ATMA stands for Android Travel Mate Application with a built in OCR suite aswell as Image Processing suite both installed in the Android device which enables Travellers and Tourists to easily capture the native country language Books pages, signboards, banners and hotel menus etc. The built-in OCR converts the text embedded in the captured image into Unicode text format. It also provides translation facility so that Tourists can translate the Native Language Unicode text into their own country language.

Panchal et al (2015) designed an Android-platform based text detection and translation application that is able to recognize text present in digital image, translate the text from English to various other languages like Arabic,German, French, Russian, Spanish etc. The app was alsoable to convert the text into speech and allows the audio file to be stored locally for later use. He and his team of researchers experimented with a set of 100 random business card pictures, captured using a smartphone camera. Findings revealed a maximum accuracy-rate that is greater than 90%, more efficient and consumes less memory rendering it a suitable app for smartphones.

J. Yang et al (2001) propose a system for translating signs written in Chinese to English. Their prototype automatically extracts sign regions from images applyingan adaptive search algorithm that uses color information and performs edge detection at different scales. The sign regions detected are segmented using Gaussian mixture models before being fed to a commercial OCR software. Their system uses Example-Based Machine Translation (EBMT) to translate the signs. This particular research above justifies the need to conduct such studies as present studies mainly focused on text recognition and translation application for smartphone targeted towards overcoming language barriers during travelling. Findings from this study will not only fill the gap in the literature, provide an insight on further enhancing existing food ingredient translator apps but also disclose the latest trends in text detection and translation system for camera based smartphone specifically in the scanning a plethora of food packaging and wrappings.

3. Methodology

3.1 Application Design

FCB application design comprises of four stages or processes which are as follows – Stage 1 Initialization, Stage 2: Data acquisition, Stage 3: Data processing and Stage 4: Data presentation. These stages are further divided into these three steps: (1) Language selection, (2) Data input and process, and (3) E-Code Extraction and presentation. These stages or processes are demonstrated via dashed blocks as illustrated in Figures 1 and 2 below. The following subsection details out these stages or processes.

Figure 1: Language Selection

3.1.1 Language Selection

Language selection is critical to identify input and output data whereby users select the language of the food code which are printed on the food packaging or wrappings, and the preferred output language (user's mother tongue or spoken language). This stage is deemed essential for processing the code and for displaying the E-code translation in the user's preferred language. Nonetheless, a default language is set to English for the purposes of handling cases involving unexpected language selections. The present version of FCB supports the following language selections - English, Malay, Chinese, Japanese, Thai, and Dutch. A GUI based feature is also designed and integrated to prompt the user for languages selection as illustrated in Figure 1 above.

3.1.2 Data Input

Data input refers to the data acquisition process. FCB implements two options for data acquisition namely textual and graphical. With these two options, the user is able to select the desired method in order to increase the flexibility of the application. The graphical option is triggered by scanning the food wrappings or packaging using the image capture resources located in the smart device in this case the camera. The process is then subsequently proceeded by two image preparation steps. First, the image should be cropped specifically to the segment where the food ingredients and E-Code are presented in the image. Secondly, the captured image is processed by an android image library to extract the image features. Then, the extracted features of the image are used as an input for an Optical Character Recognition (OCR) module of the input language.

3.1.3 E-Code Extraction and Presentation.

The E-Code extraction and presentation step is the final step which displays the obtained text for further modification by the user. The textual data acquisition dictates the user to enter the text manually. The final output of this process is the E-Code extraction and its translation in a textual format or version. Figure 2 below sums up the Application Processes of the FCB Design Application in a pictorial format.

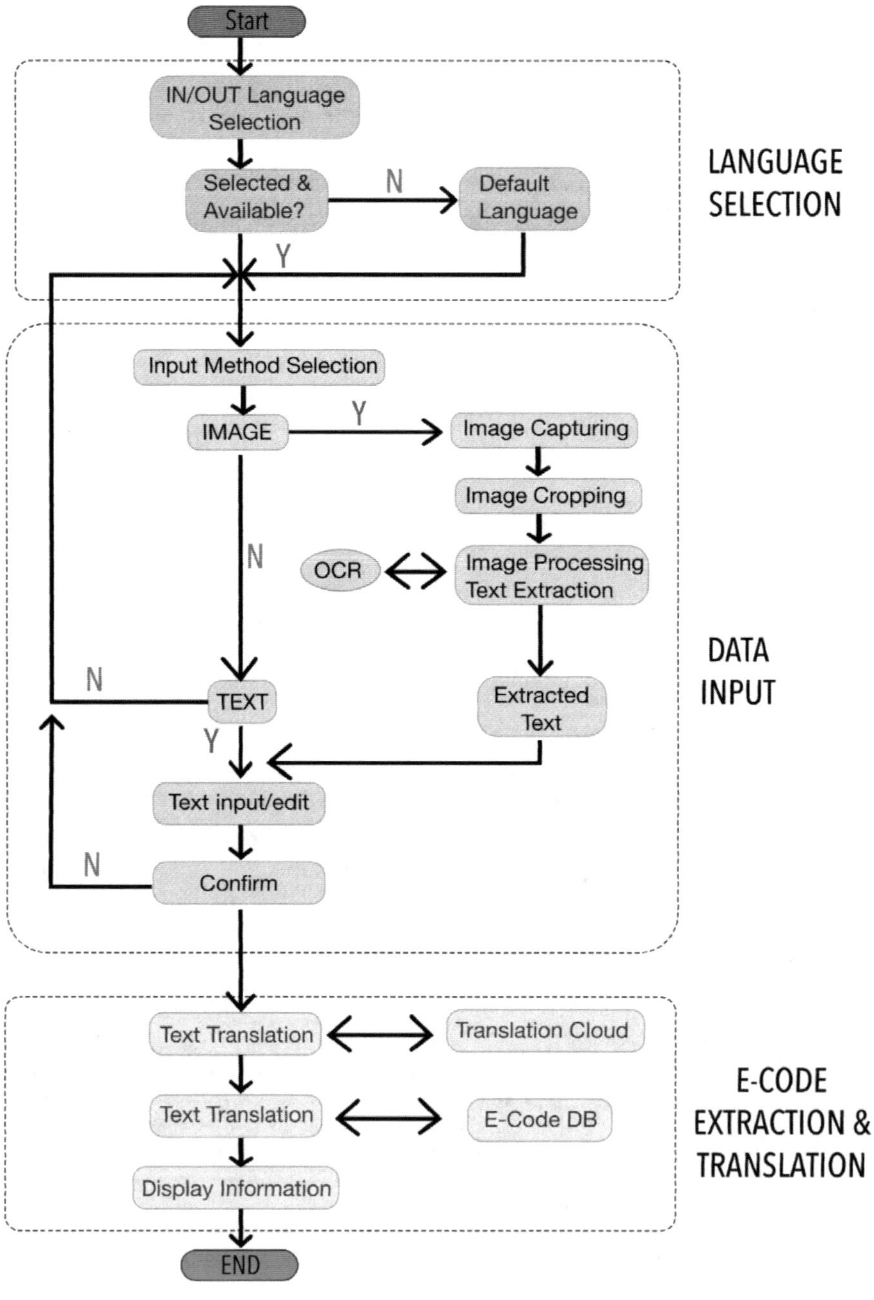

Figure 2: FCB Application Design

4. Results

4.1 FCB Application Work Process Screenshots

Figure 3 and 4 below illustrates the two versions of the FCB (formerly termed as Food Ingredient Translator [F.I.T]) work process screenshots.

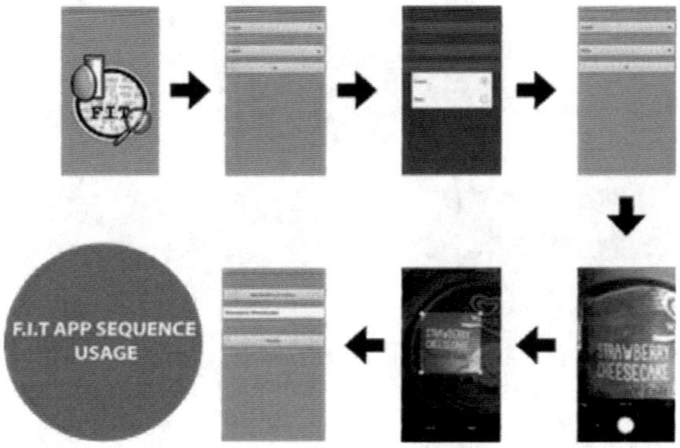

Figure 3 : Screenshots of FCB First Trial Version (Text Output)

Figure 4 below presents the upgraded version of the FCB that is able to automatically analyze the list of words (ingredients & food codes) collectively and translate them via **AUDIO OUTPUT (Say it)** or **TEXT OUTPUT (Snap it)**.

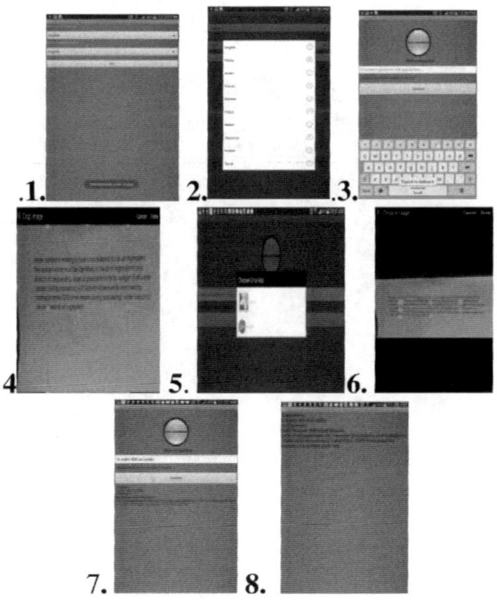

Figure 4: Screenshots of FCB's Upgraded Version (Text and Audio Output)

Figure 5 below depicts the 'SAY IT' feature of the FCB whereby users can directly speak into the microphone icon when it appears on the screen. The app will analyse users' voice input and translate the food codes and ingredients into the selected target language.

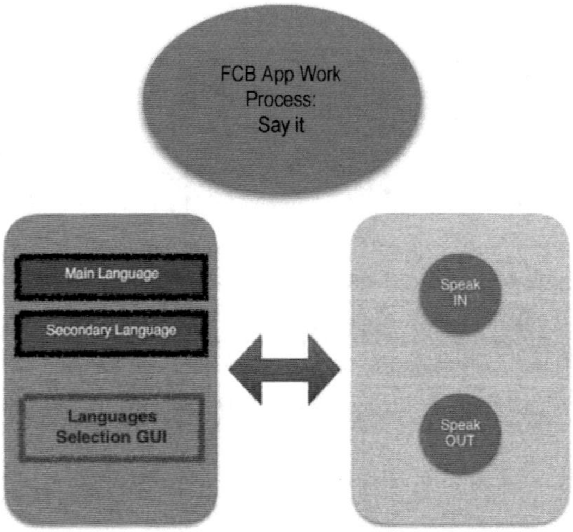

Figure 5: The 'SAY IT' Application Design of FCB

4.2 Application Evaluation Based on Similar Application Comparison

The similar application compared to the Food Code Breaker (FCB) application is the ipiit – Scan Food Ingredients application. Table 1 below displays a similar application comparison of FCB and ipiit. As illustrated in Table 1, ipiit application does not have several features provided in FCB such as 'audio output' which allows users to speak or articulate the food ingredients, 'translates food ingredients and E-Codes into other languages' which also acts as a translating device and 'multi lingual medium of instruction' which provides a multi lingual option for the user.

Table 1: Similar Application Comparison of FCB and ipiit

Feature	**Food Code Breaker (FCB)**	**ipiit – Scan Food Ingredients**
Based on Android OS	√	√
Scans bar codes of food products	√	√
Text output	√	√
Audio output	√	
Translates Food ingredients and E-Codes into other languages	√	
Multi lingual Medium of Instruction	√	
Monolingual Medium of instruction		√
Checks for allergy triggers	√	√
Checks for kosher or 'halal' status	√	
Checks for vegetarian or non- vegetarian origins	√	
Checks for Nutrition Facts		√
Provides information on food ingredients	√	√
Creates personalized preferences		√
Product comparison feature		√
Rate food product feature		√
Online mode	√	√
Offline mode		√

4.3 Application Evaluation

Section 4.3 presents the FCB application user interface evaluation based on eight rules as proposed by Shneiderman B, Plaisant C, Cohen M, dan Jacobs S. (2010) – (1) Consistency, (2) Provide Universal Usability, (3) Provide Information Feedback, (4) Design Final Dialog, (5) Simple Error Handling, (6) Internal Locus of Control, (7) Reversal Action Allowance, (8) Reduce the Information Capacity for Short-Term Memory. Analysis revealed FCB only employed 50% of the proposed eight rules:

1. Consistency
FCB keeps its consistency by using the same font type, colour, layout, menu, and menu icon in each pages. It applies to website application and mobile application.

2. Provide Universal Usability
FCB is user friendly. Its feature eases users to use the application. The 'audio output' for instance enables users with literacy or physical disabilities to use the app while the multi-lingual option caters to a wider range of users. The 'Back' button utilizes a universal icon which relates to users globally.

3. Reversal Action Allowance
FCB application integrates a simple reversal action. There is a Back button in almost all interfaces of the application.

4. Reduce the Information Capacity for Short-Term Memory FCB integrates the reduction of information capacity for short-term memory. The menu icons use universally recognized icons.

5. Conclusions and Recommendations

5.1 Conclusions

Based on the result of this research, it can be concluded that FCB has many benefits; particularly useful for travelers with dietary restrictions, whether religious, ethical, or allergy related. FCB provides an easy to use and multi-option code reader food code translation platform. FCB reads the food ingredients in either textual or image formats and is able to provide a customizable languages translation feature by which visitors to the country can quickly scan and translate the ingredients of the food they buy. It's audio output is especially appealing and user friendly as it specifically caters to users with either literacy or physical disabilities.

5.2 Recommendations

For further efficiency, this application will incorporate the remaining 50% of the eight rules in feature improvement as discussed in section 4.3 earlier. FCB can be further technological advanced to other platforms such as Blackberry and iOS and into offline mode. With this, users who use Blackberry operating system and iOS can also use FCB. Therefore, widen the target user segmentation. FCB could also be integrated with social media such as Facebook, Instagram, and Snapchat to facilitate users' in obtaining FCB application information. Further research is also strongly recommended towards further improvement of the FCB as the present accuracy rate is directly dependant on quality of input image and FCB has demonstrated difficulty to recognize cursive characters and symbols.

IOP Conf. Series: Journal of Physics: Conf. Series **1019** (2018) 012013 doi:10.1088/1742-6596/1019/1/012013

References

[1] Connolly, Thomas M., Begg, Carolyn E. (2005). Database Systems: A Practical Approach to Design, Implementation and Management. (4th edition). London: Pearson Education

[2] Garrett, Jesse J. (2010). The Elements of User Experience: User-Centered Design for the Web and Beyond. (2nd Edition). Berkeley : New Riders

[3] John Spink, Douglas C. Moyer, Hyeonho Park, Yongning Wu, Victor Fersht, Bing Shao, Miao Hong, Seung Yeop Paek, and Dmitry Edeler (2015). Introducing Food Fraud including Translation and Interpretation to Russia, Korean, and Chinese Languages. Food Chemistry 189 (2015), pp. 102 – 107.

[4] J. Yang, J. Gao, Y. Zhang, X. Chen, and A. Waibel, ``An automatic sign recognition and translation system," in Proceedings of the 2001 workshop on Perceptive user interfaces, ser. PUI '01. New York, NY, USA: ACM, 2001, pp. 1--8. [Online]. Available: http://doi.acm.org/10.1145/971478.971490.

[5] Nasser I Allheeib (2015) Mobile Application Development HALAL Food and Stores. Proceedings International Conference on Computing, Communication and Control Engineering (IC4E'2015) May 22-23, 2015 Dubai (UAE), pp. 103-108

[6] Panchal et al (2015). Design of Text Detection and Translation System for Camera Based Android Smartphone. International Journal for Scientific Research & Development| Vol. 3, Issue 1, 2015 | ISSN (online): 2321-0613

[7] Shneiderman B, Plaisant C, Cohen M, dan Jacobs S. (2010). Designing the User Interface: Strategies for Effective Human-Computer Interaction. (5th edition). New York : Addison-Wesley.

[8] Syahrul N. Junaini, & Johari Abdullah (2008) MyMobiHalal 2.0: Malaysian Mobile Halal Product Verification using Camera Phone Barcode Scanning and MMS. Proceedings of the International Conference on Computer and Communication Engineering 2008 May 13-15, 2008 Kuala Lumpur, Malaysia, pp. 528-532

[9] Somerville, Ian. (2011). Software Engineering. (9th edition). New York: Addison-Wesley.

[10] Titus al. (2015). Android Travelmate Applications with OCR & Its Language Translation. International Journal of Advanced Research in Computer Science and Software Engineering 5(3), March - 2015, pp. 1263-1270

[11] Yusuka Oda, Horoyuki Fudaba, Graham Neubig, Hideaki Hata, Sakriani Sakti, Hideaki Hata, Satoshi Nakamura (2015). Learning to Generate Pseudo-Code from Source Code Using Statistical Machine Translation. 30th IEEE/ACM International Conference on Automated Software Engineering.

IOP Conf. Series: Journal of Physics: Conf. Series **1019** (2018) 012014 doi:10.1088/1742-6596/1019/1/012014

A Secured OpenFlow Protocol Using Elliptic Curves Cryptographic for Software Defined Networks

Layth A Khalil Al Dulaimi [1], R Badlishah Ahmad[2], Naimah Yaakob[1] and Qasim Mohammed Hussein[3]

[1] Embedded Networking and Advance Computing Cluster, School of Computer & Communication Engineering, Universiti Malaysia Perlis (UniMAP) Tingkat 1, Kampus Tetap Pauh, Putra 02600 Arau, Perlis.

[2] Faculty of Informatics and Computing, Universiti Sultan Zainal Abidin, 21300 Kuala Terengganu, Terengganu, Malaysia.

[3] Control and Computer Engineering Department., Tikrit University,Tikrit - Iraq

laythadnan@studentmail.unimap.edu.my

Abstract. A Software-Defined Network (SDN) architecture was proposed to enhance the performance, flexibility and scalability of networks. However, SDN features such as centralized controlling, network programmability and virtualization present new security challenges to networks. This article will address security challenges related to SDN connection data transfer channels (i.e., control layer and infrastructure layer) and propose a novel connection data transfer channels architecture based on an OpenFlow protocol. Existing SDN OpenFlow protocols rely on higher layer secure mechanisms such as TLS (Transport Layer Security)/ SSL (Secure Sockets Layer) sessions. It is concluded that the transfer of secured data using elliptical curve encryption provides a more efficient SDN network. The proposed architecture was implemented in a testbed and its security features were analyzed.

1. Introduction

Secure Software-Defined Network data transfer between the control layer and infrastructure layer using elliptic curve cryptography is based on encryption and decryption, the proposed security mechanisms on both layers. The SDN architecture comprises three planes: Application layer, Control layer and Infrastructure layer. (APIs) Open Application Programmable Interfaces communicate between the Control and Infrastructure layers.

• Application Layer: The application layer consists of the end-user business applications and other control elements. The limit between the application layers and control layers is transferred by the northbound Open Application Programmable Interfaces (API). The SDN architecture offers different benefits such as control, flexibility, effective fragmentation, network management, network control, low operating cost, low cost plug-in devices, upon request and the provision of resources [1].

• Control Layer: The control layer consists of a centralized controller which provides unified oversight functions. Essentially, the controller supervises the packet forwarding functions of the network through an interface. Additionally, it controls all the network processes such as routing, indoctrination,

IOP Conf. Series: Journal of Physics: Conf. Series **1019** (2018) 012014 doi:10.1088/1742-6596/1019/1/012014

finishes and billing functions. The OpenFlow protocol is the most widely used control protocol in the SDN domain.

• Infrastructure layer: SDN isolates the control layer from the infrastructure layer of the network. It transfers the network intellect to a centralized controller. Consequently, the infrastructure layer now contains low end switches connected by network links. Wireless access points and the Internet are connected to the infrastructure layer. User traffic is transported through the infrastructure layer.

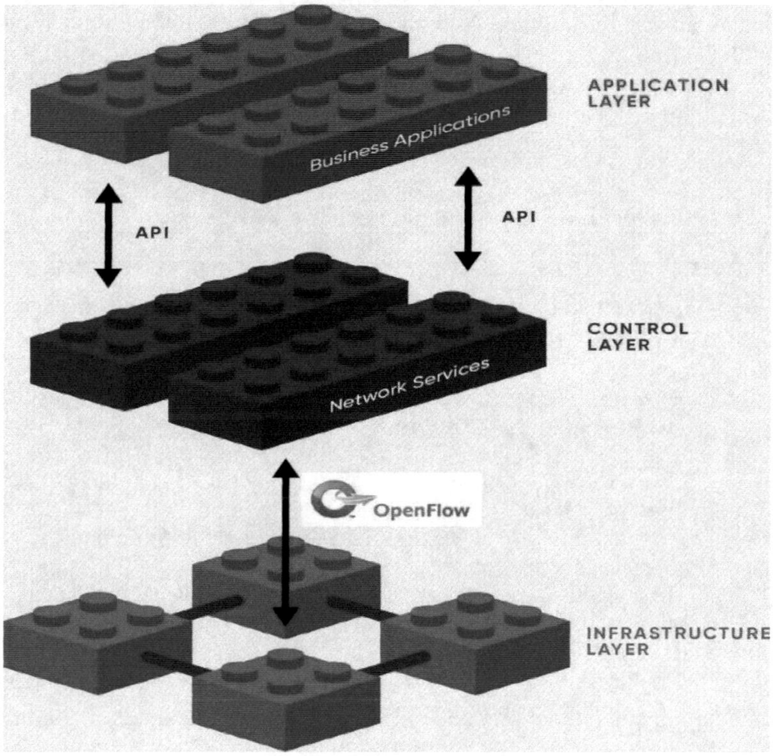

SDN Architecture.

2. *Security of SDN*

SDNs are vulnerable to security threats that can be created in different units of the network. Security threats in an SDN network can be separated into four distinct types:

1) Application layer Security,
2) Control layer Security,
3) Infrastructure layer Security
4) Layer communication Security.

SDNs contain two data transfer channels: control layer data transfer and Infrastructure layer data transfer.
Thus, the communication security threat can be divided into:
(1) Security relating to Control data transfer.
(2) Security relating to Infrastructure data transfer.

2.1. Security relating to Control data transfer

The core security issue in relation to Control data transfer is a lack of IP level security. Current SDN control protocols rely on higher layer secure mechanisms such as TLS (Transport Layer Security)/ SSL (Secure Sockets Layer). For example, the widely used OpenFlow protocol uses TLS/SSL based control data transfer [2]. However, higher layer secure mechanisms are extremely vulnerable to IP built attacks such as IP spoofing, Denial-of-service (DoS) and reset attacks [2]. Thus, higher layer defense mechanisms are not able to deliver the required level of durability and security for the Control data transfer [2].

Furthermore, a strong authentication mechanism is required between the controller and Infrastructure data transfer. If this is absent intruders can penetrate valid Infrastructure data transfer and launch security attacks on the Control data transfer. For example, the attacker can insert fake flow requests for performance-related DoS attacks [3]. However, TLS/SSL does not perform strong authentication between layers. For example, the authentication mechanism of TLS/SSL is vulnerable to IP spoofing attacks [2].

Table 1. Known attacks on OpenFlow control data transfer channels.

Attack type	description
Denial-of-service DoS Distributed denial of service DDoS	Attackers sends requests that consumes server resources to make the controller and Infrastructure unresponsive to legitimate traffic.
BEAST- Browser Exploit Against SSL/TLS attack	The attacker using cipher block chaining.
RC4 preferences in TLS	The attacker can recover the full plaintext when it is repeatedly encrypted in the same, or in several different, terms.
Reset attack	The attacker inserts a sequence of TCP reset requests to prematurely reset the communication session.
POODLE attack	The attacker's forces change the TLS sessions allowing the data to be changed using padding at the end of the block cipher. As a result, encryption codes become less secure each time it is passed.
LUCKY 13	The attacker uses a man-in-the-middle attack to recover the plaintext from a Cipher-block chaining encrypted TLS session.

The network controller is the main component of the SDN network because it comprises central intelligence and controls its functions. Consequently, attacks on the network controller represent the most serious threats to the structure of the SDN. Control data transfer is the only interface that enables a connection between the controller layer and the infrastructure layer. Thus, security in relation to data transfer control is a key factor in ensuring proper communication with the layers. A DoS attack on the SDN controller is presented in [4] in which an attacker, on an ongoing basis, sends IP packets with random headers to the controller through the control data transfer channel. This leaves the controller in an unresponsive state and unable to publish flow rules in the data transfer. TLS is used for optional menu connections in the latest open flow specification due to its complex configuration [5] Configuration is required to generate specific network site certificates and signed machine certificates corresponding to the site-level private keys for controlling and transferring data [6]. Thus, many SDN and equipment vendors omit support for TLS in the infrastructure layer. This leaves the data channel transferring control vulnerable to security attacks. The control channel must therefore be secured using other mechanisms.

2.2. Security of the SDN data channel

The current SDN traffic architecture is unencrypted which means attackers can perform the mechanisms of an "SDN scanner" to collect network information [7]. This information can then be used to perform attacks such as DoS, reset, replay and spoofing attacks [8]. Furthermore, the current SDN data transfer channel does not contain any integrity protection mechanism. An attacker could

IOP Conf. Series: Journal of Physics: Conf. Series **1019** (2018) 012014 doi:10.1088/1742-6596/1019/1/012014

therefore modify the flow of change or destroy the data without being noticed by the network operator. The rotation of data flows may result in a lower quality of service (QoS) between connection layers [7].
The SDN architecture also requires powerful mechanisms for the mutual authentication of the data transmutation channel. However, hackers can impersonate these authentication mechanisms as legitimate converters and inject fraudulent traffic flows to the data level [8]. Using this method, an attacker could exhaust the flow tables and reduce the bandwidth available for the user's traffic [2]. Furthermore, it will also affect the level of control by causing unnecessary flow requests to the controller [8].

3. Elliptic Curve Cryptosystem

Elliptic curve cryptography (ECC) is one of the public key encryption algorithms and is based on the theory of an elliptical curve on limited fields. It is used to make cryptographic keys smaller, faster and more efficient. The functions and characteristics of elliptic curves have been studied in mathematics for 150 years. [9] Their use in cryptography was suggested for the first time by Neal Koblitz and then Victor Miller in 1985. [10]. ECC has received acceptance in many accredited organizations and in security protocols since the beginning of 1990 [11].

4. Elliptic Curves

Elliptic curves are mathematical constructs that have been studied by mathematicians from the seventeenth century onwards [12].

Definition: An elliptic curve E over the finite field Fq is defined by equation (1) [13]
y 2 = x3 + ax + b,
Where a, b ∈Fq, and 4a3 + 27b2 ≠ 0 (mod q), together with a special point O, termed the point at Infinity.
The set E(Fq) consists of all points (x, y), x, y ∈Fq, which satisfy the defining equation (2), together with O.

Algebraic Elliptic Curves

The essential operation on an elliptic curve is the addition of points on the curve. To perform the addition of points on elliptic curves an algebraic formula is required. The following result provides such a formula [14].

Theorem (Algebraic Computation Law Algorithm):
let E be an elliptic curve given by equation (1) [15, 16]
Let E (Fq): y2 = x3 +ax + b,
Let Pi =(x\i , yi) ∈ E.
Let P3 = P1 ⊕ P2
Let P0=(x0,y0) ∈ E. Then: ΘP0=(x0 , -y0).
If x1= x2 and y1+ y2= 0 mod q,
then P1 ⊕ P2=*O*,
Otherwise, let

$$\lambda = \frac{y_2 - y_1}{x_2 - x_1}, \quad \text{if } x_1 \neq x_2 \qquad (2)$$

$$\lambda = \frac{3x_1^2 + a}{2y_1}, \quad \text{if } x_1 = x_2 \qquad (3)$$

P3 = P1 ⊕ P2 is given by :

$$x_3 = \lambda^2 - x_1 - x_2 \qquad (4)$$
$$y_3 = \lambda (x_1 - x_3) - y_1 \qquad (5)$$

5. OpenFlow protocol security methods

To increase the security of the network the security of the protocol is increased. The protocol is therefore responsible for protecting the transmission of data in a safe, robust and highly efficient way by using an Elliptic Curve Cryptography algorithm to transfer data. Efficiency and reliability will then be increased for each data transmission. The proposed method for protecting the protocol OpenFlow is to use an Elliptic curve cryptography algorithm that can encrypt and decrypt the data to be transported. Using an Elliptic curve cryptography that allows itself to encrypt and decrypt means a generation key is important as every public key and private key needs to be generated.

The sender will encrypt the message with the receiver's public key and the decryption device will then decrypt its key. This will increase efficiency and it is thus useful to have an elliptical curve algorithm that is effective in terms of file size data and encrypted files. This is achieved in the OpenFlow protocol where the protection used in TLS sessions has changed, increasing the security of the user protocol. This method will be useful for military intelligence as it facilitates transfer data by encrypting and decrypting data where only the source and destination can display information.

5.1 Encryption Alogrithm

A new group will be defined for encryption. To satisfy the following condition, a and b are selected:

$$a^2 + 4a^2 + 27b^2 (\text{mod } p) \neq 0$$

In this case, a and b is smaller than p. Ep (a, b) will represent a group with pair a and b.

All pairs satisfying the above conditions will be used for encryption. In general, pairs will be found using the following method.

1. $0 \leq x < p$, compute $x^3 + ax + b \ (\text{mod p})$
2. Determine whether the above result will have a square root. If it does not, x doesn't exist in Ep. If x exists, y will have two values. Pair x and y will then be used for encryption.

   Then P + Q are computed as follows: P, Q ∈ Ep (a, b);

   1. P + O = P
   2. P = (x, y) => P + (x, -y) = O. (x, -y) is reverse element of P.

      i.e. E23(1,1), P = (13, 7) => -P = (13, 16).

   3. P = (x1, y1), Q = (x2, y2), P ≠ -Q => compute P + Q = (x3, y3) as follows:

$$x3 \equiv lamda^2 - x1 - x2 \ (\text{mod p})$$
$$y3 \equiv \text{lamda}(x1 - x3) - y1 \ (\text{mod p})$$

Therefore, lamda should be computed as follows:

$$\text{lamda} = \begin{cases} \frac{y2-y1}{x2-x1} & P \neq Q \\ \frac{3x1^2 + a}{2y1} & P = Q \end{cases}.$$

5.1.1. *Encryption and Decryption*

- First, data G and group Ep(a, b) are produced.
- User B should produce security key nb and open Pb = nb * G as public.
- User B sends Pb to User A. User A will encrypt information Pm with Pb as follows:
- User A will produce random number K.
- Cm = (kG, Pm + kPb)

- User A will send Cm.
- User B will decrypt Cm with security key nb as follows:
- Pm + kPb – nb(kG) = Pm + k(nbG) – nb(kG) = Pm.

If an attacker knows G and KG, he/she can find K. This means the encryption algorithm is stronger with smaller bits.

6. Security analysis

This occurs when the algorithm has been applied in the emulator for networks NS3.25 and using the OpenFlow Protocol 1.3. It involves a single controller, two switches and two users in an SDN scenario.
After changing the TLS sessions in original protocol and using the Elliptic curve cryptography algorithm, the second table and the third table present a comparison between the current method and the proposed method through a calculation given in units of a million instructions per second (**MIPS**).

Table 2. OpenFlow Protocol using an Elliptic curve cryptography algorithm

Size of key	Computation time(MIPS)
150	$3.8 * 10^{10}$
205	$7.1 * 10^{18}$
234	$1.6 * 10^{28}$

Table 3. OpenFlow Protocol using Normal TLS

Size of key	Computation time(MIPS)
512	$3.8 * 10^{4}$
768	$2 * 10^{8}$
1024	$3 * 10^{11}$
1280	$1 * 10^{14}$
1536	$3 * 10^{16}$
2048	$3 * 10^{20}$

Thus, the algorithm is novel and strong enough to provide secure data transfer in the OpenFlow protocol used in the architecture of Software Defined Networks.

7. CONCLUSION

This paper addressed security challenges regarding connection data transfer channels in SDN mechanisms. It proposed a novel and secure method of data transfer from the control layer to infrastructure layer in the OpenFlow protocol, that of Elliptic Curve Cryptography. Reliability and efficiency were expected to increase for each transmission of data, The method increased network security by increasing the security of the protocol used in the Software Defined Network. This was achieved using an elliptic curve algorithm that allows itself to encrypt and decrypt the data that will be transferred and execute active classification. Any node between source and destination can be used to view information.

Finally, the security features and performance of the proposed architecture in NS3.25 simulation was analyzed. The results showed that the use of the proposed algorithm protected communication channels against attacks such as spoofing, DoS, replay, reset and eavesdropping attacks. However, there is a penalty incurred for security performance in terms of latency and throughput. In future, the focus will be on how to leverage resources to enhance the performance of the proposed method.

References

[1] https://www.opennetworking.org/wpcontent/uploads/2015/10/Principles_and_Practices_for_Securing_Software-Defined_Networks_applied_to_OFv1.3.4_V1.0.pdf

[2] McBride, M., Cohn, M., Deshpande, S., Kaushik, M., Mathews, M., & Nathan, S. (2013). *Sdn security considerations in the data center*. Open Networking Foundation-ONF SOLUTION BRIEF.

[3] Kreutz, D., Ramos, F. and Verissimo, P., 2013, August. Towards secure and dependable software-defined networks. In *Proceedings of the second ACM SIGCOMM workshop on Hot topics in software defined networking* (pp. 55-60). ACM.

[4] Fonseca, P., Bennesby, R., Mota, E., & Passito, A. (2012). *A replication component for resilient OpenFlow-based networking*. Paper presented at the Network Operations and Management Symposium (NOMS), 2012 IEEE.

[5] Specification, O. S. Version 1.4. 0, October 14, 2013

[6] Zerkane, S., Espes, D., Le Parc, P. and Cuppens, F., 2016, May. *Software Defined Networking Reactive Stateful Firewall.* In IFIP International Information Security and Privacy Conference (pp. 119-132). Springer International Publishing.

[7] Shin, S., & Gu, G. (2013). *Attacking software-defined networks: A first feasibility study*. Paper presented at the Proceedings of the second ACM SIGCOMM workshop on Hot topics in software defined networking.

[8] Scott-Hayward, S., O'Callaghan, G., & Sezer, S. (2013). *SDN security: A survey*. Paper presented at the Future Networks and Services (SDN4FNS), 2013 IEEE SDN For.

[9] Abbas, S. A. (2015). *Enhancing the Security of Identity and Access Management in Cloud Computing using Elliptic Curve Cryptography*. IJERMT, 4(7), 8-15.

[10] Sudha, S., & Viswanatham, V. (2013*). ADDRESSING SECURITY AND PRIVACY ISSUES IN CLOUD COMPUTING*. Journal of Theoretical & Applied Information Technology, 48(2).

[11] Nida, P., Dhiman, H., & Hussain, S. (2014). *A survey on identity and access management in cloud computing*. Int. J. Eng. Res. Technol, 3(4).

[12] Certicom Corp., 1998 *" The Elliptic Curve Cryptosystem "*, A Certicom White Paper,

[13] Fibíková, L., & Vyskoc, J. (2001). *Practical cryptography-the key size problem*: PGP after years. Paper presented at the Proceedings in Workshop "Santa's Get Together.

[14] Yan, S. Y. (2002). *Number theory for computing*: Springer Science & Business Media.

[15] Silverman, J. H. (1986). *Heights and elliptic curves Arithmetic geometry* (pp. 253-265): Springer.

[16] Araki, K., Satoh, T., & Miura, S. (1998). *Overview of elliptic curve cryptography*. Paper presented at the International Workshop on Public Key Cryptography.

Performance Analysis of Single and Multi-controller using OpenFlow Protocol for SoftwareDefined Networks

Layth A Khalil Al Dulaimi [1] R Badlishah Ahmad[2], Naimah Yaakob[1] ,Israa Sh Ahmed [3] and Mohamed Elshaikh [1]

[1] Embedded Networking and Advance Computing Cluster, School of Computer & Communication Engineering, Universiti Malaysia Perlis (UniMAP) Tingkat 1, Kampus Tetap Pauh, Putra 02600 Arau, Perlis.

[2] Faculty of Informatics and Computing, Universiti Sultan Zainal Abidin, 21300 Kuala Terengganu, Terengganu, Malaysia.

[3] Centre of Excellent Advanced Communication Engineering, School of Computer & Communication Engineering, Universiti Malaysia Perlis (UniMAP), 01000 kangar, Perlis.

laythadnan@studentmail.unimap.edu.my

Abstract. A Software Defined Network (SDN) is a relatively advanced way to implement communication networks. The decision making SDN separates the control layer, which determines where the packets are sent, from the basic infrastructure (also known as the infrastructure layer or data plan) which restores the packets to the specified destination. The newly emerging standard for enactment is the OpenFlow protocol, which contains a standard protocol for communications between the control layer and the infrastructure layer. This study analyzes the impact of OpenFlow single and multi-controllers on the performance of the OpenFlow protocol network. The research was conducted using the discrete event network simulation, omnet++, and involved analyzing the main network metrics including End to End Delay and Throughput. The results indicate that the type of controller used has a clear impact on the performance of the network.

1. Introduction

Widely used protocols in the Internet, known as TCP/IP, are a form of communication, basic contact language and protocols used to connect hosts to the Internet. The TCP/IP protocol is currently the best know network due to the successful development of Internet networks. It is therefore useful to study the behavior and efficiency of this protocol further by taking advantage of a simulation.

The need for new techniques and mechanisms requires the development of a new type of network that can be programmed for different needs and purposes by software applications based on the network operating system. It has now become easier to structure the current network and protocol designs due to the development of a new protocol called OpenFlow. OpenFlow is the first unified communications interface to be positioned between redirection and the layered structure controls of software defined network architecture. SDN is a computer network approach that enables network administrators to manage the services provided by the network by extracting higher-level functionality.

SDN has developed the OpenFlow protocol because it enables flexible control of the network devices that serve as to exchange messages between an OpenFlow controller layer and an OpenFlow switches layer [1]. Furthermore, the procedures for organizing data transfer between network elements occur at different layers in the OpenFlow protocol which can enhance performance efficiency, provide high consistency and extra connectivity, and ensure strict data protection standards.

For any new protocol in the network, there are several methods that can be used to test acceptability, evaluation and performance. One approach is to simulate a range of advantages such as cost creation, scalability, flexibility, accessibility, redundancy (for several purposes) and a simulation speed that is often faster than in real time. Easily adjustable simulation speed can be slower or faster than real time, as required. The method can run on real devices, applications and operating systems and thus offers test results that are highly realistic, however. such experiments are expensive to build. Therefore, it is very helpful to use an eraser whereby any possible scenario can be applied and, when it cannot access operation and software systems, it can be changed in any of the systems or applications as needed. Network companies have attempted to deal with distributed environments through the development of new technical solutions for routing protocols. However, there is limited information on the performance of each of these companies and there are no realistic performance comparisons available on a large scale. Consequently, very few performance ratings for OpenFlow protocols and its behavior in various domains using the OMNeT++ have been provided. Simulation tools can provide an appropriate environment for the design, construction and testing of different scenarios for users, offering practical feedback and near-reality when developing real-world systems. This allows system designers to test and determine the validity and efficiency of the design before the system is deployed. A notable feature of the simulation is that it facilitates the evaluation of different network metrics using various verification mechanisms, producing results that cannot be empirically measured on the largest geographically distributed architecture.

2. Related works

This paper focuses on the assessment and analysis of network performance and results using an OMNeT++ simulator. OMNeT++provides many advantages as a form of network simulation and has been widely used in academic research. It is capable of simulating manufacturing, airports, crowd management and estimating weather. OMNeT++ is scalable whereby all modules are implemented as C++ modules and are linked together in a single process [2]. Furthermore, OMNeT++ can modify parameters such as link bandwidth and delay as well as the configuration of network size, mobility pattern or speed for correcting performance results. When time is a concern in OMNeT++ the performance results need to be repeated for correction and accuracy. OMNeT++ supports the OpenFlow network as an extension of the INET framework including the spanning tree protocol (STP).

Regarding the OpenFlow protocol, the OMNeT++ has many limitations, one of which is that it is not new as the latest version of OpenFlow is 1.4 while, in this paper, OpenFlow switch specification 1.2 will be used [3]. The OMNeT++ uses C++ modules with its simulation engine code as well as all devices and objects, and is a user-level executable program that operates exactly like a ns-3 network simulator [4]. The NOX OpenFlow controller is a user-level program, however OMNeT++ and NOX cannot be compiled and linked together to form a single executable program [5. OMNeT++ supports several functions such as STP, however STP is not supported by ns-3 as there is no TCP connection between simulated hosts. When using a TCP connection in a real model the results are packet losses and congestion [6].

The simulated OMNeT++ module does not connect to a real OpenFlow controller and thus does not serve as an external entity for measuring the effects of the OpenFlow protocol, whereas Mininet and EstiNet can do so. The network emulator separates namespaces such as Mininet, which reduces the overheads of the system rather than acting as a single simulation process. Mininet emulated hosts use a virtualization approach and can run real applications and exchange information [7]. The network emulator gives unpredictable outcomes that yield different experimental results on each run because the emulated hosts run on CPU speed, current system activities, system load, memory bandwidth, number of emulated hosts and multiplexing over the CPU [7]. Mininet needs to run up a shell process to emulate each host using a user-space or kernel-space (OpenVswitch) to simulate each OpenFlow switch. Thus, Mininet is less scalable than EstiNet, ns-3 and OMNeT++ [8]. Mininet can only be used to study the behavior of the emulated hosts, it cannot be used to study the duration of network/application performance. Mininet GUI can be used to observe the packet playback of a simulation run, where the user needs to write Python code to set up and configure the simulation case. However, OMNeT++ has a GUI, which can be used for the observation of results whereby users need to write C++ code to set up and configure the simulation case. On balance, it is better to use OMNeT++, even it takes time and effort to create simulations, as once modules are created it is much easier to then create new ones [9].

IOP Conf. Series: Journal of Physics: Conf. Series **1019** (2018) 012015 doi:10.1088/1742-6596/1019/1/012015

3. Simulation setup for experimentation

The simulation scenarios of the OpenFlow module can be viewed in both a single controller and a multi-controller to analyze the results. This will help determine which model behaves more efficiently and more quickly

A. Simulation Scenarios

In this assessment, we use the same topology for both single controller and multi controller networks. These represent Malaysian states as shown in Figure 1 and Figure 2.

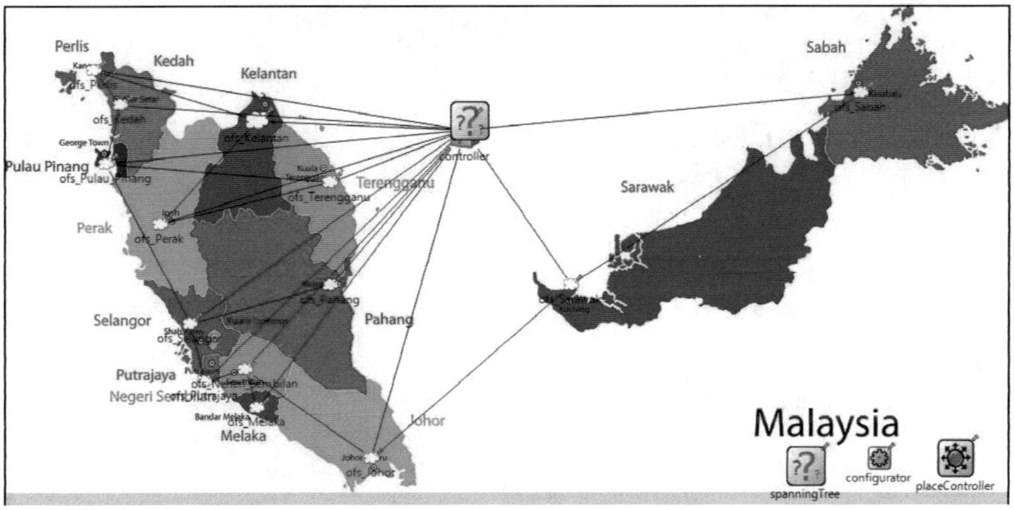

Figure 1. Single Controller

This topological diagram of Malaysia contains 14 locations, Representation of state capitals are places where, in the sole observer scenario, the OpenFlow protocol is applied in single control layer scenarios.

Where there is single control for the whole of Malaysia in the OpenFlow protocol, the protocol is connected to the infrastructure layer of each state capital. The capitals of the Malaysian states have been linked according to the calculated distance between themselves in km by using an optical cable for the simulation. The parameters of the simulation program were set up using accurate measurements of the distance between states as shown in Table 1.

Table 1 Distances between Malaysian states in Single Controller.

ofs_Perlis.gate++ <--> DistanceChannel { distance = 47km; } <--> ofs_Kedah.gate++;
ofs_Kedah.gate++ <--> DistanceChannel { distance = 171km; } <--> ofs_Pulau_Pinang.gate++;
ofs_Pulau_Pinang.gate++ <--> DistanceChannel { distance = 442km; } <--> ofs_Terengganu.gate++;
ofs_Pulau_Pinang.gate++ <--> DistanceChannel { distance = 360km; } <--> ofs_Selangor.gate++;
ofs_Terengganu.gate++ <--> DistanceChannel { distance = 367km; } <--> ofs_Perak.gate++;
ofs_Selangor.gate++ <--> DistanceChannel { distance = 277km; } <--> ofs_Pahang.gate++;
ofs_Selangor.gate++ <--> DistanceChannel { distance = 41km; } <--> ofs_Putrajaya.gate++;
ofs_Putrajaya.gate++ <--> DistanceChannel { distance = 56km; } <--> ofs_Neheri_Sembilan.gate++;
ofs_Neheri_Sembilan.gate++ <--> DistanceChannel { distance = 322km; } <--> ofs_Johor.gate++;
ofs_Neheri_Sembilan.gate++ <--> DistanceChannel { distance = 137km; } <--> ofs_Melaka.gate++;
ofs_Sarawak.gate++ <--> DistanceChannel { distance = 1038km; } <--> ofs_Sabah.gate++;
ofs_Sarawak.gate++ <--> DistanceChannel { distance = 734 km; } <--> ofs_Johor.gate++;// Direct
ofs_Perak.gate++ <--> DistanceChannel { distance = 346km; } <--> ofs_Kelantan.gate++;
ofs_Perlis.gate++ <--> DistanceChannel { distance = 410km; } <--> ofs_Kelantan.gate++;

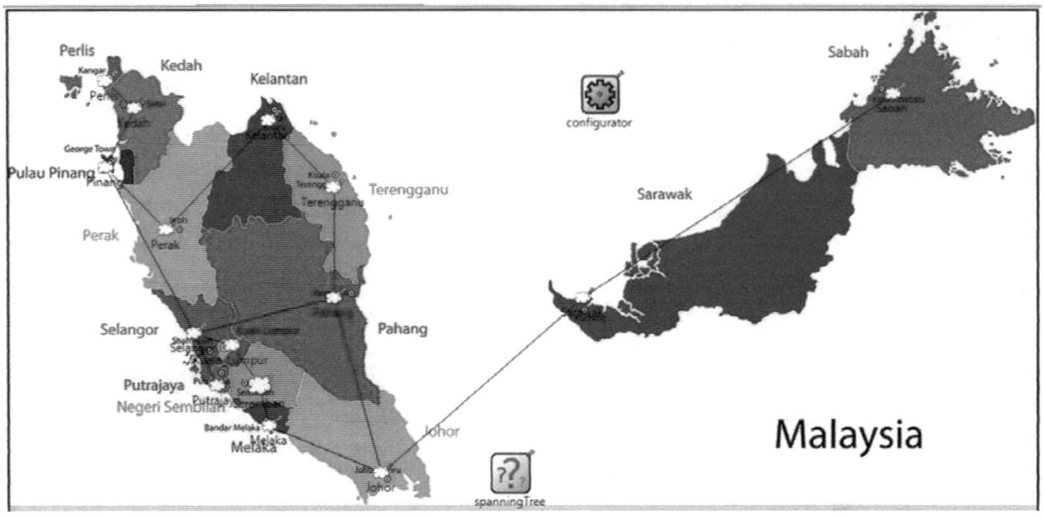

Figure 2. Multi Controller

Table 2 shows the distances between the Malaysian states in the multiple control scenario where each state is given a control and the states are associated with each other through the control units in the multiple controller layer. Each state has a structure for the SDN and the infrastructure layer is managed by a single control for the same state.

Table 2 Distances between Malaysian states in the Multi-Controller.

```
Perlis.gate++ <--> DistanceChannel { distance = 47km; } <--> Kedah.gate++;
Kedah.gate++ <--> DistanceChannel { distance = 171km; } <--> Pinang.gate++;
Pinang.gate++ <--> DistanceChannel { distance = 161km; } <--> Perak.gate++;
Perak.gate++ <--> DistanceChannel { distance = 204km; } <--> Selangor.gate++;
Perak.gate++ <--> DistanceChannel { distance = 346km; } <--> Kelantan.gate++;
Kelantan.gate++ <--> DistanceChannel { distance = 161km; } <--> Terengganu.gate++;
Selangor.gate++ <--> DistanceChannel { distance = 277km; } <--> Pahang.gate++;
Selangor.gate++ <--> DistanceChannel { distance = 41km; } <--> Putrajaya.gate++;
Selangor.gate++ <--> DistanceChannel { distance = 32km; } <--> Kuala_Lumpur.gate++;
Pahang.gate++ <--> DistanceChannel { distance = 226km; } <--> Terengganu.gate++;
Seremban.gate++ <--> DistanceChannel { distance = 86km; } <--> Melaka.gate++;
Melaka.gate++ <--> DistanceChannel { distance = 225km; } <--> Johor.gate++;
Johor.gate++ <--> DistanceChannel { distance = 323km; } <--> Pahang.gate++;
Johor.gate++ <--> DistanceChannel { distance = 734km; } <--> Sarawak.gate++;//airplan
Sarawak.gate++ <--> DistanceChannel { distance = 1038km; } <--> Sabah.gate++;
Kuala_Lumpur.gate++ <--> DistanceChannel { distance = 69km; } <--> Seremban.gate++;
Putrajaya.gate++ <--> DistanceChannel { distance = 36km; } <--> Kuala_Lumpur.gate++;
Putrajaya.gate++ <--> DistanceChannel { distance = 56km; } <--> Seremban.gate++;
Putrajaya.gate++ <--> DistanceChannel { distance = 137km; } <--> Melaka.gate++;
```

B. Performance Metric

The performance measure of the assessment is the mean End to End Delay for the nodes and Throughput in the investigated networks. In single control, the main server will be in Putrajaya to generate traffic for the network. This will be achieved by changing the size of the message length packet (0.5k, 1k, 1.5k, 2k, 2.5k) and (5k, 10k, 15k, 20k, 25k) for network efficiency in both single control and multi controller cases. The simulation was run for 300 seconds to enable measurements to be taken for the OpenFlow protocol.

4. Analyzing the performance scenario

End-to-End Delay refers to the time taken for a packet to be transmitted across a network from source to destination. It is a term commonly used in IP network monitoring, and differs from round-trip time (RTT) in that only the path from source to destination is measured [10]. 1k=1024 kilobits. Figures 3 and 4 show the End-to-End Delay for networks using a Single controller and a Multi-controller.

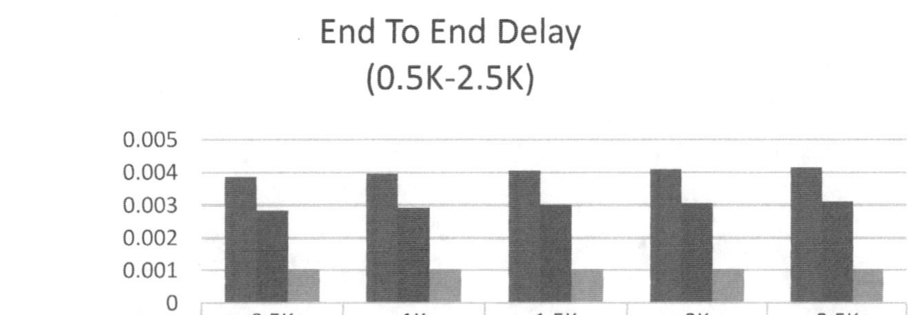

	0.5K	1K	1.5K	2K	2.5K
■ SingleController	0.003870489	0.003954435	0.00405548	0.004099805	0.00414479
■ MulitiController	0.002835145	0.002918751	0.003014788	0.003059244	0.003103124
■ Difference	0.001035345	0.001035684	0.001040692	0.001040562	0.001041666

Figure 3. End-to-End Delay for network in Single Controller and Multi Controller

The difference between the use of a single or multi-controller is extremely close, as shown in Figure 3. This behavior illustrates the nature of the network if designed using one of the mechanisms and provides useful information for designers of the network on End-to-end delay.

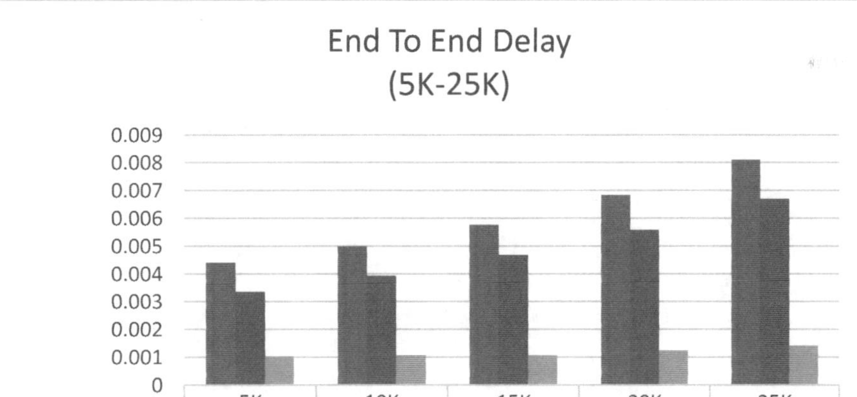

	5K	10K	15K	20K	25K
■ singleController	0.00439936	0.00500544	0.005775279	0.006841608	0.008113494
■ mulitiController	0.003359771	0.003949819	0.004694201	0.005592438	0.006704837
■ Difference	0.00103959	0.00105562	0.001081078	0.00124917	0.001408657

Figure 4. End-to-End Delay for network in Single Controller and Multi-Controller

It is also notable that, when traffic is generated in the form (5K, 10K, 15K ,20K, 25K), the behavior and nature of the network are as shown when using the OpenFlow protocol and both mechanisms are used in the simulation program.

Figures 5 and 6 illustrate the Throughput (the ratio of the total amount of data reached by a receiver from a sender to the time it takes for the receiver to obtain the last packet is referred to as Throughput) [10]. Regarding network Throughput, in using a single or multi-controller mechanism, the behavior of the network in a multi-controller design using the OpenFlow controller protocol produces an increase in Throughput. Figures 5 and 6 show that the use of a multi-controller is therefore more positive in this respect.

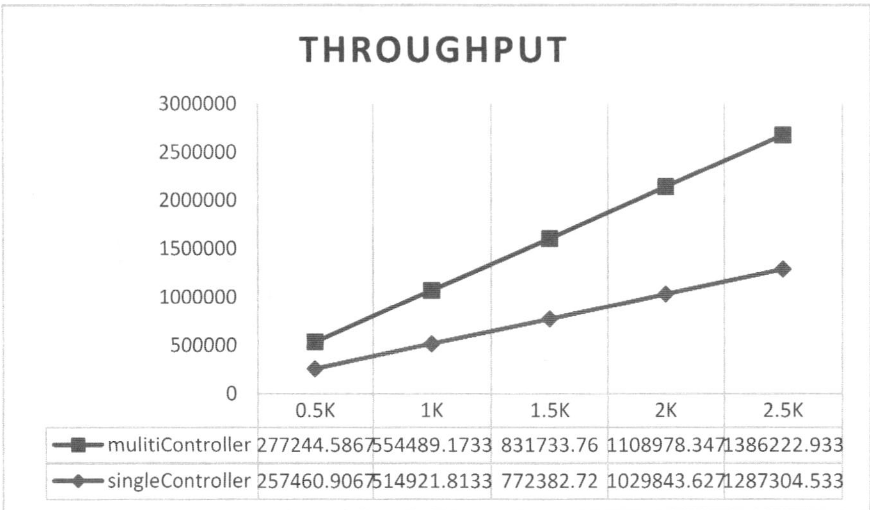

Figure 5. Throughput for network in Single Controller and Multi-Controller

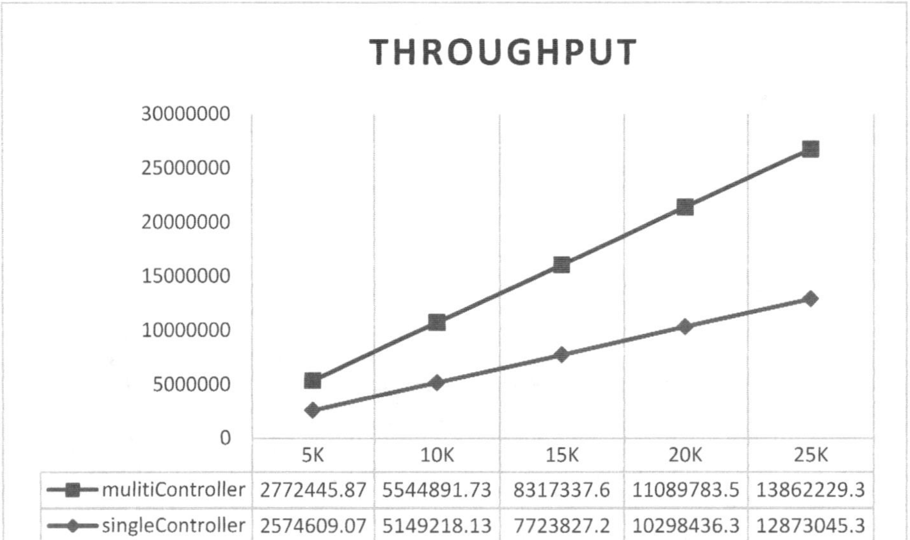

Figure 6. Throughput for network in Single Controller and Multi-Controller

5. Conclusion

This paper has introduced network metrics using single or multi-controller technology, followed by a performance evaluation of the OpenFlow protocol. The results indicate that the performance of the OpenFlow protocol in the network is affected by both single or multi-controller technology. The objective of this study is to determine the impact of using a single or multi-controller on more than one module in a network. This provides information for the designers of networks as well as facilitating study of the behavior of an OpenFlow protocol in such scenarios. Future studies can be conducted using these mechanisms for many purposes depending on the actual and virtual need for reality, and the effects can be studied to avoid any costly mistakes when real networks are designed on the ground.

IOP Conf. Series: Journal of Physics: Conf. Series **1019** (2018) 012015 doi:10.1088/1742-6596/1019/1/012015

References

[1] Kang, M., Kang, E.Y., Hwang, D.Y., Kim, B.J., Nam, K.H., Shin, M.K. and Choi, J.Y., 2013, March. Formal modeling and verification of SDN-OpenFlow. In *Software Testing, Verification and Validation (ICST), 2013 IEEE Sixth International Conference on* (pp. 481-482). IEEE

[2] Wang, S.Y., Chou, C.L. and Yang, C.M., 2013. *EstiNet openflow network simulator and emulator*. IEEE Communications Magazine, 51(9), pp.110-117.

[3] OpenFlow Switch Consortium, OpenFlow Switch Specification Version 1.0. 0.(2009). Internet: *http://www. archive. OpenFlow. org/documents/OpenFlow-spec-v1. 0.0. pdf.*

[4] Klein, D. and Jarschel, M., 2013, March. An OpenFlow extension for the OMNeT++ INET framework. In *Proceedings of the 6th International ICST Conference on Simulation Tools and Techniques* (pp. 322-329).

[5] Gude, N., Koponen, T., Pettit, J., Pfaff, B., Casado, M., McKeown, N. and Shenker, S., 2008. NOX: towards an operating system for networks. *ACM SIGCOMM* Computer Communication Review, *38*(3), pp.105-110.

[6] Fall, K. and Floyd, S., 1996. *Simulation-based comparisons of Tahoe, Reno and SACK TCP. ACM SIGCOMM* Computer Communication Review, *26*(3), pp.5-21.

[7] Lantz, B., Heller, B. and McKeown, N., 2010, October. *A network in a laptop: rapid prototyping for software-defined networks*. In Proceedings of the 9th ACM SIGCOMM Workshop on Hot Topics in Networks (p. 19).

[8] Wang, S.Y., Chou, C.L. and Yang, C.M., 2013. *EstiNet openflow network simulator and emulator.* IEEE Communications Magazine, 51(9), pp.110-117.

[9] Chamberlain, T., 2013. *Learning OMNeT++.* Packt Publishing Ltd.

[10] A. Abdou, A. Matrawy and P. C. van Oorschot, May 2015"Accurate One-Way Delay Estimation With Reduced Client Trustworthiness," in *IEEE Communications Letters*, vol. 19, no. 5, pp. 735-738,

Design and Development of Disabled Parking System for Smart City

R A J M Gining[1], S S M.Fauzi[1], A F A Hadi[1], M Z A Razak[1], and M N F Jamaluddin[1]

[1]Faculty of Computer and Mathematical Sciences, Universiti Teknologi MARA, Perlis Campus, 02600 Arau, Perlis, Malaysia.

E-mail: shukorsanim@uitm.edu.my

Abstract. Violation of the disabled rights has become one of the highly debated issues in today's local community. Nowadays, the community is less aware of the abuse that occurs in their surroundings. In disabled-parking spaces context, it has become common to violate that allocated spaces without guilty. Having this situation is a complete disaster as it now becomes a norm for the non-disabled to park at the disabled-parking spaces. This project aims to design and develop alarm system in securing the parking spaces for the disabled by using current technology which is Internet of Things (IoT). The components used are Arduino UNO, RFID RC522 reader, HC-SR04 Ultrasonic Sensor, LED and buzzer as the alarm. This product system caters the understanding on how to protect the disabled-parking spaces.

1. Introduction

Urbanization in these few decades had been influencing the way citizen live and work. The urban areas are expanding at a breakneck pace, depending on the population and business activities and most importantly the growth of the population that proliferates in these regions [1]. In 2030, United Nations predicts that our population will be increased to 1.35 billion, and thus making it to more than 7 billion people currently [2].

In conjunction with the rise of the population, an example of what Smart City architecture looks like in which operating the cities in an innovative path is related to the solution of the urbanisation crisis [3]. In this case is the crisis about the disabled parking spaces violation by the non-disabled. This research is conducted to find a solution for the issue that uses the Internet of Things (IoT) technology as part of the building blocks for the research and development.

Urban areas are covering the population that is seeking the best lifestyle and fulfilling their needs. However, the problem with the urbanisation is that the vehicle parking spaces are becoming limited due to the high-density population. As the demand for parking spaces are increasing, some people tend to break the rules and violating others rights. In this context is violating the disabled parking spaces. Not only cars that violate, but even motorcyclist also take granted of the spaces to park their motorcycle easily. It becomes a norm to the community, and if no further action or solution were taken, the disabled would be helpless in any situation.

Moreover, for the authorities to monitor the parking spaces are quite challenging as to detect illegal parkers cannot be recognised easily [1] - which made this a continuous problem for the disabled to face every day if there is no solution provided to them. The issue for the disabled parking spaces can be solved, and able-bodied can realise the importance of respecting the disabled rights. Hence, the outcome of this research can contribute for educating the people to aware the disabled place in the community.

This project intends to monitor the disabled parking spaces using the Internet of Things as a platform. The proposing project covers the usage of microcontroller platform, along with motion sensors and alarm system as to warn illegal parker. The prototype developed would picturised the situation in protecting the disabled parking spaces.

This paper outlines, introduction in the first part, which discusses on the overview and objective of the paper. Next part is related works. After that is the detailed discussion regarding the method, result and discussion. The last part refers to the final conclusion of this paper.

2. Related Works

The DisAssist System was developed to assist the disabled in finding their parking spots, at the real-time condition and also enable authentication of the parkers to be exactly legal to park at the reserved spots [1]. DisAssist System role enhanced the disabled experience in parking management system with the usage of Machine-to-Machine (M2M) technology that ease the findings of the disabled parking spaces. Furthermore, with the authentication system that enables the deserved parkers to park at their spaces (in this case is the disabled), can be monitor by the authority and further action can be taken to those who violated the rules and regulations. Another study developed locomotive alarm and parking system to monitor the locomotive using infrared technology. The function of the system is to automatically park, alarm and avoiding accidents between locomotives [4]. SmartEye was developed using infrared technology to monitor the parking spaces, and the sensors are put on the road. The information about the parking is then sent to the content management, known as SmartRep that collects the data about parking spaces. This system also comes with an application for the user to search empty parking spaces in the area. It transmits 433 MHZ signals onto the server with the use of monitoring the parking area.

Table 1 illustrates the comparison between the three of the related works. These comparison criteria are based on what technology does it use, involving the use of Radio Identification (RFID) and the law enforcement role in monitoring the system.

Table 1. Comparison between related works..

	Wireless Sensor	Infrared	RFID Tag	Law Enforcement monitoring
DisAssist System	Yes	Yes	Yes	Yes
Locomotive Alarm and Parking System	No	No	No	Yes
SmartEye	Yes	No	No	Yes

The Disabled Parking System is different from all the compared works as mentioned earlier. Instead of relying on law enforcement monitoring, the system can give alarm itself whenever violations have occurred. It is a burden for the authorities to come whenever a violation occurs. By using the alarm technology, Disabled Parking System can self-remove the violators by making them

IOP Conf. Series: Journal of Physics: Conf. Series **1019** (2018) 012016 doi:10.1088/1742-6596/1019/1/012016

like the attention to the public. Psychologically, the violators will be aware of their action, and with others also notice them, the probability is high that they will move to other parking spaces.

3. Implementation of Disabled Parking System

Figure 1 illustrates the flow of the Disabled Parking System. The process starts with a car entered the parking spot. Once it reaches a distance that is set up, the RFID reader will be activated to find any authorized RFID tag. If the RFID detected is legal, then the system will indicate that the parking is deserved to the disabled by blinking the green LED. Vice versa, the LED will blink red, and buzzer sound will be on to indicate that the violation of parking space by the non-disabled. Also noted that the red LED and buzzer will be sound as long as RFID reader cannot detect the RFID tag. By using this simulation process, the further steps can be planned especially on setting close approximation of the sensor and RFID reader.

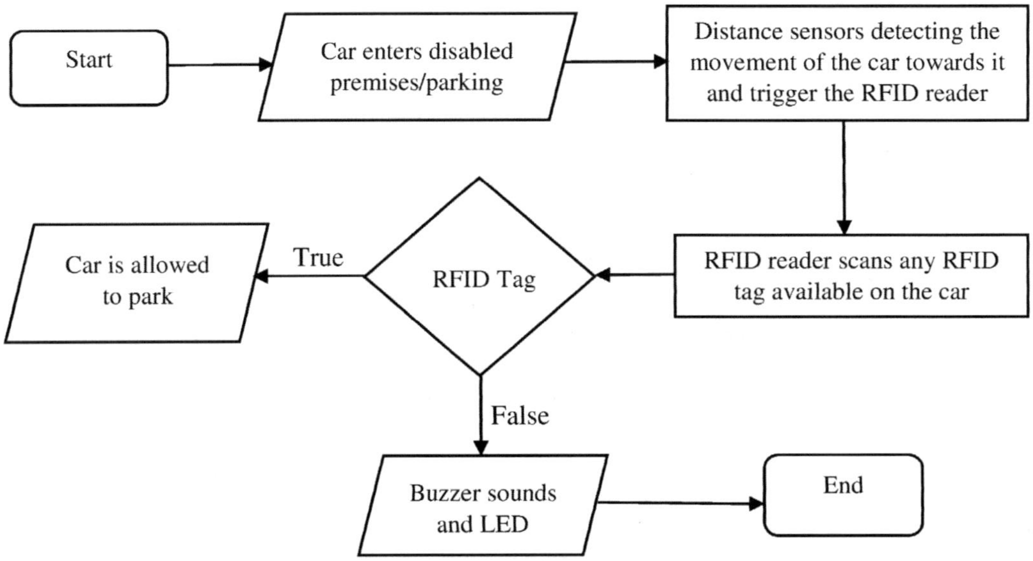

Figure 1. Flowchart for Disable Parking System working flow

The design and development of the study involve three development sub-phases aimed to design a functional prototype as illustrated in Figure 2.

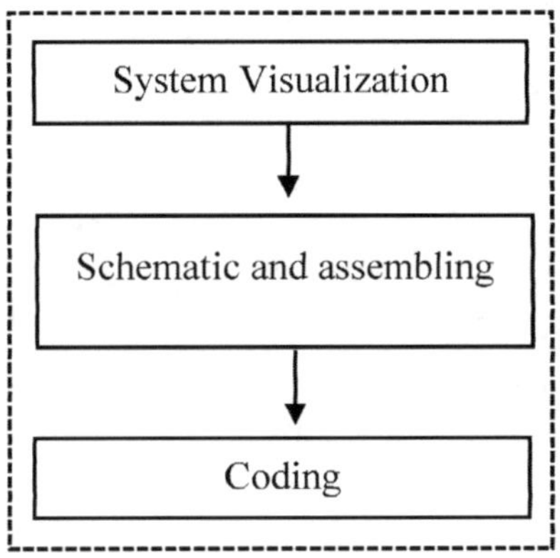

Figure 2. Phases involved in the development of Disabled Parking System

IOP Conf. Series: Journal of Physics: Conf. Series **1019** (2018) 012016 doi:10.1088/1742-6596/1019/1/012016

3.1. System Visualization

In this step, the system is simulated through visual on how it should work based on the flowchart in Figure 1. Figure 3 and 4 is the visualisation of the Disabled Parking System. From this visualisation, it illustrates the suitability of the sensor location on the parking space. The sensor resided at the in front of the parking space as to detect any car entering the parking. In Figure 3 is the illustration of what the situation like when a car with 'true' RFID tag that can be resembled as the disabled. Similar to Figure 4.4, is the opposite situation when 'false' RFID tag is entering the spaces, activating the buzzer and LED to warn the violators.

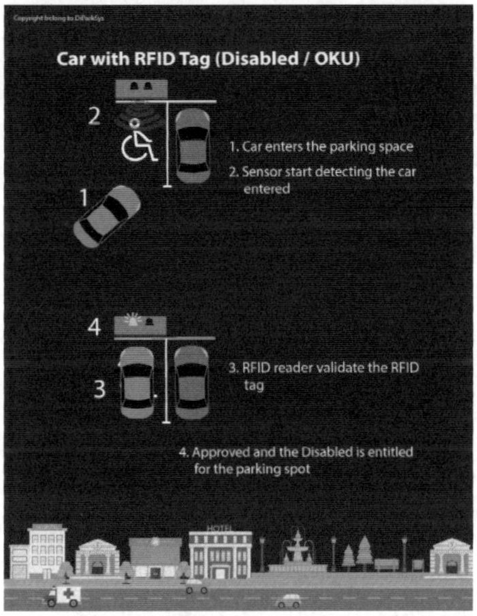

Figure 3. Disabled Parking System Visual of Car with RFID Tag

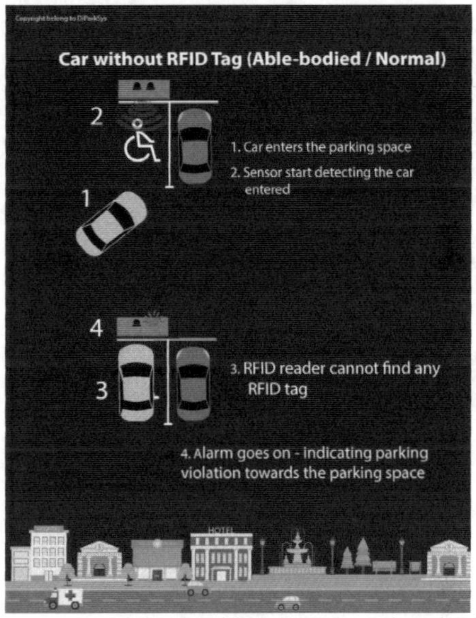

Figure 4. Disabled Parking System Visual of Car without RFID Tag

3.2. Schematic Diagram and Assembles

The next part is to draw a schematic based on the visualisation of the Disabled Parking System. Schematic is important because it supports the comprehension of the system during the assembling part. Electrical components consist of several rules and understanding. The process of wiring must be done precisely by going through the calculations of how much voltage for each component and the flow of the wiring with a connection to Arduino Uno pin. Otherwise, the components will be damaged because of the overpower supplied to them. Below is the list of components that were identified and used for this project.

1) Arduino Uno – microcontroller as the main component
2) Ultrasonic Range Sensor HC-SR04 – sensor for detecting the car movement
3) RFID RC522 Reader – Passive RFID reader
4) Passive buzzer – sound alarm for the system
5) Red and Green LED – indicator for the disabled and able-bodied
6) 220 Ohm Resistor – Resist excessive current for LED
7) Breadboard – additional pin connector for Arduino Uno
8) Jumper wire – connect components to Arduino Uno

The Fritzing software is used to draw the schematics as it was the perfect tool that most developers used for their project. In this software, there are a lot of Arduino components, sensors and other related tools that can be used. Developers also can detect any error in the connection of the components and correct it right away.

Figure 5 illustrates the first connection to the breadboard for ultrasonic sensor HC-SR04. The sensor consists of four pins. The trig pin, echo pin, integrated power supply pin or 'vcc' and ground pin or 'gnd'. The sensor works in a manner starting from echo that will spread the sonar around the area. Once it hit an object, it will reflect back to the sensor, and trig pin will read for further action. The ultrasonic sensor used 5V of power and did not need any resistor or transistor on the breadboard. The ability of the sensor in reading distances can be up to more than 70cm. For this project, the distance will be used less than 30cm for triggering RFID reader.

The second connection as in Figure 6 involves connecting the RFID RC522 Reader to the breadboard. RC522 reader consists of 8 pins in which for the project will be used only 7. Besides the 'vcc' and 'gnd' pin for power source and ground, the reader also being built with five other pins that will read and write the tags. It includes RST, IRQ MISO, MOSI, MISO and SCK. The RC522 only requires 3.3V and can be powered up directly at the 3V3 pin on Arduino Uno without using any components in the middle between them.

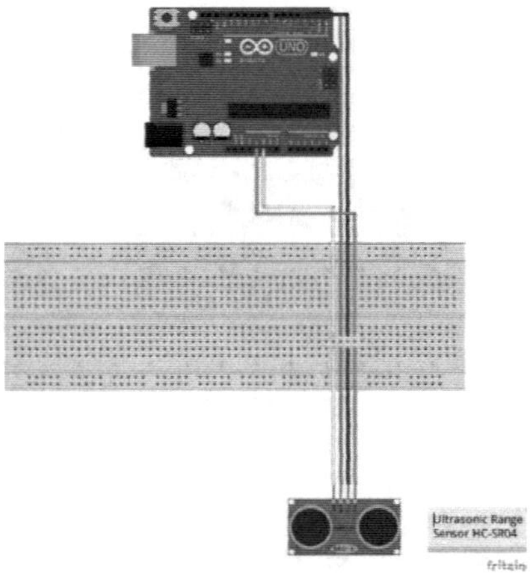

Figure 5. Ultrasonic Sensor Connection to Breadboard

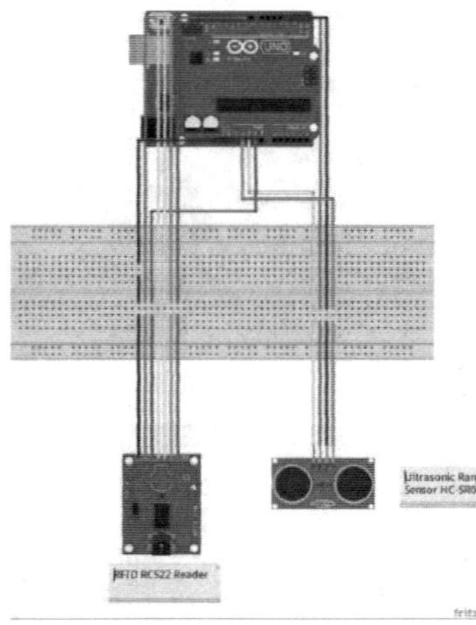

Figure 6. RFID RC522 reader connection to breadboard

As in Figure 7, the final part of this assembly was the connection between passive buzzer and also the red and green LED. The buzzer used three pins with two of the jumper wire ('vcc' and 'gnd') shared the same current flow of the RFID RC522 and Ultrasonic sensor. Meanwhile, for the LED, it requires the 220ohm resistor because the LED only consumed 2V of power. Otherwise, it will be damaged because of the overpower.

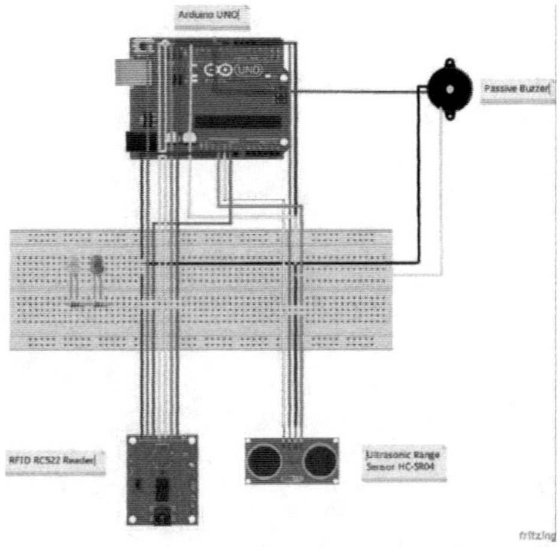

Figure 7. Passive buzzer and LED connection to breadboard

3.3. Programming the Arduino Uno

The commanding of Arduino was done after the assembled part had completed. In this development process or known as sketches, the Arduino Integrated Development Environment or Arduino IDE was used and connected to the Arduino Uno using USB cable. There is two main part in the Arduino process; the setup and loop.

3.3.1. Setup Process. Before programming the main process, the Arduino Uno had to be defined first of its pin that connected to the component. The setup segment also must have contained the library that the program referred to. Arduino has many libraries for the components. For example, the RFID RC522 has its libraries to be used. Also, different RFID module also has different libraries to be used in the loop process.

The next part was defining the pins that were used to connect the components. The void setup function starts using the libraries, initialise variables and pin modes either input or output. The setup function only run once for each power-up of the Arduino Uno. Serial begin Sets the data rate in bits per second. The higher the rate of data, the faster the process of Arduino. For this project, it was suitable to set the data rate to only 9600 bauds. Another rate could be 115200 bauds, which was the fastest processing for the board.

3.3.2. Loop Process. Loop section is the primary process for Arduino. In this part, the loop function subject for checking the pin consecutively and allows the developer to change and respond of that pin. It controls the Arduino Uno behaviour by giving the task of what it should do. The sequence of the loop started with the programming of the ultrasonic sensor that converts microseconds to centimetres or inches.

Next, the loop would process the sensor by giving it a condition to carry out a certain task. In this project, the sensor would activate the RFID RC522 reader if the distance is less than 7 inches. Along with RFID reader, the buzzer also would sound, indicating the probability of a violation of the parking. Once RFID tag was found as true, the buzzer would stop.

In Figure 8, the program segment shows the RFID 'false' process. Once it detects the 'false' state, it then sounds up the buzzer with displaying the message. In Figure 9 is the program segment of 'true' state of RFID tag. Delay for this project was put for 10,000 milliseconds which equals to 10 minutes - which results in the sensor and RFID reader taking a new reading for every 10 minutes.

```
 if (content.substring(1) == "BE 84 AE 73"){

 Serial.println("Hello dear, please remove your car. It's not yours..");
 for(i=0; i<100; i++){
 digitalWrite(buzzer,HIGH);// sound
 delay(1);//delay1ms
 digitalWrite(buzzer,LOW);//not sound
 delay(1);//ms delay
 }
}
```

Figure 8. Program segment of 'false' RFID tag

```
else
{
  content.substring(1) == "17 D9 84 AB";
  Serial.println("Hi dear, this parking is for you. Have a nice day!");
  Serial.println();

  //exit(0);
  delay(10000);

}
```

Figure 9. Program segment of 'true' RFID tag

4. Conclusion

Disabled Parking System was designed and developed as a conceptual guideline of how technology can be applied as to preserve disabled rights, specifically regarding parking rights. Through this project, the usage of Internet of Things (IoT) approach and technology can be broadened for more sophistication in concreting of disabled parking spaces protection.

References

[1] Lambrinos, L., & Dosis, A. (2013). DisAssist: An Internet of Things and Mobile Communications platform for Disabled Parking Space Management.

[2] Seto, K. C., Güneralp, B., & Hutyra, L. R. (2012). Global forecasts of urban expansion to 2030 and direct impacts on biodiversity and carbon pools. Proceedings of the National Academy of Sciences of the United States of America, 109(40), 16083–16088. https://doi.org/10.1073/pnas.1211658109

[3] Nam, T., & Pardo, T. (2011). Smart city as urban innovation: Focusing on management, policy, and context. Proceedings of the 5th International Conference on Theory and Practice of Electronic Governance, 185–194. https://doi.org/10.1145/2072069.2072100

[4] Wang, X., Miao, C., Yan, H., & Zhang, C. (2010). The design of locomotive alarm and parking system based on infrared communication. 2010 2nd Pacific-Asia Conference on Circuits, Communications and System, PACCS 2010, 1, 5–8. https://doi.org/10.1109/PACCS.2010.5627077

Development of a Wearable Device for Sign Language Recognition

Radzi Ambar[1, 2], Chan Kar Fai[1], Mohd Helmy Abd Wahab[1], Muhammad Mahadi Abdul Jamil[2] and Ahmad Alabqari Ma'radzi[2]

[1] Computational Signal, Imaging and Intelligence (CSII) Research Group, Department of Computer Engineering, Faculty of Electrical and Electronic Engineering, Universiti Tun Hussein Onn Malaysia.

[2] Biomedical Modelling and Simulation (BIOMEMS) Research Group, Department of Electronic Engineering, Faculty of Electrical and Electronic Engineering, Universiti Tun Hussein Onn Malaysia.

aradzi@uthm.edu.my

Abstract. This project describes the development of a sign language translator that converts sign language into speech and text by using wearable device. The glove-based device is able to read the movements of a single arm and five (5) fingers. The device consists of five (5) flex sensors to detect finger bending, and an accelerometer to detect arm motions. Based on the combination of these sensors, the device is able to identify any particular gestures that correspond to words and phrases in American Sign Language (ASL), and translate it into speech via a speaker and text which is displayed on an LCD screen. This article explains mainly on the hardware design of the device. Based on preliminary experimental results on nine (9) simple sign languages, the device demonstrated an average value of 0.74 s to convert a sign language into speech and text, which demonstrate the usefulness of the proposed device.

1. Introduction

Sign languages have been used since three centuries ago by humans as a medium to deliver message and communication, especially for the deaf communities. New sign languages emerge throughout the world where deaf communities exist [1]. Sign languages can consist of simultaneously combining hand shapes, orientation and movement of the hands, arms or body, and facial expressions to express a speaker's thoughts.

However, there are not many normal people understand sign language. Thus, for those who are using sign language for everyday tool of communication might find it difficult to communicate or even to express their thoughts to others. Therefore, as the results of tremendous evolution of technology, hearing aid devices have been introduced to help deaf communities to communicate with others or hear others. Hearing aid devices would help for those who didn't lost their sense of hearing completely, while the group of people who have hearing impairment can just depend on sign language to communicate between each other.

There are various types of off-the-shelf hearing aid devices which include behind-the-ear, in-the-ear and canal aid to help deafness and other communication disorder [2]. Although hearing aid devices

IOP Conf. Series: Journal of Physics: Conf. Series **1019** (2018) 012017 doi:10.1088/1742-6596/1019/1/012017

are helpful, user might experience problems such as uncomfortable feeling and hearing background noise with this kind of device. Therefore, researchers have been developing various methods that can translate sign language gestures. Basically, there are two known methods: vision-based system and wearable devices. Vision-based systems utilize image processing method by feature extraction techniques to identify hand and fingers movements [3-5]. There are many other studies on sign language translation using vision-based system explained in [6]. The advantage of vision-based systems is that the systems may not require the users to be attached with sensory devices that can be messy and uncomfortable. However, the vision-based systems are difficult to develop because it require complex and extensive computations in developing algorithms for feature and movement recognition.

Wearable devices for sign language translation usually utilize sensors attached on the user or glove-based approach. The devices can convert sign language into text or even speech. Jingqiu et al. utilized ARM processor for a data glove where the recognition results are displayed on an LCD, and also voice output via speaker [7]. They described the recognition approach using template matching algorithm which demonstrated above 91% recognition accuracy. Shukor et al. developed a data glove that utilized tilt sensor, accelerometer and flex sensors to recognize sign language [8]. The device is capable to translate 12 sign languages. Ahmed et al. presented a glove-based sign language translator that utilizes five (5) flex sensors [9]. The translated results are displayed on an LCD and prerecorded audio sound via speaker. However, there are only four (4) gestures that are designed for user input which can produce audio commands. Wearable devices for translating sign languages are indeed a very popular method among researchers. However, there are various challenges when using this method, such as comfort of wearing the device, difficulty to move hands due to data cables and costs of using the most suitable sensors to achieve the best results with limited errors.

The main objective of this work is to design a wearable device to help deaf communities communicate easily with public. The end target of this work is to develop a device by considering the comfort of use, where the wearable will be wirelessly connected to a microcontroller to eliminate the use of cables. However, this work is the continuation of the early development of the work done in [10]. The device is a glove-based sign language translation device consists of five (5) flex sensors and an accelerometer that are connected to an Arduino microcontroller via cables. Based on the combination of these sensors, the device is able to identify any particular gestures that correspond to words and phrases in American Sign Language (ASL), and translate it into speech via a speaker and text which is displayed on an LCD screen. In the next few chapters, hardware design and preliminary experimental results of the device are presented.

2. Hardware components and circuit design

Figure 1 shows the hardware components of the proposed device consists of input sensors (an accelerometer and five (5) flex sensors) and the output display (an LCD and a speaker). An Arduino Mega microcontroller is used to process the data from all sensors. As shown in the figure, the input values from flex sensors and accelerometer are conveyed to the Arduino Mega, where these values are used in a gesture recognition algorithm to translate any gestures. However, the gesture recognition algorithm will not be described in this article. The results of translation are processed and the output result will be displayed simultaneously on an LCD as well as through audio speech using a speaker.

As shown in Figure 1, an Arduino Mega 2560 receives and processes input signals from sensors, and then, provides the commands to the output component. The board contains 16 analogue input pins which is sufficient to handle a total of eight (8) analogue input from flex sensors and accelerometer. Furthermore, the device is also installed with an LCD to display the result of translation through text. The LCD requires six (6) analogue pins to be connected to the Arduino's analogue pins. An audio speaker is also used in this device to produce the appropriate speech for any sign language. The speaker is connected to an analogue pins on the Arduino boards. Therefore, the device requires 15 analogue pins on the Arduino board, where Arduino Mega is more than enough to handle all the required pins. Prior to selecting Arduino Mega 2560, a cheaper Arduino Uno board was considered for the device. However, Arduino Uno has only six (6) analogue input pins which is not enough analogue pins to cater for the project requirement.

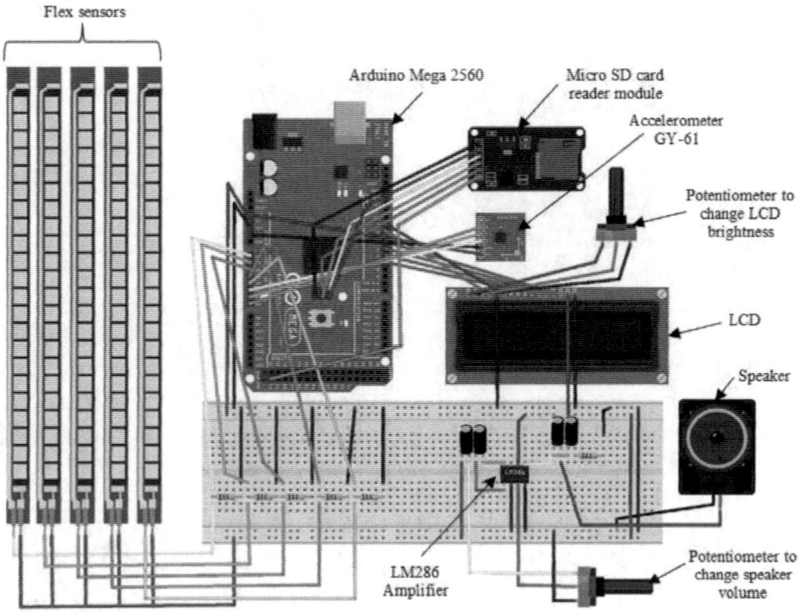

Figure 1. Circuit diagram for sign language translator device.

The developed device uses five (5) flex sensors from Spectra Symbol. Flex sensor is a type of resistor printed with a conductive particles embedded polymer ink [11]. The sensor works by increasing the resistance value when bent with the ink on the outside of the curve. The value of resistance does not change if it bends on the other side. As the sensor bend more, the resistance of the sensor increase as well. In this work, flex sensors are functioning as an angle displacement measurement components which can calculate the bending angle of five (5) fingers that help to recognize a gesture of a sign language.

Figure 2 shows a GY61 accelerometer sewed on a white cotton glove. In this work, it is used to measure the acceleration of the hand on x-, y-, z-axes to determine the roll, pitch and yaw postures. GY61 consists of an ADXL335 chip that produces acceleration values that can be used to measure the angle of the three (3) axis at a current position corresponding to the origin position. This particular sensor consists of five (5) pins which are Vcc, GND, x-, y- and z-axes pins.

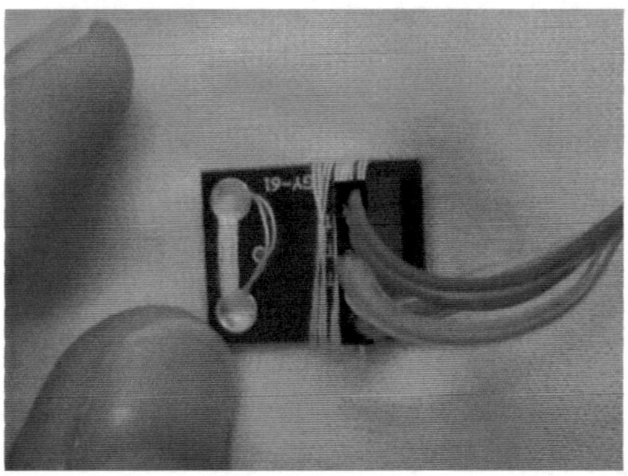

Figure 2. GY61 accelerometer sewed on white cotton glove.

Figure 3. Amplifier circuit module for changing speaker volume.

IOP Conf. Series: Journal of Physics: Conf. Series **1019** (2018) 012017 doi:10.1088/1742-6596/1019/1/012017

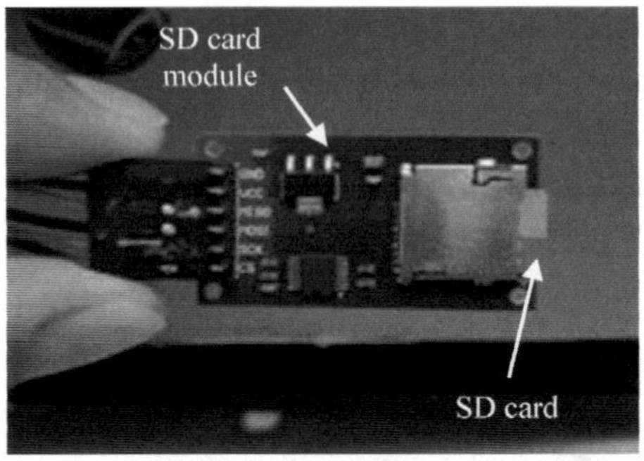

Figure 4. SD card reader module connected to Arduino's ICSP pins.

An audio speaker is used in this work to produce the audio speech which corresponds to a sign language. The 0.5 W speaker is used due to its low power consumption and optimum frequency range which can play audio between 600 Hz until 10 KHz of frequency. The speaker is connected to an LM286 amplifier circuit as shown in Figure 3. When a certain movement is translated, the corresponded audio speech will be played through the speaker. The amplifier circuit is to amplify the audio speech recorded in a micro secure digital (SD) card when played. An external 9V battery is used as the power supply for this amplifier circuit. As shown in Figure 3, a potentiometer is used to control the speech volume.

Figure 4 shows the micro SD card reader module that is used in this work to store audio speech files that correspond to a sign language. This module uses Serial Peripheral Interface (SPI) interface as the communication interface. In-Circuit Serial Programming (ICSP) pins on Arduino Mega 2560 are connected to the SD card reader which provides easy data communication between the SD card reader and microcontroller.

In order to use this module, a source code that can determine whether the connection between chip select pin and Arduino Mega is fail or not has been developed. A digital pin 53 on the Arduino Mega 2560 digital pin was set as the chip select pin for the SD card reader module. Figure 5 shows the result when the Arduino failed to read the SD card. Figure 6 shows the result when Arduino succeeded in reading the SD card. Figure 6 also shows all of the audio speech files recorded in WAV file format that are stored in a 16 GB SD card. In this work, WAV audio files are used but not MP3 files. This is because one of the Arduino Mega 2560 limitation is, it can only recognize WAV audio files.

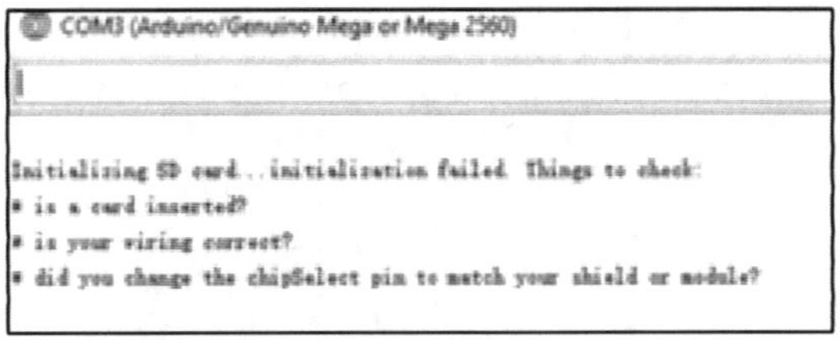

Figure 5. Arduino failed to read the SD Card.

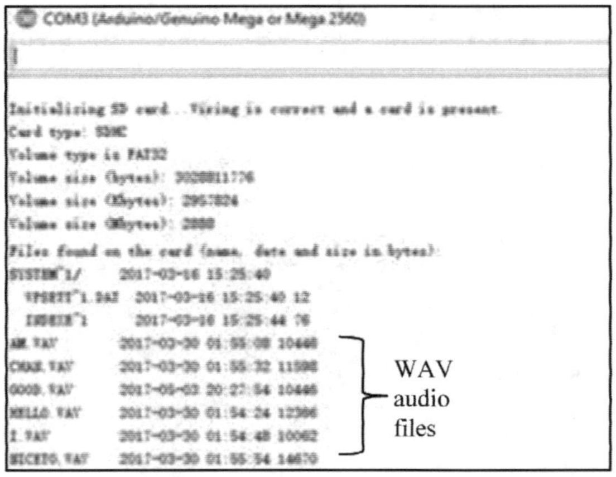

Figure 6. Arduino successfully read the SD Card.

Figure 7. Initial design of the prototype for the glove-based device.

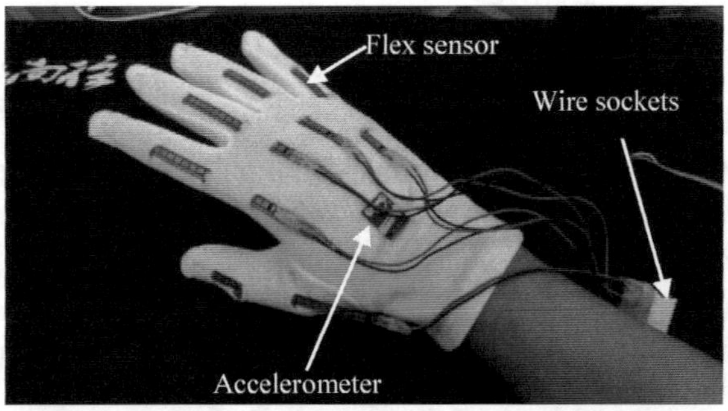

Figure 8. Current design of the prototype for the glove-based device.

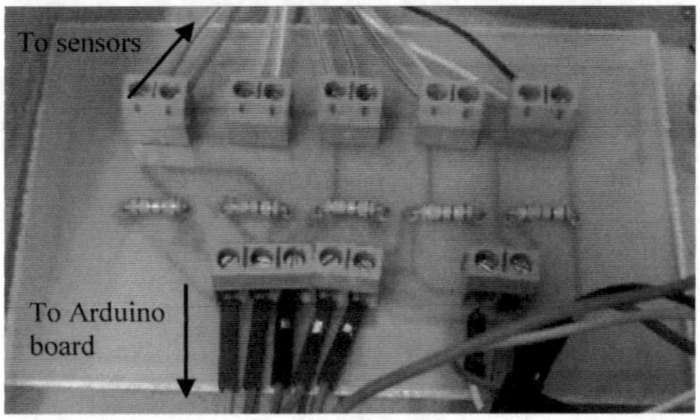

Figure 9. Printed circuit board to connect microcontroller with sensors.

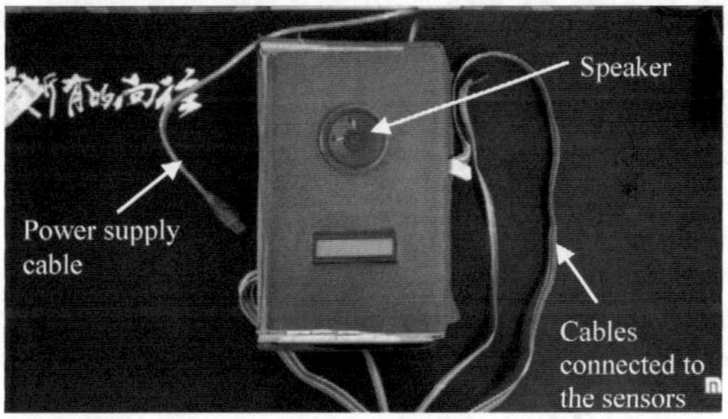

Figure 10. The main box that contains all the components.

3. Development of the prototype

Figure 7 shows the initial design of prototype for the glove-based sign language translator device developed in [10]. As shown in the figure, the flex sensors are attached on the glove using black tapes. The problem with attaching flex sensors with tapes was that the flex sensors will become loose after the glove is being used for a couple of times. Besides, the messy wires are a major problem that affected the reading of those sensors.

The problems stated above were overcome in the current prototype design. Figure 8 shows the current design of the prototype for the glove-based device. As shown in the figure, all of the flex sensors were sewed on the white cotton glove. By doing this, the readings of the sensors are not affected by the movement of glove. Furthermore, compared to the initial design, the current design makes the glove portable and more mobile, where the cables are separated by using two (2) wire sockets as shown in Figure 8. Furthermore, a printed circuit board (PCB) was also created to prevent unwanted problems such as bad connections between connecting cables. Figure 9 shows the PCB for connecting Arduino microcontroller with all sensors. For future work, in order to develop a unique device, the wearable will be wirelessly connected to the microcontroller to eliminate the use of cables.

Figure 10 shows the main box that is used to store all components such as Arduino board, an LCD and a speaker. LCD and speaker are placed on the top of the box for easier and convenience way to read the text and listen to the speech of the translated sign language. The power supply cable (blue cable) is connected to a 5 V power supply for the Arduino Mega which is connected to a power bank. Two ribbon cables (one for accelerometer and the other one for flex sensors) with socket connectors shown in Figure 10 connects Arduino board with all sensors.

IOP Conf. Series: Journal of Physics: Conf. Series **1019** (2018) 012017 doi:10.1088/1742-6596/1019/1/012017

4. Preliminary experiments and results

In this section, the steps on how to use the developed sign language translation device and also preliminary experimental results to test the device are described in detail. For preliminary experiment, a user was asked to use the device according to predetermined procedures which involve executing nine (9) sign languages.

4.1. Procedures of using the Sign Language Translation Device

First, the user needs to connect a 5 V power supply to the device with the blue cable shown in Figure 10. By doing this, the LCD will turn ON and displays "Welcome to Sign Language Translator" as shown in Figures 11 and 12.

Figure 11. LCD display "Welcome to…".

Figure 12. LCD displays "Sign Language Translator".

After a few seconds, it will start to check the condition of the SD card. If there is no SD card detected or the SD card is malfunctioning, the LCD will display "SD Card fail, check connection" as shown in Figure 13. On the other hand, if the SD card has no connection problem, it will display "SD Card is good to go." as shown in Figure 14. Then, it will go into a waiting period and displays "Waiting…" as shown in Figure 15. During this condition, user's hand and all five (5) fingers are extended as shown in Figure 16.

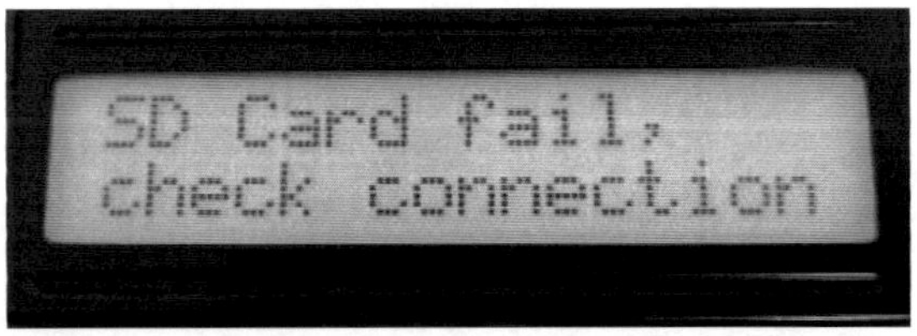

Figure 13. LCD displays "SD card fail, check connection".

Figure 14. LCD displays "SD card is ready to go.".

Figure 15. LCD displays "Waiting…".

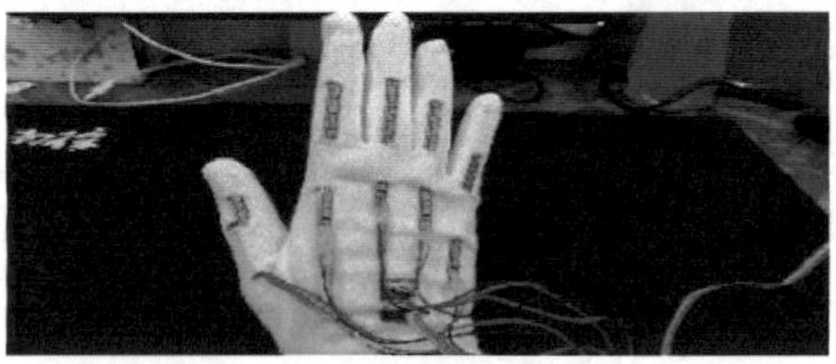

Figure 16. Sign language for "Waiting…".

Now, the user can start to move the hand and fingers, and the device is ready to translate nine (9) predetermined sign languages. Figures 17 to 22 shows the example of only three (3) sign languages done by the user being converted into texts displayed on the LCD screen, and speech is played on the speaker simultaneously.

Figure 17. LCD displays "Hello!".

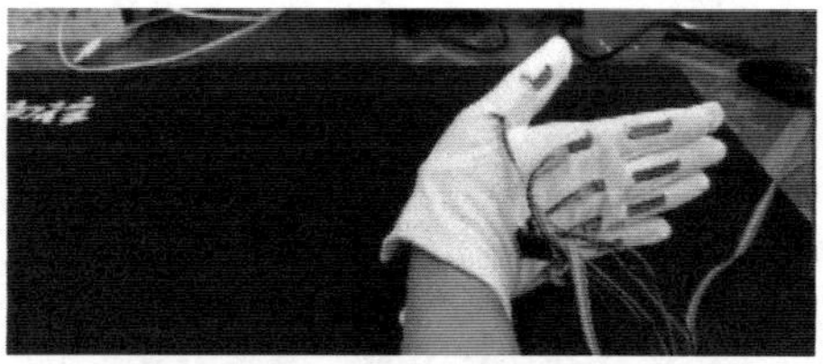

Figure 18. Sign language for "Hello!".

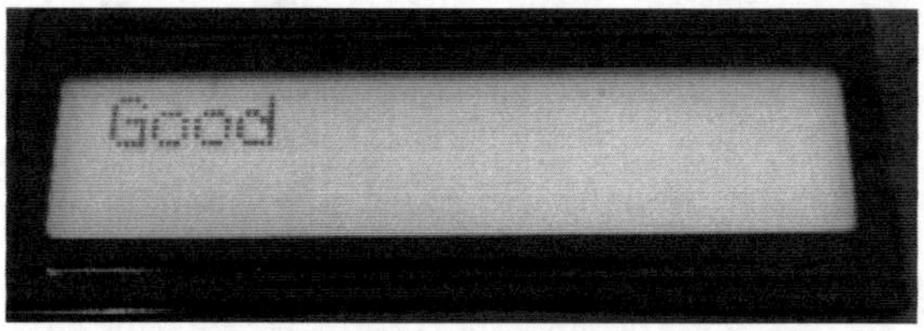

Figure 19. LCD displays "Good".

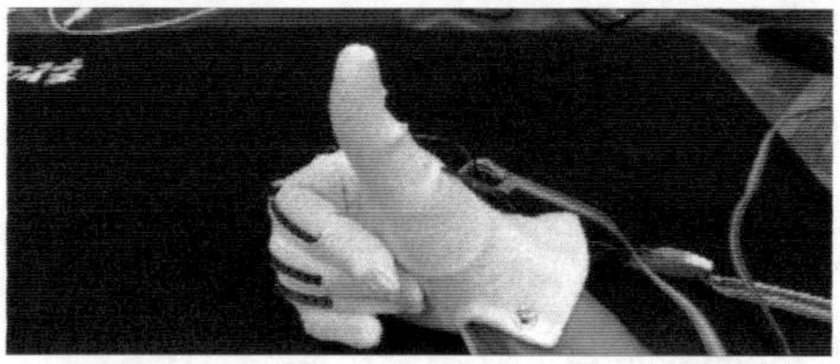

Figure 20. Sign language for "Good".

Figure 21. LCD displays "Evening".

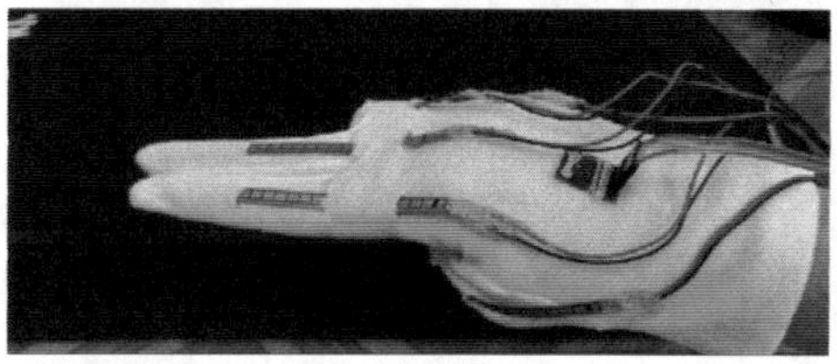

Figure 22. Sign language for "Evening".

4.2. Waiting time and translation error analysis

Figure 23 shows the average waiting time (out of 10 times) for the device to display and play the audio speech after the movement of the fingers and hand based on nine (9) predetermined sign languages. Based on these results, the device demonstrated an average value of 0.74 s to display and play audio speech of a sign language which can be assumed to be a very good time response.

Figure 24 shows the number of error occurred, when the same pattern of sign language being repeated for 10 times. This results show that majority of the sign languages are translated into words and audio speech successfully. However, for the sign language "Hello", four times error was recorded. The reason for this error might be due to the soldering defect on the thumb's flex sensor. Nevertheless, these data shows encouraging results in producing an efficient and reliable wearable device for translating sign language.

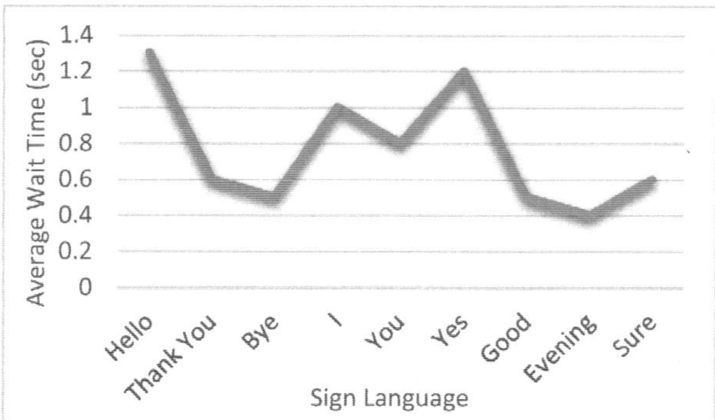

Figure 23. Average waiting time to display sign language translation.

Sign Language	Number of Error
Hello	4
Thank You	1
Bye	0
I	2
You	1
Yes	0
Good	0
Evening	0
Sure	0

Figure 24. Number of error occurred.

5. Conclusion

This paper described the design of a glove-based sign language translator device. The developed device is able to read the movements of every fingers and arm using two (2) types of sensor, an accelerometer and five (5) unit of flex sensors. In this paper, detail hardware design of the device and preliminary experimental results has been described. The results show that the device demonstrated an average value of 0.74 s to convert a sign language into speech and text, which demonstrate the usefulness of the proposed device. At the current stage, the device is connected to the microcontroller via cables. Based on preliminary testing of the device, the cables can disturb the hand movements. Therefore, in the future, wireless connection between the device and microcontroller will be implemented to solve this problem.

References

[1] Stokoe W. C. (1960) *Sign Language Structure: An Outline of the Communicative Systems of the American Deaf.* Linstock Press, Silver Spring, MD, 1960.

[2] Igari S. and Fukumura N. (2016) Recognition of Japanese Sign Language Words Represented by Both Arms using Multi-Stream HMMs. *In Proc. of IMCIC-ICSIT*, pp. 157–162, 2016.

[3] Madhuri Y., Anitha G. and Anburajan M. (2013) Vision-based sign language translation device. *In Proc. of the 2013 Int. Conf. on Information Communication and Embedded Systems (ICICES)*, pp. 565–568, 2013.

[4] Bauer B. and Kraiss K. F. (2002) Video-Based Sign Recognition Using Self-Organizing Subunit. *In Proc. of Int. Conf. on Pattern Recognition*, vol. 2, pp. 434–437, 2002.

[5] Dreuw P. et al. (2007) Speech Recognition Techniques for a Sign Language Recognition System. *InterSpeech-2007*, pp. 2513-2516, 2007.

[6] Huang T. S., and Wu Y. (1999) Vision-based gesture recognition: a review. *In Gesture Workshop*, Gif-sur-Yvette, France, vol. 1739 of LNCS, pp. 103–115, 1999.

[7] Jingqiu W. and Ting Z. (2014) An ARM-Based Embedded Gesture Recognition System Using a Data Glove. *The 26th Chinese Control and Decision Conference*, pp. 1580–1584, 2014.

[8] Shukor A. Z. et al. (2015) A New Data Glove Approach for Malaysian Sign Language Detection. *Procedia Computer Science* 76, pp. 60–67, 2015.

[9] Ahmed S. and. Islam R. (2015) Electronic Speaking System for Speech Impaired People: Speak Up. *In Proc. of 2nd Int. Conf. on Electrical Engineering and Information & Communication Technology (ICEEICT)*, 2015.

[10] Ambar R. et al. (2018) Preliminary Design of a Dual-Sensor Based Sign Language Translator Device. *In: Ghazali R., Deris M., Nawi N., Abawajy J. (eds) Recent Advances on Soft Computing and Data Mining. SCDM 2018. Advances in Intelligent Systems and Computing, Springer, Cham,* vol 700, pp. 353–362.

[11] "Flex Sensor Hookup Guide" [Online]. Available: https://learn.sparkfun.com/tutorials/flex-sensor-hookup-guide [Accessed:25-May-2017]

Investigation of Pulse electric field effect on HeLa cells alignment properties on extracellular matrix protein patterned surface

Muhammad Mahadi Abdul Jamil[1*], Mohamed Ahmed Milad Zaltum[1], Nur Adilah Abd Rahman[1], Radzi Ambar[1], Morgan C T Denyer[2], F Javed[4], Farshid Sefat[3], Masoud Mozafari[5], Mansour Youseffi[3]

[1]Biomedical Modeling and Simulation (BIOMEMS) Research Group, Department of Electronic Engineering, Faculty of Electrical and Electronics Engineering,
University Tun Hussein Onn Malaysia, 86400 Parit Raja, Batu Pahat, Johor, Malaysia
[2]Faculty of Life Sciences, [3]Faculty of Engineering and Informatics, University of Bradford, Yorkshire, University of Bradford, University of Bradford, Yorkshire, BD7 1DP, UK
[4]Pharmacy, School of Applied Sciences, University of Huddersfield, Queensgate, Huddersfield, UK
[5]Nanotechnology and Advanced Materials Department, Materials and Energy Research Center (MERC), Iran

*Author for correspondence: Muhammad Mahadi Abdul Jamil, Tel.: +60167179179,
Email: mahadi@uthm.edu.my.

Abstract
Cell behavior in terms of adhesion, orientation and guidance, on extracellular matrix (ECM) molecules including collagen, fibronectin and laminin can be examined using micro contact printing (MCP). These cell adhesion proteins can direct cellular adhesion, migration, differentiation and network formation *in-vitro*. This study investigates the effect of micro-contact printed ECM protein, namely fibronectin, on alignment and morphology of HeLa cells cultured *in-vitro*. Fibronectin was stamped on plain glass cover slips to create patterns of 25μm, 50μm and 100μm width. However, HeLa cells seeded on 50μm induced the best alignment on fibronectin pattern (7.66° ±1.55SD). As a consequence of this, 50μm wide fibronectin pattern was used to see how fibronectin induced cell guidance of HeLa cells was influenced by 100μs and single pulse electric fields (PEF) of 1kV/cm. The results indicates that cells aligned more under pulse electric field exposure (2.33° ±1.52SD) on fibronectin pattern substrate. Thus, PEF usage on biological cells would appear to enhance cell surface attachment and cell guidance. Understanding this further may have applications in enhancing tissue graft generation and potentially wound repair.

Keywords: HeLa; cells alignment; pulse electric field; micro contact printing; extra cellular matrix, cell proliferation, cell guidance

1. Introduction

Micro-contact printing (MCP) is a procedure that allows a surface to be well-designed with a material like extra cellular matrix (ECM) protein, in a clear outline [1-4]. Cell behavior in terms of adhesion, orientation and guidance, on ECM molecules such as fibronectin, collagen and laminin can be examined using MCP [5-6]. These cell adhesion proteins can guide cellular migration, adhesion and network establishment *in-vitro* [7]. These cellular characteristics can be used to deduce the cell signaling paths related with the cell interactions with the surface [8-9]. Fibronectin attach to collagen, fibroblasts in the ECM and the plasma membrane of the cells and its high molecular weight protein. This protein plays a vital role in cell adhesion and re-organiation of the ECM. The association of the cells and ECM is also facilitated by fibronectin. Fibronectin is a dimeric glycoprotein, consist of dimmers, with each around 250,000 molecular weight. Each of the subunits is doubled over three time (FN1, FN2, FN3) and contain the amino acid repeat FN1, FN2, FN3. Fibronectin can attach to collagen at FN1/2 region, whereas the FN3 region takes up contact with the respective cells [10, 7 Alternative splicing forms completely different kinds of fibronectin. For example the liver builds fibronectin that flows as a liquefied element of the serum. Nevertheless, static fibronectin is also formed in various tissues [7].

In natural skeletal tissues, the cells and the structures form spatial and oriented geometries and patterns [11]. In recent research, significant work has been done to know cell surface attachment and alignment for biomedical applications [12-13]. The prospective uses of such biomedical systems consist of the micro-textures that helps in enhancing wound closure and tissue integration [14-15], [3-4]. Meanwhile, various polymer and metals have been commercially introduced. These metals and polymers frequently show poor attachment with the proteins and cells. This makes them hard to incorporate into the body tissues. To increase the bioactivity and biocompatibility, much work have been done to deposit bioactive coatings onto implant surfaces. Contact guidance can be promoted by the use of chemical guidance cues [16]. This is a fact which take account of the cell guidance, as they extend through the micro-patterns. Another researcher [17] observed the consequences of micro-patterned on cell guidance. In microfluidic devices, BioMEMS and lab-on-a-chip applications, Polydimethylsiloxane (PDMS) has been widely used [18-20] largely because of its biocompatibility and mechanical properties [21-22]. Researchers have also analyzed the consequences of micro-patterning on cell dispersion and guidance on PDMS [23, 9]. Formation of detailed surface designs on glass, metal or polymer might allow further clarification about the elements that control cell attachment, cell growth, variation and molecular signaling paths on material surfaces [24-25]. These surface-engineered materials are expected to be able to control the shape, size cell/cell coupling of the attached cells during guided network formation.

The aims of this study are to characterize HeLa cell guidance on micro contact printed fibronectin patterns and to then determine how that guidance is modified by pulse electric field.

2. Materials and methods

2.1. Cell culture

In this study, HeLa cells were cultured in RPMI-1640 medium supplemented with 10% fetal bovine serum (FBS) and 1% penicillin-streptomycin at 37°C in a 5% humidified CO_2 incubator. When the HeLa cells reached 80-90% confluence, media was removed from the tissue culture flask, then, 3ml of phosphate buffered saline (PBS) was added in the flask to wash the cells, which was removed after few seconds. 2 ml of TryPle Express from GIBCO was added to detach the cells from bottom of the flask. TryPle Express works best in a warm surrounding, so the flask was incubated for seven minutes at 37°C. After detachment, the HeLa cells were counted by haemocytometer. Thereafter, 800µl of cells suspension at a concentration of 3.75×10^4cell/ml (equivalent to 30,000 cells) were transferred into a 4mm cuvette for electroporation.

2.2. Stamp fabrication

Main templates were made up of 25µm, 50µm and 100µm wide ridge (width) with 5µm deep channels at the University of Glasgow [26]. PDMS stamps were formed at the University of Bradford from

IOP Conf. Series: Journal of Physics: Conf. Series **1019** (2018) 012018 doi:10.1088/1742-6596/1019/1/012018

these templates. Initially, the template was coated by means of 2% dimethylchlorosilane mixed with 98% trichloroethylene which make the surface hydrophobic.
After that, 9ml of sylgard (silicon-based elastomer gel) and 1ml of curing agent were mixed with the templates and left to cure overnight. Vacuum processing was done to eliminate any air bibbles. After curing, the sylgard gel was removed from the template which enabled the generation of sylgard stamps showing a negative relief of the template to be formed. These stamps were then used in functionalizing plain glass slides and gold coated test substrates with various proteins. Figure 1 show the images of the PDMS stamps for 25μm, 50μm and 100μm width.

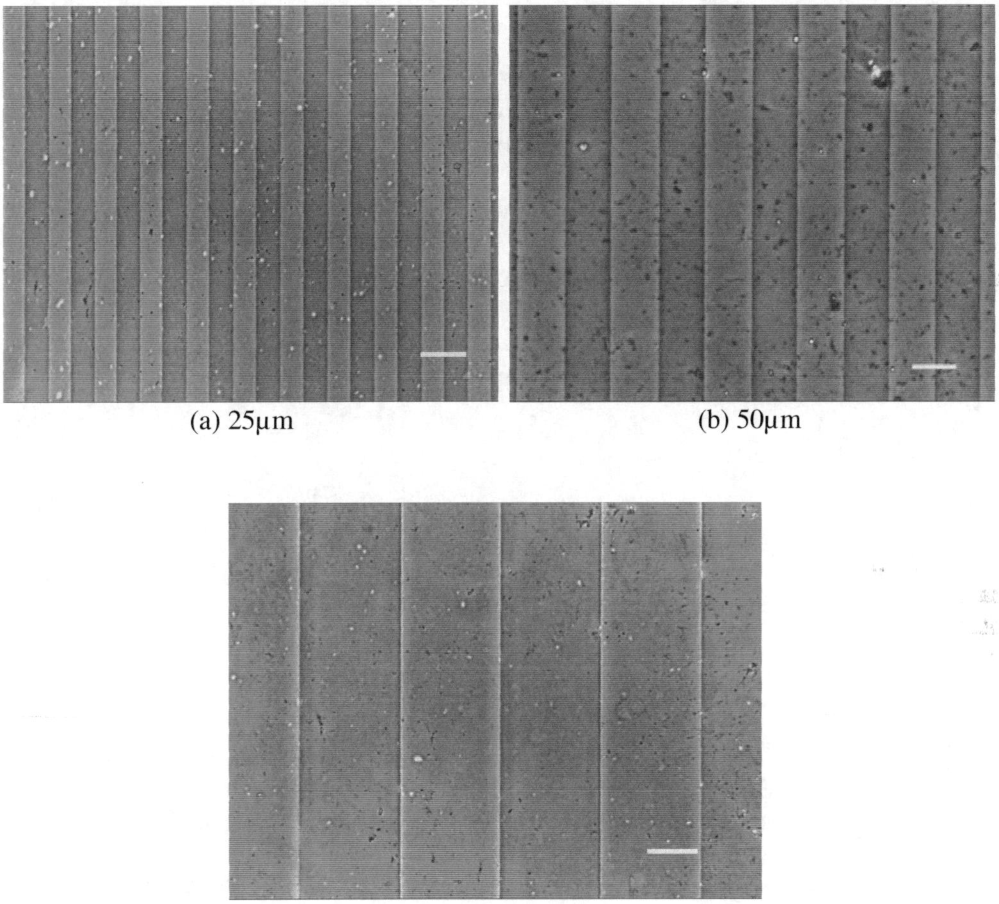

(a) 25μm (b) 50μm

(c) 100μm

Figure 1. PDMS stamps with (a) 25μm, (b) 50μm and (c) 100μm width (scale bar 50μm)

2.3. Micro contact printing (MCP) technique

All working environments (surfaces) before MCP, stamps and coverslips were sterilized with 70% ethanol and air dried in a laminar flow hood. After that plain glass coverslips were micro-contact printed with fibronectin. Inking the PDMS stamps was achieved by applying a 50μg/ml solution of fibronectin in PBS [27] to the PDMS stamps for 60 seconds [28, 4]. Following inking the stamps were air-dried for 30 seconds. The stamps were then placed immediately in contact with the substrate and pressed slightly using another glass cover slip for 50 seconds [29]. The same method was repeated for each stamp of the 25μm, 50μm and 100μm stamps. This allowed the substrate to be patterned with 25μm, 50μm and 100μm wide protein coated tracks which are separated by 25μm, 50μm and 100μm wide uncoated tracks. Figure 2 demonstrates this process, in which the MCP patterns were successfully transferred to the substrates. Though imperfections in the stamp and the use of manual pressure caused some differences in the track widths of the stamped protein.

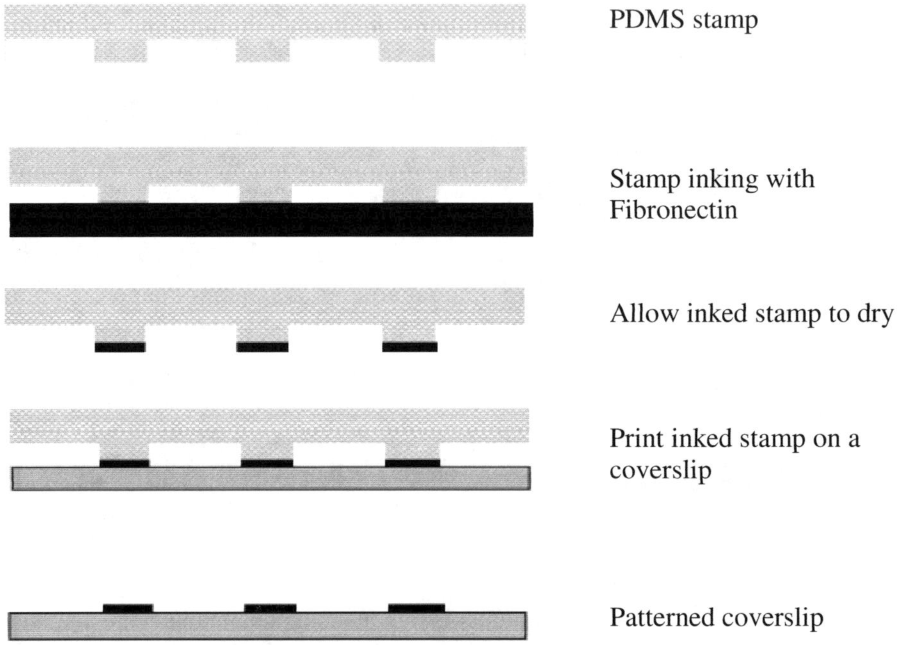

Figure 2. Illustration of the stamping process

2.4. Cell Plating

Plain glass slides and glass slides coated with un-patterned fibronectin were seeded with non-electroporated (NEP) HeLa cells at a concentration of 30,000 cells/coverslip for 18hrs as a control group. At the same time, glass slides with fibronectin micro-contact patterned printed with three different gratings (25μm, 50μm and 100μm widths) were also seeded with non-electroporated HeLa cells for 18hrs. After 18hrs, cells were then imaged with standard phase contrast light microscope. The pattern with better alignment was used for seeding electrically treated HeLa cells. This allowed the impact of single pulse electric fields on cell guidance to be determined by comparing cell guidance on the optimum fibronectin pattern with and without PEF treatment.

3. Results and discussion

3.1. Micro contact printing (MCP) technique

The stamp consisted of ridges with 25μm, 50μm and 100μm width fibronectin coated tracks separated by 25μm, 50μm and 100μm width uncoated tracks. Micro-contact printing patterned of fibronectin on the cover glass (substrates) resulted in the substrates being successfully patterned with 25μm, 50μm and 100μm width of repeated gratings as shown in Figure 3. The distances between the coated and uncoated channels have been marked by the yellow arrows.

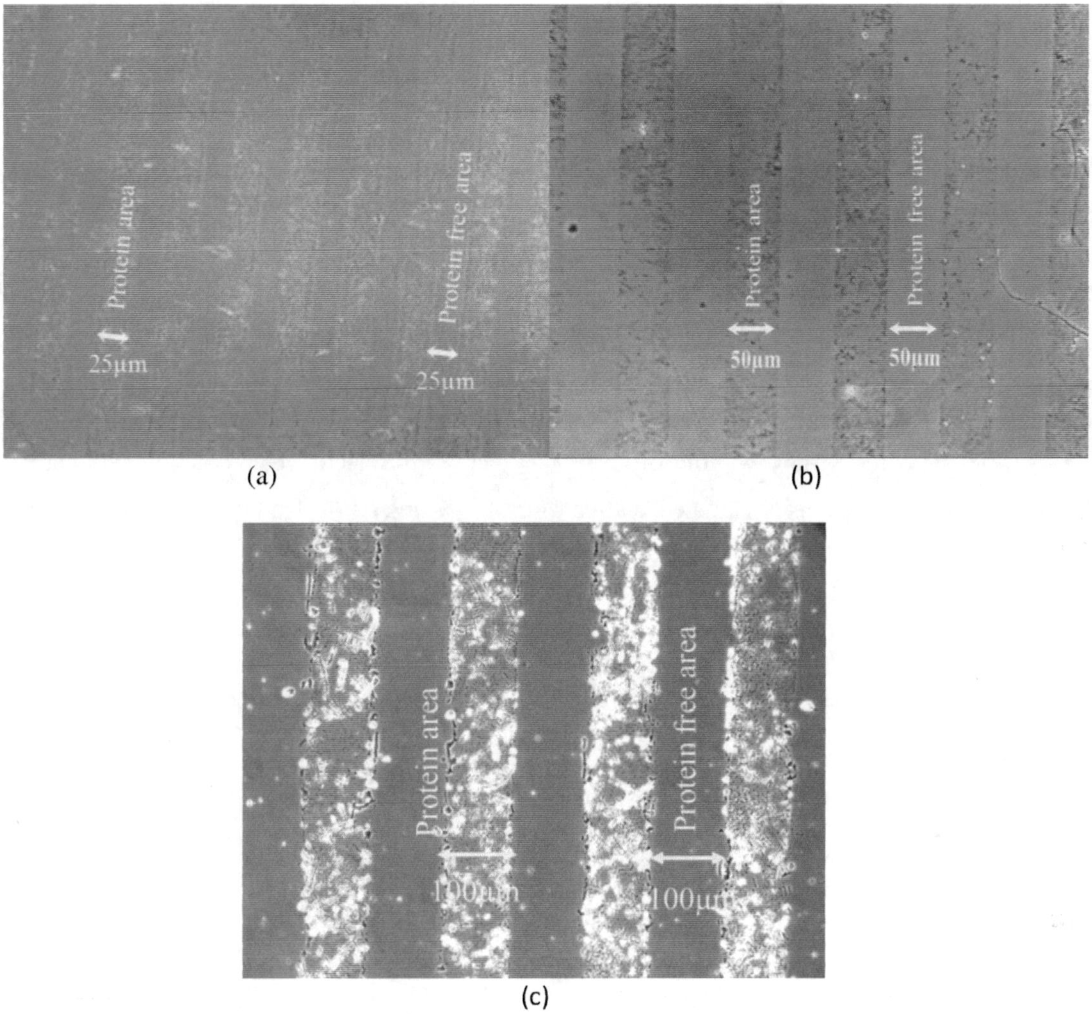

Figure 3. Actual images of (a) 25µm, (b) 50µm and (c) 100µm width protein patterns printed on glass coverslip

3.2. Hela cells interactions with micro-patterned surface

Substrate printed with chemical cues caused different cell responses like cell migration and cell alignment along the cues [1, 4, 17]. The effects of initial cell attachment on substrate were investigated by recording cell images up to 18hrs after cell culture using DinoCapture imaging software. After 18hrs in culture, the photomicrograph obtained from the plated HeLa cells showed different alignment to the MCP. It became clear that the HeLa cell seeded on uncoated and coated unpattern substrates (control group) did not show any alignment. This is obvious because even though coated but there is no pattern for the cell to follow. Figure 4 shows the morphology of HeLa cells on the two control group substrates (uncoated and fibronectin-coated but unpattern substrates). Even though the control group are identical in terms of initial cell seeding density, passage number and incubator environment, they showed different morphology and adhesion capacity when compared the pattern counterpart group. Most cell in the control group are rounded or rather more of an epithelial morphology which imply less adherence to the substrate [30].

IOP Conf. Series: Journal of Physics: Conf. Series **1019** (2018) 012018 doi:10.1088/1742-6596/1019/1/012018

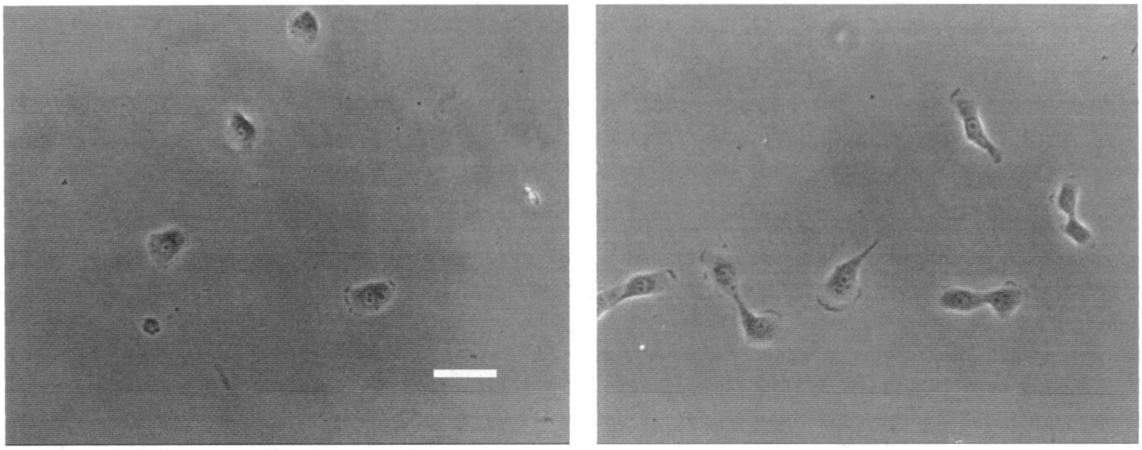

(a) Un-coated Sample without EP (b) Fibronectin coated Sample without EP

Figure 4. Photomicrographs of HeLa cells after 18hrs of seeding on (a) Un-coated coverslip glass, (b) Fibronectin coated on coverslip glass (control experiments) (scale bar 50μm)

However, the HeLa cells reacted differently to the various gratings of fibronectin patterned. HeLa cells seeded on 25μm patterned substrates induced a poor alignment (34° ±6.78SD). On the other hand, 50μm induced best alignment of the cells to the fibronectin pattern (7.66° ±1.55SD). Similarly, 100μm show good alignment (12.06° ±1.5SD) as compared to the 25μm pattern but less than the 50μm pattern. As it shown in Figure 5, we can deduce that the fibronectin pattern substrate showed a better attachment with cells elongating along the pattern especially with the 50μm and 100μm though more articulate with 50μm.

Furthermore, the 50μm induced a better alignment was used to check the behavior of HeLa cells under the influence of electric field. Hence the cells were electroporated (with 1kv/cm 100μs & single pulse) and seeded on to a 50μm fibronectin micro-patterned substrate and their morphology compared to the non-electroporated counterpart. The results as shown in Figure 6 indicates that cells aligned (2.33° ±1.52SD) on fibronectin pattern substrate especially under pulse electric field exposure: with better alignment of the cells in the EP group (2.33° ±1.52SD), when compared to NEP group (7.66° ±1.55SD) on the same 50μm fibronectin patterned substrate.

This could be that the electric field stimulated the signaling path way for cell adhesion and hence increase the cell spreading [31]. This finding is also in agreement to that of Zhang *et al.* (2011) [32]. Hence, in this research, it was also speculated that the pulse electric field could have stimulated the signaling path way, there by modulating the filopodia to align to the pattern with more pronounce elongation than the non-electrically treated counterpart. Thus, the outcome of the research can be utilized in cell assembling and guidance of cells for wound healing application and tissue regeneration for implantation, for example antibody fluorescence based experiments to measure specific integrin up-regulation.

Finally, the overall conclusion of this study is that PEF seems to enhance cell guidance on glass substrate micro-patterned with 50μm wide repeat gratings of fibronectin. This could indicate that in some way PEF and electroporation either modifies the distribution of fibronectin specific membrane bound integrin or results in their up, or down regulation. The manipulation of both may have implications in the enhancement of engineered tissue grafts technologies.

IOP Conf. Series: Journal of Physics: Conf. Series **1019** (2018) 012018 doi:10.1088/1742-6596/1019/1/012018

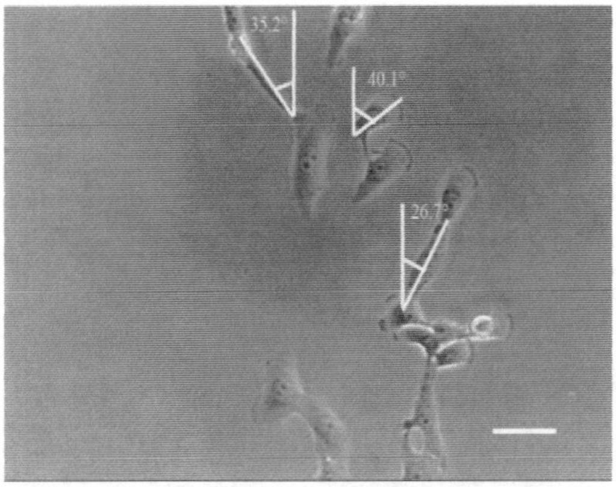

(a) 25μm micro contact printed of Fibronectin Sample

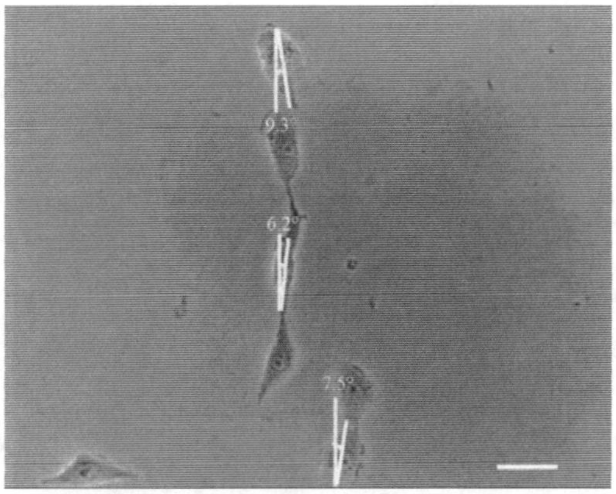

(b) 50μm micro contact printed of Fibronectin Sample

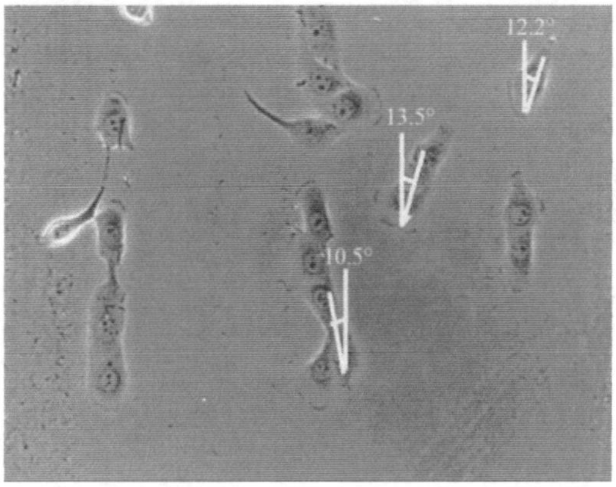

(c) 100μm micro contact printed of Fibronectin Sample

Figure 5. Photomicrographs of HeLa cells after 18hrs of seeding (a) 25μm, (b) 50μm and (c) 100μm micro contact printed of Fibronectin on coverslip glass (scale bar 50μm)

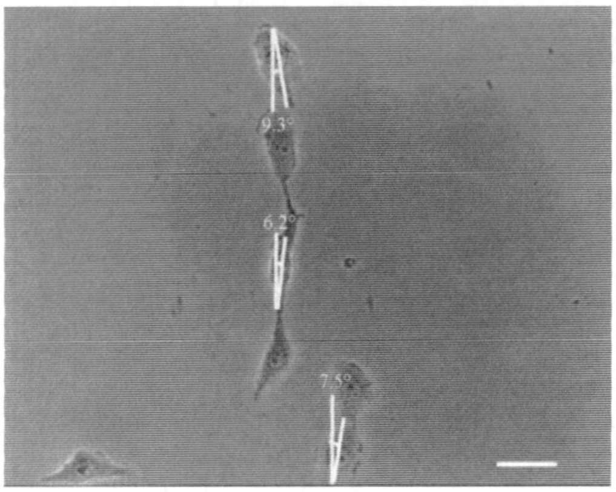

(a)

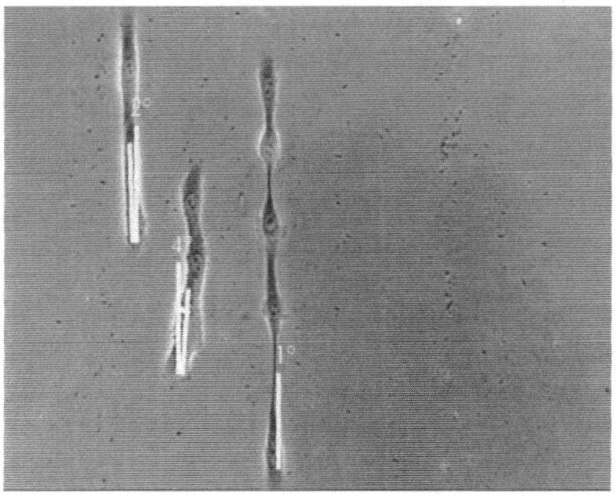

(b)

Figure 6. Photomicrographs of HeLa cells after 18hrs of seeding 50μm micro-contact printed of Fibronectin on coverslip glass (a) without EP and (b) with EP (scale bar 50μm)

4. Conclusion
This study mainly focuses on effects of PEC exposure on fibronectin pattern surface with HeLa cells for cell guidance. Stamp fabrication is performed in the first step of this experimentation. Once achieved, micro contact printing technique is achieved. This permitted the substrate to be patterned by 25μm, 50μm and 100μm width protein coated tracks. It was analyzed through experimentation that 50μm shows the most suitable for HeLa cells plating. This sample was also analyzed with PEF exposure. Results show that HeLa cells exposed to PEF show more aligned on the pattern surface when compared to control sample.

5. Acknowledgments
The researchers would like to thank Ministry of Higher Education Malaysia and UTHM for the research grant under the Tier 1 Research Grant (U865) for supporting this research work.

6. References
[1] Sefat, F., Denyer, M. C. T., & Youseffi, M. (2011). Imaging via widefield surface plasmon resonance microscope for studying bone cell interactions with micropatterned ECM proteins. *Journal of microscopy*, **241**(3), 282-290.

[2] Ricoult, S. G., Goldman, J. S., Stellwagen, D., Juncker, D., & Kennedy, T. E. (2012). Generation of microisland cultures using microcontact printing to pattern protein substrates. *Journal of neuroscience methods*, **208**(1), 10-17.

[3] Berends, R. F., Youseffi, M., Sefat, F., Khaghani, S. A., & Denyer, M. (2009). Investigating keratinocyte cell responses to ECM proteins using Microcontact printing. In *Journal of Anatomy*. Vol. **215**, No. 6, pp. 711-711.

[4] Khaghani, S., Sefat, F., Denyer, M., & Youseffi, M. (2008). Alignment of rat primary chondrocyte cells to collagen type-1, ficronectin and laminin. *In Journal of Anatomy*. **213**(3), pp. 351-351.

[5] Kumar, A., Abbott, N. L., Biebuyck, H. A., Kim, E., & Whitesides, G. M. (1995). Patterned self-assembled monolayers and meso-scale phenomena. *Accounts of Chemical Research*, **28**(5), 219-226.

[6] Kleinman, H. K., Sephel, G. C., TASHIRO, K. I., Weeks, B. S., Burrous, B. A., Adler, S. H., ... & Martin, G. R. (1990). Laminin in neuronal development. *Annals of the New York Academy of Sciences*, **580**(1), 302-310.

[7] Pankov, R., & Yamada, K. M. (2002). Fibronectin at a glance. *Journal of cell science*, **115**(20), 3861-3863.

[8] Carter, S. B. (1967). Haptotactic islands: a method of confining single cells to study individual cell reactions and clone formation. *Experimental cell research*, **48**(1), 189-193.

[9] Salber, J., Gräter, S., Harwardt, M., Hofmann, M., Klee, D., Dujic, J., & Groll, J. (2007). Influence of Different ECM Mimetic Peptide Sequences Embedded in a Nonfouling Environment on the Specific Adhesion of Human-Skin Keratinocytes and Fibroblasts on Deformable Substrates. Small, **3**(6), 1023-1031.

[10] Abdul Jamil, M. M., Youseffi, M., Britland, S.T., Liu, S., See, C.W., Somekh, M.G., Denyer, M.C.T. (2007). Widefield Surface Plasmon Resonance Microscope: A Novel Biosensor Study of Cell Attachment to Micropatterned Substrates. 3rd Kuala Lumpur International Conference on Biomedical Engineering, Biomed, 15, 334-337.

[11] Zhu, B., Lu, Q., Yin, J., Hu, J., & Wang, Z. (2005). Alignment of osteoblast-like cells and cell-produced collagen matrix induced by nanogrooves. *Tissue engineering*, **11**(5-6), 825-834.

[12] Feinberg, A. W., Wilkerson, W. R., Seegert, C. A., Gibson, A. L., Hoipkemeier-Wilson, L., & Brennan, A. B. (2008). Systematic variation of microtopography, surface chemistry and elastic modulus and the state dependent effect on endothelial cell alignment. *Journal of Biomedical Materials Research Part A*, **86**(2), 522-534.

[13] Hamilton, D. W., Wong, K. S., & Brunette, D. M. (2006). Microfabricated discontinuous-edge surface topographies influence osteoblast adhesion, migration, cytoskeletal organization, and proliferation and enhance matrix and mineral deposition in vitro. *Calcified tissue international,* **78**(5), 314-325.

[14] Dalby, M. J., Riehle, M. O., Yarwood, S. J., Wilkinson, C. D., & Curtis, A. S. (2003). Nucleus alignment and cell signaling in fibroblasts: response to a micro-grooved topography. *Experimental cell research*, **284**(2), 272-280.

[15] Khang, D., Sato, M., Price, R. L., Ribbe, A. E., & Webster, T. J. (2006). Selective adhesion and mineral deposition by osteoblasts on carbon nanofiber patterns. *International journal of nanomedicine*, **1**(1), 65.

[16] Clark, P., Connolly, P., Curtis, A. S., Dow, J. A., & Wilkinson, C. D. (1990). Topographical control of cell behaviour: II. *Multiple grooved substrata. Development*, **108**(4), 635-644.

[17] Su, W. T., Liao, Y. F., & Chu, I. M. (2007). Observation of fibroblast motility on a micro-grooved hydrophobic elastomer substrate with different geometric characteristics. *Micron*, **38**(3), 278-285.

[18] Whitesides, G. M. (2006). The origins and the future of microfluidics. *Nature*, **442**(7101), 368-373.

[19] Cecchini, M., Girardo, S., Pisignano, D., Cingolani, R., & Beltram, F. (2008). Acoustic-counterflow microfluidics by surface acoustic waves. *Applied Physics Letters*, **92**(10), 104103.

[20] Yea, K. H., Lee, S., Choo, J., Oh, C. H., & Lee, S. (2006). Fast and sensitive analysis of DNA hybridization in a PDMS micro-fluidic channel using fluorescence resonance energy transfer. *Chemical Communications*, **14**, 1509-1511.

[21] Peterson, S. L., McDonald, A., Gourley, P. L., & Sasaki, D. Y. (2005). Poly (dimethylsiloxane) thin films as biocompatible coatings for microfluidic devices: cell culture and flow studies with glial cells. *Journal of Biomedical Materials Research Part A*, **72**(1), 10-18.

[22] Griscom, L., Degenaar, P., LePioufle, B., Tamiya, E., & Fujita, H. (2002). Techniques for patterning and guidance of primary culture neurons on micro-electrode arrays. *Sensors and Actuators B: Chemical*, **83**(1), 15-21.

[23] Chaw, K. C., Manimaran, M., Tay, F. E. H., & Swaminathan, S. (2007). Matrigel coated polydimethylsiloxane based microfluidic devices for studying metastatic and non-metastatic cancer cell invasion and migration.Biomedical microdevices, **9**(4), 597-602.

[24] Ogaki, R., Alexander, M., & Kingshott, P. (2010). Chemical patterning in biointerface science. *Materials today*, **13**(4), 22-35.

[25] Diener, A., Nebe, B., Lüthen, F., Becker, P., Beck, U., Neumann, H. G., & Rychly, J. (2005). Control of focal adhesion dynamics by material surface characteristics. *Biomaterials*, **26**(4), 383-392.

[26] Abdul Jamil, M. M., Youseffi, M., Twigg, P. C., Britland, S. T., Liu, S., See, C. W., & Denyer, M. C. (2008). High resolution imaging of bio-molecular binding studies using a widefield surface plasmon microscope. *Sensors and Actuators B: Chemical*, **129**(2), 566-574.

[27] Shen, K., Qi, J., & Kam, L. C. (2008). Microcontact printing of proteins for cell biology. *Journal of Visualized Experiments,* JoVE. **22**, e1065-e1065.

[28] Tan, J. L., Liu, W., Nelson, C. M., Raghavan, S., & Chen, C. S. (2004). Simple approach to micropattern cells on common culture substrates by tuning substrate wettability. *Tissue engineering*, **10**(5-6), 865-872.

[29] Abdul Jamil, M. M. (2007). Application of a novel high resolution widefield surface plasmon microscope in cell engineering, wound healing and development of new binding assays. PhD thesis submitted to the University of Bradford.

[30] Sefat, F. (2013). Cell engineering of human bone monolayers and the effect of growth factors and microcontact printed ECM proteins on wound healing. The role of ECM proteins, TGF¿-1, 2 and 3 and HCl/BSA in cellular adhesion, wound healing and imaging of the cell surface interface with the widefield surface plasmon microscope (Doctoral dissertation, University of Bradford).

[31] Mamman, H. B., & Jamil, M. M. (2015). Investigation of electroporation effect on HT29 cell lines adhesion properties. 2015 2nd International Conference on Biomedical Engineering (ICoBE), IEEE. 1-4.

[32] Zhang, X., Kang, L. G., Ding, L., Vranic, S., Gatalica, Z., & Wang, Z. Y. (2011). A positive feedback loop of ER-α36/EGFR promotes malignant growth of ER-negative breast cancer cells. *Oncogene*, **30**(7), pp. 770-780.

IOP Conf. Series: Journal of Physics: Conf. Series **1019** (2018) 012019 doi:10.1088/1742-6596/1019/1/012019

Lighting system design using green energy from living plants

M.F. Lee[1], M.N.N. Mohd Zain[2] and C.S. Lai[3]

[1,2,3]Faculty of Technical & Vocational Education, Universiti Tun Hussein Onn Malaysia, 86400 Parit Raja, Batu Pahat, Johor

mflee@uthm.edu.my, nnaimzain@gmail.com, lcsern@uthm.edu.my

Abstract. Green energy is the energy generated from nature such as water, wind, solar and wave that aimed to reduce environmental damage or destruction, sustain the use of energy and natural resources as well as promote the use of renewable energy. This study aimed to analyse the electrical energy generated by the three types of living plants selected at the parking area of the Faculty of Technical and Vocational Education and design a lighting system for the parking area by using living plants as the source of energy generation. This study consisted two phases, Phase I: Analysing the energy generated from the living plants; and Phase II: designing the lighting system for the parking area phase using living plants. The result of the analysis of the energy produced by living plants showed that the selected living plants are able to produce energy with the sufficient amount needed by the lighting system at the parking area. Based on the analysis of the generation of energy from living plants, the lighting system was designed by using step-up circuit and the charging circuit. In conclusion, living plants should be fully utilized as part of the energy resources to generate energy.

1. Introduction

Most of the publics' knowledge of green technology is a technology which produces an electric power from resources such as water, wind, sunlight, and waves. According to [1], green technology is the use of products, equipment and systems that are intended to preserve nature and reduce the negatively impact on life. Basically, green technology is divided into several criteria, namely:

i. Reduce destruction or damage to nature.
ii. Has a low level of gas emissions
iii. Preserving energy and natural resources
iv. Promote the use of renewable resources

Green energy is a technology application in power generation system and minimize the adverse effects resulting from human daily activities to the environment [2]. In this rapid development society in Malaysia, the use of electricity is utmost important and become a must in daily life. This growing progress requires constant energy supply without any interruption. At the same time, the increase in electricity tariffs has also increased over time. Universities in Malaysia are also impressed and burdened by the increasing in electricity tariffs.

The previous research findings noted that the current problem facing by all the country around world is no sustainable source of energy that does not interfere with the country's ecosystem [4]. According to a study by [5], 68% electricity generation nowadays relies on the fuel-burning which have long-term effects on the country's ecosystem. Hence, it reinforces the reason why research on the generate electricity by using green resource has to be carried out and taken seriously as a source of energy-producing without severe side-effects. Many users are unaware of the importance and many consumers do not know about the advantages of living plants that can produce good and sustainable electricity. Therefore, a technical studies on harvesting electric energy from living plants was conducted. Harvesting energy form living plants is not a new things but only a small portion of people know about this technology. Living plants is one of the energy generator that do not require any high cost. Also, it is a green energy generation system without disturbing any ecosystem besides helping in the improvement of fresh air system at the particular area [3]. The previous studies in determining the electricity generated by living plants only being done on pulai tree, banana tree and aloe vera. The findings showed that these trees have produced voltage of 0.8V 0.913V and 0.945V each [6]. Basically, this study was aimed to:

i. Analyze the electricity generated by living plants around the parking area at FPTV.
ii. Design the lighting system for parking area in FPTV using green energy from living plants.

2. Experiment Details

This study was conducted to determine the level of electricity generated by the living plants found around the parking area at Faculty of Technical & Vocational Education, Universiti Tun Hussein Onn Malaysia (refer Figure 1). Based on the gathered information, a green energy-efficient lighting system was designed. There are three types of living plants around the parking were *garcinia subelliptica*, *gliricidia sepium* and *murraya paniculata* as showed in Figure 2.

Figure 1 : Parking Area at FPTV, UTHM

Figure 2(a) : *Garcinia Subelliptica* Figure 2(b): *Gliricidia Sepium* Figure 2(c) : *Murraya Paniculata*

In order to determine the voltage and current generate by the living plants, copper rods and carbon steel rods were used as an electric absorber metal located around the roots of the tree. These two rods were then connected to a multimeter that proposed by [6] as showed in Figure 3. There were 13 times of data collection on the voltage and current readings for each types of the plants regardless the types of soil. In this study, factors that may influence the voltage and current generated by the living plants only focus on the temperature and the weather. Figure 4 showed an example of experiment being done to obtain the voltage and current generated by living plant.

IOP Conf. Series: Journal of Physics: Conf. Series **1019** (2018) 012019 doi:10.1088/1742-6596/1019/1/012019

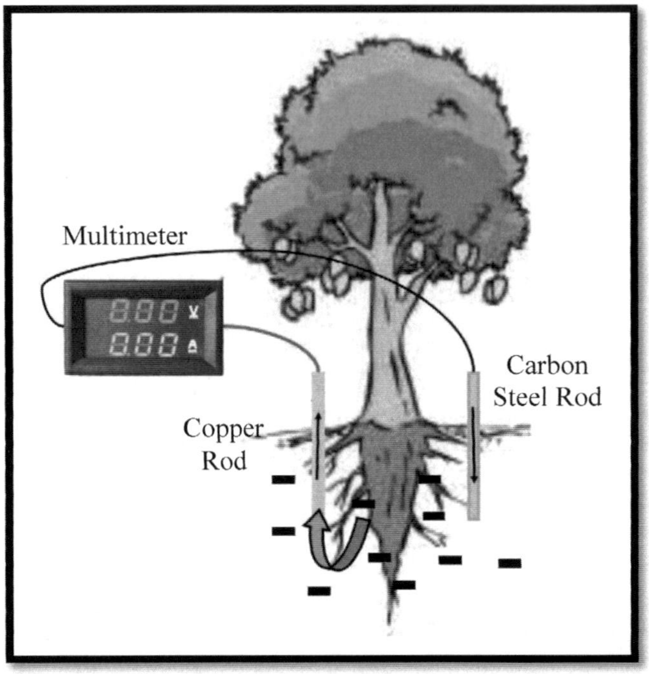

Figure 3: Method to determine voltage and current from living plant

Figure 4: Example of experiment

3. Results

There were three types of living plants that found at the parking area of FPTV. The result of voltage and current generated by these three different types of plants were discussed in the following subtopics.

3.1. Voltage and current harvested from garcinia subelliptica

Table 1: Voltage and current generated by *garcinia subelliptica*

NO	Time	Temperature	Voltage	Current	Weather
1.	Morning	25°c	0.621V	0.01A	Sunny
2.	Morning	28 °c	0.570V	0.016A	Cloudy
3.	Noon	32°c	0.632V	0.010A	Raining
4.	Noon	27 °c	0.687V	0.011A	Raining
5.	Morning	28 °c	0.690V	0.010A	Sunny
6.	Noon	32 °c	0.691V	0.002A	Sunny
7.	Morning	25 °c	0.676V	0.013A	Sunny
8.	Morning	28 °c	0.677V	0.003A	Sunny
9.	Noon	31 °c	0.678V	0.001A	Cloudy
10.	Noon	31 °c	0.676V	0.013A	Sunny
11.	Evening	28 °c	0.668V	0.002A	Sunny
12.	Morning	22 °c	0.602V	0.005A	Cloudy
13.	Evening	28 °c	0.583V	0.007A	Cloudy

Table 1 shows that the average voltage generated by *garcinia subelliptica* was 0.652V, and the average of current generated was 0.007A.

3.2. Voltage and current harvested from gliricidia sepium

Table 2: Voltage and current generated by *gliricidia sepium*

NO	Time	Temperature	Voltage	Current	Weather
1.	Morning	25°c	0.510V	0.003A	Sunny
2.	Morning	28°c	0.489V	0.007A	Cloudy
3.	Noon	32°c	0.680V	0.010A	Raining
4.	Night	27°c	0.681V	0.002A	Raining
5.	Morning	28°c	0.681V	0.004A	Sunny
6.	Noon	32°c	0.682V	0.013A	Sunny
7.	Morning	24°c	0.679V	0.001A	Sunny
8.	Morning	28°c	0.679V	0.011A	Sunny
9.	Noon	30°c	0.681V	0.013A	Cloudy
10.	Noon	31°c	0.683V	0.009A	Sunny
11.	Evening	28°c	0.680V	0.011A	Sunny
12.	Morning	22°c	0.674V	0.001A	Cloudy
13.	Evening	28°c	0.678V	0.009A	Cloudy

Table 2 shows that the average voltage generated by *gliricidia sepium* was 0.581V, and the average of current generated was 0.007A.

IOP Conf. Series: Journal of Physics: Conf. Series **1019** (2018) 012019 doi:10.1088/1742-6596/1019/1/012019

3.3. Voltage and current harvested from murraya paniculata

Table 3: Voltage and current generated by *murraya paniculata*

NO	Time	Temperature	Voltage	Current	Weather
1.	Morning	25°c	0.574V	0.007A	Sunny
2.	Morning	26 °c	0.380V	0.006A	Cloudy
3.	Noon	31 °c	0.424V	0.004A	Raining
4.	Night	27 °c	0.724V	0.001A	Raining
5.	Morning	28 °c	0.728V	0.016A	Sunny
6.	Noon	32 °c	0.730V	0.012A	Sunny
7.	Morning	25 °c	0.731V	0.014A	Sunny
8.	Morning	28 °c	0.731V	0.012A	Sunny
9.	Noon	31 °c	0.734V	0.011A	Cloudy
10.	Noon	31 °c	0.728V	0.003A	Sunny
11.	Evening	28 °c	0.726V	0.004A	Sunny
12.	Morning	22 °c	0.717V	0.009A	Cloudy
13.	Evening	28 °c	0.713V	0.011A	Cloudy

Table 3 shows that the average voltage generated by *murraya paniculata* was 0.664V, and the average of current generated was 0.008A.

4. Lighting system design

Based on the data gathered on three types of living plants at parking area FPTV, *murraya paniculata* generated the highest voltage compare to another two types of plants. The existing lamp type at parking area are the conventional lamp which each of them need 60 watt and 240volt to operate. The lamps will switch on from 7pm to 7am each day. The energy consumed by each lamp is 0.66kWh (60watt x 11 hour). To allow the AC lamp to be switched on using the electricity produce by tree, the existing circuit needs to be adjusted. Figure 5 shows the block diagram for lighting system at parking area FPTV using *murraya paniculata* as power supply. Figure 6 shows the proposed wiring for the lighting system for the parking area. Meanwhile, Figure 7 shows the designed circuit for voltage booster and battery charger in order to generate the needed electricity by existing lamps at parking area FPTV.

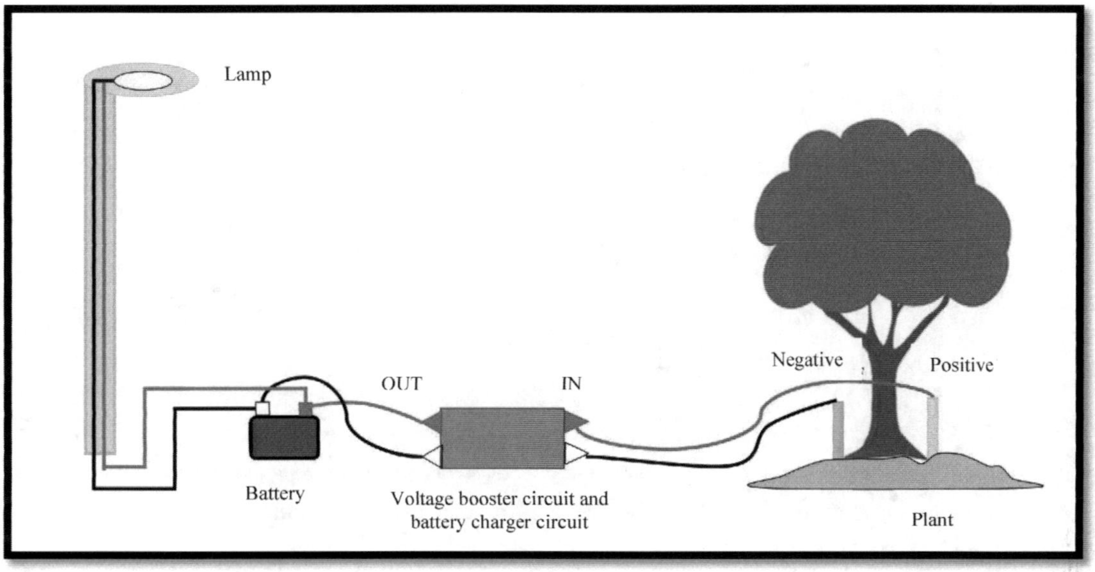

Figure 5 : Block diagram for lighting system at parking area FPTV using living plants as power supply

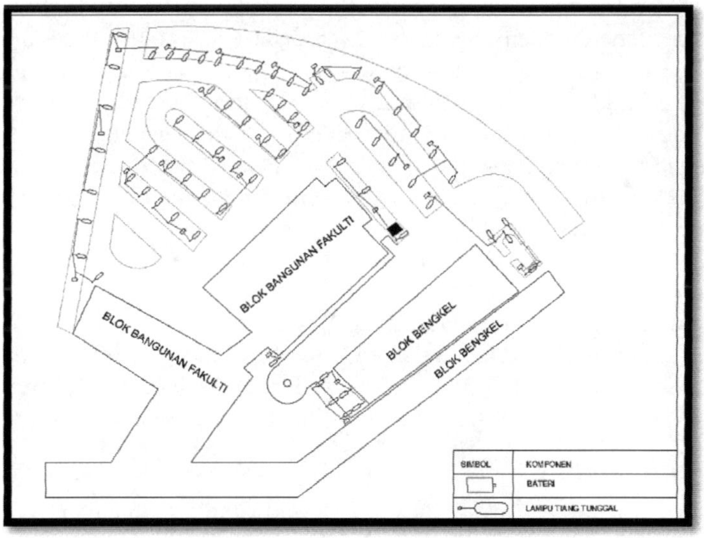

Figure 6: Proposed wiring for the lighting system for the parking area

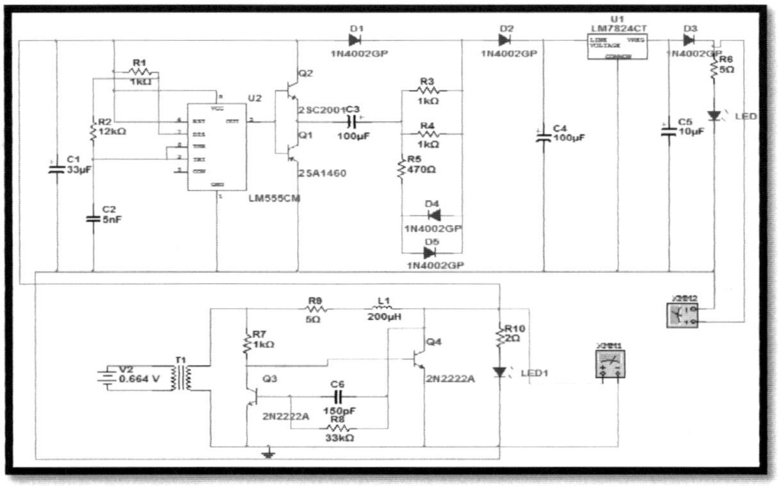

Figure 7: Voltage booster and battery charger circuit

5. Conclusion

In conclusion, studies related to the harvesting electric energy from living plants should be increased to create green energy consumption society. Finding of this study showed that living plants are able to produce high voltage and current as supply to the lighting system at the parking area in FPTV. However, the voltage generated among the trees are depend on the type of plants, temperature, weather and time [7]. During cloudy weather and raining, the voltage and current reading is lower than sunny weather. Further study on testing the designed lighting system for parking area at FPTV is recommended in order to investigate the functionality of the circuit. This study also showed an example and reference to researchers in the future especially those who are interested in studying green energy as well as to expose the public on generating electricity using living plants. Future research will be done is to test the functional of the proposed lighting system.

Acknowledgement

The publication of this work was supported by the FRGS grant (No. Vot: 1554) awarded by Ministry of Higher Education Malaysia (MOHE) and the Office for Research, Innovation, Commercialization and Consultancy Management (ORICC) of Universiti Tun Hussein Onn Malaysia. The authors would also like to thank to those who graciously gave their time to participate in this study.

References

[1] Bhardwaj, M. (2015). The Advantages and Disadvantages of Green Technology, Journal of Basic and Applied Engineering Research, 2(22), 1957–1960.

[2] Bokhari, M., Abdullah, A. N., Syed, S. N., Jano, Z., & Saadan, R. (2014). Perspektif Etnik Di Negeri Melaka, Journal of Human Capital Development , 7(2), 33–46.

[3] Bird, L., & Cardinal, K. (2005). Trends in Utility Green Pricing Programs (2004) Technical Report. NREL/TP-640-40777 October 2005

[4] Howells, M., & Alexander, R. (2012). Perspectives on Sustainable Energy for the 21, (May). Retrived from https://sustainabledevelopment.un.org/content/documents/1131Energy_SD21.pdf

[5] Timmers, R. (2012). Electricity generation by living plants in a plant microbial fuel cell. Retrived from http://edepot.wur.nl/209871

[6] Choo, Y. Y., & Dayou, J. (2013). A Method to Harvest Electrical Energy from Living Plants, Journal of Science and Technology, 79–90.

[7] Sarkar, J., & Bhattacharyya, S. (2008). Short Communication Green electricity production with living plants and bacteria in a fuel cell, (September 2007). Retrived from https://doi.org/10.1002/er

Performance Analysis of DC-DC Buck Converter for Renewable Energy Application

N H Baharudin, T M N T Mansur, F A Hamid, R Ali, M I Misrun
School of Electrical System Engineering, Universiti Malaysia Perlis, Arau, 02600, Perlis, Malaysia
E-mail: norhanisahbaharudin@gmail.com

Abstract. This paper presents the analysis of DC-DC buck converter for renewable energy application. Renewable energy resources mostly produce DC voltages that are used DC-DC converter to convert the DC voltage from one level to another level according to different applications. This DC-DC buck converter is used to step down the high level DC voltage to low level DC voltage. By varying the duty cycle, the performance of buck converter in term of its output voltage can be obtained and analysed. Simulations have been performed using PSIM and validated with hardware implementation. All of the results are recorded and showed that very low voltage can be produced using this topology.

1. Introduction

Since most renewable energy produce DC voltage, power electronic circuit such as DC-DC converters are used to convert a DC voltage to a different DC voltage level [1]. Nowadays these converters are widely used with modern electronic application [1]. In photovoltaic (PV) application, maximum power point tracking (MPPT) with buck, boost, buck-boost, cuk, sepic and zeta converters are used commonly for peak power tracking [1][2]. In buck converter, the output voltage is always lower than the input voltage while in boost converter, the output voltage is always higher than input voltage [3]. For buck-boost converter, the output can be either lower than or higher than the input voltage depending on their duty cycle (D). If the duty cycle is higher than 0.5, the output voltage is greater than input and becoming boost converter. On the other hand, if the duty cycle is less than 0.5, the output voltage is lower than the input voltage and becoming buck converter. Figure 1 show the DC-DC converter in PV application with MPPT. Prior to transfer maximum power from the PV panel to the load, load matching must be achieved by varying the duty cycle of the DC-DC converter [4].

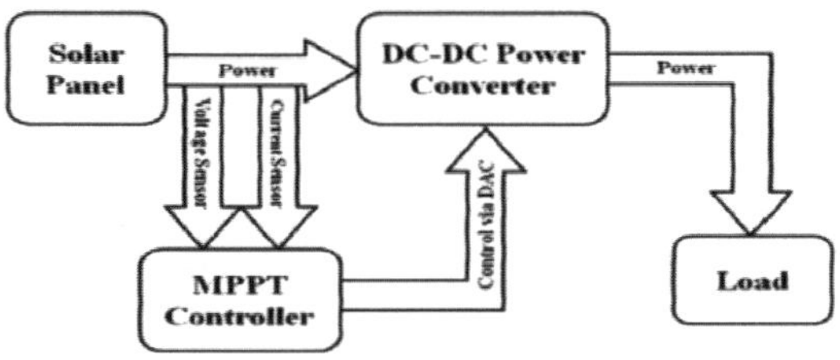

Figure 1. PV system with DC-DC converter and MPPT [4]

2. Topology of DC-DC buck converter

Basically a DC-DC buck converter consists of power switches such as MOSFET or BJT that will act as switches to control pulses. The circuit topology as shown in Figure 2 consists of MOSFET as power switch and this circuit will step down high level DC voltage to low level DC voltage. The output voltage for this circuit is always smaller than the input voltage [1]. The purpose of this circuit is to produce the output that is purely DC. So, the filter such as LC low pass filter was added in the basic circuit to produce the DC output voltage. When the switch is ON, the diode will become reverse bias and provide energy to the load and inductor. When the switch is OFF, diode will become forward bias and inductor current will flow through the diode. Some of its stored energy will be transferred to the load. This circuit topology can be connected to low load from high level voltage [3] and also can be used in high range step down converter [6].

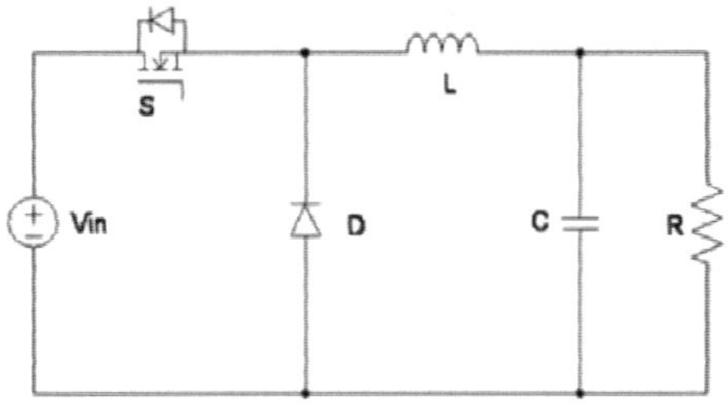

Figure 2. Buck converter topology [5]

3. Mathematical analysis of DC-DC buck converter

In this part, the relationship of input voltage, V_s and duty cycle in producing low output voltage related with buck converter are observed. The value of output voltage, V_o depends on the duty cycle. Equation (1) can be used in calculating the output voltage based on the duty cycle, D. Equation (2) and (3) are used to calculate the value of inductor while designing the buck converter. In calculating the value of inductor, it must be 25% greater than the minimum value of inductor, L_{min} obtained in Equation (2) to make sure that the circuit will be operated continuously. Furthermore, the maximum inductor current, I_{max} and minimum inductor current, I_{min} can be calculated by using Equation (4), (5) and (6). The value of I_{min} must be positive in order to sustain the continuous current. The value of capacitor,C can be calculated by using Equation (7) with the output ripple voltage which can be defined as a fraction of the output voltage.

$$D = \frac{V_o}{V_s} \quad (1)$$

$$L_{min} = \frac{(1-D)R}{2f} \quad (2)$$

$$L = 1.25L_{min} \quad (3)$$

$$I_L = \frac{V_o}{R} \quad (4)$$

$$I_{max} = I_L + \frac{\Delta i_L}{2} \quad (5)$$

$$I_{min} = I_L - \frac{\Delta i_L}{2} \quad (6)$$

$$C = \frac{1-D}{8L(\frac{\Delta V_O}{V_O})f^2} \tag{7}$$

4. Methodology

DC-DC buck converter is designed to step down 5V DC voltage source to 2V and 3V DC output voltages. The selected switching frequency, f is 25 kHz which is more than 20 kHz to avoid audio noise with 10 ohm load resistance, R and 10uF capacitor, C. The duty cycle, D of 0.4 and 0.6 are tested to produce 2 V and 3V DC voltage respectively. The circuit is simulated using PSIM software by using calculated parameters as shown in Equation (1) – (7) and the output waveforms will be analysed further. Low pass filter is included in this circuit to produce purely DC output voltage.

For the hardware implementation, components selection depend upon the maximum current and voltage that components can withstand according to their specification such as MOSFET, inductor, capacitor and resistor. Considering the power switches such as MOSFET, the designer should consider the maximum current and voltage stress during on and off condition that can be handled by the MOSFET and diode. Furthermore, the designer should also consider the maximum current that the inductor can support and the value of its operating frequency to avoid components break down.

The designed hardware is tested, analysed and measured using oscilloscope and digital multimeter. By differentiating the duty cycle of 0.4 and 0.6, the expected result of 2 V and 3 V DC voltages can be produced with this designed circuit of buck converter. MOSFET is triggered using gate driver to switch ON and OFF of the pulses. Similar parameters of inductor, capacitor and resistor which are 150uH, 10uF and 10Ω tested respectively for simulation and hardware testing.

5. Analysis of the simulated results and its hardware implementation.

An analysis of DC buck converter from the simulation results and hardware implementation will be presented. All of the data are recorded and tabulated for further discussion.

5.1. Simulation results

Figure 3 shows the circuit design for DC-DC buck converter that can convert high input of DC voltage to low output DC voltage that had been run in PSIM. Voltmeters that labeling with V_{ind}, V_{diode}, V_{cap} and V_{out} are used to measure the voltage drop across the inductor, diode, capacitor and load resistor respectively.

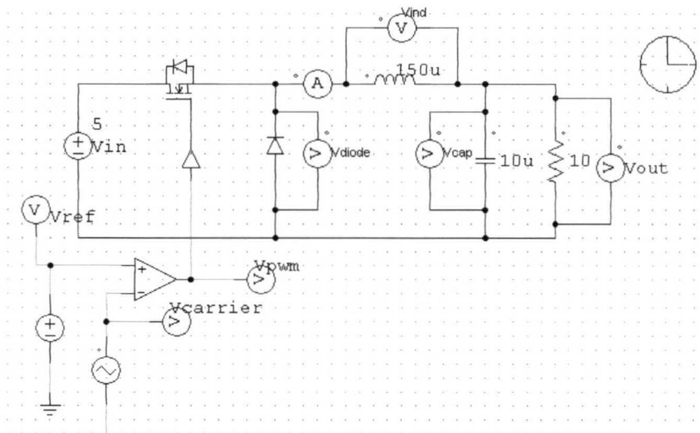

Figure 3. The designed of DC buck converter.

Figure 4 shows the output waveform when the duty cycle is 0.4 while Figure 5 shows the output waveform when the duty cycle is 0.6. The output voltage waveform is equal to 2 V for the duty cycle of 0.4 and 3 V for the duty cycle of 0.6. The comparison data was recorded in Table 1 for the duty cycle of 0.4 and 0.6 respectively. Based on simulation results, the value of output voltage depends on the duty cycle proportionally. When the duty cycle is smaller, the output voltage produced will be lower.

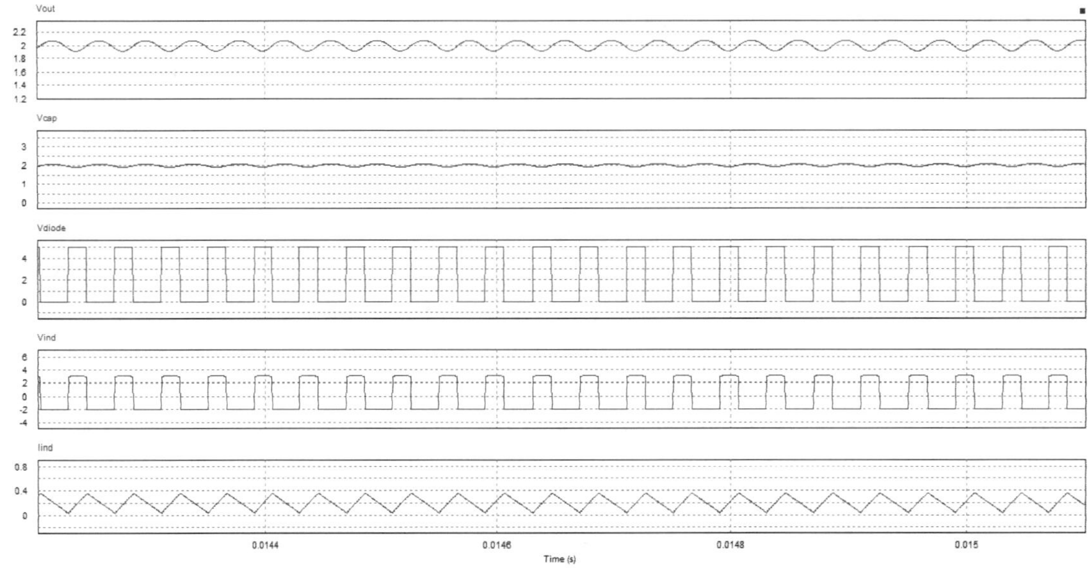

Figure 4. Output waveforms when the duty cycle is 0.4

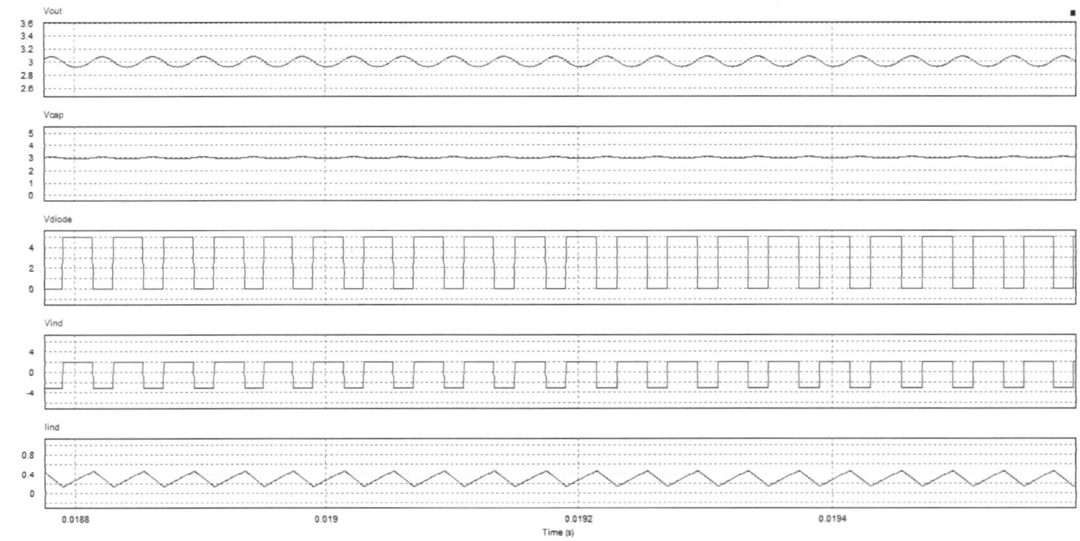

Figure 5. Output waveforms when the duty cycle is 0.6

5.2. Hardware prototype results.

For hardware prototype, the output waveforms and the reading of voltage across the components as well as the current flow through the components will be recorded for further analysis. Table 1 presented that the output voltage across the load is reduced to half as compared to the simulation and calculated value due to some losses from the high voltage drops across the diode and harmonic issues. This problem can be fixed by replacing the diode with second MOSFET that has low voltage drop and reduce the losses. Figure 6 – 10 show the output waveform across the DC buck converter such as capacitor, diode, inductor, load resistor as well as the inductor current.

Table 1. Comparison of hardware results for different duty cycle.

Parameters	Duty Cycle, D = 0.4	Duty Cycle, D = 0.6
V_{out}	0.970 V	1.760 V
V_{ind}	0.330 V	0.250 V
V_{cap}	1.090 V	1.810 V
V_{diode}	1.040 V	1.910 V
I_{ind}	0.019 A	0.035 A

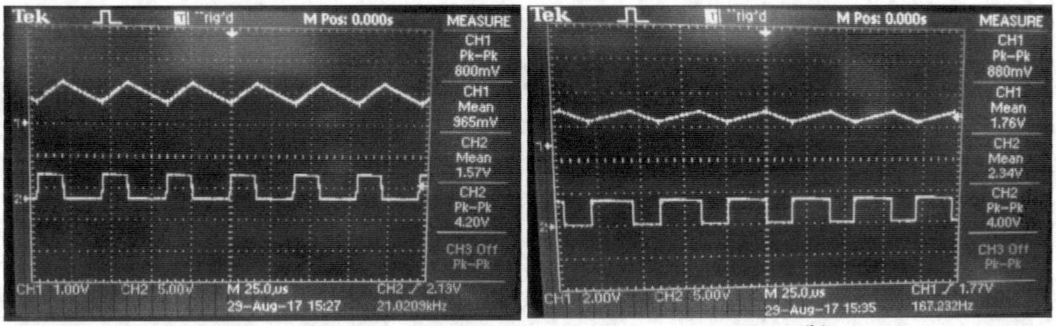

(a) (b)

Figure 6. Load resistor output voltage at (a) duty cycle, D of 0.4 (b) duty cycle, D of 0.6.

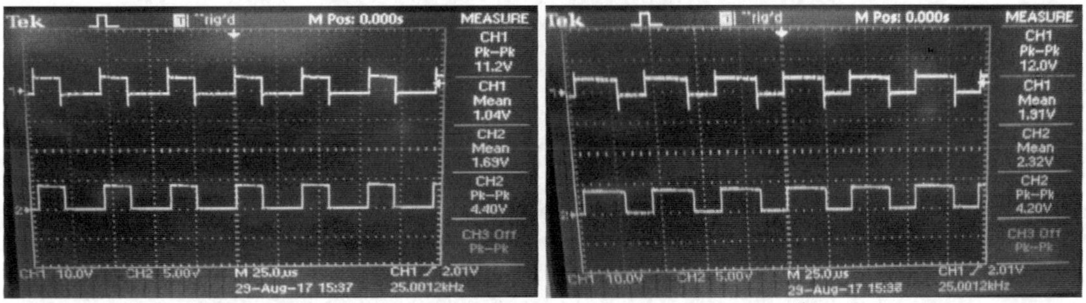

(a) (b)

Figure 7. Diode voltage at (a) duty cycle, D of 0.4 (b) duty cycle, D of 0.6.

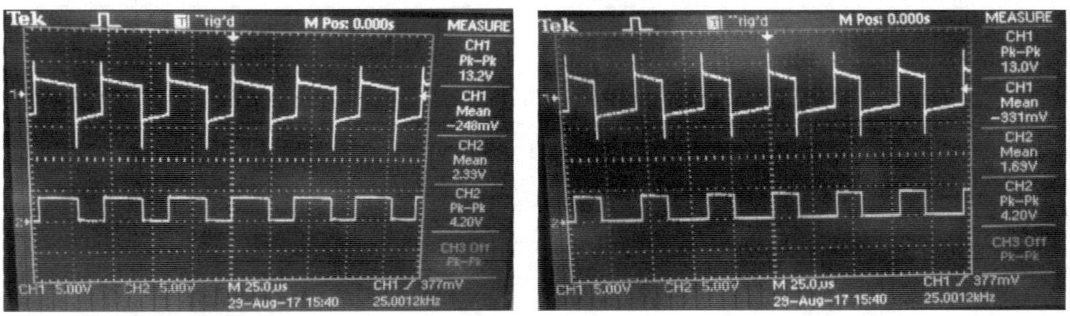

(a) (b)

Figure 8. Inductor voltage at (a) duty cycle, D of 0.4 (b) duty cycle, D of 0.6.

IOP Conf. Series: Journal of Physics: Conf. Series **1019** (2018) 012020 doi:10.1088/1742-6596/1019/1/012020

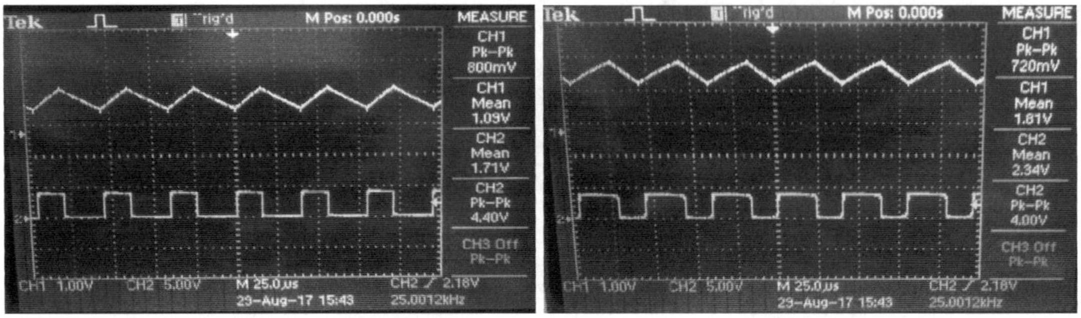

(a) (b)

Figure 9. Capacitor voltage at (a) duty cycle, D of 0.4 (b) duty cycle, D of 0.6.

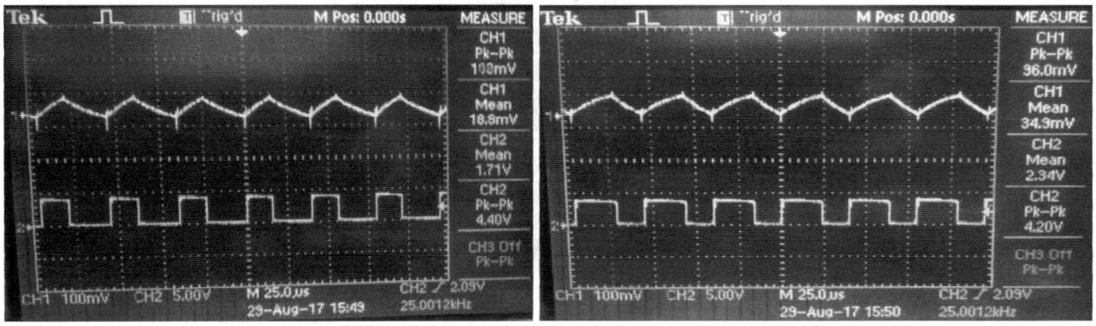

(a) (b)

Figure 10. Inductor current (a) duty cycle, D of 0.4 (b) duty cycle, D of 0.6.

5.3. Comparison between simulation and hardware implementation results.
Table 2 shows the comparison between calculation, simulation and hardware prototype results with different duty cycle of the DC buck converter. The experiment varies the duty cycle to analyze the output voltages. The output voltage and current obtained from hardware prototype are different compared to calculation and simulation result by half due to some losses occur during high voltage drops across the diode and harmonic issues

Table 2. Summary results of the designed DC buck converter.

Parameters	Duty Cycle, D = 0.4			Duty Cycle, D = 0.6		
	Calculation	Simulation	Hardware	Calculation	Simulation	Hardware
V_{out}	2.0 V	2.0 V	0.965 V	3.0 V	3.0 V	1.760 V
I_{ind}	0.2 A	0.2 A	0.019 A	0.3 A	0.3 A	0.035 A

6. Conclusion
This paper presents the analysis of DC buck converter that can be applied for renewable energy application that can be reduced or step down its input voltage to a lower level depending upon the load requirement. By varying the duty cycle of the DC buck converter, the output voltage can be varied proportionally. Even though the hardware implementation shows the output voltage is reduced by half from the simulation and calculated results, this problem can be solved with including a second MOSFET in replacing the diode to reduce the losses and lower voltage drop as compared to diode.

Acknowledgments
The author would like to acknowledge the support from the Fundamental Research Grant Scheme (FRGS) Project code: 9003-00559 from the Ministry of Higher Education Malaysia.

References

[1] G. Dileep and S. N. Singh, "Selection of non-isolated DC-DC converters for solar photovoltaic system," *Renew. Sustain. Energy Rev.*, vol. **76**, no. March, pp. 1230–1247, 2017.

[2] U. S. Ali, "Z-source DC-DC Converter with Fuzzy Logic MPPT Control for Photovoltaic Applications," *Energy Procedia*, vol. **90**, no. December 2015, pp. 163–170, 2016.

[3] M. H. Taghvaee, M. A. M. Radzi, S. M. Moosavain, H. Hizam, and M. Hamiruce Marhaban, "A current and future study on non-isolated DC-DC converters for photovoltaic applications," *Renew. Sustain. Energy Rev.*, vol. **17**, pp. 216–227, 2013.

[4] D. G and S. N. Singh, "Selection of non-isolated DC-DC converters for solar photovoltaic system," *Renew. Sustain. Energy Rev.*, vol. **76**, no. August 2016, pp. 1230–1247, 2017.

[5] P. K. Jain, W. Thermal, P. Station, G. State, E. Corporation, and A. Kumar, "Interleaved DC to DC Buck Converter for Low Power Application," pp. 0–4, 2015.

[6] M. S. Ali, S. K. Kamarudin, M. S. Masdar, and A. Mohamed, "An Overview of Power Electronics Applications in Fuel Cell Systems : DC and AC Converters," *Sci. World J.*, vol. 2014, pp. 1–9, 2014.

[7] M. K. Hossain and M. H. Ali, "Transient stability augmentation of PV/DFIG/SG-based hybrid power system by parallel-resonance bridge fault current limiter," *Electr. Power Syst. Res.*, vol. **130**, pp. 89–102, 2016.

IOP Conf. Series: Journal of Physics: Conf. Series **1019** (2018) 012021 doi:10.1088/1742-6596/1019/1/012021

Diabetic Retinopathy (DR) on Retinal Image: A Pilot Study

Wan Azani Mustafa [1,*] **,Aimi Salihah Abdul-Nasir** [2] **, Haniza Yazid** [3]

[1,2] Faculty of Engineering Technology, Universiti Malaysia Perlis, UniCITI Alam Campus, Sungai Chuchuh, 02100 Padang Besar, Perlis, Malaysia

[3] School of Mechatronic Engineering, University Malaysia Perlis, 02600 Ulu Pauh, Arau, Perlis, Malaysia

[1,*] *wanazani@unimap.edu.my,* [2] *aimisalihah@unimap.edu.my,* [3] *haniza.yazid@gmail.com*

Abstracts. Eyes are part of an organ in a human which is important for vision especially retinal. Retinal images are used to detect and diagnose many diseases such as diabetes or hypertension. Diabetic retinopathy is the name given to disease of the retina due to diabetes. The eyes have to be examined by an optometrist to detect the lesion. Diabetic retinopathy symptoms may include blurred vision, difficulty reading and also vision loss. Generally, DR patients perceive no symptoms until visual loss develops which happens usually in the later disease stages when the treatment is less effective. DR is not a curable disease, but laser treatment can prevent major vision loss if detected in the early stages. In this paper, a fundamental of DR was studied included stages, causes, symptoms, and risks. This study will give the direction for a researcher in order to propose a new DR detection method.

1. Introduction

In human eye consist of three main layers of membrane that each layer performs a different function. The outer layer of the eye is defined as a fibrous tunic that composed of the cornea that admits and refracts the light, and a sclera function is to provide structure, strength and protects the inner parts of the eye [1], [2]. Otherwise, the middle membrane is known as a vascular tunic which is distributed to three components include the iris, ciliary body, and choroid. In details, the purpose of the iris is to regulate the amount of light that enters the eyeball. Meanwhile, ciliary body aqueous humor and alter the shape of the lens for near or far vision and the choroid provides blood supply and absorbs the scattered light. On the other hand, in the anatomical structure of the eye which has the inner sensory located at the innermost layer of the membrane that also called the nervous tunic [3]. This inner sensory is known as a retina which contains information and plays the vital role for the visualization in human.

In this paper, the fundamental about Diabetic Retinopathy (DR) was studied. The objective of this review is to study and explored the Diabetic Retinopathy (DR) on the retinal images such as symptoms, stages, causes, and risk. Section 2 discussed an overview of the Diabetic Retinopathy (DR). Section 2.1 and section 2.2 explained details about stages and causes. The symptoms and risks were described in section 2.3 and 2.4. Finally, section 3.0 described the conclusion of this research. Figure 1 shows the structures of the eye. The details function of the structure was described as below [2], [4];

- Conjunctiva: Supply nutrients to the cell and remove the waste product. It also acts as preventing reflection of light within the eyeball.
- Cornea: Help to maintain the shape of the anterior chamber of the eyeball.

- Sclera: Help to control the amount of light entering the eye. Enough light enters to allow a person to see.
- Iris: Just contains only cone cell. Responsible for good visual acuity. Part of the retina that is directly opposite the pupil.
- Lens: Curved structure, flexible and is transparent. The function is to focus incoming light rays.
- Retina: Acts as a suspension for the lens. Helps to maintain the shape of the posterior chamber of the eyeball.

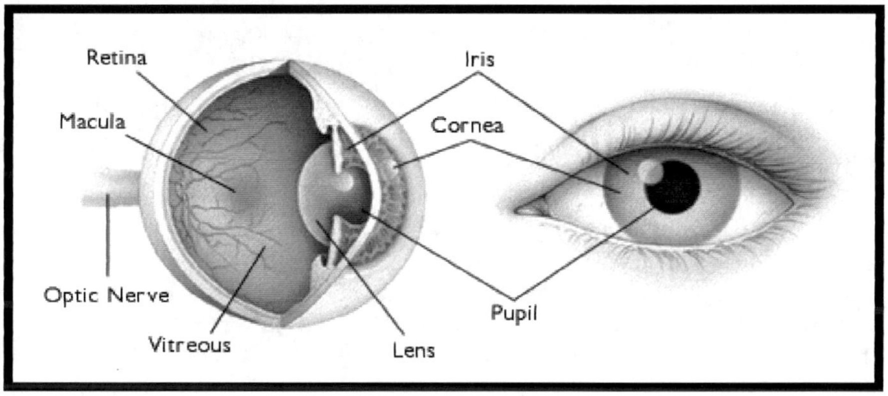

Figure 1: Structure of eye [5].

2. Diabetic Retinopathy (DR)

The retinal is a specialized part of eye organ in a human which contains the network of blood vessels [6], [7]. In a clinical study, the retinal image is very important in diagnostic tools and helps the ophthalmologists which assist by computer-aided analysis to perform diagnosis, treatment and screening of various epidemic eye diseases, such as glaucoma, age related macular degenerative myopia and Diabetic Retinopathy (DR) [8]–[10]. All these diseases were diagnosed from the information of blood vessel which is the most crucial features in the retina. For information, the first ranking of ophthalmic pathologic that contribute to the blindness among people is Diabetic Retinopathy (DR). In 2000, the estimation of all age groups whose suffered from diabetes is 2.8% and rise up to 4.4% in 2030 thus shows the aggressively increase of 171 million in 2000 to 366 million of diabetes patient in 2030 [11], [12].

Diabetic retinopathy is one of the major eye diseases that affected the retina. This conditions caused by complications of diabetes, which can eventually lead to blindness. The abnormal blood glucose level elevation is the main cause of DR. The first manifestations of DR are tiny capillary dilations known as microaneurysms [13], [14]. On the other hand, DR advancement will cause neovascularization, macular edema and exudates and also the cotton wool spot where at the later phase causes retinal segregation. Figure 1.2 shows the differences between normal retina images with DR retina image.

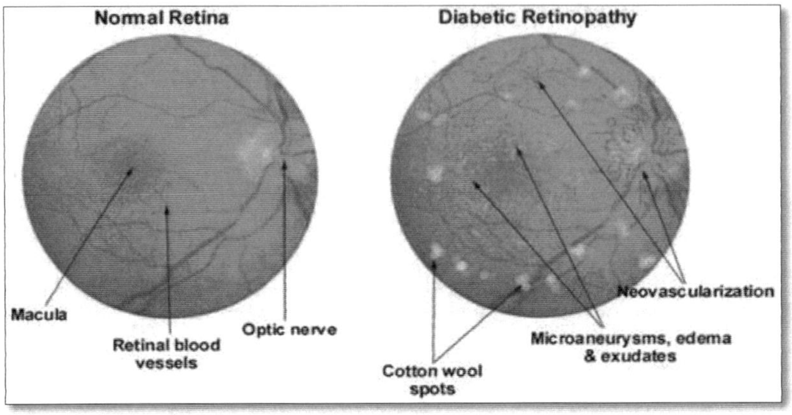

Figure 2: Normal Retina Image and Diabetic Retinopathy Retina Image [15], [16].

2.1 Stages of Diabetic Retinopathy

Diabetic retinopathy consists of four stages. The four stages include [17]–[20]:

1. Mild Nonproliferative Retinopathy: At this earliest stage, microaneurysms occur. They are small areas of balloon-like swelling in the retina's tiny blood vessels.
2. Moderate Nonproliferative Retinopathy: As the disease progresses, some blood vessels that nourish the retina are blocked.
3. Severe Nonproliferative Retinopathy: Many more blood vessels are blocked, depriving several areas of the retina with their blood supply. These areas of the retina send signals to the body to grow new blood vessels for nourishment.
4. Proliferative Retinopathy: At this advanced stage, the signals sent by the retina for nourishment trigger the growth of new blood vessels. This condition is called proliferative retinopathy. These new blood vessels are abnormal and fragile. They grow along the retina and along the surface of the clear, vitreous gel that fills the inside of the eye. By themselves, these blood vessels do not cause symptoms or vision loss. However, they have thin, fragile walls. If they leak blood, severe vision loss and even blindness can result.

2.2 Causes of the Diabetic Retinopathy in Retinal

Besides, blood vessels damaged from diabetic retinopathy can cause vision loss in two ways [21]–[27]:

1. Fragile, abnormal blood vessels can develop and leak blood into the center of the eye, blurring vision. This is proliferative retinopathy and is the fourth and most advanced stage of the disease.
2. Fluid can leak into the center of the macula, the part of the eye where sharp, straight-ahead vision occurs. The fluid makes the macula swell, blurring vision. This condition is called macular edema. It can occur at any stage of diabetic retinopathy, although it is more likely to occur as the disease progresses. About half of the people with proliferative retinopathy also have macular edema.

2.3 Symptoms of Diabetic Retinopathy

Diabetic retinopathy often has no early warning signs. Even macular edema, which may cause vision loss more rapidly, may not have any warning signs for some time. In general, however, a person with macular edema is likely to have blurred vision, making it hard to do things like reading or drive. In some cases, the vision will get better or worse during the day.

In the first stage which is called non-proliferative diabetic retinopathy (NPDR), there are no symptoms, it is not visible to the naked eye and patients will have 20/20 vision [28], [29]. The only way to detect

NPDR is by fundus photography, in which microaneurysms can be seen. If there is reduced vision, fluorescein angiography can be done to see the back of the eye. Narrowing or blocked retinal blood vessels can be seen clearly and this is called retinal ischemia means lack of blood flow. Macular edema may occur in which blood vessels leak contents into the macular region can happen at all stages of NPDR. The macular edema symptoms are blurring, darkening or distorted images with not the same between two eyes. 10 percent of diabetic patients will get vision loss related to macular edema.

On the second stage, neovascularization form at the back of the eye as a part of proliferative diabetic retinopathy (PDR), they can burst, bleed and blur vision, because the new blood vessels are weak [30], [31]. The first time this happens, it may not be very severe. In most cases, it will leave just a few specks of blood, or spots, floating in a person's visual field, though the spots often go away after a few hours. On funduscopic exam, a doctor will see cotton wool spots, flame hemorrhages, and dot-blot hemorrhages.

2.4 Risk factors for Diabetic Retinopathy Disease

All people with diabetes mellitus are at risk, those with Type I diabetes and those with Type II diabetes [32]. The longer a person has diabetes, the higher the risk of developing some ocular problem. Between 40 to 45 percent of Americans diagnosed with diabetes have some stage of diabetic retinopathy. Several factors can influence the development and severity of diabetic retinopathy, including [33]–[35]:

1) Blood Sugar: Controlling your blood sugar is the key risk factor that you can affect. Lower blood sugar levels can delay the onset and slow the progression of diabetic retinopathy.
2) Blood Pressure: A major clinical trial demonstrated that effectively controlling blood pressure reduces the risk of retinopathy progression and visual acuity deterioration. High blood pressure damages your blood vessels, raising the chances for eye problems. Target blood pressure for most people with diabetes is less than 130/80 mmHg.
3) Duration of diabetes: The risk of developing diabetic retinopathy or having your disease progress increases over time. After 15 years, 80 percent of Type 1 patients will have diabetic retinopathy. After 19 years, up to 84 percent of patients with Type 2 diabetes will have diabetic retinopathy.
4) Blood lipid levels (cholesterol and triglycerides): Elevated blood lipid levels can lead to greater accumulation of exudates, protein deposits that leak into the retina. This condition is associated with a higher risk of moderate visual loss.
5) Ethnicity: While diabetic retinopathy can happen to anyone with diabetes, certain ethnic groups are at higher risk because they are more likely to have diabetes. These include African Americans, Latinos and Native Americans.
6) Pregnancy: Being pregnant can cause changes to your eyes. If you have diabetes and become pregnant, your risk for diabetic retinopathy increases. If you already have diabetic retinopathy, it may progress. However, some studies have suggested that with treatment these changes are reversed after you give birth and that there is no increase in long-term progression of the disease.

3. Conclusion

In the medical field, the retinal image is very important in diagnostic tools and help ophthalmologist which assist by computer-aided analysis to perform diagnosis, treatment, and screening of various epidemic eye diseases. Diabetic retinopathy is one of the major eye diseases that affected the retina. This conditions caused by complications of diabetes, which can eventually lead to blindness. DR propagation also causes neovascularization, hemorrhages, macular edema and in later stages, retinal detachment. Moreover, the vascular peculiarity is one of the DR emergencies, thus the automated segmentation of blood vessels plays a necessary part in the inspection programs for early detection of eye disease especially in the diabetic retinopathy for further diagnose and treatment.

4. Acknowledgment

This work was supported by Short Term Grant Scheme (Grant No. 9001-00514).

5. Reference

[1] J. Marshall, T. W. Cronin, and S. Kleinlogel, "Stomatopod eye structure and function: A review," *Arthropod Structure and Development*, vol. 36, no. 4. pp. 420–448, 2007.

[2] W. M. Haschek, C. G. Rousseaux, M. A. Wallig, L. Teixeira, and R. R. Dubielzig, "Eye," in *Haschek and Rousseaux's Handbook of Toxicologic Pathology*, 2013, pp. 2095–2185.

[3] James Garrity, Whitney, and Betty MacMillan, "Structure and Function of the Eyes - Eye Disorders." [Online]. Available: http://www.merckmanuals.com/home/eye-disorders/biology-of-the-eyes/structure-and-function-of-the-eyes. [Accessed: 28-Aug-2017].

[4] M. Hayhoe and D. Ballard, "Eye movements in natural behavior," *Trends in Cognitive Sciences*, vol. 9, no. 4. pp. 188–194, 2005.

[5] E. H. Land and J. J. McCann, "Lightness and Retinex Theory," *J. Opt. Soc. Am.*, vol. 61, no. 1, pp. 1–11, 1971.

[6] W. A. Mustafa, H. Yazid, and S. Yaacob, "Illumination Correction of Retinal Images Using Superimpose Low Pass and Gaussian Filtering," in *International Conference on Biomedical Engineering (ICoBE)*, 2015, pp. 1–4.

[7] W. A. Mustafa, H. Yazid, and S. Yaacob, "A Review : Comparison Between Different Type of Filtering Methods on the Contrast Variation Retinal Images," in *IEEE International Conference on Control System, Computing and Engineering*, 2014, pp. 542–546.

[8] T. Wong and P. Mitchell, "The eye in hypertension," *Lancet*, vol. 369, no. 9559. pp. 425–435, 2007.

[9] N. R. Miller and N. J. Newman, "The eye in neurological disease," in *Lancet*, 2004, vol. 364, no. 9450, pp. 2045–2054.

[10] W. A. Mustafa, H. Yazid, S. Yaacob, and S. Basah, "Blood vessel extraction using morphological operation for diabetic retinopathy," *IEEE Reg. 10 Symp.*, no. 3, pp. 208–212, Apr. 2014.

[11] D. A. Antonetti, R. Klein, and T. W. Gardner, "Diabetic retinopathy.," *N. Engl. J. Med.*, vol. 366, no. 13, pp. 1227–39, 2012.

[12] B. Ramasubramanian and G. Anitha, "An Efficient Approach for the Detection of New Vessels in Diabetic Retinopathy Images," *Int. J. Eng. Innov. Technol.*, vol. 2, no. 3, pp. 240–244, 2012.

[13] A. A. Alghadyan, "Diabetic retinopathy - An update," *Saudi Journal of Ophthalmology*, vol. 25, no. 2. pp. 99–111, 2011.

[14] L. M. Aiello, "Perspectives on diabetic retinopathy," *American Journal of Ophthalmology*, vol. 136, no. 1. pp. 122–135, 2003.

[15] W. A. Mustafa, H. Yazid, and N. Mazlan, "An enhancement method on illumination images : A survey," *J. Adv. Rev. Sci. Res.*, vol. 28, no. 1, pp. 33–41, 2016.

[16] P. V. Priya, A. Srinivasarao, and J. V. C. Sharma, "Diabetic Retinopathy - Can Lead To Complete Blindness," *Int. J. Sci. Invent. Today*, vol. 2, no. 4, pp. 254–265, 2013.

[17] P. H. Scanlon, "Diabetic retinopathy," *Medicine*, vol. 38, no. 12. pp. 656–660, 2010.

[18] D. S. Fong, L. Aiello, T. W. Gardner, G. L. King, G. Blankenship, J. D. Cavallerano, F. L. Ferris, and R. Klein, "Diabetic retinopathy," *Diabetes Care*, vol. 26, no. 1. pp. 226–229, 2003.

[19] C. P. Wilkinson, F. L. Ferris, R. E. Klein, P. P. Lee, C. D. Agardh, M. Davis, D. Dills, A. Kampik, R. Pararajasegaram, J. T. Verdaguer, and F. Lum, "Proposed international clinical diabetic retinopathy and diabetic macular edema disease severity scales," *Ophthalmology*, vol. 110, no. 9, pp. 1677–1682, 2003.

[20] A. N. Kollias and M. W. Ulbig, "Diabetic retinopathy: Early diagnosis and effective treatment.,"

Dtsch. Ärzteblatt Int., vol. 107, no. 5, pp. 75–83; quiz 84, 2010.

[21] J. Tang and T. S. Kern, "Inflammation in diabetic retinopathy," *Progress in Retinal and Eye Research*, vol. 30, no. 5. pp. 343–358, 2011.

[22] C. M. Morello, "Etiology and natural history of diabetic retinopathy: An overview," in *American Journal of Health-System Pharmacy*, 2007, vol. 64, no. 17 SUPPL.

[23] A. Clark, W. H. Morgan, S. Kain, H. Farah, K. Armstrong, D. Preen, J. B. Semmens, and D. Y. Yu, "Diabetic retinopathy and the major causes of vision loss in Aboriginals from remote Western Australia," *Clin. Exp. Ophthalmol.*, vol. 38, no. 5, pp. 475–482, 2010.

[24] D. S. Fong, F. L. 3rd Ferris, M. D. Davis, and E. Y. Chew, "Causes of severe visual loss in the early treatment diabetic retinopathy study: ETDRS report no. 24. Early Treatment Diabetic Retinopathy Study Research Group.," *Am. J. Ophthalmol.*, vol. 127, no. 2, pp. 137–141, 1999.

[25] M. W. Stewart, "Pathophysiology of diabetic retinopathy," in *Diabetic Retinopathy: Evidence-Based Management*, 2010, pp. 1–30.

[26] D. S. Fong, F. L. F. Iii, M. D. Davis, and E. Y. Chew, "Causes of severe visual loss in the early treatment diabetic retinopathy study: {ETDRS} report no. 24," *Am. J. Ophthalmol.*, vol. 127, no. 2, pp. 137–141, 1999.

[27] C. J. Bulpitt, M. A. Sleightholm, B. Hunt, A. E. Fletcher, A. Palmer, and E. M. Kohner, "Causes of death and risk factors in young and old diabetic patients referred to a retinopathy clinic," *J. Diabetes its Complicat.*, vol. 10, no. 3, pp. 160–167, 1996.

[28] J. Tam, K. P. Dhamdhere, P. Tiruveedhula, B. J. Lujan, R. N. Johnson, M. A. Bearse, A. J. Adams, and A. Roorda, "Subclinical Capillary Changes in Non-Proliferative Diabetic Retinopathy," *Optom. Vis. Sci.*, vol. 89, no. 5, pp. E692–E703, 2012.

[29] R. C. Silvia and R. Vijayalakshmi, "Detection of Non-Proliferative Diabetic Retinopathy in fundus images of the human retina," in *2013 International Conference on Information Communication and Embedded Systems (ICICES)*, 2013, vol. 1, pp. 978–983.

[30] R. Klein, B. E. Klein, and S. E. Moss, "Epidemiology of proliferative diabetic retinopathy.," *Diabetes Care*, vol. 15, no. 12, pp. 1875–1891, 1992.

[31] G. S. Ramlugun, V. K. Nagarajan, and C. Chakraborty, "Small retinal vessels extraction towards proliferative diabetic retinopathy screening," *Expert Syst. Appl.*, vol. 39, no. 1, pp. 1141–1146, 2012.

[32] T. S. T. Smith, J. Szetu, and R. R. A. Bourne, "The prevalence and severity of diabetic retinopathy, associated risk factors and vision loss in patients registered with type 2 diabetes in Luganville, Vanuatu," *Br. J. Ophthalmol.*, vol. 91, no. 4, pp. 415–419, 2007.

[33] N. Cheung, J. J. Wang, S. L. Rogers, F. Brancati, R. Klein, A. R. Sharrett, and T. Y. Wong, "Diabetic retinopathy and risk of heart failure," *J Am Coll Cardiol*, vol. 51, no. 16, pp. 1573–1578, 2008.

[34] M. B. Sultan, C. Starita, and K. Huang, "Epidemiology, risk factors and management of paediatric diabetic retinopathy.," *Br. J. Ophthalmol.*, vol. 96, no. 3, pp. 312–7, 2012.

[35] A. Dizdarevic, A. Alikadic-Husovic, and V. Jusufovic, "Risk factors and diabetic retinopathy," *Med. Glas. Ljek. Komore Zenickodobojskog Kantona*, vol. 9, no. 1, pp. 104–106, 2012.

Binarization of Document Image Using Optimum Threshold Modification

Wan Azani Mustafa[1*], Mohamed Mydin M. Abdul Kader[1]
[1]Faculty of Engineering Technology, Universiti Malaysia Perlis, UniCITI Alam Campus, Sungai Chuchuh, 02100 Padang Besar, Perlis, Malaysia

wanazani@unimap.edu.my

Abstract. Document image binarization is one important processing step, especially for data analysis. In this paper, a new binarization based local thresholding technique 'WAN' was presented. The proposed algorithm is known as 'WAN' after the first name of the author in this paper. WAN has been inspired from the Sauvola's binarization method and exhibits its robustness and effectiveness when evaluated on low quality document images. The objective of the WAN method is to improve the Sauvola method and achieve a better binarization results, specifically, for non-uniform document images. The results of the numerical simulation indicate that the WAN method is the most effective and efficient (f-measure 72.274 and NRM = 0.093) compared to the Sauvola method, Local Adaptive method, Niblack method, Feng Method, and Bernsen method.

1. Introduction

There are many challenges addressed in handwritten document image binarization, such as faint characters, bleed-through and large background ink stains [1]–[3]. Document image binarization is the process that segments the document image into the text and background by removing any existing degradations [1], [4]. Recently, many document image binarization methods have been proposed in the literature [5]–[8]. However, selecting the most optimum threshold for binarization is a difficult task due to the presence of a variety of degradations in document images [9]–[10].

Previous studies concentrated on proposing a new method or algorithm to solve the degradation of document images. In 2008, Nikolaos and Dimitrios reviewed a few enhancement and binarization techniques to find the best approach for the future research [11]. They summarized that combination of pre-processing and binarization algorithm able to improve and finally provide the new method. Many researchers agree that it's very difficult to propose a perfect algorithm since the document image in badly condition dealing with many information such as text and structure [11]–[13]. Gatos *et al.* [12] discussed the challenges and strategies to improve the document image binarization based on a combination of Multiple Binarization Techniques and Adapted Edge Information. This approach has a number of advantages: firstly, (i) combining the binarization results of several state-of-the-art methodologies; (ii) incorporating the edge map of the grey scale image; and (iii) applying efficient image post-processing based on mathematical morphology for the enhancement of the final result. Research finding by Shijian *et al.* [14] also points towards the edge information to propose a new method. However, they concentrated on the surface and stroke edge. The result is more effective compared to the Gatos method [12], Sauvola

IOP Conf. Series: Journal of Physics: Conf. Series **1019** (2018) 012022 doi:10.1088/1742-6596/1019/1/012022

method [15] and Otsu method [16]. In 2009, Reza and Mohamed published a paper in which they described the new model of a low quality document image using virtual diffusion processes [17]. This technique focuses on the shadow- through and bleed-through problem.

Besides, Laurence *et al.* [18] mentioned the significant relationship between Total Variation regularization and Non-local Means filtering in order to binarize a degrade document image. This approach consists two steps; (1) in order to reduce the effect of background noise, they proposed to apply the Total Variation framework and the result considered as a mask image. (2) The Non-local means was performed to clean the image from noise and bleed-through. However, Shi *et al.* [19] found differences suggesting that the technique based on shape feature is more effective compared to the technique based on background normalization [18], [20]. The threshold selection is based on the stroke width consistency and the result performance is superior compared to the existing binarization technique such as Otsu method [16], Niblack method [21], and Multistage Adaptive Thresholding (MAT) method [22].

Several studies investigating document binarization based on Otsu modification [23]–[25]. Starting 2010, Nina *et al.* [23] mentioned the significant combination between the recursive extension of Otsu thresholding and selective bilateral filtering of scanned handwritten images. This approach considered background estimation before applying the post- processing stage [23], [25]. The above findings contradict the study by Zhang and Wu [24]. They examined the modification algorithm based on the Adaptive Otsu method. The proposed method is based on three main steps; (1) applied the Wiener filter in order to eliminate noise, (2) improved adaptive Otsu's method, and (3) dilation and erosion operators were performed to preserve stroke connectivity and fill possible breaks, gaps, and holes. The advantage of this approach is that faster processing time compared to Recursive Otsu Thresholding Method [23] and AdOtsu method [25]. Similarly, Reza and Mohamed also proposed a new novel method based on the Otsu modification known as AdOtsu method [25]. The main idea of this technique was considered parameterless behavior such as average stroke width and the average line height. A positive result was achieved compared to Sauvola method [15], Otsu Method [16], and Lu and Tan method [14].

In this paper, a new binarization method based on the maximum threshold was discussed. The proposed method inspired from Sauvola method and is known as 'WAN' method. The proposed method experiments on 14 non-uniform document images. A few image quality assessment such as f-measure, sensitivity, NRM (Negative Rate Metric), and Peak Signal Noise Ratio (PSNR) was performed in order to compare the effectiveness the method. Summary, this paper is organized in the following sections: Section 2.0 describes the proposed binarization methods. Section 3.0 presents the analysis of results and Section 4.0 explains the conclusion of this work.

2. Proposed Approach

The proposed algorithm is inspired by the Sauvola method. In this paper, we put forward our proposition of calculating the binarization threshold which is likely to work better for many (if not all) types of degraded and noisy documents. The Sauvola method able to solve the problem of black noise depending on the impact on the standard deviation value by using a range of grey level values in the images [15], [26]–[27]. However, the Sauvola method failed to segment if the contrast between the foreground and background is small or if the text is in thin pen stroke text. So, the WAN method was proposed to overcome this problem by obtaining the maximum threshold value. The main advantage of the proposed method over Sauvola is that it considerably improves binarization for "lost" detail images by shifting up the binarization threshold. The Sauvola algorithm is denoted as follows;

$$T = m\left(1 - k\left(1 - \frac{\sigma}{R}\right)\right) \tag{1}$$

Where, R is the grey level (128), m is the mean value, σ is the standard deviation, and k was set is 0.2 (default value). This method outperforms Niblack algorithm in images where the text pixels have near 0 grey value and the background pixels have near 255 grey values. However, in images where the grey values of text and non-text pixels are close to each other, the results degrade significantly. Based on research, the mean value m will give a high effect on the threshold value as shown in figure 1. According to figure 1, example, if the mean value less than average (m -20) the result was blurred and more information details were lost. Otherwise, if the mean value more than the average (m +20) the result is better and improved. In this paper, the specific value was found to replace the normal mean value. The aim is to find the specific value more than the normal average.

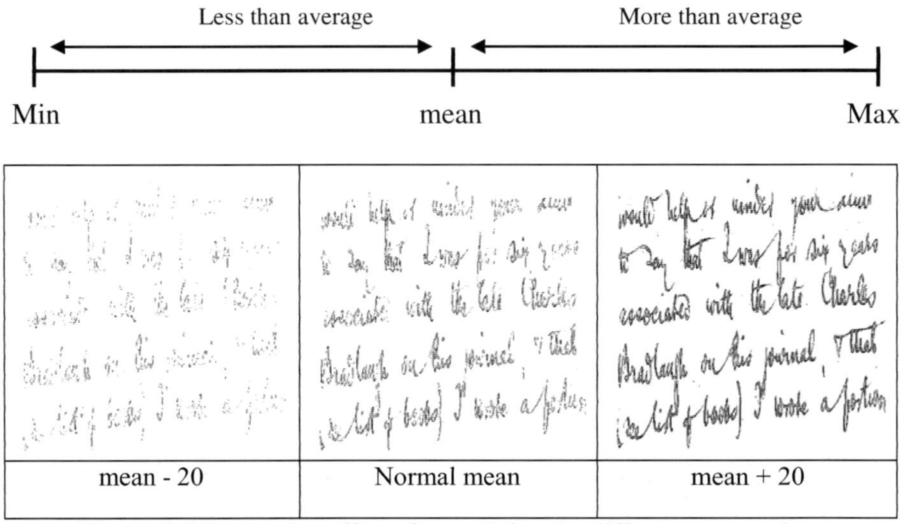

Figure 1. Binarization effect after applying the different mean value.

Actually, the Sauvola method failed to binarize the low contrast region because the threshold value is low. Therefore, the proposed method tends to increase the threshold value to segment the information in the low contrast region. However, if the threshold value is higher, it's will introduce noise and artefact on the resulting image. So the specific and the maximum threshold value needs to be proposed. In this paper, the maximum mean is calculated in order to replace the original mean. The maximum mean equation is depicted as follows;

$$m_{\max} = \frac{\max(x, y) + mean}{2} \quad (2)$$

Where, $\max(x, y)$ is the maximum intensity of input image and $mean$ is the original mean for the whole image. The average between the highest intensity and mean image was calculated. The main target is to improve the lost details on binarization result and at the same time to reduce the noise and any artefact. The final proposed algorithm is;

$$T = \frac{\max(x, y) + mean}{2}\left[1 - k\left(1 - \frac{\sigma}{R}\right)\right] \quad (3)$$

where, k and R value used a default value form Sauvola method. From this algorithm, the low contrast problem can be solved and automatically increased the binarization result. In order to evaluate the proposed method and compare the results with a few local methods, 14 document images were tested and the results are given in the following section.

3. Experimental Result

In this paper, 14 document images from Handwritten Document Image Binarization Contest (H-DIBCO) [28] dataset experimented. The images contain various degradations such as shadows, non-uniform illumination, stains, smudges, bleed-through and faint characters [1]. All the processed images are in greyscale images and the size of each image is 400 × 400 pixels, 72 dpi, and 8-bit depth. All the programs were written in MATLAB from an Asus laptop with AMD Athlon™ II P320 Dual-Core Processor 2.10GHz and 3.00GB RAM. Three examples of document images were illustrated in figure 2. The first row shows the original image with degradation and shadowing problem and follows by binarization methods from Sauvola method and lastly by the proposed method (WAN). Based on visual criteria, the proposed algorithm seems to outperform the other methods with respect to image quality and preservation of meaningful textual information. Besides, the Sauvola result shows poor binarization compared to the proposed method. The problem of Sauvola method already overcome using the proposed algorithm.

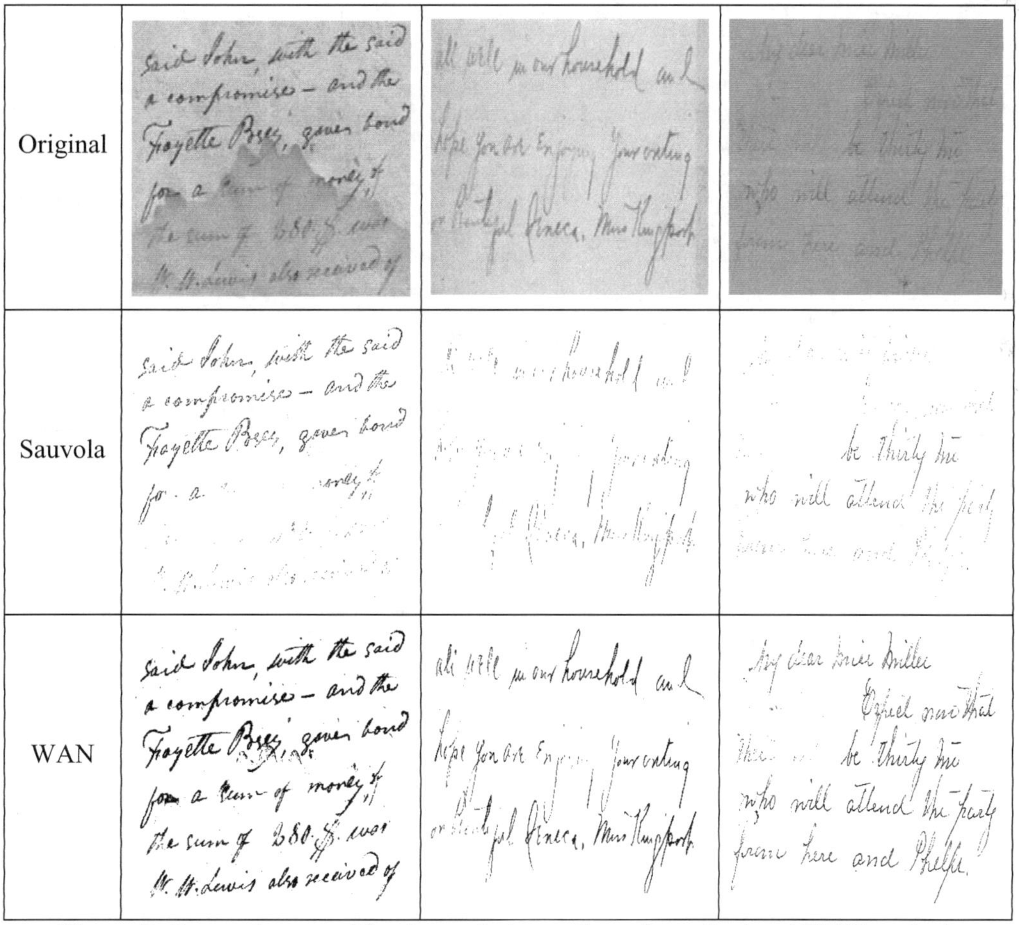

Figure 2. Comparison resulting image between Sauvola method and WAN method.

Next, a few image quality assessment (IQA) was calculated to compare with the Sauvola method and proposed method. In this paper, the evaluation based on the F-measure, Sensitivity, NRM (Negative Rate Metric), and PSNR (Peak Signal to Noise Ratio) was obtained. All the assessment equation can be referred in H-DIBCO [29]–[30]. The highest F-measure, Sensitivity, and PSNR while lowest NRM value represents a good binarization result. Table 1 provides the comparison result between Sauvola method and the proposed method (WAN). Average, in term of f-measure and sensitivity, the proposed method achieved the highest result (f-measure = 72.274 and sensitivity = 87.676) compared to the Sauvola method (f-measure = 56.321 and sensitivity = 45.343). Besides, a comparison based on NRM again shows that the performance of the proposed method obtained 0.093, which is lower than the Sauvola methods. However, based on PSNR the proposed method slightly lower (13.614) compared to the Sauvola method (14.612).

Table 1: The comparison based on a few objective measurements on Sauvola method and WAN method.

	Sauvola				**WAN**			
	F-measure	Sensitivity	NRM	PSNR	F-measure	Sensitivity	NRM	PSNR
H1	83.153	74.797	0.128	17.631	76.402	94.907	0.042	14.765
H2	47.349	31.301	0.344	12.281	83.990	86.119	0.078	15.545
H3	8.148	4.251	0.479	12.145	88.065	86.392	0.071	18.266
H4	84.884	75.039	0.125	18.871	86.658	94.114	0.035	18.519
H5	82.120	75.949	0.123	17.065	34.259	99.138	0.130	6.262
H6	77.328	66.711	0.168	15.906	52.439	98.400	0.070	9.314
H7	52.074	36.028	0.321	13.581	76.918	71.639	0.147	15.463
H8	85.501	78.754	0.108	17.225	69.182	99.462	0.036	12.017
H9	52.726	36.014	0.320	11.310	90.638	90.816	0.052	16.678
H10	51.646	35.338	0.324	11.799	52.334	94.531	0.124	7.471
H11	20.827	11.676	0.442	11.171	56.766	92.480	0.100	9.166
H12	31.078	18.442	0.408	13.285	74.132	61.438	0.194	16.091
H13	29.484	17.560	0.413	13.918	70.233	60.740	0.199	16.043
H14	82.176	72.942	0.136	18.375	73.823	97.293	0.029	14.989
Average	**56.321**	**45.343**	**0.274**	**14.612**	**72.274**	**87.676**	**0.093**	**13.614**

Besides that, in order to prove the improvement of binarization, the increment percentages was calculated. The result of increment based on f-measure, sensitivity, NRM, and PSNR was illustrated in figure 3.

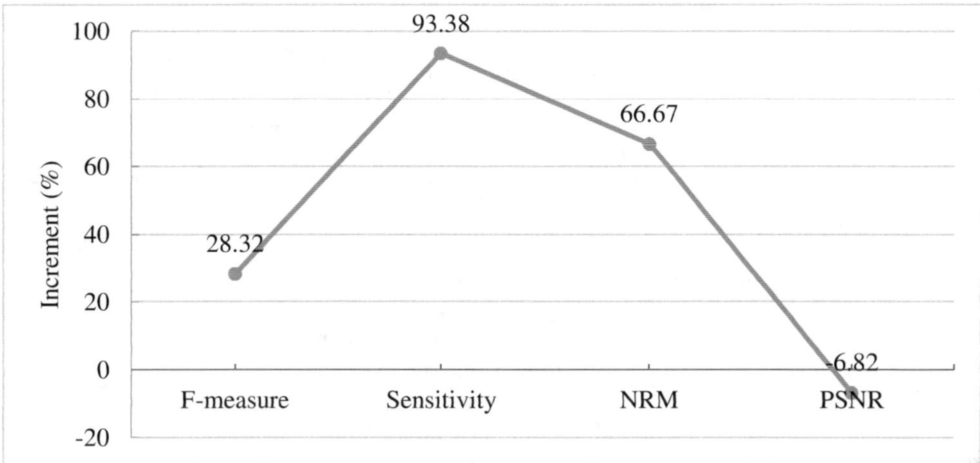

Figure 3: The increment after applying the WAN method.

The histogram in figure 3 indicates the increment (%) of binarization after employing the proposed method. Based on figure 3, the evaluation result based on sensitivity shows the highest improvement which is 93.38% and f-measure obtained lower increment which is 28.32% compared to the Sauvola method. However, in term of PSNR, the proposed method slightly decreased (6.82%) compared to the Sauvola method.

4. Conclusion

Document binarization is an important application in vision processing. In this paper, a new binarization method was proposed for low quality document image known as 'WAN' method. The aim of this paper is to improve the Sauvola method and achieved a better binarization compared to a few existing binarization methods. The proposed method performs better than the contemporary methods, especially when the input image has very few or no text (white image) and also when the intensity variations between the text and background are extremely low. Based on the result performance, the proposed method achieved a good result in term of F-measure, Sensitivity and NRM compared to the original Sauvola method. In the near future, we will propose a new algorithm which will use the more reliable methodology to enhance the work.

5. Reference

[1] K. Ntirogiannis, B. Gatos, and I. Pratikakis, "A combined approach for the binarization of handwritten document images," *Pattern Recognit. Lett.*, vol. 35, pp. 3–15, Jan. 2014.

[2] B. M. Singh, R. Sharma, D. Ghosh, and A. Mittal, "Adaptive binarization of severely degraded and non-uniformly illuminated documents," *Int. J. Doc. Anal. Recognit.*, pp. 393–412, 2014.

[3] D. Rivest-Hénault, R. Farrahi Moghaddam, and M. Cheriet, "A local linear level set method for the binarization of degraded historical document images," *Int. J. Doc. Anal. Recognit.*, vol. 15, no. 2, pp. 101–124, 2012.

[4] W. A. Mustafa and H. Yazid, "Illumination and Contrast Correction Strategy using Bilateral Filtering and Binarization Comparison," *J. Telecommun. Electron. Comput. Eng.*, vol. 8, no. 1, pp. 67–73, 2016.

[5] P. K. More and D. D. Dighe, "A Review on Document Image Binarization Technique for Degraded Document Images," *Int. Res. J. Eng. Technol.*, pp. 1132–1138, 2016.

[6] X. Chen, L. Lin, and Y. Gao, "Parallel nonparametric binarization for degraded document images," *Neurocomputing*, vol. 189, pp. 43–52, 2016.

[7] G. Vara Lakshmi & P. Kamala, "Improving Degraded Document Images Using Binarization Techniques," *Int. J. Electron. Commun. Eng.* , vol. 5, no. 2, pp. 1–8, 2016.

[8] M. Soua, R. Kachouri, and M. Akil, "Improved Hybrid Binarization based on Kmeans for Heterogeneous document processing," in *9th International Symposium on Image and Signal Processing and Analysis, ISPA 2015*, 2015, pp. 210–215.

[9] R. C. Gonzalez and R. E. Woods, *Digital Image Processing*, Third. Upper Saddle River, NJ, USA: Prentice Hall, 2008.

[10] H. Tanaka, "Threshold correction of document image binarization for ruled-line extraction," in *Proceedings of the International Conference on Document Analysis and Recognition, ICDAR*, 2009, pp. 541–545.

[11] N. Ntogas and V. Dimitrios, "A Binarization Algorithm for Historical Manuscripts," in *International Conference on Comunications*, 2008, pp. 41–51.

[12] B. Gatos, I. Pratikakis, and S. J. Perantonis, "Improved document image binarization by using a combination of multiple binarization techniques and adapted edge information," *19th Int. Conf. Pattern Recognit.*, pp. 1–4, Dec. 2008.

[13] P. Vinicius, K. Borges, J. Mayer, and E. Izquierdo, "Document Image Processing for Paper Side Communications," in *IEEE Transactions On Multimedia*, 2008, vol. 10, no. 7, pp. 1277–1287.

[14] S. Lu, B. Su, and C. L. Tan, "Document image binarization using background estimation and stroke edges," *Int. J. Doc. Anal. Recognit.*, vol. 13, no. 4, pp. 303–314, Oct. 2010.

[15] J. Sauvola and M. Pietikäinen, "Adaptive document image binarization," *Pattern Recognition*, vol. 33. pp. 225–236, 2000.

[16] N. Otsu, "A Threshold Selection Method from Gray-Level Histograms," in *IEEE Transactions On Systrems, Man, And Cybernetics*, 1979, vol. 20, no. 1, pp. 62–66.

[17] R. F. Moghaddam and M. Cheriet, "Low quality document image modeling and enhancement," *Int. J. Doc. Anal. Recognit.*, vol. 11, no. 4, pp. 183–201, Feb. 2009.

[18] L. Likforman-Sulem, J. Darbon, and E. H. B. Smith, "Enhancement of historical printed document images by combining Total Variation regularization and Non-local Means filtering," *Image Vis. Comput.*, vol. 29, no. 5, pp. 351–363, Apr. 2011.

[19] J. Shi, N. Ray, and H. Zhang, "Shape based local thresholding for binarization of document images," *Pattern Recognit. Lett.*, vol. 33, no. 1, pp. 24–32, Jan. 2012.

[20] W. A. Mustafa, H. Yazid, and S. Yaacob, "Illumination Normalization of Non-Uniform Images Based on Double Mean Filtering," in *IEEE International Conference on Control Systems, Computing and Engineering*, 2014, pp. 366–371.

[21] W.Niblack, *An Introduction to Digital Image Processing.* Englewood Cliffs: Prentice-Hall, 1986.

[22] F. Yan, H. Zhang, and C. R. Kube, "A multistage adaptive thresholding method," *Pattern Recognit. Lett.*, vol. 26, pp. 1183–1191, 2005.

[23] O. Nina, B. Morse, and W. Barrett, "A Recursive Otsu Thresholding Method for Scanned Document Binarization," in *IEEE Workshop on Applications of Computer Vision (WACV)*, 2010, pp. 307–314.

[24] Y. Zhang and L. Wu, "Fast Document Image Binarization Based on an Improved Adaptive Otsu ' s Method and Destination Word Accumulation," *J. Comput. Inf. Syst.*, vol. 6, no. 7, pp. 1886–1892, 2011.

[25] Reza Farrahi Moghaddamn Mohamed Cheriet, "AdOtsu: An adaptive and parameterless generalization of Otsu's method for document image binarization," *Pattern Recognit.*, vol. 45, pp. 2419–2431, Jun. 2012.

[26] B. Bataineh, S. N. H. S. Abdullah, and K. Omar, "An adaptive local binarization method for document images based on a novel thresholding method and dynamic windows," *Pattern Recognit.*

Lett., vol. 32, pp. 1805–1813, 2011.

[27] J. Sauvola, T. Seppanen, S. Haapakoski, M. Pietikiiinen, M. Vision, M. P. Group, and I. Oulu, "Adaptive Document Binarization," in *International Conference on Document Analysis and Recognition*, 1997, pp. 147–152.

[28] K. Zagoris, "H-DIBCO '16," 2016. [Online]. Available: http://vc.ee.duth.gr/h-dibco2016/. [Accessed: 01-Jan-2016].

[29] B. Gatos, K. Ntirogiannis, and I. Pratikakis, "ICDAR 2009 Document Image Binarization Contest (DIBCO 2009)," *2009 10th Int. Conf. Doc. Anal. Recognit.*, no. Dibco, pp. 1375–1382, 2009.

[30] W. A. Mustafa and H. Yazid, "Background Correction using Average Filtering and Gradient Based Thresholding," *J. Telecommun. Electron. Comput. Eng.*, vol. 8, no. 5, pp. 81–88, 2016.

IOP Conf. Series: Journal of Physics: Conf. Series **1019** (2018) 012023 doi:10.1088/1742-6596/1019/1/012023

Binarization of Document Images: A Comprehensive Review

Wan Azani Mustafa[1], Mohamed Mydin M. Abdul Kader[1]
[1]Faculty of Engineering Technology, Universiti Malaysia Perlis, UniCITI Alam Campus, Sungai Chuchuh, 02100 Padang Besar, Perlis, Malaysia

wanazani@unimap.edu.my

Abstract. Document image binarization is one important pre-processing step, especially for data analysis. Extraction of text from images and its recognition may be challenging due to the presence of noise and degradation in document images. In this paper, seven (7) types of binarization method were discussed and tested on Handwritten Document Image Binarization Contest (H-DIBCO 2012). The aim of this paper is to provide comprehensive review methods in order to binary document images in the damaging background. The results of the numerical simulation indicate that the Gradient Based method most effective and efficient compared to other methods. Hopefully, the implications of this review give future research directions for the researchers.

1. Introduction

There are many challenges addressed in handwritten document image binarization, such as faint characters, bleed-through and large background ink stains [1]–[4]. Document image binarization is the process that segments the document image into the text and background by removing any existing degradations [1]. Many document image binarization methods have been proposed in the literature. However, selecting the most optimum method for binarization is a difficult task due to the presence of a variety of degradations in document images.

Previous studies have primarily concentrated in order to propose a new method or algorithm due to solve the degradation of document images. In 2008, Nikolaos and Dimitrios review a few enhancement and binarization techniques to find the best approach for the future research [5]. They summarize that combination of pre-processing and binarization algorithm able to improve and finally provide the extensive method. Many researchers agree that it's very difficult to propose a perfect algorithm because the document image in badly condition dealing with many information such as text and structure [5], [6]. Gatos *et al.* discussed the challenges and strategies to improve the document image binarization based on a combination of Multiple Binarization Techniques and Adapted Edge Information [6]. This approach has a number of advantages: firstly, (i) combining the binarization results of several state-of-the-art methodologies; (ii) incorporating the edge map of the grey scale image; and (iii) applying efficient image post-processing based on mathematical morphology for the enhancement of the final result. Research finding by Shijian *et al.* also points towards the edge information to propose a new method [7]. However, they more concentrate on the surface and stroke edge. The result finding more effective compared to the Gatos method [6], Sauvola method [8] and Otsu method [9]. In 2009, Reza and Mohamed published a paper in which they described the new model of a low quality document image using virtual diffusion processes [10]. This technique focuses on the shadow- through and bleed-through problem.

IOP Conf. Series: Journal of Physics: Conf. Series **1019** (2018) 012023 doi:10.1088/1742-6596/1019/1/012023

Several studies investigating document binarization based on Otsu modification [11]–[13]. Starting 2010, Nina *et al.* mention the significant combination between the recursive extension of Otsu thresholding and selective bilateral filtering of scanned handwritten images [11]. This approach considers background estimation before applied the post- processing stage [11], [13]. The above findings contradict the study by Zhang and Wu. They examined the modification algorithm based on the Adaptive Otsu method [12]. The proposed based on three main steps; (1) applied the Wilner filter in order to eliminate noise, (2) improved adaptive Otsu's method, and (3) dilation and erosion operators were performed to preserve stroke connectivity and fill possible breaks, gaps, and holes. The advantage of this approach is that faster processing time compared to Recursive Otsu Thresholding Method [11] and AdOtsu method [13]. Similarly, Reza and Mohamed also proposed a new novel method based on the Otsu modification known as AdOtsu method [13]. The main idea of this technique was considered parameterless behavior such as average stroke width and the average line height. The positive result was achieved compared to Sauvola method, Otsu Method, and Lu and Tan method.

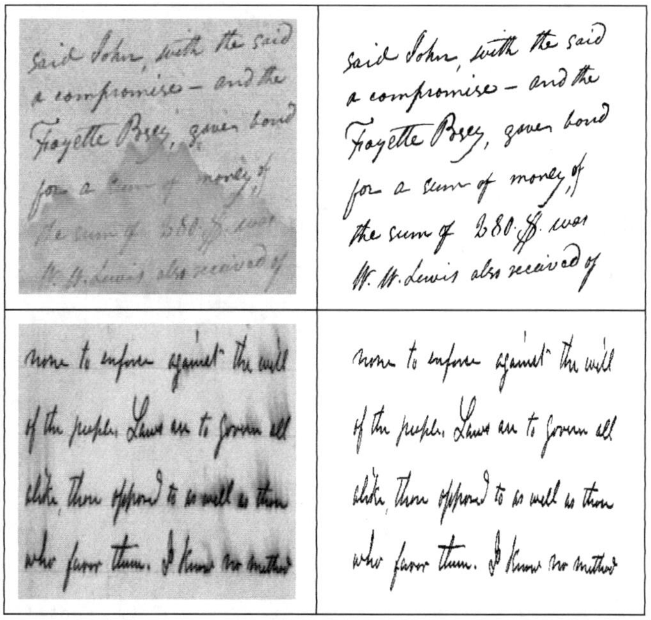

Figure 1. The document image; (left) the original image and (right) the benchmark image.

Figure 1 (left) has shown the input original image for binarization and figure 1 (right) has shown the binary image for the same. Document image binarization is generally performed in the pre- processing phase of distinctive archive picture handling related requisitions, for example, optical character distinguishment (OCR) and report picture recovery.

In this paper, a comprehensive review of 7 types of binarization methods such as Otsu method, Niblack method, Bernsen method and Gradient Based method was discussed. A few image quality assessment such as Misclassification Error (ME) and Peak Signal Noise Ratio (PSNR) was performed in order to compare the effectiveness of every method. Summary, this paper is organized in the following sections: Section 2 describes the review comparison binarization methods. Section 3 presents the analysis of results and Section 5 gives the conclusion of the results.

2. Reviews of Binarization Methods

This section explains a few popular binarization methods that were used in this study. Normally, a poor image quality caused by noise, illumination, artefacts from the camera, and degradation of the image affected the binarization process [14]–[16].

2.1 Gradient Based Thresholding

Haniza and Arof presented a new binarization method using gradient based thresholding by constructing a threshold surface [17]. This method contains three important phases: first, construct the inverse image $T(i, j)$, second obtaining the k value between -255 to 255, and finally applying binarization to separate object and background.

$$result(i, j) = \begin{cases} 0, (object) & if \;\; I(i, j) < T(i, j) + k_0 \\ 255, (background) & if \;\; I(i, j) > T(i, j) + k_0 \end{cases} \quad (1)$$

where $I(i, j)$ is the intensity of the original image and k_0 is the minimum sum of absolute difference intensity. Based on ME, the proposed binarization produced better and effective results compared to other methods.

2.2 Niblack Method

The main purpose of Niblack method is to set the threshold value based on local standard deviation and local mean. The threshold for each pixel was determined by [18]:

$$T(x, y) = m(x, y) + k\delta(x, y) \quad (2)$$

where, standard deviation $\delta(x, y)$ and local mean $m(x, y)$ were determined by 80 x 80 windowing size [19] and standard k value is - 0.2. This method does not work correctly if the image suffers from non-uniform illumination.

2.3 Otsu Method

The technique obtained the threshold value automatically based on global variance and between-class variance. In the non-uniform image, Otsu assumes the image contains two areas: dark and bright in order to purpose final algorithm [9]. Finally, Otsu thresholding is determined by:

$$k = \frac{\sigma^2 B}{\sigma^2 G} \quad (3)$$

where, k a threshold value, $\sigma^2 B$ are a global variance of the entire image, and $\sigma^2 G$ between-class variance.

2.4 Nick Method

Khurshid *et al.* proposed new strategies to improve the Niblack method by shifting the thresholding value downward [20]. The threshold is obtained based on the following equation:

$$T(x,y) = m + k\sqrt{\frac{(I^2 - m^2)}{N}} \tag{4}$$

The k factor value is similar with Niblack and the windowing size is defined as 15 x 15, while I and m represent the intensity pixel and mean of grey scale image. N represents the image size.

2.5 Bradley Method
By using the integral image as the input image, this method is an improvement of Wellner's method [21] and it is robust to illumination changes within the image. The key idea of the algorithm is that every image's pixel is set to black if its brightness is T percent lower than the average brightness of surrounding pixels in the window of the specified size, otherwise it is set to white. The default windowing size is 15 by 15 and T is 10 [22].

$$T = m\left(1 - \frac{k}{100}\right) \tag{5}$$

2.6 Bernsen Method
The Bernsen algorithm is based on the estimation of a local threshold value for each pixel. This value is assigned as the local threshold value only if the difference between the lowest and the highest grey level value is bigger than a threshold k . Otherwise, it is assumed that the window region contains pixels of one class (foreground or background). The default windowing size (w) is 3-by-3 and k is 15 [23]. The final equation as follows;

$$T(x,y) = \frac{Z_{\max} + Z_{\min}}{2} \tag{6}$$

where, $Z_{\min}$ and $Z_{\max}$ are the lowest and highest grey level pixel values.

2.7 Local Adaptive Thresholding
Local Adaptive Thresholding is a basic and simple algorithm to separate the foreground from the background with non-uniform illumination. For each pixel in the image, a threshold has to be calculated. If the pixel value is below the threshold, it is set to the background value, otherwise, it assumes the foreground value. The default local windowing size (w) is 15 by 15 and local threshold (T) is a 0.05 [14].

$$T = \frac{\max + \min}{2} \tag{7}$$

3. Result

In this paper, 14 document images from Handwritten Document Image Binarization Contest (H-DIBCO) [24] dataset experimented. The images contain various degradations such as shadows, non-uniform illumination, stains, smudges, bleed-through and faint characters [1]. All the processed images are in greyscale images and the size of each image is 400 × 400 pixels, 72 dpi, and 8-bit depth. All the programs were written in C programming and ran using Ubuntu with a Linux 3.5 from an Asus laptop with AMD Athlon™ II P320 Dual-Core Processor 2.10GHz and 3.00GB RAM. The seven (7) types of binarization were tested and shown in figure 2. Three examples of document images were illustrated in figure 2. The first row shows the original image with degradation and shadowing problem and follows by binarization methods such Otsu, Local Adaptive, Niblack, Bernsen, Bradley, Nick, and Gradient Based method. Based on observation, the resulting image from Bradley and Gradient Based is satisfied compared to other methods.

Original	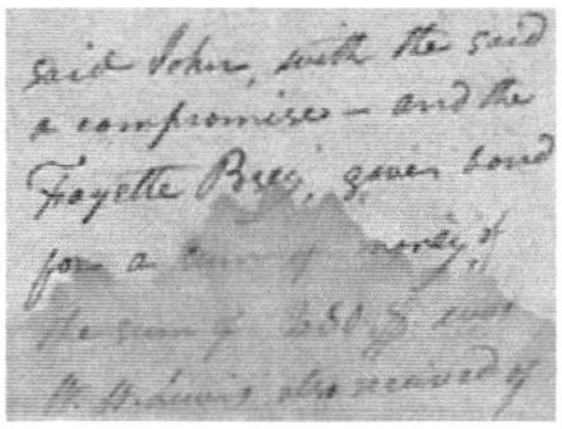	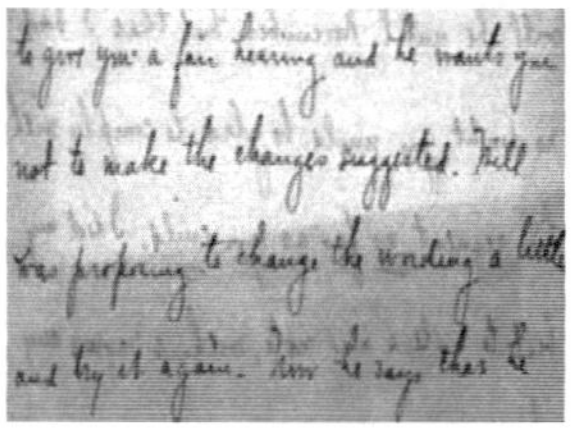	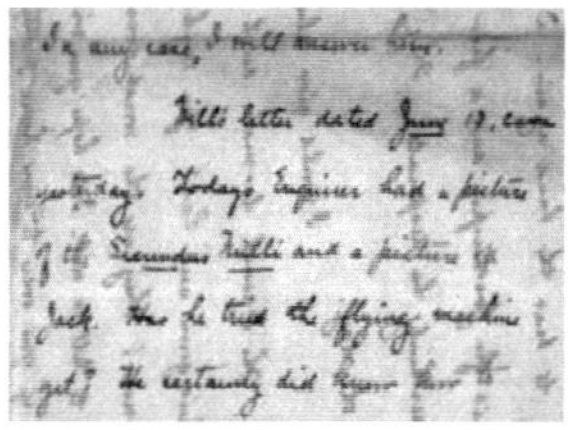
Otsu	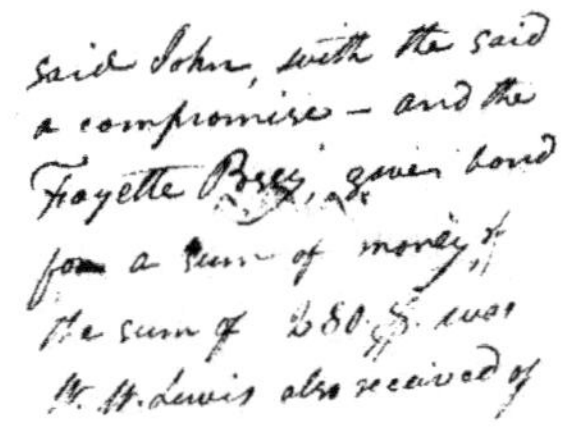	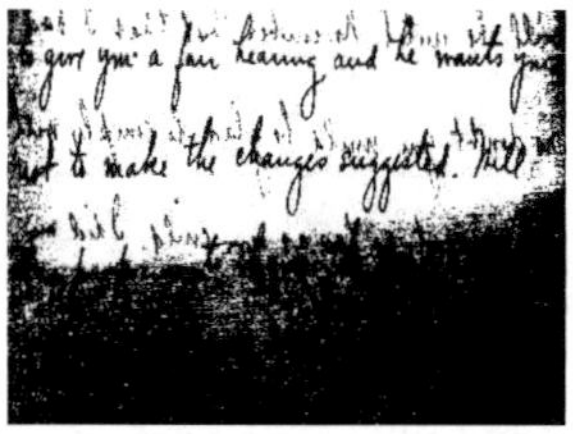	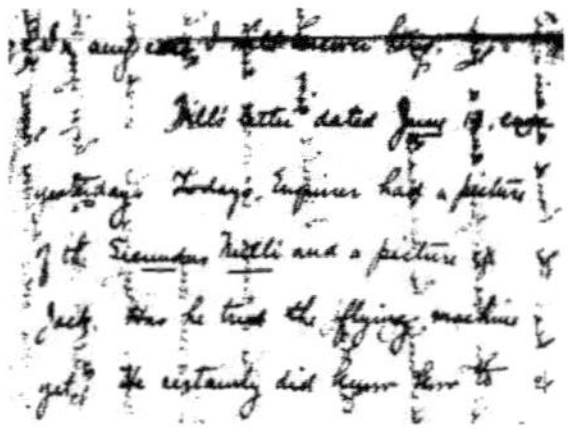
Local Adaptive	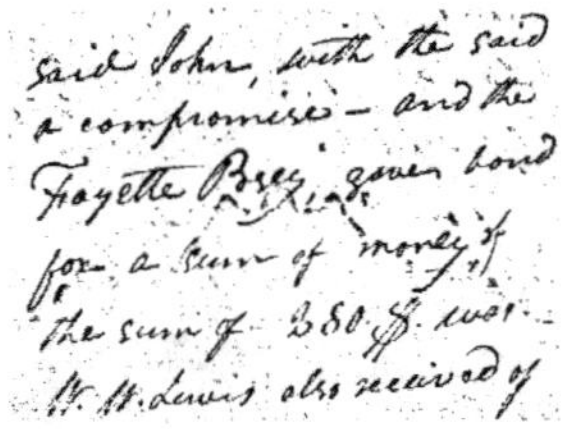	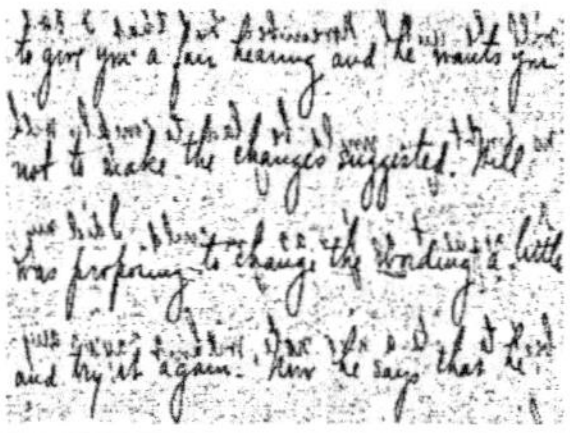	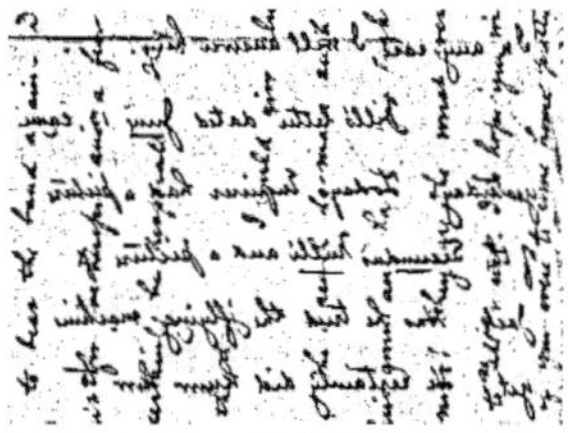
Niblack	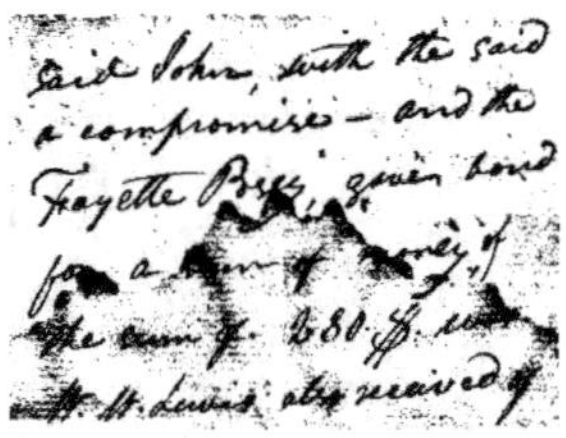	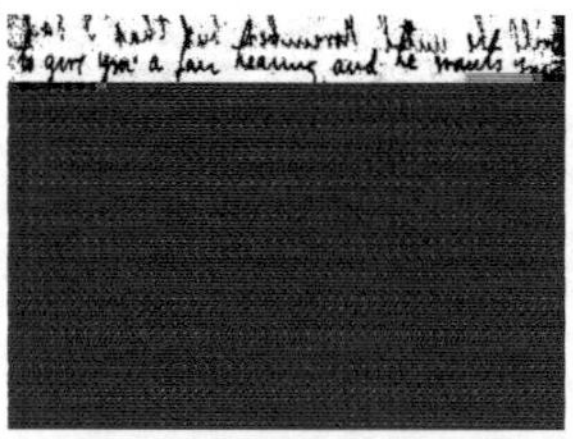	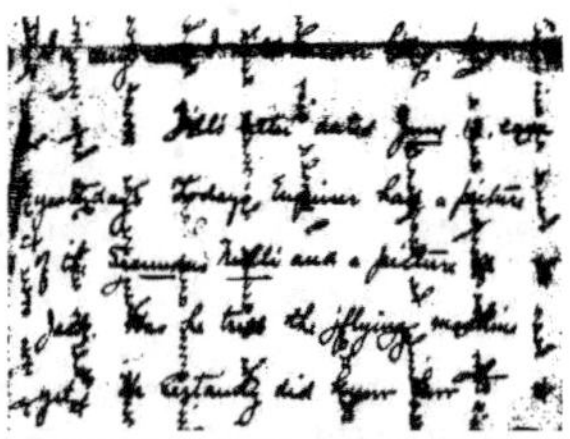
Bernsen	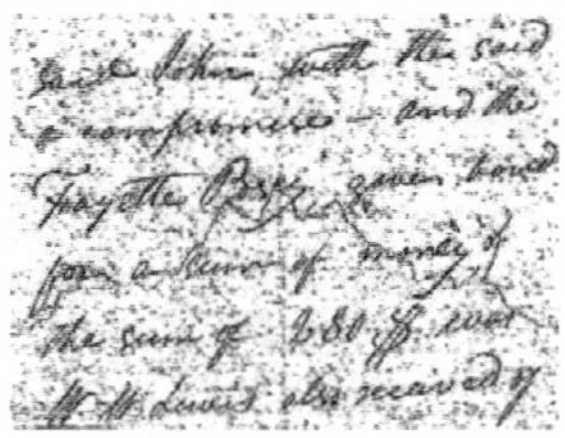		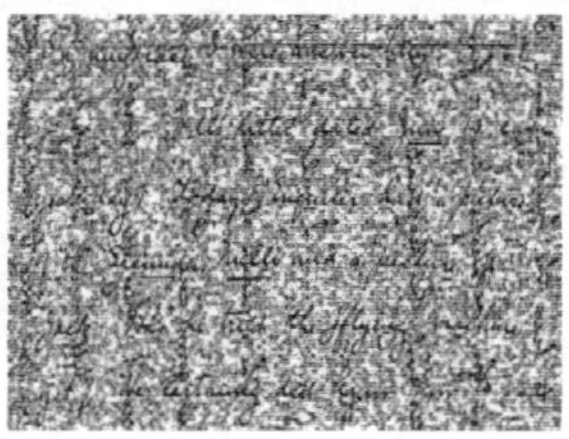

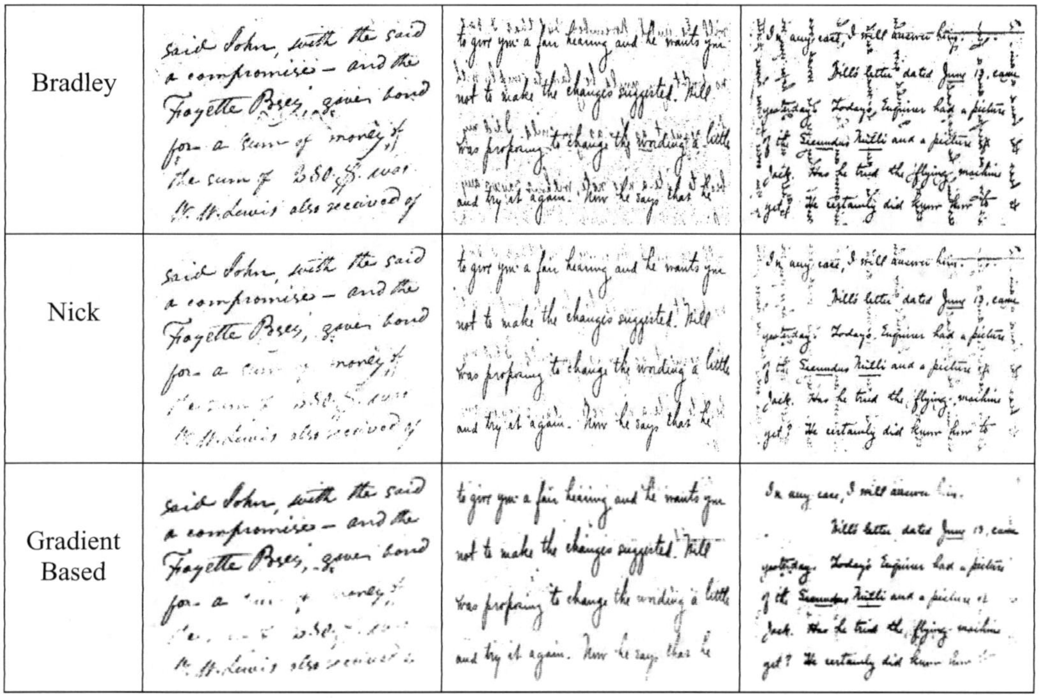

Figure 2. The comparison of binarization method on document images.

Next, a few image quality assessment (IQA) was calculated to find the best binarization methods. In this paper, the evaluation based on ME (Misclassification Error), PSNR (Peak Signal to Noise Ratio) and MPM (Misclassification Penalty Metric) was obtained. All the assessment equation can be referred on H-DIBCO [25], [26]. The lower of ME and MPM, while higher of PSNR value represent the good binarization result. Table 1 provides the comparison result of 7 binarization methods. The lowest value of ME indicates a good performance in case of binarization as displays by the Bradley method (ME = 0.0285) and follow by Gradient Based method (ME = 0.0292) respectively. Opposite, the higher value in term of PSNR present the best binarization was come from the Nick method (PSNR = 24.8349) compared to others. Besides, comparisons based on MPM show that the performance of the Local Adaptive method obtained 0.2376, which is lower than the other methods. The highest value of MPM came from the Otsu method (24.1048) since it produces a degraded and non-uniform image.

IOP Conf. Series: Journal of Physics: Conf. Series **1019** (2018) 012023 doi:10.1088/1742-6596/1019/1/012023

Table 1. A few evaluation techniques

Method	ME	PSNR	MPM (10^{-3})
Otsu	0.0886	16.5185	24.1048
Local Adaptive	0.0390	16.0604	0.2376
Niblack	0.1196	9.7405	7.2171
Bernsen	0.1073	12.1823	3.2518
Bradley	0.0285	19.3170	0.9831
Nick	0.0301	24.8349	1.7177
Gradient Based	0.0292	19.6308	1.5587

Method	ME	PSNR	MPM (10^{-3})
Otsu	0.0886	16.5185	24.1048
Local Adaptive	0.0390	16.0604	0.2376
Niblack	0.1196	9.7405	7.2171
Bernsen	0.1073	12.1823	3.2518
Bradley	0.0285	19.3170	0.9831
Nick	0.0301	24.8349	1.7177
Gradient Based	0.0292	19.6308	1.5587

Method	ME	PSNR	MPM (10^{-3})
Otsu	0.0886	16.5185	24.1048
Local Adaptive	0.0390	16.0604	0.2376
Niblack	0.1196	9.7405	7.2171
Bernsen	0.1073	12.1823	3.2518
Bradley	0.0285	19.3170	0.9831
Nick	0.0301	24.8349	1.7177
Gradient Based	0.0292	19.6308	1.5587

In order to prove the sensitivity of each binarization methods, the evaluation based on the F-measure and the Accuracy was obtained. Normally, the higher on F- measure and Accuracy shows the effectiveness of binarization method [7], [27]. The result is presented in figure 3. As shown in the figure, Local Adaptive method achieved the highest result which is (F-measure = 87.678) and (Accuracy = 98.729) compared to other methods. While, the lowest value in term of F-measure and Accuracy came from Niblack method (F-measure = 52.657) and (Accuracy = 88.037).

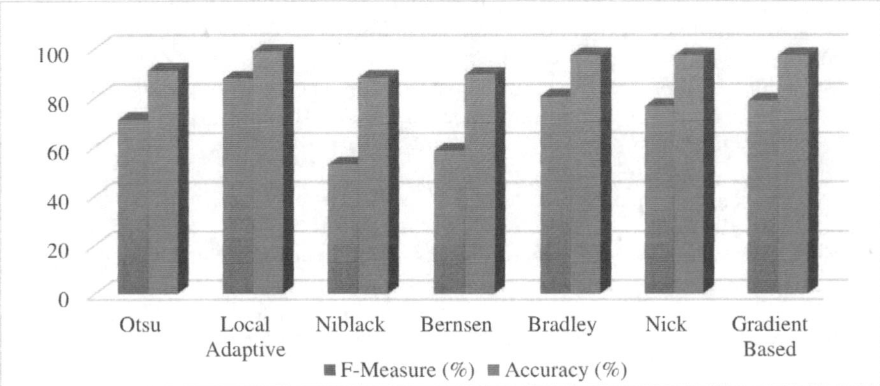

Figure 3. Comparison of seven types binarization methods based on F-measure and Accuracy

4. Gaps in Literature

Many techniques have been proposed so far for document binarization as shown in literature survey. It has been concluded from the existing research is that no technique is perfect for every case. Therefore still some research is required in this field of image binarization. Following are the main limitations of this research work:-

1. Many researchers have used image filters to reduce the noise from the image, but the use of the guided filter (best edge preserving filter) is not found. It may increase the accuracy of the available binarization methods
2. In the most of the techniques, the contrast enhancement is either done by traditional methods or not done. So, adaptive contrast enhancement is required.
3. Most of the methods have neglected the use of edge map which has the ability to map the exact character in an efficient manner.

5. Conclusion

This paper has focused on the seven (7) types of binarization method and experimented on degraded document image. Document binarization is an important application of vision processing. The main objective of this paper is to evaluating the shortcomings of algorithms for degraded image binarization. It has been found that each technique has its own benefits and limitations; no technique is best for every case. The main limitations of existing workers are found to be noisy and low intensity images. In near future, we will propose a new algorithm which will use the more reliable methodology to enhance the work. We will propose a new algorithm which will use nonlinear enhancement as a pre-processing technique to improve the results further.

6. References

[1] K. Ntirogiannis, B. Gatos, and I. Pratikakis, "A combined approach for the binarization of handwritten document images," *Pattern Recognit. Lett.*, vol. 35, pp. 3–15, Jan. 2014.
[2] W. A. Mustafa, "A proposed optimum threshold level for document image binarization," *J. Adv. Res.*

Comput. Appl., vol. 1, no. 1, pp. 8–14, 2017.

[3] W. A. Mustafa and H. Yazid, "Illumination and Contrast Correction Strategy using Bilateral Filtering and Binarization Comparison," *J. Telecommun. Electron. Comput. Eng.*, vol. 8, no. 1, pp. 67–73, 2016.

[4] W. A. Mustafa and H. Yazid, "Background Correction using Average Filtering and Gradient Based Thresholding," *J. Telecommun. Electron. Comput. Eng.*, vol. 8, no. 5, pp. 81–88, 2016.

[5] N. Ntogas and V. Dimitrios, "A Binarization Algorithm for Historical Manuscripts," in *International Conference on Comunications*, 2008, pp. 41–51.

[6] B. Gatos, I. Pratikakis, and S. J. Perantonis, "Improved document image binarization by using a combination of multiple binarization techniques and adapted edge information," *19th Int. Conf. Pattern Recognit.*, pp. 1–4, Dec. 2008.

[7] S. Lu, B. Su, and C. L. Tan, "Document image binarization using background estimation and stroke edges," *Int. J. Doc. Anal. Recognit.*, vol. 13, no. 4, pp. 303–314, Oct. 2010.

[8] J. Sauvola and M. Pietikäinen, "Adaptive document image binarization," *Pattern Recognition*, vol. 33. pp. 225–236, 2000.

[9] N. Otsu, "A Threshold Selection Method from Gray-Level Histograms," in *IEEE Transactions On Systrems, Man, And Cybernetics*, 1979, vol. 20, no. 1, pp. 62–66.

[10] R. F. Moghaddam and M. Cheriet, "Low quality document image modeling and enhancement," *Int. J. Doc. Anal. Recognit.*, vol. 11, no. 4, pp. 183–201, Feb. 2009.

[11] O. Nina, B. Morse, and W. Barrett, "A Recursive Otsu Thresholding Method for Scanned Document Binarization," in *IEEE Workshop on Applications of Computer Vision (WACV)*, 2010, pp. 307–314.

[12] Y. Zhang and L. Wu, "Fast Document Image Binarization Based on an Improved Adaptive Otsu ' s Method and Destination Word Accumulation," *J. Comput. Inf. Syst.*, vol. 6, no. 7, pp. 1886–1892, 2011.

[13] Reza Farrahi Moghaddamn Mohamed Cheriet, "AdOtsu: An adaptive and parameterless generalization of Otsu's method for document image binarization," *Pattern Recognit.*, vol. 45, pp. 2419–2431, Jun. 2012.

[14] R. C. Gonzalez and R. E. Woods, *Digital Image Processing*, Third. Upper Saddle River, NJ, USA: Prentice Hall, 2008.

[15] M. D. Vlachos and E. S. Dermatas, "Non-uniform illumination correction in infrared images based on a modified fuzzy c-means algorithm," *J. Biomed. Graph. Comput.*, vol. 3, no. 1, pp. 6–19, Nov. 2012.

[16] M. A. Bakhshali, "Segmentation and enhancement of brain MR images using fuzzy clustering based on information theory," *Soft Comput.*, vol. 20, pp. 1–8, 2016.

[17] H. Yazid and H. Arof, "Gradient based adaptive thresholding," *J. Vis. Commun. Image Represent.*, vol. 24, no. 7, pp. 926–936, Oct. 2013.

[18] W.Niblack, *An Introduction to Digital Image Processing.* Englewood Cliffs: Prentice-Hall, 1986.

[19] J. Shi, N. Ray, and H. Zhang, "Shape based local thresholding for binarization of document images," *Pattern Recognit. Lett.*, vol. 33, no. 1, pp. 24–32, Jan. 2012.

[20] K. Khurshid, I. Siddiqi, C. Faure, and N. Vincent, "Comparison of Niblack inspired Binarization methods for ancient documents," *Proc. SPIE-IS&T Electron. Imaging*, vol. 7247, pp. 1–9, 2009.

[21] Wellner, "Adaptive thresholding for the digital desk.," *Tech. Rep. Eur.*, pp. 93–110, 1993.

[22] D. Bradley and G. Roth, "Adaptive Thresholding Using the Integral Image," *J. Graph. GPU, Game Tools*, vol. 12, no. 2, pp. 13–21, 2011.

[23] J. Bernsen, "Dynamic thresholding of grey-level images.," in *Proceedings of the Eighth International Conference on Pattern Recognition*, 1986, pp. 1251–1255.

[24] K. Zagoris, "H-DIBCO '16," 2016. [Online]. Available: http://vc.ee.duth.gr/h-dibco2016/. [Accessed: 01-Jan-2016].

[25] W. A. Mustafa, H. Yazid, and S. Yaacob, "Illumination Normalization of Non-Uniform Images Based on Double Mean Filtering," in *IEEE International Conference on Control Systems, Computing and Engineering*, 2014, pp. 366–371.

[26] W. A. Mustafa, H. Yazid, M. Jaafar, M. Zainal, A. S. Abdul-, and N. Mazlan, "A Review of Image Quality Assessment (IQA): SNR, GCF, AD, NAE, PSNR, ME," *J. Adv. Res. Comput. Appl.*, vol. 7, no. 1, pp. 1–7, 2017.

[27] S. Thilagamani and N. Shanthi, "Gaussian and Gabor Filter Approach for Object Segmentation," *J. Comput. Inf. Sci. Eng.*, vol. 14, pp. 1–7, 2014.

IOP Conf. Series: Journal of Physics: Conf. Series **1019** (2018) 012024 doi:10.1088/1742-6596/1019/1/012024

Document Image Database (2009 - 2012): A Systematic Review

Wan Azani Mustafa[1,*], **Mohamed Mydin M. Abdul Kader**[1]
[1]Faculty of Engineering Technology, Universiti Malaysia Perlis, UniCITI Alam Campus, Sungai Chuchuh, 02100 Padang Besar, Perlis, Malaysia

[1,*] *wanazani@unimap.edu.my*, [2] *mohamedm@unimap.edu.my*

Abstract. Document image binarization contributes significantly to the success of the document image analysis and recognition challenging tasks. Image quality can play an important role in addressing the issue of binarization effectiveness. In this paper, a comprehensive review of document database was presented. Review based on image from Document Image Binarization Contest (DIBCO) 2009 to 2012 consists handwritten and printed image. The best algorithm for each year is discussed and analysed. Implications of the review give the direction for future binarization approach developments.

1. Introduction

There are many challenges addressed in handwritten document image binarization, such as faint characters, bleed-through and large background ink stains [1]–[3]. Document image binarization is the process that segments the document image into the text and background by removing any existing degradations [1]. Many document image binarization methods have been proposed in the literature. However, selecting the most optimum method for binarization is a difficult task due to the presence of a variety of degradation in document images.

In 2009, Reza and Mohamed published a paper in which they described the new model of a low quality document image using virtual diffusion processes [4]. This technique focuses on the shadow- through and bleed-through problem. Several studies investigating document binarization based on Otsu modification [5]–[7]. Starting 2010, Nina *et al.* mention the significant combination between the recursive extension of Otsu thresholding and selective bilateral filtering of scanned handwritten images [5]. This approach considers background estimation before applied the post-processing stage [5], [7]. The above findings contradict the study by Zhang and Wu. They examined the modification algorithm based on the Adaptive Otsu method [6]. The proposed based on three main steps; (1) applied the Wilner filter in order to eliminate noise, (2) improved adaptive Otsu's method, and (3) dilation and erosion operators were performed to preserve stroke connectivity and fill possible breaks, gaps, and holes. The advantage of this approach is that faster processing time compared to Recursive Otsu Thresholding Method [5] and AdOtsu method [7]. Similarly, Reza and Mohamed also proposed a new novel method based on the Otsu modification known as AdOtsu method [7]. The main idea of this technique was considered parameterless behavior such as average stroke width and the average line height. The positive result was achieved compared to Sauvola method, Otsu Method, and Lu and Tan method.

IOP Conf. Series: Journal of Physics: Conf. Series **1019** (2018) 012024 doi:10.1088/1742-6596/1019/1/012024

In this paper, a comprehensive review of document image database from 2009 to 2016 was presented. The database consists two types of image which is handwritten and printed image. The specification in terms of size and resolution also was discussed. Furthermore, the best binarization method for each year from the first rank to the third rank was explained. The rest of this paper is organized as follows: Section 2 describes an overview of the document image. Sub-section 2.1 to 2.4 described the database image. Finally, Section 3 explained the conclusion of this work.

2. Literature Of Database

Document Image Binarization Contest (DIBCO) is established and popular database image. Many researchers used this database in their research and published in the local and international journal [8][9][10]. This research concentrated on DIBCO 2009, DIBCO 2010, DIBCO 2011, and DIBCO 2012 based on a number of images, size, and resolution. The three best methods for each year also was described.

2.1 DIBCO 2009

In International Document Image Binarization Contest (DIBCO 2009), thirty five (35) research groups have participated in the competition with forty three (43) different algorithms. The database of DIBCO 2009 contains two types which are; five (5) handwritten images and five (5) printed image. This database image can be reached at http://www.iit.demokritos.gr/~bgat/DIBCO2009/benchmark. Figure 1 shows the example of an image in the DIBCO 2009 database. The database image present different of resolution (Pixels Per Inch) and size of the pixel as shown in Table 1. The handwritten image labeled 'H0' and printed image 'P0'.

Figure 1. Image from DIBCO Database 2009

Table 1. Specification of image database

Images	Size (pixels)	Resolution (ppi)
H01	2025 x 426	300
H02	946 x 1366	200
H03	582 x 492	72
H04	1091 x 581	72
H05	1341 x 713	72
P01	1268 x 263	96
P02	1223 x 310	96
P03	1153 x 493	96
P04	1849 x 357	96
P05	1218 x 259	300

Based on the contest at 10th International Conference on Document Analysis and Recognition [11], the first ranks winner comes from Institute for Infocomm Research, Singapore. They proposed a new binarization techniques based on four steps; (1) document background extraction, (2) stroke edge detection, (3) local thresholding, and (4) post-processing [12]. Fabrizio and Marcotegui from Université Pierre, France achieved second place with finding a new algorithm using toggle mapping operator [9]. They focused on the morphological operator such as erosion and dilation in order to divide the input image into three classes; foreground, background and homogeneous. The third rank is method proposed by Henault et al. based on local probabilistic models and the calculus of variation [13], [14].

2.2 DIBCO 2010

The document database 2010 consist ten (10) handwritten images with different size and specification. All the sample images are non-uniform and degraded caused by large background ink stains, shadow and other. The database image, benchmark and evaluation tool can be download at http://www.iit.demokritos.gr/~bgat/H-DIBCO2010/benchmark. Example document images as illustrated in Figure 2. Based on observation Figure 2, image show very low-quality with many noise and non-uniform background. Non-uniform illumination and noise can come from various sources such as aging filaments, sunlight, camera sensors or a noisy environment [15]. Many researchers have agreed that very complicated and difficult to segment the document image in low-grade condition [16]–[19]. The specification in terms of size and resolution are presented in Table 2. All the images show the different size with the large size came from 'H02' image (1570 x 841).

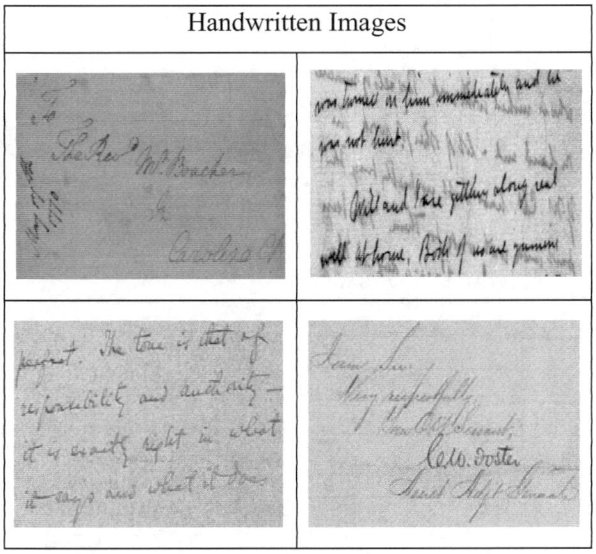

Figure 2. Image from DIBCO Database 2010

Table 2. Specification of image database

Images	Size (pixels)	Resolution (ppi)
H01	1489 x 380	300
H02	1570 x 841	72
H03	786 x 423	96
H04	935 x 537	96
H05	1726 x 391	72
H06	945 x 366	300
H07	1742 x 467	72
H08	2280 x 326	72
H09	1158 x 637	300
H10	1768 x 624	72

The binarization contest using document database 2010 was done and published at 12th International Conference on Frontiers in Handwriting Recognition. The objective is to find the advanced and effective algorithm in term of measurement criteria such as F-Measure, Negative Rate Metric (NRM) and Misclassification penalty metric (MPM). According to Pratikakis et al. [20], Sixteen (16) distinct research groups have participated in the competition with seventeen (17) different algorithms. The winner is research by Yosef et al. [21] from Ben-Gurion University, Computer Science department, Israel. The proposed method based on adaptive technique and a few pre-processing steps such as removed unwanted objects and accurate local region-based active contour. Interestingly, this rank was shared with the method proposed by B. Su et al. [12]. Second rank achieved by T. Lelore and F. Bouchara from South University of Toulon-Var, France. This method quite complicated to estimate the edge and text location before performing the clustering algorithm [22]. Howe from Smith College, MA, USA at the third rank [23].

They suggest the combination between Laplacian operator and Canny edge detection to detect the text region. This approach is simple however very effective

2.3 DIBCO 2011

Database 2011 consists eight (8) handwritten images and eight (8) printed images with different pixel size, however same resolution (72ppi). The database can be accessed at http://utopia.duth.gr/~ipratika/DIBCO2011/benchmark. A few sample images are shown in Figure 3. Based on observation, the major problem came from the non-uniform background. Table 3 shows the specification of the database image, where the handwritten image labeled 'HW' and printed image 'PR'.

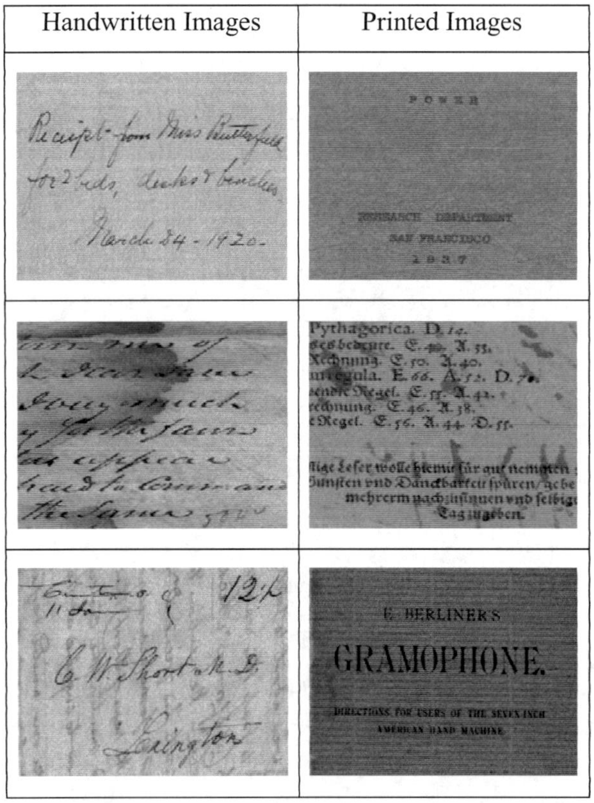

Figure 3. Image from DIBCO Database 2011

Table 3. Specification of image database

Images	Size (pixels)	Resolution (ppi)
HW1	645 x 743	72
HW2	1218 x 781	72
HW3	1870 x 511	72
HW4	469 x 597	72
HW5	1623 x 261	72
HW6	787 x 687	72
HW7	982 x 657	72
HW8	998 x 410	72
PR1	1381 x 368	72
PR2	1180 x 371	72
PR3	1203 x 363	72
PR4	1838 x 798	72
PR5	690 x 682	72
PR6	1315 x 1069	72
PR7	600 x 564	72
PR8	859 x 323	72

In 2010, Pratikakis et al. published a paper in which they described the advanced and winner in document binarization using document database 2011 [24]. Sixteen (17) research groups have participated in the competition with eighteen (18) different algorithms. Based on four types of measurement calculation, the best algorithm is proposed by T. Lelore and F. Bouchara from South University of Toulon-Var, France [22]. In 2010 competition, this approaches achieved the second place, however after applied slightly improvement on method became more successful. The improvement is applied a median filtering of the input image and then upscale it using linear interpolation. Furthermore, binarization method by B.Su et al. [12] from National University of Singapore drop to the second rank. The details method was explained in section 2.1. The third rank again achieved by Howe from Smith College, MA, USA [23]. The key element of this approaches is the minimized a global energy function for efficient exact computation. The details method was explained in section 2.2.

2.4 DIBCO 2012

In contrast, the document database 2012 provided 14 handwritten images with the same resolution (72ppi). However, every image shows the different pixel size and the database can be download at http://utopia.duth.gr/~ipratika/HDIBCO2012/benchmark. Table 4 described the specification of database image in terms of size and resolution. The largest size is 'H01' (2245 x 1317) and the smallest is 'H04' (961 x 854). Figure 4 illustrated a few sample images from database 2012. Based on Figure 6, the major problem images is caused by the fade ink and degrade of background.

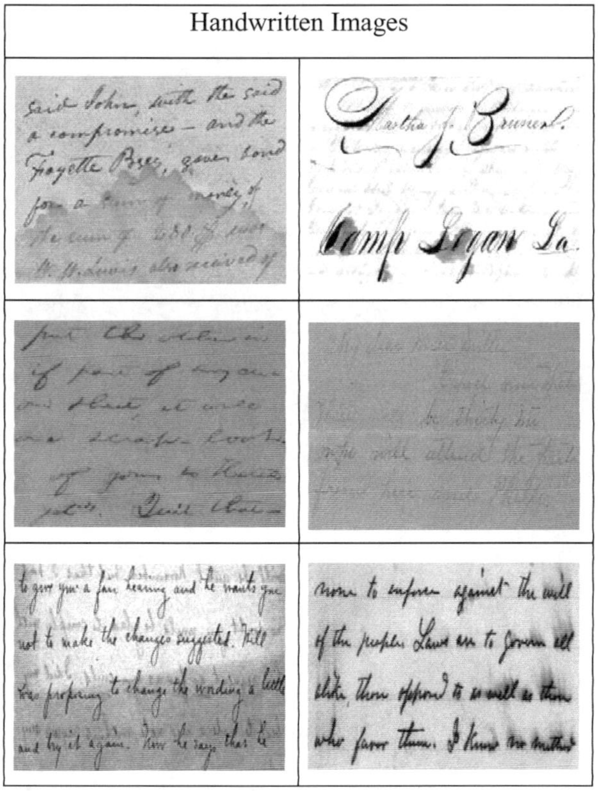

Figure 4. Image from DIBCO Database 2012

Table 4. Specification of image database

Images	Size (pixels)	Resolution (ppi)
H01	2245 x 1317	72
H02	1285 x 843	72
H03	1709 x 1371	72
H04	961 x 854	72
H05	1661 x 633	72
H06	1499 x 939	72
H07	1221 x 297	72
H08	1645 x 453	72
H09	1389 x 527	72
H10	1735 x 1021	72
H11	1623 x 945	72
H12	1841 x 433	72
H13	1703 x 847	72
H14	2800 x 768	72

Pratikakis et al. [25] indicated that nineteen (19) distinct research groups have participated in the competition with twenty four (24) different algorithms because certain participating research groups have submitted more than one algorithm. After the evaluation process, the first rank belonging to Howe from Smith College, Department of Computer Science, Northampton (MA), USA. Last year, this method placed at third ranking. The key is to optimize a global energy function based on the Laplacian image and dynamically setting the regularization coefficient [23]. Thibault Lelore and Frédéric Bouchara from South University of Toulon- Var, La Garde, France achieved the second rank [22]. The input image divided into three group image (Text, Background and Unknown). The details explanation method can be referred in section 2.2 and 2.3. The third rank is method proposed by B. Su et al. [12] from Institute for Infocomm Research, & School of Computing, National University of Singapore. This method described in Section 2.1.

3. Conclusion
Binarization plays a vital role in the document recognition and analysis. The low quality image such as degraded and non-uniform background will effect on the post-processing stage. This study set out to review in detail the available information on document image database. The review on DIBCO 2009 to DIBCO 2012 in terms pixel size and resolution. The best binarization method for each database image also was discussed. The results of this review indicate that binarization technique using background estimation and stroke edges is best and efficient for this database. Further research might explore the comparison effect between handwritten and printed image on binarization field.

4. Acknowledgment
This work was supported by Short Term Grant Scheme (Grant No 9001-00576)

5. References
[1] K. Ntirogiannis, B. Gatos, and I. Pratikakis, "A combined approach for the binarization of handwritten document images," *Pattern Recognit. Lett.*, vol. 35, pp. 3–15, Jan. 2014.
[2] B. M. Singh, R. Sharma, D. Ghosh, and A. Mittal, "Adaptive binarization of severely degraded and non-uniformly illuminated documents," *Int. J. Doc. Anal. Recognit.*, pp. 393–412, 2014.
[3] D. Rivest-Hénault, R. Farrahi Moghaddam, and M. Cheriet, "A local linear level set method for the binarization of degraded historical document images," *Int. J. Doc. Anal. Recognit.*, vol. 15, no. 2, pp. 101–124, 2012.
[4] R. F. Moghaddam and M. Cheriet, "Low quality document image modeling and enhancement," *Int. J. Doc. Anal. Recognit.*, vol. 11, no. 4, pp. 183–201, Feb. 2009.
[5] O. Nina, B. Morse, and W. Barrett, "A Recursive Otsu Thresholding Method for Scanned Document Binarization," in *IEEE Workshop on Applications of Computer Vision (WACV)*, 2010, pp. 307–314.
[6] Y. Zhang and L. Wu, "Fast Document Image Binarization Based on an Improved Adaptive Otsu ' s Method and Destination Word Accumulation," *J. Comput. Inf. Syst.*, vol. 6, no. 7, pp. 1886–1892, 2011.
[7] Reza Farrahi Moghaddamn Mohamed Cheriet, "AdOtsu: An adaptive and parameterless generalization of Otsu's method for document image binarization," *Pattern Recognit.*, vol. 45, pp. 2419–2431, Jun. 2012.
[8] A. Shaus, B. Sober, E. Turkel, and E. Piasetzky, "Beyond the ground truth: Alternative quality measures of document binarizations," in *Proceedings of International Conference on Frontiers in Handwriting Recognition, ICFHR*, 2017, pp. 495–500.

[9] I. Ben Messaoud, H. Amiri, H. El Abed, and V. Märgner, "New binarization approach based on text block extraction," in *Proceedings of the International Conference on Document Analysis and Recognition, ICDAR*, 2011, pp. 1205–1209.

[10] T. H. N. Le, T. D. Bui, and C. Y. Suen, "Ternary entropy-based binarization of degraded document images using morphological operators," in *Proceedings of the International Conference on Document Analysis and Recognition, ICDAR*, 2011, pp. 114–118.

[11] B. Gatos, K. Ntirogiannis, and I. Pratikakis, "ICDAR 2009 Document Image Binarization Contest (DIBCO 2009)," *2009 10th Int. Conf. Doc. Anal. Recognit.*, no. Dibco, pp. 1375–1382, 2009.

[12] S. Lu, B. Su, and C. L. Tan, "Document image binarization using background estimation and stroke edges," *Int. J. Doc. Anal. Recognit.*, vol. 13, no. 4, pp. 303–314, Oct. 2010.

[13] R. Farrahi Moghaddam and M. Cheriet, "RSLDI: Restoration of single-sided low-quality document images," *Pattern Recognit.*, vol. 42, no. 12, pp. 3355–3364, 2009.

[14] D. Rivest-Hénault and M. Cheriet, "Image Segmentation Using Level Set and Local Linear Approximations," in *Image Analysis and Recognition: 4th International Conference, ICIAR 2007, Montreal, Canada, August 22-24, 2007. Proceedings*, M. Kamel and A. Campilho, Eds. Berlin, Heidelberg: Springer Berlin Heidelberg, 2007, pp. 234–245.

[15] W. A. Mustafa and H. Yazid, "Illumination and Contrast Correction Strategy using Bilateral Filtering and Binarization Comparison," *J. Telecommun. Electron. Comput. Eng.*, vol. 8, no. 1, pp. 67–73, 2016.

[16] Y. Yang and H. Yan, "An adaptive logical method for binarization of degraded document images," *Pattern Recognition*, vol. 33. pp. 787–807, 2000.

[17] J. G. Kuk and N. I. Cho, "Feature Based Binarization of Document Images Degraded by Uneven Light Condition," *2009 10th Int. Conf. Doc. Anal. Recognit.*, pp. 748–752, 2009.

[18] W. A. Mustafa, "A proposed optimum threshold level for document image binarization," *J. Adv. Res. Comput. Appl.*, vol. 1, no. 1, pp. 8–14, 2017.

[19] B. Su, S. Lu, and C. L. Tan, "Robust document image binarization technique for degraded document images," *IEEE Trans. Image Process.*, vol. 22, pp. 1408–1417, 2013.

[20] I. Pratikakis, B. Gatos, and K. Ntirogiannis, "H-DIBCO 2010 - Handwritten Document Image Binarization Competition," *12th Int. Conf. Front. Handwrit. Recognit.*, pp. 727–732, Nov. 2010.

[21] I. Bar-Yosef, N. Hagbi, K. Kedem, and I. Dinstein, "Fast and Accurate Skew Estimation Based on Distance Transform," *2008 Eighth IAPR Int. Work. Doc. Anal. Syst.*, pp. 402–407, 2008.

[22] T. Lelore and F. Bouchara, "Document image binarisation using Markov field model," in *Proceedings of the International Conference on Document Analysis and Recognition, ICDAR*, 2009, pp. 551–555.

[23] N. R. Howe, "A Laplacian energy for document binarization," in *Proceedings of the International Conference on Document Analysis and Recognition, ICDAR*, 2011, pp. 6–10.

[24] I. Pratikakis, B. Gatos, and K. Ntirogiannis, "ICDAR 2011 Document Image Binarization Contest (DIBCO 2011)," *Int. Conf. Doc. Anal. Recognit.*, pp. 1506–1510, Sep. 2011.

[25] K. Ntirogiannis, B. Gatos, and I. Pratikakis, "ICFHR2012 Competition on Handwritten Document Image Binarization (H-DIBCO 2012)," *Int. Conf. Front. Handwrit. Recognit.*, pp. 809–813, 2012.

IOP Conf. Series: Journal of Physics: Conf. Series **1019** (2018) 012025 doi:10.1088/1742-6596/1019/1/012025

Contrast Enhancement Based on Fusion Method: A Review

Wan Azani Mustafa[1,*], Mohamed Mydin M. Abdul Kader[2]
[1]Faculty of Engineering Technology, Universiti Malaysia Perlis, UniCITI Alam Campus, Sungai Chuchuh, 02100 Padang Besar, Perlis, Malaysia

[1] *wanazani@unimap.edu.my,* [2] *mohamedm@unimap.edu.my*

Abstract. Image enhancement is a fundamental step and plays an important role in image processing, pattern recognition, and computer vision. Image enhancement can sharpen the edges of objects in an image, making it easier to extract objects and attain more information from the enhanced image. In this paper, a few popular image enhancement methods based Fusion Method was studied. This paper presents an exhaustive review of these studies and suggests a direction for future developments of image enhancement methods. Each method shows the owned advantages and drawbacks. In future, this work will give the direction to other researchers in order to propose new advanced enhancement techniques.

1. Introduction

Image enhancement can be considered as one of the fundamental processes in image analyses. The goal of contrast enhancement is to improve the quality of an image to become more suitable for a particular application. Till today, numerous image enhancement methods have been proposed for various applications and efforts have been directed to further increase the quality of the enhancement results and minimize the computational complexity and memory usage. In this paper, a few popular enhancement approaches such as histogram equalization and fusion method was reviewed. The advantages and drawback of the method were elaborated deeply and systematic. This paper is organized in the following sections: Section 2 describes the method using fusion techniques and Section 3 gives the conclusion of the work.

2. Fusion Method

A large and growing body of literature has investigated about enhancement methods. Image enhancement deals with improving the quality of images, where the objective is to emphasize wanted features and make them less obscured [1], [2]. The area of image enhancement is very appealing, where many fundamental image enhancement techniques are developed based on a very simple concept [3]–[6]. In the recent decades, image enhancement based on contrast and luminosity correction has been focused [7]–[12].

In 2010, Al-amri *et al.* [13] published a paper that reviewed two types of contrast enhancement techniques, namely linear contrast and non-linear contrast. The linear type represented by three methods such as Max-Min contrast, Percentage contrast, and Piecewise contrast. Four methods of the non-linear technique namely Histogram equalization, Adaptive histogram equalization, Homomorphic Filter, and Unsharp Mask. After evaluation, the Piecewise contrast and the Homomorphic Filter technique produced the best result compared to the other methods. According to Vu and Caplier [14], they found that the combination of two non-linear functions and the difference of Gaussians filter were efficient to solve the contrast and illumination problem. In this experiment, this approach obtained a very high recognition rate even for the challenging illumination conditions. The proposed algorithm also has a low computational complexity. In contrast to [14], a study by Cheng *et al.* [15] argued that 2D Gaussian Filtering technique is more effective. This technique focuses on contrast stretching between dark and bright areas. The proposed method was effective to eliminate the luminosity and enhance the contrast compared to the method by [14]. This view is supported by Cheng *et al.* [16] that discovered a combination of quotient image and different smoothing filter knows as Different Smoothing filters Quotient Image (DSFQI). The Gaussian distribution in 2D was performed in order to smooth the contrast variation image. This technique was more effective for illumination normalization, compared to Self-Quotient Image (SQI) [17] and Dynamic Morphological Quotient Image (DMQI). The finding based on the smoothing filter is consistent with the research by Ambule *et al.* [18] which introduced an adaptive smoothing filter. However, this method focuses on the median filter. The adaptive median filter able to eliminate the noise and enhance the image quality. Furthermore, this technique is simple and has fast processing time. An *et al.* [17] proposed pre-processing techniques by using a combination of Lambertian reflectance model and Self-Quotient Image (SQI) to enhance the contrast variation images under changing lighting conditions. This paper used the TV_L model for finding the low frequency and high frequency component. The approach is very effective and suitable for a real-time application. The research study by Wang *et al.* [19] also found the Lambertian reflectance function for contrast enhancement. They performed experimental investigations on face images by applying Weber's law method in order to eliminate the unwanted light and automatically enhanced the contrast variation. This technique consists of two steps: (1) by using the Laplace operator to measure the intensity variation: (2) by applying an algorithm from Lambertian reflectance as a way to produce the final result.

In recent years, there has been an increasing amount of literature on image enhancement method and the main purpose is to improve the quality of the image by performing an adjustment on bright or dark intensity [20], [21]. Rivera *et al.* [22] proposed a new image enhancement algorithm known as Content-Aware Algorithm. Normally. The images were captured in abnormal lighting (dark conditions) and producing an image with low levels of brightness. The aim of this method was to preserve the flat region's information by smoothing and gradient (edge) sharpening. This method enhanced the image by adaptively creating the mapping functions and produced the *ad hoc* transformation for every individual image. The *ad hoc* transformation comes from the contrast in the boundaries and the texture regions. The finding is consistent with the finding by Jha *et al.* [23], which proposed a new method known as Non-linear Non-Dynamic Stochastic Resonance. The proposed method isable to enhance the dark original image. Generally, images are degraded due to noise, blurring, and occlusion. Besides that, the artefacts have the ability to produce a degradation in the abnormal lighting conditions. By using the low contrast information, the sub-threshold was obtained and applied to enhance the contrast. The result of the experiment was better compared to the existing spatial domain techniques. In 2011, Fang *et al.* [24] suggested a new technique using image fusion. A fusion explained as changing the properties of one entity according to the properties of another entity. Fusion plays a very important role in evaluating the information from more than one image by fusing it, especially in the satellite image processing and medical image processing where analysing the information is crucial. For example, when the doctor does not understand the condition of a patient by using one medical image, then they fuse the two images to get the accurate information. In this work, different fusion techniques along with different enhancement techniques are compared based on standard deviation, contrast level, and average gradient. In 2012, the above finding supported the study by Saleem *et al.* [25], which reported that contrast enhancement based on image fusion. In this technique, a multi-channel property was determined using the Laplacian pyramid. A simple model was developed to represent saturation, contrast variation, and image brightness. The main advantage of this approach is, it is effective to enhance local and global contrast image without affecting the colour balance. Based on the result performance, this method is successfully compared to HE and

CLAHE. Figure 1 illustrates the comparison result between a few methods. However, this method is not suitable and less efficient for a real-time system

A number of the study examined the relationship between global and local effect. The global and local method has their own advantages and drawbacks. In order to remain the advantages and at the same time minimize the drawbacks, a method based on a combination of the global and the local technique was proposed by Garvert and Gollisch [26] and Celik [27]. The objectives of both studies are to enhance the contrast and improved the quality images without any distortions and noise. Garvert and Gollisch [26] proposed a method based on contrast adaptation, while a method presented by Celik [27] is known as Spatial Entropy-based Contrast Enhancement in DCT (SECEDCT). Another algorithm for illumination correction was proposed by Glaister *et al.* [28] which calculated the underlying illumination through Multi-stage Illumination Modelling. This paper was divided into three parts. The first part deals with estimating the illumination map using non-parametric, second, by determining the quadratic surface reflectance through parametric modelling, and finally, calculating the final illumination estimation and used it to remove illumination variation.

Figure 1. Comparison of classical enhancement algorithm: (a) the original image, (b) CLAHE, (c) imadjust, (d) HE, and (e) Saleem method [25].

As highlighted by Wu *et al.* [29], a novel algorithm using tone mapping was presented. This technique is based on brightness adjustment to improve the contrast image without increasing the noise. The advantages of this approach are, it can reduce the computational complexity and easily implemented in the hardware. In 2012, the findings by Hasikin and Mat Isa [30] contradicted the study by Wu *et al.* [29], where they found that tone mapping correction of low contrast images was not successful in producing better enhanced images. According to an investigation by Hasikin and Mat Isa [30] and Zhou *et al.* [31], they set local and global adaptive information as the main parameters in order to enhance the low brightness and low contrast in the non-uniform images. Hasikin and Mat Isa [30] studied the effect of low contrast images and they proposed a new effective enhancement method based on fuzzy set theory. The proposed work used the fuzzy set theory by maximizing the fuzzy measurements in the image. The fuzzy set offers low complexity and high performance while enhancing the image even in the bad contrast conditions. The finding is consistent with findings by Zhou *et al.* [31], which focuses on local and global details. They proposed a novel method namely Overall Brightness and Local Contrast Adaptive Enhancement (OBLCAE) and experimented with colour images. Unfortunately, the proposed algorithm by Zhou *et al.* [31] is not suitable for very low illumination image enhancement because of inherent defects caused by the brightness non-linear mapping function. Besides, the finding by Hasikin and Mat Isa [30] supported by the finding Vlachos and Dermatas [32] investigated a method based on Fuzzy C-Means (FCM). However, this technique focuses on the modification of FCM algorithm that able to reduce the non-uniform illumination effects significantly and does not introduce brightness variations if the background is uniform. However, the weakness of this technique is the random selection of the prototypes of classes in the first step of the modified FCM algorithm.

Recently, image processing and computer vision approach are growing progressively in all areas of medical imaging, especially in modern ophthalmology. The high quality of retinal images is important for precise detection and diagnosis either manually or automatically. The aim of the pre-processing stage is to improve the diagnostic in fundus images for visual assessment and also for the computer-aided segmentation. Based on the previous study, many researchers agreed that retrospective correction techniques were effective to enhance the illumination in medical images, especially, retinal image [33], [34]. This view is supported by H. Lee and Kim [35] who agreed that retrospective correction is a better way in order to improve image quality. They suggested a new approach for bi-level images based on penalized non-linear least squares to measure the roughness of illumination. Major advantages of this approach are, it does not require tuning of design parameters, not suffering from a trivial minimizer, and effective optimization.

In 2010, Kubecka *et al.* [33] presented a B-spline shading model for non-uniform illumination modification following Shannon's entropy. It applied two models which were a parametric local bias model for the formulation of criterion function and its derivatives and multiplicative model of non-uniform illumination. The example of the resulting images is shown in Figure 2.

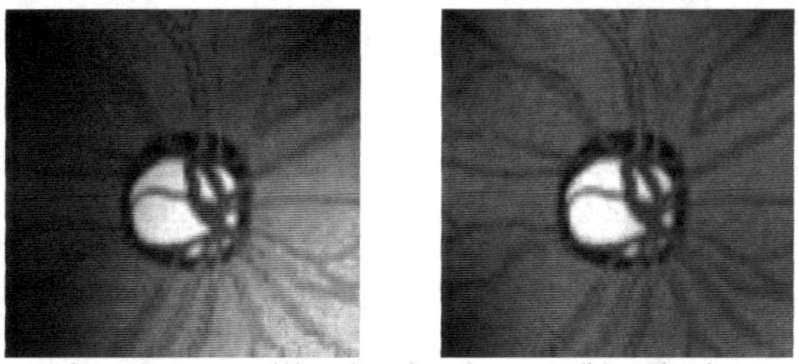

Figure 2. (a) Original non-homogenous illumination image. (b): The image after processing by the presented algorithm [33].

However, based on the same application, Zheng *et al.* [34] proposed a method based on Gradient Distribution Sparsity to solve the bias field problem in retinal fundus images. The illumination inhomogeneity was automatically calculated using Parametric Model of Bias Field and Gradient Distribution Sparsity. The experiment result showed that this technique was efficient to eliminate the bias effect and automatically improved the blood vessel segmentation. In the same years, an easy technique for illumination normalization was proposed by Leahy *et al.* [36]. The approach was developed by obtaining the illumination profile using multiplicative image formation models and Laplace interpolation. The proposed method was applied to the retinal image. The latest research by Rasta *et al.* [37] produced comprehensive reviews on contrast and illumination enhancement on retinal image. The main objective of all techniques is to enhance the contrast and improve the segmentation result for retinal vasculature. This review involved eight pre-processing techniques in order to correct uneven illumination and low contrast which appeared in the retinal images. Based on the analyses, the quotient based and homomorphic filtering was an effective technique to reduce the variation of illumination. However, this technique eliminated some features such as the macula.

3. Conclusion

One major area of digital image processing is image enhancement. The main objective of image enhancement tis to improving the quality of images emphasize wanted features and make them less obscured. In this paper, we present an overview of the background and related work in the area of image enhancement that focusses on Fusion method. The most obvious finding to emerge from this study is that Fusion method technique is popular because it is easy to implement and fast processing. In addition, many types of Fusion method are based on the global technique. However, these global processing techniques are found to be insufficient to overcome variations due to illumination changes. Therefore, the fusion method is effective compared to the histogram equalization, however, a few methods show the complicated algorithm and slow process. This study was done to find the gaps in the existing research and possible solutions to overcome these gaps in the future

4. Acknowledgment

This work was supported by Short Term Grant Scheme (Grant No. 9001-00576) from University Malaysia Perlis, Malaysia

5. References

[1] W. A. Mustafa and H. Yazid, "Illumination and Contrast Correction Strategy using Bilateral Filtering and Binarization Comparison," *J. Telecommun. Electron. Comput. Eng.*, vol. 8, no. 1, pp. 67–73, 2016.

[2] W. A. Mustafa and H. Yazid, "Background Correction using Average Filtering and Gradient Based Thresholding," *J. Telecommun. Electron. Comput. Eng.*, vol. 8, no. 5, pp. 81–88, 2016.

[3] M. Kaur, J. Kaur, and J. Kaur, "Survey of Contrast Enhancement Techniques based on Histogram Equalization," *Int. J. Adv. Comput. Sci. Appl.*, vol. 2, no. 7, pp. 1–5, 2011.

[4] B. Pamornnak, S. Limsiroratana, and M. Chongcheawchamnan, "Color correction scheme for different illumination and camera device conditions," in *IEEE TENCON Spring Conference*, 2013, pp. 430–434.

[5] X. Zhang, F. Ding, Z. Tang, and C. Yu, "Salt and pepper noise removal with image inpainting," *AEU - Int. J. Electron. Commun.*, vol. 69, no. 1, pp. 307–313, Jan. 2015.

[6] S. Jenifer, S. Parasuraman, and A. Kadirvelu, "Contrast enhancement and brightness preserving of digital mammograms using fuzzy clipped contrast-limited adaptive histogram equalization algorithm," *Appl. Soft Comput. J.*, vol. 42, pp. 167–177, 2016.

[7] W. A. Mustafa, H. Yazid, and S. Yaacob, "Illumination Normalization of Non-Uniform Images Based on Double Mean Filtering," in *IEEE International Conference on Control Systems, Computing and Engineering*, 2014, pp. 366–371.

[8] W. A. Mustafa, H. Yazid, and S. Yaacob, "Illumination Correction of Retinal Images Using Superimpose Low Pass and Gaussian Filtering," in *International Conference on Biomedical Engineering (ICoBE)*, 2015, pp. 1–4.

[9] A. Konieczka, J. Balcerek, A. Chmielewska, and D. Adam, "Approach to local contrast enhancement," in *Signal processing, algorithms and application*, 2015, pp. 16–19.

[10] D. J. Chung, K. Sung, F. C. Osuagwu, H. H. Wu, C. Lassman, and D. S. K. Lu, "Contrast enhancement patterns after irreversible electroporation: Experimental study of CT perfusion correlated to histopathology in normal porcine liver," *J. Vasc. Interv. Radiol.*, vol. 27, no. 1, pp. 104–111, 2016.

[11] M. Kahani, A. Kamali-Asl, H. Ghadiri, and S. Hashemi, "A method for material decomposition in dual-energy contrast enhancement digital mammography," *Meas. J. Int. Meas. Confed.*, vol. 88, pp. 87–95, 2016.

[12] X. xi Meng, Y. qiu Zhang, H. qiang Liao, H. chao Liu, H. lin Jiang, S. jun Ke, and W. hua Dong, "Dynamic contrast-enhanced MRI for the assessment of spinal tumor vascularity: correlation with angiography," *European Spine Journal*, pp. 1–10, 2016.

[13] S. S. Al-amri, N. V Kalyankar, and S. D. Khamitkar, "Linear and Non-linear Contrast Enhancement Image," *J. Comput. Sci.*, vol. 10, no. 2, pp. 139–143, 2010.

[14] N. Vu and A. Caplier, "Illumination-Robust Face Recognition Using Retina Modeling," in *International Conference Image Processing*, 2009, pp. 3289–3292.

[15] Y. Cheng, Z. Jin, and C. Hao, "Illumination Normalization based on 2D Gaussian illumination model," in *International Conference on Advanced Computer Theory and Engineering (ICACTE) Illumination*, 2010, pp. 451–455.

[16] Y. Cheng, Z. Jin, and C. Hao, "Illumination normalization based on different smoothing filters quotient image," in *International Conference on Intelligent Networks and Intelligent Systems*, 2010, pp. 28–31.

[17] G. An, J. Wu, and Q. Ruan, "An illumination normalization model for face recognition under varied lighting conditions," *Pattern Recognit. Lett.*, vol. 31, no. 9, pp. 1056–1067, 2010.

[18] V. Ambule, M. Ghute, K. Kamble, and S. Katre, "Adaptive Median Filter for Image Enhancement," *Int. J. Eng. Scicen Innov. Technol.*, vol. 2, no. 1, pp. 318–323, 2013.

[19] B. Wang, W. Li, W. Yang, and Q. Liao, "Illumination normalization based on weber's law with application to face recognition," *IEEE Signal Process. Lett.*, vol. 18, pp. 462–465, 2011.

[20] S. C. F. Lin, C. Y. Wonga, M. A. Rahman, G. Jiang, S. Liu, N. Kwoka, H. Shi, Y.-H. Yuc, and T. Wu, "Image enhancement using the averaging histogram equalization (AVHEQ) approach for contrast improvement and brightness preservation," vol. 46, pp. 356–370, 2015.

[21] S. Singh and S. Sharma, "A Survey of Image Enhancement Techniques," *Int. J. Comput. Sci.*, vol. 2, no. 5, pp. 1–5, 2014.

[22] A. R. Rivera, B. Ryu, and O. Chae, "Content-aware dark image enhancement through channel division," *IEEE Trans. Image Process.*, vol. 21, no. 9, pp. 3967–3980, 2012.

[23] R. K. Jha, R. Chouhan, and P. K. Biswas, "Contrast enhancement of dark images using stochastic resonance," *IET Image Process.*, vol. 6, no. September 2010, pp. 230–237, 2012.

[24] X. Fang, J. Liu, W. Gu, and Y. Tang, "A method to improve the image enhancement result based on image fusion," *2011 Int. Conf. Multimed. Technol.*, pp. 55–58, 2011.

[25] A. Saleem, A. Beghdadi, and B. Boashash, "Image fusion-based contrast enhancement," *EURASIP J. Image Video Process.*, vol. 10, no. 1, pp. 1–17, 2012.

[26] M. M. Garvert and T. Gollisch, "Local and global contrast adaptation in retinal ganglion cells," *Neuron*, vol. 77, no. 5, pp. 915–928, 2013.

[27] T. Celik, "Spatial Entropy-Based Global and Local," *IEEE Trans. Image Process.*, vol. 23, no. 12, pp. 5298–5308, 2014.

[28] J. Glaister, A. Wong, and D. A. Clausi, "Illumination correction in dermatological photographs using multi-stage illumination modeling for skin lesion analysis," in *Annual International Conference of the IEEE Engineering in Medicine and Biology Society (EMBC)*, 2012, pp. 102–105.

[29] Y. Wu, Z. Liu, Y. Han, and H. Zhang, "An image illumination correction algorithm based on tone mapping," *Int. Congr. Image Signal Process.*, pp. 645–648, Oct. 2010.

[30] K. Hasikin and N. A. Mat Isa, "Adaptive fuzzy contrast factor enhancement technique for low contrast and nonuniform illumination images," *Signal, Image Video Process.*, vol. 8, no. 8, pp. 1591–1603, 2012.

[31] Z. Zhou, N. Sang, and X. Hu, "Global brightness and local contrast adaptive enhancement for low illumination color image," *Optik (Stuttg).*, vol. 125, no. 6, pp. 1795–1799, 2014.

[32] M. D. Vlachos and E. S. Dermatas, "Non-uniform illumination correction in infrared images based on a modified fuzzy c-means algorithm," *Int. J. Emerg. Technol. Adv. Eng.*, vol. 3, no. 1, 2013.

[33] L. Kubecka, J. Jan, and R. Kolar, "Retrospective illumination correction of retinal images," *Int. J. Biomed. Imaging*, vol. 2010, no. 1, pp. 1–10, 2010.

[34] Y. Zheng, B. Vanderbeek, R. Xiao, E. Daniel, D. Stambolian, M. Maguire, J. O'Brien, and J. Gee, "Retrospective illumination correction of retinal fundus images from gradient distribution sparsity," *Biomedical Imaging (ISBI), 2012 9th IEEE International Symposium on*. IEEE, pp. 972–975, 2012.

[35] H. Lee and J. Kim, "Retrospective correction of nonuniform illumination on bi-level images," vol. 17, no. 26. Optical Society of America, pp. 23880–23893, 2009.

[36] C. Leahy, A. O'Brien, and C. Dainty, "Illumination correction of retinal images using Laplace interpolation.," *Appl. Opt.*, vol. 51, no. 35, pp. 8383–9, Dec. 2012.

[37] S. H. Rasta, M. E. Partovi, and A. Javadzadeh, "A comparative study on preprocessing techniques in diabetic retinopathy retinal images: illumination correction and contrast enhancement," *J. Med. Signals Sens.*, vol. 5, no. 1, pp. 40–48, 2015.

A Review of Histogram Equalization Techniques in Image Enhancement Application

Wan Azani Mustafa[1,*], Mohamed Mydin M. Abdul Kader[2]
[1]Faculty of Engineering Technology, Universiti Malaysia Perlis, UniCITI Alam Campus, Sungai Chuchuh, 02100 Padang Besar, Perlis, Malaysia

[1] *wanazani@unimap.edu.my,* [2] *mohamedm@unimap.edu.my*

Abstract. Image enhancement can be considered as one of the fundamental processes in image analysis. The goal of contrast enhancement is to improve the quality of an image to become more suitable for a particular application. Till today, numerous image enhancement methods have been proposed for various applications and efforts have been directed to further increase the quality of the enhancement results and minimize the computational complexity and memory usage. In this paper, an image enhancement methods based on Histogram Equalization (HE) was studied. This paper presents an exhaustive review of these studies and suggests a direction for future developments of image enhancement methods. Each method shows the owned advantages and drawbacks. In future, this work will give the direction to other researchers in order to propose new advanced enhancement techniques.

1. Introduction

The image enhancement technique is to make the digital picture more appealing to our eyes, for example, making the images smooth or sharp. This is an important topic in digital image processing. It can help humans and computer vision algorithms obtaining accurate information from the enhanced images. The visual quality and certain image properties, such as brightness, contrast, signal-to-noise ratio, resolution, edge sharpness, and color accuracy were improved through the enhancement process [1], [2]. Recently, many image enhancement methods have been developed based on various digital image processing techniques and applications. They can be developed in the spatial domain or spatial-frequency domain. The enhanced image will provide useful information for post-processing, especially in segmentation stage. This paper is organized in the following sections: Section 2 describes the related work of the studied using HE and Section 3 gives the conclusion of the work.

2. Histogram Equalization

A large and growing body of literature has investigated about enhancement methods. Image enhancement deals with improving the quality of images, where the objective is to emphasize wanted features and make them less obscured [3], [4]. The area of image enhancement is very appealing, where many fundamental image enhancement techniques are developed based on a very simple concept [5]–[8]. In the recent decades, image enhancement based on contrast enhancement has been focused [9]–[14].

IOP Conf. Series: Journal of Physics: Conf. Series **1019** (2018) 012026 doi:10.1088/1742-6596/1019/1/012026

Many researchers argued that Histogram equalization (HE) is a simple and an easy method to enhance the contrast and improve the image quality [15]–[17]. Since 1997, Yeong Kim [18] raised several concerns about contrast problem and suggested Brightness preserving Bi-Histogram Equalization (BBHE) in order to enhance the contrast. The average intensity value was applied as a separating point to differentiate between a dark area and bright area. The above finding contradicts the study by Wang *et al.* [19]. The author presented that a median intensity value is more accurate as the separating point compared to the average intensity. These results were contradicted that suggested the minimum mean brightness between original and output image as the separating point is more specific and accurate compared to the BBHE and Dualistic Sub Image Histogram Equalization (DSIHE) [20]. Research conducted by Ooi and Isa [21] proposed a new improvement in histogram equalization known as Quadrant Dynamic Histogram Equalization (QDHE). The first step in this technique was to divide the histogram into four sub-quadrant histograms based on the median value of the original image. After normalizing each sub-histogram, finally, the image was equalized. A major advantage of QDHE is that it's enhanced the image without any intensity saturation, noise amplification, and over-enhancement. In 2010, Ooi *et al.* [22] presented a new method based on Plateau level equation, namely Bi-Histogram Equalization with a Plateau Level (BHEPL). The main objective of this paper is to improve the BBHE technique in term of processing time. The process of this method also involved mean brightness preserving histogram equalization method with a clipped histogram equalization method. However, interestingly, this is contrary to a study conducted by Sengee *et al.* [23]. They suggest an extension method of BBHE based on the Neighbourhood Metric. This method involved a few steps: First, a large histogram was divided into the sub-region using Neighbourhood Metric. Second, based on mean, the histogram of the original image was separated into two sub-regions and process independently. The results enhanced the local contrast and preserved the brightness of the original image. The comparison of result performance is illustrated in Figure 1.

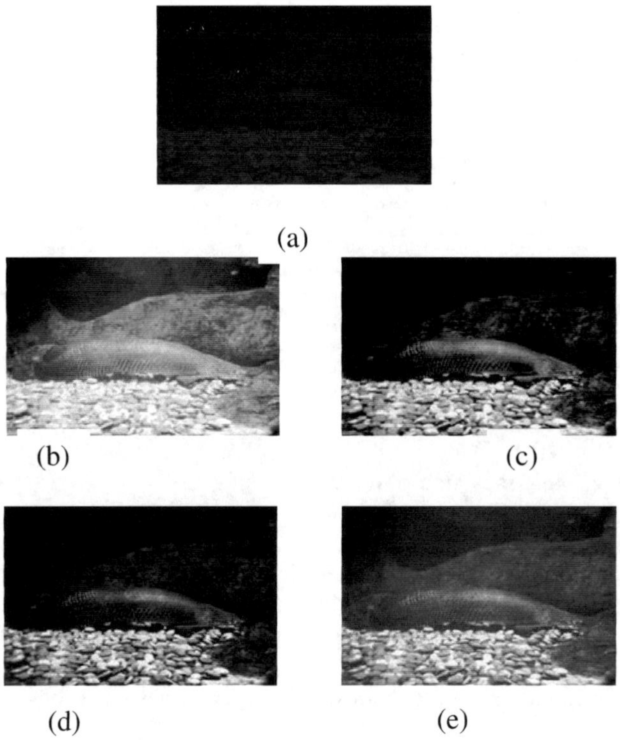

IOP Conf. Series: Journal of Physics: Conf. Series **1019** (2018) 012026 doi:10.1088/1742-6596/1019/1/012026

Figure 1. The comparison of enhancement result based on histogram modification: (a) Original image, (b) HE, (c) BBHE [23], (d) BHEPL [22], and (e) QDHE [21].

In a different study, Salah *et al.* [24] explained a new approach to solving the illumination problem on the face images using Histogram Equalization (HE). The technique is based on the combination of gamma correction and the Retinal filter's compression function namely GAMMA-HM-COMP. The Retinal filter is a new enhancement method, and the result was effective compared to the three conventional enhancement methods which are histogram equalization [1], gamma correction [25] and log transformation [1]. In another study, Tan *et al.* [26] proposed a Background Brightness Preserving Histogram Equalization (BBPHE) method based on non-linear Histogram equalization (HE). Based on the background and non-background level techniques, the original image was separated into three interval histograms: (1) low grey level, (2) medium grey level, and (3) high grey level. The objective of this method is to enhance the object contrast and maintaining the background brightness. Similarly, Moniruzzaman et al. [27] proposed a modification of Brightness Preserving Bi-Histogram Equalization (BPBHE) using the edge pixels data. In order to prove the effectiveness, the Average Mean Brightness Error (AMBE) was calculated and the result was presented in table 1. The lowest of AMBE shows the high quality image and good performance technique.

Table 1. The AMBE result on different histogram technique.

Image	HE	BBHE	BPBHE
A	24.3350	29.7404	3.4737
B	42.0428	35.4111	2.5079
C	53.9887	13.6073	13.6073
D	11.386	8.5763	3.3188

Hashemi *et al.* [28] proposed a novel enhancement method based on Genetic Algorithm using a simple chromosome structure and corresponding operator. The method was tested on the image that has a low dynamic range. In 2013, research finding by Chaudhary and Patil [29] also suggested a simple method based on Genetic Algorithm. The advantages of both methods are fast processing time, efficient, and produce a high-quality image. Besides that, they also produced a comprehensive comparison between a BBHE, DSIHE, MMBBHE, MPHE, and RMSHE. The analyses were done based on PSNR and contrast ratio.

In another study, Shome *et al.* [30] examined a method using Contrast Limited Adaptive Histogram Equalization (CLAHE) in order to normalize the contrast variation in the retinal image. CLAHE is an adaptive extension of Histogram Equalization followed by thresholding, which helps in the dynamic preservation of the local contrast features of an image. This proposed method used a non-mean based approach to improve the quality of the Diabetic Retinopathy (DR) image while preserving the sharpness and minutes of the details. The method also increased the local contrast pixels. The above view was supported studied by Sundaram *et al.* [31] where a CLAHE technique with a slight modification was suggested in order to enhance the mammogram images. The method was known Histogram Modified

Contrast Limited Adaptive Histogram Equalization (HM-CLAHE). The optimization technique was used to adjust the level of strong contrast and the local details for more relevant interpretation. Based on Enhancement Measure (EME) result, this method was better compared to the HE, Unsharp masking (USM), and CLAHE. Tan *et al.* [26] introduced a modification of HE based on non-linear technique. The aim is to enhance the image contrast, while preserving the background brightness for images with well-defined background brightness. The original image was divided into three sub-images by using the proposed algorithm. Finally, only the problem region was normalized. The correction process based on sub-images technique also was supported by Shanmugavadivu and Balasubramanian [32]. They proposed a new method called as Thresholded and Optimized Histogram Equalization (TOHE). The main process has divided the histogram using the Otsu thresholding. Based on the result performance, this approaches is successful compared to the HE, BBHE, Range-Limited Bi-Histogram Equalization (RLBHE). Figure 2 shows the comparison result between TOHE and a few histogram methods. Circle in figure 2 represent the improvement and drawback of each method.

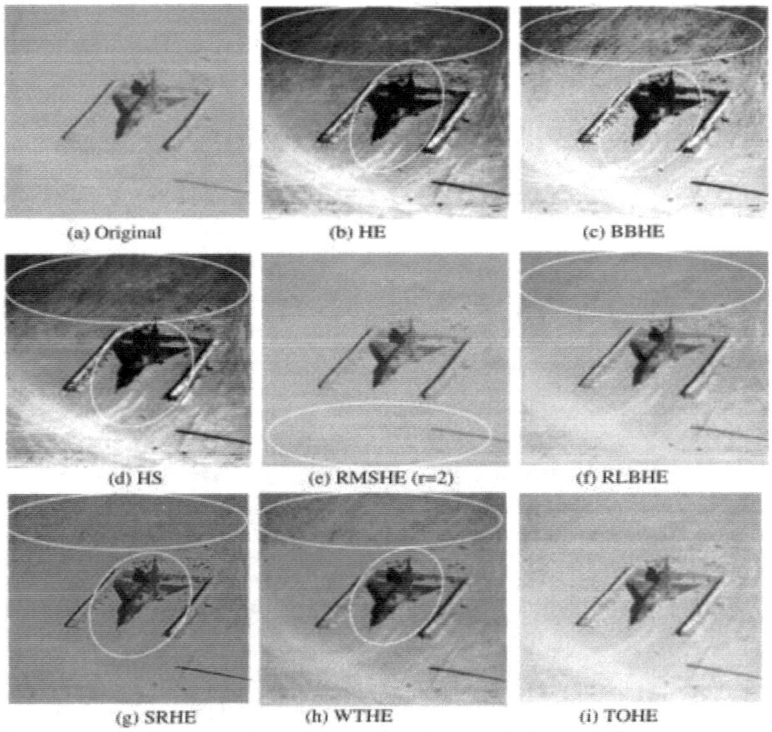

Figure 2. Comparison of the result performance after applying a different histogram process.

On the other side, Lin *et al.* [33] suggested a new HE method was applied to the colour image known as Averaging Histogram Equalization (AVHEQ). The technique also separates the original image into sub-images and equalized independently. A new mathematical algorithm in order to determine the optimal averaging threshold was proposed. The result is better compared to the conventional methods such as BBHE, DSIHE, and BHEPL.

Summary, the Histogram Equalization (HE) technique popular because is easy to implement and fast processing [34][23][35]. However, this technique produces many drawbacks such as it adds noise to the output image, increasing the contrast of its background and the signal gets distorted [15]. The HE may produce over enhancement result and saturation artefacts due to the stretching of the grey levels over the full grey level range. In addition, many types of HE are based on the global technique. However, these global processing techniques of image processing are found to be insufficient to overcome variations due to illumination changes [36]. In non-uniform illumination, variation is still difficult to deal with using these global processing techniques [37], [38].

3. Conclusion

One major area of digital image processing is image enhancement. The main objective of image enhancement is to improve the quality of images emphasize wanted features and make them less obscured. This study has discussed an overview of the background and related work in the area of image enhancement using HE. The HE technique is simple and easy to apply. Recently, many modifications on HE was presented in order to find the best normalization technique. In future, research on the mathematical algorithm in HE should be explored.

4. Acknowledgment

This work was supported by Short Term Grant Scheme (Grant No. 9001-00576) from University Malaysia Perlis, Malaysia

5. References

[1] R. C. Gonzalez and R. E. Woods, *Digital Image Processing*, Third. Upper Saddle River, NJ, USA: Prentice Hall, 2008.

[2] S. Sharma, *Fundamentals of Digital Image Processing*, 4th ed. New Delhi: S.K. Kataria & Sons, 2013.

[3] W. A. Mustafa and H. Yazid, "Illumination and Contrast Correction Strategy using Bilateral Filtering and Binarization Comparison," *J. Telecommun. Electron. Comput. Eng.*, vol. 8, no. 1, pp. 67–73, 2016.

[4] W. A. Mustafa and H. Yazid, "Background Correction using Average Filtering and Gradient Based Thresholding," *J. Telecommun. Electron. Comput. Eng.*, vol. 8, no. 5, pp. 81–88, 2016.

[5] M. Kaur, J. Kaur, and J. Kaur, "Survey of Contrast Enhancement Techniques based on Histogram Equalization," *Int. J. Adv. Comput. Sci. Appl.*, vol. 2, no. 7, pp. 1–5, 2011.

[6] B. Pamornnak, S. Limsiroratana, and M. Chongcheawchamnan, "Color correction scheme for different illumination and camera device conditions," in *IEEE TENCON Spring Conference*, 2013, pp. 430–434.

[7] X. Zhang, F. Ding, Z. Tang, and C. Yu, "Salt and pepper noise removal with image inpainting," *AEU - Int. J. Electron. Commun.*, vol. 69, no. 1, pp. 307–313, Jan. 2015.

[8] S. Jenifer, S. Parasuraman, and A. Kadirvelu, "Contrast enhancement and brightness preserving of digital mammograms using fuzzy clipped contrast-limited adaptive histogram equalization algorithm," *Appl. Soft Comput. J.*, vol. 42, pp. 167–177, 2016.

[9] W. A. Mustafa, H. Yazid, and S. Yaacob, "Illumination Normalization of Non-Uniform Images Based on Double Mean Filtering," in *IEEE International Conference on Control Systems, Computing and Engineering*, 2014, pp. 366–371.

[10] W. A. Mustafa, H. Yazid, and S. Yaacob, "Illumination Correction of Retinal Images Using Superimpose Low Pass and Gaussian Filtering," in *International Conference on Biomedical Engineering (ICoBE)*, 2015, pp. 1–4.

[11] A. Konieczka, J. Balcerek, A. Chmielewska, and D. Adam, "Approach to local contrast enhancement," in *Signal processing, algorithms and application*, 2015, pp. 16–19.

[12] D. J. Chung, K. Sung, F. C. Osuagwu, H. H. Wu, C. Lassman, and D. S. K. Lu, "Contrast enhancement patterns after irreversible electroporation: Experimental study of CT perfusion correlated to histopathology in normal porcine liver," *J. Vasc. Interv. Radiol.*, vol. 27, no. 1, pp. 104–111, 2016.

[13] M. Kahani, A. Kamali-Asl, H. Ghadiri, and S. Hashemi, "A method for material decomposition in dual-energy contrast enhancement digital mammography," *Meas. J. Int. Meas. Confed.*, vol. 88, pp. 87–95, 2016.

[14] X. xi Meng, Y. qiu Zhang, H. qiang Liao, H. chao Liu, H. lin Jiang, S. jun Ke, and W. hua Dong, "Dynamic contrast-enhanced MRI for the assessment of spinal tumor vascularity: correlation with angiography," *European Spine Journal*, pp. 1–10, 2016.

[15] N. Longkumer, M. Kumar, and R. Saxena, "Contrast Enhancement Techniques using Histogram Equalization : A Survey," *Int. J. Curr. Eng. Technol.*, vol. 4, no. 3, pp. 1561–1565, 2014.

[16] N. S. P. Kong, H. Ibrahim, and S. C. Hoo, "A Literature Review on Histogram Equalization and Its Variations for Digital Image Enhancement," *Int. J. Innov. Manag. Technol.*, vol. 4, no. 4, pp. 386–389, 2013.

[17] R. Jaiswal, A. G. Rao, and H. P. Shukla, "Image Enhancement Techniques Based on Histogram Equalization," *Int. J. Adv. Electr. Electron. Eng.*, vol. 1, no. 2, pp. 69–78, 2010.

[18] Yeong-Taeg Kim, "Contrast enhancement using brightness preserving bi-histogram equalization," *IEEE Trans. Consum. Electron.*, vol. 43, no. 1, pp. 1–8, 1997.

[19] Y. Wang, Q. Chen, and B. Zhang, "Image enhancement based on equal area dualistic sub-image histogram equalization method," *IEEE Trans. Consum. Electron.*, vol. 45, no. 1, pp. 68–75, 1999.

[20] S. D. Chen and A. R. Ramli, "Minimum mean brightness error bi-histogram equalization in contrast enhancement," *IEEE Trans. Consum. Electron.*, vol. 49, no. 4, pp. 1310–1319, 2003.

[21] C. H. Ooi and N. A. M. Isa, "Quadrants dynamic histogram equalization for contrast enhancement," *IEEE Trans. Consum. Electron.*, vol. 56, no. 4, pp. 2552–2559, 2010.

[22] C. H. Ooi, N. P. Kong, and H. Ibrahim, "Bi-histogram equalization with a plateau limit for digital image enhancement," *IEEE Trans. Consum. Electron.*, vol. 55, no. 4, pp. 2072–2080, 2010.

[23] N. Sengee, A. Sengee, and H. K. Choi, "Image contrast enhancement using bi-histogram equalization with neighborhood metrics," *IEEE Trans. Consum. Electron.*, vol. 56, no. 4, pp. 2727–2734, 2010.

[24] A. Salah-eldin, K. Nagaty, and T. Elarif, "An Enhanced Histogram Matching Approach Using the Retinal Filter ' s Compression Function for Illumination Normalization in Face Recognition," in *International Conference on Image Analysis and Recognition*, 2008, pp. 873–883.

[25] S. Shan, W. Gao, B. Cao, and D. Zhao, "Illumination Normalization for Robust Face Recognition Against Varying Lighting Conditions," in *IEEE International Workshop on Analysis and Modeling of Faces and Gestures (AMFG'03)*, 2003, pp. 157–164.

[26] T. L. Tan, K. S. Sim, and C. P. Tso, "Image enhancement using background brightness preserving histogram equalisation," *Electron. Lett.*, vol. 48, no. 3, p. 155, 2012.

[27] M. Moniruzzaman, M. Shafuzzaman, and M. F. Hossain, "Brightness Preserving Bi-Histogram Equalization Using Edge Pixels Information," in *International Conference on Electrical Information and Communication Technology (EICT)*, 2013.

[28] S. Hashemi, S. Kiani, N. Noroozi, and M. E. Moghaddam, "An image contrast enhancement method based on genetic algorithm," *Pattern Recognit. Lett.*, vol. 31, no. 13, pp. 1816–1824, 2010.

[29] C. Chaudhary and M. K. Patil, "Review of Image Enhancement Techniques Using Histogram," *Int. J. Appl. or Innov. Eng. Manag.*, vol. 2, no. 5, pp. 343–349, 2013.

[30] S. K. Shome, S. Ram, and K. Vadali, "Enhancement of Diabetic Retinopathy Imagery Using Contrast Limited Adaptive Histogram Equalization," *Int. J. Comput. Sci. Inf. Technol.*, vol. 2, no. 6, pp. 2694–2699, 2011.

[31] M. Sundaram, K. Ramar, N. Arumugam, and G. Prabin, "Histogram modified local contrast enhancement for mammogram images," *Appl. Soft Comput. J.*, vol. 11, no. 8, pp. 5809–5816, 2011.

[32] P. Shanmugavadivu and K. Balasubramanian, "Thresholded and Optimized Histogram Equalization for contrast enhancement of images q," *Comput. Electr. Eng.*, vol. 40, no. 3, pp. 757–768, 2014.

[33] S. C. F. Lin, C. Y. Wonga, M. A. Rahman, G. Jiang, S. Liu, N. Kwoka, H. Shi, Y.-H. Yuc, and T. Wu, "Image enhancement using the averaging histogram equalization (AVHEQ) approach for contrast improvement and brightness preservation," vol. 46, pp. 356–370, 2015.

[34] R. Garg, B. Mittal, and S. Garg, "Histogram Equalization Techniques For Image Enhancement," *Int. J. Electron. Commun. Technol.*, vol. 2, no. 1, pp. 107–111, 2011.

[35] K. Singh and R. Kapoor, "Image enhancement using Exposure based Sub Image Histogram Equalization," *Pattern Recognit. Lett.*, vol. 36, no. 1, pp. 10–14, 2014.

[36] V. P. Vishwakarma, S. Pandey, and M. N. Gupta, "Adaptive histogram equalization and logarithm transform with rescaled low frequency DCT coefficients for illumination normalization," *Int. J. Recent Trends Eng. Technol.*, vol. 1, no. 1, pp. 318–322, 2009.

[37] Y. Li, X. Liu, and Z. Gao, "Shadow determination and compensation for face recognition," *Int. J. Mach. Learn. Cybern.*, vol. 5, no. 4, pp. 599–605, 2014.

[38] L. N. V. Carneiro and G. Cámara-Chávez, "Logarithm Discrete Cosine Transform Domain and Discrimination Power Analysis for Illumination Invariant Face Recognition," in *Proceedings of the International Conference on Image Processing, Computer Vision, and Pattern Recognition (IPCV)*, 2012, pp. 1–7.

PSO-Based PI Controller for Speed Sensorless Control of PMSM

A.M. Nazelan, M. K. Osman, A.A.A. Samat and N. A. Salim
Faculty of Electrical, Universiti Teknologi Mara
13500 Permatang Pauh, Pulau Pinang, Malaysia
E-mail: amnazelan@gmail.com

Abstract. In the controlling three phase AC motor especially permanent magnet synchronous motor (PMSM), good controller is required to achieved the desired target of speed. This paper presents a comparison among different technique of speed and position estimation by using Proportional and Integral controller (PI) tune by using Heuristic, MATLAB PID auto tuner and Particle Swarm Optimization (PSO) method of sensorless control for PMSM. The reason of this study is to identify which method can produce better result. Simulation that have been conducted by using PSO optimization technique h produced better result than the conventional controller in terms of system parameter such as rise time (Tr), settling time (Ts), percent overshoot (%Os), percent undershoot (%Us) and root mean square error (RMSE) of PMSM.

Keywords: Permanent Magnet Synchronous Motor, PI controller, Particle Swarm Optimization, Artificial Neural Network, Model Reference Adaptive Control.

1. Introduction

Permanent Magnet Synchronous Motor (PMSM) has been widely used for various industries of technologies in robotics, electrical and automation. PMSM are well known because of its high power density, large torque to inertia ratio, high efficiency and simple structure. It consists of two primary parts which is permanent magnet that rotates called rotor and static core surrounded by three equally space winding called stator. There have two type of rotor. First is surface mounted permanent magnet where the magnets are attached at the surface of the rotor and second is interior mounted permanent magnet where the magnets are attached inside the rotor. Permanent magnets on the motor generate rotor flux. So the electrical rotor position is same with the rotor flux position [1]. For controlling the PMSM, it is necessary to have the information of speed and position of the rotor and normally encoder has been used to identify it [2]. Since the mechanic sensor is too expensive, increase the cost and decrease the stability, sensorless methods are implemented. It can save some space because installation of encoder requires bigger space [3].

Nowadays, sensorless technique for Permanent Magnet Synchronous Motor (PMSM) has various types. Generally, sensorless techniques have two methods which is back electromagnetic force (EMF) estimation and signal injection method [4], [5]. Back EMF method offer a good performance at high speed revolution per minute (RPM) for PMSM. But, at low speed and zero speed, the estimation accuracy will be decrease because the parameter reading is too small compare to the noise [6]. For signal injection method, their offer good performances at all state include zero speed. But, disadvantage of this method is requiring extra hardware and delay problem due to digital implementation of the digital filters and control algorithm [7].

The demand for variable speed drive in both low and high power applicants has resulted in great variety of products from different researcher, each offering a great variety of features [8]. A variety of techniques and strategies have been developed to overcome all this problem and weakness such as Sliding Mode Observer (SMO), Extended Kalman Filter (EKF), Artificial Neural Network (ANN) and Model Referenced Adaptive System (MRAS) [2], [9]–[11].

In this paper focus is on comparing the MRAS algorithm with same adaptive mechanism but different tuning method in order to identify which method can provide better result for speed and position tracking. Conventional method that have been propose using Proportional and Integral controller (PI)[6]. But, it was reported that low speed region still hunger with accurate controller. Therefore, this research is intend to improve the existing controller in term of decreasing the time to reach reference speed of PMSM with another type of tuning method. Particle Swarm Optimization (PSO) method was being prove the effectiveness for best parameter tuning in many application before.

Model reference adaptive system (MRAS) consist of two functional block which is reference model and adjustable model [12]–[15]. Main idea of this system is comparing the reference parameter (current i_d & i_q for this case) between actual parameter. Error of this signal will be use to estimate the speed and position of the PMSM. Then, the signal is fed into the adaptation algorithm to estimate the quantity of which is use to tune the adjustable model. This method is simple and required less computation [6].

I. PMSM Model

In this section, PMSM mathematical model are describe to understand the behavior. The PMSM can be modeled in stationary reference (α, ß) and rotor reference (d, q) as shown in Fig. 1.

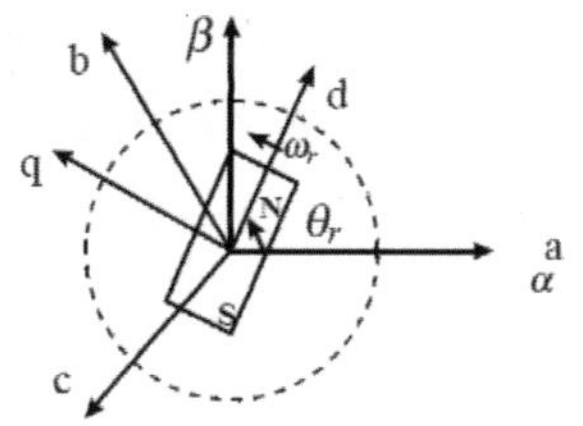

Figure 1: The stationary and synchronous frame [2]

The PMSM equivalent circuit as shown in fig. 2 are considered in order to determine the mathematical model of the PMSM.

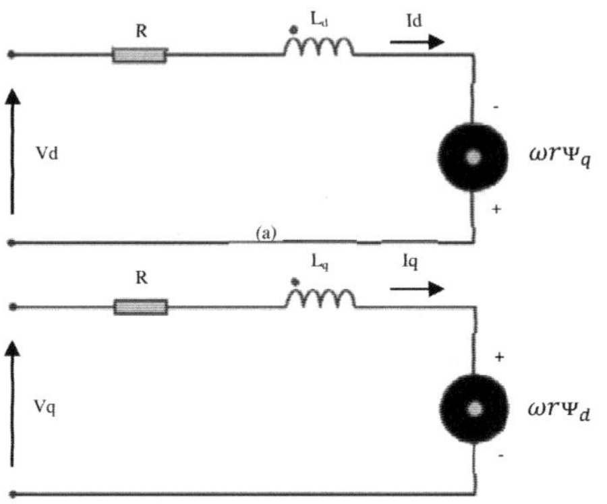

Figure 2. The PMSM equivalent circuit

Fig. 2 (a) and (b) is the equivalent circuit for voltage at d-axis and q-axis. From fig.2 the voltage equation can be derived as:

$$V_d = R(i_d) + L_d \frac{di_d}{dt} - \omega_r \psi_q \quad (1)$$

$$V_q = R(i_q) + L_q \frac{di_q}{dt} - \omega_r \psi_d \quad (2)$$

where R is the phase resistor winding, ωr is the electrical angular velocity of the rotor and ψ_d, ψ_q is flux for d and q. Flux equation be derived as:

$$\psi_d = L_d(i_d) + \psi_m \quad (3)$$

$$\psi_q = L_q(i_q) \quad (4)$$

In the equation of flux above, ψ_m is the flux that generated by magnetic pole and stator.

2. POSITION AND ESTIMATION METHOD

A. Model Reference Adaptive System

Model reference adaptive control system (MRAS) scheme was first proposed by Professor Witark of the United States in 1958 [16]. The general view and idea MRAS is a closed loop system that contains reference model and adjustable model [12], [15]. This system will compare the output of plant (PMSM) with the desired response from reference model. The parameters from the plant will update into the system and then, the system will determine the error between reference models. If the parameters have an error, adjustable model will automatically tune and counter the error to match with reference model. The reference model is independent from the rotor speed, calculation of variable from the terminal voltage and current. Meanwhile, for the adaptive model is dependent on it.

The difference between these state variables is then used to drive an adaptation mechanism which generates the estimated speed ($\hat{\omega}$) [17]. Generalized of this system is shown in Fig. 2. Input for reference and adjustable model is same which is voltage in stationary form (*ud & uq*) that project from three phase (*ua, ub & uc*).

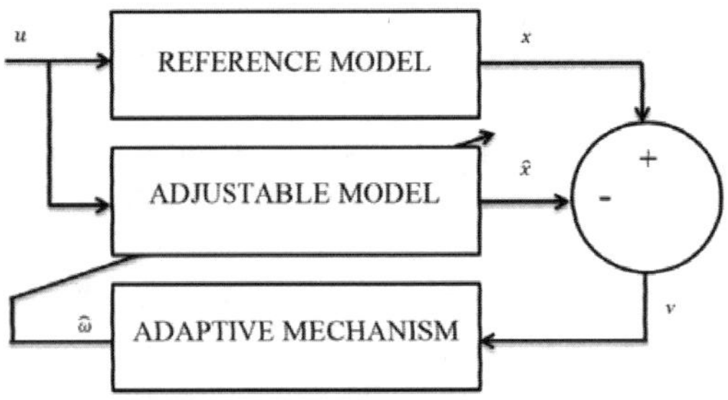

Figure 3. Block diagram of MRAS control scheme

In fig. 2, x and $\hat{x}$ are the state vector of the reference model and adjustable model, respectively. Reference model is develop in ideal condition where output of the model x will be used to compare with the measured performance of $\hat{x}$. The difference vector of v will be used as input for the adaptive mechanism to modify the parameters of the adjustable model.

B. Model Reference Adaptive System on PI Controller

Mainly, MRAS sensorless control is based on model of the motor at the rotating reference plane. In the rotor rotating reference frame, the model of the PMSM stator current describe as follows:

$$\frac{d}{dt}[i_d] = -\frac{R}{L_d} i_d + \frac{\omega_r L_q}{L_d} i_q + \frac{V_d}{L_d} \tag{5}$$

$$\frac{d}{dt}[i_q] = -\frac{\omega_r L_q}{L_d} i_d - \frac{R}{L_d} i_q + \frac{V_q}{L_q} - \frac{\omega_r \psi_r}{L_d} i_d \tag{6}$$

where i_d, i_q is dq-axis stator current, u_d, u_q is dq-axis stator voltage, R is stator resistance, L_d, L_q is stator inductance and ψr is rotor permanent magnet flux. By estimating dq-axis current, estimated speed can be formulate as follows [16]:

$$\hat{\omega} = \int_0^t k_1 [i_d \hat{i}_q - i_q \hat{i}_d - \frac{\psi f}{L}(i_q - \hat{i}_q)] d\tau$$

$$+ k_2 [i_d \hat{i}_q - i_q \hat{i}_d - \frac{\psi f}{L}(i_q - \hat{i}_q)] + \hat{\omega}(0) \tag{7}$$

where $\hat{\omega}$ is estimated rotor angular speed, $\widehat{\imath d}, \widehat{\imath q}$ is dq-axis stator estimated current, k1, k2 is PI regulator coefficients. The equation (2) and (3) will transform to matrix equation:

$$\frac{d}{dt}\begin{bmatrix} i_d + \frac{\psi_m}{L_d} \\ i_q \end{bmatrix} = \begin{bmatrix} \frac{-R}{L_d} & \frac{\omega_r L_q}{L_d} \\ \frac{-\omega_r L_q}{L_d} & \frac{-R}{L_d} \end{bmatrix} \begin{bmatrix} i_d + \frac{\psi_m}{L_d} \\ i_q \end{bmatrix}$$

$$\begin{bmatrix} \frac{1}{L_d} & 0 \\ 0 & \frac{1}{L_d} \end{bmatrix} \begin{bmatrix} V_d + \frac{R\psi_m}{L_d} \\ V_q \end{bmatrix} \tag{8}$$

In order to simplify matrix in (8), assume:

$$\left.\begin{aligned} i'_d &= i_d + \frac{\psi_m}{L_d} \\ i'_q &= i_q \\ V'_d &= V_d + \frac{R\psi_m}{L_d} \\ V'_q &= V_q \end{aligned}\right\} \tag{9}$$

Substitute equation (9) into (8) to simplify the matrix equation of (8)

$$\frac{d}{dt}[i'_d] = -\frac{R}{L_d} i'_d + \frac{\omega_r L_q}{L_d} i'_q + \frac{V'_d}{L_d} \tag{10}$$

$$\frac{d}{dt}[i'_q] = -\frac{\omega_r L_q}{L_d} i'_d - \frac{R}{L_d} i'_q + \frac{V'_q}{L_q} \tag{11}$$

Then, the value of i'_d, i'_q, V'_d and V'_q in the equation (10) and (11) must be expand back based on equation at (8):

$$\frac{d}{dt}[i'_d] = -\frac{R}{L_d} i_d + \frac{\omega_r L_q}{L_d} i_q + \frac{V_d}{L_d} \tag{12}$$

$$\frac{d}{dt}[i'_q] = -\frac{\omega_r L_q}{L_d} i_d + \frac{\omega_r L_q}{L_d}\frac{\psi_m}{L_d} - \frac{R}{L_d} i_q + \frac{V_q}{L_q} \tag{13}$$

While,

$$\omega e = \frac{\omega_r L_q}{L_d} \tag{14}$$

Then, substitute (14) into (12) and (13):

$$\frac{d}{dt}[i'_d] = -\frac{R}{L_d} i_d + \omega e(i_q) + \frac{V_d}{L_d} \tag{15}$$

$$\frac{d}{dt}[i'_q] = -(\omega e) i_d + \frac{\omega_r L_q}{L_d}\frac{\psi_m}{L_d} - \frac{R}{L_d} i_q + \frac{V_q}{L_q} \tag{16}$$

To generalized the error:

$$e = i - \hat{i} \tag{17}$$

where e is the error between reference and adjustable model, i and $\hat{i}$ is reference and adjustable current. The simple matrix can be form below:

$$\frac{d}{dt} e = Ae - (\ddot{A} - A) i' \tag{18}$$

$$\frac{d}{dt}\begin{bmatrix} e_d \\ e_q \end{bmatrix} = \begin{bmatrix} \frac{-R}{L_d} & \frac{\omega'_r L_q}{L_d} \\ \frac{-\omega'_r L_q}{L_d} & \frac{-R}{L_d} \end{bmatrix} \begin{bmatrix} e_d \\ e_q \end{bmatrix} + J[\omega e - \omega\hat{e}] \begin{bmatrix} i_d \\ i_q \end{bmatrix}$$

From the previous formula:

$$i = \begin{bmatrix} i_d \\ i_q \end{bmatrix}, \hat{i} = \begin{bmatrix} i'_d \\ i'_q \end{bmatrix} \tag{19}$$

Substitute the formula in (19) into (17)

$$e = \begin{bmatrix} i_d - i'_d \\ i_q - i'_q \end{bmatrix}$$

$$J = \begin{bmatrix} 0 & -1 \\ 1 & 0 \end{bmatrix}$$

$$\frac{d}{dt} e = Ae - W \tag{20}$$

$$W = (\ddot{A} - A)\hat{x}$$

The adaptive law shown in (17) can be obtain by implement PI Controller.

$$\hat{\omega} = [k_p + k_i]\hat{i}_s * i_s$$

$$\hat{i}_s * i_s = \zeta\omega$$

$$\zeta\omega = e^{i'} J^{i} = \hat{i}_s * i_s$$

$$e = i'_d i_q - i'_q i_d - \frac{\psi_t}{L}(i'_q - i_q) \tag{21}$$

Substitute the equation (18) into (19):

$$\hat{\omega} = \left[k_p + k_i\right]\left[i'_d i_q - i'_q i_d - \frac{\psi_t}{L}(i'_q - i_q)\right] + \hat{\omega}_{re}(0) \tag{19}$$

$$\hat{\theta} = \int_0^t \hat{\omega}_{re}$$

C. Particle Swarm Optimization

Particles Swarm Optimization (PSO) is a optimization technique developed by Dr. Eberhart and Dr. Kennedy in 1995 that inspired by social behavior of bird flocking or fish schooling. Due to many advantage can be found by using this technique including simplicity and easy implementation, this algorithm can be used widely in many field such as classification, neural network training, machine study, signal procession and etc. This algorithm is also appropriate for parameter optimization in continues search space in many dimensions [18].

PSO concept is using a number of agent (particles) that constitute a swarm moving around in the search space looking for the best solution. Each particles in search space adjust its 'flying' according to its own flying experience well as the flying experience of other particles and the best are called (pbest). Individual particles also have the best knowledge of values by the group (gbest). Each agent uses the information connecting to: current position (x,y), current velocity (vx,vy), the distance between the current position and gbest group to change the position. Velocity and position of each agent modified by equation (18) and (19) below.

$$e = v_i^k + c_i * rand_1 * \left(pbest_i - s_i^k\right) + c_2 * rand_2 * \left(gbest - s_i^k\right) \tag{22}$$

$$s_i^{k+1} = s_i^k v_i^{k+1} \tag{23}$$

Where:

v_i^k = current velocity of agent i at iteration k
v_i^{k+1} = new velocity of agent i at iteration k
s_i^k = current position of agent i at iteration k
s_i^{k+1} = denotes the position of agent I at the next iteration k+1
$pbest_i$ = personal best agent i
$gbest$ = global best of the population
c_1 = adjustable cognitive acceleration constant (self-confidence)
c_2 = adjustable cognitive acceleration constant (swarm confidence)
$rand_{1,2}$ = random number between 0 and 1

For this paper, PSO is ;use to optimize the value of PI controller that can minimum the error and overshoot of the PMSM sensorless control. There are four step for this algorithm optimize the value of PI that illustrate in Fig. 4 below.

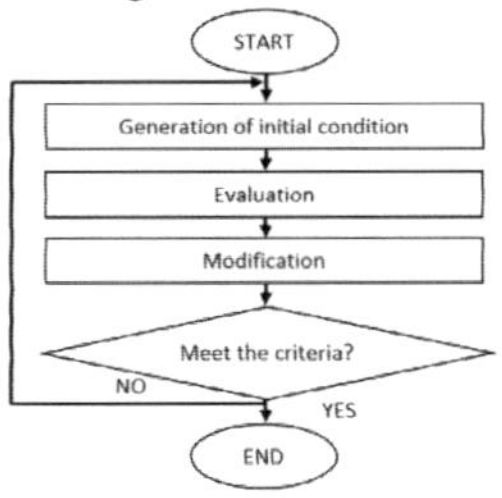

Figure 4: Flowchart for the steps of PSO

Step 1: Generation of initial conditions of each agent

For the first stage, the initial input and speed of each agent will be choose randomly. Each agent will set the current search point to pbest and at the same time, gbest also will determine the best value of pbest. Then, the best value for the number agent is stored for evaluation process.

Step 2: Evaluation of searching point of each agent

Second stage is evaluation for the stored value of agent where output of the system will produce an error and overshoot. The data will be analyze and new agent will be replace to see the output of system behavior. If the agent is better than pbest at this time, it will replace current value. Also, if the value of pbest is better than gbest, it will replace current value that stored inside gbest. Then, the best agent will stored.

Step 3: Modification of each searching point

By yusing the equation (19)(20), the value of agent will be declared to find the point.

Step 4: Checking to exit condition

In this step, several test were carried out before the end of the process. If the system not reach the desired output, the process will repeated again from step 1 and if otherwise the process will stopped.

Table 1: Parameter of PSO

Variable	Value	Description
n	25	Size of the swarm "no. of birds"
Bird step	10	Max. no. of "bird steps"
dim	2	Dimension of the problem
c1	1.5	Velocity constant
c2	0.15	Velocity constant
w	0.9	Momentum of inertia
minKp	10	Min. value of agent for Kp
maxKp	60	Max. value of agent for Kp
minKi	0	Min. value of agent for Ki
maxKi	5	Max. value of agent for Ki

In this research, controller that being use for adaptation method for MRAS is Heuristic-PI, MATLAB PI and PSO-PI controllers.

Heuristic-PI Controller

Proportional Integral (PI) controller is a common feedback loop consist of direct and integrator gain that the gain value is set based on try and error (manually tune). This method generally used in industrial control system as shown in Fig. 5.

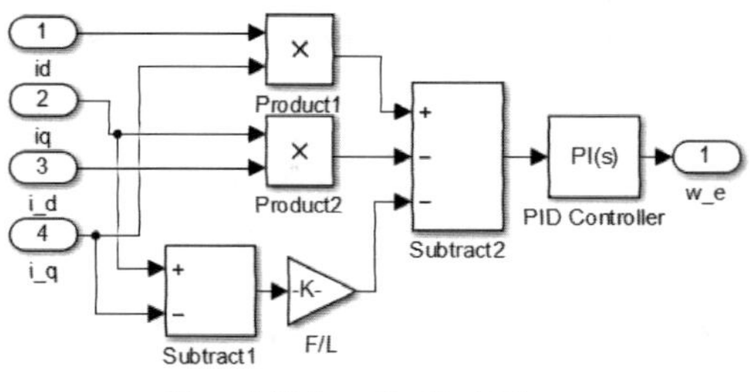

Figure 5: PI Controller Block Diagram

MATLAB-PI Controller

MATLAB-PI (M-PI) controller is the PI controller that tuned by using MATLAB tuner. It much simpler method where user only set the response time and transient behavior before the system automatically develop the gain value

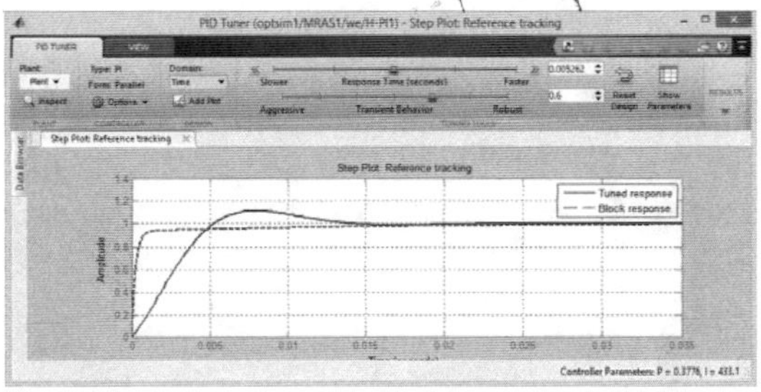

Figure 6: PID tuner that provide by MATLAB Simulink block

PSO-PI controller

PSO-PI controller is the PI controller that gain of the system is tuned by using Particle Swarm Optimization. Based on PSO-PI controller block diagram in Fig. 7, the simulation was carried out.

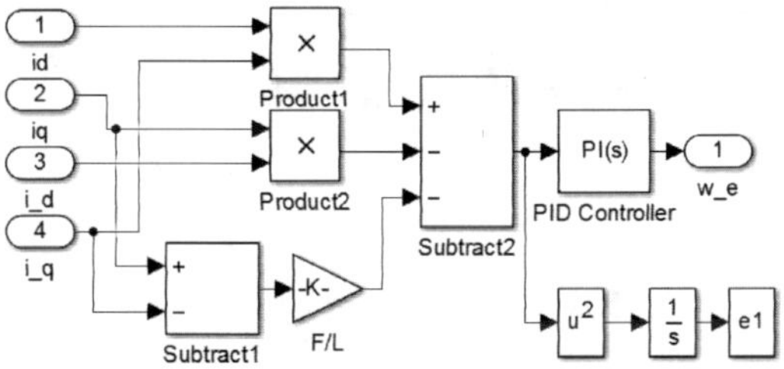

Figure 7: PSO-PI Controller Block Diagram

3. Simulation Result

First, the simulation results are using Heuristic-PI (H-PI) controller, then MATLAB Autotune PI (M-PI), and lastly is Particles Swarm Optimization PI (PSO-PI) controller. Simulation is perform in the MATLAB Simulink and time is kept 0.3sec. Result are based on system performance metric such as rise time (Tr), settling time (Ts), percent overshoot (%Os), percent undershoot (%Us) and root mean square error (RMSE) where its verified under four (4) different conditions.

1) Speed and load both are constant.
2) Speed is constant and load varied.
3) Speed is varied and load is constant.
4) Speed and load both are varied.

For condition 1- Speed and load torque are constant [speed is 1500r/min and load torque is 2Nm]

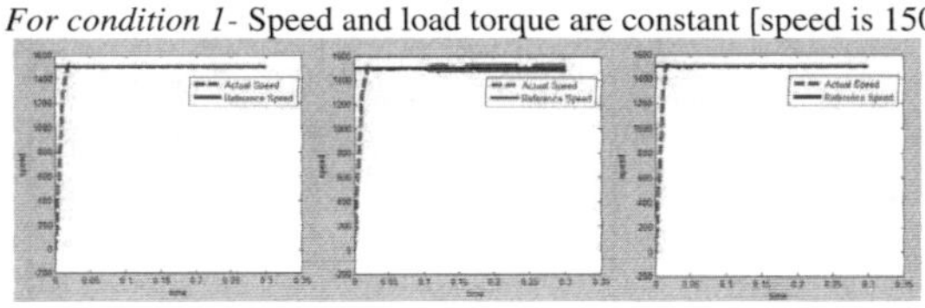

(a) (b) (c)

Figure 8. Reference and actual speed of (a) H-PI, (b) M-PI and (c) PSO-PI method for condition 1

For condition 1, PSO-PI produce better result where Tr is 0.026751s, Ts is 0.036654s, Os is 0.9431% and RMSE is 309.7595.

For condition 2- Speed is constant and load torque is varied [speed is 1500r/min and load torque is 0-0.075sec (2Nm), 0.075-0.15sec (1Nm), 0.15-0.225sec (2Nm) and 0.225-0.3sec (1Nm)].

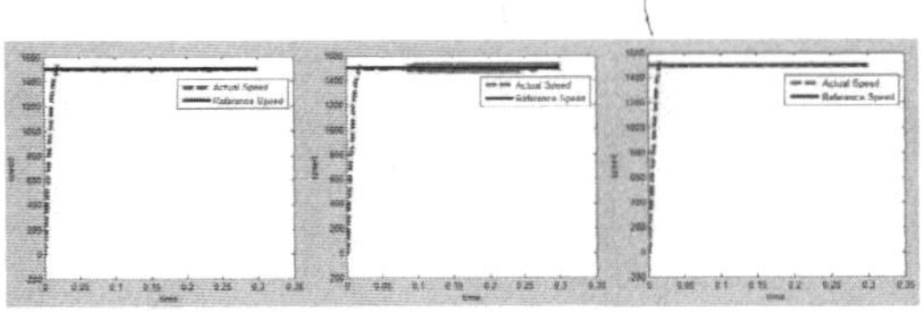

(a) (b) (c)

Figure 9. Reference and actual speed of (a) H-PI, (b) M-PI and (c) PSO-PI method for condition 2

For condition 2, PSO-PI produce better result where Tr is 0.026782s, Ts is 0.036713s, Os is 0.8816% and RMSE is 309.8039.

For condition 3- Speed is varied and load torque is constant [speed is 0-0.1sec (0-1500r/min), 0.1-0.18sec (1500r/min), 0.18-0.25sec (800r/min) and 0.25-0.3sec (1500r/min) and load torque is 2Nm].

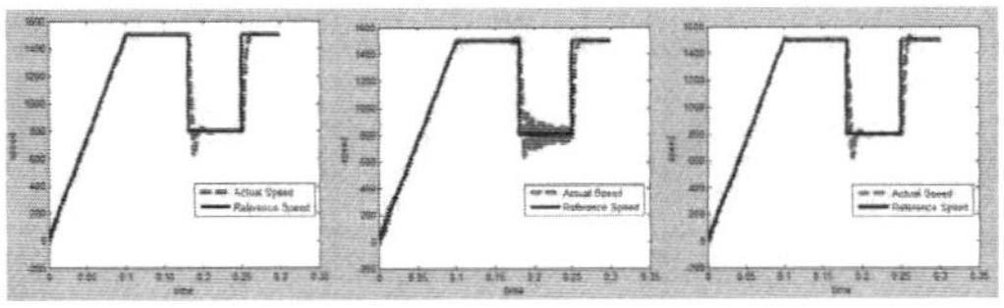

(a) (b) (c)

Figure 10. Reference and actual speed of (a) H-PI, (b) M-PI and (c) PSO-PI method for condition 3

For condition 3, PSO-PI produce better result where Tr is 0.088513s, Ts is 0.263733s, Os is 0.9559% and RMSE is 76.6674.

For condition 4- Speed and load torque are varied [speed is 0-0.1sec (0-1500r/min), 0.1-0.18sec (1500r/min), 0.18-0.25sec (800r/min) and 0.25-0.3sec (1500r/min) and load torque is 0-0.075sec (2Nm), 0.075-0.15sec (1Nm), 0.15-0.225sec (2Nm) and 0.225-0.3sec (1Nm)].

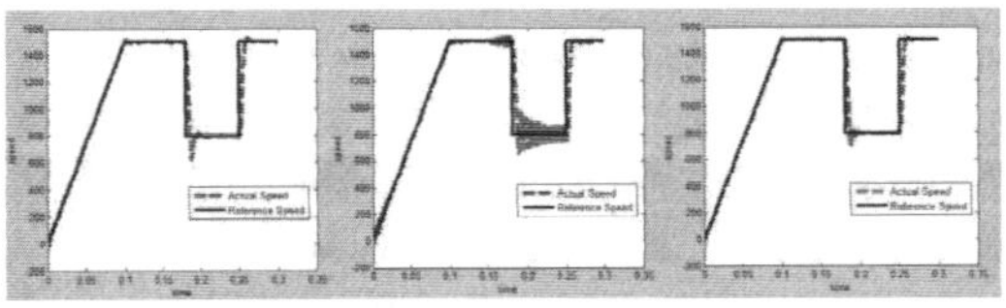

(c)

(a) (b)

Figure 11. Reference and actual speed of (a) H-PI, (b) M-PI and (c) PSO-PI method for condition 4

For condition 4, PSO-PI produce better result where Tr is 0.087982s, Ts is 0.0s263733, Os is 1.1356% and RMSE is 75.3531. Table 2 to 4 shows the overall result that produce from different method that have being used.

Table 2: Result by using H-PI controller for condition 1 to 4

	Cond. 1	Cond. 2	Cond. 3	Cond. 4
Tr	0.027678	0.027509	0.09316	0.092181
Ts	0.039906	0.039693	0.26455	0.262614
%Os	1.5386	1.4781	1.1247	1.9413
RMSE	311.5337	310.4529	78.2006	77.1321

Table 3: Result by using M-PI controller for condition 1 to 4

	Cond. 1	Cond. 2	Cond. 3	Cond. 4
Tr	0.03434	0.034835	0.10394	0.104045
Ts	0.29863	0.298435	0.264172	0.263029

%Os	2.7949	2.7285	1.6554	1.9325
RMSE	345.1328	346.5095	80.5208	89.2568

Table 4: Result by using PSO-PI controller for condition 1 to 4

	Cond. 1	Cond. 2	Cond. 3	Cond. 4
Tr	0.026751	0.026782	0.088513	0.087982
Ts	0.036654	0.036713	0.263733	0.262141
%Os	0.9431	0.8816	0.9559	1.1356
RMSE	309.7595	309.8039	76.6674	75.3531

4. Conclusion

In this paper a new design method is proposed to determine stability of step response in term of rise time, settling time, maximum overshoot and root mean square error. Amongst this three controllers that were tested, PSO-PI was found to be producing better results compared to H-PI and M-PI. It shows that this method can improve the dynamic performance of the system in a better way.

ACKNOWLEDGMENT

This paper was supproted by the Fundamental Research Grant Scheme (FRGS) (FRGS/1/2015/TK04/UITM/02/12) funded by the Ministry of Education and Faculty of Electrical Engineering Universiti Teknologi MARA.

References

[1] Z. Zheng, Y. Li, and M. Fadel, "Sensorless Control of PMSM Based on Extended Kalman Filter," *Eur. Conf. Power Electron. Appl.*, pp. 1–8, 2007.

[2] M. Elbuluk, "A sliding mode observer for sensorless control of permanent magnet synchronous motors," *Conf. Rec. 2001 IEEE Ind. Appl. Conf. 36th IAS Annu. Meet. (Cat. No.01CH37248)*, vol. 2, no. C, pp. 1273–1278, 2001.

[3] Z. Qiao, T. Shi, Y. Wang, Y. Yan, C. Xia, and X. He, "New sliding-mode observer for position sensorless control of permanent-magnet synchronous motor," *IEEE Trans. Ind. Electron.*, vol. 60, no. 2, pp. 710–719, 2013.

[4] E. M. Fernandes, a. C. Oliveira, M. a. Vitorino, E. C. Dos Santos, and W. R. N. Santos, "Speed sensorless PMSM motor drive system based on four-switch three-phase converter," *Proceedings, IECON 2014 - 40th Annu. Conf. IEEE Ind. Electron. Soc.*, pp. 902–906, 2014.

[5] Z. Q. Zhu and L. M. Gong, "Investigation of effectiveness of sensorless operation in carrier-signal-injection-based sensorless-control methods," *IEEE Trans. Ind. Electron.*, vol. 58, no. 8, pp. 3431–3439, 2011.

[6] A. Mishra, V. Mahajan, P. Agarwal, and S. P. Srivastava, "MRAS based estimation of speed in sensorless PMSM drive," *2012 IEEE 5th Power India Conf. PICONF 2012*, 2012.

[7] F. Cupertino, a. Guagnano, a. Altomare, and G. Pellegrino, "Position estimation delays in signal injection-based sensorless PMSM drives," *3rd IEEE Int. Symp. Sensorless Control Electr. Drives, SLED 2012*, pp. 1–6, 2012.

[8] P. Kumar, M. Kumar, and S. Dahiya, "Sensorless Speed Control of PMSM using SVPWM Technique Based on MRAS Method for Various Speed and Load Variations," *Proc. World Congr. Eng. 2015*, vol. I, no. 1, pp. 1–6, 2015.

[9] D. Fodor, J. Vass, Z. Juhász, and K. Bíró, "Extended Kalman Filter Based Speed Sensorless Ac Motor Control," p. 6.

[10] R. K. R. Kumar, R. a. Gupta, and a. K. Bansal, "Identification and Control of PMSM Using Artificial Neural Network," *2007 IEEE Int. Symp. Ind. Electron.*, pp. 30–35, 2007.

[11] Z. Wang, Q. Teng, and C. Zhang, "Speed identification about PMSM with MRAS," *2009 IEEE 6th Int. Power Electron. Motion Control Conf. IPEMC '09*, vol. 3, pp. 1880–1884, 2009.

[12] G. Wu and X. Xiao, "Speed controller of servo system based on MRAS Method," *Proc. IEEE Int. Conf. Ind. Technol.*, 2009.

[13] P. Vaclavek and P. Blaha, "Synchronous Machine Drive Observability Analysis and Sensorless Control Design," *2008 Ieee 2Nd Int. Power Energy Conf. Pecon, Vols 1-3*, no. PECon 08, pp. 265–270, 2008.

[14] H. M. Kojabadi and M. Ghribi, "MRAS-based adaptive speed estimator in PMSM drives," *Int. Work. Adv. Motion Control. AMC*, vol. 2006, no. 1, pp. 569–572, 2006.

[15] K. Jinsong, Z. Xiangyun, W. Ying, and H. Dabing, "Study of position sensorless control of PMSM based on MRAS," *Proc. IEEE Int. Conf. Ind. Technol.*, no. 3, pp. 3–6, 2009.

[16] W. Gao and Z. Guo, "Speed sensorless control of PMSM using model reference adaptive system and RBFN," *J. Networks*, vol. 8, no. 1, pp. 213–220, 2013.

[17] M. Dursun, A. F. Boz, M. Kale, and M. Karabacak, "Sensorless Speed Control of Permanent Magnet Synchronous Motor With Hybrid Speed Controller Using Model Reference Adaptive," vol. 3, no. 1, pp. 24–37, 2014.

[18] W. F. Tarmizi, I. Elamvazuthi, N. Perumal, K. Nurhanim, M. K. A. A. Khan, S. Parasuraman, and a. V. Nandedkar, "A Particle Swarm Optimization-PID controller of a DC Servomotor for Multi-Fingered Robot Hand," *2016 2nd IEEE Int. Symp. Robot. Manuf. Autom. ROMA 2016*, pp. 1–6, 2017.

The Emergence Themes of Staff and Course Evaluation at UniMAP: A Grounded Theory Approach.

Hanif Suhairi Abu Bakar, Tengku Kastriafuddin Shah Tengku Yaakob, Zuraidi Ahmad Mokhdzar and Nor Azmi Hj Johari

School of Human Development and Techno-communication (iKOM), Universiti Malaysia Perlis, Perlis, Malaysia

hanifsuhairi@unimap.edu.my

Abstract. This paper concentrates on the emergence factors and themes that surfaced as result of paper and online course evaluation activities among students at Universiti Malaysia Perlis (UniMAP) that have never been reported prior to this. Before the online course and lecturer evaluation practices every semester, students need to fill out the form manually in order to evaluate the courses and lecturers, and this brings a lot of disadvantages and problems to the university. School of Human Development and Technocommunication (iKOM) has initiated the online course and evaluation approach, and is now being utilized by all schools at UniMAP. This research and innovation has also been awarded gold at the 11th Innovative and Creative Convention held at Universiti Sains Islam Malaysia (USIM) in 2015. The actual research attempted to assist the university to improve its way of handling and processing information that occurs every semester or biannually by focusing on factors that contribute to its predicament for staff and students at UniMAP. In addition, the research seeks to understand the antecedents, contexts, and prevailing conditions that influence it as well as the phenomenon, the coping strategies, and the consequences resulting from the coping strategies. Using a grounded theory methodology, in-depth interviews were conducted on students and staff from different schools at UniMAP. The data were analyzed using NVivo 10 by methodically coding and categorizing the data in open, axial, and selective coding to arrive at a model development. Working hypotheses were obtained based on informants' individual comments and experiences that uniquely contribute to the body of knowledge on this new phenomenon at UniMAP. The involved elements derived from the model are the antecedents, the phenomenon itself, the coping strategies by informants, and the consequences resulting from these strategies. The significance of this study lies in its attempt to provide a working framework for reducing the dilemma when the time comes to execute staff and course evaluation that will eventually add up to time and money lost. This study is hoped to benefit all the parties: staff, students, the university, and other stakeholders to maximize the benefits of online evaluation practices. At a time when most assessments are done online around the world, this research is both timely and necessary.

Keywords: Online Evaluation; Grounded Theory; Facebook; Course and Staff Evaluation

1. Introduction

The main intention of the current research was to construct a grounded theory of the online evaluation phenomenon on the basis of staff and students' report about their own experience on previously manually executed course and lecturer assessment.

Second, it is expected that this study can offer a more profound descriptive account of the phenomenon. Nevertheless, the most fundamental ground to the researchers is to come out with a working hypothesis or process model of online evaluation. The researchers conducted the present study to examine the process by which online evaluation occurs and to propose a preliminary paradigm model (Strauss & Corbin, 1998) that can be tested in future research. The researchers preferred grounded theory methodology as it is preferably suited to construct a data-based theory that can be utilized as a foundation for future studies (Creswell, 1998; Strauss & Corbin, 1998).

Given the prevailing nature of online assessment, the researchers questioned how staff and students perceive it and whether their experiences are aligned with the existing views in the literature. The researchers believed that a grounded theory analysis would facilitate researchers to ascertain other characteristics of online course and staff evaluation that might justify its practice and make rooms for further improvements.

2. Literature Review

2.1. The Validity and Reliability of Course and Lecturer Evaluations

Researchers commonly believe student evaluations (both online and paper evaluations) of instructors to be extremely trustworthy (Aleamoni, 1999; Centra, 1993; Hobson & Talbot, 2001) in contrast with evaluations done by colleagues or trained observers that have not been found to be reliable, hence not applicable (Centra, 1993).

Commonly, student assessments of lecturers have been established to be associated with ratings of instructor's skills in course organization, relationship with students, and fair grading; variance in organizational skill elucidated most inconsistency in student evaluations (Jirovec, Chathapuram, & Rosegrant-Alvarez, 1998).

As soon as lecturers collect mid-semester feedback from students and have a truthful conversation about it with supervisor, it leads to higher evaluations at the end of the semester as well as higher final exam scores, providing facts that satisfactory evaluations can lead to better teaching (Overall & Herbert, 1979). Even though carry marks do have some consequences on how students rate instructors (Johnson & Valen, 2002), its effect is reasonably low (Gigliotti & Buchtel, 1990) and can be statistically attuned for (Greenwald & Gillmore, 1997). Carry marks do not have as large of an effect as do how much students feel they have learned (Bard, 1987), how much they felt stimulated by the class (Remedios & Lieberman, 2008), and whether the class was appropriately difficult (Centra, 2003). Differing to the "retaliation" theory, students who do poorly in a class are equally or less likely than those who do well to complete course evaluations (Liegle & McDonald, 2004).

2.2. Comparison of Online and Paper Evaluation

The main drawback of online course evaluation compared to paper evaluation is the low response time (Anderson, Cain, & Bird, 2005; Avery, Bryant, Mathios, Kang, & Bell, 2006). However, utilizing online media to remind students can drastically boost the response time (Norris & Conn, 2005). Noticeably, evaluation scores do not change via online evaluation in contrast to paper evaluations (Avery et. al, 2006; Donovan, Mader, & Shinsky, 2006). According to Anderson et al. (2005), Donovan et al. (2006), and Kasiar, Jennifer, Schroeder, and Holstad (2002), students place more and informative comments on online evaluations compared to the traditional paper evaluation. What is more, students and staff viewed online evaluation more optimistically in contrast to its predecessor (Anderson et al., 2005; Donovan et al., 2006).

IOP Conf. Series: Journal of Physics: Conf. Series **1019** (2018) 012028 doi:10.1088/1742-6596/1019/1/012028

Students' comments on the online forms were more frequent and longer than the paper forms. They liked the online method better and believed that they might present more beneficial responses online (Anderson et al., 2005).

Ballantyne (2003) suggested that to increase the response rate, there needs to be approaches such as providing computer access, having school encouragement for the system, and allowing students see how their responses are being utilized for the better. For further improvement, the online evaluation needs to be planned so that students cannot rate the same course and lecturer more than once while the anonymity is still protected (Ballantyne, 2003). According to Ballantyne (2004), there are types of students that will likely respond more via online, namely male students, younger students, undergraduates, and full-time students. Students responding online were less likely to comment, but online comments were lengthier than paper comments (Ballantyne, 2004).

Regarding the expenditure, Bothell and Henderson (2003) compared the costs of an online evaluation system to paper evaluations and they found that it saved them $235,000 a year when Brigham Young University switched to online evaluations. The decreases come from a reduction in printing costs, personnel assistance for distribution and collection, processing, reporting, and fewer time taken away from lecturers in the classroom (Bothell & Henderson, 2003).

3. Method

3.1 Informants

Sixty academic and non-academic staff as well as students (n = 60) from various departments at UniMAP participated in this research. Data were collected over a period of three months right after final examinations in order to get fresh experiences from the informants. All informants were chosen by their first-hand encounter with the manual and online staff and course evaluation procedure in order to get more accurate information.

All the researchers also have the experience of real situation handling the two types of evaluations. The researchers also interviewed the Head of Academic Management Department whom at that particular moment has been given the green light by the university to purchase two new Optical Mark Recognition (OMR) machines.

3.2 Procedure

Data were gathered using grounded theory processes described in Strauss and Corbin (1990, 1998). Creswell (1998) suggested that a standard grounded theory research will comprise 20–30 interviews that collectively saturate the categories that emerge during analysis.

Data accumulated in four phases that differ with regard to purpose and data collection strategies (Harry, Sturges, & Klingner, 2005) are summarised in Table 1.

Table 1. Four Phases in Data Collection

Phase	Coding	Purpose	Informants
1	Open	Categorize codes within categories for advanced analysis	20 individuals and 5 in focus groups
2	Axial	Specify codes in detail; relay codes to one another to create themes	5 individuals and 5 in focus groups
3	Selective	Create paradigm model and discuss themes in relation to model; establish plot that integrates paradigm model	15 in focus groups
4	Selective	Test, certify, and explain paradigm model until saturated; recognize surfacing principles constant with paradigm model; perform member checks.	10 in focus groups

Data analysis of current study was completed by using NVivo 10 as to meet the terms of ‘trustworthiness’, ‘rigorousness’, or ‘quality’ of the data; therefore, it is important that these are carried out in a thorough and transparent manner (Crawford, Leybourne, & Arnott, 2000; Creswell, 1998; Kirk & Miller, 1986; Lincoln & Guba, 1985; Miles & Huberman, 1994; Seale, 1999). Thus, using software in the data analysis process has been thought by some to add rigor to qualitative research (Richards & Richards, 1994). Phase 1 data collection incorporated individual interviews and focus group interviews that allowed the researchers to generate a list of codes within five categories in the paradigm model, specifically antecedents, definitions of the phenomenon, context and conditions that affect the phenomenon, coping strategies, and consequences of the coping strategies taken. The main reason of open coding is to distinguish significant topics creditable of closer analysis and explanation (Strauss & Corbin, 1998). In all, 25 staff and students were interviewed individually, and 35 in focus groups. The interviews at this stage were to implore preliminary thoughts and impressions about online evaluation in a setting in which informants could react to each other (Krueger, 1994). All interviews were based on both open-ended responses to the following questions:

(1) How would you describe manual and online course and lecture’s evaluation?
(2) What are the factors that influence you to use manual and then online evaluation?
(3) Which would you prefer most; manual or online evaluation? Why?
(4) What are some positive and negative aspects of manual and online evaluation?
(5) How do you cope with the negative aspects of manual and online evaluation?

These questions are consistent with the major elements of the paradigm model described by Strauss and Corbin (1998). Table 2 shows 98 preliminary codes obtained from the interview sessions with each code representing a significant topic of discussion related to online and paper course, and staff evaluation.

Table 2. Initial Categories and Codes in Phase 1

Category	Code
Antecedents	1. Last-minute work.
	2. Do not care much.
	3. Not paying attention.
	4. Consider as less important.
	5. Assuming that lecturer(s) would not read their comments.
	6. Taking paper evaluation for granted.
	7. Consider as less important.
	8. Assuming management will take for granted whether it is done or not.
	9. Lecturers are afraid if the students will tell the truth.
	10. Lecturers are afraid if the students are not telling the truth.
	11. Had to do it because of instruction by immediate supervisor.
	12. Sense of duty.
	13. Need to inform the staff's bad performance.
	14. Need to praise the staff's excellent performance.
	15. Had to do it because of the carry marks.
	16. Responsibility.
	17. To further improve his or her own teaching techniques.
	18. To know his or her own weaknesses.
	19. Had to do it because of instructions from the dean.
	20. Duty.
Definitions	21. To change the culture to support online student evaluation processes.
	22. To identify the strengths and weaknesses of lecturers while performing their duties.
	23. To further improve the course's content.
	24. To further improve the lecturer's style of delivery.
Contexts and Conditions that Influence the Antecedents	25. The improvement of Internet facilities at UniMAP.
	26. Technology and Internet cannot be separated from students.
	27. Facebook and Google are free.
	28. Facebook and Google are common.
	29. Can be completed anywhere and everywhere.
	30. Can be completed anytime.
	31. Obligations to the university.
	32. To improve teaching and learning.
	33. Being instructed to do so.
	34. Had to be done for staff evaluation by the head of department.
	35. Free application such as Facebook where students hang around often.
Phenomenon	36. Can easily remove the unpleasant students' evaluation.
	37. Can easily instruct the students face-to-face.
	38. Can easily explain to students face-to-face.
	39. The improvement of staffs' knowledge of Internet applications.
	40. The enhancement of students' awareness of Internet applications.
	41. Free application such as Google Doc.
	42. Responding to student concerns for privacy and anonymity.
	43. Immediate availability of data for analysis and reporting.
	44. More extensive qualitative responses to the open-ended questions.
	45. Enforce uniformity for the evaluation of all courses.
	46. Providing faculty with the flexibility to add items specific to their courses

	to 30 and address accreditation requirements. 47. Having the opportunity to complete evaluations during free time. 48. The surveys cannot be misplaced. 49. There are no papers to be wasted. 50. Delivered OMR sheets arrive in an unscannable form because of damaged forms. 51. Incorrect printing. 52. Incorrect information being inputted. 53. Academic audit report is not completed. 54. Records are not updated. 55. OMR reporting is behind schedule. 56. Inaccurate data input. 57. OMR sheets are misplaced. 58. OMR sheets are not collected. 59. Delayed data collection. 60. Prolonged OMR processing time. 61. Copied OMR sheets are blurry. 62. The dissimilarities of items in OMR sheets are not compatible with some schools. 63. Staff insufficient. 64. Inappropriate for team-teaching subjects. 65. No minutes of meetings. 66. Numerous and tedious OMR processing procedures. 67. OMR machines are always out of order. 68. OMR machines are expensive. 69. The maintenance of OMR machines is expensive. 70. Distant OMR processing location. 71. Students fill out the forms for granted – yielding unreliable results. 72. Expensive OMR forms. 73. Had to queue in order use the OMR machines. 74. Schools gets a non-conformant status for failing to produce academic evaluation reports. 75. Qualitative data cannot be interpreted by OMR machines. 76. Results are often ambiguous and fail to provide action-oriented solutions. 77. Paper surveys are regularly delivered to the wrong place. 78. Paper surveys are not being delivered. 79. Summaries are often ambiguous and fail to provide action-oriented solutions. 80. Copied OMR forms cannot be read by the OMR machines. 81. Evaluations are the basis for highly stressful decisions rather than a tool for improving teaching and learning. 82. Paper surveys are regularly delivered to the wrong place. 83. Paper surveys are not being delivered. 84. Students know that their comments will not be read. 85. Hard to ensure that students will evaluate only once. 86. Hard to ensure sure that only students will evaluate, not staff themselves. 87. Hard to make sure that only students will evaluate, not staff. 88. Participation from students is not 100%.
Coping Strategies	89. Using copied OMR forms to save cost.

	90. Lecturers fill out the OMR forms themselves.
	91. Lecturers discard OMR forms that yield negative comments.
	92. Carry on with copied OMR forms or sheets even though many were rejected by OMR machines.
Consequences	93. Improve quality of administrative work.
	94. Improve quality of feedback or information.
	95. Improve quality of teaching and learning.
	96. Cut costs.
	97. Less time-consuming.

Many staff and students for instance, mentioned characteristics of online technology such as social networking are facilitating factors to online evaluation. The researchers thus concluded that the focus group and individual interviews saturated the codes that were necessary to understand the occurrence of online evaluation after similar codes emerged during the interviews. Phase 2 that exercised axial coding encompassed of five individual interviews and five in a focus group, which aimed to elucidate codes in detail. Phase 2 clusters codes into themes and patterns related to online evaluation. Axial coding allowed the researchers to structure an advanced understanding of each of the five main components in the paradigm model (Strauss & Corbin, 1998).

A structured interview was used in which each individual responded to three questions and follow-up probes. Question 1 was intended to familiarize individuals to a typical online evaluation situation. Questions 1 and 2 were based on the five questions presented in Phase 1 of data collection. Follow-up probe questions were added to focus in more detail on prominent categories mentioned during Phase 1.

(1) What are the positive and negative consequences of paper and online evolutions? Probes: quality of life and quality of academic.

(2) What do you do to cope with paper and online evaluation's inadequacy? Probes: active, aggressive, avoidance, cognitive.

(3) What are the impacts of paper and online evaluations? Probes: emotional, behavior, social, physical, academic.

The researchers identified a range of preliminary themes on the foundation of five categories and codes as illustrated in Table 2. The researchers merged themes in accordance with the interviews. Phase 2 provides the researchers with the possibility to organize and label themes that could be tested in Phase 3. Therefore, macrothemes and themes identified in Phase 2 were viewed as provisional, subject to revision and removal in Phase 3.

Selective coding was intended for Phase 3. According to Strauss and Corbin (1998), selective coding is the process of assessing the macrothemes and themes identified in Phase 2, linking the categories to one another, and developing a plot that integrates the paradigm model. Fifteen informants participated in three focus groups interviews to serve several purposes such as to duplicate crucial codes and themes that were identified in Phases 1 and 2 (Strauss & Corbin, 1998).

Second selective coding was used to delve into detail the connection among macrothemes within each category (Strauss & Corbin, 1998). A third objective was to request informants to discuss the five categories in the paradigm model related to each other; for example, informants discussed the relationship between antecedents and contextual conditions of the phenomenon in order to understand the process of online and paper evaluations (Strauss & Corbin, 1998).

This information facilitated the researchers to construct a credible paradigm model that could be authenticated in Phase 4. The final objective was to create a storyline, which consists of a descriptive story about the central phenomenon of this research. This phase consists of two major outcomes. The first outcome distinguished the range of themes that were considered within macrothemes in each of the five categories. Altogether as shown in Table 2, 98 themes were identified and were grouped into 5 macrothemes. The second outcome was a constructed preliminary paradigm model that has been validated in Phase 4 of data collection and illustrated in detail in the following section. The preliminary paradigm model consisted of the macrothemes and themes acknowledged so far.

Finally, Phase 4 was projected to validate and explain the paradigm model constructed in Phase 3. Ten informants in four focus groups were interviewed and asked to respond to a paradigm model based on the responses in Phase 3. The central idea was to perform member checks on the preliminary paradigm model. Informants were asked whether the themes and macrothemes were conceivable to them. A second purpose was to ask for comments and reactions concerning the reliability of the paradigm model.

Informants in Phase 4 were shown preliminary diagrams of the paradigm model and asked to comment on whether the model is sensible to them and in line with their own experience. A third purpose was to encourage the informants to refine or add to any of the categories, macrothemes, or themes in the paradigm model. According to Maxwell (1996), this last stage facilitated to guarantee that the final paradigm model was fully saturated, reliable, and credible.

4. Findings

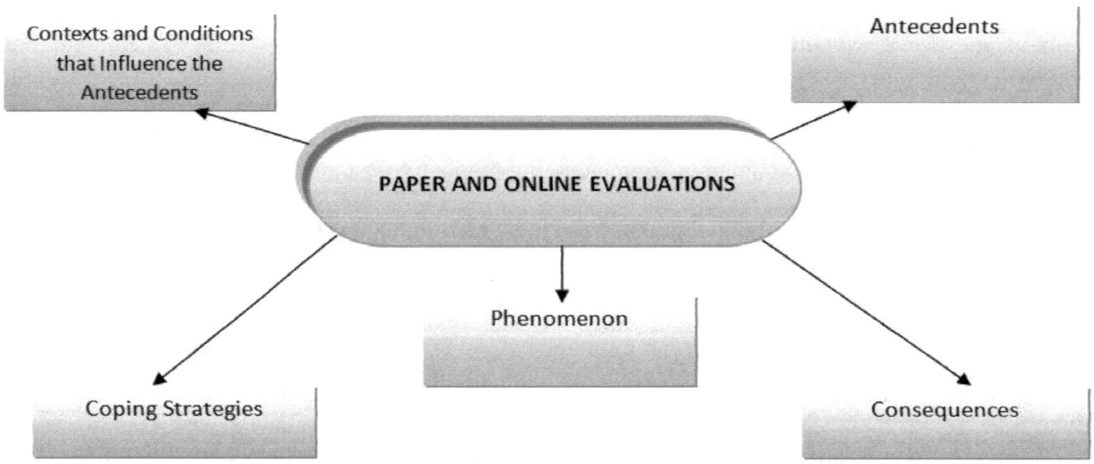

Figure 1. The emergence themes of staff and online evaluation phenomenon

4.1 Emerging Themes of Staff and Course Evaluation Phenomenon

The researchers proposed that the principal theme that emerged from the open coding investigative process which the in-depth interviews provided was that staff, especially the academic staff, desired to gain 'a sense of fulfillment' over their lives and they found that this concept could be easily achieved by their use of online media applications.

The fundamental categories that emerged from these interviews were (1) the antecedents of the phenomenon, (2) the contexts and conditions that influence the antecedents, (3) the central phenomenon, (4) the coping strategies, and (5) the consequences resulting from the coping strategies.

The identification of these five categories is the major element of the paradigm model of this particular study and should be beneficial to add to the existing literature and for the stakeholders to have a better understanding on the phenomenon.

All factors emerged evidently are interpretive in nature and entail advanced exploration; however, there was sound consistency among the informants in this research. After all, most informants illustrated these factors impulsively and when asked truthfully to assess these factors in Phase 4 of this study using peer member checks, all informants (n = 10) strongly agreed with them.

4.2 Contexts and Conditions that Influence Antecedents

For this perspective, only staff are being interviewed as it is obligatory for them to meet the standards established by university. According to the informants (M = 65%, F = 35%), the obligations to the university is the main reason for academic and non-academic staff perform the paper and online evaluations every semester. However, for some (M = 34%, F = 66%), they made it clear that they did it as it was being instructed by their head of department and immediate supervisor. Nonetheless, there are lecturers (M = 25%, F = 75%) that have desire to be included in the exercise for they need to improve their performance in class.

The improvement of the Internet facilities is the key factor that encourages the staff to switch the old and obsolete technique to newer and improved networking methods. Additionally, in this day and age these Internet technologies are impossible to be separated from staff and students. With applications such as Facebook and Google, staff and students are able to communicate, store, collaborate, and manage information anywhere and anytime.

4.3 Antecedents

Findings in the current study showed that staff and students attributed staff and course evaluations to three types of antecedents. These include the academic staff's behavior, non-academic staff's behavior, and students' behavior. Results show that laziness has been a major issue for all three types of informants as some informants (M = 70%, F = 30%) said that there are many things to achieve rather than concentrating on lecturer's evaluation, which previously has no positive impact whatsoever. Many students and staff saw that paper evaluation is actually a waste of time and resources; they had to do it only because of their carry marks. Some students voiced that they assume lecturers would not read their comments at all. For the lecturers, many of the informants (M = 69%, F = 31%) expressed that they are afraid if the students write the truth about their bad performance or because of personal factors students will tell lies regarding their teaching approach. Thus, many informants consider it as less important things to do and they will do it just to abide the instruction from the university.

On the other hand, there are informants who are totally enthusiastic about the evaluation. For students, the informants (M = 36%, F = 64%) suggest that the practice is important and it is their responsibility to enlighten the academic staff's performance whether it is an excellent or a poor ability during teaching and learning process. Being a lecturer, it is a good exercise to realize their weaknesses and thus can improve his or her teaching techniques.

4.4 Central Phenomenon

The researchers have become aware of the fundamental theme that surfaces from the interviews (during the open coding investigative process), namely staff and students achieve "a feeling of fulfillment" over their lives through the utilization of both paper and online evaluations. However, this feeling of fulfillment has produced advantages and disadvantages of the two approaches and this in turn will generate several coping strategies in order for them to regain yet another sense of satisfaction in their lives. Informants in the current study described a variety of disadvantaged aspects. However, apart from the drawback, the researchers found that there are benefits of both methods.

4.4.1 Advantages of Paper Evaluation

To academic staff, according to the informants (M = 90%, F = 10%), the most vital characteristic of paper evaluation is that they can easily remove the unpleasant students' assessment by discarding the OMR sheets. Other criteria are that they (M = 55%, F = 45%) can easily instruct and explain and administer the process of evaluation face-to-face in order to minimize mistakes.

4.4.2 Advantages of Online Evaluation

Facebook and Google Apps indirectly have improved staff and students' knowledge and awareness of Internet applications. These applications are at no cost and maintenance-free, plus they are ready to be used anywhere and anytime.

Many informants (M = 60%, F = 40%) admitted that online evaluation responds to student concerns for privacy and anonymity. Students can anonymously evaluate their lecturer with a peace of mind, that they can critique them if they are unhappy with the way their lecturer handles lectures, et cetera.

According to administrative staff, the availability of data for analysis and reporting is instantaneous in online evaluation. As soon as students input their responses, the system immediately analyzes and findings can be generated for viewing in an instant. In addition, more extensive qualitative responses to the open-ended questions can now be viewed and scrutinized with ease. Other than that, according to informants (M = 34%, F = 66%), online evaluation has enforced uniformity for the evaluation of all courses, that prior to this while exercising paper evaluation, there are difficulties when collecting and analyzing data for social science schools as the OMR sheets and analyzing machines are tuned toward engineering subjects.

Finally, the researchers found that online evaluation has provided faculties with the flexibility to add items specific to their courses to address the accreditation requirements. In other words, it provides the faculties with the total freedom to fine-tune all questions according to the course content and expertise of the lecturers.

4.4.3 Disadvantages of Paper Evaluation

There are numerous information regarding the drawback of paper evaluation from the informants (M = 80%, F = 20%) interviewed. All informants have raised the same concern and that data saturation has been detected at informant number seven during Phase 1 data collection. According to informants, since the original OMR sheets are expensive, many schools resorted to utilizing copied sheets and thus many of them cannot be analyzed by the OMR machines. Furthermore, because of this, many OMR sheets arrived in damaged state and cannot be scanned by the OMR machines. Additionally, OMR machines are always out of order plus the maintenance fees are expensive and the machines cost the university a fortune to replace them.

Another downside of paper evaluation is that qualitative data cannot be interpreted by OMR machines, albeit it is very crucial to be interpreted as all data are genuinely from the students themselves. Results and summaries of the analysis also are often ambiguous and fail to provide action-oriented solutions in view of the fact that many students fill out the forms for granted, yielding unreliable outcome. To make matters worse, the dissimilarities of items in the OMR sheets are not compatible with Social Science schools as they are arranged to meet the standards of Engineering-related subjects.

In terms of practical downsides, feedback from the informants (M = 60%, F = 40%) was that the completed OMR sheets were often misplaced or not collected at all. This frequently happened during the queuing process of data analysis. Informants also confirmed that after tedious analyzing process has been done, which in some cases took an exceptionally long time, some of the OMR sheets are regularly being delivered back to the wrong place or worse, not being delivered at all. This is due to staff that administer the overall process are insufficient in some schools that handle too many students. This leads to incomplete academic audit reports and schools are getting a non-conformant status for failing to produce academic evaluation reports.

Succinctly, informants (M = 75%, F = 25%) suggested that paper evaluations are the basis for extremely stressful decisions rather than an instrument for improving teaching and learning for the academicians.

4.4.4.Disadvantages of Online Evaluation

According to the informants (M = 66%, F = 34%), especially academic staff, it is very difficult to ensure that the students will evaluate only once and the accountability and reliability of the process is also very hard to measure as at the moment it can be certain that the staff themselves pretend to be students in order to try to evaluate with positive comments. The participation from students is of course not 100% and it is possible to generalize the population.

4.5 Coping Strategies

There are several coping strategies being enlightened by the informants (M = 52%, F = 48%) in order to deal with the weaknesses of staff and course evaluation, predominantly paper evaluation. Informants mentioned that they have done the following, for example by using copied OMR sheets to save cost and to meet the demand of high population of students. Some schools, according to the informants, will carry on with copied sheets even though many were rejected by the machines.

For some academic staff (M = 90%, F = 10%), in order to have good outcome for paper evaluation, they resorted by filling out the OMR sheets themselves or simply by discarding one that yields negative comments. For online evaluation, these lecturers will straightforwardly log in and disguise themselves as students and input very positive comments.

4.6 Consequences

By looking at the drawbacks of paper evaluation that are way too many compared to online evaluation, the university has decided to implement online evaluation for all schools beginning 2015. This accomplishment has improved the quality of administrative work significantly. There are no papers to be wasted and costs are reduced drastically. Every semester, the researchers' school has managed to report the findings to all lecturers within a time frame given and this achievement has improved the quality of teaching and learning a great deal.

5. Conclusions

The majority of educators would coincide that course evaluations are vital not only for tenure and promotion but also to give the university feedback to improve teaching and give students an opportunity for constructive input into their own learning. If the findings of this research are correct, the blend of administrative convenience, lecturers' desire for timely feedback, and course improvement through formative evaluation will cease the utilization of traditional paper evaluations and make online evaluations the method of preference.

Staff and course evaluations in any universities are likely to shape professional and career advancement, promotion, tenure, and of course the course content itself to be improved. Not many spheres in higher education become sources for more anxiety than staff and course evaluations, and few areas have been studied more for validity and reliability (Wachtel, 1998). From this current research and in line with many other studies, the researchers conclude that online staff and course evaluations bring the advantages of saving time, resources and many more over the traditional paper method of evaluation. As a consequence, academic staff may be required to utilize them in place of the more burdensome paper method.

References

[1] Anderson, H. M., Cain, J., & Bird, E. (2005). Online student course evaluations: Review of literature and a pilot study. American Journal of Pharmaceutical Education, 69(1), article 5 (2005).

[2] Anderson, H., Cain, J., & Bird, E. (2011). Online student course evaluations: Review of literature and a pilot study. American Journal of Pharmaceutical Education 2005, 69(1), Article 5.

[3] Aleamoni, L. M. (1999). Student rating myths versus research facts from 1924 to 1998. Journal of Personnel Evaluation in Education, 13(2), 153–166.

[4] Avery, R. J., Bryant, W. K., Mathios, A., Kang, H., & Bell, D. (2006). Electronic course evaluations: Does an online delivery system influence student evaluations? Journal of Economic Education, 37(1), 21–37.

[5] Ballantyne, C. S. (2003). Online evaluations of teaching: An examination of current practice and considerations for the future. In D. L. Sorenson & T. D. Johnson (Eds.), Online student ratings of instruction: New directions for teaching and learning, No. 96, Winter 2003, Jossey-Bass.

[6] Ballantyne, C. S. (2004). Online or on paper: An examination of the differences in response and respondents to a survey administered in two modes. Paper presented to the Australasian Evaluation Society Annual Conference, Adelaide, South Australia, 13–15 October, 2004. http://www.aes.asn.au/conference2004/index.htm#fri

[7] Bard, J. S. (1987). Perceived learning in relation to student evaluation of university instruction. Journal of Educational Psychology, 79(1), 90–91.

[8] Bothell, T. W., & Henderson, T. (2003). Do online ratings of instruction make sense? In D. L. Sorenson & T. D. Johnson (Eds.), Online student ratings of instruction: New directions for teaching and learning, No. 96, Winter 2003, Jossey-Bass.

[9] Centra, J. A. (2003). Will teachers receive higher student evaluations by giving higher grades and less coursework? Research in Higher Education, 44(5), 495–518.

[10] Centra, J. (1993). Reflective faculty evaluation: Enhancing teaching and determining faculty effectiveness (Jossey Bass higher and adult education series). San Francisco: Jossey-Bass Inc Pub.

[11] Crawford, K. H., Leybourne, M. L., & Arnott, A. (2000). How we ensured rigour in a multi-site, multi-discipline, multi-researcher study. Forum: Qualitative Social Research. 1, Art. 12. Retrieved from http://www.qualitative-research.net/fqs

[12] Creswell, J. W. (1998). Qualitative inquiry and research design: Choosing among five traditions. Thousand Oaks, CA: Sage.

[13] Donovan, J. E., Mader, C., & Shinsky, J. (2006). Constructive student feedback: Online vs. traditional course evaluations. Journal of Interactive Online Learning, 5(3), 283–296.

[14] Gigliotti, R. J., & Buchtel, F. S. (1990). Attributional bias and course evaluations. Journal of Educational Psychology, 82(2), 341–351.

[15] Greenwald, A. G., & Gillmore, G. M. (1997). Grading leniency is a removable contaminant of student ratings. American Psychologist, 52(11), 1209–1217.

[16] Harry, B., Sturges, K. M., & Klingner, J. K. (2005). Mapping the process: An exemplar of process and challenge in grounded theory analysis. Educational Researcher, 34, 3–13.

[17] Hobson, S. M., & Talbot, D. M. (2001). Understanding student evaluations: What all faculty should know. College Teaching, 49(1), 26–31.

[18] Jirovec, L., Chathapuram, S. R., & Rosegrant-Alvarez, A. (1998). Course evaluations: What are social work students telling us about teaching effectiveness? Journal of Social Work Education, 34(2), 229–236.

[19] Johnson, V. E. (2002). Teacher course evaluations and student grades: An academic tango. Chance, 15(3), 9–16.

[20] Kasiar, J. B., Schroeder, S. L., & Holstad, S. G. (2002). Comparison of traditional and Web-based course evaluation processes in a required, team-taught pharmacotherapy course. American Journal of Pharmaceutical Education, 66, 268–270.

[21] Kirk, J., & Miller, M. L. (1986). Reliability and validity in qualitative research. London: Sage.

[22] Krueger, R. A. (1994). Focus groups: A practical guide for applied research (2nd ed.). Thousand Oaks, CA: Sage.

[23] Liegle, J. O., & McDonald, D. S. (2004). Lessons learned from online vs. paper-based computer information students' evaluation system. Proceedings from the Information Systems Education Conference 2004, v 21 (Newport): §2214. ISSN: 1542-7382.

[24] Lincoln, Y. S., & Guba, E. G. (1985). Naturalistic inquiry. Newbury Park, CA: Sage.

[25] Maxwell, J. A. (1996). Qualitative research design: An interactive approach. Thousand Oaks, CA: Sage.

[26] Miles, M. B., & Huberman, M. A. (1994). Qualitative data analysis: An expanded sourcebook (2nd ed.). Beverly Hills, CA: Sage.

[27] Norris, J., & Conn, C. (2005). Investigating strategies for increasing student response rates to online delivered course evaluations. Quarterly Review of Distance Education, 6(1), 13–29.

[28] Overall, J. U., & Marsh, H. W. (1979). Midterm feedback from students: Its relationship to instructional improvement and students' cognitive and affective outcomes. Journal of Educational Psychology, 71(6), 856–865.

[29] Richards, L., & Richards, T. (1994). From filing cabinet to computer. In A. Bryman & R. G. Burgess (Eds.), Analysing qualitative data (pp. 146–172). London: Routledge.

[30] Remedios, R., & Lieberman, D. A. (2008). I liked your course because you taught me well: The influence of grades, workload, expectations and goals on students' evaluations of teaching. British Educational Research Journal, 34(1), 91–115.

[31] Seale, C. (1999). The quality of qualitative research. London: Sage.

[32] Strauss, A., & Corbin, J. (1998). Basics of qualitative research: Techniques and procedures for developing grounded theory (2nd ed.). Thousand Oaks, CA: Sage.

[33] Strauss, A., & Corbin, J. (1990). Basics of qualitative research: Grounded theory procedures and techniques. Newbury Park, CA: Sage.

[34] Wachtel, H. K. (1998). Student evaluation of college teaching effectiveness: A brief review. Assessment and Evaluation in Higher Education, 23(2), 191–211.

IOP Conf. Series: Journal of Physics: Conf. Series **1019** (2018) 012029 doi:10.1088/1742-6596/1019/1/012029

Design and Performance Analysis of Solar PV DC Power System for Disaster Relief Centre using PVsyst

T M N T Mansur[1,2], N H Baharudin[1] and R Ali[1]

[1]School of Electrical System Engineering, Universiti Malaysia Perlis, 02600 Arau, Perlis, MALAYSIA

[2]Centre of Excellence in Renewable Energy (CERE), Universiti Malaysia Perlis, 02600 Arau, Perlis, MALAYSIA

tunkunizar@unimap.edu.my

Abstract. One of essential elements at disaster relief camps is electricity supply. Categorized under disaster preparedness, it is required to support humanitarian aids, provide security and convenient for the victims involved. The objective of this project is to design a standalone DC power system for a disaster relief camp at Kota Bharu, a capital for Kelantan State located at east coast of Peninsular Malaysia from renewable energy resource which is solar PV. The methodology of the project are configuring DC load demand, sizing PV array and battery bank from solar energy resource and load demand and lastly simulation of the design system by using PVsyst. In general, a relief camp will consist of basic electrical appliances such as lightings, fans and hand-phone chargers while in common area could be equipped with flood lightings, air-conditioning unit and refrigerator to store foods and medicine. The solar PV array, battery bank and DC bus regulator power rating are designed accordingly to the local meteorological data and load profile based on desired operating schedule. The system operates in stand-alone mode in which the power is fully supplied in DC voltage. The energy generated from solar array will be stored in the battery bank while the operation and management of the system is controlled and coordinated by local controller to ensure energy optimization system. The amount of user's load demand is 8,820Wh per day. Based on solar energy resource, the optimum PV array size is 3.0kWp while the battery bank size is 900Ah at 48V which is designated for 4 days of autonomy. The system could meet up to 99% of load demand throughout the year except for December with 62.3% performance ratio. The loss to the system is contributed by the temperature effect to the PV module ohmic wiring loss, unused energy because of battery full, converter efficiency and battery efficiency.

1. Introduction

In 2014, Malaysia was hit by one of the worst flood disaster in history where more than 100,000 victims have been moved to evacuation centre. In Kelantan area, there are extreme cases where victims have to live without food, clean water and electricity for more than 40 hours due to the severity of floods that prevented delivery of aids [1].

Electricity supply is one of elements required to support humanitarian aids such as disaster relief centre or camps. So far, electricity technology for disaster relief centre is not given serious attention since government's building are normally been used as a centre where electricity is already available. In the meantime, outdoor camps only focused on traditional fuel such as wood, biomass, agricultural waste and conventional fossil fuels for heating, cooking, lighting and fuel driven electric generators for power [2]. However, based on the extreme case during 2014 flood disaster, backup power supply especially portable that could support basic loads such as for lighting and communication devices should be considered.

Motivated by environmental and economic conditions, renewable energy resources such as from solar photovoltaic (PV) is currently popular technologies used especially for remote and rural electrification. In recent years, tremendous improvements in efficiency and cost effectiveness of solar PV technology have been made [3]. In addition, there have been intensive efforts to utilize DC in distribution system to supplement the traditionally used AC system [4]. Moreover, many of today's consumer loads are available at low voltage DC such as LED lightings and hand phone's charger [5]. The advancement achieved in power electronics has made the DC voltage regulation a simpler task and better efficiency. Today's solid-state switching DC converters have a power conversion efficiency in the range of 95% [6]. Compared to AC system, the energy conversion losses of DC system is minimized [7] since there is no reactive power [8].

2. Methodology

The objective of this project is to design a standalone DC power system for a disaster relief camp from renewable energy resource which is solar PV. The methodology of the project are configuring DC load demand, sizing PV array and battery bank from solar energy resource and load demand and lastly simulation of the design system by using PVsyst. PVsyst is professional software used to design and simulate the performance of the system related to solar PV.

The project site that has been chosen is Kota Bharu (6.17°N, 102.28°E), a capital for Kelantan State which located at the east cost of Peninsular of Malaysia. This area is severely affected by the heavy flood in 2014. This research is expected to give benefit to the authorities involved in disaster management in the affected area.

2.1. Solar Energy Resource

Information on solar energy for Kota Bharu is obtained from Meteonorm 7.1 which provided monthly meteorological data that has been preloaded in PVsyst. The annually global solar irradiation is 1,705kWh/m^2 and the average ambient temperature is 27.2°C. Figure 1 below shows the monthly global solar irradiation and average ambient temperature for Kota Bharu.

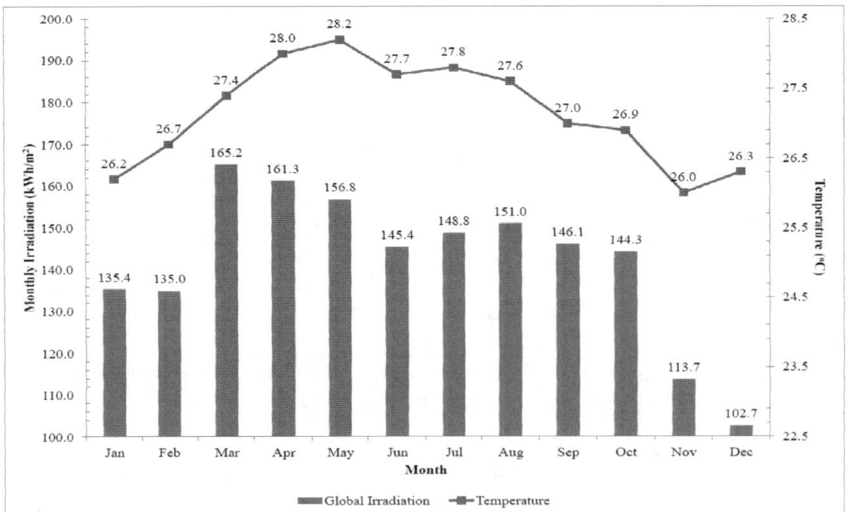

Figure 1. Global solar irradiation and average temperature for Kota Bharu.

2.2. Load Demand

In general, a relief camp will consist of basic electrical appliances such as lightings, fans and hand phones charger while a common area could be equipped with flood lightings, air-conditioning unit and refrigerator to store foods and medicine. These loads will be divided according to their power rating where the high power loads will operate at higher voltage level which is 48V while the low power appliances operates at low voltage level which is 12V. List of loads with their respective power rating and their daily energy requirement is shown in Table 1 while Figure 2 shows the hourly energy demand from the loads. From there we can summarize that loads will operate at night only for 6 hours with total daily energy demand is 8,820 Wh.

Table 1. Daily Energy Consumption for Disaster Relief Centre.

Appliances	Quantity (Unit)	Power (W)	Hours (hr)	Energy (Wh)
LED lighting	10	12	6	720
Fan	10	12	6	720
Charger	10	5	6	300
Flood lighting	10	30	6	1,080
DC Air-conditioner Unit	1	1,000	6	6,000

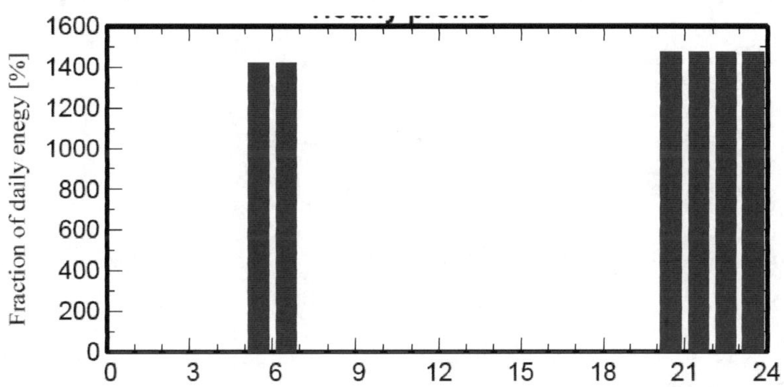

Figure 2. Hourly Energy Demand for Disaster Relief Centre.

2.3. Sizing of PV Array

Solar PV array is sized to meet the required energy demand. The output energy of PV array is a function of peak sun shine hours and temperature. The estimation of energy generated from PV array is simplified and calculated based on Equation (1) below [9].

$$E_{array} = P_{array} \times PSH_{period} \times f_{temp} \quad (1)$$

where E_{array} is PV array yield in kWh, PSH_{period} is Peak Sun Hour and f_{temp} is temperature de-rating factor. The temperature de-rating factor, f_{temp} is given by the Equation (2)

$$f_{temp} = 1 + \left[\left(\frac{\gamma_{pmp}}{100} \right) \times \left(T_{cell\_ave} \times T_{stc} \right) \right] \quad (2)$$

where γ_{pmp} is temperature coefficient for P_{mp} in % per °C, T_{cell_ave} is average daily maximum cell temperature and T_{stc} is cell temperature at standard test condition which is 25°C . The average cell temperature, T_{cell_ave} is given by Equation (3)

$$T_{cell\_ave} = T_{amb\_ave\_\max} + \left[\left(\frac{NOCT - 20}{800} \right) \times G_{amb\_ave\_\max} \right] \quad (3)$$

where $T_{amb\_ave\_max}$ is average daily maximum ambient temperature in °C, NOCT is nominal operation cell temperature in °C, $G_{amb\_ave\_max}$ is average daily maximum solar irradiance in Wm^{-2}.

The PV module used is Yingli Solar YL250P-29b which is preloaded in the PVsyst database. Selected parameter and their value for the PV module is shown in Table 2 below.

Table 2. PV Module specification

Parameter	Value at Standard Test Condition (STC)
Power output, P_{max}	250.0 Wp
Voltage at Pmax, V_{mpp}	30.23 V
Current at Pmax, I_{mpp}	8.27 A
Open-circuit voltage, V_{oc}	37.73 V
Short-circuit current, I_{sc}	8.83 A
Module efficiency	15.40 %
Temperature coefficient at Pmax, γ_{mpp}	-0.42 % / °C

2.4. *Sizing of Battery Bank*

The amount of charge required is based on Equation (4) below

$$C_{Ah\_req} = \frac{E_{req}}{V_{sys}} \times \frac{k_{storage}}{DOD_{max}} \times T_{aut} \quad (4)$$

where C_{Ah_req} is charge storage capacity required in Ah, V_{sys} is the system voltage, $k_{storage}$ is the battery efficiency, DOD_{max} is maximum depth of discharge allowed and T_{aut} is number of autonomy days.

The system voltage is set to 48V, battery efficiency is set to 97%, number of autonomy days is set to 4 days and maximum depth of discharge is set to 80%.

3. Result

PVsyst software is used to simulate the performance of the system. Summary of the system configuration is shown in Figure 3.

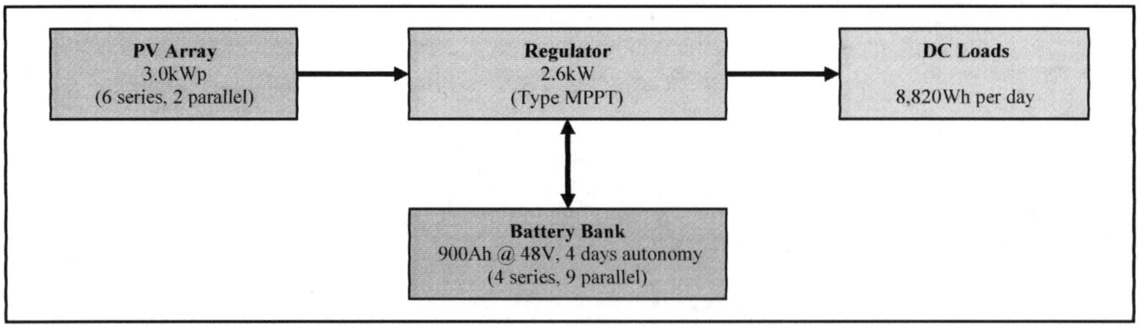

Figure 3. Summary of the system configuration.

Table 3 summarized the system performance throughout the year. In general, the system could meet the loads demand for every month except in December due to the lowest solar irradiation compared to the other months. The annual energy requirement is 3,183kWh while the system could supply up to 3,152 kWh which is 99% of load demand.

Table 3. Summary of the system performance.

	GlobHor kWh/m²	GlobEff kWh/m²	E Avail kWh	EUnused kWh	E Miss kWh	E User kWh	E Load kWh	SolFrac
January	135.4	143.5	342.8	46.0	0.00	270.3	270.3	1.000
February	135.0	137.5	330.1	71.4	0.00	244.2	244.2	1.000
March	165.2	161.4	386.5	105.0	0.00	270.3	270.3	1.000
April	161.3	149.5	355.4	68.7	0.00	261.6	261.6	1.000
May	156.8	138.4	332.0	45.3	0.00	270.3	270.3	1.000
June	145.4	126.7	305.8	31.2	0.00	261.6	261.6	1.000
July	148.8	130.8	313.8	36.0	0.00	270.3	270.3	1.000
August	151.0	137.1	327.7	32.2	0.00	270.3	270.3	1.000
September	146.1	139.9	333.5	57.2	0.00	261.6	261.6	1.000
October	144.3	143.9	343.7	55.4	0.00	270.3	270.3	1.000
November	113.7	117.0	280.4	27.1	0.00	261.6	261.6	1.000
December	102.7	105.4	254.6	0.0	30.89	239.4	270.3	0.886
Year	1705.6	1631.0	3906.1	575.5	30.89	3151.9	3182.8	0.990

Figure 4 summarized the performance ratio and solar fraction of the system and Figure 5 summarized the system's energy loss throughout the year. The nominal PV array energy at STC is 4,893 kWh but only 3, 563 kWh is available as output from it. The losses are due to the temperature effect to the PV module (11.6%), ohmic wiring loss (2.3%) and unused energy because of battery full (13.9%). From PV array output, only 3,152 kWh has been supplied to the users where the main losses are from converter efficiency (6.5%) and battery efficiency (4.5%). The overall performance ratio is 62.3%.

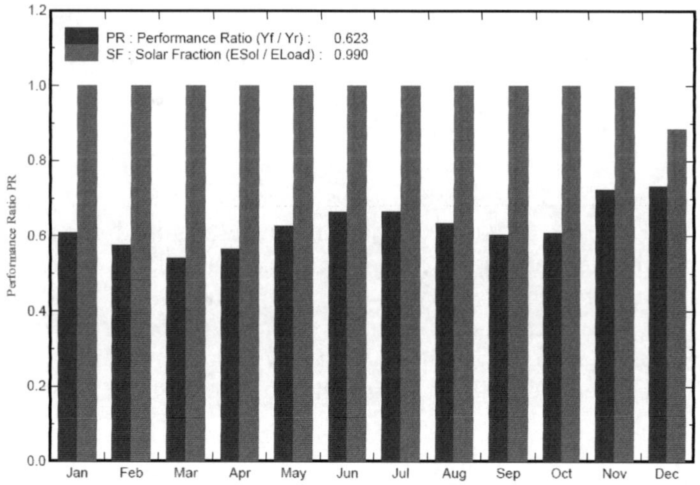

Figure 4. Performance Ratio and Solar Fraction of the system

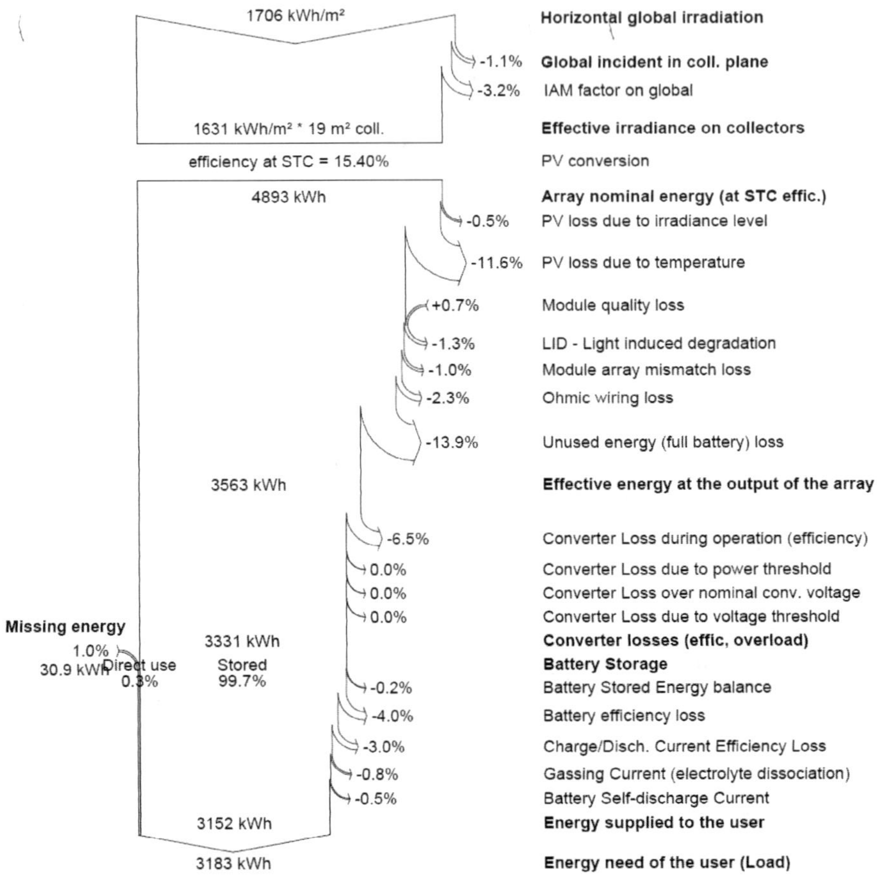

Figure 5. Loss diagram of the system.

4. Conclusion

The DC power system for disaster relief camp powered by solar PV in Kota Bharu has been designed and simulated by using PVsyst. The amount of user's load demand is 8,820Wh per day. Based on solar energy resource, the optimum PV array size is 3.0kWp while the battery bank size is 900Ah at 48V which is designated for 4 days of autonomy. The system could meet up to 99% of load demand throughout the year except for December with 62.3% performance ratio. The loss to the system is contributed by the temperature effect to the PV module ohmic wiring loss, unused energy because of battery full, converter efficiency and battery efficiency.

References

[1] Malaymail Online, "Kuala Krai flood victims trapped without food, water or electricity at shelter.," 26-Dec-2014.

[2] J. Franceschi, J. Rothkop, and G. Miller, "Off-grid Solar PV Power for Humanitarian Action: From Emergency Communications to Refugee Camp Micro-grids," *Procedia Eng.*, vol. 78, pp. 229–235, 2014.

[3] K. Shah, S. Member, P. Chen, A. Schwab, and K. Shenai, "Smart Efficient Solar DC Micro-grid."

[4] S. Sidopekso and M. Taufik, "The DC House Project for Sustainable Rural Electrification," no. November 2012, pp. 2–6, 2013.

[5] A. T. Ghareeb, A. a. Mohamed, and O. a. Mohammed, "DC microgrids and distribution systems: An overview," *IEEE Power Energy Soc. Gen. Meet.*, vol. 119, pp. 407–417, 2013.

[6] K. Shenai, K. Shah, and S. Member, "Smart DC Micro-grid for Efficient Utilization of Distributed Renewable Energy."

[7] M. Amin, Y. Arafat, S. Lundberg, and S. Mangold, "An efficient appliance for low voltage DC house," pp. 334–339, 2011.

[8] E. R. Diaz, X. Su, M. Savaghebi, J. C. Vasquez, M. Han, and J. M. Guerrero, "Intelligent DC Microgrid living Laboratories - A Chinese-Danish cooperation project," *2015 IEEE First Int. Conf. DC Microgrids*, pp. 365–370, 2015.

[9] A.M. Omar, S. Shaari, S.I. Sulaiman, *Grid-Connected Photovoltaic Power Systems Design*. Sustainable Energy Development Authority Malaysia, 2012.

Green Skills for Green Industry: A Review of Literature

Lai Chee Sern[1], Adib Farhan Zaime[1], Lee Ming Foong[1]

[1] Faculty of Technical and Vocational Education, Universiti Tun Hussein Onn Malaysia, 86400 Parit Raja, Batu Pahat Johor, Malaysia

E-mail: lcsern@uthm.edu.my, adibfarhanzaime@gmail.com, mflee@uthm.edu.my

Abstract. Apart from technical and generic skills, employers start looking for manpower with green skills which are much needed for promoting sustainable development in social, economic as well as environment. This article aimed at finding out the major green skills demanded by the green industry based on the existing literature. In the present study, document analysis was employed. The published articles and technical reports related to green skills in various disciplines were identified from electronic databases that includes ScienceDirect, Google Scholar, ResearchGate, and Academia. The findings from literature review revealed that there are ten most common green skills demanded by various industrial sectors, namely, design skill, leadership skill, management skill, energy skill, city planning skill, landscaping skill, communication skill, waste management skill, procurement skill, and financial skill. The role and directions of skill training institutions should be aligned with the needs and wants of the industrial sectors. Therefore, it is imperative for the skill training institutions to revise and revamp their curriculum in order to produce graduates with green skills to cater for the demand of manpower market.

1. Introduction

Two major types of skill have been demanded by the employers, namely, technical skills and generic skills. In general, technical skills are regarded as the skills that are related to occupation, such as programming, repairing car, and it is hands-on oriented [1]. Whereas generic skills are the abilities for dealing with job and persons such as communication skills, team work skills, problem solving and critical thinking [2, 3]. Both technical skills and generic skills are important for getting a job. However, this scenario has changed. Someone who has technical skills and generic skills are no more sufficient for him/her to secure a job [4].

Apart from the above-mentioned skills, employers start looking for manpower with green skills which are much needed for promoting sustain able development in social, economic as well as environment. To date, most economic activities do not contribute to sustainable development of social, economic and environment due to low technical and technological levels [5]. Countries around the world, especially the developing ones, have paid a very high cost on environmental pollution and climate change, which in turn will affect the social and economic activities [6]. Environmental pollution and climate change are negatively affecting the sustainability of various economical activities across the world with Malaysia being one of them. The changes of climate and environmental issues will bring for changing weather pattern, more extreme droughts or monsoons, rising sea levels and so on [7]. These consequences will cause devastating impacts on social as well as economy if effective measures and pragmatic actions are not taken to mitigate the problems.

2. The Impacts of Climate Change on Economy

One of the major challenges of today is to prevent the climate change. According to Omar [8], major human activities that cause climate change are as follows;

i. Industries - factory smoke emissions such as carbon monoxide and harmful gas causing the greenhouse effect; depletion of ozone due to the use of chlorofluorocarbons (CFCs) in aerosol product and air conditioners.
ii. Transportation - the burning of petrol causing the greenhouse effect.
iii. Logging - deforestation cause the unbalanced air content (carbon dioxide vs oxygen) in the air.
iv. Urbanization and agriculture - increase the temperature due to over-development for building construction and agriculture.

From the economy perspective, the developing countries such as Thailand and the Philippine have been affected with climate change [9]. Specifically, in 2007, Thailand has experienced economy losses of about RM17 billions due to the drought event. The worst affected sectors were agriculture and livestock [10]. Likewise, the Philippines has also encountered enormous losses due to climate change such as typhoon has caused RM24 billion, earthquake caused RM22 billion, volcano caused RM10 billion, and flood caused RM19 billion losses [11]. According to Ueda *et al.* [12], the climate change has negatively impacted on the economy of almost every country in Asia. Table 1 shows the probability of damage in Gross Domestic Product (GDP) due to climate change.

Table 1. Percentage of gross domestic product (GDP) by probability of climate change occurrence

Developing Asian Countries	GDP 2011 (USD billion)	Probability of Damage				
		5.00%	1.00%	0.50%	0.10%	0.01%
Afghanistan	16.70	0.33	1.27	1.51	1.73	1.79
Bangladesh	108.55	0.26	0.93	1.09	1.24	1.28
Cambodia	12.56	0.75	1.78	2.23	3.27	4.75
People's Republic of China	6,254.76	0.26	0.70	0.80	0.88	0.90
India	1,587.83	0.17	0.28	0.30	0.32	0.32
Indonesia	767.75	0.17	0.89	1.10	1.30	1.35
Kazakhstan	148.51	0.11	0.20	0.22	0.23	0.23
Laos	6.67	0.95	4.08	4.89	5.66	5.85
Malaysia	241.22	0.16	0.41	0.46	0.50	0.51
Maldives	1.98	1.24	2.48	4.96	14.89	100.59
Mongolia	7.55	2.00	5.96	9.94	11.93	16.03
Myanmar	48.03	0.20	2.19	2.94	3.73	3.93
Nepal	16.58	1.00	3.53	4.14	4.70	4.83
Pakistan	175.61	0.98	4.01	4.79	5.51	5.69
Philippines	193.31	0.19	0.30	0.32	0.33	0.34
Sri Lanka	53.84	0.60	1.44	1.60	2.10	3.03

Tajikistan	5.76	8.05	11.19	11.66	15.67	18.22
Thailand	323.51	0.34	1.18	2.02	3.51	3.51
Tonga	0.35	2.38	8.89	12.58	14.35	14.78
Vietnam	103.65	0.54	0.69	0.71	0.72	0.73
Total	10,184.09	0.26	0.74	0.87	1.02	1.0

Refers to Table 1, China was recorded as the country that has the highest GDP (USD6254.76 billion). On the other hand, Tonga has only USD0.35 billion, which was the lowest compared to other Asian countries. If the probability of damage is 1.00% for both countries, China was faced with USD 62.5476 billion loss of money, whereas Tonga USD0.0035 billion. This statistical data reflects that a country needs a large amount of capital to stimulate and recover the economy which many countries cannot afford to execute the economic plan due to lack of capital [12].

3. Lack of Awareness

One of the main factors that causes climate change is the lack of awareness on environmental protection among people. Previous research revealed that many people have scarce knowledge about climate change and, worse still, they did not know about the risk of climate change [15]. Similar findings have been reported by National Environmental Education and Training Foundation (NEETF) [16] and the Asia Foundation [18] which showed that majority of people do not know causes of pollution happened and do know how to react to climate change. Likewise, Vynne, and Doppelt [17] have also discovered that 97% of respondents involved in their survey did not considered the climate change preparation as their main priority. Although government of many countries have come out with policies and programmes to deal with the issues related to environmental pollution, many people are not aware of those policies and programmes [19,21,22,23]. These appalling findings reflects the fact that many people around the world are not aware of the severity of the impacts caused by climate change.

In Malaysia, the target has been set in which the Malaysian government aims to reduce carbon dioxide intensity by up to 40% by the year 2020 [25]. In line with this aim, Malaysian government had come out with 11th Malaysian Plan (11th MP) which puts part of focuses on sustainable development. There are six strategies thrusts under 11th MP, namely, (i) inclusiveness, (ii) improving wellbeing for all, (iii) accelerating human capital development for an advanced nation, (iv) pursuing green growth for sustainability and resilience, (v) strengthening infrastructure to support economic expansion, and (vi) re-engineering economic growth for greater prosperity [26]. All of these thrusts are to ensure the visions of becoming a sustainable economy and developed nation is achieved by 2020 [26]. Specifically, the fourth thrusts emphasizes on green growth focusing on four key areas such as strengthening the enabling environment for green growth, adopting the sustainable consumption and production concept, conserving natural resources for present and future generations, and strengthening resilience against climate change and natural disasters [27]. In this context, green growth means to achieve a resilient, low-carbon, and resource-efficient economy model that leads to higher quality of living and promote well-being of the citizens.

4. Objective of Study

Some countries especially the developed countries start taking precautions and steps to deal with the environmental issues by shifting current economy model to green economy [13]. Green economy means manufacture, commercial, service and other economic activities are more to environmental friendly which produce lower carbon emissions, and consume lesser unrenewable energy that lead to improved human well-being and social equity [14]. The shift to green economy has increased the pace of change in manpower market and skills demanded by the green industries [13]. The green industries that underpins green economy requires workers with green skills to perform the tasks.

IOP Conf. Series: Journal of Physics: Conf. Series **1019** (2018) 012030 doi:10.1088/1742-6596/1019/1/012030

Although workers with green skills are needed by the green industries [20], it has not been given emphasise by the higher learning institutions, especially those focus on Technical and Vocational Education and Training (TVET). The training institutions must go beyond for equipping graduates with technical and soft skills because green skills is equally important and they must have been embedded in all levels of education. In this aspect, TVET institutions have an essential role to play in produce green skills workers since TVET is closely and directly connected with the development of economy and society. This crucial role of TVET is in line with the Malaysia Education Blueprint 2015-2015 (Higher Education) which puts the emphasizes on talent excellence and quality TVET graduates as two out of its 10 shifts [24]. It is meaningless and useless to talk about quality if the skills mastered by the graduates do not meet the needs of the green industries.

To date, it remains unclear concerning the types of green skills demanded by the industries due to varied nature of work in different industries. Therefore, the current study was conducted to find out what type of green skills are actually sought by the green industry.

5. Definition of Green Skills

A long time ago, only the gardeners have green skills, but today green skills are going to be an important skills to people. The term of green skills appeared since the advent of green technologies [29]. Green skills are needed to operate and develop green technologies. In Malaysia, the Ministry of Energy, Green Technology and Water is responsible to plan and formulate policies and program related to green technologies [30]. This reflects the fact that Malaysian government pays attention to deal with environmental pollution and climate change. Although green skills are important for green industry, it is still not unclearly defined. Most of people only know about generic skills, technical skills and employability skills, but their knowledge on green skill is very limited [31].

In general, green skills are regarded as skills for sustainability which are related to the technical skills, knowledge, values and attitudes needed in the workforce to develop and support sustainable social, economic and environmental outcomes in business, industry and the community [32]. Likewise, CEDEFOP [33] defines green skills as the knowledge, abilities, values and attitudes needed to live in, develop and support a society which reduces the impact of human activity on the environment. Taken together, green skills is generally composed of three dimensions, namely, knowledge (cognitive dimension), skills/abilities (psychomotor dimension), and attitudes/values (affective dimension) needed by workers to promote sustainable development in social, economy, and environment. From the cognitive dimension, the knowledge concerning environmental protection can be regarded as an element of green skills. From psychomotor perspective, green skill refer to the ability to, for instance, minimise energy consumption, or reduce greenhouse gases. Green skills also refer to affective aspect, for example, motivation of an individual to conserve natural resources.

6. Method

In order to explore the type of green skills demanded by green industry, document analysis was employed. It is vital to retrieve all relevant research, because loss of information might be leading to bias in the study. However, it is not feasible to obtain every single study on the subject because some of them are not published and, therefore, those articles may not be searchable. In the present study, published articles and technical reports related to green skills in various disciplines were identified from electronic database that includes ScienceDirect, Google Scholar, ResearchGate, and Academia.

7. Findings and Discussion

A total of ten green skills are identified from the existing literature. The skills are commonly demanded by various sectors depending on the duties and tasks. The identified green skills are as follows:

- i) Design skill
- ii) Leadership skill
- iii) Management skill
- iv) City planning skill

v) Landscaping kill
vi) Energy skill
vii) Financial skill
viii) Procurement skill
ix) Waste management skill
x) Communication skill

Specifically, design skill is regarded as an important green skill as the element of design is applied in many sectors, such as building design, machine design, and circuit design. The designer of today should be able to integrate green elements into their design in order to produce an idea that is friendly to the environment [34].

Apart from that, leadership skill and management skill are equally important because these skills are needed to change the organizational structure, function, and operation in order to support green activities, such as lean production or life-cycle management [35].

City planning skill and landscaping skill are also considered as one of the important green skills currently and in future. This is because many parts of the world are going through urbanization and the existing metropolitans are evolving to become smart cities that aims to generate a more convenient and modern places to live in [36]. These processes need proper planning and landscaping in order to make the cities livable and sustainable in long run.

Energy skill is seen as another critical green skill. To date, energy productions in the world are mostly from fossil fuel, such as coal, natural gas and petroleum. Based on the data from World Bank [37], the total consumption of fossil fuels was 86% compared to renewable and alternative resources (14%) in 2015. The by-products of fossil fuels consumption during energy generation process are very harmful to the environment [38]. Therefore, it is imperative to train workers with energy skills that help reduce the use of non-renewable resource in energy production and consumption, and at the same time replace those non-renewable resources with the ones that are more environmental friendly and safe to use.

Another green skills needed by the industry are financial skill and procurement skill. These skills are considered as green skills because they are important for economic responsibility that contributes to sustainable development. Financial skill is needed to control the expenditure of an organisation in order to balance up the revenue and responsibility for environmental conservation [39].

Likewise, procurement professionals will have to deal many internal departments of an organization as well as external agencies to manage, coordinate and purchase materials. Within green industrial context, procurement skill is very much needed to ensure the materials purchased are environmental friendly in order to minimise the environment impact during their life cycle [40].

Waste management skill is becoming increasingly significant skill for green industry. Waste management skill means the ability to reduce, reuse, and recycle waste through proper planning, implementation, and coordination of waste management system [41]. Waste management skill is highly demanded nowadays by the waste management sector which contributes enormously to the sustainability of environment and prevention of pollution.

Apart from the above-mentioned green skills, communication skill is another skill that is categorised as green skill and it is equally important for the environmental sustainability. Within the sphere of green industry, communication skill is not only referred to as the skill needed for verbal and non-verbal communication, but it also includes technological skills for communication which minimise energy consumption and more towards to environmental friendly type of communication [42].

8. Conclusion

There are a great demand of workers with green skill in green industrial sectors. These skill are needed to perform tasks that bring in profits without jeapordising our environment and ecosystem and to ensure sustainable economic development. The afore-mentioned green skills are the common skills needed by various industrial sectors. The role and directions of skill training institutions should be aligned with the needs and wants of the industrial sectors. Therefore, it is imperative for the skill training institutions to revise and revamp their curriculum in order to produce graduates with green skills to cater for the demand of manpower market. We strongly believe that there are many more skills that are not mentioned in this paper. Further studies should be more focused by determining specific green skills needed in certain sectors, such as computer manufacturing sector, construction sector, tourism, waste management industry, and so on. The expected findings might be very useful for skill training institution to redesign and restructure their curriculum based on the industrial needs.

Acknowledgement
The authors would like to express their gratitude to the Ministry of Higher Education Malaysia for supporting the research project (Vot 1554) under the Fundamental Research Grant Scheme (FRGS).

References
[1] Laker, D.R., & Powell, J.L. (2011). The differences between hard and soft skills and their relative impact on training transfer. Human Resource Development Quarterly. 22(1), pp. 111-122.
[2] Zhang, A. (2012). Peer Assessment of Soft Skills and Hard Skills. Journal of Educational Technology Education. vol 11, pp.155-168.
[3] Schulz, B. (2008). The Importance of Soft Skills: Education Beyond Academic Knowledge. Journal of Language and Communication. 2(1), pp. 146-154.
[4] Abdullah, Z., Mohamed, A., Sabran, M.S., Muthusamy, P., & Amin, S.M. (2009). *Kemahiran Berkesan*, UPM Press, Serdang, Selangor.
[5] Beder, S. (1994). The Role of Technology in Sustainable Development, Technology and Society, Vol. 13(4), pp.14-19.
[6] Chinowsky, P., Hayles, C., Schweikert, A., Strzepek, N., Strazepek, K., & Schlosser, C.A. (2011). Climate change: comparative impact on developing and developed countries. The Engineering Project Organization Journal. 1, pp. 67–80.
[7] Mousavi, M.E., Irish, J.L., Frey, A.E., Olivera, F., & Edge, B.L. (2011). Global warming and hurricanes: the potential impact of hurricane intensification and sea level rise on coastal flooding. Climatic Change, 104(3-4), pp. 575-597.
[8] Omar, R. (2010). *Perubahan Iklim* (Climate Change). Universiti Malaysia Perlis. Retrieved from http://dspace.unimap.edu.my/dspace/bitstream/123456789/10422/1/PERUBAHAN%20IKLIM2.pdf
[9] Rajoo, A., (2006). *Ekonomi Asia Tenggara Bakal Terjejas Akibat Perubahan Iklim.* Retrived from: http://www.bernama.com/bernama/v3/bm/news_lite.php?id=208688
[10] Naruchaikusol, S. (2016). Climate Change and Its Impact in Thailand: A Short Overview on Actual and Potential Impacts of The Changing Climate in Southeast Asia. Retrieved from http://www.transre.org/files/3114/6522/5151/Climate_Change_in_Thailand_TransRe_Fact_Sheet_No.2.pdf
[11] Rincon, G. M. F., & Virtucio, F. K. (2010). Climate Change in the Philippines: A Contribution to the Country Environmental Analysis. Retrieved from http://siteresources.worldbank.org/INTPHILIPPINES/Resources/PhilippineCEACC1July.pdf
[12] Ueda, T., Brubaker, A., Foerster, J., Eunkyung, K., Ocenar, L., & Labayen, S. (2013). The Rise of Natural Disasters in Asia and The Pacific: Learning from ADB's experience Mandaluyong City, Philippines: Asian Development Bank. ISBN 978-92-9254-194-1 (Print), 978-92-9254-195-8 (PDF).
[13] European Centre for the Development of Vocational Training, CEDEFOP. (2010). Skills for Green Jobs: European Synthesis Report. Luxembourg: Publications Office of the European Union.
[14] United Nations Environment Programme, UNEP. (2011). UNEP Year Book: Emerging Issues in Our Global Environment. Retrieved from www.unep.org/yearbook/2011

[15] Oyero, O., Oyesomi, K., Abioye, T., Ajiboye, E., & Adedeji, T.K. (2015). Behaioural Practices and Climate Change Awareness in Ado Odo/Ota, Ogun State, Nigeria: Implications for Communication and Development Agenda. International Conference on African Development Issues (CU-ICADI) 2015: Social and Economic Models for Development Track, pp. 359-364.

[16] National Environmental Education and Training Foundation, NEETF. (2001). The Ninth Annual National Report Card on Environmental Attitudes, Knowledge and Behaviour. Washington DC: NEETF.

[17] Vynne, S., & Doppelt B. (2009). Climate Change Health Preparedness in Oregon: An Assessment of Awareness, Preparation and Resource Needs for Potential Public Health Risks Associated with Climate Change. Retrieved from https://scholarsbank.uoregon.edu/xmlui/bitstream/handle/1794/10739/ORPHSurveyReportFinal.pdf

[18] The Asia Foundation. (2012). Climate Change Perception Survey. Retrieved from https://asiafoundation.org/resources/pdfs/ClimateChangeperceptionsurvey.pdf

[19] Al Buloshi, A.S., & Ramadan E. (2015). Climate Change Awareness and Perception amongst the Inhabitants of Muscat Governorate, Oman. American Journal of Climate Change, 4, pp. 330-336

[20] Olga, S., Christine, H., Mercedes, D. H., & Shinyoung, J. (2011). Skills for Green Jobs a Global View. Synthesis Report Based on 21 Country Studies. Geneva. International Labour Office

[21] Benton, T. (1994). Biology and Social Theory in The Environmental Debate, in M. Redclift and T. Benton (eds), Social Theory and The Global Environment, London and New York: Routledge.

[22] Agboola, O. S., & Emmanuel, M. (2016). Awareness of Climate Change and Sustainable Development among Undergraduates from Two Selected Universities in Oyo State, Nigeria. World Journal of Education, Vol. 6 (3), pp. 70-81.

[23] Pruneau, D., Liboiron, L., Vrain, E., Gravel, H., Bourque, W., & Langis, J. (2001) People's Ideas About Climate Change : A Source of Inspiration for The Creation of Educational Programs, Canadian Journal of Environmental Education, 6, pp.121-138.

[24] Ministry of Education, MoE. (2015). Executive Summary: Malaysia Education Blueprint 2015-2015 (Higher Education). Retrieved from http://hes.moe.gov.my/event/docs/4.%20Executive%20Summary%20PPPM%202015-2025.pdf

[25] Ministry of Energy, Green Technolog, and Water, Malaysia (2011). Low Carbon Cities Framework & Assessment System. ISBN: 978-967-5893-06-3

[26] Economic Planning Unit, Prime Minister's Department, Malaysia (2015). Eleventh Malaysia Plan 2016-2020. Anchoring Growth on People. Percetakan Nasional Malaysia Berhad. Kuala Lumpur.

[27] Asohan, A. (2015, Mei 21). 11th Malaysia Plan: Broadband Gets Some Love. Digital News Asia. Retrieved from https://www.digitalnewsasia.com/digital-economy/11th-malaysia-plan-broadband-gets-some-love

[28] Sharom, A. (2002). Malaysian Environmental Law: Ten Years After Rio. Singapore Journal of International & Comparative Law, vol 6, pp. 855-890.

[29] Buntat, Y., & Othman, M. (2012). *Penerapan Kemahiran Insaniah 'Hijau'* (Green Soft Skills) *Dalam Pendidikan Teknik dan Vokasional di Sekolah Menengah Teknik*, Malaysia. Journal of Social Science, 5. 32-41. pp. 32-41. ISSN 2231-7333.

[30] Ministry of Energy, Green Technology, and Water Malaysia, KeTTHA., (2015). National Green Technology Master Plan.

[31] Zolkifli, H., Kamin, Y., Abdul Latif, A., Buntat, Y., & Awang, Z. (2016). Generic Green Skills: Industry and Perspectives on Technical Education and Vocational Training (TVET). In: TVET@Asia, 6, pp. 1-13.

[32] McDonald, G., Condon, L. & Riordan, M. (2012). The Australian Green Skills Agreement: Policy and Industry Context, Institutional Response and Green Skills Delivery. Australia, NSW: TAFE

[33] CEDEFOP (2012). *Green skills and environmental awareness in vocational education and training: Synthesis Report.* Luxembourg: Publications Office of the European Union

[34] Ragheb, G., El-Shimy,H., Ragheb, A. (2015). Green architecture: a concept of sustainability Soc. Behav. Sci., 2, pp. 324-333

[35] UNEP (2012), *Measuring progress towards a green economy*, United Nations Environment Programme Report.

[36] Adhya, A., Plowright, P. & Stevens, J., (2010). "Defining sustainable urbanism: Towards a responsive urban design," in *Proceedings of the Conference on Sustainability and the Built Environment*, 2010, pp. 17–38.

[37] BP (2016). BP Statistical Review of World Energy June 2016, Retrieved 25th July 2017 from https://www.bp.com/content/dam/bp/pdf/energy-economics/statistical-review-2016/bp-statistical-review-of-world-energy-2016-full-report.pdf

[38] Patel, N. (2014). Environmental and economical effects of fossil fuels, *J. Recent Res. Engg. Tech.*, 1, 2349–2260

[39] Krechovská, M. (2015). Financial Literacy as a Path to Sustainability. *Business Trend*, 2, 3-12

[40] Bohari, A. A. M., & Xia, B. (2015). Developing green procurement framework for construction projects in Malaysia. In The Proceedings of the 6th International Conference on Engineering, Project, and Production Management (EPPM2015) (pp.282-290). Association of Engineering, Project, and Production Management (EPPM)

[41] Bozkurt, Ö & Stowell, A. (2016). Skill in the green economy: recycling promises in the UK e-waste management sector, New Technology, Work, and Employment, 31(2), 146-160

[42] Buntat, Yahya and Othman, Mustafa (2012) Penerapan kemahiran insaniah 'hijau' (green soft skills) dalam pendidikan teknik dan vokasional di Sekolah Menengah Teknik, Malaysia. *Journal of Social Science*, 5 . pp. 32-41.

Sigma-Delta ADC Topology Implementation Based on Partial Discharge Detection using Rogowski Coil Sensor

M N K H Rohani[1], M Isa[1], M Syahril[1], C C Yii[1], A S Rosmi[1], B Ismail[1]
[1]Centre of Excellence for Renewable Energy (CERE), School of Electrical System Engineering, Pauh Putra Main Campus, Universiti Malaysia Perlis, 02600 Arau, Perlis, Malaysia.
E-mail: khairulhafizi@unimap.edu.my

Abstract. Partial discharge (PD) diagnostic on high voltage (HV) power cables is a modern technology in order to investigate the insulation health. The trending technology such as wireless sensor network (WSN) and internet of thing (IoT) are mostly process in digital signal by microprocessor where looking toward on automatic system. The high performance of analog to digital converter (ADC) in term of sampling rate which is much expensive in order to convert the high frequency of analog PD signal that captured by Rogowski coil (RC) sensor. Thus, this paper propose the implementation of the Sigma-Delta ($\sum\Delta$) ADC topology in Altera DE0-Nano development board as an effective method in this work to replace the conventional ADC chip and the cost is minimal. Then, peak PD detection technique is applied after the conversion of the digital PD signal. The Verilog hardware description language (VHDL) is used in Quartus II software which corporate with Altera DE0-nano development board for the prototype development. In order to achieve the accurate digital sampling of high frequency PD signal, the N-bit of voltage resolution is paramount.

1. Introduction

Nowadays, early detection on PD signal is useful for maintenance purposes or to schedule arrangement for replacing the HV power cable. The PD detection methods has been divided into four section, which optical detection, acoustic detection, chemical detection and electrical detection [1-2]. However, the electrical detection method is the most favored method in HV power cable for investigation between acoustic, optical and chemical detection due to on capturing the electrical signal [3-4]. The characteristic of the PD signal is low in amplitude and short duration current pulse within a nanosecond to microsecond [5-6]. The information about the insulation degradation and type of PD signal can be determined from the shape of the current pulse within alternating current (AC) cycle when it is occurring. Various types of the conventional current sensor are suggested to detect the PD signals either off-line or on-line technique such as shunt current sensor, HFCT, capacitive couple sensor, RC sensor, hall effect current sensor, giant magneto resistive (GMR) sensor and magneto impedance (MI) sensor [7]. The characteristic comparison between current sensors type are listed in

Table 1. The selection of air-cored RC sensor as a PD sensor is based on the advantages of sensors which are wide bandwidth, lightweight, linearity, fast response, no hysteresis losses, low cost and easy to construct in the laboratory [8-9].

However, the new trending of technology such as using wireless sensor network (WSN) and internet of thing (IoT) are increasing popular over the world as a smart detection technique [10-13]. Based on previous works in [14-17], the researcher using the external of ADC component in order to convert the analog of PD signal. This paper present the design and implementation of ∑Δ ADC topology in Altera DE0-Nano development board. Figure 1 shows the block diagram of PD measurement setup in this work. The details of ∑Δ ADC topology design and implementation is describe in Section 2, while the PD detection technique is proposed in Section 3. The experimental work is conducted in order for testing and validate the performance for ∑Δ ADC by using the peak detection technique of PD signal captured by RC sensor. Normally, the PD signal range is from 0.5 MHz up to 20 MHz.

Table 1. Comparison between existing current sensors characteristic [7].

Type of sensor	Shunt	CT	RC	Hall	GMR	MI
Cost	Low	Medium	Low	High	Medium	Medium
Bandwidth	DC ~ 10 MHz	0.1 Hz ~ 100 MHz	0.1 Hz ~ 100 MHz	<1 MHz	DC ~ 5MHz	DC ~ 30 GHz
DC capability	Yes	No	No	Yes	Yes	Yes
Linearity	Very Good	Fair	Very Good	Poor	Fair	Fair
Operating temperature	-55 ~ 125 °C	-50 ~ 150 °C	-20 ~ 100 °C	-40 ~ 125 °C	-40 ~ 150 °C	-40 ~ 150 °C
Integra ability	Excellent	Good	Excellent	Fair	Excellent	Excellent
Material Technology	Simple	Simple	Simple	Complicated	Very complicated	Very complicated

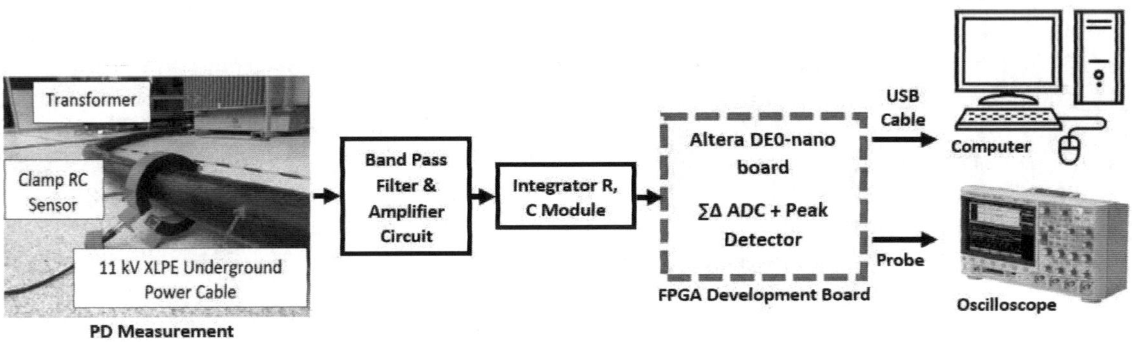

Figure 1. Block diagram of PD measurement setup

2. ∑Δ ADC Topology Implementation

The design and implementation of ∑Δ ADC using FPGA technology for PD signal output from the RC sensor is conducted in this work. Altera DE0-Nano development board with Cyclone IV chip is used as FPGA board and Quartus II software is a platform for programming and simulation. High speed ADC device is needed to sample the high frequency of PD signal. However, the conventional ADC in the market is commonly is much more expensive in high speed rate. Thus, the ∑Δ ADC topology is selected which has the advantages of using minimum number of small component, low cost and can be design depends on the sampling frequency required. The ∑Δ ADC topology is possible to implement into the FPGA by taking advantages of the low voltage differential signalling (LVDS) receiver inside the FPGA. The LVDS inside the FPGA is used as a comparator which can reduce the overall external components. Figure 2 shows the basic layout of the ∑Δ ADC implementation on FPGA which divided into two sections, off-FPGA and on-FPGA. The off-FPGA components includes system clock, signal input from PD measurement signal and integrator circuit while, LVDS comparator, Phase-Locked loop (PLL), D flip-flop, summation, edge counter, accumulator, buffer and digital output are implemented digitally in the FPGA chip.

An external resistor, *R* and capacitor, *C* network known as integrator is required in the feedback system of ∑Δ ADC topology. The main purpose of the *R* and *C* network act as low pass filter to eliminate the out of band noise during bit stream generated from the sampling process while tracking the signal input. The optimal value of *R* and *C* can be determined by following Equation 1 based on the sampling frequency, *fs* used in the ∑Δ ADC topology development. The value of *C* is fixed to 1 nF. Thus, the calculated value of *R* is 320 Ω for 4-bit resolution. By using the ∑Δ ADC topology, high precision component is not required and it has 5 % tolerance for each component. Thus, 300 Ω is selected for the *R* value in the integrator component.

$$R = \frac{2^N}{f_s \text{ x } C} \tag{1}$$

The Δ part in the yellow box as shown is Figure 2 is operated by the LVDS comparing between the input signal and close loop system from D flip-flop. The input signal from the RC sensor is connected to the positive LVDS comparator input, while the negative LVDS is connected to the integrator. The output of the D flip-flip consist of 1bit bitstream of data which represents the signal input and used for signal tracking through the integrator. To convert this 1bit bitstream of data into useable N-bit digital output, a decimation process is required. Decimation is a process of trading data rates for higher resolution. A uniform decimation process known as averaging decimation is used in the ∑ part to demonstrate the effectiveness of the ADC topology. The uniform decimator consist of counter, accumulator and summers. The counter is depends on the number of $2^{N\text{ -bit resolution}}$. However, the analysis on different N-bit resolution such as 4, 5, 6, 7 and 8 bits are set in the Verilog HDL of ∑Δ ADC topology based on the selected switch on the FPGA. The summers in the topology will add the accumulated bitstream samples up to the sample number. For example, to convert the bitstream data to a 4-bit output, 16 samples are required and accumulated to produce the output. In this case, the output data rate is reduced by a factor of 16. Thus, if the output resolution is further increase, the output data rate will decrease.

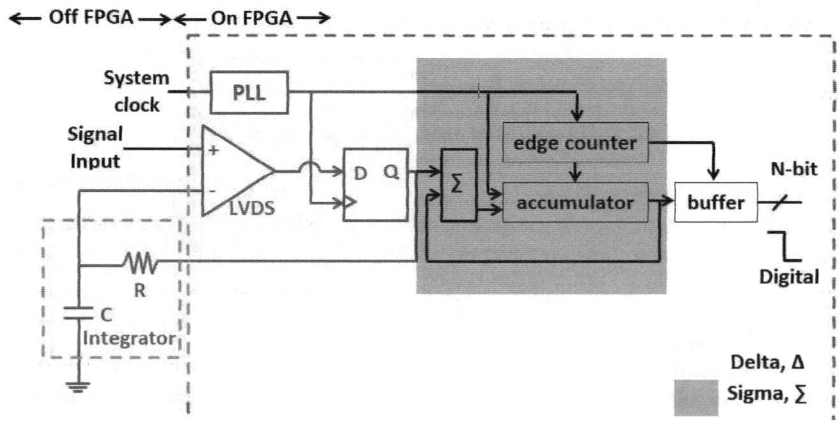

Figure 2. Implementation of Sigma-Delta ADC in FPGA.

PLL is used to increase the 50 MHz on-board system clock to a higher clock frequency required by ∑Δ ADC sampling. MegaWizard® plug-in manager in the Quartus II software is purposely used to generate the PLL for this system. 320 MHz is the maximum frequency that can be generated in the PLL manager for higher sampling signal. The conversion time of signals depends on the number of N-bit resolution. The N-bit resolution means the unique number of vertical levels of signal divided. The higher number of N-bit resolutions will increase the precision. However, the conversion time will be increased with the same PLL clock. The determination of conversion time, *t* is identified by following Equation 2, while the number of samples per 1 μs as follows in Equation 3.

$$Conversion\ time, t = \frac{1}{PLL\ Clock} \text{ x } 2^{N-Bit\ resolution} \tag{2}$$

$$Number\ of\ sample = \frac{1\ us}{Conversion\ time, t} \tag{3}$$

3. PD Detection Method

The PD detection method using a peak detector is introduced in this section. Figure 3 shows the process flow of peak detection technique that has been implemented in Verilog HDL using Quartus II software. The programming starts with the initialization of PD detection counter. The purpose of PD counter is to count the same peak signal in the certain time such as in 0.1 ms. Once the peak signal over than the threshold is detected, the FPGA will check the next peak signal within 0.1 ms. If it has the same peak signal over than the threshold voltage, thus, it is recognized as PD signal. Then, the FPGA will transmit a notification to the receiver at the main unit. If no peak signal is detected for the second time, this system will check for a new detection.

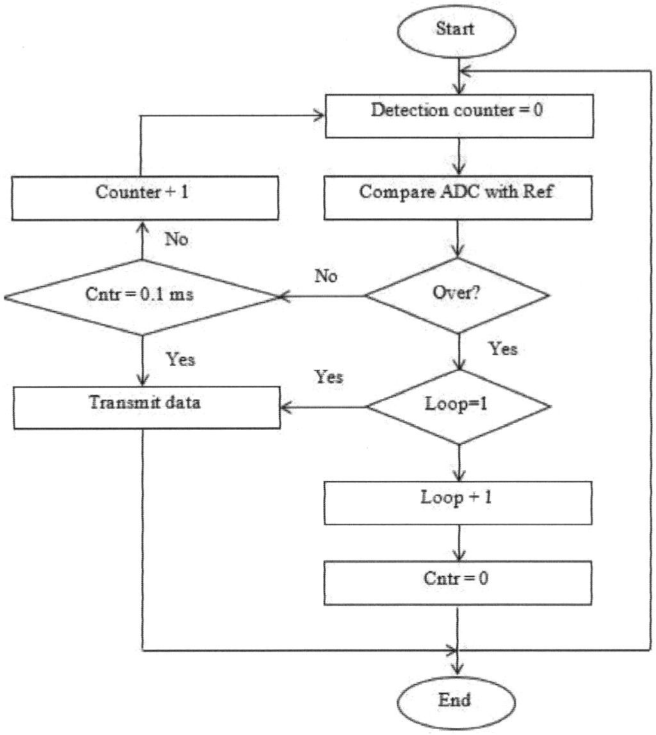

Figure 3. Process flow for PD detection method.

4. Experimental Setup
Altera DE0-Nano development board with Cyclone IV chip is used as FPGA development board that connected with the RC sensor directly as shown in Figure 1. The dip switch, LEDs, general purpose input output (GPIO) and on-board 50 MHz crystal clock (clkin) is used in ∑Δ ADC implementation. The device setup should be properly assigned the I/O port for this implementation. The compiled VHDL for ∑Δ ADC only shown in Quartus II software. For hardware implementation, the assigned pin of I/O port using the pin planner, reflects the hardware I/O pin number. Four dip switches (SW) are used to trigger the number of N-bit resolution where, N is equal to 4, 5, 6, 7 and 8. The inLVDS and inLVDS (n) have been assigned for positive and negative input respectively, while eight LEDs (out8b) are used as an indicator of the converted signal from analog to digital.

Then, the FPGA development board is connected with computer (SignalTap II Logic Analyser) and DSO-X 3024A oscilloscope. The SignalTap II Logic Analyser is used as debugging tool in Quartus II software that captured and displays signal in circuit in circuits designed for implementing in FGPA development board. The SignalTap II analyser utilizes a universal serial bus (USB) cable to communicate with the FPGA devices. In order to analyse the output from the FPGA, the SignalTap II is used to ensure the digital output is same with the value shows in the oscilloscope.

5. Measurement Results and Discussion
The conversion of the PD signal measured by RC sensor to digital is a key of success in this section. Then, the peak signal is recognized. The assessment begins with the hardware implementation of the basic ∑Δ ADC structure programmed in the Altera DE-0 Nano board with the external op-amp, active filter and RC sensor in PD pulse measurement. The analysis on the number of sample in 1 μs duration is evaluated based on the number of N-bit resolution as calculated in Table 2 based on Equations 2 and 3. The selection of N-bit resolution value is important to get the accurate peak voltage result as shown in Figure 4 which is 1.026 V.

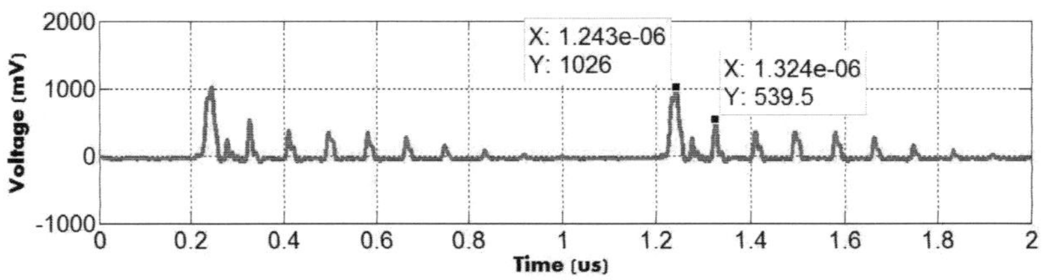

Figure 4. Amplify and filtering implementation on RC sensor output results for PD measurement.

Table 2. Calculated Value for Sampling Time and Sampling Number in 1 μs PD Signal.

N-Bit resolution	Conversion time, *t* (ns)	Number of samples
4	50	20.00
5	100	10.00
6	200	5.00
7	400	2.50
8	800	1.25

Figure 5 shows the digital output with a maximum peak of PD signal is 1.026 V at the input of the ADC with the different of resolutions captured by SignalTap II analyser. The aim of this research to implement the ∑Δ ADC topology to replace the conventional ADC chip and to determine the efficiency of high frequency PD sampling signal accuracy.

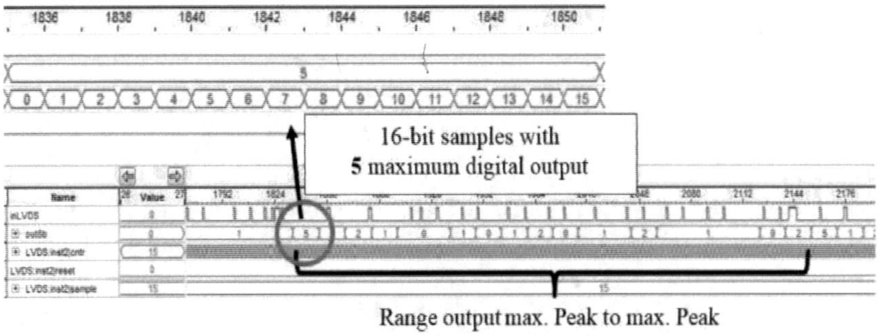
16-bit samples with
5 maximum digital output
Range output max. Peak to max. Peak

(a)

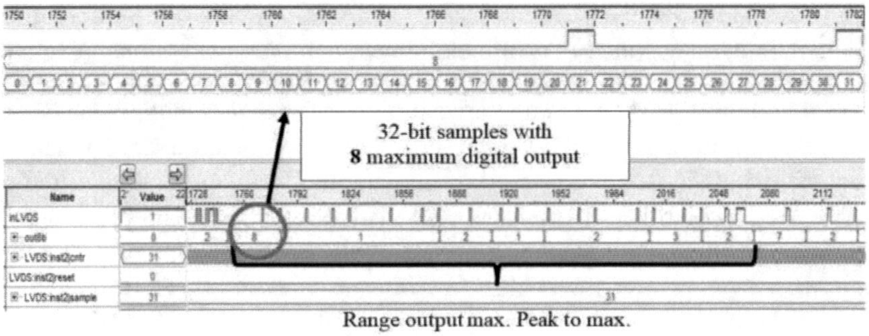
32-bit samples with
8 maximum digital output
Range output max. Peak to max.

(b)

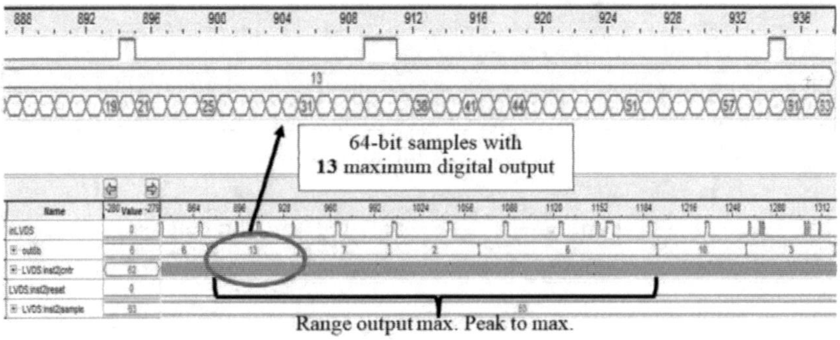
64-bit samples with
13 maximum digital output
Range output max. Peak to max.

(c)

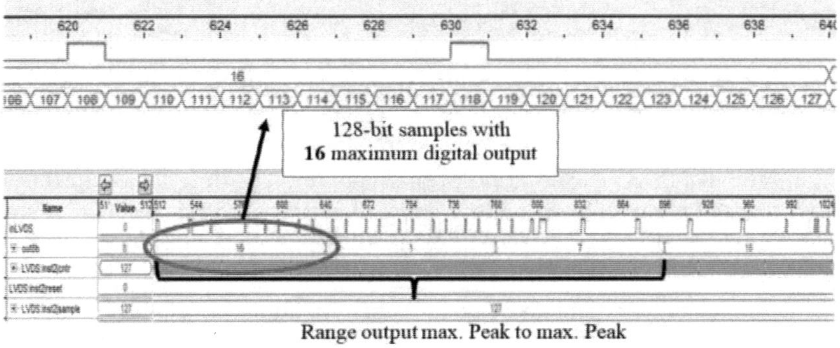
128-bit samples with
16 maximum digital output
Range output max. Peak to max. Peak

(d)

IOP Conf. Series: Journal of Physics: Conf. Series **1019** (2018) 012031 doi:10.1088/1742-6596/1019/1/012031

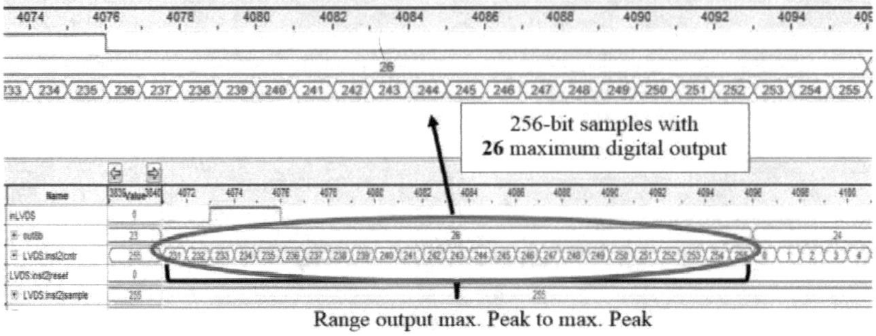

(e)

Figure 5. Sampling signal for PD measurement using ∑Δ ADC topology in SignalTap II analyzer, (a) 4-bit resolutions, (b) 5-bit resolutions, (c) 6-bit resolutions, (d) 7-bit resolutions, (e) 8-bit resolutions.

The analog signal and digital signal is compared in term of the maximum peak signal by following Equation 4 where, the output of Altera DE0-nano board is 3.3 V. The output voltage, *Vout* for digital output signal is summarized in Table 3. Based on the results, it is conclude that, the low resolutions has higher number of samples which more accurate in term of the sampling signal. Thus, 4-bit resolutions is selected in this research as the best resolution for the ∑Δ ADC to sampling the PD measurement output. However, the thresholing method is challenging to decide due to the less of number resolutions. The minimum voltage for 4-bit resolutions is 0.206 V, compare with 8-bit resolutios is 0.013 V. Although, the higher resolutios has a better in minimum digital voltage, but it is difficult to identify the greatest peak voltage as shown in Figure 5 (e) due to the value is really close, compared with 16-bit voltage resolutions.

$$Output\ voltage, V_{out} = \frac{Output\ of\ FPGA\ port}{Resolutions} \text{ x } Maximum\ digital\ output \qquad (4)$$

Table 3. Digital Output Voltage in Different Resolutions.

N-Bit	Samples	Maximum digital output	Output voltage, *Vout*
4	16	5	1.031
5	32	8	0.845
6	64	13	0.670
7	128	16	0.413

6. Conclusion

FPGA technology is an attractive device to implement digital signal processing technique and applications. The applications of ∑Δ ADC topology is feasible to develop in the FPGA development board in order to convert high frequency of PD signal propagate on the power lines from analog to digital signal captured by RC sensor. The 16-bit voltage resolutions is selected due to the optimum sampling PD signal which have 20 sampling number in 1 μs. Hence, the ∑Δ ADC topology is a novel by implementing in wireless RC sensor development based on PD detection technique and become the preferred solution. The wireless sensor applications can reduce the cost of installation where the wiring is not necessary and no need for manually data collecting in PD monitoring system.

7. References

[1] Li J, Xutao H, Liu Z, and Yao X 2016 A Novel GIS Partial Discharge Detection Sensor with Integrated Optical and UHF Methods *IEEE Trans. Power Deliv.* 1-2

[2] Mohammadi H and Haghjoo F 2016 Feasibility study for partial discharge detection and defective region identification in power transformers by using distributed sensors *IET Gener. Transm. Distrib.* **10** 4006-23

[3] Rohani M N K H, Yii C C, Isa M, Hassan S I S, Ismail B, Adzman M R, and Shafiq M 2016 Geometrical Shapes Impact on the Performance of ABS-Based Coreless Inductive Sensors for PD Measurement in HV Power Cables *IEEE Sens. J.* **16** 6625-32

[4] Shafiq M, Hussain G A, Kütt L, Elkalashy N L, and Lehtonen M 2015 Partial discharge diagnostic system for smart distribution networks using directionally calibrated induction sensors *Electr. Power Syst. Res.* **119** 447-61

[5] Xu Y, Gu X, Liu B, Hui B, Ren Z, and Meng S 2016 Special requirements of high frequency current transformers in the on-line detection of partial discharges in power cables *IEEE Electr. Insul. Mag.* **32** 8-19

[6] Mulroy P, Hurtado A, and Badetz D 2012 On-line partial discharge monitoring system for distribution networks *IEEE Int. Conf. Cond. Monit. Diagnosis*, 542-45

[7] Xiao C, Zhao L, Asada T, Odendaal W G, and Van Wyk J D 2003 An Overview of Integratable Current Sensor Technologies *Ind. Appl. Conf. 2003. 38th IAS Annu. Meet.* 1251-58

[8] Rohani M N K H, Yii C C, Isa M, Hassan S I S, Mukhtaruddin A, and Yusof N A 2016 Evaluation of Rogowski Coil Sensor Performance Using EMTP-ATP Software *IEEE 3rd Int. Conf. Electron. Des.* 446-51

[9] Isa M, Elkalashy N I, Hashmi G M and Lehtonen M 2012 Experimental Evaluation of Rogowski Coil Performance for Locating PD in Energized Overhead Covered-Conductor Feeder *J. Energy Power Eng.* **6** 949-59

[10] Chen J 2016 Model based compensation for overcoming sensor breakdown in a piezo-based wireless sensor network *Optik (Stuttg).* **127** 3138-42

[11] Dener M 2017 WiSeN: A new sensor node for smart applications with wireless sensor networks *Comput. Electr. Eng.* 1-15

[12] Li W and Kara S 2017 Methodology for Monitoring Manufacturing Environment by Using Wireless Sensor Networks (WSN) and the Internet of Things (IoT) *Procedia CIRP* **61** 323-28

[13] Sisinni E, Depari A, and Flammini A 2016 Design and implementation of a wireless sensor network for temperature sensing in hostile environments *Sensors Actuators, A Phys.* **237** 47-55

[14] De Souza Neto J M R, De Macedo E C T, Da Rocha Neto J S, Queiroz Vieira M F, Glover I A, Judd M D, Atkinson R, and Soraghan J 2013 Plausibility of incoherent detection for radiometric monitoring of insulation integrity in HV substations *Loughbrgh. Antennas Propag. Conf. LAPC* 494-99

[15] Chakrabarty C K, Ramasamy A K, and Ghani A B A 2011 Matlab and VHDL model of real time partial discharge detection using FPGA technology *IEEE Conf. Open Syst.* 389-94

[16] Emilliano, Chakrabarty C K, Ghani A B A, and Ramasamy A K 2010 Online partial discharge counting system using microcontroller PIC 16F877A and FPGA technology *Proceeding, IEEE Student Conf. Res. Dev. - Eng. Innov. Beyond,* 314-19

[17] Emilliano, Chakrabarty C K, Ghani A B A, and Ramasamy A K 2011 VHDL implementation for measurement of the distance test distribution pattern of the Tri-Axial magnetic probe for the PD detection circuit system by using 3 GHz ADC and FPGA *Proc. Int. Conf. Util. Exhib. Power Energy Syst. Issues Prospect. Asia,* 1-6

IOP Conf. Series: Journal of Physics: Conf. Series **1019** (2018) 012032 doi:10.1088/1742-6596/1019/1/012032

A Preliminary Review of Building Informatics for Sustainable Energy Management

S C J Lim[1] and B T Tee[2]

[1] Department of Engineering Education, Faculty of Technical & Vocational Education, Universiti Tun Hussein Onn Malaysia, Parit Raja, 86400, Johor, Malaysia
[2] Centre of Advanced Research on Energy (CARe), Faculty of Mechanical Engineering, Universiti Teknikal Malaysia Melaka, Hang Tuah Jaya, Durian Tunggal, 76100, Melaka, Malaysia

E-mail: [1]johnson@uthm.edu.my,[2]tee@utem.edu.my

Abstract. Building sector is a major energy consumer that represents a significant area of research in energy sustainability efforts. Building system is essentially a complex system that can be viewed from a number of perspectives, including occupants, thermal conditions, climate, engineering systems, and management practices. In this paper, we have conducted a preliminary review on the research issues, challenges and opportunities in the field of sustainable energy management, from the perspectives of building informatics. Based on literature survey, we have reviewed the status-quo of existing practice and related research challenges in building energy management. We then present the existing building information collection and processing strategy, follow up with a discussion on how informatics is being applied for sustainable energy management. Based on our discussion, we have summarized four essential research areas that are worth investigating for future research, with in-depth discussion on the rationale, main issues and existing efforts in respective area.

1. Introduction

Building is the fastest growing energy consuming sector [1]. Building sector (residential, commercial and public) accounts for 20%-40% of final energy consumption [2]. In a modern office environment, building systems such as lighting, HVAC and office equipments are the main energy consumers. According to US Green Building Council [3], buildings consume about 70% of final energy consumption via HVAC system and artificial lighting, and contributed approximately 39% of CO_2 emissions. Another survey by US Energy Information Administration also showed that total energy usage in commercial buildings has increased 7% since 2003, where electricity consumption in buildings has doubled from 38% in 1979 to 61% in 2012 [4]. Locally, in the context of Malaysia, approximately 48% of all available electrical energy are consumed in commercial and office buildings [5].

The aforementioned facts indicate that building energy consumption and management is a pressing issue and a key sector worth investigating for energy sustainability. Understanding the use of energy in buildings requires an insight into the amounts of energy consumed and the different aspect of the usage. For energy users, energy management can be described as the sum of measures planned and carried out to achieve the objective of using the minimum possible energy while the comfort levels (in offices or residential) and the production rates (in factories) are maintained [6]. The challenge is particularly difficult in the area of energy management, where rising utility costs, a lack of discounted tariffs and restrictive legislations are creating major issues in the process. With the prolonged global economic downturn, energy management is seen as a concept limited to monitoring and controlling utility consumption and reducing leakage. From the implementation perspective, building information management (BIM) system is often used to provide a common platform for information sharing that can be used by all stakeholders in design, construction, and facility management after a building is completed and occupied. A building management system (BMS), on the other hand, offers intelligent decision making process to allow integration and automation of all building technologies to optimize energy consumption. A BMS provide intelligent functionality to respond to the energy demand and comfort of office building environments for normal daily operation. Nevertheless, few have successfully identified emerging energy management opportunities beyond utility consumption [7].

With that in regard, building informatics (BI) has become an emerging research area in the building research community in order to achieve an overall operational and energy sustainability. BI is an emerging area of research where it involves the collection, processing, storage and retrieval of building related information, such as occupants data, building operation data, etc. for the purpose of building operational analysis, knowledge discovery and decision making. In the academia, building informatics is an active research area with progress in fields such as smart home [8], intelligent buildings [9], heating, ventilating and air-conditioning (HVAC) system monitoring [1] and occupants activity analysis [10]. Nevertheless, the exploration on knowledge discovered from collected building information and the application of such knowledge for optimized building operation is still less emphasized. Given the latest development in building activity sensing technology and data analytics, we reckoned that there is a need to review the latest works in this area for a better understanding of research issues, challenges and opportunities.

In this paper, we wish to discuss on the state-of-the-art research and development in the field of sustainable building energy management, from the perspective of BI. The rest of this paper is as follows: next section (Section 2) elaborates on the state-of-the-art research in building energy management. We explain the recent advances of building informatics in Section 3, and further explained this field of study in two sub-sections (Section 3.1 and Section 3.2). Lastly, we discuss the research issues pertaining this emerging field of study in Section 4 with brief discussion on future works, and conclude our work in Section 5.

2. Building Energy Management

The performance of the energy management in building might be carried out in two terms which is applying predicted method using building energy modeling software where it relies on certain assumption that the building services is perfectly efficient [11]. Another method is performing frequent energy audit in term of the building energy management and control, thermal comfort and building quality besides detecting the unregulated load by the facility managers [12]. This is relevant as the term Post Occupancy Evaluation (POE) which is the evaluation method designed for facility managers to identify and evaluate the building performance of the building could be sustained. From this perspective, several assessments can be conducted to determine the performance of buildings either it met the user needed resulting with a suggestion of measures necessary to improve overall building design, performance, and fitness [13].

In broad view, the scope of energy management should not only be limited to utility consumption by air conditioning and mechanical ventilation (ACMV) systems in facilities or IT infrastructure. In the recent years, the growing development of building energy efficiency technologies around the world has increased the interest of implementing building energy management concept in Malaysia. The Malaysian government has demonstrated the commitment towards this concept among the government buildings which include Diamond Building at Putrajaya and Low Energy Office Building of Ministry of Energy, Water & Communications. Both buildings have been designed with energy efficiency features, such as a comprehensive Energy Management System (EMS), and can achieve a Building Energy Index (BEI) less than 150 $kWh/m^2/year$ [14].

It has become imperative for organization to build dedicated teams and departments to implement processes and technologies that can help reduce consumption. However, with numerous avenues of energy consumption to be considered, it has been a challenge for such teams to measure and control it. By taking this as an opportunity, some IT service providers, environmental consultants, energy consultants and technology manufacturers dealing in energy consumption monitoring services are developing software products and proposing process reforms that can help users monitor and control their energy usage [7]. However, these service providers deliver only basic consumption data without any detailed analysis that traces the patterns of usage based upon various influencing factors, such as the impact of the weather or the scale of operations. Currently, there is still a gap in obtaining approximate solution that addresses the entire aspect of energy management [7].

3. Building Informatics

In current situation, smart or intelligent buildings are highly adaptable to changing environmental conditions. They have automated systems, including wireless sensor monitoring, to facilitate energy efficient, comfortable and cost-effective environments by optimizing structure, systems, services, building management and their interrelationships [15]. In the context of Internet of Things (IoT), Intelligent Building Energy Management Systems (BEMS) can be considered part of a much larger information system. This system is used by facilities managers in buildings to manage energy use and energy procurement and to maintain buildings systems. It is based on the infrastructure of the existing Intranets and the Internet, and therefore utilizes the same standards as other IT devices. Within this context reductions in the cost and reliability of wireless sensor networks are transforming building automation, by making the maintenance of energy efficient healthy productive work spaces in buildings increasingly cost effective [16]. Figure 1 shows an overview of how building informatics can be applied for sustainable building applications. From the figure, in general there are three main parts: data and information collection, data analytics and sustainable application. We shall elaborate in detail the related aspects of building informatics in the following sections.

3.1. Data and Information Collection

Gathering information on building environment and building systems is the essential first step for overall operational assessment. In the practical sense, sensor technology is a widely adopted in terms of building environmental parameters sensing. Typically, sensors are used for measuring parameters such as temperature, relative humidity, air quality, movement and luminance [17]. There are also multiple state change sensors (e.g., presence or activity sensor) that are used for occupancy detection [18]. From the perspective of building energy, automated meter reading technology [19] is widely implemented at some developed countries for building monitoring, where hourly reading can be automatically logged.

In relation to this, smart meters is a more advanced sensing device that is connected to the Internet. Besides monitoring energy usage, it can also be used to monitor unusual usage patterns, temperature and pressure values [20]. Such a meter can be either connected through normal wired or wireless Internet connection. In the context of building operations, Wu and Clements-Croome [21] stated that the signal of the WSNs provides a dynamic scenario of the state of indoor environment, enabling real-time diagnostics of system and component anomalies as well as off-line analysis. They indicated that the benefits of utilizing such a network is two-

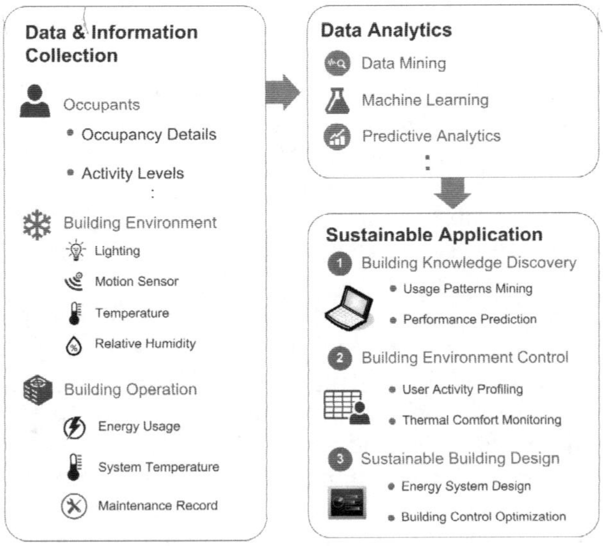

Figure 1. Building Informatics for Sustainable Building Applications

fold: energy consumption can be managed by using sensors to monitor occupant behavior, and occupants' well-being can be measured by various environmental comfort indexes.

With the advent of the IoT and connected devices paradigm, the introduction of relatively low-cost, wireless electronic communication devices not only serve as an affordable mean to gather building information, it is also useful for gathering occupant related details, such as user activity. The analysis of user activity is essential to create thermal comfort conditions that suite different thermal requirements that relates to their level of activity. In terms of user activities, it can be logged and recognized in two ways: either actively or passively. Active monitoring involves the use of personal devices such as mobile phones, wearable devices (e.g., smart watch), smart thermostat and voice assistant (e.g., Amazon Echo/Google Home), with user's consent. Wearable sensors enable high accuracy activity recognition. User activities can be accurately tracked via analyzing accelerometer sensor data [22, 23] that is present in some of these devices. In relation to this, latest software technology such as Google's Platform Activity Recognition API have enabled precise recognition of user activities such as walking, jogging, cycling or driving using mobile phones or wearable devices (e.g. smart watch). Besides, voice recognition feature of these devices also enable users to provide active feedback to the environmental control system. For instance, user can either indicate their preference on the thermal environment (e.g., temperature and lighting conditions are automatically adjusted when user indicate "relax mode") or issue direct request via natural language (e.g., "I feel warm"). On the other hand, passive monitoring involves non-intrusive infra-red (PIR) sensor where it is triggered by user's motion. The advantage of using infrared sensors over other types of sensing techniques is the privacy preserving property, where occupants' identity cannot be identified, albeit at the cost of sensing precision. On the other hand, video recording can also be used to monitor user activities as a less-privacy preserving option.

3.2. Application in Sustainable Energy Management

As conventional building loads such as HVAC and artificial lighting are related to the way how a building is utilized by occupants, occupants detection and activity profiling are important building information. Such an informatics can be used to optimize the environmental control system, so that office conditions are adjusted in line with the individual preference [10]. Basically, the first step is occupancy detection, which aimed at determining the number of occupants in a building. Langensiepen et al. [10] proposed a way for worker presence detection using a number of installed sensory devices. Dong et al. [24] used Hidden Markov Models (HMMs) to perform occupancy detection in an intelligent office environment setup, and achieved 73% of detection accuracy. Besides motion detection, their occupancy detection approach also includes acoustic and CO_2 detection for improved accuracy in an open office plan setting. Lim et al. [25] proposed a simple classification approach to predict occupancy patterns based on data collected from wireless sensor network. These are important for human-centered environmental control and energy usage in buildings.

Nevertheless, detecting the number of occupants only serve as one perspective of occupancy analysis. Occupant activity monitoring and recognition is one of the actively researched dimension of building research. Its primary aim is to build user profile based on recognized occupants' activities to control indoor environmental parameters. A study by Akhlaghinia et al. [26] have found that occupants tend to have a more systematic behavior pattern that can be predicted using prior behavioral monitoring. Thus, it can be used as a know-how to better manage building energy resources and also ensure occupants safety. For instance, Luhr et al. [18] proposed a novel data mining approach named EFP-Tree to detect emergent behaviors of building occupants. Their mining algorithm is able to detect the presence of new, possibly anomalous behavior via comparison with mined rules from sensor event log. Such a characteristic is proven useful for helping elderly building occupants. From the literature of computer vision and sensor analytics, human activities can be detected recognized by using a number of ways, such as: (1) sensor-based techniques (i.e., infrared sensing [27] or mobile wearable sensors [28, 22, 23]); or (2) vision-based techniques (i.e., using video camera recording) [29]. In relation to this, a number of computational techniques that can be effective for activity recognition such as Hidden Markov Models (HMMs) [30], neuro-fuzzy inference [31] and Bayesian classifiers [22].

Next, building informatics is applied for discovering building related knowledge (e.g., know-how) for an overall efficient building operation. For instance, Xiao and Fan [32] proposed an applicable framework for mining building automation system (BAS) database. Their approach consists of five major steps, including data preparation, clustering analysis, association rule mining, post-mining and application of discovered knowledge in two application context, which is proven useful in ensuring energy efficiency of building operation. Doukas et al.[17] proposed a rule-based expert system in BEMS that incorporates historical records and expert rules for efficient building energy control. The system is able to evaluate current building loads with desirable ones (from historical data), perform decision making, and link such a decision with building actuators for optimized building control. Another perspective of research involves the prediction of building operational performance through informatics. Such a prediction is important for building energy planning. For example, Li et al.[1] focused on hourly cooling load in a building. They have applied artificial neural network (ANN) and support vector machine (SVM) to predict hourly cooling load in a building with high accuracy. Chou et al.[20] predicted the coefficient of performance of a refrigeration system using data mining techniques. Based on the data collected via smart meters, various data mining techniques including artificial neural network (ANN), SVM and classification and regression tree (CART) are used to effectively predict performance.

With the application of computational techniques, building informatics also serve as an important source of knowledge to assist designers in proposing energy efficient building design. Ahmed et al.[33] proposed the use of data mining algorithms on real-time sensed weather data, coupled with internal daylight analytical tools, for proposing best design of building lighting. For this purpose, they have applied support vector machine (SVM) to estimate indoor daylight levels and decision tree algorithm for evaluation of multiple building design with different daylight methods and procedures, to assist building designers with a reliable and simple approach in lighting assessment and decision making. On another related research work, They have also proposed a data mining model that incorporates thermal comfort standards and indoor daylight procedures, to observe correlation between weather conditions, building characteristics and low-energy comfortable rooms [34]. They have conducted their study at a low-energy building with many sustainable energy features such as solar panels, geothermal heat pumps and heat recovery system, monitored by a building management system that is complete with wireless and wired sensors in every room. Their research indicates that Naive-Bayes classifier outperformed SVM and decision tree in thermal comfort prediction, and such a result is used to optimize energy savings.

4. Discussion

From our discussion on previous sections, we reckoned that building informatics is a significant research area for ensuring an energy-efficient and sustainable building operation. Nevertheless, we do observe a few research issues that are worth for further investigation, and these may serve as future research directions. We discuss each issues as follows:

(i) **Energy-Efficient Monitoring:** We notice that building related research often neglect the energy consumption aspects of monitoring devices, or regard these as insignificant with respect to the energy consumption of building operational equipments. With the increasing number of sensing or monitoring devices installed in buildings, the sensing activity itself can be a significant energy consuming factor. Aiming at minimizing the overall energy consumption of such an activity, WSN is deemed a promising solution. However, how to best configure WSN for prolonged operation is not a trivial task. To the best of our knowledge, there are only a few related research. For instance, Liu et al. [35] proposed a cluster-based architecture design to minimize the energy dissipation for gathering sensor node data in a wireless mobile sensor network. They have proposed algorithms for cluster-head election and cluster formation, aimed for low power consumption wireless sensor network design. Ploennigs et al. [36] proposed a simple and scalable model-based virtual sensor for building heat consumption analysis using simple temperature sensors. Virtual sensors are used to estimate a projected sensor value from other measurement. They have suggested a relational model which resulted in reasonable estimations for heat consumption estimation.

(ii) **Data Design and Cleaning:** This is a typical issue in informatics field. Typically in the field of building informatics, there are heterogeneous building systems involved (e.g., different BAS and BEMS installed) from different solution providers that might be using different communication protocol and standards. Consequently, different application programming interfaces (APIs) are required to obtain building data, inconsistency in data structure definition and description (e.g., "RH" vs "Humidity") may present, causing difficulties in data integration efforts for comprehensive, overall building analysis. Besides, sensory data are normally very noisy and at times incomplete. Among the techniques used to ensure data validity and consistency is clustering algorithm (e.g., k-means) for outlier detection [21]. We do suggest that additional research involving the techniques for building related data cleaning are essential to ensure high quality input for data mining activity at later stage.

(iii) **Knowledge Discovery and Modeling:** From the literature, there are multiple techniques (e.g., in data mining) used for knowledge discovery and modeling schemes proposed for knowledge reuse, with the aim of decision making. Some of these studies [32, 17] have indicated that we are able to discover significant knowledge patterns from building related data. While these are useful in the perspective of existing building, previously collected operational data may not be always useful for designing future buildings due to design changes. This is known as the "cold start problem", where the required dataset is not always present. In relation to this, there are researcher that proposed a building simulation approach in training machine learning model for building energy management [19]. Their studies showed that the model trained with such an approach has comparable internal consistency with model trained using actual field data in the energy consumption prediction task. On the other hand, another issue worth noting is how can we incorporate existing human know-how into the decision making mechanisms or systems. Human inputs can be important where actual building operational constraints can be better included. We believe that a synergy between human and machine-assisted knowledge can be very useful in completing the knowledge discovery and modeling process.

(iv) **Conflict Resolution:** With the increasing number of building occupants, ensuring overall thermal comfort of occupants is a challenging issue. From previous studies, it is well-understood that dissatisfaction with indoor thermal conditions is the most common source of occupant complaints in office buildings [37]. There are also numerous studies that have indicated how a poor indoor thermal condition can impact health, productivity and occupants' behavior [38]. With regards to building energy consumption, achieving a balance between occupant's comfort and building energy efficiency is challenging. This is mainly due to heterogeneous thermal preference of building occupants, and resolving conflicts between occupants' needs is one of the key towards achieving such a balance. Research area related to such an adaptive building control is emerging. For instance, Camacho et al. [39] proposed an ontology-based approach towards realizing an adaptive home and building automation system (HBAS). Their proposed solution is able resolve conflict among building occupants' in terms of multiple needs of environmental needs, by using pre-defined ontology of user environmental preference as knowledge support. Their case study showed that using such a system energy efficiency can be maximized while ensuring the comfort of occupants.

5. Conclusion

In this paper, we have conducted a preliminary review on building informatics for sustainable energy management. We have discussed the status-quo, issues and challenges of existing building energy management practice, and how building informatics are being applied in modern practice. In the perspective of building informatics, we have discussed the aspects of building information collection and processing, and also how information processing and knowledge management are being applied for enabling sustainable building energy management. Based on these review, we have summarized and discussed four essential research issues for future research. We do believe that these are important areas that require further research focus to realize sustainability in building energy management.

References

[1] Li Q, Meng Q, Yoshino H and Mochida A 2009 *Energy Conversion and Management* **50** 90–6
[2] Prez-Lombard L, Ortiz J and Pout C 2008 *Energy and Buildings* **40** 394–398 ISSN 0378-7788 URL `http://www.sciencedirect.com/science/article/pii/S0378778807001016`
[3] USGBC Buildings and climate change URL `http://www.eesi.org/files/climate.pdf`
[4] CBECS 2012 2012 commercial buildings energy consumption survey: Energy usage summary URL `http://www.eia.gov/consumption/commercial/reports/2012/energyusage/`
[5] Noranai Z and Md Yusof M 2011 *Proceedings of World Academy of Science, Engineering and Technology* vol 77 pp 745–751
[6] Machinchick T 2013 An introduction to energy management systems URL `http://www.creativecarbonscotland.com`
[7] Gehani A 2012 The future of energy management Tech. rep. Cognizant
[8] WSU 2014 Casas smart home project URL `http://ailab.wsu.edu/casas/datasets.html`
[9] MERL 2009 Merlsense dataset URL `https://sites.google.com/a/drwren.com/wmd/`
[10] Langensiepen C, Lotfi A and Puteh S 2014 *IEEE International Conference on Fuzzy Systems*

[11] Menezes A C, Cripps A, Bouchlaghem D and Buswell R 2012 *Applied Energy* **97** 355 – 364 ISSN 0306-2619 energy Solutions for a Sustainable World - Proceedings of the Third International Conference on Applied Energy, May 16-18, 2011 - Perugia, Italy URL http://www.sciencedirect.com/science/article/pii/S0306261911007811
[12] Brady L and Abdellatif M 2017 *Energy and Buildings* **149** 142 – 150 ISSN 0378-7788
[13] Akashah M, Ali F W and Zahari A S 2015 *Jurnal Teknologi* **75** 27–37
[14] Aliagha G, M H, O S A and Ali K 2013 *Journal of Energy Technologies and Policy* **3**
[15] Grondzik W T, Kwok A G, Stein B and Reynolds J S 2010 *Mechanical and Electrical Equipment for Buildings* 11th ed (John Wiley and Sons)
[16] Grindvoll H, Vermesan O, Crosbie T, Bahr R, Dawood N and Revel G M 2012 *Journal of Information Technology in Construction* **17** 43
[17] Doukas H, Patlitzianas K D, Iatropoulos K and Psarras J 2007 *Building and Environment* **42** 3562 – 3569 ISSN 0360-1323 URL http://www.sciencedirect.com/science/article/pii/S036013230600312X
[18] Luhr S, West G and Venkatesh S 2007 *Pervasive and Mobile Computing* 95–116
[19] de Wilde P, Martinez-Ortiz C, Pearson D, Beynon I, Beck M and Barlow N 2013 *Advanced Engineering Informatics* **27** 457 – 465 ISSN 1474-0346 URL http://www.sciencedirect.com/science/article/pii/S1474034613000505
[20] Chou J S, Hsu Y C and Lin L T 2014 *Expert Systems with Applications* **41** 2144 – 2156 ISSN 0957-4174 URL http://www.sciencedirect.com/science/article/pii/S0957417413007446
[21] Wu S and Clements-Croome D 2007 *Energy and Buildings* **39** 1183–1191
[22] Yang S I and Cho S B 2008 *Proceedings of IEEE International Conference on Multisensor Fusion and Integration for Intelligent Systems (MFI)* pp 100–105
[23] Ravi N, Dandekar N, Mysore P and Littman M 2005 *Proceedings of the Seventeenth Conference on Innovative Applications of Artificial Intelligence (IAAI/AAAI 2005)*
[24] Dong B, Andrews B, Lam K P, Hoynck M, Zhang R, Chiou Y S and Benitez D 2010 *Energy and Buildings* **42** 1038–1046
[25] Lim S, Puteh S and Goh K 2016 *IEEE International Conference on Industrial Engineering and Engineering Management* (Bali, Indonesia)
[26] Akhlaghinia M, Lotfi A, Langensiepen C and Sherkat N 2008 *Journal of Internet Technology* **9** 333–338
[27] Shankar M, Burchett J B, Hao Q, Guenther B D and Brady D J 2006 *Optical Engineering* **45** 106401–106401–10 ISSN 0091-3286 URL http://dx.doi.org/10.1117/1.2360948
[28] Hong J H, Yang S I and Cho S B 2010 *Expert Systems with Applications* **37** 4680–4686
[29] Shih H C 2014 *Energy and Buildings* **77** 270–280
[30] Madokoro H, Honma K and Sato K 2012 *Proceedings of International Joint Conference on Neural Networks* pp 1–8
[31] Ahad M, Tan J, Kim H and Ishikawa S 2008 *Proceedings of International Conference on Control, Automation and Systems (ICCAS)* pp 1896–1901
[32] Xiao F and Fan C 2014 *Energy and Buildings* **75** 109 – 118 ISSN 0378-7788 URL http://www.sciencedirect.com/science/article/pii/S0378778814001169
[33] Ahmed A, Otreba M, Korres N E, Elhadi H and Menzel K 2011 *Advanced Engineering Informatics* **25** 364 – 379 ISSN 1474-0346 information mining and retrieval in design URL http://www.sciencedirect.com/science/article/pii/S1474034610000959
[34] Ahmed A, Korres N E, Ploennigs J, Elhadi H and Menzel K 2011 *Advanced Engineering Informatics* **25** 341 – 354 ISSN 1474-0346 information mining and retrieval in design URL http://www.sciencedirect.com/science/article/pii/S147403461000100X
[35] Liu C M, Lee C H and Wang L C 2007 *Journal of Parallel and Distributed Computing* **67** 1187 – 1200 ISSN 0743-7315 URL http://www.sciencedirect.com/science/article/pii/S0743731507001207
[36] Ploennigs J, Ahmed A, Hensel B, Stack P and Menzel K 2011 *Advanced Engineering Informatics* **25** 688 – 698 ISSN 1474-0346 special Section: Advances and Challenges in Computing in Civil and Building Engineering URL http://www.sciencedirect.com/science/article/pii/S1474034611000528
[37] Federspiel C 1998 *ASHRAE Transactions* **104** 912–923
[38] Rupp R F, Vsquez N G and Lamberts R 2015 *Energy and Buildings* **105** 178–205
[39] Camacho R, Carreira P, Lynce I and Resendes S 2014 *Expert Systems with Applications* **41** 6161 – 6173 ISSN 0957-4174 URL http://www.sciencedirect.com/science/article/pii/S0957417414002140

Acknowledgments
The work described in this paper was supported by a research grant by University Tun Hussein Onn Malaysia (Grant Ref: MDR U098).

IOP Conf. Series: Journal of Physics: Conf. Series **1019** (2018) 012033 doi:10.1088/1742-6596/1019/1/012033

An Optimization of Electrical Output Power for Piezoelectric Energy Harvester using Different Micro-Cantilever Beam Geometries

A S Rosmi[1], M Isa[1], B Ismail[1], M N K H Rohani[1], Y Wahab[2],

[1]Centre of Excellence for Renewable Energy (CERE), School of Electrical System Engineering (SoESE), Universiti Malaysia Perlis, Malaysia.
[2] Advanced Multi-Disciplinary MEMS-Based Integrated NCER Centre of Excellence (AMBIENCE), School of Microelectronic Engineering (SoME), Universiti Malaysia Perlis, Malaysia.
E-mail: effa_kh@yahoo.com

Abstract. A MEMS-based piezoelectric energy harvester is capable in producing the electrical output at certain level. The investigation on ability to successfully generating the electrical output power is highly recommended. Accordingly, an optimization on electrical output power based on micro-cantilever beam geometry is presented in this paper via simulation work using COMSOL Multiphysics software. Thus, four different geometries named as PPG 1, PPG 2, PPG 3 and PPG 4 are individually characterized. The simulation result shows the superior performances from PPG 2 which generating 393 mV of output voltage with 39 nA output current, while the output power is 15 nW at 10 MΩ optimal load and 1g vibration amplitude. The excitation frequency for PPG 2 is as low as 60 Hz. Therefore, the modification of micro-cantilever geometry is able to increase the production of electrical output power.

1. Introduction

The present of alternative energy sources from the ambient such as wind, solar, thermal, radio frequency and mechanical vibration offering space to green source electrical power generation [1-3]. Harvesting the mechanical vibration using piezoelectric energy harvester is mostly reported by the previous researchers either in experimental or simulation studies [4-6]. The mechanical vibration is a good candidate for alternative energy because it is said as clean, always present and sustainable energy [7-9]. However, the direct electrical power produced by the piezoelectric harvester is stated as too low [10]. Thus, an enhancement of electrical output power for piezoelectric harvester is obviously required before connected to the power management circuitry. A selection of different piezoelectric material and operating modes are the most famous approach for the output optimization rather than micro-cantilever beam geometry modification technique which is less complicated. Thus, there are four different models named as PPG 1, PPG 2, PPG 3 and PPG 4 are designed and simulated. The Zinc Oxide (ZnO) material is mainly used in this paper due to its high piezoelectric coupling, environmental friendly and flexibility in processing [11-14].

2. Piezoelectric Power Generator (PPG) 3D modelling

The micro-cantilever beam PPG sensor 3D model is designed using COMSOL Multiphysics software approach. Figure 1 gives a picture of major PPG sensor components in 3D model geometry

IOP Conf. Series: Journal of Physics: Conf. Series **1019** (2018) 012033 doi:10.1088/1742-6596/1019/1/012033

including the piezoelectric material, bottom electrode, top electrode, proof mass, and supporting beam. There are four different geometries are proposed to investigate the mechanical to electrical performances in terms of electrical output power. The geometric shapes are rectangular with narrow proof mass (PPG 1), rectangular with big proof mass (PPG 2), trapezoidal (PPG 3) and wide-fixed trapezoidal (PPG 4) are designed.

Figure 2 portrays the geometries with the corresponding 3D model using COMSOL Multiphysics software. The PPG sensor model is made up of Si as the supporting beam with a thin layer of ZnO deposited on top of the Si beam. A proof mass from Si material is attached at the beam free end to reduce the frequency resonance and easily match with the target vibration frequency for maximum electrical output power generation. The frequency target should be less than 200 Hz, the existence frequency in the ambient [10]. The proof mass is designed to have a dimension of 6000 x 6000 x 420 μm^3 except for the PPG 1 and PPG 4. The proof mass dimension for PPG 1 is 6000 x 3000 x 420 μm^3 meanwhile for PPG 4 is 4000 x 4000 x 200 μm^3. The proof mass different value is purposely given for structural firmness control to avoid the beam from broken during operation. The length dimension of PPG sensor is varied in range of 1000 μm to 5500 μm. The thickness dimension for micro-cantilever beam PPG sensor based on previous experimental work to ensure the fabrication reliability which is 0.3 μm for ZnO and 30 μm for Si [12]. The Body Load, F_v boundary is assigned to the whole structure as input for a reason the PPG sensor start to vibrate and simultaneously inducing a mechanical strain. The fixed end of the PPG sensor is terminated by a load resistance in the range of 10^2 up to 10^{10} for demonstration in the real applications. Thus, the optimal value of load resistance could be found and also to investigate the effect of the resistive load value to the electrical output power. This is done by adding the electrical circuit interface. The Floating Potential boundary is applied for the ZnO layer upper face, while the Ground boundary for the lower face for the electrical behavior investigation. The model is meshed to a group of simpler finite element bricks before FEM analysis conducted.

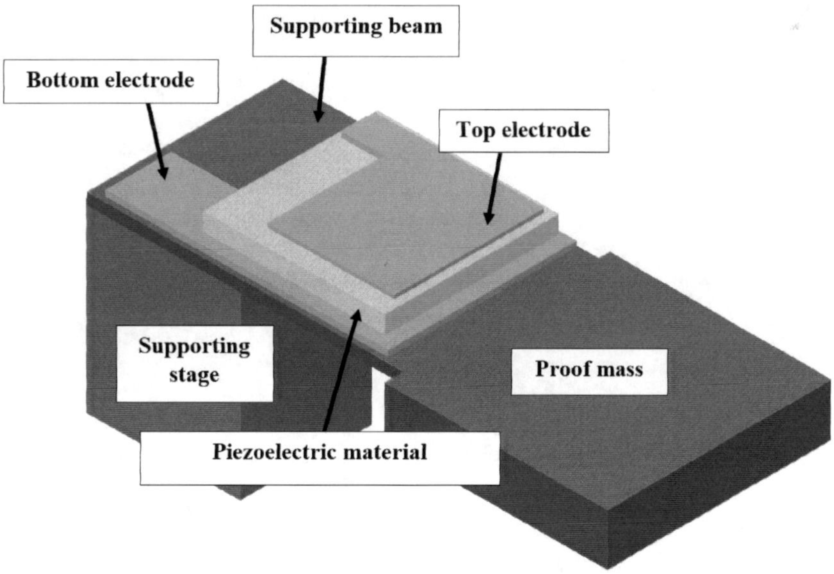

Figure 1. 3D model geometry.

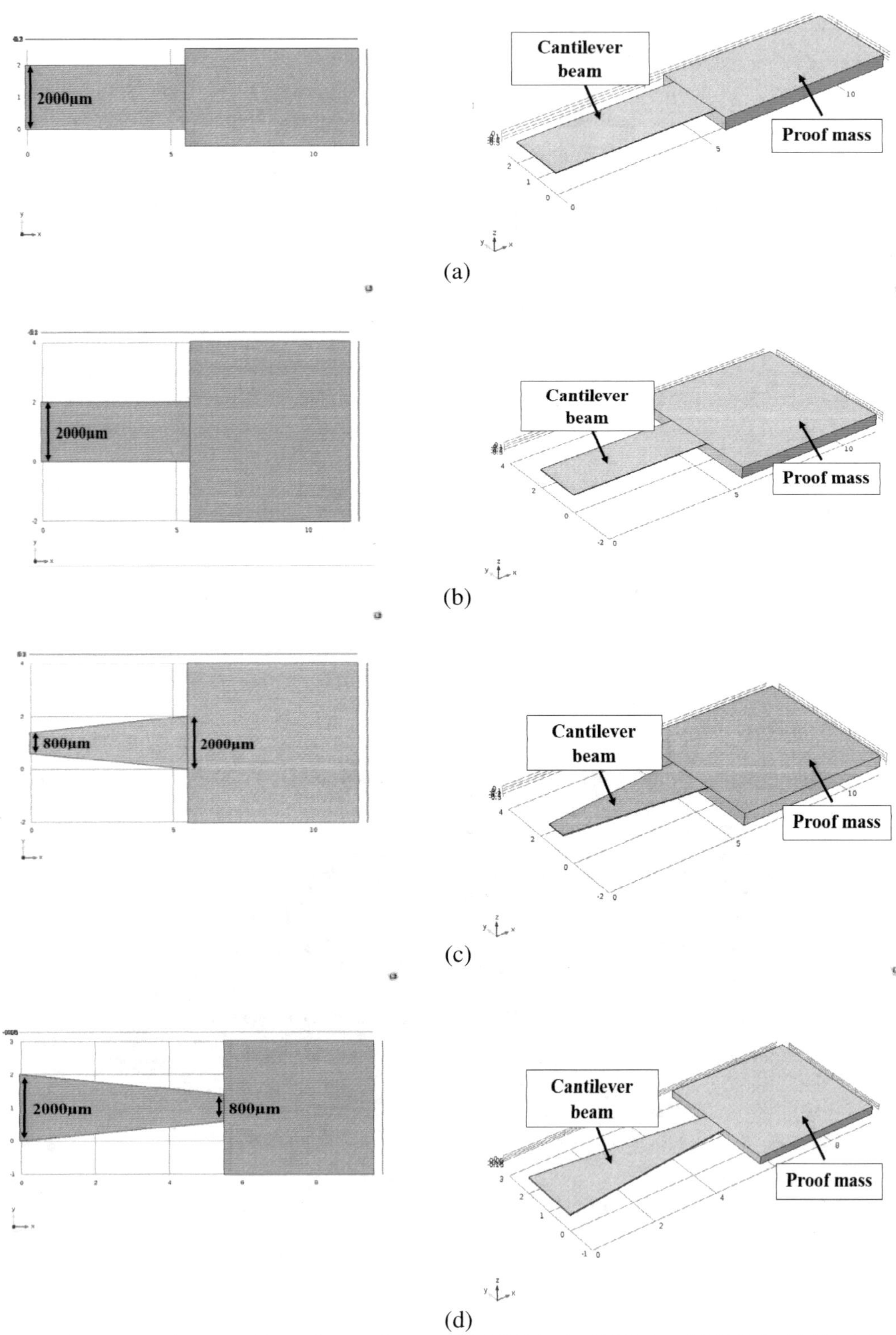

Figure 2. Micro-cantilever beam shapes: (a) PPG 1; (b) PPG 2; (c) PPG 3; (d) PPG4

IOP Conf. Series: Journal of Physics: Conf. Series **1019** (2018) 012033 doi:10.1088/1742-6596/1019/1/012033

3. Result and discussion

This section is divided into three section which are frequency resonance, resistive load, and electrical output result. The corresponding result for PPG 1, PPG 2, PPG 3 and PPG 4 are discussed in details.

3.1 Frequency Resonance Result

The proposed PPG sensor is purposely designed with different micro-cantilever beam geometries in order to determine the lower frequency resonance. Furthermore, a modification to length dimension is made as the change to width dimension has no significant effect on frequency resonance [5]. There are ten samples of length dimension taken into consideration starting from 1000 μm to 5500 μm with 500 μm separation. Hence, Figure 3 depicts the influence of different length dimension to PPG sensor frequency resonance.

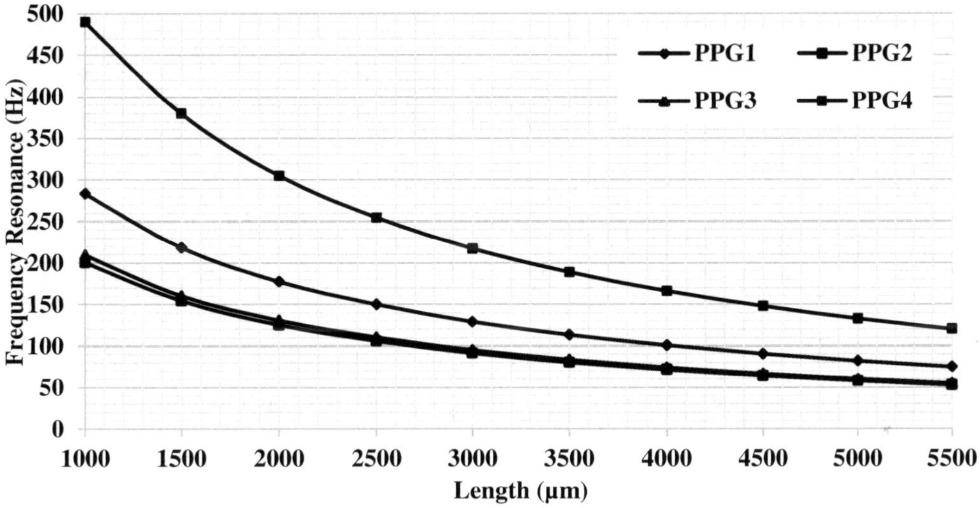

Figure 3. The influence of length dimension against frequency resonance.

From the plotted result, the frequency resonance for the PPG sensor designs are matched to the frequency target which is less than 200 Hz at the specific length dimension. Both PPG 2 and PPG 3 are discovered to achieve the match frequency at 1500 μm, while the PPG 1 and PPG 4 at 2000 μm and 3500 μm respectively. The frequency resonance is revealed to decrease when the length dimension is increasing.

The PPG 2 tends able to operate at the lowest frequency resonance as low as 52.77 Hz at the longest length dimension compared to other designs. The PPG 3 offers second lower frequency resonance which is 54.70 Hz followed by PPG 1 with 74.59 Hz, while PPG 4 indicates the higher frequency resonance which is 120.17 Hz. Fortunately, the frequency resonance for all four proposed PPG sensor are less than 200 Hz. The main factor contributes to this condition is longer length dimension resulting lower beam stiffness. The beam stiffness is measured based on the induced stress at the beam fixed-end where the lower induced stress indicates the beam at lower stiffness. The lower stiffness allow the beam to swing easily when subjected to enough energy at resonance. Moreover, the low frequency resonance is easily reachable when the beam stiffness is low. As to strengthen these arguments, an analysis on length dimension against induced stress is made and the result clearly plotted in Figure 4. Thus, from the result, the lowest induced stress is obtained at the longest length dimension. Therefore, both results in Figures 3 and 4 prove the longer length dimension resulting a lower induced stress that offers lower frequency resonance.

IOP Conf. Series: Journal of Physics: Conf. Series **1019** (2018) 012033 doi:10.1088/1742-6596/1019/1/012033

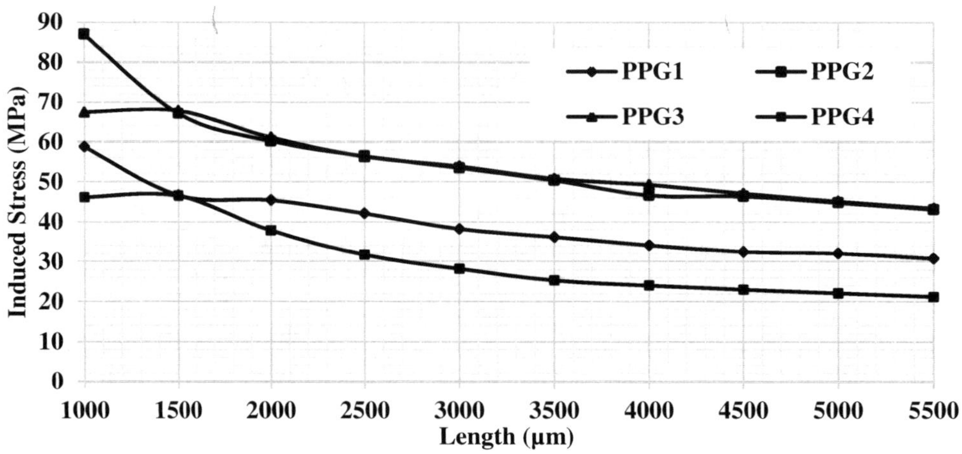

Figure 4. The influence of length against induced stress.

Consequently, a new excitation frequency is required as to avoid any failure due to PPG sensor reaching the maximum strength when vibrates exactly at the resonance. Table 1 is listing the obtained PPG sensor frequency resonance and the corresponding arbitrarily adjusted frequency.

Table 1. The PPG sensor adjusted excitation frequency.

PPG sensor	Frequency resonance, Hz	Adjusted frequency, Hz
PPG 1	74.59	80
PPG 2	52.77	60
PPG 3	54.70	65
PPG 4	120.17	130

3.2 Resistive Load Result

An analysis on determining the optimal load resistance, R_{Load} is conducted in order to demonstrate the proposed PPG sensor to the real application. Based on the result plotted in Figure 5, PPG 2 precedes with 393 mV followed by PPG 1 with 376 mV. Then, the output voltage produced by PPG 3 is 127 mV, while 158 mV from PPG 4. There is obviously seen the output voltage for each PPG sensor is remain constant starting from 10 MΩ.

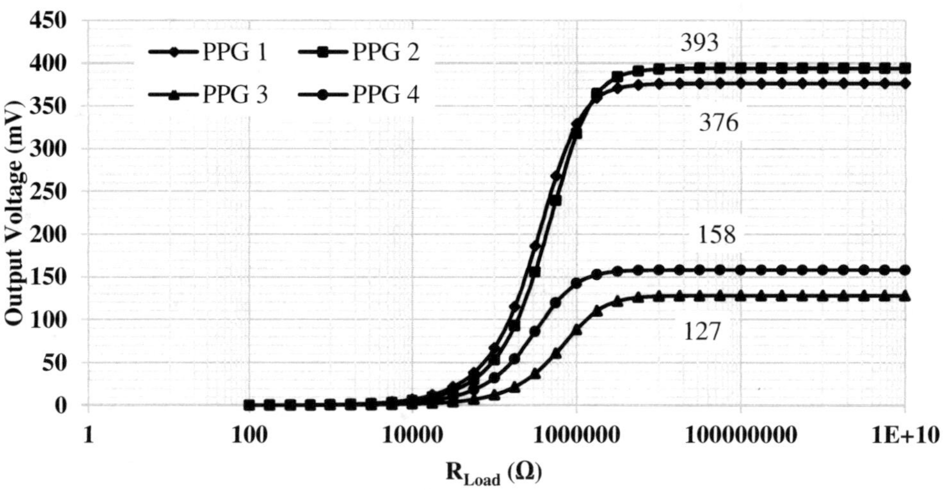

Figure 5. The corresponding resistive load for each PPG sensor designs.

The Kirchhoff's Voltage Law (KVL) is applied to describe the finding on output voltage remains constant starting from 10 MΩ although the R_{Load} value is continue increased. The R_{Load} is parallel connected with PPG sensor. The KVL states that the total voltage in a closed loop is equal to zero as simply derived in Equations (1) and (2) based on the circuit in Figure 6.

$$-V_{in} + V_{Load} = 0 \quad (1)$$

Thus,

$$V_{in} = V_{Load} \quad (2)$$

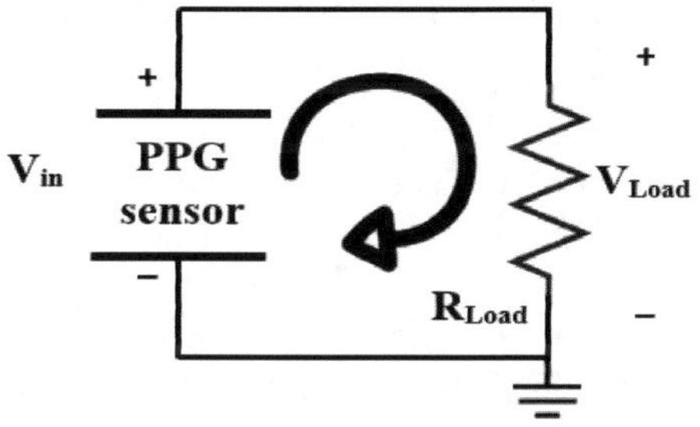

Figure 6. The load resistance, R_{Load} connection.

Therefore, based on Equation (2), the PPG sensor accumulated voltage from harvested vibration is revealed when the R_{Load} at the optimal value.

3.3 Electrical output power result

Figures 7 and 8 depict the electrical output results due to vibration amplitude dependence at 10 MΩ, the optimal R_{Load} value. As clearly seen, the relationship between acceleration and produced electrical output is directly proportional.

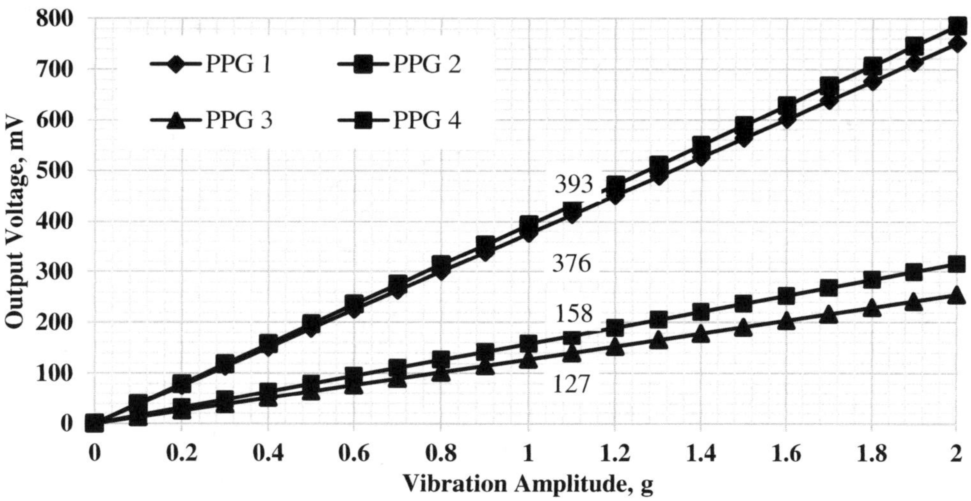

Figure 7. The output voltage for each PPG sensor.

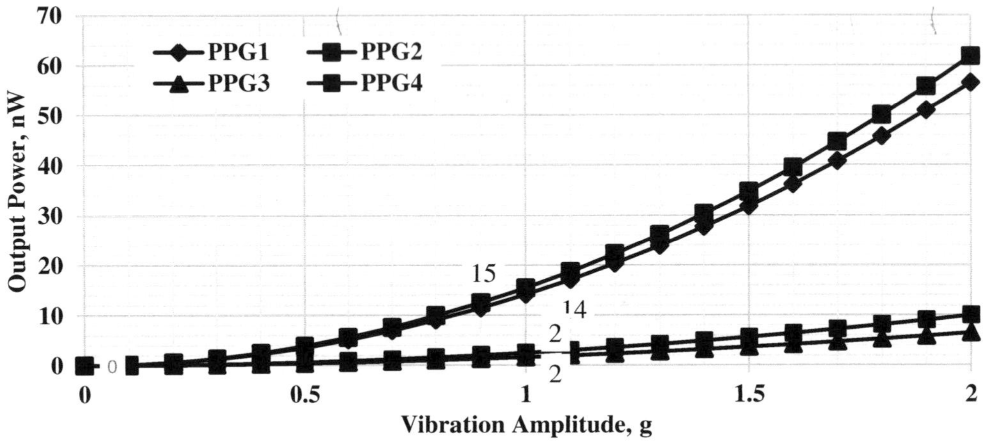

Figure 8. The output power for each PPG sensor.

The produced electrical output is increasing when the acceleration value is further increased. Hence, the obtained result at 1 g vibration amplitude is referred in order to describe the electrical output performance for each PPG sensor. 1 g is equal to the acceleration from gravity which is 9.81 m/s^2. Therefore, the PPG 1 is producing 376 mV of output voltage, while 14 nW for the corresponding output power. Then, PPG 2 produces 393 mV of output voltage, while the output power is 15 nW. The next proposed PPG sensor namely as PPG 3 is able to generate 127 mV, and 2 nW of output voltage, and output power correspondingly. The fourth PPG sensor which is PPG 4 is successfully producing 158 mv of output voltage, while 2 nW for the output power.

4. Conclusion

This research paper deals with electrical output power optimization by the simple and less complicated micro-cantilever beam modification. The main focus is to design the PPG sensor with lower vibration frequency to easily match with the target ambient vibration frequency. The different micro-cantilever beam geometry produced different performance outputs based on the obtained results. From the frequency resonance analysis, all the proposed PPG sensor tend able to operate at lower frequency which is less than 200 Hz. From the finding, the proposed geometries are suitable to be used as ambient vibration harvester. However, the PPG 2 sensor gives a superior performances with producing the higher electrical output power at lowest frequency.

5. References

[1] Bhatia D, Sharma H, Meena R S and Palkar V R 2016 A Novel Zno Piezoelectric MicrocantileverEnergy Scavenger: Fabrication And Characterization *J. Sens. Bio-Sensing Res.* 45-52 **9** 45–52

[2] Hassan H F, Hassan S I S and Rahim R A 2014 Acoustic Energy Harvesting Using Piezoelectric Generator for Low Frequency Sound Waves Energy Conversion *Int. J. Eng. Technol.* **5** 4702–07

[3] Lallart M, Wu Y and Guyomar D 2012 Switching Delay Effects on Nonlinear Piezoelectric Energy Harvesting Techniques *IEEE Trans. Ind. Electron.* **59** 464–72

[4] Lin H C, Wu P H, Lien I C and Shu Y C 2013 Analysis of An Array of Piezoelectric Energy Harvesters Connected in Series *Smart Mater. Struct.* **22** 1–11

[5] Littrell R and Grosh K 2012 Modeling and Characterization of Cantilever-Based MEMS Piezoelectric Sensors and Actuators *J. Microelectromechanical Syst.* **21** 406–13

[6] Liu H, Lee C, Kobayashi T, Jui C and Quan C 2012 Physical Piezoelectric MEMS-Based Wideband Energy Harvesting Systems using a Frequency-Up-Conversion Cantilever Stopper *Sensors Actuators A. Phys.* **186** 242–48

IOP Conf. Series: Journal of Physics: Conf. Series **1019** (2018) 012033 doi:10.1088/1742-6596/1019/1/012033

[7] Nordin A N and Salleh H 2010 Simulation of a MEMS Piezoelectric Energy Harvester *Symp. Des. Test Integr. Packag. MEMS/MOEMS* 7–11
[8] Park J C, Park J Y and Lee Y 2010 Modeling and Characterization of Piezoelectric d_{33}-Mode MEMS Energy Harvester *J. Microelectromechanical Syst.* **19** 1215–22
[9] Park J, Lee S and Kwak B M 2012 Design Optimization of Piezoelectric Energy Harvester Subject to Tip Excitation *J. Mech. Sci. Technol.* **26** 137–43
[10] Rahim R A, Hassan S I S, Malek F and Nordin J M 2013 Harmonics Suppression Single-fed Dual-Circularly Polarized Microstrip Patch Antenna for Future Wireless Power Transmission *Int. J. Eng. Technol.* **5** 4423–30
[11] Rosmi A S, Saadon S, Hassan S I S and Wahab Y 2017 Preliminary Results on Shape Optimization of Ambient- Based Piezoelectric Energy Harvester *AIP Conf. Proc.* **1808** 1–5
[12] Selvaraj S, Mano S, Sugumaran P, Meenaa J and Jayaraman P 2014 Multiphysics Modeling and Simulation of Nanoscale Cantilever Array Sensor for the Detection of HIV Antigens *J. Nanosci. Nanotechnol.* **2** 15–18
[13] Wang Z. *et al.* 2013 Large Power Amplification of a Piezoelectric Energy Harvester Excited by Random Vibrations *IEEE 26th Int. Conf. MicroElectroMechanical Syst.* 106–09
[14] Yang S M, Lee T and Jeng C A 2009 Physical Development of a Thermoelectric Energy Harvester with Thermal Isolation Cavity by Standard CMOS Process *Sensors Actuators A Phys. J.* **153** 244–50

Multi-Attribute Decision Making with VIKOR Method for Any Purpose Decision

Dodi Siregar[1], Heri Nurdiyanto[2], S Sriadhi[3], Diana Suita[4], Ummul Khair[1], Robbi Rahim[5*], Darmawan Napitupulu[6], Achmad Fauzi[7], Abdurrozzaq Hasibuan[8], M Mesran[9] and Andysah Putera Utama Siahaan[4]

[1]Department of Informatics, Universitas Harapan Medan, Indonesia
[2]Department of Informatics, STMIK Dharma Wacana, Indonesia
[3]Department of Electrical Engineering, Universitas Negeri Medan, Indonesia
[4]Department of Civil Engineering, Universitas Harapan Medan, Indonesia
[5]School of Computer and Communication Engineering, Universiti Malaysia Perlis, Malaysia
[6]Research Center for Quality System and Testing Technology, Indonesian Institute of Sciences, Indonesia
[7]Department of Computer Engineering, STMIK Kaputama, Indonesia
[8]Department of Industrial Engineering, Universitas Islam Sumatera Utara, Indonesia
[9]Department of Informatics, STMIK Budi Darma, Indonesia

*usurobbi85@zoho.com

Abstract. Implementation of Decision Support System for various purposes now can facilitate policy makers to get the best alternative from a variety of predefined criteria, one of the methods used in the implementation of Decision Support System is VIKOR (Vise Kriterijumska Optimizacija I Kompromisno Resenje), VIKOR method in this research got the best results with an efficient and easily understood process computationally, it is expected that the results of this study facilitate various parties to develop a model any solutions.

1. Introduction

Decision support systems utilize private resources in a manner with computer skills to improve decision results, so this is a computer-based support system for decision-making that deals with semi-issues structured [1] [2]. Decision-making is always correlated with the uncertainty of the results of decisions taken, to reduce this uncertainty factor, the decision requires valid information about the conditions that have been, and may occur, then processed the information into several alternative problems solving as a material consideration in deciding the steps to be implemented, so that the decision taken is expected to provide maximum benefits [3].

The decision support system that will present in this research uses VIKOR method which aims to gather information about all data related to multiple attributes and multiple-criteria [4]. The use of VIKOR method is used because the method it can choose highly effective and efficient criteria for determining decision outcomes with multiple attributes and multiple-criteria [4] [5] [6]. VIKOR is a multiple attribute decision-making method used to solve problems in discrete space [7]. Therefore, in MADM is usually used to perform assessment or selection of several alternatives in a limited number,

IOP Conf. Series: Journal of Physics: Conf. Series **1019** (2018) 012034 doi:10.1088/1742-6596/1019/1/012034

the process of multiple attribute decision making in this paper shows the calculation of VIKOR method gradually with 25 alternatives and five criteria to determine the accuracy of the process of VIKOR.

2. Methodology

Some previous studies which discussed the VIKOR method is Papathanasiou [7], Papathanasiou makes web-based application of TOPSIS and VIKOR MCDM, the policy makers can choose 2 (two) different decision result from application, and also another researcher Nisel [8] analyzed the application of VIKOR method to determine the rank of graduation from business program students, and many other studies discussing experimental results with VIKOR and its combination, from several types of research it was observed that VIKOR could arrange for many alternatives, so it can be assumed that the VIKOR algorithm is still relevant to use and in this research VIKOR method is calculated gradually to facilitate the calculation process for many criteria, many attributes and many alternatives.

The VIKOR method focuses on ranking and chooses from a set of samples with different criteria, which can help decision makers to get a final determination [4].

VIKOR is a method for optimizing/optimizing multiple criteria in a complex system. VIKOR's basic concept is to rank the current samples by looking at the results of the corresponding values or regrets (R) of each sample. The VIKOR method has applied by some researchers in the case of MCDM [7]. The formula of the VIKOR method is as follows:

1) Normalization of the matrix

   $R_{ij} = \frac{(X*j - Xij)}{(X*J - X'J)}$

   Information:
   Xij = Value of sample i data criteria j
   (I = A, B, C, D, E)
   (J = 5 criteria)
   X*j = Best value in one criterion
   X'j = the worst value in one criterion
2) Calculating the S and R-Value

   Si= $\sum_{j=1}^{n} wj\ x\ (Rij)$

   Wj = weighting criteria
   The value of S is achieved from the sum of the result of the multiplication of the criteria weights by the data in each sample
   Ri= Max j [wj x Rij], largest value from [Wj x Rij]
   The value of R is the largest value of the multiplication of the weight of the criteria with the normalized data of each sample
3) Calculates the VIKOR index

   Formula= $[\frac{Si - S'}{S* - S'}]$ x V + $[\frac{Ri - R'}{R* - R'}]$ x (1-V)

   S'= smallest S value (the best value)
   S*= the largest S value
   R'= the smallest R value
   R*= the largest R value

3. Result and Discussion

The analysis of the VIKOR method can be seen gradually in the following test process gradually, for the first step is to determine the criteria to be given the weight value for VIKOR calculation process, Table 1 below are the criteria that are used in this paper:

Table 1. Criteria and Weight

No	Criteria	Weight (%)
1	C1	30
2	C2	25
3	C3	17
4	C4	20
5	C5	8

The next process is to determine alternatives with values for each criterion:

Table 2. Alternative and Criteria Value

No	Alternative	Criteria				
		C1	C2	C3	C4	C5
1	A1	0	21	0	15	0
2	A2	11	23	5	13	3
3	A3	12	10	0	12	0
4	A4	10	15	0	12	0
5	A5	5	9	16	8	0
6	A6	9	8	13	5	0
7	A7	15	14	0	7	0
8	A8	24	12	0	18	0
9	A9	17	17	16	20	0
10	A10	5	8	17	5	6
11	A11	9	7	17	12	8
12	A12	10	10	5	20	0
13	A13	17	22	0	11	0
14	A14	0	12	0	11	0
15	A15	5	8	10	6	7
16	A16	5	8	7	4	0
17	A17	0	0	9	4	0
18	A18	10	14	0	11	0
19	A19	7	5	0	8	0
20	A20	11	15	0	5	0
21	A21	13	9	0	4	0
22	A22	17	20	0	4	0
23	A23	23	20	0	5	0
24	A24	0	0	0	0	0
25	A25	0	0	0	0	0

From table value of each criterion will be normalized data, the result can see as below, with example value Criteria C1

$$R\ (A1),\ C1 = \frac{(24\text{-}0)}{(24\text{-}0)} = \frac{24}{24} = 1$$

$$R\ (A2),\ C1 = \frac{(24-11)}{(24-0)} = \frac{13}{24} = 0.54$$

R (A1) and R (A2) are samples of normalization calculation of first criterion matrix with alternative 1 and alternative 2, and matrix normalization process is executed for all criteria and alternatives, the final result of matrix normalization process could be seen in Table 3 below:

IOP Conf. Series: Journal of Physics: Conf. Series **1019** (2018) 012034 doi:10.1088/1742-6596/1019/1/012034

Table 3. Normalization Matrix and Weight

No	Alternative	Criteria				
		C1	C2	C3	C4	C5
1	A1	1	0,08	1	0,25	1
2	A2	0,54	0	0,71	0,35	0,62
3	A3	0,5	0,56	1	0,4	1
4	A4	0,58	0,35	1	0,4	1
5	A5	0,79	0,61	0,06	0,6	1
6	A6	0,62	0,65	0,23	0,75	1
7	A7	0,37	0,39	1	0,65	1
8	A8	0	0,48	1	0,1	1
9	A9	0,29	0,26	0,06	0	1
10	A10	0,79	0,65	0	0,75	0,25
11	A11	0,62	0,69	0	0,4	0
12	A12	0,58	0,56	0,71	0	1
13	A13	0,29	0,04	1	0,45	1
14	A14	1	0,48	1	0,45	1
15	A15	0,79	0,65	0,41	0,7	0,12
16	A16	0,79	0,65	0,58	0,8	1
17	A17	1	1	0,47	0,8	1
18	A18	0,58	0,39	1	0,45	1
19	A19	0,71	0,78	1	0,6	1
20	A20	0,54	0,35	1	0,75	1
21	A21	0,46	0,61	1	0,8	1
22	A22	0,29	0,13	1	0,8	1
23	A23	0,04	0,13	1	0,75	1
24	A24	1	1	1	1	1
25	A25	1	1	1	1	1

After the results obtained from the normalization of criteria and alternatives, the next is to multiply the value of normalization and weight so that the performance as table 4 below:

Table 4. Result Normalization x Weight

No	Alternativ e	Criteria				
		C1	C2	C3	C4	C5
1	A1	30	2	17	5	8
2	A2	16,2	0	12,07	7	4,96
3	A3	15	14	17	8	8
4	A4	17,4	8,75	17	8	8
5	A5	23,7	15,25	1,02	12	8
6	A6	18,6	16,25	3,91	15	8
7	A7	11,1	9,75	17	13	8
8	A8	0	12	17	2	8
9	A9	8,7	6,5	1,02	0	8
10	A10	23,7	16,25	0	15	2
11	A11	18,6	17,25	0	8	0
12	A12	17,4	14	12,07	0	8
13	A13	8,7	1	17	9	8
14	A14	30	12	17	9	8
15	A15	23,7	9,75	6,97	14	0,96
16	A16	23,7	16,25	9,86	16	8
17	A17	30	25	7,99	16	8

18	A18	17,4	9,75	17	9	8
19	A19	21,3	19,5	17	12	8
20	A20	16,2	8,75	17	15	8
21	A21	13,8	15,25	17	16	8
22	A22	8,7	3,25	17	16	8
23	A23	1,2	3,25	17	15	8
24	A24	30	25	17	20	8
25	A25	30	25	17	20	8

Based on the formula VIKOR method that has been described and from the value of table 3 and table 4 and got the value of Q (VIKOR index) by using formula 3 of VIKOR, the results index value in table 5 below:

Table 5. Index VIKOR

No	Alternative	Q Value
1	A1	0,9325
2	A2	0,8135
3	A3	0,4435
4	A4	0,4347
5	A5	0,5879
6	A6	0,4799
7	A7	0,4229
8	A8	0,292
9	A9	**0**
10	A10	0,78
11	A11	0,3595
12	A12	0,8211
13	A13	0,323
14	A14	0,8416
15	A15	0,5576
16	A16	0,6792
17	A17	0,707
18	A18	0,4478
19	A19	0,6492
20	A20	0,4632
21	A21	0,4968
22	A22	0,384
23	A23	0,3279
24	A24	1
25	A25	1

From the table above obtained the data that the sample (A9) has the smallest index value, and A9 is the best ranking, from the data table 5 above obtained graph of the process of using the VIKOR method for each criterion and alternatives that exist.

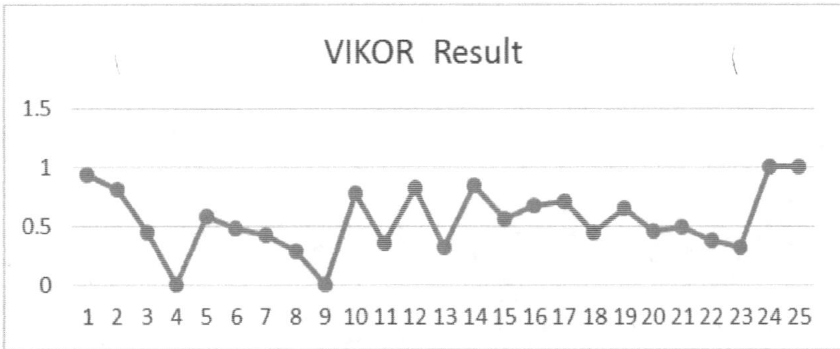

Figure 1. VIKOR Result

Based on the above graph shows the results of calculations VIKOR method meet proper distribution with a uniform process.

4. Conclusion

The experiment of the VIKOR method can help to complete effective decision-making because the concept is straightforward and easy to understand and the computation process is efficient and can measure the relative performance of various decision alternatives

5. References

[1] Risawandi and R. Rahim, "Study of the Simple Multi-Attribute Rating Technique For Decision Support," *International Journal of Scientific Research in Science and Technology (IJSRST)* , vol. 2, no. 6, pp. 491-494, 2016.

[2] Syamsudin and R. Rahim, "Study Approach Technique for Order of Preference by Similarity to Ideal Solution (TOPSIS)," *International Journal of Recent Trends in Engineering & Research,* vol. 3, no. 3, pp. 268-285, 2017.

[3] Mesran, G. Ginting, R. Rahim and Suginam, "Implementation of Elimination and Choice Expressing Reality (ELECTRE) Method in Selecting the Best Lecturer (Case Study STMIK BUDI DARMA)," *International Journal of Engineering Research & Technology (IJERT),* vol. 6, no. 2, pp. 141-144, 2017.

[4] C. T. Sasanka and K. Ravindra, "Implementation of VIKOR Method for Selection of Magnesium Alloy to Suit Automotive Applications," *International Journal of Advanced Science and Technology,* vol. 83, pp. 49-58, 2015.

[5] C.-H. Wang and C.-T. Pang, "Using VIKOR Method for Evaluating Service Quality of Online Auction under Fuzzy Environment," *International Journal of Computer Science Engineering and Technology,* vol. 1, no. 6, pp. 307-314, 2011.

[6] M. Xue, X. Tang and N. Feng, "An Extended VIKOR Method for Multiple Attribute Decision Analysis with Bidimensional Dual Hesitant Fuzzy Information," *Mathematical Problems in Engineering,* vol. 2016, pp. 1-16, 2016.

[7] J. Papathanasiou, N. Ploskas, T. Bournaris and B. Manos, "A Decision Support System for Multiple Criteria Alternative Ranking Using TOPSIS and VIKOR: A Case Study on Social Sustainability in Agriculture," in *International Conference on Decision Support System Technology*, Belgium, 2016.

[8] S. Nisel, "An Extended VIKOR Method for Ranking Online Graduate Business Programs," *International Journal of Information and Education Technology,* vol. 4, no. 1, pp. 103-107, 2014.

IOP Conf. Series: Journal of Physics: Conf. Series **1019** (2018) 012035 doi:10.1088/1742-6596/1019/1/012035

Mobile Application Detection of Road Damage using Canny Algorithm

G Gunawan[1], Heri Nuriyanto[2], S Sriadhi[3], Achmad Fauzi[4], Ari Usman[5], F Fadlina[6], Haida Dafitri[5], Janner Simarmata[7], Andysah Putera Utama Siahaan[8,9] and Robbi Rahim[8*]

[1]Department of Computer Engineering and Informatics, Politeknik Negeri Medan, Indonesia
[2]Department of Informatics, STMIK Dharma Wacana, Indonesia
[3]Department of Electrical Engineering, Universitas Negeri Medan, Indonesia
[4]Department of Computer Engineering, STMIK Kaputama, Indonesia
[5]Department of Informatics, Universitas Harapan Medan, Indonesia
[6]Department of Informatics Management, AMIK STIEKOM Sumatera Utara, Indonesia
[7]Universitas Negeri Medan, Indonesia
[8]School of Computer and Communication Engineering, Universiti Malaysia Perlis, Malaysia
[9]Faculty of Computer Science, Universitas Pembangunan Panca Budi, Indonesia

*usurobbi85@zoho.com

Abstract. Edge detection is one of the most frequent processes in digital image processing for various purposes, one of which is detecting road damage based on crack paths that can be checked using a Canny algorithm. This paper proposed a mobile application to detect cracks in the road and with customized threshold function in the requests to produce useful and accurate edge detection. The experimental results show that the use of threshold function in a canny algorithm can detect better damage in the road.

1. Introduction

One of the image processing techniques used is edge detection [1]. Edge detection is standard in digital image processing because it is one of the first steps in segmenting the image, which aims to present the objects contained in the picture [1] [2] [3] [4]. Edge detection serves to identify the boundary of an object over an overlapping image. So that if the edge line in the picture could accurately identify, all objects could found and the fundamental properties such as area, shape, and size of the object can be measured [5] [6]. Image edge is the position where the pixel intensity of the picture changes from low value to high value or vice versa [1] [5]. Currently, several algorithms could use for edge detection, for example, Sobel [7], Canny [7], Prewitt [7], frei-chen [7] and Laplacian algorithms [8]. In this research taken canny algorithm, Edge is a change of intensity value of gray degree quickly or suddenly in a short distance [5] [8]. The purpose of detecting the edges themselves is to group the objects in the image, and also be used to analyze the picture further [9] [10] [11] [12].

In addition to the use of canny algorithm, this research also uses threshold function in application which is a parameter to determine the level of image clarity of edge detection results done, the crack examination on the road is very important to detect the condition of the road that is used because of the effect on the safety of the street drivers, fracture on roads could be taken seriously by repairing the road.

Chitra [4] applies the gradient calculation of magnitude and gradient of direction to improve edge detection in the image, then Biswas [13] enhances edge detection by applying fuzzy sets to reduce non-noise in the picture so that the accuracy level is higher, in principle improvisation edge detection canny algorithm is possible done, only by using the threshold function applied to the mobile application contained in the experimental results in this paper.

2. Methodology

The canny algorithm is an algorithm that performs the examination with the minimum error rate to produce the optimum edge image, but the canny algorithm process is not sufficient to reduce the picture that has many edges or strokes which is really an edge or not, for that canny algorithm in Improvised by adding a threshold function so that edge detection errors on the image can be reduced well by providing the minimum and maximum values at the time of edge detection processing.

Results of this research was conducted to detect cracks in the road which is the beginning of damage to the road, to facilitate the process of detecting road cracks directly designed a mobile-based applications to detect road damage with processors using canny algorithms and threshold functions, with applications designed policy makers can take a direct solution to fix the road or just do maintenance only.

The selection of canny algorithm as edge detection process because ease of implementation in the language of various programming language especially Java besides also easy to be developed by applying various function especially threshold function.

There are some of the most optimum edge detection criteria that can be met by canny algorithm [3] [10] [11]:

1. Detect well, Ability to put and mark all existing edges under the selection of convolution parameters performed. It also provides very high flexibility regarding determining the level of edge thickness detection as desired.
2. Localize well with Canny; it is possible to produce a minimum distance between the edges detected with the original edge.
3. The response is clear, there is only one response to each edge so easily detected and does not cause confusion on subsequent image processing.

There is five step in performing canny edge detection [5] [7] [14]:

1. Eliminate the noise present in the picture by implementing the Gaussian filter. This process will produce an image that looks a bit blurry, and this is intended to get the edge of the actual picture. If not performed then fine lines will also be detected as an edge. Here is one example of a Gaussian filter with $\sigma = 1.4$.

$$\frac{1}{159}\begin{matrix} 2 & 4 & 5 & 4 & 2 \\ 4 & 9 & 12 & 9 & 4 \\ 5 & 12 & 15 & 12 & 5 \\ 4 & 9 & 12 & 9 & 4 \\ 2 & 4 & 5 & 4 & 2 \end{matrix}$$

2. Perform edge detection by searching horizontally (Gx) and vertically (Gy).
3. Determining the edge direction found using the following formula.

$$\theta = \arctan \frac{Gy}{Gx}$$

Next, divide into four colors so that the lines with different directions and have different colors. The division is 0 - 22.5 and 157.5 - 180 degrees yellow, 22.5 - 67.5 are green, and degrees 67.5 - 157.5 are red

4. Minimizes the edge line that appears by applying non-maximum suppression to produce a slimmer edge line.
5. The final step is binary by applying two thresholds, the following figure will show the image shape before processing (a) and after processing (b). The image used is a grayscale image with threshold value 0.05.

Thresholding is used to adjust the amount of gray degree present in the picture. By using Thresholding the gray degree can be changed as desired, e.g. desired using the level of gray 16, then live to divide the value of a gray degree with 16. The process of Thresholding is a process of changing the quantization of the image, so to do Thresholding with the gray degree can be used the formula:

$$X \frac{W}{B}$$

Where:
W is the gray degree value before Thresholding
B is the desired number of degrees of gray
X is the gray degree value after Thresholding

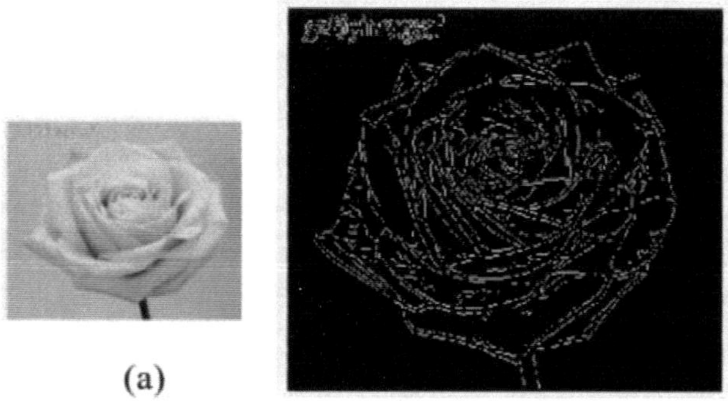

(a)

(b)

Figure 1. Before and After Edge Detection

The results of this research is a smartphone application that can be used to detect cracks in roads with canny algorithms and threshold functions, apps designed using Android Studio and Android API 24 SDK

3. Result and Discussion

The Canny algorithm is one of the operators used for edge detection, and this operator is like a Sobel operator. Convolution canny algorithm involving canny kernel and convolution is perform with pixel value of the image that to be processed, figure 2 below will be test for edge detection.

IOP Conf. Series: Journal of Physics: Conf. Series **1019** (2018) 012035 doi:10.1088/1742-6596/1019/1/012035

Figure 2. Crack damage in the road

For the process of detecting cracks in the road could be seen in the following pseudo code, the code is:

```
public void setLowThreshold(float threshold) {
        if (threshold < 0) throw new IllegalArgumentException(); lowThreshold = threshold;
   }

public float getHighThreshold() {return highThreshold;}

public void setHighThreshold(float threshold) {
        if (threshold < 0) throw new IllegalArgumentException();
highThreshold = threshold; }

Arrays.fill(data, 0);
int offset = 0;
for (int y = 0; y < height; y++) {
for (int x = 0; x < width; x++) {
if (data[offset] == 0 && magnitude[offset] >= high) {
follow(x, y, offset, low);
}
offset++;
} }

private void follow(int x1, int y1, int i1, int threshold) {
int x0 = x1 == 0 - x1 : x1 - 1;
int x2 = x1 == width – 1 - x1 : x1 + 1;
int y0 = y1 == 0 - y1 : y1 - 1;
int y2 = y1 == height - 1 - y1 : y1 + 1;

data[i1] = magnitude[i1];
 for (int x = x0; x <= x2; x++) {
  for (int y = y0; y <= y2; y++) {
    int i2 = x + y * width;
     if ((y != y1 || x != x1)
       && data[i2] == 0
       && magnitude[i2] >= threshold) {
          follow(x, y, i2, threshold);
            return;} } }
}
```

IOP Conf. Series: Journal of Physics: Conf. Series **1019** (2018) 012035 doi:10.1088/1742-6596/1019/1/012035

The above pseudo code are some critical processes for detecting crack paths with canny algorithms and threshold functions. The result of edge detection process from the figure 2 by using the canny algorithm could be seen in figure 3 below:

Figure 3. Edge Detection Result

Based on the process of the canny algorithm that has been described and implement in mobile based edge detection application, the results can see in figure 4 below:

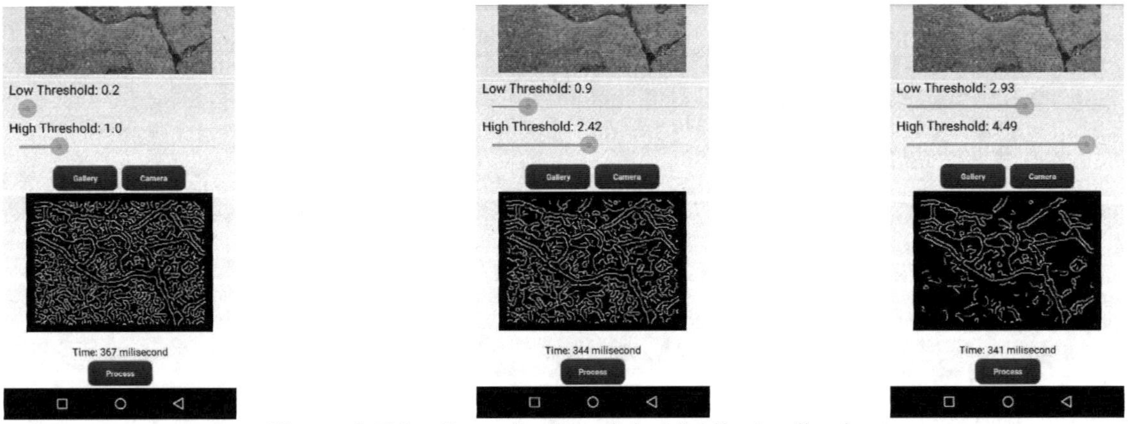

Figure 4. Edge Detection Result in Mobile Application

In Figure 4 there are some results of the edge detection process with the same object but have different results when the threshold value is replaced, so it can be concluded that the edge detection results in the image with the canny algorithm can affect when the threshold value on the picture is change, and this process also affects the time of edge detection process.

The Result of detection road damage by using canny algorithm and threshold function produce fast enough when using application because every object of image changed into grayscale image, and then by determining the minimum and maximum threshold values can reduce the detection error against objects that are not part of the road damage.

4. Conclusion

The Canny algorithm applied to a mobile application can be used to detect damage on the road very well with accurate results, but also by adding threshold function to the image can know the cracks in the image more precisely because it will eliminate the smaller edge of pixel intensity of the edge pixel image that had crack.

5. References

[1] D. Apdilah, M. Y. Simargolang and R. Rahim, "A Study of Frei-Chen Approach for Edge Detection," *International Journal of Scientific Research in Science, Engineering and Technology,* vol. 3, no. 1, pp. 59-62, 2017.

[2] R. Rahim, T. Afriliansyah, H. Winata, D. Nofriansyah, Ratnadewi, and S. Aryza, "Research of Face Recognition with Fisher Linear Discriminant," IOP Conf. Ser. Mater. Sci. Eng., vol. 300, p. 012037, 2018.

[3] H. Banouni, B. Faiz, D. I. E. H. Ouacha, M. Boutaib and M. Derra, "The Canny Edge Detection Method vs. the Radius of Curvature Method for Determining the Time of Flight on Ultrasound," *International Journal of Signal System Control and Engineering Application,* vol. 9, no. 3, pp. 48-54, 2016.

[4] A. D. Chitra and P. Ponmuthuramalingam, "An Approach for Canny Edge Detection Algorithm for Face Recognition," *International Journal of Science and Research (IJSR),* vol. 4, no. 11, pp. 24-27, 2015.

[5] S. Vijayarani and M. Vinupriya, "Performance Analysis of Canny and Sobel Edge Detection Algorithms in Image Mining," *International Journal of Innovative Research in Computer and Communication Engineering,* vol. 1, no. 8, pp. 1760-1767, 2013.

[6] V. Saini and R. Garg, "A Comparative Analysis on Edge Detection Techniques Used in Image Processing," *IOSR Journal of Electronics and Communication Engineering (IOSRJECE),* vol. 1, no. 2, pp. 56-59, 2012.

[7] S. Bhardwaj and A. Mittal, "A Survey on Various Edge Detector Techniques," *Procedia Technology,* vol. 4, pp. 220-226, 2012.

[8] R. Muthukrishnan and M. Radha, "EDGE DETECTION TECHNIQUES FOR IMAGE SEGMENTATION," *International Journal of Computer Science & Information Technology (IJCSIT),* vol. 3, no. 6, pp. 259-265, 2011.

[9] Y. Ramadevi, T. Sridevi, B. Poornima and B. Kalyani, "SEGMENTATION AND OBJECT RECOGNITION USING EDGE DETECTION TECHNIQUES," *International Journal of Computer Science & Information Technology (IJCSIT),* vol. 2, no. 6, pp. 153-161, 2010.

[10] A. Abbaszadeh and S. Daneshvar, "A Review Of Different Type," *JOURNAL OF CURRENT RESEARCH IN SCIENCE,* vol. 1, pp. 871-875, 2016.

[11] N. Ahmadi and G. Akbarizadeh, "Iris Recognition System based on Canny and LoG Edge Detection Methods," *Journal of Soft Computing and Decision Support Systems,* vol. 2, no. 4, pp. 26-30, 2015.

[12] D. Nofriansyah et al., "A New Image Encryption Technique Combining Hill Cipher Method, Morse Code and Least Significant Bit Algorithm," J. Phys. Conf. Ser., vol. 954, no. 1, p. 012003, 2018.

[13] R. Biswas and J. Sil, "An Improved Canny Edge Detection Algorithm Based on Type-2 Fuzzy Sets," in *2nd International Conference on Computer, Communication, Control and Information Technology*, India, 2012.

[14] G. Yang and F. Xu, "Research and analysis of Image edge detection algorithm Based on the MATLAB," *Procedia Engineering,* vol. 15, pp. 1313-1318, 2011.

Breadth First Search Approach for Shortest Path Solution in Cartesian Area

Robbi Rahim[1*], Dahlan Abdullah[2], Saiful Nurarif[3], Mukhlis Ramadhan[3], Badrul Anwar[3], Muhammad Dahria[4], Surya Darma Nasution[5], Tengku Mohd Diansyah[6] and Mufida Khairani[6]

[1]School of Computer and Communication Engineering, Universiti Malaysia Perlis, Malaysia
[2]Department of Informatics, Universitas Malikussaleh, Jl. Medan-Banda Aceh, Aceh Utara, Indonesia
[3]Department of Computer System, STMIK Triguna Dharma, Jl. Jenderal Abdul Haris Nasution No.73, Medan 20219, Indonesia
[4]Department of Information System, STMIK Triguna Dharma, Jl. Jenderal Abdul Haris Nasution No.73, Medan 20219, Indonesia
[5]Department of Informatics, STMIK Budidarma, Medan, Indonesia
[6]Department of Informatics, Universitas Harapan Medan, Medan, Indonesia

*usurobbi85@zoho.com

Abstract. Determining the shortest path is one problem that is much discussed using some algorithm like Djikstra, Floyd Warshall and in this research an algorithm Breadth First Search are used, Breadth First Search algorithms in this study is used to determine the shortest route and optimal from a Cartesian field, the best and optimal route search experiment of cartesian areas using Breadth First Search algorithm can be perform very well and gets some route options from the best to the longest route.

1. Introduction

The development of computer-based technology is currently experiencing rapid development, and the development has also covered various aspects of the field of life with the aim to facilitate the completion of tasks charged, Artificial intelligence is the most developed field today.

Shortest Path is a pretty much solved problem with algorithms like Djikstra, Depth First Search and Breadth First, Yan ZHOU[1] conducted a shortest path study using the Breadth First Search algorithm by approaching the search from the starting point to the destination point by examining the layer by layer against all the possibilities that can be passed and remove all the paths that are not possible to pass. Similar to Yan ZHOU this experiment try to finding the shortest path that can pass through the Cartesian field. This issue could be explained as follows; it is identified that an object will move from the center point (0,0) to point A (m, n). The object should only bend at the grid points and always step in parallel with the x-axis or y-axis. The object must not cross the path that has been passed and should not cross the point that has been crossed. After that provided a set of provisions that limit the movement of the object. The question is how the trajectories that the object can pass through using the terms defined above.

Breadth First Search is one solution that can be used to solve the above problems, Breadth-first search is an algorithm that performs a wider search by visiting the node preorder i.e. visiting a node and then visiting all the nodes adjacent to the node first[2][3]. Next, unvisited nodes and neighbors of the previously visited nodes, so it is doubtful that the visited node will be revisited [5][6].

This study attempts to show some of the path solutions from the initial location to the destination location, using the Breadth First Search algorithm. The given solution can consist of many solutions with the path meet the criteria of the problem already mentioned, Research conducted by previous research Yan ZHOU, this research not only provides the shortest path solution but also provides another form of optimal search if there are obstacles that block the process of the path being traversed.

2. Theory

A. Breadth first search

In breadth first search method, all nodes at level n will be visited first before visiting nodes at level n+1 (next level). The search starts from the root node and continues to the 1st level from left to right, then moves to level-2 and so on[2][3][7]. As an illustration of the first search breadth process could be seen in the following Tree process until the discovery of the solution [8].

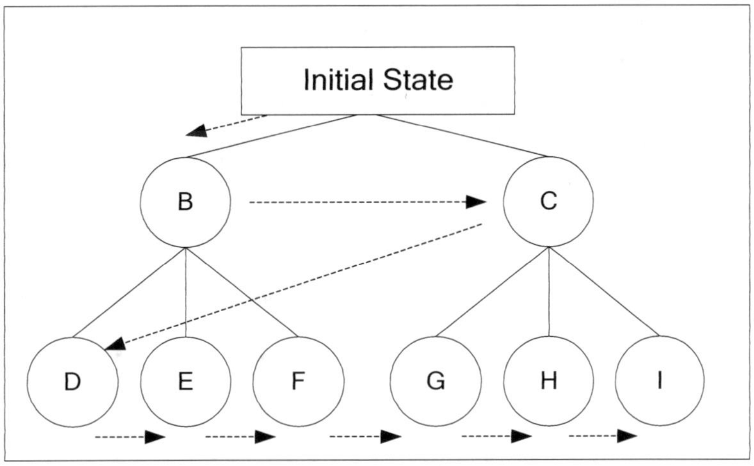

Figure 1. Procedure Breadth First Search

The advantage of using this algorithm is that it will not get stuck, and if there is more than one solution then the breadth-first search will find all the solutions, and if there is more than one solution then the minimum solution will be found[9][10]. While the weakness of this algorithm pretty much requires more memory, because it stores all the nodes in the tree and requires quite a long time because it will test all the nodes on level n to get a solution the level (n + 1)[11][12][13], for pseudo code Breadth First Search in this article use is below:

```
For N = 0 To UBound(Solution(Row).Step)
        Step.Matrix(N + 1, 0) = N
        Step.Matrix(N + 1, 1) = "(" & _
        Solution(Row).Step(N).X & "," & _
        Solution (Row).Step(N).Y & ")"
        'Ant Movement
        If N > 0 Then
           'Initial
           LineS(N).X1 = ConvX(Solution(Row).Step(N - 1).X)
           LineS(N).Y1 = ConvY(Solution(Row).Step(N - 1).Y)
           'Goal
           LineS(N).X2 = ConvX(Solution(Row).Step(N).X) + 15
```

```
        LineS(N).Y2 = ConvY(Solution(Row).Step(N).Y) + 15

      End If
Next N
```

B. Cartesian field trajectory

One of the issues that can put into the topic of discussion of AI is the search for trajectories in the field of Cartesian [14][15][16]. This problem applies the tracking tree method using the Breadth First Search algorithm to find possible solutions.

This issue is described with an object moving from the center point (0,0) to point an (m, n) on the Cartesian plane provided that it must move parallel to the x-axis or y-axis (horizontal or vertical) and should not cross the path Which he has been through and should not cross the point he has been through.

Implementation of the tracking tree is as follows, and we first set the initial state as the root node (root node). Next, do the development of child nodes that may lead to the destination situation by complying with and not violating existing rules.

For example, the object's destination point is the point (m, n) = (3, 2). Rules, objects may only move within quadrant I (positive x-axis and positive y-axis), the movement of the object should not exceed the maximum limit, x = 3 and y = 2, then with the tree obtained the possible movement toward the state of destination is as follows:

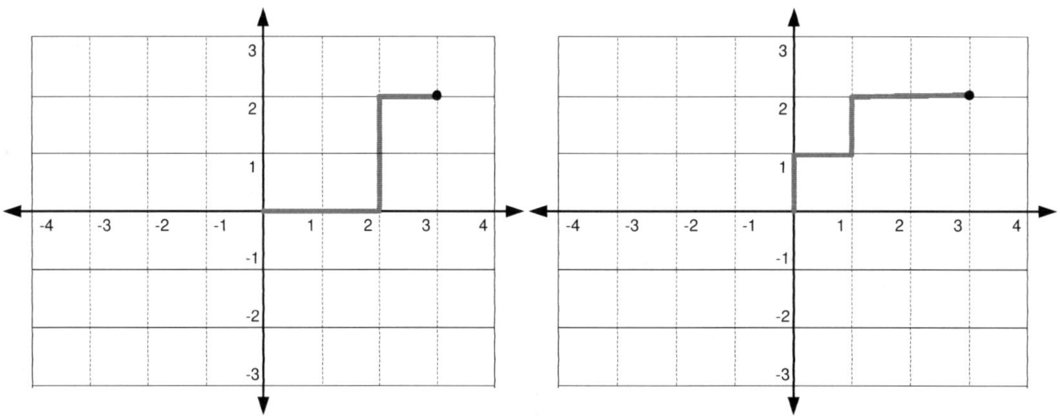

Figure 2. (a) First Solution, (b) Second Solution

3. Result and Discussion

The path search that can pass by an object in the Cartesian field could be accomplished with the help of a tracking tree, then at the point to go through a series of rules that restrict the movement of the object. Restrictive provisions are the maximum permissible limit of movement, the position of the obstacle to which the object should not pass, the position where the object must pass and cross only the allowed quadrant to pass, starting from the initial position as a root node, the BFS algorithm then finds a solution by developing root nodes to the next levels, all possible moves, not violating the terms and producing new conditions as fully developed. Search ends when no new nodes or conditions could be revealed. All the nodes that are the destination position are the solutions. The solution of the shortest path finding solution using BFS method in this research is different from the previous research, the search process is perform in Cartesian area and every point in cartesian area will be calculated as the possibility of the passing path so the search process will be more optimal and from the traceable solution can be known. See the movement process of objects that are allowed are:

a. Move right (x + 1).
b. Move left (x - 1).
c. Move up (y + 1).

d. Move down (y - 1).
e. All object movements above are legal if they comply with the following provisions:
   1) Not crossing the point of obstacle positions.
   2) Not crossing the post of a point that has been pass before.
   3) Does not exceed the maximum movement limit of the object (x or y).
   4) Not crossing quadrants that are not allowed to pass. The quadrant in question is quadrant I (where x is positive, and y is positive), quadrant II (where x is negative, and y is positive), quadrant III (where x is negative, and y is negative) and quadrant IV (where x is positive, and y is positive).

The process of searching the solution search using the Breadth First Search algorithm are performed in the following Cartesian case:

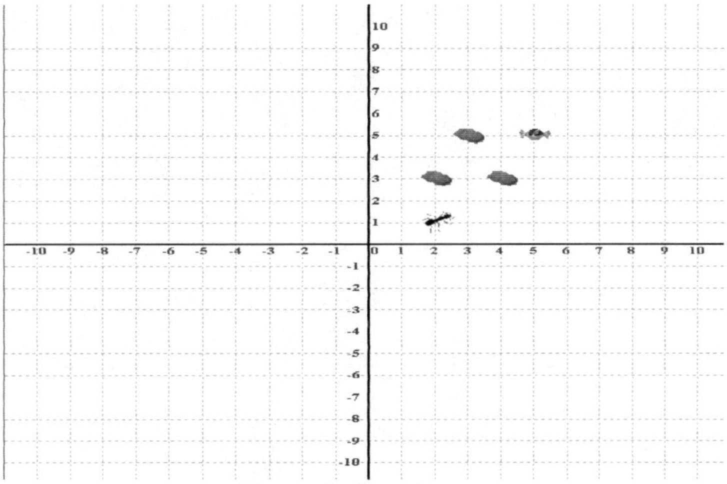

Figure 3. Sample case

Based on tests carried out, several solutions can be taken to achieve the desired goal, here is a solution obtained:

Table 1. Solutions obtained

No	Steps	Few Solution Step		
1	7 Step Found 8 Solution	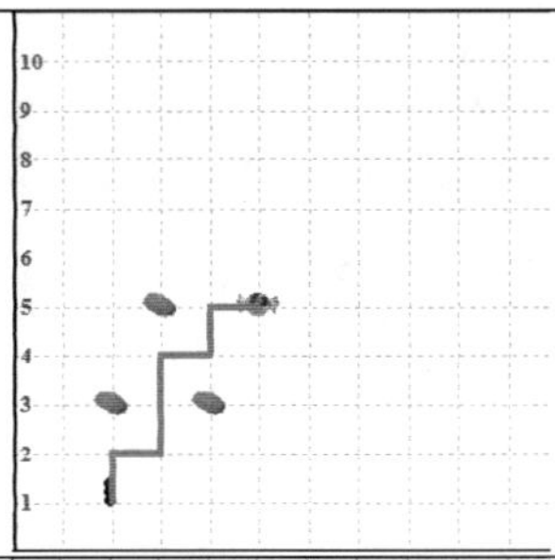	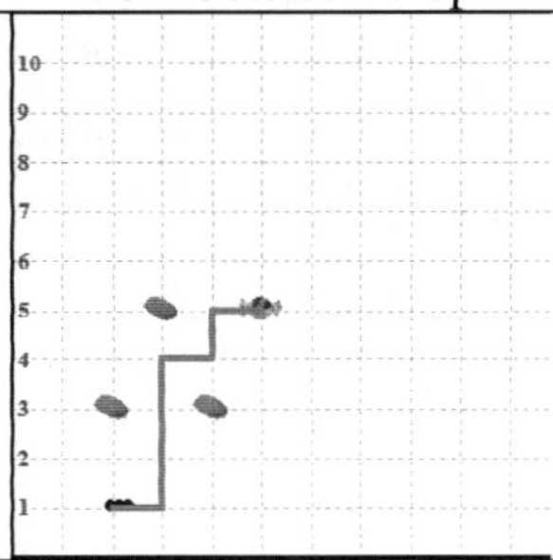	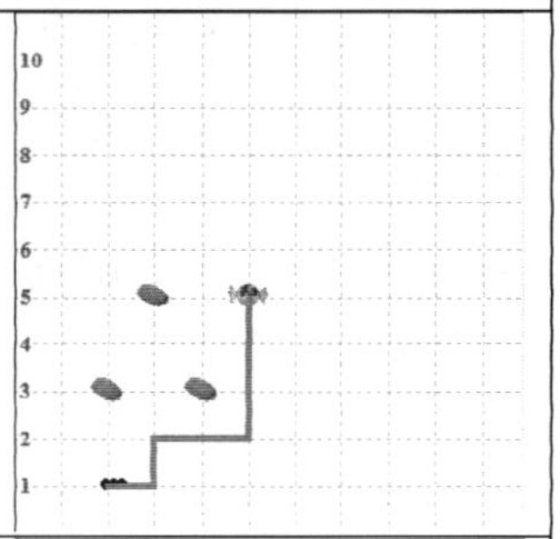
2	9 Step Found 61 Solution	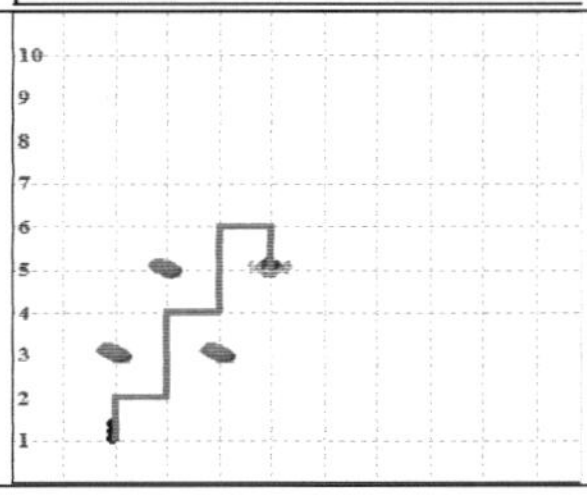	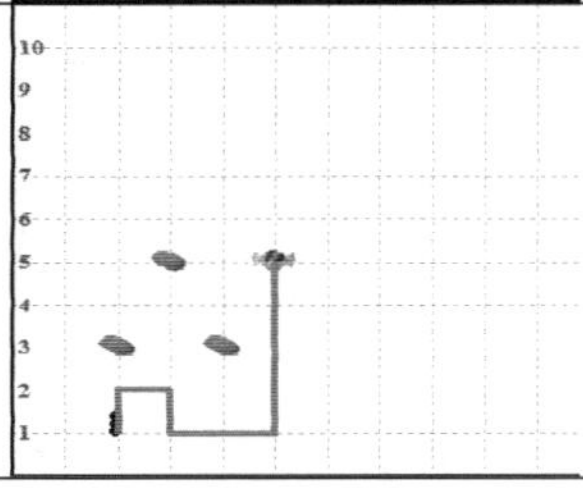	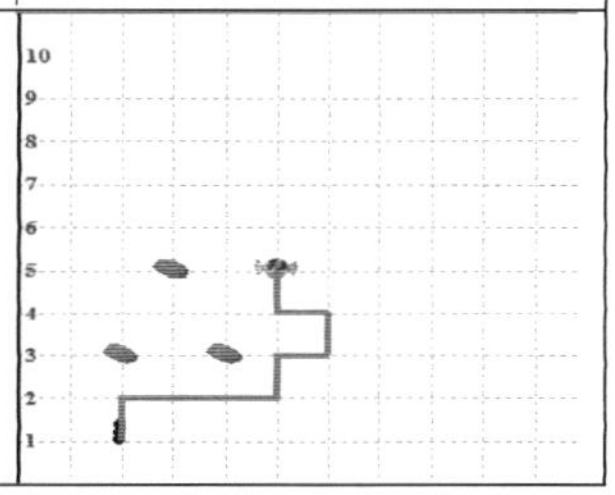

IOP Conf. Series: Journal of Physics: Conf. Series **1019** (2018) 012036 doi:10.1088/1742-6596/1019/1/012036

		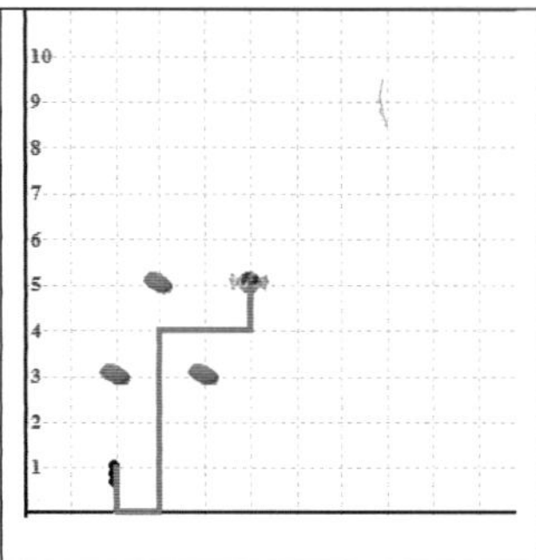	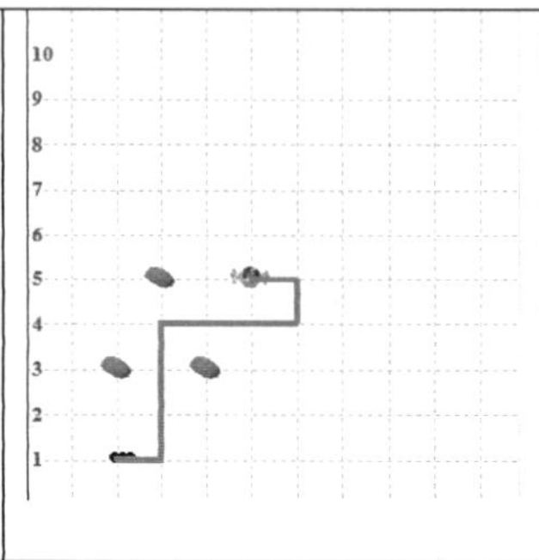	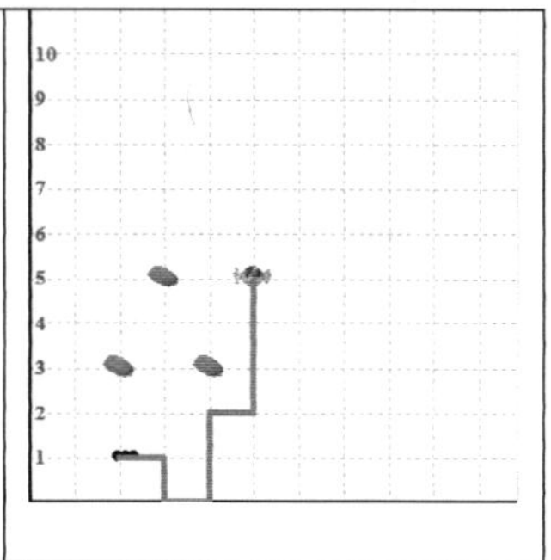
3	11 Step Found 236 Solution	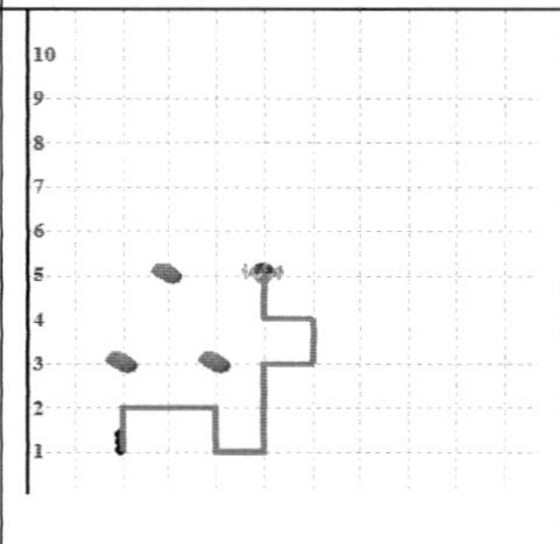	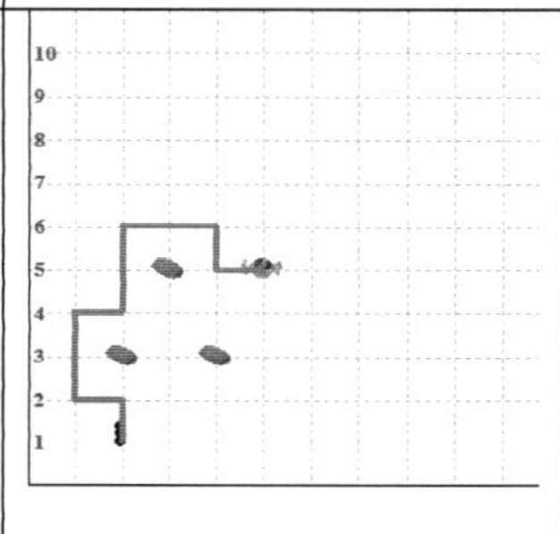	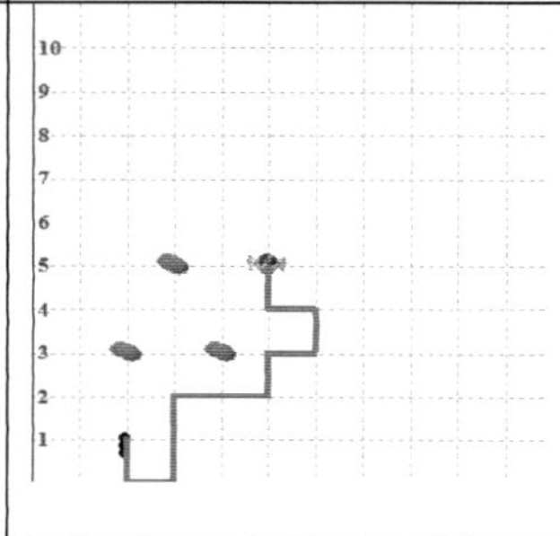
		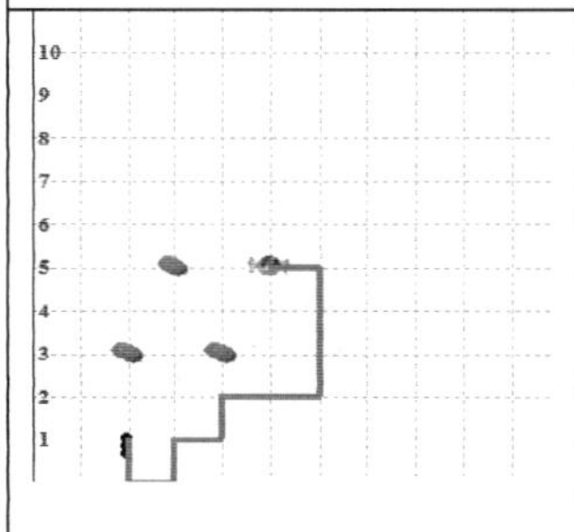	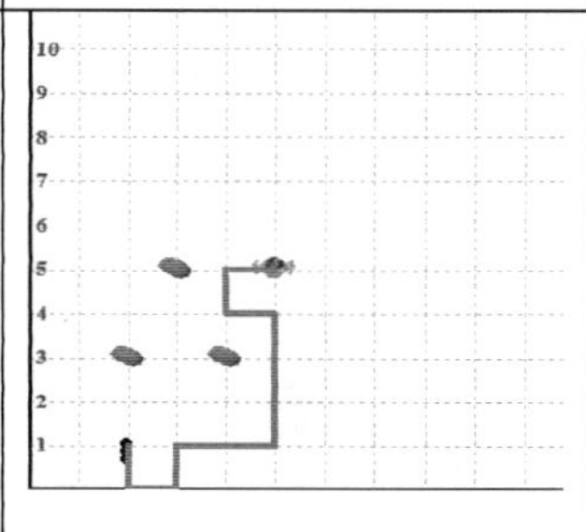	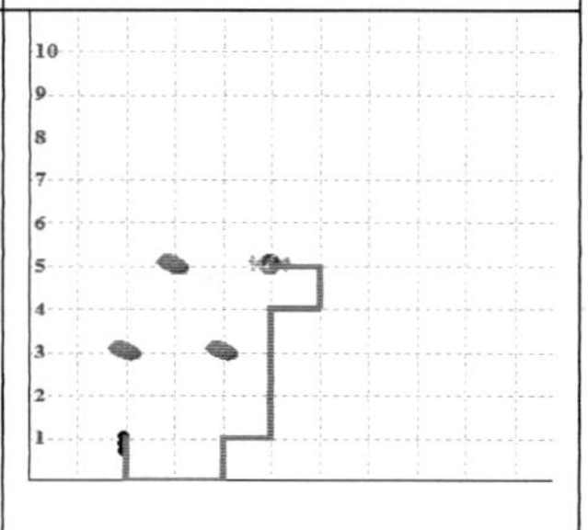
		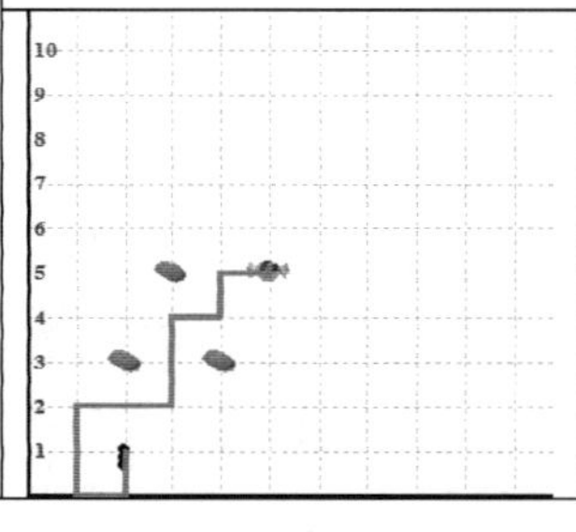	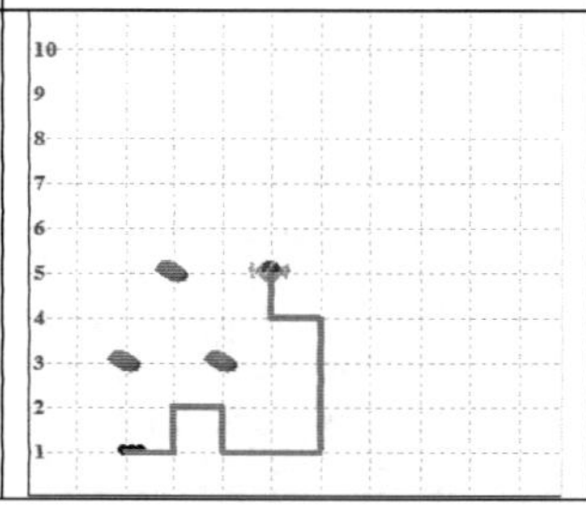	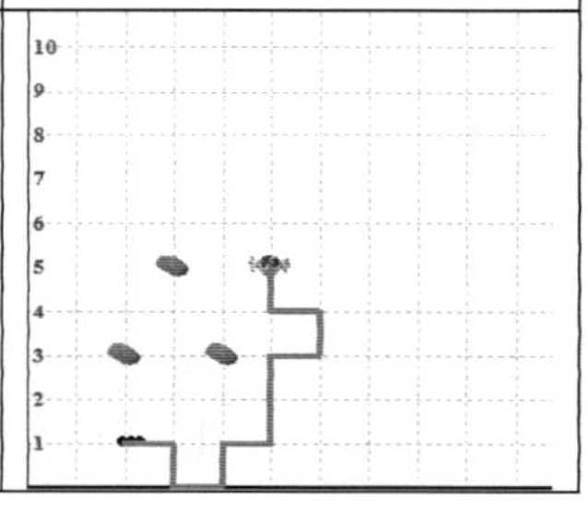

Testing was accomplished in the area of Cartesian to locate the path through which the object produces some options that could be passed, with the most optimal solution consists of 7 steps.

4. Conclusion

The Breadth First Search algorithm applied to determine the path from the original location to the destination location in the Cartesian field yields several alternative solutions based on the tests performed. Thus the Breadth First Search algorithm applied to the Cartesian field can produce some optimal solution (shortest path) and solution which is non-optimal.

5. References

[1] Y. Zhou, W. Wang, D. He, and Z. Wang, "A fewest-turn-and-shortest path algorithm based on breadth-first search," *Geo-Spatial Inf. Sci.*, vol. 17, no. 4, pp. 201–207, 2014.

[2] J. Silvela and J. Portillo, "Breadth-first search and its application to image processing problems," *IEEE Trans. Image Process.*, vol. 10, no. 8, pp. 1194–1199, 2001.

[3] P. Ossona de Mendez and P. Rosenstiehl, "Coding Properties of Breadth-First Search Orderings," *Electron. Notes Discret. Math.*, vol. 5, pp. 253–255, 2000.

[4] S. Beamer, K. Asanovic, and D. Patterson, "Searching for a Parent Instead of Fighting Over Children: A Fast Breadth-First Search Implementation for Graph500," *Tech. Rep. UCB/EECS-2011-117, EECS Dep. Univ. California, Berkeley*, pp. 1–9, 2011.

[5] D. Ajwani, R. Dementiev, V. Osipov, and U. Meyer, "Breadth first search on massive graphs," *shortest path Probl. ninth DIMACS implemenation Chall.*, vol. 74, pp. 291–307, 2009.

[6] A. S. M. Lumenta, "Perbandingan Metode Pencarian Depth-First Search, Breadth-First Search Dan Best-First Search Pada Permainan 8-Puzzle," *e-journal Tek. Elektro dan Komput.*, 2014.

[7] C. Rhee, Y. D. Liang, S. K. Dhall, and S. Lakshmivarahan, "Efficient algorithms for finding depth-first and breadth-first search trees in permutation graphs," *Inf. Process. Lett.*, vol. 49, no. 1, pp. 45–50, 1994.

[8] R. Zhou and E. A. Hansen, "Breadth-first heuristic search," *Artif. Intell.*, vol. 170, no. 4–5, pp. 385–408, 2006.

[9] I. Papoutsakis, "On approximating tree spanners that are breadth first search trees," *J. Comput. Syst. Sci.*, vol. 82, no. 5, pp. 817–825, 2016.

[10] O. G. Attia, T. Johnson, K. Townsend, P. Jones, and J. Zambreno, "CyGraph: A reconfigurable architecture for parallel breadth-first search," in *Proceedings of the International Parallel and Distributed Processing Symposium, IPDPS*, 2014, pp. 228–235.

[11] Y. Zhang and E. A. Hansen, "Parallel Breadth-First Heuristic Search on a Shared-Memory Architecture," in *AAAI Workshop on Heuristic Search, Memory-Based Heuristics and Their Applications*, 2006.

[12] S. Beamer, A. Buluc, K. Asanovic, and D. Patterson, "Distributed memory breadth-first search revisited: Enabling bottom-up search," in *Proceedings - IEEE 27th International Parallel and Distributed Processing Symposium Workshops and PhD Forum, IPDPSW 2013*, 2013, pp. 1618–1627.

[13] A. V. Goldberg, S. Hed, H. Kaplan, R. E. Tarjan, and R. F. Werneck, "Maximum flows by incremental breadth-First search," in *Lecture Notes in Computer Science (including subseries Lecture Notes in Artificial Intelligence and Lecture Notes in Bioinformatics)*, 2011, vol. 6942 LNCS, pp. 457–468.

[14] K. L. Wright, J. I. Hamilton, M. A. Griswold, V. Gulani, and N. Seiberlich, "Non-Cartesian parallel imaging reconstruction," *Journal of Magnetic Resonance Imaging*, vol. 40, no. 5. pp. 1022–1040, 2014.

[15] F. Porikli, "Integral histogram: A fast way to extract histograms in Cartesian spaces," in *Proceedings - 2005 IEEE Computer Society Conference on Computer Vision and Pattern Recognition, CVPR 2005*, 2005, vol. I, pp. 829–837.

[16] J. Wang, J. Wang, G. Zeng, R. Gan, L. Shipeng, and B. Guo, "Fast neighborhood graph search using cartesian concatenation," in *Multimedia Data Mining and Analytics: Disruptive Innovation*, 2015, pp. 397–417.

Knowledge Management Organization to Gain Performance in Private University

Elizabeth Haloho[1*], Lailan Tawila[1], Idawati Purba[1] and Robbi Rahim[2]

[1]Department of Management, Universitas Sari Mutiara Indonesia, Indonesia
[2]School of Computer and Communication Engineering, Universiti Malaysia Perlis, Malaysia

*elisabeth.haloho@gmail.com

Abstract. The purpose of this reasearch to develop quality of education Private Universities in Medan. Competitive advantage of each Private university could be different one another but in this reasearch will give insight on how using internal resource such as knowledge management will be benefitting the management. Higher education institution is aimed to become the center of development of education through research, teaching and outreach programs, technology development, art and social culture which will be expected to enhance the quality of life of people. There is increment number of students who finished Senior High School in Medan every year. With this reason, private universities are encourage to compete amongs them to atract new students to register. Therefore each of them must have a competitive advantage to become known by potential students and win the competition. The research is located in Medan, North Sumatera, with 100 respondents are Head of Study Programs from 11 universities in Medan. The research used Descriptive Statistic and Inferential in the data analysis. The analysis model is using Structural Equation Modelling (SEM) and calculated by AMOS. The result showed that Knowledge Management has significant relation with Competitive advantage and university performance.

1. Introduction

Universities are the centers of higher education whereas the maintenance, and development of science[1], technology[2] and/or arts in the hope of improving the quality of life in society, nation and country[3], [4]. Quality of education in Indonesia is still relatively lower if compared to other countries in Southeast Asia[5]. The position of Indonesian universities at International level can be seen from the list of the world's best universities issued by the Times Higher Education Supplement (THES) which is still out of top 100 best universitis in the world.

From the rankings of Best universities in Indonesia there is still no university from North Sumatra included in the top 10 ranking of the best Universities in Indonesia. Two universities in North Sumatera which ranked in the top 100 are Universitas Sumatera Utara (USU) ranked 19th, UNIMED ranked 21th (www.kemenristekdikti.org). Some of the factors leading to the Bankruptcy of the universities are lack of interest in prospective students who want to register. As a result, the university is not able to finance education operations which rely on public funds. Other problems are faced by the majority of private universities in Indonesia, of which related to mismanagement of finance, lack of facilities and infrastructure, bad leadership, lack of human resources and bad image of the institution which leading to unhealthy organization and sometimes lawsuit on court by the founder[6].

IOP Conf. Series: Journal of Physics: Conf. Series **1019** (2018) 012037 doi:10.1088/1742-6596/1019/1/012037

Development and change occur in a geographical manner. This developments known as globalization[7]. Such conditions demanding each private universities to be able digging and developing sources of competitive advantage in order to survive and win competition. University Performance can be formed from competitive advantage. This is in accordance with research by Agha[7] that stated Competitive Advantage can improve Organizational Performance. In order for the company to remain competitive and gain competitive advantage, managers can try to improve organizational performance by managing each of the core competencies dimensions of shared vision; cooperation and empowerment[8], [9]

However, in managing competitive advantage, it must be done well in order to achieve superior performance. Ccompetitive advantage will not boost high performance if 1) a company that has competitive advantage but failed to develop it into a compound advantage, 2) the company failed to take advantage of that potential, 3) if it has some competitive advantage but failed to make the right combination, 4) the company deliberately sacrificed the competitive advantage [10]. Another study by Diana, 2009[11] says that sustainable competitive advantage cannot always be decisive in competition because of the effects of rapid development such as technology and market changes forcing competitive advantage must also change. The unique advantages which were difficult to imitate cannot be sustained because of technological developments that eventually can be duplicated as well.

Competitive advantage is arguable as a topic discussed in strategic management[12]–[14]. Barney[15] further suggests that a company has the potential to produce a competitive advantage. But it must have four attributes: (a) having values; (b) rare; (c) cannot be imitated and (d) there is no substitute. The internal resources that can increase the competitive advantage is by implementing knowledge management[12], [16], [17]. Therefore, this research will see the effect knowledge management on competitive advantage and university Performance.

2. Methodology

The research method used in this study is descriptive survey research method and explanatory survey method. The type of investigation used is causality, which is the type of research that states a causal relationship. The unit of analysis of this study is the Private University that operates in Medan city, while the observation unit is the Head of Study Program at private universities in Medan city. The research model that will be used as an analytical tool in this study is Partial Least Square-Path Modeling (PLS-PM)

The population of this research is the Head of Study Program registered at the private University in Medan City. Based on data from *Kopertis Sumatera Utara*, currently the number of private universities which were active in Medan is as many are 23 universities, but not all of the existing private universities will be the object of research. From the data obtained, private universities included in the criteria are 11 private universities located in Medan. Then to choose head of Program Studies at selected private universities, this research is using incidental sampling with a proportional number of as many as 100 respondents.

3. Result and Discussion

Confirmatory Factor Analysis Result for exogenous constructs used to test the the dimensions that built up the latent variables. Table 1 presents P value of the model which results have significant loading with the Goodness of fit, so the model is acceptable. The significance level of 0.320 indicates that the null hypothesis has no differences between the sample covariance matrix and the estimated population covariance matrix. Therefore this exogenous construct is accepted, please see table 1 below:

Table 1. Confirmatory Factor Analysis

Goodness of Fit Index	Cut-off Value	Result Analysis	Evaluation Model
$\chi 2$ – *Chi-square*	expected to be small (df=82)≤ 87.437	87.437	Good
Probability	≥ 0.05	0.320	Good
RMSEA	≤ 0.08	0.026	Good
GFI	≥ 0.90	0.899	Marginal
AGFI	≥ 0.90	0.852	Marginal
TLI	≥ 0.95	0.989	Good
CFI	≥ 0.95	0.991	Good

After a confirmatory analysis and a fit model are obtained, each variable can be used to define a latent construct so the full SEM analysis can be done. The result for the structural model is presented on Figure 1 below

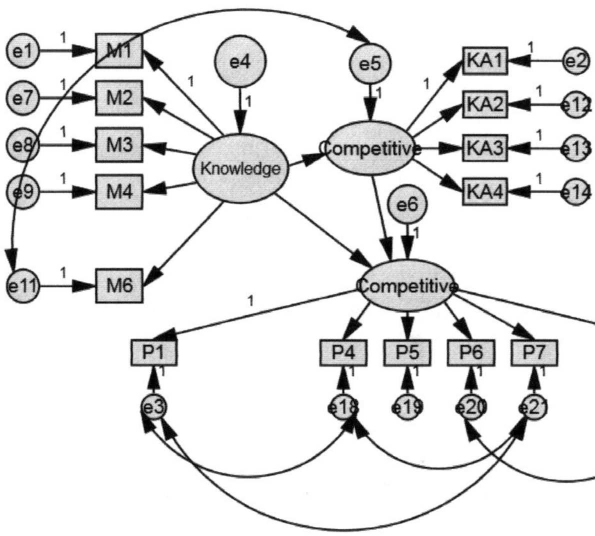

Figure 1. Structural Equation Model

The result of coefficient correlation on regression presented at Table 2 has the best fit correlation

Table 2. Regression Weight SEM

			Estimate	S.E.	C.R.	P	Label
Competitive	<---	knowledge	.465	.126	3.704	***	par_1
Performance	<---	Competitive	.922	.392	2.351	.019	par_9
Performance	<---	knowledge	.848	.384	2.208	.027	par_20
M1	<---	knowledge	1.000				
KA1	<---	Competitive	1.000				
P1	<---	Performance	1.000				
M2	<---	knowledge	1.069	.170	6.274	***	par_2
M3	<---	knowledge	1.036	.172	6.026	***	par_3
M4	<---	knowledge	1.164	.185	6.301	***	par_4
M6	<---	knowledge	.676	.176	3.830	***	par_5

IOP Conf. Series: Journal of Physics: Conf. Series **1019** (2018) 012037 doi:10.1088/1742-6596/1019/1/012037

			Estimate	S.E.	C.R.	P	Label
KA2	<---	Competitive	1.509	.259	5.832	***	par_6
KA3	<---	Competitive	1.162	.254	4.581	***	par_7
KA4	<---	Competitive	1.040	.215	4.832	***	par_8
P4	<---	Performance	8.683	4.932	1.761	.078	par_10
P5	<---	Performance	8.332	4.817	1.730	.084	par_11
P6	<---	Performance	5.895	3.416	1.726	.084	par_12
P7	<---	Performance	3.335	2.284	1.461	.144	par_13
P9	<---	Performance	4.829	2.914	1.657	.097	par_14

Table 2 indicates that not all variables analyzed has correlation and effect to other variables because they are at the significant level expected since CR is more than 2,00 and value of probability is more than 0,05

Hypothesis Experiment I

H_o = Knowledge management has no positive significant effect on competitive advantage
H_1 = Knowledge management has a positive and significant effect on competitive advantage
As shown at Table 2 that $P<0.05$ and $CR >2$ thus H_o rejected and H_1 is accepted. So the result of this research is knowledge management has significant level on competitive advantage

Hypothesis Experiment II

H_o = Competitive advantage has no positive and significant effect on organization performance
H_2 = Competitive advantage has positive and significant effect on organization performance
As shown at Table 2 that $P<0.05$ and $CR <2$ thus H_o rejected and H_2 is accepted. So the result of this research competitive advantage has no significant level on performance

Hypothesis Experiment III

H_o = Knowledge has no positive and no significant effect on organization performance
H_3 = Knowledge has positive and significant effect on organization performance
Based on Table 2 it can concluded that H_o is rejected and H_3 is accepted because P value is less than 0.05 and CR is more than 2. So the result of this research is knowledge management has significant effect on organization performance

Knowledge management is a process of human activity related to knowledge, but does not deal with the special nature of different types of knowledge, or the relative importance to different knowledge within an organization. Knowledge must be balanced within an organization because it is very important. The University's strategic resources can also be a competitive advantage. The strategic resource could be the knowledge management. In achieving competitive advantage, knowledge management is important in assisting organizations in developing their innovative product. It is recommended that knowledge management used as an essential Business tool to gain competitive advantage and in turn encourage economic growth.

We are not only in the new millennium, but also in a new era: the era of knowledge. Sustainable competitive advantage depends on building and utilizing core competencies. The company's resource-based view (RBV) defines strategic assets as rare, valuable, inimitable and non-replaceable assets. Knowledge is a potential strategic asset to be a source of competitive advantage for an organization. Knowledge management can be used to create a competitive advantage an achieve organization performance.

4. Conclusion

Knowledge management has a positive and significant effect on competitive advantage, and also has a significant influence on organizational performance, so in this study can be concluded that exogenous variables that have significant influence only knowledge management, if the higher education want to achieve competitive advantage, they should be improve their knowledge management in order to compete with other universities.

References

[1] J. Suyono, A. Sukoco, M. I. Setiawan, Suhermin, and R. Rahim, "Impact of GDP Information Technology in Developing of Regional Central Business (Case 50 Airports IT City Development in Indonesia)," *J. Phys. Conf. Ser.*, vol. 930, no. 1, p. 12045, Dec. 2017.

[2] R. Rahim, "Man-in-the-middle-attack prevention using interlock protocol method," *ARPN J. Eng. Appl. Sci.*, vol. 12, no. 22, pp. 6483–6487, 2017.

[3] R. Rahim, Mesran, A. Putera, U. Siahaan, and S. Aryza, "Composite performance index for student admission," *Int. J. Res. Sci. Eng.*, vol. 3, no. 3, pp. 68–74, 2017.

[4] Mesran, G. Ginting, Suginam, R. Rahim, M. Aan, and R. Rahim, "Implementation of Elimination and Choice Expressing Reality (ELECTRE) Method in Selecting the Best ... Implementation of Elimination and Choice Expressing Reality (ELECTRE) Method in Selecting the Best Lecturer (Case Study STMIK BUDI DARMA)," *Int. J. Eng. Res. Technol.*, vol. 6, no. 2, NaN-2017, pp. 141–144, 2017.

[5] P. harliana and R. Rahim, "Comparative Analysis of Membership Function on Mamdani Fuzzy Inference System for Decision Making," *J. Phys. Conf. Ser.*, vol. 930, no. 1, p. 12029, Dec. 2017.

[6] S. Khotimah, "PERUMUSAN STRATEGI BAGI PERGURUAN TINGGI SWASTA (PTS) UNTUK MERAIH KEUNGGULAN BERSAING (STUDI KASUS PADA UNIVERSITAS MERDEKA MALANG) Siti Khotimah," *SPREAD*, vol. 2, no. April, pp. 73–84, 2012.

[7] S. Agha, L. Alrubaiee, and M. Jamhour, "Effect of Core Competence on Competitive Advantage and Organizational Performance," *Int. J. Bus. Manag.*, vol. 7, no. 1, p. 192, Dec. 2011.

[8] A. Ismadi Ismail, R. C. Rose, H. Abdullah, and J. Uli, "THE RELATIONSHIP BETWEEN ORGANISATIONAL COMPETITIVE ADVANTAGE AND PERFORMANCE MODERATED BY THE AGE AND SIZE OF FIRMS," *Asian Acad. Manag. J.*, vol. 15, no. 2, pp. 157–173, 2010.

[9] D. Siregar, D. Arisandi, A. Usman, D. Irwan, and R. Rahim, "Research of Simple Multi-Attribute Rating Technique for Decision Support," *J. Phys. Conf. Ser.*, vol. 930, no. 1, p. 12015, Dec. 2017.

[10] A. Balkytė and M. Tvaronavičienė, "PERCEPTION OF COMPETITIVENESS IN THE CONTEXT OF SUSTAINABLE DEVELOPMENT: FACETS OF " SUSTAINABLE COMPETITIVENESS ""

[11] D. Bratić, "Knowledge And Knowledge Management As A Competitive Advantage Author," *Knowl. Knowl. Manag. acta Graph.*, vol. 20, pp. 1–4, 2009.

[12] M. E. Porter, "TECHNOLOGY AND COMPETITIVE ADVANTAGE," *J. Bus. Strategy*, vol. 5, no. 3, pp. 60–78, Jan. 1985.

[13] S. E. South, "COMPETITIVE ADVANTAGE: THE CORNERSTONE OF STRATEGIC THINKING," *J. Bus. Strategy*, vol. 1, no. 4, pp. 15–25, Jan. 1981.

[14] L. Sołoducho-Pelc, "Competitive Advantage: The Courage in Formulating Objectives and Expansiveness of a Strategy," *Procedia - Soc. Behav. Sci.*, vol. 150, pp. 271–280, Sep. 2014.

[15] J. Barney, "Firm Resources and Sustained Competitive Advantage," *J. Manage.*, vol. 17, no. 1, pp. 99–120, Mar. 1991.

IOP Conf. Series: Journal of Physics: Conf. Series **1019** (2018) 012037 doi:10.1088/1742-6596/1019/1/012037

[16] O. R. Mahdi, M. K. Almsafir, and L. Yao, "The role of knowledge and knowledge management in sustaining competitive advantage within organizations: A review," *African J. Bus. Manag.*, vol. 5, no. 23, pp. 9912–9931, 2011.
[17] D. Kaveh, E. M. Bamipour, M. S. Far, and M. S. Far, "The Relationship between Knowledge Management and Competitive Advantage: A Case Study in Small and Medium-Sized Companies in the Packaging Industry of Khorasan Razavi," *Eur. Online J. Nat. Soc. Sci. Spec. Issue New Dimens. Econ.*, vol. 4, no. 1, 2015.

Breadth First Search Approach for Shortest Path Solution in Cartesian Area

Robbi Rahim[1*], Dahlan Abdullah[2], Saiful Nurarif[3], Mukhlis Ramadhan[3], Badrul Anwar[3], Muhammad Dahria[4], Surya Darma Nasution[5], Tengku Mohd Diansyah[6] and Mufida Khairani[6]

[1]School of Computer and Communication Engineering, Universiti Malaysia Perlis, Malaysia
[2]Department of Informatics, Universitas Malikussaleh, Jl. Medan-Banda Aceh, Aceh Utara, Indonesia
[3]Department of Computer System, STMIK Triguna Dharma, Jl. Jenderal Abdul Haris Nasution No.73, Medan 20219, Indonesia
[4]Department of Information System, STMIK Triguna Dharma, Jl. Jenderal Abdul Haris Nasution No.73, Medan 20219, Indonesia
[5]Department of Informatics, STMIK Budidarma, Medan, Indonesia
[6]Department of Informatics, Universitas Harapan Medan, Medan, Indonesia

*usurobbi85@zoho.com

Abstract. Determining the shortest path is one problem that is much discussed using some algorithm like Djikstra, Floyd Warshall and in this research an algorithm Breadth First Search are used, Breadth First Search algorithms in this study is used to determine the shortest route and optimal from a Cartesian field, the best and optimal route search experiment of cartesian areas using Breadth First Search algorithm can be perform very well and gets some route options from the best to the longest route.

1. Introduction

The development of computer-based technology is currently experiencing rapid development, and the development has also covered various aspects of the field of life with the aim to facilitate the completion of tasks charged, Artificial intelligence is the most developed field today.

Shortest Path is a pretty much solved problem with algorithms like Djikstra, Depth First Search[1] and Breadth First Search[2], Yan ZHOU[3] conducted a shortest path study using the Breadth First Search algorithm by approaching the search from the starting point to the destination point by examining the layer by layer against all the possibilities that can be passed and remove all the paths that are not possible to pass[4]–[6]. Similar to Yan ZHOU this experiment try to finding the shortest path that can pass through the Cartesian field. This issue could be explained as follows; it is identified that an object will move from the center point (0,0) to point A (m, n). The object should only bend at the grid points and always step in parallel with the x-axis or y-axis. The object must not cross the path that has been passed and should not cross the point that has been crossed. After that provided a set of provisions that limit the movement of the object. The question is how the trajectories that the object can pass through using the terms defined above.

Breadth First Search is one solution that can be used to solve the above problems, Breadth-first search is an algorithm that performs a wider search by visiting the node preorder i.e. visiting a node and then visiting all the nodes adjacent to the node first. Next, unvisited nodes and neighbors of the previously visited nodes, so it is doubtful that the visited node will be revisited[7], [8].

This study attempts to show some of the path solutions from the initial location to the destination location, using the Breadth First Search algorithm. The given solution can consist of many solutions with the path meet the criteria of the problem already mentioned, Research conducted by previous research Yan ZHOU, this research not only provides the shortest path solution but also provides another form of optimal search if there are obstacles that block the process of the path being traversed.

2. Theory

A. Breadth First Search

In breadth first search method, all nodes at level n will be visited first before visiting nodes at level n+1 (next level). The search starts from the root node and continues to the 1st level from left to right, then moves to level-2 and so on. As an illustration of the first search breadth process could be seen in the following Tree process until the discovery of the solution[9], [10].

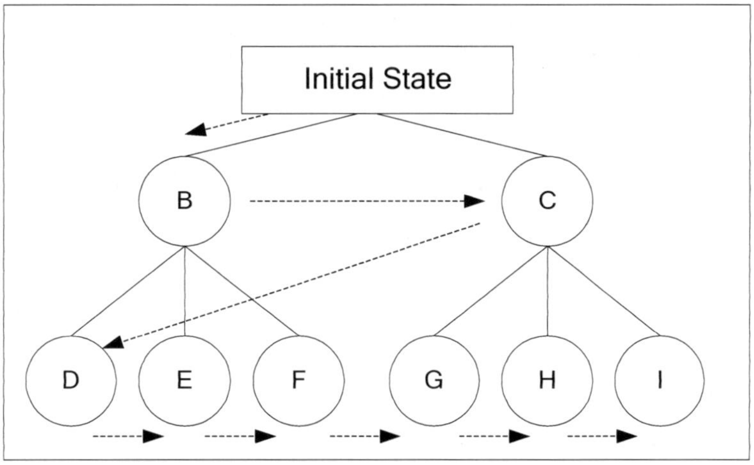

Figure 1. Procedure Breadth First Search

The advantage of using this algorithm is that it will not get stuck, and if there is more than one solution then the breadth-first search will find all the solutions, and if there is more than one solution then the minimum solution will be found[11][12]. While the weakness of this algorithm pretty much requires more memory, because it stores all the nodes in the tree and requires quite a long time because it will test all the nodes on level n to get a solution the level (n + 1)[10], [12], for pseudo code Breadth First Search in this article use is below:

```
For N = 0 To UBound(Solution(Row).Step)
        Step.Matrix(N + 1, 0) = N
        Step.Matrix(N + 1, 1) = "(" & _
        Solution(Row).Step(N).X & "," & _
        Solution (Row).Step(N).Y & ")"
        'Ant Movement
        If N > 0 Then
           'Initial
           LineS(N).X1 = ConvX(Solution(Row).Step(N - 1).X)
```

```
        LineS(N).Y1 = ConvY(Solution(Row).Step(N - 1).Y)
        'Goal
        LineS(N).X2 = ConvX(Solution(Row).Step(N).X) + 15
        LineS(N).Y2 = ConvY(Solution(Row).Step(N).Y) + 15

    End If
Next N
```

B. Cartesian Field Trajectory

One of the issues that can put into the topic of discussion of AI is the search for trajectories in the field of Cartesian [13][14][15]. This problem applies the tracking tree method using the Breadth First Search algorithm to find possible solutions.

This issue is described with an object moving from the center point (0,0) to point an (m, n) on the Cartesian plane provided that it must move parallel to the x-axis or y-axis (horizontal or vertical) and should not cross the path Which he has been through and should not cross the point he has been through. Implementation of the tracking tree is as follows, and we first set the initial state as the root node (root node). Next, do the development of child nodes that may lead to the destination situation by complying with and not violating existing rules.

For example, the object's destination point is the point (m, n) = (3, 2). Rules, objects may only move within quadrant I (positive x-axis and positive y-axis), the movement of the object should not exceed the maximum limit, x = 3 and y = 2, then with the tree obtained the possible movement toward the state of destination is as follows:

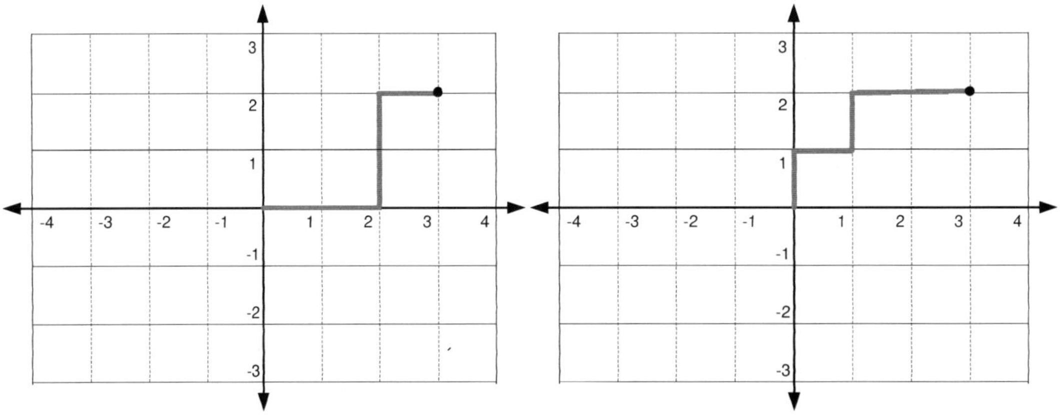

Figure 2. (a) First Solution, (b) Second Solution

3. Result and Discussion

The path search that can pass by an object in the Cartesian field could be accomplished with the help of a tracking tree, then at the point to go through a series of rules that restrict the movement of the object. Restrictive provisions are the maximum permissible limit of movement, the position of the obstacle to which the object should not pass, the position where the object must pass and cross only the allowed quadrant to pass, starting from the initial position as a root node, the BFS algorithm then finds a solution by developing root nodes to the next levels, all possible moves, not violating the terms and producing new conditions as fully developed. Search ends when no new nodes or conditions could be revealed. All the nodes that are the destination position are the solutions. The solution of the shortest path finding solution using BFS method in this research is different from the previous research, the search process is perform in Cartesian area and every point in cartesian area will be calculated as the possibility of the passing path so the search process will be more optimal and from the traceable solution can be known. See the movement process of objects that are allowed are:

a. Move right (x + 1).
b. Move left (x - 1).
c. Move up (y + 1).
d. Move down (y - 1).
e. All object movements above are legal if they comply with the following provisions:
   1) Not crossing the point of obstacle positions.
   2) Not crossing the post of a point that has been pass before.
   3) Does not exceed the maximum movement limit of the object (x or y).
   4) Not crossing quadrants that are not allowed to pass. The quadrant in question is quadrant I (where x is positive, and y is positive), quadrant II (where x is negative, and y is positive), quadrant III (where x is negative, and y is negative) and quadrant IV (where x is positive, and y is positive).

The process of searching the solution search using the Breadth First Search algorithm are performed in the following Cartesian case:

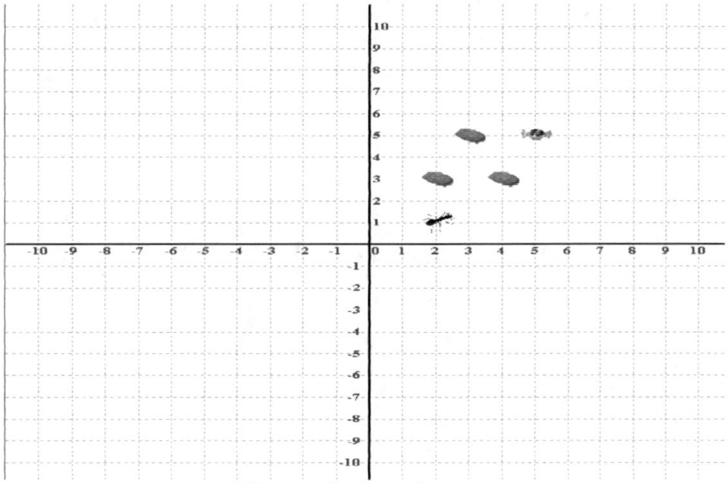

Figure 3. Sample case

Based on tests carried out, several solutions can be taken to achieve the desired goal, here is a solution obtained:

Table 1. Solutions obtained

No	Steps	Few Solution Step		
1	7 Step Found 8 Solution	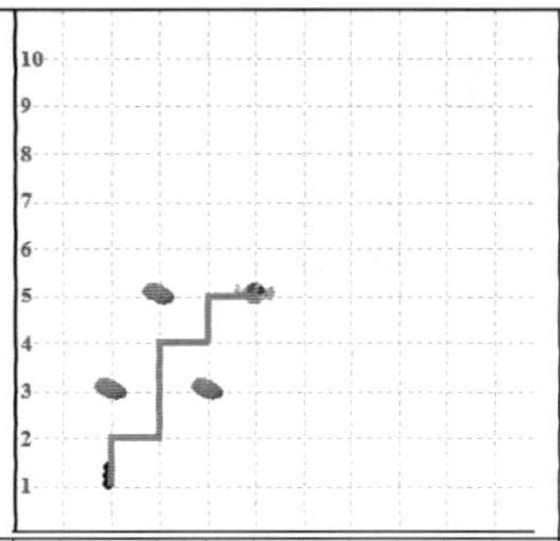	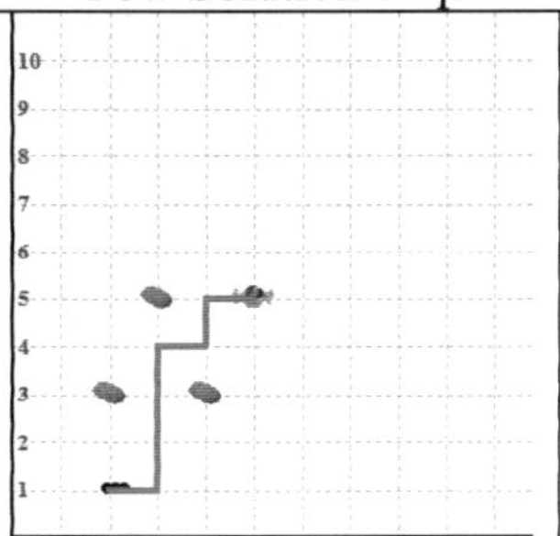	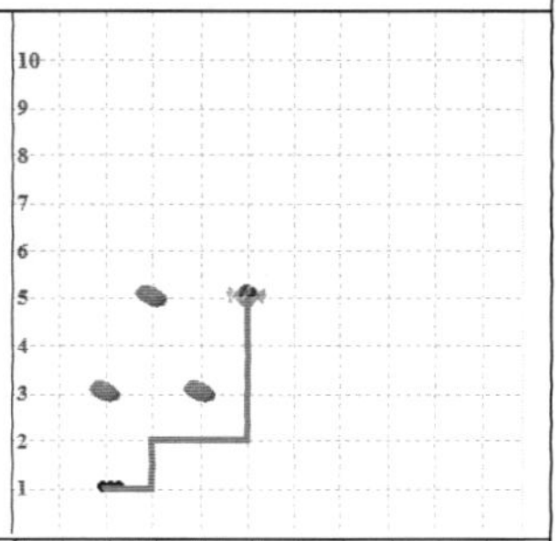
2	9 Step Found 61 Solution	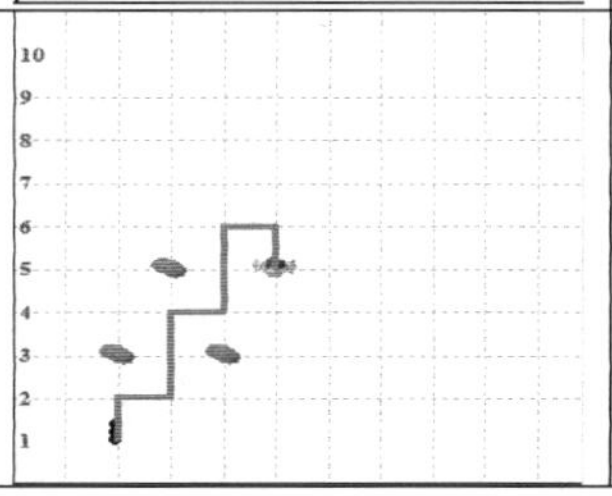	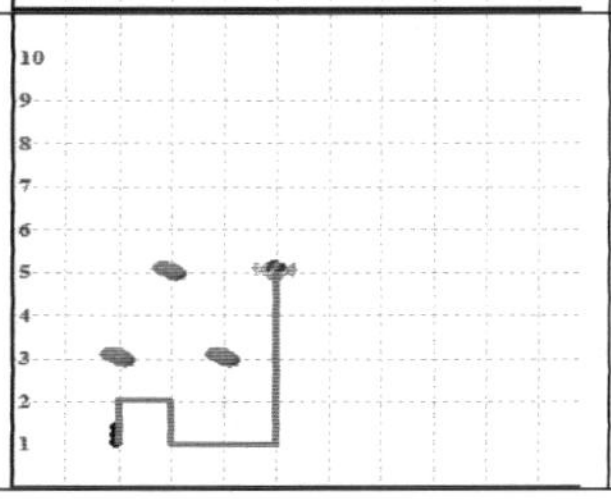	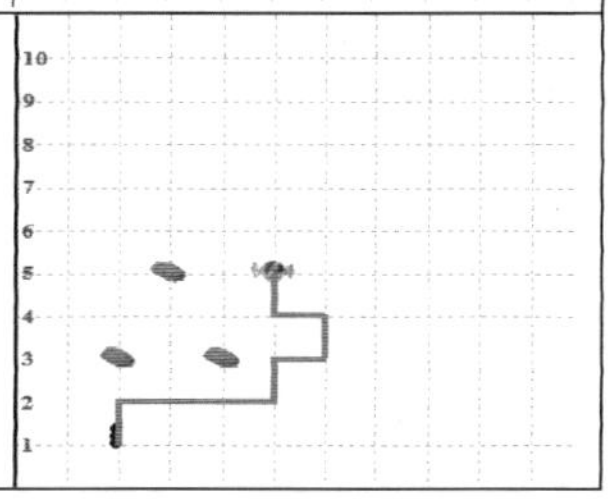

		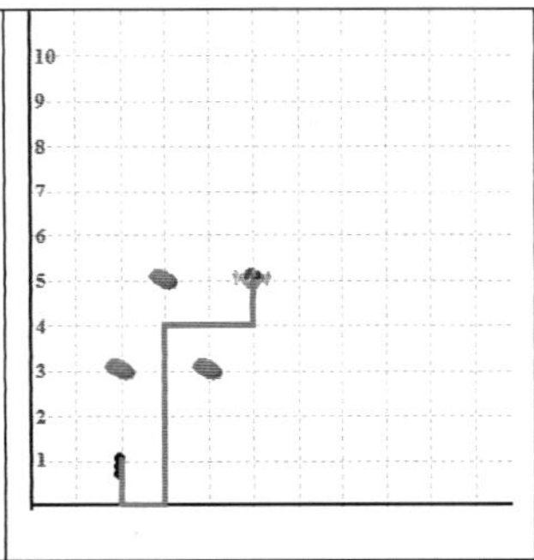	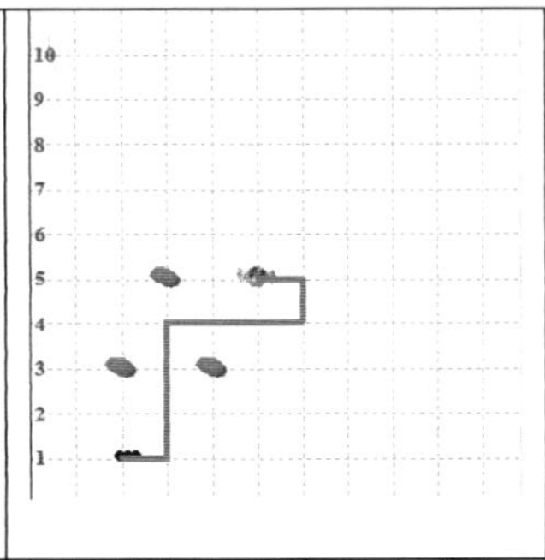	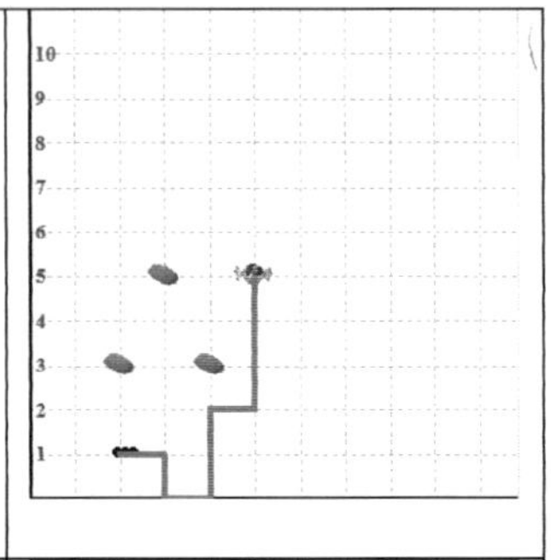
3	11 Step Found 236 Solution	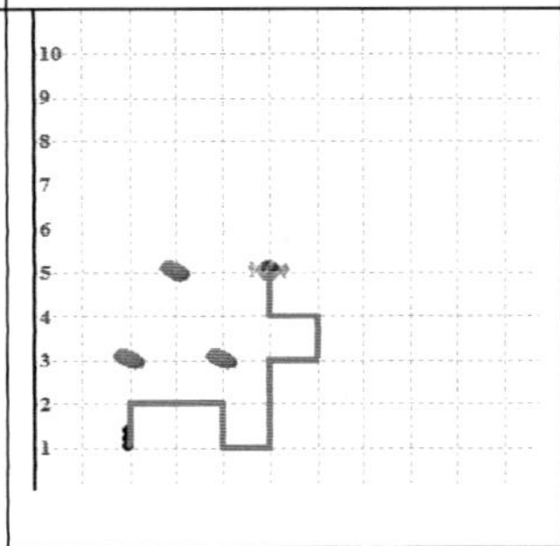	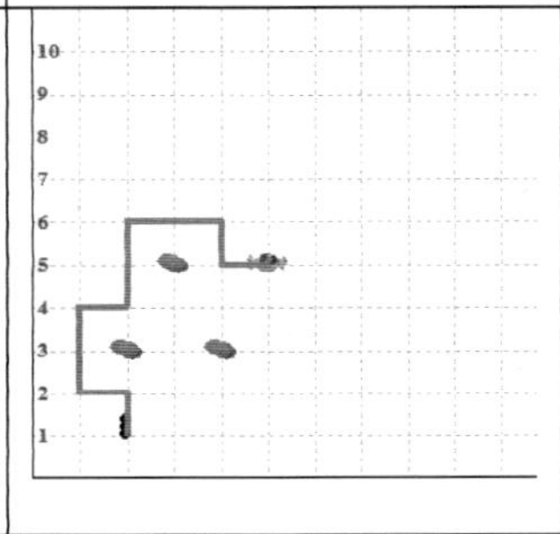	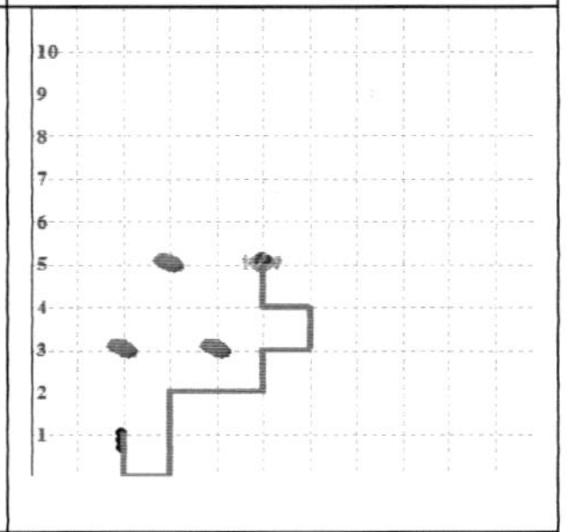
		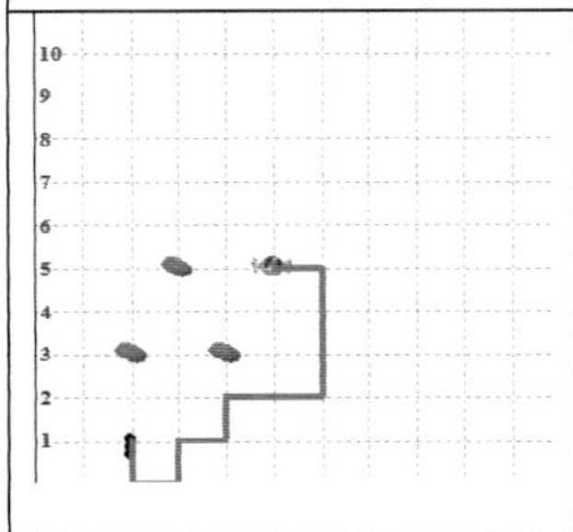	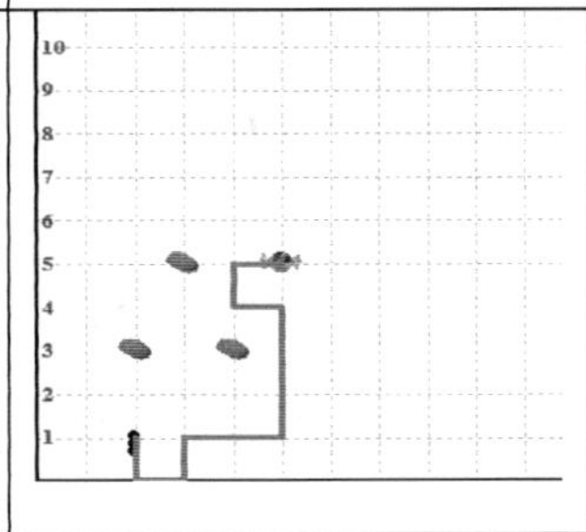	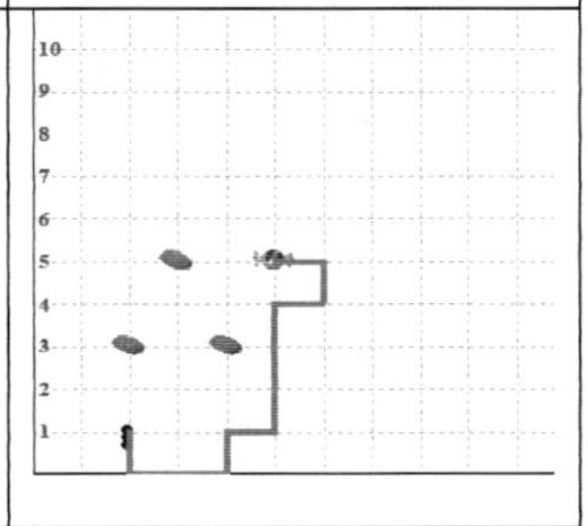
		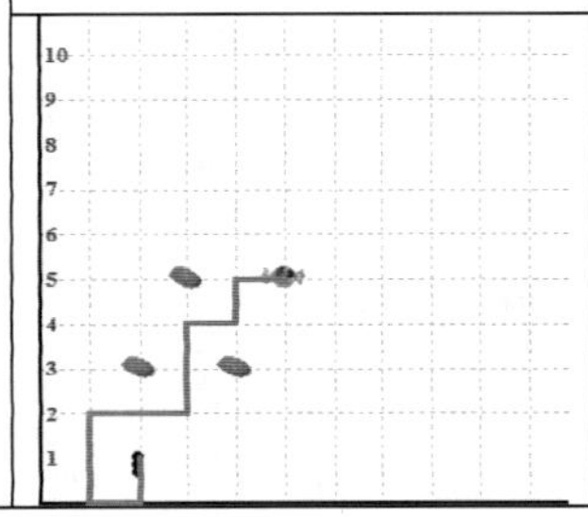	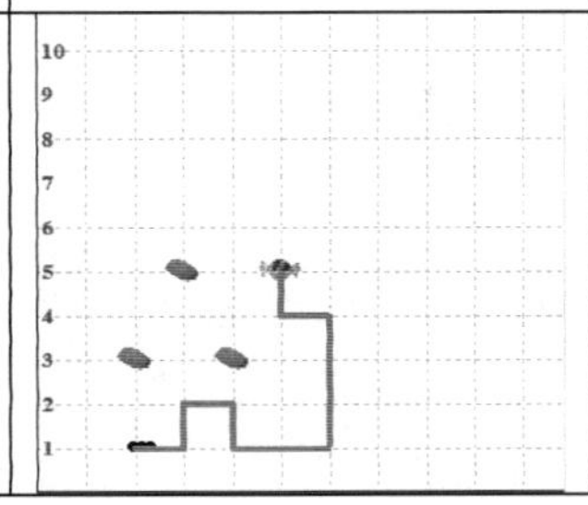	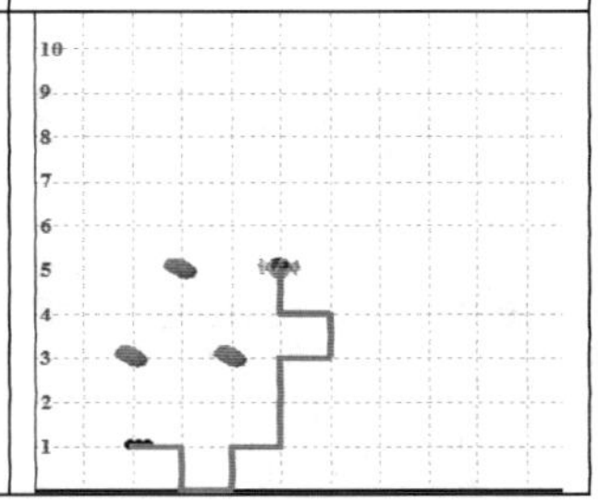

Testing was accomplished in the area of Cartesian to locate the path through which the object produces some options that could be passed, with the most optimal solution consists of 7 steps.

4. Conclusion

The Breadth First Search algorithm applied to determine the path from the original location to the destination location in the Cartesian field yields several alternative solutions based on the tests performed. Thus the Breadth First Search algorithm applied to the Cartesian field can produce some optimal solution (shortest path) and solution which is non-optimal.

References

[1] Rahim R *et al.*, 2018 Block Architecture Problem with Depth First Search Solution and Its Application *J. Phys. Conf. Ser.* **954**, 1 p. 12006.

[2] Ratnadewi R Sartika E M Rahim R Anwar B Syahril M and Winata H, 2018 Crossing Rivers Problem Solution with Breadth-First Search Approach in *IOP Conference Series: Materials Science and Engineering* **288**, 1.

[3] Zhou Y Wang W He D and Wang Z, 2014 A fewest-turn-and-shortest path algorithm based on breadth-first search *Geo-Spatial Inf. Sci.* **17**, 4 p. 201–207.

[4] Rahim R Nurarif S Ramadhan M Aisyah S and Purba W, Dec. 2017 Comparison Searching Process of Linear, Binary and Interpolation Algorithm *J. Phys. Conf. Ser.* **930**, 1 p. 12007.

[5] Rahim R Ahmar A S Ardyanti A P and Nofriansyah D, Dec. 2017 Visual Approach of Searching Process using Boyer-Moore Algorithm *J. Phys. Conf. Ser.* **930**, 1 p. 12001.

[6] Rahim R Zulkarnain I and Jaya H, 2017 A review: search visualization with Knuth Morris Pratt algorithm in *IOP Conference Series: Materials Science and Engineering* **237**, 1 p. 12026.

[7] Zhou R and Hansen E A, 2006 Breadth-first heuristic search *Artif. Intell.* **170**, 4–5 p. 385–408.

[8] Rhee C Liang Y D Dhall S K and Lakshmivarahan S, 1994 Efficient algorithms for finding depth-first and breadth-first search trees in permutation graphs *Inf. Process. Lett.* **49**, 1 p. 45–50.

[9] Ajwani D Dementiev R Osipov V and Meyer U, 2009 Breadth first search on massive graphs *shortest path Probl. ninth DIMACS implemenation Chall.* **74** p. 291–307.

[10] Goldberg A V. Hed S Kaplan H Tarjan R E and Werneck R F, 2011 Maximum flows by incremental breadth-First search in *Lecture Notes in Computer Science (including subseries Lecture Notes in Artificial Intelligence and Lecture Notes in Bioinformatics)* **6942 LNCS** p. 457–468.

[11] Papoutsakis I, 2016 On approximating tree spanners that are breadth first search trees *J. Comput. Syst. Sci.* **82**, 5 p. 817–825.

[12] Attia O G Johnson T Townsend K Jones P and Zambreno J, 2014 CyGraph: A reconfigurable architecture for parallel breadth-first search in *Proceedings of the International Parallel and Distributed Processing Symposium, IPDPS* p. 228–235.

[13] Wright K L Hamilton J I Griswold M A Gulani V and Seiberlich N, 2014, Non-Cartesian parallel imaging reconstruction, *Journal of Magnetic Resonance Imaging*, **40**, 5. p. 1022–1040.

[14] Porikli F, 2005 Integral histogram: A fast way to extract histograms in Cartesian spaces in *Proceedings - 2005 IEEE Computer Society Conference on Computer Vision and Pattern Recognition, CVPR 2005* **I** p. 829–837.

[15] Wang J Wang J Zeng G Gan R Shipeng L and Guo B, 2015, Fast neighborhood graph search using cartesian concatenation, in *Multimedia Data Mining and Analytics: Disruptive Innovation*, p. 397–417.

Super-Encryption Cryptography with IDEA and WAKE Algorithm

Dahlan Abdullah[1], Robbi Rahim[2*], Andysah Putera Utama Siahaan[2], Ananda Faridhatul Ulva[1], Zahratul Fitri[1], M Malahayati[3] and H Harun[3]

[1]Department of Informatics, Universitas Malikussaleh, Aceh, Indonesia
[2]School of Computer and Communication Engineering, Universiti Malaysia Perlis, Malaysia
[3]Department of Law, Universitas Malikussaleh, Aceh, Indonesia

*usurobbi85@zoho.com

Abstract. The security and confidentiality of information becomes an important factor in communication, the use of cryptography can be a powerful way of securing the information, IDEA (International Data Encryption Algorithm) and WAKE (Word Auto Key Encryption) are some modern symmetric cryptography algorithms with encryption and decryption function are much faster than the asymmetric cryptographic algorithm, with the combination experiment IDEA and WAKE it probable to produce highly secret ciphertext and it hopes to take a very long time for cryptanalyst to decrypt the information without knowing the key of the encryption process.

1. Introduction

Encryption is a standard way used to scramble data, so it is unreadable by unidentified parties[1]–[4], there are many cryptographic algorithms that can be used such as Triple DES, RSA, Blowfish, Twofish, AES[5] commonly used to secure data, there are also other symmetric algorithms such as IDEA (International Data Encryption Algorithm)[6] and WAKE (Word Auto Key Encryption)[7] which has a pretty good security process.

Super-Encryption[8] is one technique that can be used to secure data by combining 2 (two) type or more symmetric cryptographic algorithms, this process is commonly used to re-encrypt the encrypted ciphertext result from cryptographic algorithm[9]–[11], the combination of cryptographic algorithms[12] can produce safer ciphertext from various cryptanalyst attacks that want to know data or information from ciphertext, super-encryption in this paper using IDEA and WAKE algorithm.

The IDEA and WAKE algorithm used one by one, the first ciphertext (C1) is generated from the IDEA algorithm encryption process and then re-encrypted using the WAKE algorithm to generate the new Ciphertext (C2), the encryption and decryption process of the IDEA and WAKE algorithms use different keys, using different keys makes cryptanalyst take a longer time to know the data or information from the ciphertext, the use of cryptography does not guarantee data or information free from attacks but can make irresponsible parties be longer in getting the data.

2. Methodology

Super-Encryption in this paper perform step by step where plaintext encrypted with IDEA algorithm and then re-encrypted with WAKE algorithm, for process IDEA algorithm as in the following function:

a. *Key Formation*

This formation process begins by splitting 128-bit keys into eight pieces of 16-bit sub-key, key rotated 25 bits to the left and divided into eight subkeys again[13], [14].

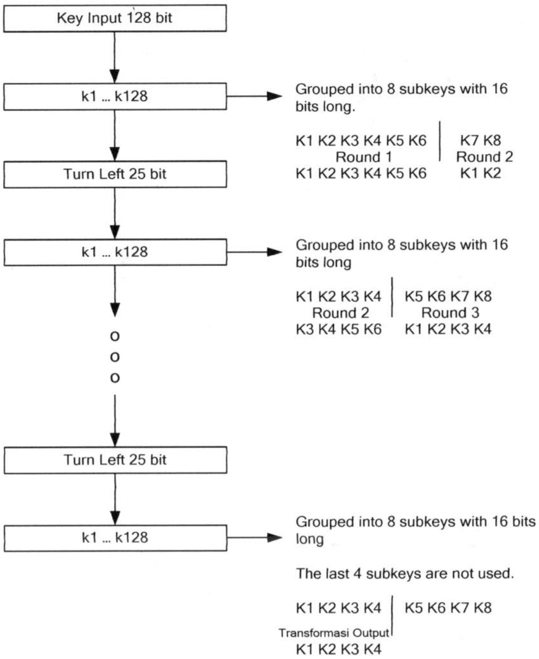

Figure 1. Key Formation

b. *Encryption Process*

64-bit plaintext is divided into four sub-blocks of 16 bits used for the first iteration. At each iteration, four sub-blocks are XOR-aligned, added, multiplied by another and with six pieces 16-bit sub-keys, next is the sequence of the IDEA algorithm process:

1. Multiply X1 with K1 mod (216 + 1).
2. Add X2 with K2 mod 216.
3. Add X3 with K3 mod 216.
4. Multiply X4 with K4 mod (216 + 1).
5. XOR results from step 1 and 3.
6. XOR results from step 2 and 4.
7. Multiply the result from step 5 with K5 mod (216 + 1).
8. Add the result from step 6 and 7 mod 216.
9. Multiply the result from step 8 with K6 mod (216 + 1).
10. Add the result from step 7 and 9.
11. XOR results from step 1 and 9.
12. XOR results from step 3 and 9.
13. XOR results from step 2 and 10.
14. XOR results from step 4 and 10.

c. *Decryption Process*

The decryption process is precisely the same as the encryption process. The difference lies only in the rules of the sub-key. The encryption process reverses the order of the sub-key, and the sub-key is inverse. The sub-key on the output transformation step of the encryption process is invoked and used as a sub-key in round 1 of the decryption process [11].

Next is how WAKE algorithm works on the encryption and decryption process [6]:
a. A process of forming table S-Box (Substitution Box).
b. A process for forming the key.
c. Encryption and decryption process.

The core of the WAKE method lies in the process of forming the S-Box table and the critical building process. The S-Box table of the WAKE method is flexible and different for each round. Super-encryption IDEA and WAKE algorithm can be seen in Figure 2 below:

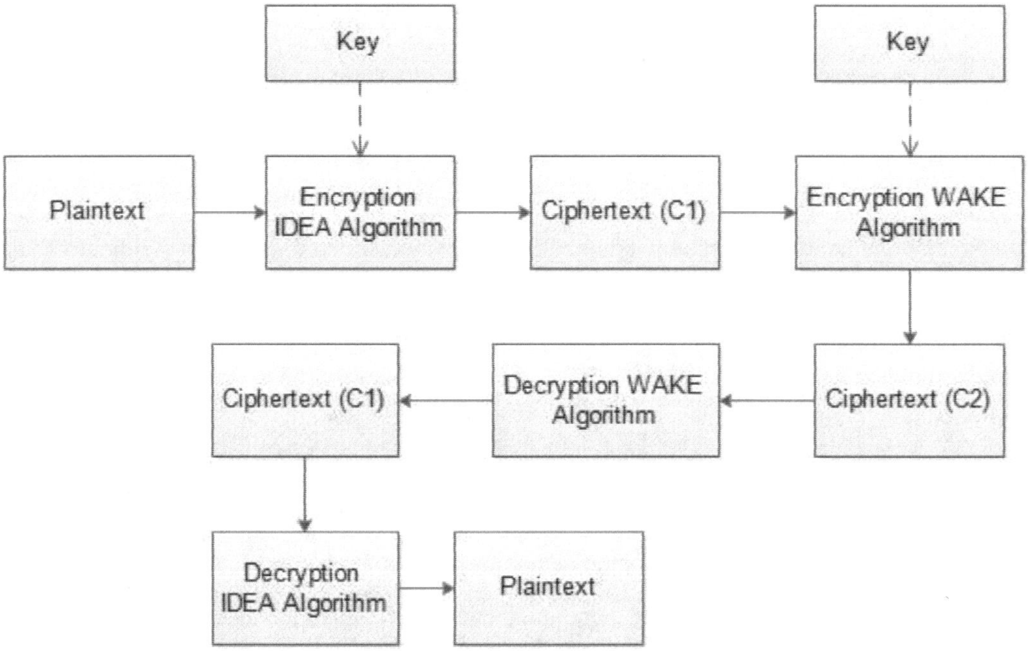

Figure 2. Super-Encryption IDEA and WAKE Algorithm

3. Result and Discussion

The experiment performed by step by step, for the first stage determines the plaintext and key as below:
Plaintext = DAHLANRR
Key = InternationalCon
The first step is to do the process of forming the key with the IDEA algorithm based on the process of figure 1 so the results obtained as follows:
Key = InternationalCon = 195C9B985D1A5BDB985B10DBDB925B9D
Encryption by using IDEA algorithm using a key that has been formed got a result as follows:
Ciphertext = IðŽÁ¿ I
the ciphertext of the IDEA algorithm results is then re-encrypted with the WAKE algorithm by using different keys as follows:
Plaintext = IðŽÁ¿ I
Key = Dahlanabdullahok
Encryption process from WAKE algorithm can seen below:
ASCII from 'I' = 49
ASCII from ' ' = 20
ASCII from 'ð' = F0
ASCII from 'Ž' = 8E
ASCII from 'Á' = C1
ASCII from '¿' = BF

ASCII from ' ' = 09
ASCII from 'I' = 49
Plaintext in hexadecimal = 4920F08EC1BF0949

Key in hexa = 77BD76EE

Cipher Text = Plaintext XOR Key
49 XOR 77 = 3E = '>'
20 XOR BD = 9D = '• '
F0 XOR 76 = 86 = '†'
8E XOR EE = 60 = '`'
C1 XOR 77 = B6 = '¶'
BF XOR BD = 02 = '_'
09 XOR 76 = 7F = '• '
49 XOR EE = A7 = '§'

Ciphertext = >□†`¶_□§

Super-encryption algorithm IDEA and WAKE can be done well in the experiments performed, the combination of these two algorithms is very suitable because the ciphertext of the IDEA algorithm in the form of ASCII can be re-encrypted with WAKE algorithm that changes the plaintext into the new hexadecimal form and then re-encrypted, for more details of the combination of these two algorithms can be seen in table 1 below:

Table 1. Super-Encryption IDEA and WAKE Algorithm

No	Plaintext	Key	IDEA	WAKE	Combination
1	ICCAI USU	uzumakinarutooke	Û".w)žœ	p¾_¥pÝ_·l	âßZÊNÔÊx
2	Conference Medan	uzumakinarutooke	'\ƒØt6• á	z':,\□1ŠZ˜t©\™5Š	«¡×<MËÙ_
3	Kolaborasi Publikasi	uzumakinarutooke	‡¥!g"~6	r'8…['&…J"t´LŸ8□Rœ'	_zñÅ^iHš_
4	Scopus Conference	scopusindexingok	]Ëä}Gl	Ùû¹>ÿëöåö°+øý¸-ï	†Å_ª›å‵"
5	GWA Scopus Yes	scopusindexingok	&O–Þ3®	ÍÏ—nÙû¹>ÿëö_ïë	¬Ÿ™Ø•Fåà

The result of the combination of IDEA and WAKE algorithm produces different ciphertext, with this combination is probably make cryptanalyst difficult to know the plaintext of ciphertext.

4. Conclusion

The IDEA and WAKE algorithm used in this paper can be combined as well, the use of different keys on each algorithm can also be done, the next development of the encryption process can be a mixed mode where the encryption process is done simultaneously for two different algorithms or more materials.

References

[1] Rahim R and Ikhwan A, 2016 Cryptography Technique with Modular Multiplication Block Cipher and Playfair Cipher *Int. J. Sci. Res. Sci. Technol.* **2**, 6 p. 71–78.
[2] Putera A Siahaan U and Rahim R, Aug. 2016 Dynamic Key Matrix of Hill Cipher Using Genetic Algorithm *Int. J. Secur. Its Appl.* **10**, 8 p. 173–180.
[3] Rahim R and Ikhwan A, Nov. 2016 Study of Three Pass Protocol on Data Security *Int. J. Sci.*

Res. **5**, 11 p. 102–104.

[4] Rahim R, 2017 Man-in-the-middle-attack prevention using interlock protocol method *ARPN J. Eng. Appl. Sci.* **12**, 22 p. 6483–6487.

[5] Tange H and Andersen B, 2013 Attacks and countermeasures on AES and ECC in *16th International Symposium on Wireless Personal Multimedia Communications (WPMC)* p. 1–5.

[6] Chen J Xue D and Lai X, 2008 An analysis of international data encryption algorithm(IDEA) security against differential cryptanalysis *Wuhan Univ. J. Nat. Sci.* **13**, 6 p. 697–701.

[7] Legito L and Rahim R, 2017 SMS Encryption Using Word Auto Key Encryption *Int. J. Recent Trends Eng. Res.* **3**, 1 p. 251–256.

[8] Iswahyudi C Setyaningsih E and Widyastuti N, 2012 Pengamanan kunci enkripsi citra pada algoritma super enkripsi menggunakan metode end of file *Pros. Semin. Nas. Apl. Sains Teknol. Periode III* November p. 278–285.

[9] Kartikadarma E Listyorini T and Rahim R, 2018 An Android mobile RC4 simulation for education *World Trans. Eng. Technol. Educ.* **16**, 1 p. 75–79.

[10] Rahim R Hartama D Nurdiyanto H Ahmar A S Abdullah D and Napitupulu D, 2018 Keylogger Application to Monitoring Users Activity with Exact String Matching Algorithm *J. Phys. Conf. Ser.* **954**, 1 p. 12008.

[11] Nurdiyanto H Rahim R and Wulan N, Dec. 2017 Symmetric Stream Cipher using Triple Transposition Key Method and Base64 Algorithm for Security Improvement *J. Phys. Conf. Ser.* **930**, 1 p. 12005.

[12] Yu L Wang Z and Wang W, 2012 The Application of Hybrid Encryption Algorithm in Software Security *Fourth Int. Conf. Comput. Intell. Commun. Networks* p. 762–765.

[13] Ludwig G A and Alegre P, 2007 International Data Encryption Algorithm (IDEA) **8**, 1 p. 6–11.

[14] Daemen J Govaerts R and Vandewalle J, 1993, Weak Keys for IDEA, in *Advances in Cryptology — CRYPTO' 93*, (Berlin, Heidelberg: Springer Berlin Heidelberg), p. 224–231.

IOP Conf. Series: Journal of Physics: Conf. Series **1019** (2018) 012040 doi:10.1088/1742-6596/1019/1/012040

Application of Web Based Book Calculation using Deterministic Dynamic Programming Algorithm

Dahlan Abdullah[1*], Robbi Rahim[2], Dedy Hartama[3], A Abdisyah[4], Z Zulmiardi[5], Syahril Efendi[6]

[1]Department of Informatics, Universitas Malikussaleh, Aceh, Indonesia
[2]School of Computer and Communication Engineering, Universiti Malaysia Perlis, Perlis, Malaysia
[3]Department of Information System, STIKOM Tunas Bangsa, Pematang Siantar, Indonesia
[4]Faculty of Management, Universitas Sumatera Utara, Medan, Indonesia
[5]Department Chemical Engineering, Universitas Malikussaleh, Aceh, Indonesia
[6]Department of Computer Science and Information Technology, Universitas Sumatera Utara, Medan, Indonesia

* dahlan@unimal.ac.id

Abstract. The library procurement materials is one of the technical services activities in a library as an effort to provide information needed by librarians. in doing the calculation of the procurement book (stock) of in library Universitas Malikussaleh still not automated and computerized and common error occurred on the the given amount of procurement the library because calculation was carried out using the tools count, with the application calculation book of stock it would make simply in calculation stock in library of Universitas malikussaleh. This application calculation stock of book is using deterministic dynamic programming algorithm for calculation process, deterministic dynamic programming itself is a method with gradually solutions and every stage will get an optimal solutions. By using this method it would know whether are the amount of stock of book has been Optimal. This Research conducted by the writer in unit. The library of Universitas Malikussaleh are taking total which start in 2012-2016 and the result by using dynamic programming deterministic algorithm is 270,779 as the optimal solutions in 2016.

1. Introduction

The rapid development of web applications since the advent of internet technology is very helpful in the ease and speed of delivery, delivery and acceptance of information[1]–[3]. Starting from companies, schools, colleges, and other institutions or organizations has been a lot of web applications in sales, promotions, learning activities and other activities where required delivery, dissemination and acceptance of information[4]–[7] so as to provide convenience for users (users) who need Web applications or can be called also with web-based software has grown rapidly both in terms of use, size, language used and the complexity[1], [8]. Web apps were originally only static and navigated oriented websites, and more used as product brochures or online company profiles. At this time web applications have a lot of dynamic, interactive and task oriented for use in information systems, telecommunications, commerce, banking and others

In doing the calculation of the procurement (stock) of books in the Library of Malikussaleh university has been computerized but not yet systematized often occur inefficiency in procurement of library materials calculated considering the calculation is still using Microsoft Excel, and this is not comfortable for book calculation.

Looking at the problems that sometimes occur, a thought arises about how we can create something with internet-based technology to help solve the problem of stock book calculation, problem solving is done by deterministic dynamic programming approach[9].

2. Methodology

The optimization algorithm[10] can be defined as an algorithm or numerical method to find the value of x such that it yields f (x) of value as small as (or as large) as possible for a given function f, which may be accompanied by some limitation on x. here, x can be a scalar or vector of continuous or discrete values. The optimization algorithm is slightly different from the search algorithm or search algorithm[10]–[12]. In the search algorithm, there is a certain criterion which states whether an element xi, is a solution or not. In contrast, in the optimization algorithm there may be no such criteria but only objective functions that describe whether or not a given configuration is good[10]. Because the objective function can provide a more general definition of a problem, the optimization algorithm can be said to be a generalization of the search algorithm[9], [10], [12].

Based on the method of operation the optimization algorithm can be divided into two large classes, such as deterministic and probabilistic algorithms[10]; and in this study using deterministic algorithm. Dynamic Programming[13] is one of the algorithm design techniques developed to solve very complex problems by breaking the problem into many sub-problems[13]. Dynamic programming is one of the algorithms used to find the optimal value of a problem. In dynamic programming, the solution of a problem is divided into several steps until the solution of a problem can be viewed as a series of interrelated decisions[9], [13]. The use of deterministic dynamic programming in this study can be seen in Figure 1 below:

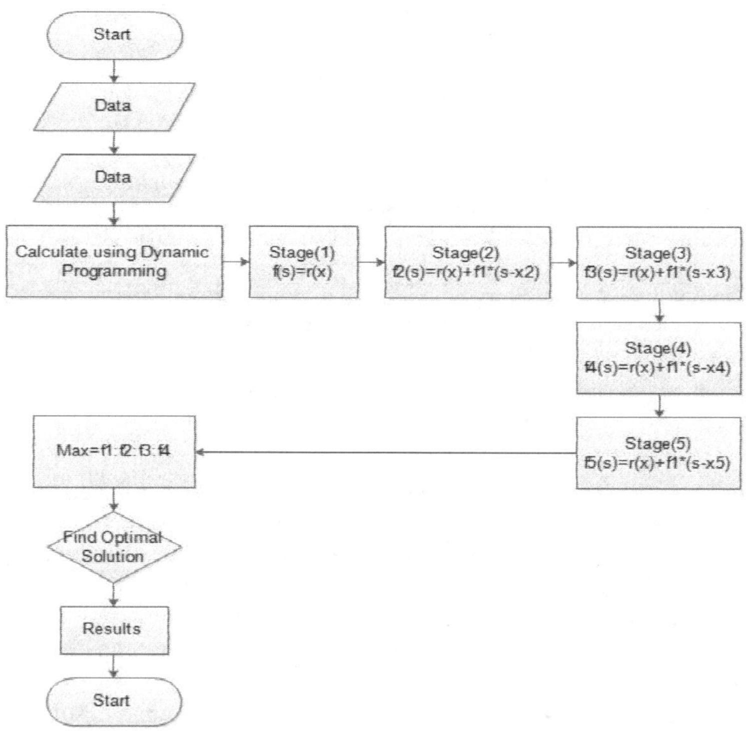

Figure 1. Flowchart Deterministic Dynamic Programming

IOP Conf. Series: Journal of Physics: Conf. Series **1019** (2018) 012040 doi:10.1088/1742-6596/1019/1/012040

3. Result and Discussion

Dynamic programming algorithm is tasked to describe the steps of the number of book stock in a few years starting from 2012-2016 to determine the optimal data in the year for data processing engineer own research field study at the Library University Malikussaleh In dynamic programming is divided into several stage then each stage will be compared with the decision value and maximization of data on the previous stage after it gets up to the nth stage then the step to determine the optimum is to combine some decision values (x) and maximize stock (fn) of each stock and then find the most optimum value in the decision (x), for example data used note table 1 below:

Table 1. Book data each year

Year	Total Stock
2012	60,985
2013	64,715
2014	66,039
2015	73,415
2016	87,824

The above data is the total of summed data from the General Collection and book data of the course of 2012-2016, with calculated formula as:

$f_n(S_{n,}X_n)= r_s(X_n); f_{n-1*}(S_{n,-1})$: n= 2, 3
n= 5
Xn= 5(decisiom)
SN=5(Corresponding data)

Table 2. Initiation data

F0	60,985
F1	64,715
F2	66,039
F3	73,415
F4	87,824

Therefore, the application of stock calculation of this book only calculate only do research on 5 years only then step from calculation of stock book in dynamic programming algorithm also as much as 5, for step 1 as in table 3 below:

Table 3. First step

	0	1	2	3	4	f_1	X_1
0	60,985					60,985	0
1	60,985	62,750				64,715	1
2	60,985	62,750	66,039			66,039	2
3	60,985	62,750	66,039	73,415		73,415	3
4	60,985	62,750	66,039	73,415	87,824	87,824	4

It is a series of early stages that will be a benchmark on the next data the formula f (s) = r (x)
f (0) = (60,985) = 60,985
f (1) = (60,985) f1 (1) = (62,750) (1)
f (2) = (60,985) f2 (1) = (62,750) (1) f2 (2) = (66,039) (2)
f3 (2) = (62,750) f3 (3) = (73,415) (3) f3 (4) = (60,985)
f4 (1) = (62,750) (1) f4 (2) = (66,039) (2) f4 (3) = (73,415) (3) f4 (4) = (87,824) (4)

Phase 1 is equal to the original data that is f (s) = r (x) with the number of optimum stock obtained that is 87,824 at decision 4

Table 4. Second step

	0	1	2	3	4	f_2	X_2
0	121,970					121,970	0
1	123,735	123,735				123,700	0,1
2	127,024	125,500	127,024			127,024	0,2
3	134,400	128,789	128,789	134,400		134,400	0,3
4	148,809	136,165	132,078	136,165	148,809	148,809	0,4

Step 2 has a relationship with step 1 where each value will be added based on decision table after deducted in this first step For more details can be seen in the formula below:
Formula f2 (s) = r (x) + f1 * (s-x2)
f2 (0) = r (60,985) x (0) + f1 (60,985) s-x2 (0-0) = 121,970
f2 (1) = r (60,985) x (0) f1 (62,750) s-x2 (1-0) = 123,735
f1 (1) = r (62,750) (1) + f1 (60,985) S-X2 (1-1) = 123,735
f2 (2) = r (60,985) x (0) + f1 (66,039) s-x2 (2-0) = 127,024
f2 (2) = r (62,750) (1) + f1 (62,750) s-x2 (2-1) = 125,500
f2 (2) = r (66,039) x (2) + f1 (60,985) s-x2 (2-2) = 127,024
f2 (3) = r (60,985) x (0) + f1 (87,824) s-x2 (3-0) = 148,809
f2 (3) = r (62,750) (1) + f1 (66,039) s-x2 (3-1) = 123,735
f2 (3) = r (66,039) x (2) + f1 (62,750) s-x2 (3-2) = 127,024
f2 (3) = r (87,824) x (3) + f1 (60,985) s-x2 (3-3) = 148,809
f2 (4) = r (60,985) x (0) + f1 (87,824) s-x2 (4-0) = 148,809
f2 (4) = r (62,750) x (1) + f1 (73,415) s-x2 (4-1) = 136,165
f2 (4) = r (66,039) x (2) + f1 (66,039) s-x2 (4-2) = 132,078
f2 (4) = r (73,415) x (3) + f1 (62,750) s-x2 (4-3) = 136,165
f2 (4) = r (87,824) x (4) + f1 (60,985) s-x2 (4-4) = 148,809

the next process is the third step, the value as follows:

Table 5. Third step

	0	**1**	**2**	**3**	**4**	f_3	X_3
0	182,955					182,955	0
1	184,720	184,720				184,720	0,1
2	188,009	186,485	188,009			188,009	0,2
3	195,385	189,774	189,774	195,385		195,385	0,3
4	209,794	197,150	193,063	197,150	209,794	209,794	0,4

Step 3 is the final step of the search for its values obtained initial data summed with the optimum value of the data in the previous step (step 2)

Function f3 (s) = r (x) + f1 * (s-x2)

f3(0)=r(60,985)x(0)+f1(121,970)s-x2(0-0)= 182,955
f3(1)=r(60,985)(0)+f1(123,735,970)S-X2(1-0)=184,720
f3(1)=r(62,750)(1)+f1(121,970)S-X2(1-1)=184,720
f3(2)=r(60,985)x(0)+f1(127,024)s-x2(2-0)=188,009
f3(2)=r(62,750)(1)+f1(123,735)s-x2(2-1)=186,485

IOP Conf. Series: Journal of Physics: Conf. Series **1019** (2018) 012040 doi:10.1088/1742-6596/1019/1/012040

f3(2)=r(66,039)x(2)+f1(121,970)s-x2(2-2)=188,009
f3(3)=r(60,985)x(0)+f1(134,400)s-x2(3-0)=195,385
f3(3)=r(62,750)(1)+f1(127,024)s-x2(3-1)=189,774
f2(3)=r(66,039)x(2)+f1(123,735)s-x2(3-2)=189,774
f2(3)=r(73,415)x(3)+f1(121,970)s-x2(3-3)=195,385
f2(4)=r(60,985)x(0)+f1(182,955)s-x2(4-0)= 209,794
f2(4)=r(62,750)x(1)+f1(134,400)s-x2(4-1)=197,150
f2(4)=r(66,039)x(2)+f1(127,024)s-x2(4-2)=193,063
f2(4)=r(73,415)x(3)+f1(121,970)s-x2(4-3)=197,150
f2(4)=r(87,824)x(4)+f1(121,970)s-x2(4-3)=209,794
the next process is the fourth step, the value can see in table VI:

Table 6. Fourth step

	0	1	2	3	4	f_4	X_4
0	243,940					243,940	0
1	245,705	245,705				245,705	0,1
2	248,994	247,470	248,994			248,994	0,2
3	256,370	250,759	250,759	256,370		256,370	0,3
4	270,779	258,135	254,048	258,135	270,779	270,779	0,4

Step 4 is the final step of the search for its values obtained initial data summed with the optimum value of the data in the previous step (step 3)

Function f3(s)=r(x)+f1*(s-x2)
f3(0)=r(60,985)x(0)+f1(182,955)s-x2(0-0)= 243,940
f3(1)=r(60,985)x(0)f1(184,720)s-x2(1-0)=245,705
f3(1)=r(62,750)(1)+f1(182,955)S-X2(1-1)=245,705
f3(2)=r(60,985)x(0)+f1(188,009)s-x2(2-0)=248,994
f3(2)=r(62,750)(1)+f1(184,720)s-x2(2-1)=247,470
f3(2)=r(66,039)x(2)+f1(182,955)s-x2(2-2)=248,994
f3(3)=r(60,985)x(0)+f1(195,385)s-x2(3-0)=256,370
f3(3)=r(62,750)(1)+f1(188,009)s-x2(3-1)=250,759
f2(3)=r(66,039)x(2)+f1(184,720)s-x2(3-2)=250,759
f2(3)=r(73,415)x(3)+f1(182,955)s-x2(3-3)=256,370
f2(4)=r(60,985)x(0)+f1(182,955)s-x2(4-0)= 270,779
f2(4)=r(62,750)x(1)+f1(195,385)s-x2(4-1)=258,135
f2(4)=r(66,039)x(2)+f1(188,009)s-x2(4-2)=254,048
f2(4)=r(73,415)x(3)+f1(184,720)s-x2(4-3)=258,135
f2(4)=r(87,824)x(4)+f1(182,955)s-x2(4-4)= 270,779

the fifth step is the last dynamic programming process to determine the optimum value, note the following table 7:

Table 7. Fifth step

	0	1	2	3	4	f_5	X_5
0	304,925					304,925	0
1	306,690	306,690				306,690	0,1
2	309,979	308,455	309,979			309,979	0,2
3	317,355	311,744	311,744	317,355		317,355	0,3
4	331,764	319,120	315,033	319,120	331,764	331,764	0,4

After the search to stage 5, the last step is to compare each stage from step 1, step 2, step 3, and step 4 so that found the optimum value is = 331,764 at the decision to 0 and 4.

4. Conclusion

The experiment process of calculating book stock, Dynamic programming algorithm deterministic is a method with the completion of a gradual process of data year 2012-2016, the data is described into smaller stages of the problem and dynamic programming to make the problem of the problem of the process of the problem becomes more clear to be known, and also the use of dynamic programming algorithm on stock book calculation tends to be more flexible than other optimization techniques but It takes a long time to solve the problem because it describes the solution in several stages, from the results used using dynamic programming algorithm in the can that the optimum stock is 331,764 for 2016.

References

[1] S. Al-Fedaghi, "Developing web applications," *Int. J. Softw. Eng. its Appl.*, vol. 5, no. 2, pp. 57–68, 2011.

[2] J. Suyono, A. Sukoco, M. I. Setiawan, S. Suhermin, and R. Rahim, "Impact of GDP Information Technology in Developing of Regional Central Business (Case 50 Airports IT City Development in Indonesia)," in *Journal of Physics: Conference Series*, 2017, vol. 930, no. 1.

[3] D. Siregar, D. Arisandi, A. Usman, D. Irwan, and R. Rahim, "Research of Simple Multi-Attribute Rating Technique for Decision Support," *J. Phys. Conf. Ser.*, vol. 930, no. 1, p. 012015, Dec. 2017.

[4] R. Rahim *et al.*, "INA-Rxiv: The Missing Puzzle in Indonesia's Scientific Publishing Workflow," *J. Phys. Conf. Ser.*, vol. 1007, no. 1, p. 012032, Apr. 2018.

[5] K. Adiyarta, D. Napitupulu, R. Rahim, D. Abdullah, and M. Setiawan, "Analysis of e-learning implementation readiness based on integrated elr model," *J. Phys. Conf. Ser.*, vol. 1007, no. 1, p. 012041, Apr. 2018.

[6] E. Kartikadarma, T. Listyorini, and R. Rahim, "An Android mobile RC4 simulation for education," *World Trans. Eng. Technol. Educ.*, vol. 16, no. 1, pp. 75–79, 2018.

[7] D. Napitupulu *et al.*, "Analysis of Student Satisfaction Toward Quality of Service Facility," *J. Phys. Conf. Ser.*, vol. 954, no. 1, p. 012019, Jan. 2018.

[8] M. Elazony, M. Al-kholy, A. Khalifa, S. Nouh, M. Hussein, and G. Tharwat, "Improve information retrieval and e-learning using mobile agent based on semantic web technology," *Int. J. Web Semant. Technol.*, vol. 6, no. 3, 2015.

[9] S. Kavitha and N. P. Ratchagar, "A Deterministic Dynamic Programming Approach for Optimization Problem with Quadratic Objective Function and Linear Constraints," *Int. J. Math. Comput. Phys. Electr. Comput. Eng.*, vol. 7, no. 3, pp. 565–569, 2013.

[10] H. Aguiar e Oliveira Junior, L. Ingber, A. Petraglia, M. Rembold Petraglia, and M. Augusta Soares Machado, "Global Optimization and Its Applications," Springer, Berlin, Heidelberg, 2012, pp. 11–20.

[11] M. J. HARLEY and T. R. E. CHIDLEY, "DETERMINISTIC DYNAMIC PROGRAMMING FOR LONG TERM RESERVOIR OPERATING POLICIES," *Eng. Optim.*, vol. 3, no. 2, pp. 63–70, Jan. 1978.

[12] M. Karamouz and M. H. Houck, "COMPARISON OF STOCHASTIC AND DETERMINISTIC DYNAMIC PROGRAMMING FOR RESERVOIR OPERATING RULE GENERATION," *J. Am. Water Resour. Assoc.*, vol. 23, no. 1, pp. 1–9, Feb. 1987.

[13] G. de Cooman and M. C. M. Troffaes, "Dynamic programming for deterministic discrete-time systems with uncertain gain," *Int. J. Approx. Reason.*, vol. 39, no. 2–3, pp. 257–278, Jun. 2005.

IOP Conf. Series: Journal of Physics: Conf. Series **1019** (2018) 012041 doi:10.1088/1742-6596/1019/1/012041

Lip Reading Analysis of English Letters as Pronounced by Filipino Speakers Using Image Analysis

Cruz, Hans Miguel [1,*], Puente, Jofet Kane T [1,*], Santos, Christian[1] ,Vea Larry A.[1], Rajendaran Vairavan[2]

[1]School of Information Technology, Mapua University, 333 Sen. Gil Puyat Ave., Makati City 1200, Philippines
[2]School of Microelectronic Engineering, University Malaysia Perlis (UniMAP), Kampus Alam, Pauh Putra, 02600 Arau, Perlis, Malaysia.

Abstract. Lip reading or speech reading is a skill of understanding speech by watching the speaker's actions. Main factors that will lead to a good speech reading, and these are: lip reading experience, good language knowledge, normal vision, good verbal short term memory, and familiarity with the speaker. Hence, this study aims to develop a model that will recognize the spoken English letters through lip reading. This study focuses on the analysis of lip reading through video processing. The video that was used in the experiment was limited to pre-recorded videos of the subjects pronouncing each letter of the English language. The test subjects are limited to a total of thirty Filipino English speakers, composed of 15 men and 15 women, to be specific. Out of the thirty subjects, two of them was an expert in speaking the English language. The test subject does not have a moustache and beard, also that the test subjects does not have lip deformation, and lastly the test subjects are limited to do extra unnecessary movements. The authors created their own image database that was used in the experiments. They pursued the experiment in a controlled environment consisting only of supplemental lighting. The camera that was used to record the video is a DSLR camera using a 640x424 image resolution running at 24fps. Only the movement of the exterior lip region is included in the experiment without regarding to the movement of the tongue and teeth. Overall results showed that the system demonstrated an overall precision accuracy of 45.62%.

1 Introduction

Lip reading or speech reading is a skill of understanding speech by watching the speaker's actions. The speaker's actions are said to be the following: speaker's mouth movements, head, face, eye and torso movements. These are main factors that will lead to a good speech reading, and these are: lip reading experience, good language knowledge, normal vision, good verbal short term memory, and familiarity with the speaker. These factors are the basis that lead to speech reading. It is also stated that the important aspects of visual image capture for speech reading are lightning, view, color, distance, look, and image compression [1,2 & 3].

There are also conflicts on lip reading, It is stated that these are factors that make lip reading difficult such as; speech is designed to be heard and not viewed, most movements of the tongue and teeth that are not visible ,not visible actions of tongue, lips, teeth (m/p, f/v, th), distinctive mouth shapes ('a','i','u') , most consonants that are liable to error, due to lips look alike ('d','t','k'), and lastly the consonants that easily distinguish (m , n, th). The largest part of information is derived from the movements of the mouth region [4]. Lip movements contributes significantly to speech information where the audio signal is absent or of low quality. A major drawback in dealing with visual cues is dealing with the large amount of visual data. This leads to storage problems and additional computational effort; thereby, leading to increase in processing time [5]. Visual speech information through lip reading is very useful for human speech recognition [6]. Much of the research on lip reading uses frontal facial images. A frontal facial image contains much useful information [7].

As Summerfield (1987) has remarked: any comprehensive account of how speech is perceived should encompass audio-visual speech perception. The ability to see as well as to hear has to be integral to the design, not merely a retro-fitted after thought" [8]. Study regarding lip reading by [9] was done by utilizing Using 3D on Hidden Markov Model to analyze visual speech information, such as lip movement. A study by [10] demonstrated a novel technique which modelled visual speech movements by employing fuzzy neural network to analyze the behavior of images derived from video film with particular focus on input coding. A work by [11] proposed lip information extraction by employing

* Corresponding author: puentejofet@yahoo.com

IOP Conf. Series: Journal of Physics: Conf. Series **1019** (2018) 012041 doi:10.1088/1742-6596/1019/1/012041

fusion of geometric and motion feature analysis. A research by [12] demonstrated application of kohonen algorithm for development of lip recognition software.

Thus, this study aims to develop a model that will recognize the spoken English letters through lip reading. This study demostartes the Test and validate done on the developed model for the lip reading analysis of English letters pronounced by the Filipino speakers. This study focuses on the analysis of lip reading through video processing. The video that was used in the experiment was limited to pre-recorded videos of the subjects pronouncing each letter of the English language.

2 Methodology

The study applies a controlled-type laboratory experiment consisting of male and female participants. The participants pronounced the English letters that consists from letter A-Z. The set of letters used are stated in the variables. The data gathered are processed under MATLAB to format the video to sequence of images .AVI, segment the lip area and consists of initial eight (8) key feature points. Comparison was done between 8 key feature points, 12 key feature points and 16 key feature points. The data were analysed under certain variables and the authors decided to select thirty (30) students or any person that speaks English letters properly. The test subjects consists of fifteen (15) female and fifteen (15) male and the test was held in an enclosed lighted room that supports the quality of the test video, and the room that was used is the same for all test subjects. The distance of the camera from the speaker/subject is one meter and the height of the camera (from the floor up to the camera lens) is 3 meters. The initial and final state of the test subject's lips is closed. Figure 1 shows the setup for data gathering. The distance of the camera from the test subject is a meter away, and the camera is focused at the face of the test subject .The authors gathered the test subject's data by recording a video using a Nikon D3100 camera with the lens specification 18mm-55mm. The subjects pronounced a letter from the English alphabet which is used as a part of the data. The numerical data gathered from the subjects was used to train the model.

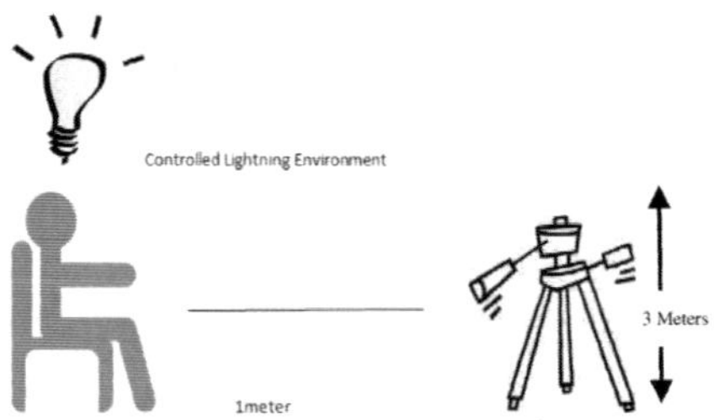

Fig. 1. Data gathering Setup

2.1 Converting Video to image sequence using MATLAB integrated in Java

In this phase, the authors converted the pre-recorded video to image sequences for the image analysis.MATLAB was utilized.The authors developed a process in MATLAB that will automatically create the folder and convert the video into images. Twelve (12) frames of images for processing were taken. The authors converted the images in .jpg format for the pictures.

2.2 Lip detection using Viola-Jones Algorithm and point plotting using Point distribution model tracking KLT Algorithm

In Viola-Jones technique is used based on exploring the input image by means of sub window capable of detecting features [13]. Viola Jones algorithm has 4 stages: Haar Feature Selection, Creating an Integral Image, Adaboost Training and Cascading classifiers.Haar features are like simple square waveforms which detects "darker" blocks adjacent to "lighter" blocks also known as "features". The feature value is calculated by the sum of the pixel intensities in the light rectangles subtracted from the sum of the pixels in the dark rectangle. In order to improve the summing of the intensities of the pixels in a given rectangle, Integral image was calculated at every point. Integral image is computed by the sum of the pixels of the rectangle from upper then to the left corner of the given point; after calculating the integral image of every point the area of any rectangle can be now calculated. The formula in solving for the area of the rectangle is D+A-(B+C). Adaboost Training combines the "weak" classifiers to come up with a "strong" classifier. The last step is to cascade classifiers, each sub window of the original image will now be tested against a series of classifiers and if the sub window fails at any stage of the testing, the sub window is rejected and if it passes all through the classifiers it is classified as a face.

After the images are pre-processed, the next step is the process on how the lip segmentation works. The original mask image is the result of inversed output during the detection of the flesh. The image is converted into CIELAB color space and the flesh is detected, after the flesh is detected the output is being inversed so that the red color (lips) is detected. After wards, the image goes through the Sobel mask detection which helps to enhance the mask that will be created. After all that process was done, the last part is to combine the output of the previous phases.

Point Distribution Model (PDM) and Kanade Lucas Tomasi (KLT) tracking algorithm template is created using sixteen point. This model describes the lip contour by a set of points. Figure 2 is the representation of Point Distribution model for Lip point plotting in MATLAB works. Comparison was done between 8 key feature points, 12 key feature points and 16 key feature points. The comparison was done by clustering the several alphabets into 8 clusters and 5 frames (F2 to F6) . This was then repeated again with second set of 8 clusters and 5 frames (F7 to F11). The difference between these two major clusters is the arrangement of several alphabets in each respective frames. Figure 3 shows the key feautres for the lips. The max number of points consist of 16 feature in relation of the horizontal and vertical distance of the lips. the 16 points are labelled as:left corner, right corner, top left, top right, top left left, top left right, top right left, top right right, bottom left,bottom right, bottom left left, bottom left right, bottom right left, bottom right right, bottom left and bottom right feature set. This is done speaker-wise. The normalized values are used for feature selection and classification.

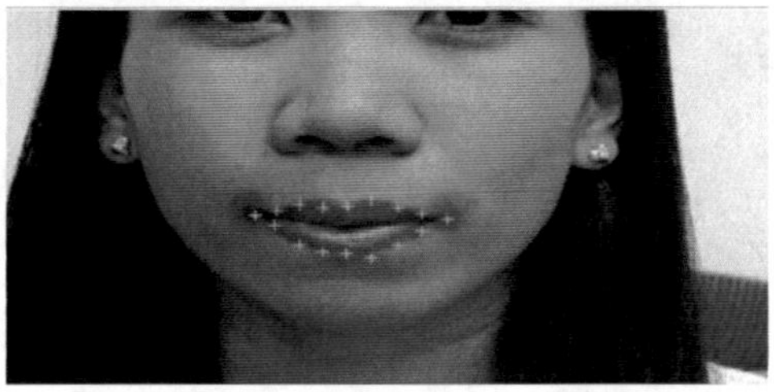

Fig. 2. Image sequence point plotting

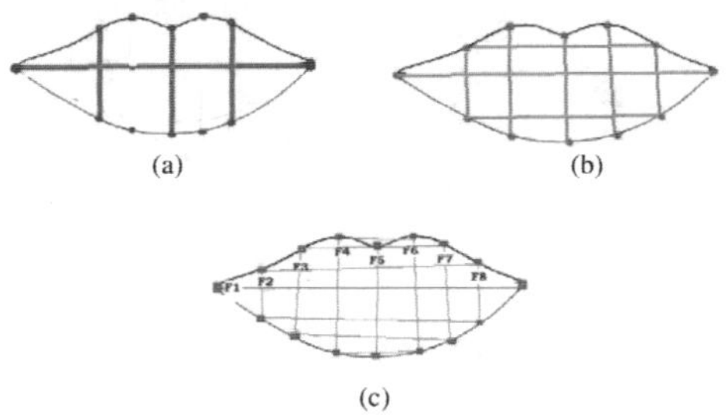

Fig. 3. Key features of the lips (a) 8 key features (b) 12 key features (c) 16 key features.

2.3 Lip Feature Extraction And Selection Using Matlab Works

This is done by defining new important lip features from selecting twelve (12) relevant images that will be used for image processing and will produce raw data sets. The feature extraction phases are done under MATLAB works by means of the extracting data through point plotting to produce the raw data sets of the distance of each feature points. After that process, the raw data are converted in to decimal points that pertains to the distance of the lips while moving according to image sequence; all of the data was automatically saved in an excel file, and it is converted to .mod extension file by using the prototype integrated program in java '.mod' converter that is done by the authors and the .mod file is used as the model data in rapid miner. From the feature extracted, the authors selected twelve relevant frames for each letter. The selected twelve relevant frames are used for point plotting that is consisted of sixteen feature points that are used for Feature Extraction. For the feature selection phase, it is done by analysing the decision tree of the model in the rapid miner and they analyse all the extracted data and focus on the selected frames which consists of frames 2, 4, 6, 8, and 10.

2.4 Lip Feature Extraction And Selection Using Matlab Works

After the extraction and selection of data, a model was created based on the results of the classification. Decision tree is a structure where it mutually partitions the inputted dataset and labels it to identify the data points.The authors used a total of thirty test subjects with two tries per English letter that was used for establishing the trained data sets, and for the test sets they used different test subjects that consists only of one try per letter. The model that is created in Rapid Miner using decision tree J48 algorithm was embedded in the prototype, which was also the same program that was used for pre-classification.

2.5 Model Testing and Analysis

For the model testing, the authors used another 10 test person to establish data for the test sets per letters that will consist of 1 try only. After the authors established the data sets, it is saved as CSV file and it is loaded to the prototype. Once the test data is loaded to the prototype, the software will start to recognize the test sets to the trained model to see how accurate is the model.The authors used Java to create a prototype, with the displayed process, and used the model built by Rapid Miner and tested it by developing a prototype, by using decision tree J48 algorithm [14].

3 Result and discussion

Table 1 shows the results for Key features of lips according to its respective key features. Based on Table 1 percentage by having a sixteen key features and it is used to establish the model training data. From Frame 2, it has a 75.09% average because it's where the lips is starting to open and most of the letters has the same starting phase of the lips. Then in Frame 3, it increases to 76.46% because in this Frame the lips is starting to move from its starting phase. In frame 4 it continue to rise to 78.21%, Frame 5 has a percentage of 76.92. Compare from frame 4 frame 5 was lower than Frame 4 because some of the letters depends on how it was pronounced by the speaker. Frame 6 has the value of 77.15%, Frame 7 has a percentage of 76.99. Frame 8 and has a same value of percentage which is 77.45%. Frame 10 has 77.60% and lastly Frame 11 which is the closing state of the lips has a percentage of 76.31%. Based on Table 1, it is observed that 16 key features point showed best percentage per frame for the letter clustering compared to the 12 key feature points and 8 key feature points. These results are based from the experiment that is done by the authors to see which key features has the higher percentage per frame that is used to establish the model for this study. Those percentage per frame is based on the distance of the key feature points. Thus 16 key feature points is established for the prototype testing as it shows better evaluation points on the lip movement. The results of the prototype testing is shown in Table 2. The results is based on the precision of the prototype to distinguish each letter as it is being pronounced.

Table 1. Results for Key features of lips

8 key features		12 key features		16 key features	
Frame	(%)	Frame	(%)	Frame	(%)
F2	71.53	F2	73.50	F2	75.09
F3	73.73	F3	75.25	F3	76.16
F4	72.44	F4	76.23	F4	78.21
F5	74.03	F5	75.17	F5	76.92
F6	72.67	F6	76.17	F6	77.15
F7	74.11	F7	75.40	F7	76.99
F8	75.63	F8	75.63	F8	77.45
F9	73.80	F9	75.40	F9	77.75
F10	72.51	F10	76.23	F10	77.60
F11	71.53	F11	76.77	F11	76.31

Table 2. Prototype testing results

Letter	Precision	Letter	Precision
A	25%	N	41%
B	46%	O	32%
C	72%	P	23%
D	81%	Q	48%
E	58%	R	44%
F	61%	S	31%
G	32%	T	42%
H	50%	U	47%
I	42%	V	31%
J	35%	W	28%
K	51%	X	34%
L	77%	Y	49%
M	51%	Z	55%
Total Accuracy		45.26%	

The overall results will be discussed in terms of per letter analysis.Based on the results, it indicates that letter A has a precision rate of 25%; upon reviewing the results the authors found out that mostly of the frames has a precision rate below 50% and the only frame that has a higher precision rate is Frame 11.Letter B has a precision rate of 46%. By reviewing the results of each frames, the authors observe that it is difficult to predict in the middle stages of the frames (4,5,6), because those frames are the peak opening of the lips that causes some of letters clustered with B is similar with each other.

Letter C has a high precision rate of 72% and Letter D has a high precision rate of 81% and letter E showed a precision rate of 58%. Letter F showed a good precision rating with 61%. Upon reviewing the results it shows that almost all the time that this letter is correctly predicted.Letter G has a precision rate of 32%. The authors noticed that each frames has a precision rate that is ranging from 20-30% and the highest value of precision rate is 80%. These are Frames 7 and 10. Letter H gave a fifty percent (50%) precision rate during the testing phase. Letter I has a value of fourty two percent (42%). Based from the assessment, frame 11 has a zero percent (0%) precision and frames 4 and 5 have the precision value of fourty percent (40%).Letter J had a precision rate of 41%. Based from the assessment, Frame 8 has a precision rate of zero percent (0%) while the rest of the frames had a precision rate value below fifty percent (50%).

Letter K gave a precision rate of fifty one percent (51%) during the testing phase. Letter L showed a seventy seven percent (77%) precision rate during the testing phase. Letter M gave a precision rate of seventy seven percent (77%). Upon reviewing the result it showed that almost all of the time, this letter is correctly predicted and if not, usually it is predicted when it's clustered with letter L during the Frames 4 and 8; this two (2) letters has almost identical movements.Letter N has a precision rate of fourty one (41%), upon reviewing the results, the authors found out that most of the frames have a precision rate below fifty percent (50%) and the only frame that has a higher precision rate is Frame 3.Letter O had a precision rate of 25%; upon reviewing the results the authors found out that the value of the frames is ranging from 10-50%. Letter P has a low precision rate of 23%. Letter R had a precision rate of fourty four (44%), upon observing the result, the authors found out that some of the frames has a precision value below 50% and frame 8 has the higher precision value which is 70%.Letter S had a precision rate of thirty one (31%).

Based from the assessment, Frame 8 and 9 had a lowest precision rate while the rest of the Frames has a precision rate value below 50%. The highest precision rate is Frame 10.Letter T had a precision rate of thirty one percent (31%). Based from the reviewing of the result, Frame 10 and 11 has a lowest precision rate while the rest of the Frames has a precision rate value below fifty percent (50%). Letter U had a precision rate of fourty seven (47%) during the testing phase. Letter V had a precision rate of thirty one percent (31%). Based from the reviewing of the result, Frame 10 and 11 has a lowest precision rate while the rest of the Frames has a precision rate value below fifty percent (50%).Letter X had a precision rate of thirty one percent (31%). Based from the reviewing of the result, Frame 10 and 11 has a lowest precision rate while the rest of the Frames has a precision rate value below fifty percent (50%).Letter Y had a precision rate of fourty nine percent (49%) during the testing phase. Letter Z gave a precision rate of fifty five percent (55%).Hence, overall the system showed an overall precision accuracy of 45.62%.

Thus, in this work, the sixteen (16) Key Points consists of the following: Upper center lips, Lower center Lips, Left corner Lips, Right corner lips, Upper left corner, Upper right corner, Lower right corner, and the Lower left corner of the lips. Was set Out of this points the Authors come up with the following features: F1(L-R), F2(TOPLL-BOTLL), F3(TOPL-BOTL), F4(TOPLR-BOTLR), F5(TOP-BOT), F6(TOPRL-BOTRL), F7(TOPR-BOTR), F8(TOPRR-BOTRR), F9(TOPLL-TOPRR), F10(TOPL-TOPR),F11(TOPLR-RL), F12(BOTLL-BOTRR), F13(BOTL-BOTR), F14(BOTLR-BOTRL).Based on this, it was found out that the most useful and distinguishable feature movements are the vertical distance of features 4(TOPLR-BOTLR), 5(TOP-BOT), 6(TOPRL-BOTRL) and the horizontal distance of feature 1.

The model is correctly recognized by means of conducting a test that will have a good performance accuracy rate that established a good model data sets. If the test data sets are tested to the model data it brings a good recognition rate from the model. By using some of the algorithms in RapidMiner it established a model in batch cross fold validation and has a relevant recognition rate between the model and test sets. Using the KNN algorithm, this is applied for making the model of lip reading analysis because it recognizes a random data and does not give a better result.

From the results taken from Decision tree J48 batch algorithm method in RapidMiner, using this algorithm for letter recognition this gives the most efficient way compared from the other algorithm. It consists of sixty two percent (62%) correctly classified instances and has a percentage of error thirty eight percent (38%). Most of the letters have an error in classification because of phonemes, because some letters has the same way on how to pronounce but different letter. As the results some of the testing has a different result because of that error.

As the authors could not find concrete evidence to show on how to improve and on how to prevent detecting words that have the same pronunciation but different letters. The authors recommend future research for its future study. The Authors concluded that the key features of the lips are the vertical distances especially Feature 4, 5, 6 which is the middle vertical feature in the lips and the horizontal Feature which is Feature 1.

They also recommend the following to the future authors of this study. In order to achieve a better and more accurate lip segmentation, they can use lipsticks on the subjects to easily discriminate the skin from the lips. Another benefit from this is that it can lower the processing time of the method because the detection of the flesh and inversing its result to detect the red/lips can be removed from the process. They showed a deeper and thorough studies to fully understand and differentiate the phoneme letters with each other; restrict the train subjects to professional English speakers such as call center agents, English linguist, and latter. to have a more accurate model.It is better to use a controlled light such as face lights and reflectors aside from the room lights to acquire better video quality.

4 Conclusion

The objective of the study was to detect letter that has been said by the speaker thru video processing to determine the highest recognition rate in lip reading analysis by using the 16 key points for point plotting, and also to develop a model that recognize the English letters from A-Z by taking video of college students that are Filipino English speakers. Thus, the objective set were achieved. In order to achieve a better and more accurate lip segmentation in the future , lipsticks on the subjects is suggested to easily discriminate the skin from the lips. This research is deemed significant to serve as a stepping stone for the future researches related to the topic. It will also help in translating the conversation of the people in a video to a subtitle which will be helping the authorities in case of reviewing CCTV footage from a crime scene.

References

1. 1.WenJuan, Y.; YaLing, L.; MingHui, D. In *Advanced Computer Theory and Engineering (ICACTE), 2010 3rd International Conference on*; IEEE, 2010; Vol. 6, pp V6-363.
2. M.Z. Ibrahim, D.J. Mulvaney, *Geometrical-based lip-reading using template probabilistic multi-dimension dynamic time warping*, In *Journal of Visual Communication and Image Representation*, Volume 30, 2015, Pages 219-233.
3. Sunil S. Morade, Suprava Patnaik, *Comparison of classifiers for lip reading with CUAVE and TULIPS database*, In Optik - *International Journal for Light and Electron Optics*, Volume 126, Issue 24, 2015, Pages 5753-5761.
4. Saitoh, T.; Konishi, R. In *Pattern Recognition (ICPR), 2010 20th International Conference on*; IEEE, 2010; pp 1356–1359.
5. Huang, Y.; Liang, J.; Pan, B.; Fan, X. In *Systems Man and Cybernetics (SMC), 2010 IEEE International Conference on*; IEEE, 2010; pp 2402–2405.
6. Okubo, M.; Watanabe, T. In *Automatic Face and Gesture Recognition, 1998. Proceedings. Third IEEE International Conference on*; IEEE, 1998; pp 187–192.
7. Mahmoud, H. A.; Muhaya, F. Bin; Hafez, A. In *Future Information Technology (FutureTech), 2010 5th International Conference on*; IEEE, 2010; pp 1–4.
8. Summerfield, Q. **1987**.
9. Min, K. Y.; Zuo, L. H. In *Electronics and Optoelectronics (ICEOE), 2011 International Conference on*; IEEE, 2011; Vol. 1, pp V1-115.
10. Bothe, H.-H. In *Proceedings of the 1995 ACM symposium on Applied computing*; ACM, 1995; pp 450–454.
11. Yingjie, M.; Haiyan, Z.; Yingjie, H.; Jinyang, L. In *Fuzzy Systems and Knowledge Discovery (FSKD), 2011 Eighth International Conference on*; IEEE, 2011; Vol. 4, pp 2186–2190.
12. Demetz Clement (2003). Lip recognition Software using a Kohonen Algorithm for image Compression. Computer-Aided Engineering, University of Wisconsin-Madison.
13. Khan, I.; Abdullah, H.; Zainal, M. S. Bin. *Int. J. Eng.***2013**, *3* (4), 8269.
14. Kaur, G., & Chhabra, A. (2014). Improved J48 classification algorithm for the prediction of diabetes. *International Journal of Computer Applications*, *98*(22).

IOP Conf. Series: Journal of Physics: Conf. Series **1019** (2018) 012042 doi:10.1088/1742-6596/1019/1/012042

A Review of Reconfigurable Frequency Switching Technique on Micostrip Antenna

M. S. Shakhirul[1], M. Jusoh[2], Y. S. Lee[3] and C. R. Nurol Husna[4]

[1, 3, 4] Faculty of Engineering Technology (FTK), Department of Electronic Engineering Technology, Universiti Malaysia Perlis (UniMAP), Unicity Alam Campus, 02100 Padang Besar, Perlis, Malaysia.
[2] Bioelectromagnetics Research Group (BioEM), School of Computer and Communication Engineering, (UniMAP), 02600 Pauh Putra, Perlis, Malaysia.

Email: shakhirulsalleh@unimap.edu.my

Abstract. This paper is briefly describes review of techniques design a reconfigurable frequency on microstrip antenna. There are several methods have been consequential or developed used to provide as well as improve the switching frequencies technique in a single terminal antenna by researchers. This paper investigates different techniques undergoes by some of the research that has been performed to achieved various operating frequencies.

Keywords: reconfigurable frequency, switching technique, microstrip antenna.

1. Introduction

Reconfigurable frequency are gaining popularity among researcher due to constraint design improvements and their need in many wireless communication. Reconfigurable antenna can be defined as the antenna with the capability to reconfigure certain characteristics such as frequency, polarization and pattern. The reconfiguration is not limited to a single characteristic, but it can be a combination of characteristics depending to the application. The reconfigurable antenna idea was first introduced in early 1930 [1]. However the practical demonstration started from 1999. Since then, microstrip antenna has undergone significant improvement and been used as a platform to design reconfigurable antenna [2-6].

In designing reconfigurable antenna, the recent state of RF switches used is FET switches, PIN diodes and MEMS switches. The switch is a device to act as either a circuit maker or circuit breaker in order to configure the antenna parameters. For a reconfigurable antenna, there are several types of suggested RF switch such as PIN diode or micro-electro mechanical system (MEMS) to design. Generally, the RF switch performs as a circuit breaker. The principle of this switch is to regulate the number of the RF current to flow along the desired direction. Hence, to achieve recongfigurability, the position of the RF switch is crucial and has to be optimized in the antenna design. The common approach to realize switchable frequency is using PIN diodes due to the high handling competency and cost efficient if compared to the MEMS [7-8].

The PIN diode switch capable to perform switching in fast times and relatively high current handling, it`s very common in microwave circuit applications capabilities. Due to contact potential and inertial effects, old-fashioned electromechanical RF switching elements is naturally slow response devices [9-12]. The advantage of PIN diode switches are it`s speed of operation and ease of packaging. Typical PIN diode switches has less than 100 ns switching speeds. An important characteristic of PIN diode is it consist of nearly pure resistance at microwave frequencies. For biasing purpose, the resistance could range from a value of 1 Ω to 10k Ω. Normally, 10mA bias current required for ON state operation.

2. Literature Review

The reconfigurable frequency antenna can control the antenna performance in terms of different operating frequencies in a single terminal. Such operation is intended for performance improvement, especially to minimize interference with other wireless system and maximizing throughput [13]. In general, the resonant of antenna is determined by the effective length of the radiator. The antenna includes the type of dipole, monopole, loop, slot and microstrip antennas. The effective length plays a significant role to determine the operating frequency of the antenna hence, by controlling the effective length, frequency can be configured. By several methods of switching, frequency reconfiguration can be achieved. As an example; varying the patch size; reconfigurable matching network; changing the current flow; mechanically configuring using metasurface; and varying the length of the slot. All this switching technique is explained in the following subsection.

2.1. Frequency Switching Method: Varying the Patch Size

A reconfigurable monopolar patch antenna has been reported in [14], which capable to reconfigure up to eight different frequency bands using four PIN diodes. The PIN diodes are used to connect main patch to four different sizes of small patches. The resonant frequency can be tuned from 1.82 GHz to 2.48 GHz by controlling the total patch size. Figure 1 shows the biasing circuit of the proposed antenna [14].

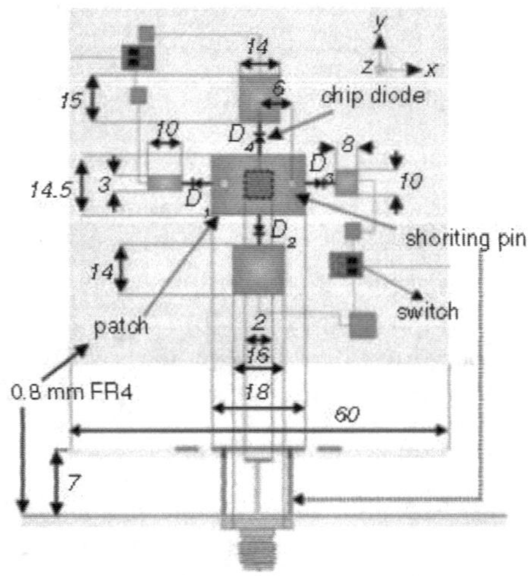

Figure 1: Frequency reconfigurable monopolar patch antenna [14].

Another design that used the similar method has been done by [15]. The proposed tunable antenna made of a slotted rectangular patch loaded by a number of posts close to the patch edge. Three posts are short circuited to the ground plane connected with a set of PIN diode switches. Reconfigurability is obtained by selecting the appropriate post to be shorted thus provides four different frequency reconfiguration. The geometry of the proposed switchable antenna is shown in Figure 2. This technique can easily achieved the desired operating frequency by analysed the total size of patches. However, the drawback of this method is complicated antenna design and biasing circuit.

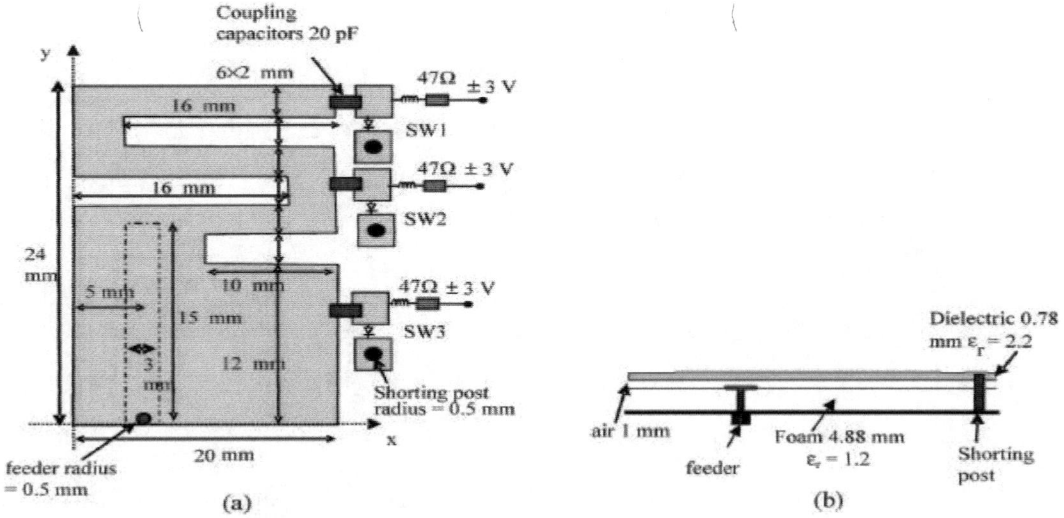

Figure 2: The geometry of the proposed tunable antenna: (a) top view and (b) vertical view [15]

2.2. Frequency Switching Method: Reconfigurable Matching Network (Stub Tunner)

Antenna with the reconfigurable matching network has been presented by [16], where three different frequency bands can be switched using two PIN diodes. The antenna is matched at 5.2 GHz by forward biasing the diode on the left and by forward biasing on the right, the antenna is matched to 6.4 GHz. Meanwhile, when neither of two diodes is biased, the antenna is matched at 5.8 GHz. The described matching network is illustrated in Figure 3. This design offers a simple antenna geometry, but gives limited frequency reconfiguration.

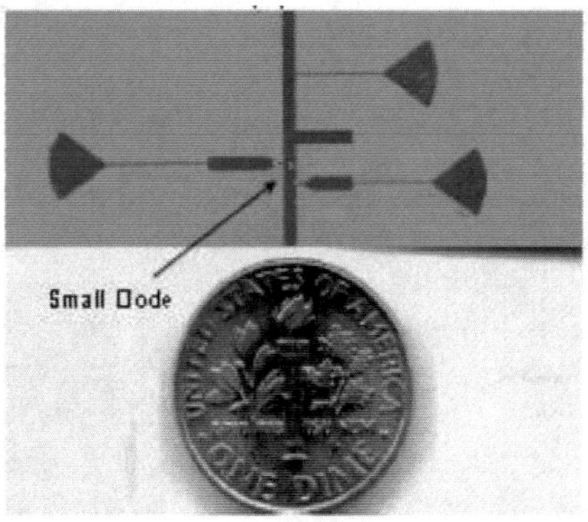

Figure 3: Reconfigurable matching network antenna [16]

2.3. Frequency Switching Method: Changing the Current Flow

A reconfigurable cedar-shaped microstrip antenna for wireless applications are presented by [17], invented the switching frequency method by changing the current flow of the patch. The author used is switched to reconfigure four different frequency bands. Slits are introduced at the edge of the cedar shaped patch. The proper configuration of the switches alters the current flow and changes the operating frequencies. The geometry of proposed antenna shown in Figure 4 (a). The current flow of cedar-shaped antenna at a different state of configuration is illustrated in Figure 4 (b). Meanwhile, the resonance frequency gain results of the antenna are tabulated in Table 1.

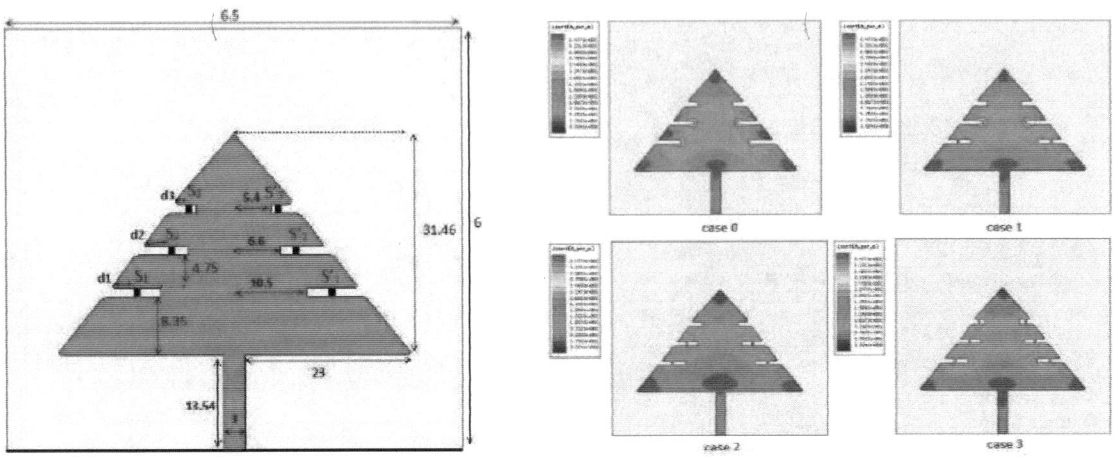

(a) (b)

Figure 4: (a) Geometry of Cedar-shape microstrip antenna, (b) Current distribution for switching case 0, 1, 2 and 3 presented in Table 1 [17].

Table 1: Switch states and the antenna performance results [17]

Case	(S_1, S_1’)	(S_2, S_2’)	(S_3, S_3’)	Resonance Frequency (GHz)	Gain (dB)
0	OFF	OFF	OFF	2.075	0.2798
1	ON	OFF	OFF	2.145	1.0373
2	ON	ON	OFF	2.295	2.1824
3	ON	ON	ON	2.325	2.5098

On the other hand, this paper used the copper strip to represent a switch. Hence, in order to implement PIN diodes into the proposed antenna becomes difficult as the biasing circuit will become more complicated. The biasing circuit requires components such as capacitor, inductor and PIN diode to be properly located so the antenna is not becomes a short circuit. Moreover, the antenna needs to add small slot between switches to provide different state of configuration. Furthermore, the changing structure of the antenna will affect the operating frequency and antenna performances.

2.4. Frequency Switching Method: Mechanically Configure using Metasurface

Another method to configure frequency is by using metasurface instead of the embedded switching circuit. Zhu *et al.* [18] invented frequency-reconfigurable antenna using metasurface. The proposed antenna consist of a simple circular patch antenna, design on Rogers substrates RO4350B with full ground plane. The metasurface of the proposed antenna is designed on a single-side substrate and composing of a number of rectangular-loop unit cells as shown in Figure 5.

The metasurface of the proposed antenna constructed by placing rotating the metasurface around the center and relative to the patch antenna. As the rotation angle of metasurface increase from 10°, 25°, 35°, 55° and 80°, the resonant frequency shifted to 4.77 GHz, 4.9 GHz, 5.07 GHz, 5.31 GHz and 5.5 GHz, respectively. This kind of method categories is mechanically reconfigurable antennas, where the parts of the antenna structures (metasurface) consist of movable parts for turning the frequency. The drawback of such designs was the difficulties of the fabricated process to ensure the moveable part can be still attached to the antenna and the same time can be flexible tuneable.

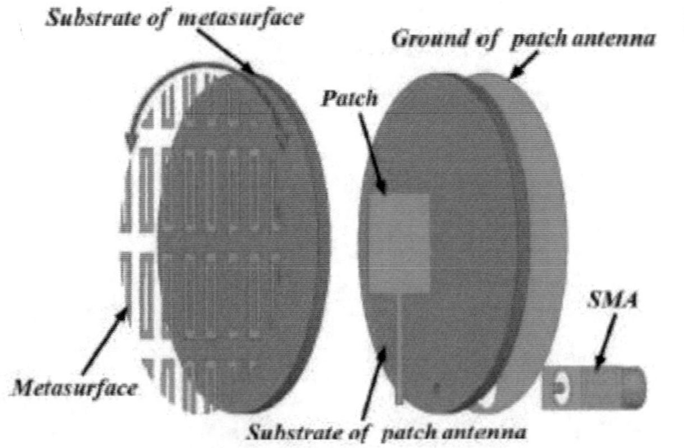

Figure 5: Assembly schematic of proposed antenna [18]

2.5. Frequency Switching Method: Varying the Length of the Slot
Micostrip slot antenna has the potential to be a better candidate for frequency reconfiguration because it offers a wide frequency range turning and conveniences of the resonant frequency turning with switches of varactors across the slot. A PIN diodes are used to vary the slotted size of the antenna [19]. Four PIN diodes used along the slotted of the antenna are capable to configure up to five different band frequencies. Figure 6 shows the switchable frequency structure on the ground plane of the proposed antenna. The different size and structure of the slot was produce when PIN diode is on and off. Hence, the frequency can vary from 2.11 GHz up to 10.92 GHz.

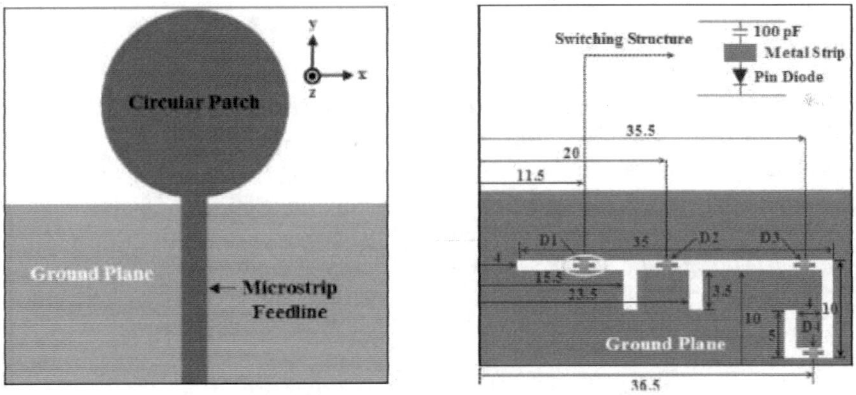

(a) (b)

Figure 6: Configuration of the UWB monopole antenna structure: (a) front view and (b) back view [19]

Similar approach for configurable frequency using the slotted technique on the ground plane has been reported in [20]. Five PIN diodes were placed in the slot to tune the length of the slot, thus provides operating frequency from 1.98 GHz to 3.59 GHz. The antenna is capable to switch up to nine frequency bands by turning on and off five switches at different configuration in order to varying the length of the slot. Six small slots with a width of 0.3 mm are introduced to separate between switches for biasing proposed. Since the small slots separation effected the ground plane efficiency, a surface mount RF capacitor with a value of 100 pF is used to provide RF wave connection throughout the ground plane. Figure 7 illustrates the proposed reconfigurable frequency antenna.

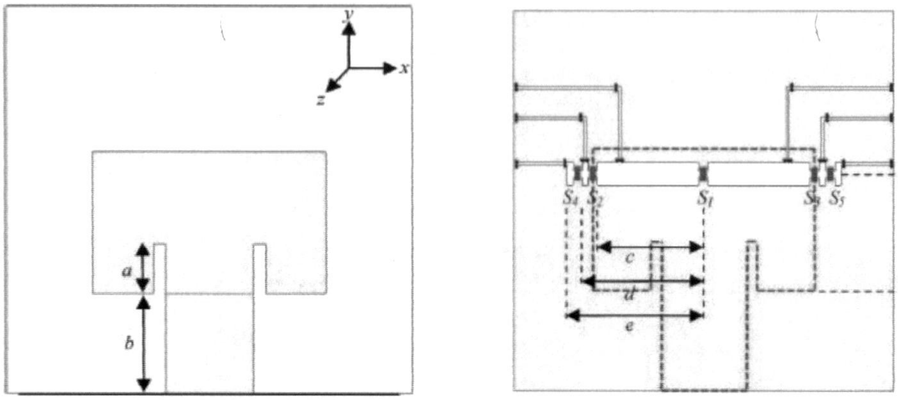

(a) (b)

Figure 7: Frequeny reconfigure microstrip patch slot antenna: (a) front view and (b) back view [20]

3. Conclusions

A several technique to design reconfigurable frequency antenna has been accomplished in this study. The reconfigurable frequency include varying the patch size; reconfigurable matching network; changing the current flow; mechanically configuring using metasurface; and varying the length of the slot is presented. The understanding on the microstrip antenna properties and their frequency response behaviour are necessary in order to improve the antenna performances without affected other parameters.

References

[1] Haupt, R. L., & Lanagan, M. (2013). Reconfigurable antennas. *IEEE Antennas and Propagation Magazine*, *55*(1), 49–61. doi:10.1109/MAP.2013.6474484

[2] Bernhard, J. T., Wang, R., Clark, R., & Mayes, P. (2001). Stacked Reconfigurable Antenna Elements for Space-Based Radar Applications. *Antennas and Propagation Society International Symposium, 2001. IEEE*, *1*, 158–161. doi:10.1109/APS.2001.958816

[3] Feldner, L. M., Nordquist, C. D., Christos, G., & Laboratories, S. N. (2005). RF MEMS Reconfigurable Triangular Patch Antenna. *Antennas and Propagation Society International Symposium, 2005 IEEE*, *2A*, 388–391. doi:10.1109/APS.2005.1551824

[4] Lai, M., Hsieh, T. W. J., & Jeng, C. W. S. (2009). Design of reconfigurable antennas based on an L-shaped slot and PIN diodes for compact wireless devices, (April 2008). doi:10.1049/iet-map

[5] Yang, X., Wang, B., Wu, W., & Introduction, I. (2005). Pattern Reconfigurable Patch Antenna with Two Orthogonal Quasi-Yagi Arrays. In *Antennas and Propagation Society International Symposium, 2005 IEEE* (pp. 617–620). doi:10.1109/APS.2005.1552087

[6] Sor, J., Qian, Y., Chang, M. F., & Itoh, T. (2000). Multi-mode microstrip antennas for reconfigurable aperture. In IEEE Antennas and Propagation Society International Symposium. Transmitting Waves of Progress to the Next Millennium. 2000 Digest. Held in conjunction with: USNC/URSI National Radio Science Meeting (Cat. No.00CH37118) (Vol. 1, pp. 318–320). IEEE. doi:10.1109/APS.2000.873825

[7] M.F.Ismail, M. K. A. R., & H. A. Majid. (2012). Wideband Frequency Reconfiguration using PIN diode. *Microwave Optical Technology Letter, 54*, 1407-1412.

[8] Y. Yu, J. X., H. Li, & S. He. (2011). An electrically small frequency re-configurable antenna with a wide tuning range. *IEEE Antennas Wireless Propagation Letter, 10*, 103–106.

[9] G. Monti, R. De Paolis, & Tarricone, L. (2009). Design of 3-State Reconfigurable CRLH Transmission Line Based on MEMS Switches. *Progress In Electromagnetics Research (PIER), 95*, 283-297

[10] J. C. Maloney, M. P. Kesler, L. M. Lust, L. N. Pringle, T. L. Fountain, P. H. Harms, & G. S. Smith, S. f. a. a., " in Antennas and Propagation Society International Symposium, 2000. IEEE, 2000, pp. 310-313 vol.1. (2000). *Switched fragmented aperture antennas,.* Paper presented at the IEEE Antennas and Propagation Society International Symposium.
[11] Varlamos, P. K., & C. N. Capsalis. (2002). Electronic Beam Steering Using Switched Parasitic Smart Antenna Arrays. *Progress In Electromagnetic Research, 36*, 101-119
[12] W. H. Weedon, W. J. Payne, & G. M. Rebeiz. *MEMS-switched reconfigurable antennas.* Paper presented at the IEEE Antennas and Propagation Society International Symposium.
[13] Christodoulou, C.G.; Tawk, Y.; Lane, S.A.; Erwin, S. R. (2012). Reconfigurable Antennas for Wireless and Space Applications. *Proceedings of the IEEE*, *100*(7), 2250 – 2261. doi:10.1109/JPROC.2012.2188249
[14] Han, T.-Y., & Huang, C.-T. (2010). Reconfigurable monopolar patch antenna. *Electronics Letters*, *46*(3), 199. doi:10.1049/el.2010.3242.
[15] Sheta, A., Mahmoud, S. F., & Member, L. S. (2008). A Widely Tunable Compact Patch Antenna, *7*, 40–42
[16] Nikolaou, S., Kim, B. K. B., & Vryonides, P. (2009). Reconfiguring antenna characteristics using PIN diodes. *Antennas and Propagation, 2009. EuCAP 2009. 3rd European Conference on*, 3748–3752
[17] M. A. Madi, M. Al-Husseini, A. H. Ramadan, K. Y. Kabalan, and A. E.-H. (2012). A Reconfigurable Cedar-Shaped Microstrip Antenna for Wireless Applications. *Progress In Electromagnetics Research C*, *25*, 209–221. doi:10.2528/PIERC11101204
[18] Zhu, H. L., Liu, X. H., Cheung, S. W., Member, S., & Yuk, T. I. (2014). Frequency-Reconfigurable Antenna Using Metasurface. *IEEE Transactions on Antennas and Propagation*, *62*(1), 80–85.
[19] Aboufoul, T., Member, S., Alomainy, A., & Parini, C. (2012). Reconfiguring UWB Monopole Antenna for Cognitive Radio Applications Using GaAs FET Switches. *IEEE Antenna And Wireless Propagation Letters*, *11*, 392–394.
[20] Majid, H. A., Abdul Rahim, M. K., Hamid, M. R., Murad, N. A., & Ismail, M. F. (2013). Frequency-Reconfigurable Microstrip Patch-Slot Antenna. *IEEE Antennas and Wireless Propagation Letters*, *12*, 218–220. doi:10.1109/LAWP.2013.2245293

IOP Conf. Series: Journal of Physics: Conf. Series **1019** (2018) 012043 doi:10.1088/1742-6596/1019/1/012043

Steady State versus Transient Signal for Fault Location in Transmission Lines

M.N.Hashim, M.K.Osman, M.N.Ibrahim, A.F.Abidin

Faculty of Electrical, Universiti Teknologi Mara
13500 Permatang Pauh, Pulau Pinang, Malaysia

E-mail: miminurzilah@gmail.com

Abstract. This paper focuses on determination of suitable time interval of features extraction either during steady state signal or transient signal of impedance based method to estimate the location of fault using Artificial Neural Network (ANN). Simulation studies have been carried on power system model using MATLAB & SIMULINK. The faulted current signal taken at single-ended of transmission lines are used as input in the proposed scheme. Different types of faults at different operating condition have been considered to this study. This studies extracted five features that are delta minimum to maximum, amplitude, standard deviation, energy and mean of the signal for both condition. The proposed studies is evaluated by Mean Square error (MSE) from ANN. Most researchers using steady state signal as their input and there also have researchers used transient signal as the input but there have no study or reason for choosing the signal at that condition. However, by carrying out this studies, it shows that not all features are suitable to extract during one time signal condition.

1. Introduction

Transmission Line need to deliver energy from power supply to consumer but physically it exposed to the environment, therefore it would facing high possibility to fault due to several causes such as tree falls, lightning, storms and insulation breakdown. At the meantime the requirement on providing a continuous power supply without significant cost urging to have modern technology of power system protection and fault location considered as crucial needed. The restoration can be expedited if the location of the fault is either known or can be estimated with reasonable accuracy.

When a fault occurs on a transmission line it causes a sudden change in the current and voltage signals as well as measured impedances at the relay location(1). Basically fault location method are broadly classified as traveling wave technique and impedance based technique as shown in Figure 1. Traveling wave based on the arrival time of fault signal using the reflected wave measurement meanwhile the impedance based technique used the current or voltage signal measurement (2–4).

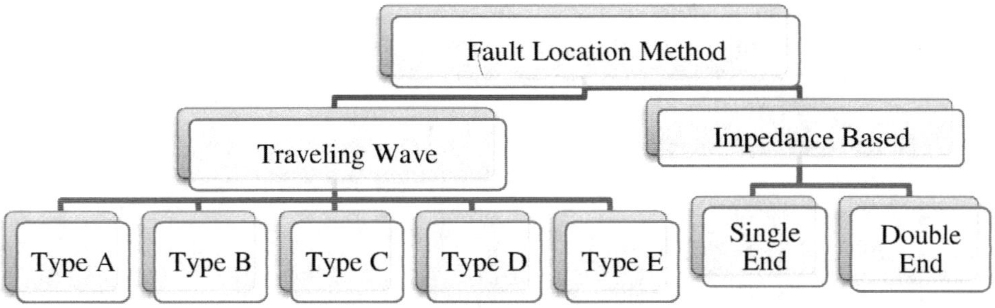

Figure 1 Types of Fault Location Technique

Figure 2 presents the time interval for pre-fault, fault and post fault and one can distinguish. Pre-fault signal is signal recorded before fault occurs or at normal signal without fault. It recorded within the pre-fault interval. The steady state signal called as post-fault signal is a signal recorded during the fault occurs up to detected steady state signal instant or within the fault interval meanwhile transient signal called as fault signal is signal recorded after the fault occurs or within the post-fault interval.(5)

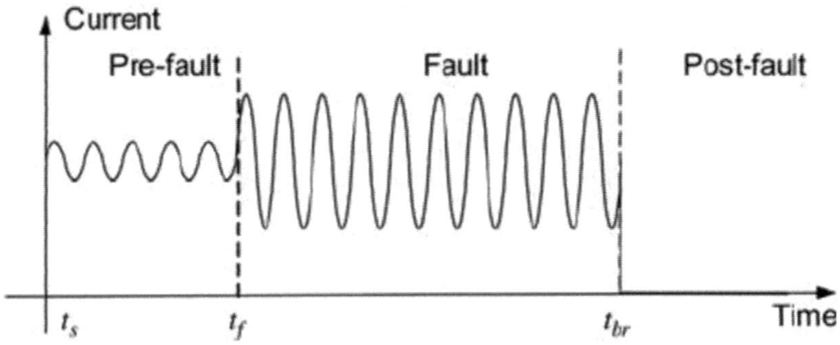

Figure 2 Pre-fault, Fault and Post-fault time interval

Nowadays fault location in transmission line using the intelligence system becoming a trending to many researchers due to capability of the system to estimate fault location with high accuracy. The selection input data is important for ANN to estimate the location correctly. Therefore, the data collection approach would influence the performance of system developed. Iżykowski *et al.* said there is another possibility which concerns post-fault analysis program with fault location algorithm included. It is shows post-fault data have been used for estimate fault location(6). Most researchers have chosen to use post-fault data for their study on fault location in transmission line such as (7–9). Ray and Mishra also uses post-fault of current waveform to estimate fault location and develop a system with error less than 0.29%(10) while Ghimire used post-fault voltage and current from two end method to estimate fault location(11) with less than 3.9% for impedance based method. In contrast, researcher (12–14) studies fault location in transmission line using faulted current signal. There are no significant reason of choosing post-fault signal or fault signal. Moreover, there have no study on the suitable selection time interval to collect the data for fault location.

This paper aims to determine the appropriate time to extract features of impedance based technique for fault location in transmission lines using ANN. Thus, comparing the result to have the suitable features for fault location in transmission line. This paper are limited to 10 types of faults using single-ended of impedance based method. The 900 experimental data generated using MATLAB SIMULINK for data training, validation and testing as input for ANN. All the data have been developed under MATLAB environment. The length of transmission line is 300km with voltage of 500KV, 50Hz.

IOP Conf. Series: Journal of Physics: Conf. Series **1019** (2018) 012043 doi:10.1088/1742-6596/1019/1/012043

2. Methodology

Figure 3 describe the chronology of the work carried out for this study. After simulation of transmission lines, two types of data collection method will be explored which are during fault condition and post-fault condition. Then the data will be normalized. Next features will be extract as input into ANN to perform fault location and lastly evaluate the performance of the system.

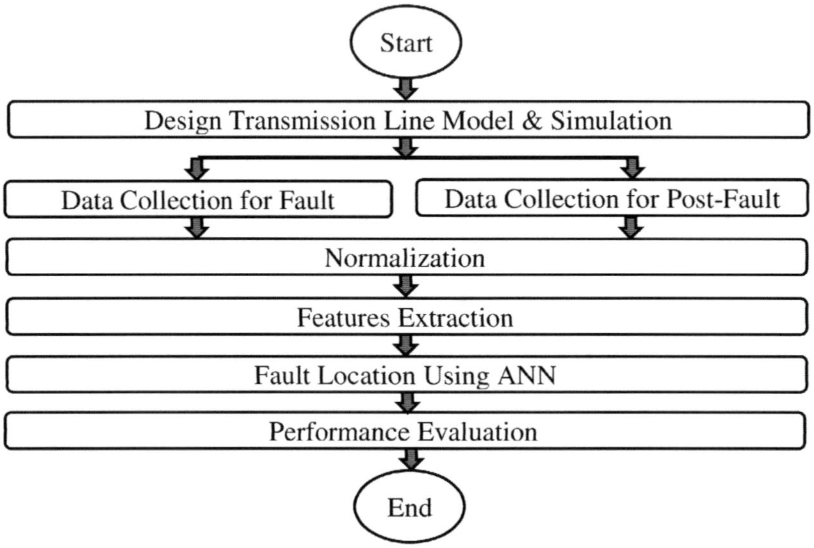

Figure 3 Flow Chart of the Study.

2.1. Transmission Line Model

This study generate the data by using type single-ended of impedance based technique to perform the simulation of fault in transmission line. The transmission line model considered is single line diagram with total length of 300km as shown in Figure 4. The model was developed using MATLAB/SIMULINK. . The study is tested with varying fault location, fault types, fault resistance and inception angle as mentioned in Table 1.

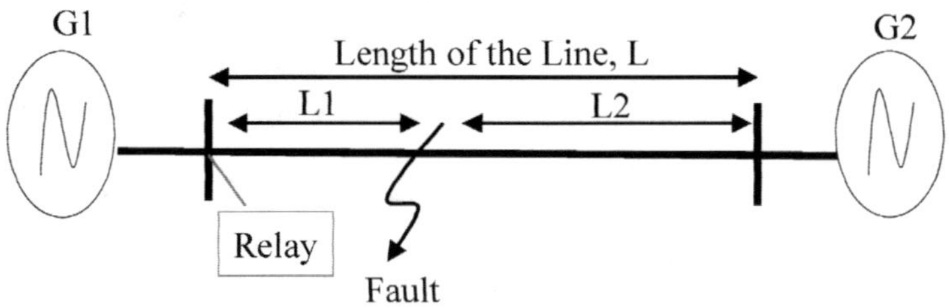

Figure 4 Single Line Diagram for Single Ended Transmission Line Model

The fault is created at 0.023s with sampling frequency of 1.25 MHz and the parameter setting for:

$$R1 = 0.01273\ \Omega/km;\ R0 = 0.3864\ \Omega/km;$$
$$L1 = 0.9337e\text{-}3\ H/km;\ L0 = 4.1264e\text{-}3\ H/km;$$
$$C1 = 12.74e\text{-}9\ F/km;\ C0 = 7.751e\text{-}9\ F/km$$

Table 1 Varies Parameters Value

Parameters	Value
Fault Location, L (km)	0-500
Fault Resistance, Rf (Ω)	0.01, 1, 10, 50, 100, 200
Fault inception angle, Ft(degree)	0, 90

2.2. Data Collection Method

This paper conducted 10 types of fault that are three types of single line to ground fault (AG, BG, CG), three types of double line fault (AB, AC, BC), three types of double line to ground fault (ABG, ACG, BCG) and lastly a three phase fault (ABCG). This fault location scheme using one faulted current signal as data collection reference. The current signal was chosen because current is sensitive to any changes especially during fault. The data was collected for one cycle of fault interval and post-fault interval for next process.

Figure 5 presents the example waveform of three-phase current signal recorded using scope in SIMULINK for fault occurs at 187km. Fault incipience is the starting time to collect the data for fault interval while fault clearance is the starting time to collect the data for post-fault.

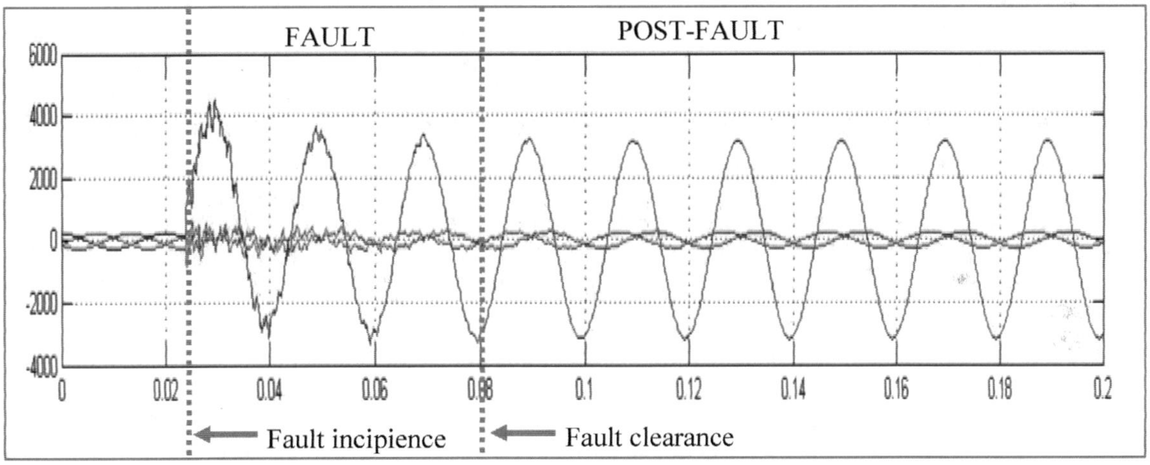

Figure 5 Specification of time intervals according to position for 187km fault

2.3. Features Extraction

This study were extracted five features from both interval condition that are delta minimum to maximum, amplitude, standard deviation, energy and mean of the current signal. The features extracted use as input into ANN to estimate the location of fault. Features is an important element to ANN determine the output of the system. The suitability of the features are considered to have better fault location system.

Delta minmax is peak to peak amplitude of waveform while the amplitude is the maximum point of the waveform from origin. The features extracted using formulae of mean by equation (1)

$$\bar{x} = \frac{1}{n}\sum_{i=1}^{n} x_i \quad (1)$$

Formulae of standard deviation by equation (2)

$$\sigma = \sqrt{\frac{1}{n}\sum_{i=1}^{n}(x_i - \bar{x})^2} \tag{2}$$

Formulae of energy by equation (3)

$$E = \frac{1}{n}\sum_{i=1}^{n}|x_i|^2 \tag{3}$$

2.4. *Fault Location using ANN*

There have several process for ANN to determine the fault location as shown in Figure 6. Since the values of data was too high, normalization process is needed to avoid ANN taking too much time to calculate and estimate the output. Next several features need to be extracted as input into ANN. Lastly, ANN will estimate the location based on data given.

Figure 6 Main Process toward ANN

Basic algorithms used in fault locators are intended to make distance to fault calculation as accurate as possible. A structure of ANN called MLP is developed that consists of three layers which are input layer (Features extracted from faulted current), hidden layer and output layer (fault location). The NN are perform independently based on types of fault as in Figure 7. The MLP network was selected because the model display an efficient learning environment where each layer is connected with weight and bias for minimizing error between target and obtained value. The process of training, validation and testing are performed using neural network toolboxes in MATLAB with suitable number of hidden layer. Lavenberg-Marquardt backpropagation is used as training algorithm. The transfer function of the hidden layer and output layer are respectively sigmoid and linear function. The suitable number of hidden layer used is within 1 to 20 hidden layers.

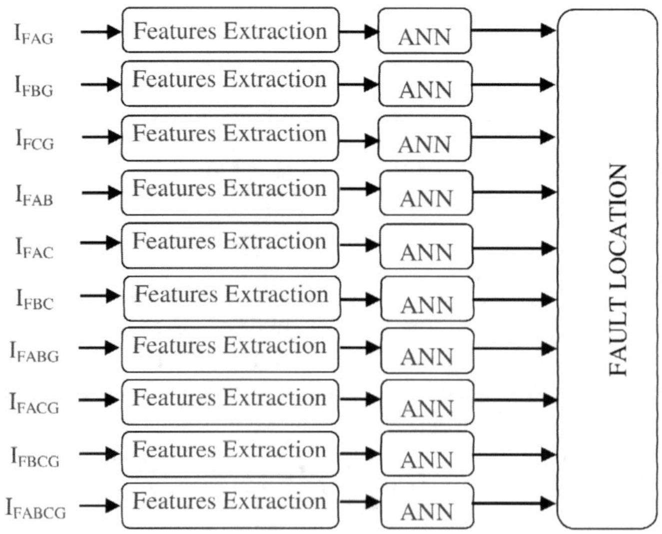

Figure 7 ANN structure

2.5. Performance Evaluation
The performance of system are evaluated by calculating performance indicator. Performance indicator chosen is MSE of ANN for testing data. The Mean Square Error, MSE is an indicator for the quality of the estimated data from targeted output. Best results of MSE is 0. The formulae is on equation (4).

$$MSE = \frac{1}{n}\sum_{i=1}^{n}(Output\ ANN_i - Output\ Target_i)^2 \qquad (4)$$

Where n represent the number of data

3. Results and Discussion

3.1. Transmission Line Model
Figure 8 shows the design of single-ended of single line diagram using SIMULINK MATLAB. All the setting parameters followed the setting describe in methodology.

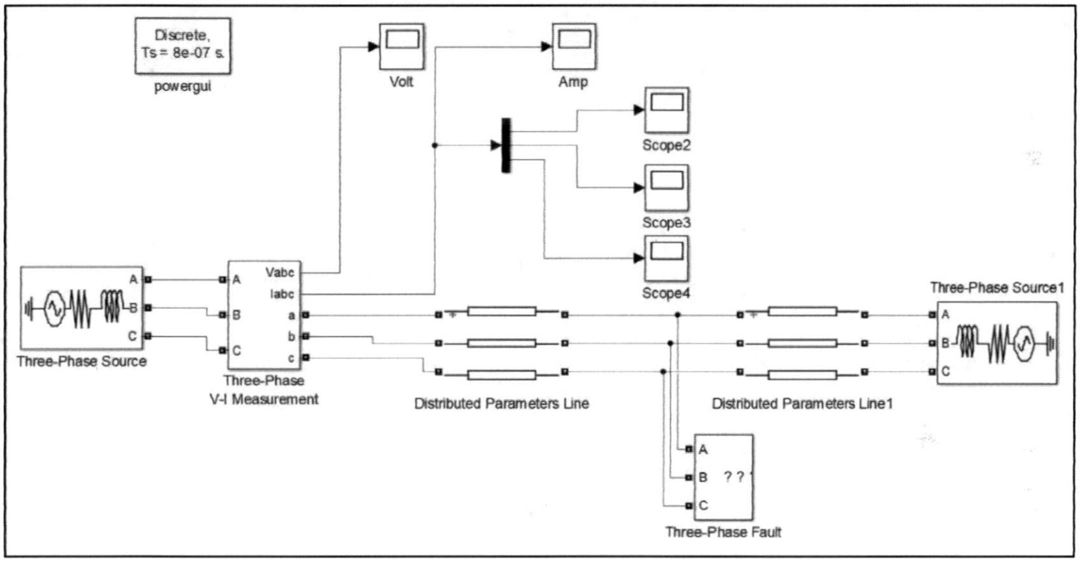

Figure 8 Transmission Line Model using MATLAB SIMULINK

3.2. Features Extraction
Figure 9 shows the example of training distribution data from features extraction of AG fault for fault condition and post-fault condition. Each figure illustrated the graph of normalized current, I_{norm} against fault location. The features extracted into five features that are delta minmax, amplitude, mean, standard deviation and energy from both condition. The features were used as input for ANN to estimate the location. As can be seen, when the fault location is increase, the I_{norm} will decrease exponentially. From the figure, the pattern of data for fault condition and post-fault condition is almost the same except for mean features.

IOP Conf. Series: Journal of Physics: Conf. Series **1019** (2018) 012043 doi:10.1088/1742-6596/1019/1/012043

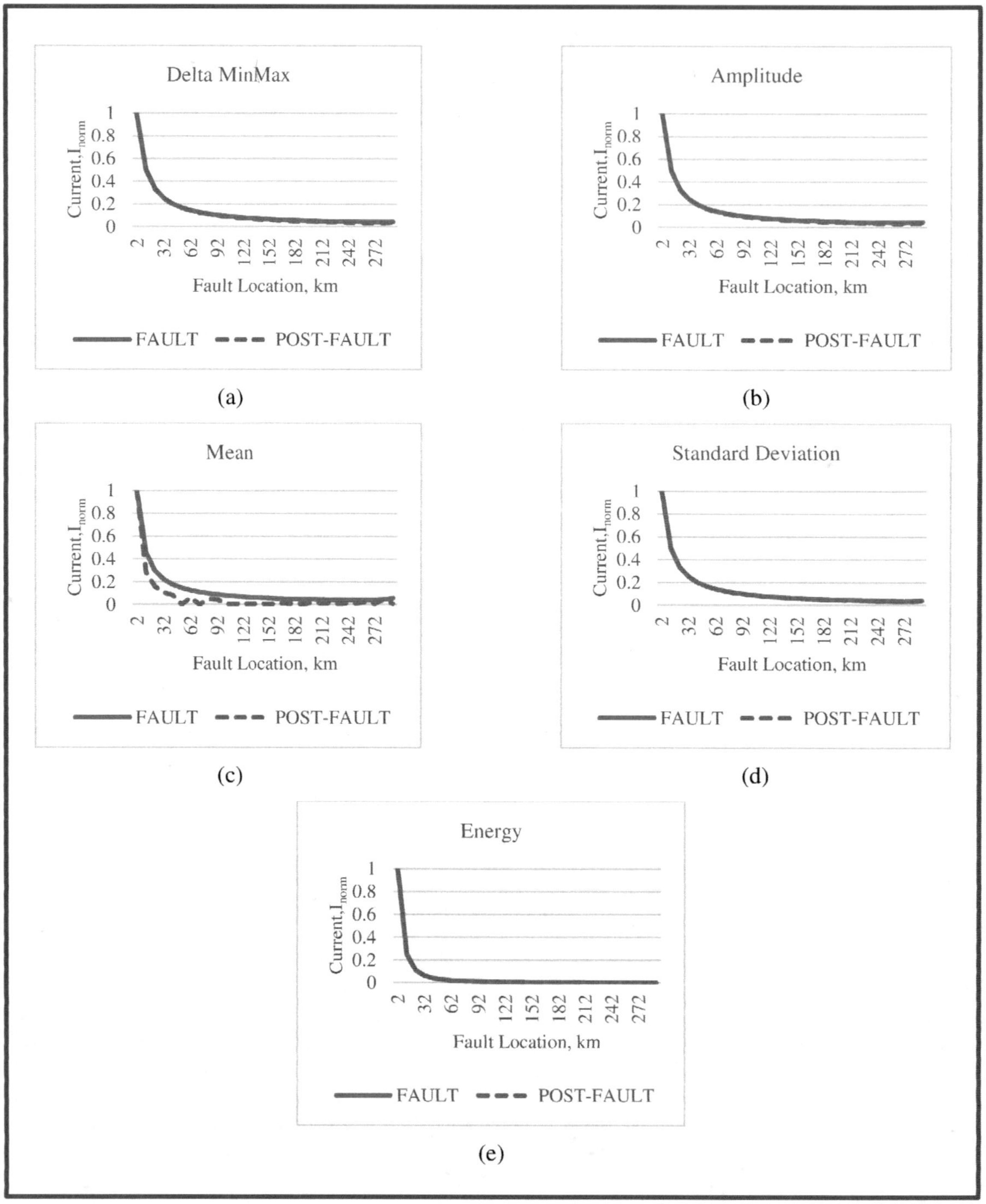

Figure 9 Data Training of AG Fault

(a)Delta MinMax (b) Amplitude (c) Mean (d) Standard Deviation (e) Energy

3.3. Performance of Features Extracted

All the five features performance at the fault condition and post-fault condition was determined based on MSE results and the results are presented in Table 2.

Table 2 MSE Results for all features

Types of Fault	Delta MinMax		Amplitude		Mean		Standard Deviation		Energy	
	Fault	Post-Fault	Fault	Post-Fault	Fault	Post-Fault	Fault	Post-Fault	Fault	Post-Fault
AG	27.79	0.56	56.14	4.41	1450.30	3322.78	0.3	0.37	0.16	0.4
BG	10.09	2.61	22.35	17.11	538.66	2193.29	0.24	4.25	1.57	0.92
CG	6.27	2.43	36.42	1.81	943.78	4658.53	0.28	1.63	1.14	0.97
AB	2.95	0.55	4.29	2.41	2576.78	692.99	0.7	0.84	0.65	4.43
AC	1.57	0.35	2.20	3.91	0.031	1127.81	1.01	0.68	0.68	4.47
BC	18.58	6.22	1.71	5 1.15	0.10	2150.63	0.74	0.51	0.43	36.48
ABG	5.36	2.55	33.10	3.28	1944.51	1684.38	0.98	1.01	4.35	1.98
ACG	15.03	1.20	2.95	10.95	0.639	555.94	0.98	0.89	0.72	2.24
BCG	15.13	8.43	6.73	6.24	0.864	2464.03	0.25	0.77	0.6	31.09
ABCG	9.41	3.42	2.38	1.06	0.739	602.63	0.79	0.87	0.61	9.91
Average	27.79	0.56	16.83	5.12	745.640	1945.30	0.3	0.37	1.09	9.29

In view of the results obtained, the MSE value for post-fault condition are lower for delta minmax and amplitude but higher for mean, standard deviation and energy based on the average results. Delta minmax and amplitude were reduced about 27.23 and 11.71 each for MSE value by choosing post-fault condition compared to fault condition. In contrast, MSE value for mean, standard deviation and energy were reduced by 1200, 0.07 and 8.2 each by choosing fault condition compared to post-fault condition. Among all the features, mean features was found to be highest of MSE and have significant reduction for fault condition and post-fault condition. It can be seen contribution of features extraction data has the greatest influence to the results of MSE.

The table was interpreted in figure as in Figure 10. The figure shows the MSE results from testing data ANN against the types of faults on both cases for all the features that are (a) delta minmax, (b) amplitude, (c) mean, (d) standard deviation of the signal and, (e) energy. From the figure, it can be seen that each types of fault have different criteria for MSE results for each features. The ANN shows more difficult to estimate fault location for single phase to ground fault since MSE values is higher during this types of fault except for energy features.

IOP Conf. Series: Journal of Physics: Conf. Series **1019** (2018) 012043 doi:10.1088/1742-6596/1019/1/012043

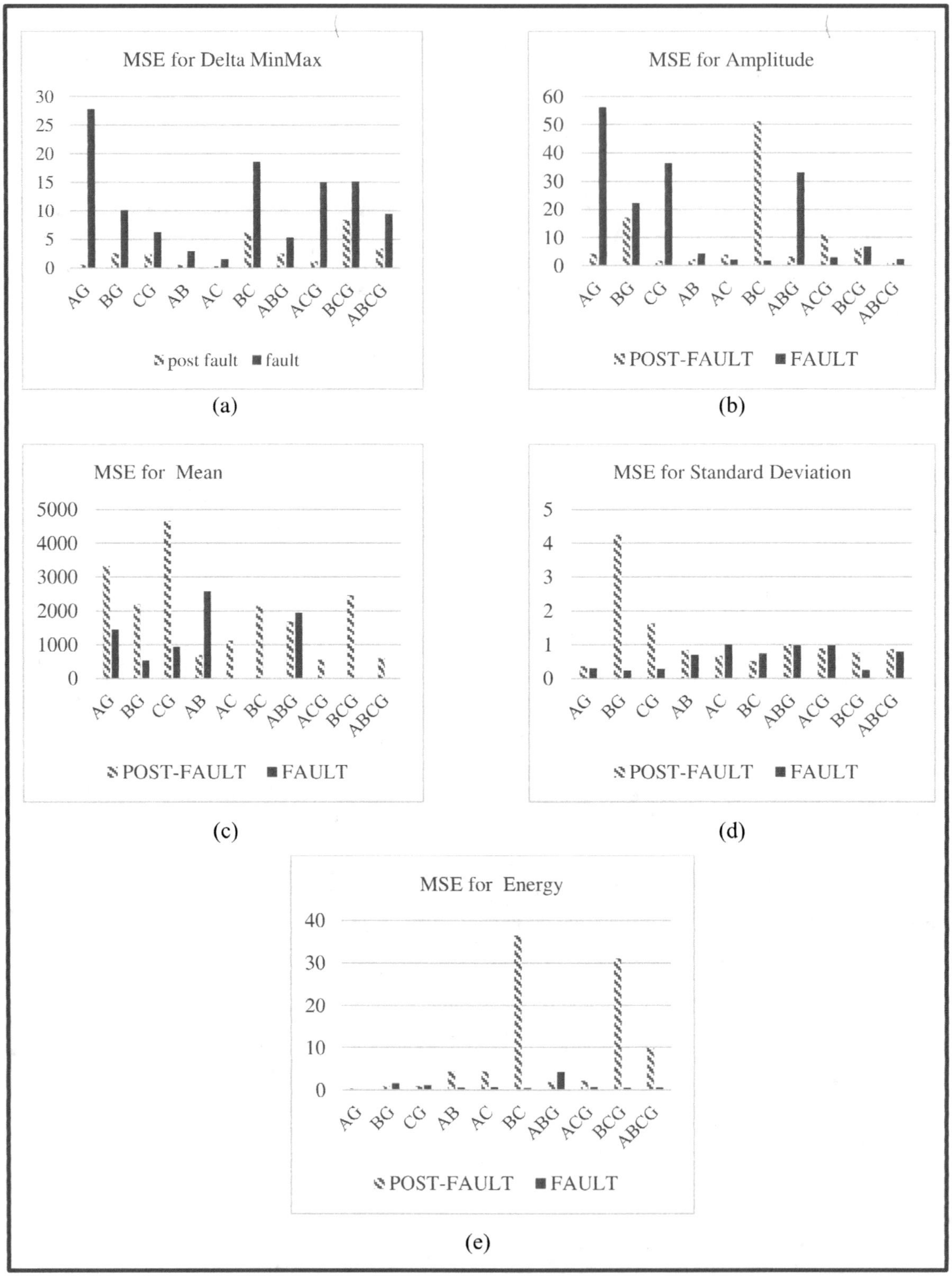

Figure 10 MSE results for all features

(a) Delta minmax (b) Amplitude (c) Mean (d)Standard Deviation (e) Energy

4. Conclusion

In this paper, fault location is calculated using single-ended impedance based method. From the results, certain features are suitable collected during fault condition such as standard deviation, energy and mean of signal meanwhile the rest is suitable collected during post-fault condition such as delta minimum to maximum and amplitude. The performance evaluation is based on the MSE results of each features. Therefore, the suitable features for fault location by rank is standard deviation of fault condition (MSE=0.299), delta minimum to maximum of post-fault condition (MSE=0.560), energy of fault condition (MSE=1.091), amplitude of post-fault condition (MSE=5.118) and lastly mean of fault condition (MSE=745.640).

Acknowledgments

The authors would like to thank to all parties that support and involve directly or indirectly into this research especially UiTM Penang. This research is supported by the Fundamental Research Grant Scheme (FRGS), Grant No: FRGS/1/2015/TK04/UITM/03/1 from the Ministry of Higher Education, Malaysia.

References

1. Rao Jonnalagadda A, Hagos G. Review of Performance of Impedance Based and Travelling Wave Based Fault Location Algorithms in Double Circuit Transmission Lines. J Electr Electron Eng. 2015;3(4):65.
2. Krzysztof G, Kowalik R, Rasolomampionona D. Travelling wave fault location in hv lines. Acta Energ. 2011;3(8):5–12.
3. Schweitzer EO, Guzman A, Mynam M V., Skendzic V, Kasztenny B, Marx S. Protective Relays with Traveling Wave Technology Revolutionize Fault Locating. IEEE Power Energy Mag. 2016;14(2):114–20.
4. Zhu Y, Fan X. Fault location scheme for a multi-terminal transmission line based on current traveling waves. Int J Electr Power Energy Syst. 2013;53:367–74.
5. M.M Saha, J. Izykowski ER. Fault Location on Power Networks. Springer. 2010. 425 p.
6. Iżykowski J, Wydawnicza O, Wrocławskiej P. Fault location on power transmission lines. 2008. 221 p.
7. Hussain S, Osman A. Fault location on series-compensated lines using unsynchronized measurements. IEEE Power Energy Soc Gen Meet [Internet]. 2013;78:277–84.
8. Tahar B. Fault Location in EHV Transmission Lines using Artificial Neural Networks. Int J Appl Math Comput Sci. 2004;14(1):69–78.
9. Hessine M Ben, Jouini H, Chebbi S. Neural Network Approach to Fault Location for High Speed Protective Relaying of Transmission Lines. In: Computational Intelligence Applications in Modeling and Control. 2015. p. 315–42.
10. Ray P, Mishra DP. Support vector machine based fault classification and location of a long transmission line. Eng Sci Technol an Int J. 2016;19(3):1368–80.
11. Sushma Ghimire. Analysis of Fault Location Methods on Transmission Lines. MSc Thesis, Dept. Elect. Eng. ,University Of New Orelans; 2014.
12. Patel M, Patel RN. Fault analysis in transmission lines using neural network and wavelets. In: 2015 2nd international Conference on Signal Processing and Integrated Networks (SPIN). 2015. p. 719–24.
13. Jamil M, Kalam A, Ansari AQ, Rizwan M. Wavelet-FFNN Based Fault Location Estimation of a Transmission Line. Electr Eng Res. 2013;1(July):77–82.
14. Jamil M, Kalam A, Ansari AQ, Rizwan M. Generalized neural network and wavelet transform based approach for fault location estimation of a transmission line. Appl Soft Comput J. 2014;19:322–32.

Content-Based Image Retrieval in Medical Domain: A Review

Nor Asma Mohd Zin[1], Rozianiwati Yusof[1], Saima Anwar Lashari[2], Aida Mustapha[2], Norhalina Senan[2], Rosziati Ibrahim[2]

[1]Faculty of Computer Science and Mathematics,
Universiti Teknologi MARA, Bukit Ilmu, 18500 Machang, Kelantan, Malaysia
{norasma, rozian696}@kelantan.uitm.edu.my
[2] Faculty of Computer Science and Information Technology,
Universiti Tun Hussein Onn Malaysia, Parit Raja, 86400 Batu Pahat, Johor, Malaysia
{saima, aidam, halina, rosziati}@uthm.edu.my

Abstract. Content-based Image Retrieval (CBIR) aids radiologist to identify similar medical images in recalling previous cases during diagnosis. Although several algorithms have been introduced to extract the content of the medical images, the process is still a challenge due to the nature of the feature itself where most of them are extracted in low level form. In addition to the dimensionality reduction problem caused by the low-level features, current features are also insufficient to convey the semantic meaning of the images. This paper reviews the recent works in CBIR that attempts to reduce the semantic gap in extracting the features from medical images, precisely for mammogram images. Approaches such as the use of relevance feedback, ontology as well as machine learning algorithms are summarized and discussed.

1. Introduction

In recent times, the usage of digital images is becoming more popular across different sectors including medical, scientific experiments, educational and so forth. Hospitals and medical institutions are generating a great number of digital images such as x-ray, mammogram and magnetic resonance imaging (MRI) as part of their daily routine. Interpreting the medical images is certainly a complex task which requires extensive knowledge. In order to assist radiologists in interpreting the medical images, researchers have developed support systems such as Computer Aided Diagnosis (CAD) system and Content Based Image Retrieval (CBIR) system. CAD would help radiologists in diagnosis and also serve as a second opinion [1]. On the other hand, Content-based Image Retrieval (CBIR) uses visual content to help users browse, search and retrieve similar medical images from a database based on the user's interest [2-3].

CBIR systems such as QBIC, PhotoBook or VisualSeek which help retrieve similar images would assist radiologists in analysing previous similar cases [4]. Researches in CBIR were initiated back in early 1990s to support the emergence of large image collection. The research area becomes more active with the advancement of the availability of multimedia technologies in the medical industry as well as the advancement in image processing and medical informatics [5]. Researchers study the retrieval of images in two parts: text- and visual-based. To retrieve image that is text-based, keywords are used to annotate the images. This manual annotation process is done by human and it has the potential of offering very precise information when images are well-named or annotated [1]. However, this process becomes much more difficult to execute as the size of image collection increases. Thus, it becomes labor-intensive and time consuming. In addition, some images cannot be annotated because they contained too much information. The other difficulty is because of the subjectivity of human perception during annotation [6]. In overcoming the problems of the retrieval of text-based image, researchers have shifted the focus to visual-based image retrieval.

Two main components in visual-based are feature extraction and feature reduction with application of certain similarity measurement. In CBIR systems, image content includes the colour, texture, and shape. They are extracted for subsequent uses during indexing and retrieval process. These features are considered low level and are usually presented in numerical forms and assigned in feature vectors. Nonetheless, the majority of users are always more interested in specific regions rather than the entire image. Thus, most of the current CBIR systems are region-based where the features are extracted only from the Regions of Interest (ROI). Representation of the images at regional level is identified to be closer to human perception system [7-8].

In the past, the extraction of these low- level features were done by different algorithms. It is then followed by features reduction process to form a good feature representation from larger collection of image features. However, these algorithms are not able to adequately represent the semantic information or high-level features (concept) including keywords and text descriptions that are commonly used by human during interpretation. The algorithms are somehow limited when dealing with larger numbers of rich content images in the database. Researchers, in overcoming these issues, have provided resolutions to reduce the semantic gap in retrieving the images via (1) ontology high-level concepts are defined by ontology, (2) machine learning tools associating low-level features with query concepts, with machine-learning tools and (3) relevance feedback (RF) to promote continuous learning of user interest.

Besides semantic annotation, another key challenge in CBIR research is the similarity measurement. By calculating the distance between their corresponding feature descriptors (feature vectors) in the featured space, the similarity between any two images can be determined, However, there are different ways to measure image features similarity, For example, using the distance metrics (i.e. Euclidean distance) where the features are represented as a vector, or via graph matching where the objects in the images are arranged in the form of relationships among them. In addition, statistical classifiers can also be trained to categories the query image into respective known classes. This approach would overcome the semantic gap by training the similarity of measurement on known and labeled data [9-10].

Section 2 of this paper presents a summary of research in CBIR which includes the process of image segmentation, low-level image extraction and feature selection. This section would also detail out the similarity measurement as well as the performance measurement. Section 3 reviews the related works of CBIR in medical domain and Section 4 provides the discussion on reducing the semantic gap. Finally, Section 5 ends the paper by providing future directions in CBIR research.

2. Content-Based Image Retrieval

Most of the Content-based Image Retrieval (CBIR) systems are based on image similarity, whereby a user inputs a query image and the system responds by providing the most similar image based on a certain similarity measurement. The results of the similar images are then displayed in descending rank. The fundamental bases of CBIR system are feature extraction and feature selection, classification and similarity measure as illustrated in Figure 1.

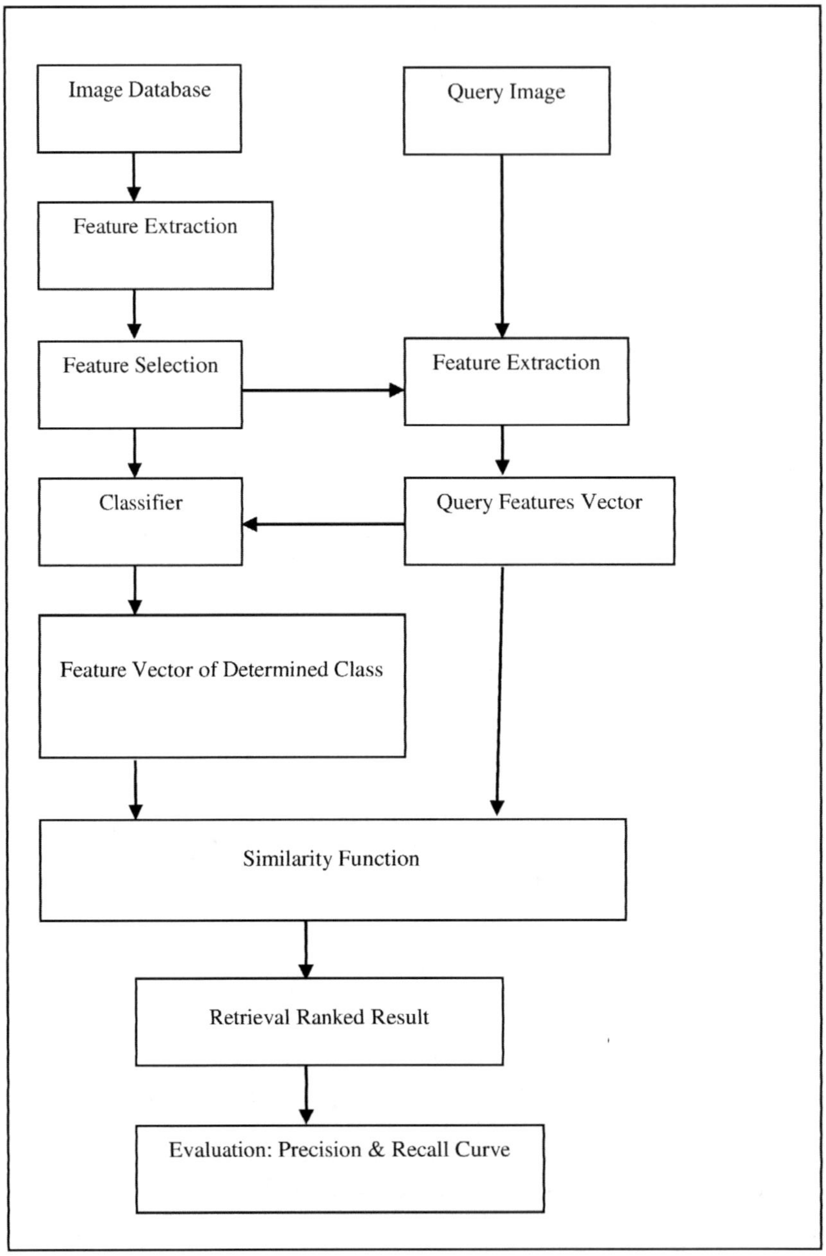

Figure 1: Framework for a CBIR system.

Features such as colour, texture and shape are often used to describe the content of the image similarity. Many algorithms have been proposed to extract image features. Image features can be extracted from the entire image or from certain regions. Thus, image segmentation process is necessary prior to feature extraction for Region-based Image Retrieval (RBIR) [11].

2.1. Image Segmentation

The selection of the types of features, global (involving the colour and texture of the histograms and the entire image layout) and local (the colour, texture and shapes of sub-images, segmented regions and interest points) features, significantly affect CBIR research. The explanations of these features mentioned are as follows:

Colour: Colour features are designed based on colour space. Colour representations that are commonly used in RBIR are RGB, LAB and LUV, HSV (HSL), YCrCb, and Hue-Min-Max-Difference (HMMD). All of these are designed to be closer to human perception. They are represented using colour-covariance matrix, colour histogram, colour correlogram, dominant colour descriptor, colour moments, colour coherent vector or colour co-occurrence matrix. Colour features are reported to be very stable and robust due to their resilience towards rotation, translation and changes during scaling [12]. However, most medical images are in grayscale except for colour photographs that are used for diagnosis in ophthalmology, pathology and dermatology [13][14]. Bunte et al. [3] used elective colour features to retrieve similar skin lesions when comparing to an actual case to verify the diagnosis or confer with similar symptoms. The colour feature representations used are limited to rank matrix learning vector quantisation (LiRaMLVQ) and a Large Margin Nearest Neighbor (LMNN) approach. They performed a comparison on the retrieval rates achieved with extracted and original features for eight different standard colour spaces, and the research has achieved a significant result for each colour space.

Shape: Shape features, in comparison to colour and texture features, have not been used commonly. Past research have shown that it is challenging to apply shape features as compared to textual and colour features as they may be inaccurate during the segmentation process [15]. Shape features are useful in cases where images comprise objects that can evidently be separated from the background. Shape features include aspect ratio, circularity, Fourier descriptor, moment invariants and Zernike moments. In retrieving medical images, shape features are the best descriptors in detecting disease, lesion or mass. Wei et al. [16] applied features of shape and margin using Zernike moments to identify mass lesions in the mammogram. The result showed that the retrieval system was capable of achieving the highest precision among all mass lesion types when retrieving round-shaped and circumscribed margin masses,

Texture: Texture refers to the visual patterns that contain homogeneity properties and is not a result of the presence of only a single colour or intensity. Texture features are represented by using the gray-level co-occurrence matrix, Tamura feature, wavelet coefficients, gabor filter and Haralick's feature. For Tamura feature, six visual texture properties which include coarseness, contrast, directionality, line likeness, regularity, and roughness are used. In Haralick's approach, the texture features are represented using the co-occurrence matrix based on the orientation and distance between image pixels. Meaningful statistics are then extracted from the matrix as the texture representation [19]. In medical domain, researchers usually use texture features because organs and tissue anomalies are well characterised by texture properties [13][16]. Tissues are expected to have consistent and homogeneous texture. Thus, it is the best option to discriminate among organ tissues in CBIR system, for intances,texture features have been used for brain tumor diagnosis [17] and mass lesion detection in mammogram images [18].

2.2. Feature Selection

For each image in CBIR systems, a feature vector is processed and stored in a visual feature database. When a query is made by a user, the feature vector for this query is first computed. By using a similarity criterion, all the vectors in the visual feature database are compared to this query vector. The images that are most similar to the query image are returned to the user. It is obvious that the CBIR systems strongly depend on the characteristics extracted from the images. Features extracted from images are represented in the form of feature vector. Then, using these features, CBIR system will index and find the most similar and relevant images.

Intuitively, a single feature is not able to describe the image visual content very well. Therefore, researchers have proposed a combination of several features so that the content of the images can be well-characterised. However, a substantial amount of features could affect the efficiency of the retrieval process. Large number of features normally would result in feature redundancy and irrelevant features. This would then lead to high dimensionality and eventually affect the efficiency of the retrieval process. There would be a considerable loss of the discriminative power of each feature when the number of features is added. This calls for the need for an elective feature selection algorithms to reduce the feature dimensionality. One of the key challenges in optimising the CBIR systems is feature selection. Feature selection can be defined as the process of selecting the combination of features among a given larger set that describes a particular data collection best. Feature selection has been applied as pre-processing technique in data mining, pattern recognition, classification of medical images, genomic data analysis, speech recognition and others. The function of feature selection is to decrease the features in order to generate precise results. It is therefore critical to choose the optimal features [19][20].

Three categories of feature selection methods are filter-, wrapper- and hybrid-based. Filter-based methods utilise general characteristics of the data independently from the classifier for the evaluation process. The evaluation process is classifier-dependent in wrapper methods. Finally, hybrid models use both filtering and wrapping methods to improve the performance of the selection process. Filter methods also do not include any induction algorithms. Thus, they are not computationally expensive. There is possibility that subsets of features that are not compatible with the algorithm in the users' application are selected [21].

In learning to rank approach, feature selection is used, whereby only parts of raw features are chosen. Importance and similarity are two properties of each feature which are defined. In the computation of feature importance, each dimension of the feature is used to rank the database images. The ranking performance such as mean average precision (mAP) or normalised discounted cumulative gain (NDCG) is taken as its importance. To calculate feature similarity between two features, the first step is to rank the database images individually. Subsequently, the correlation of these two-ranking list would be taken as the feature similarity. Based on these two properties, the loss function of feature selection is defined as choosing the most important (high importance) and representative (low similarity) features [22].

De Oliveira *et al.,* [23] implemented a two-dimensional principal component analysis (PCA) to overcome the existing PCA as it is claimed to be simpler and more straightforward to use for the characterisation of breast density texture. The retrieval process was carried out using support vector machine (SVM) which solved varied learning, classification, and prediction problems. On the other hand, Marinakis *et al.* [24] used a set of evaluation functions for CBIR, which is also known as the Fitness coach (Fc). The researchers, focusing on improving the quality of each query answer by retrieving a substantial amount of pertinent images during ranking, inserted feature selection function in the GA-based algorithm.

2.3. Similarity Measures

Similarity measures are used to identify the similarity between query image and database images. Thus, a good similarity measure would need to be employed in order to retrieve the most similar and relevant images. Many similarity measures have been introduced by researchers to improve the effectiveness and efficiency in CBIR. The measurement required for the similarity comparison is determined by the selected feature vectors. The extraction of the multi-dimensional points' features from the images allows the calculation of the distances between the multi-dimensional points. Popular metrics that are utilised to gauge the distance between two points in multi-dimensional space are Euclidean distance, Weighted Euclidean distance, Manhattan distance, Cross Correlation distance, Minimum Mean distance rule and Statistical distance [2].
Another similarity measure is performed using graph-based approach. Graph representations have received more attention recently because it is an elective tool to represent relational information. Sharma *et al.* [25] proposed a new method to determine the similarity between histological images through graph-theoretic description and matching in retrieving histological images from larger databases. In contrast to the above methods that directly measure the similarity in terms of image

information alone, Classifier-based similarity measures do not directly measure similarity in terms of image information as compared to the method mentioned earlier. It uses the classification of a query image based on a fixed set of predetermined labels to assess similarity [26].

3. Previous Works in CBIR

This section examines current practices and prospects of CBIR systems. The emphasis is placed on the summarization of major advanced CBIR approaches used for improving information retrieval. A considerable amount of literature has been published on CBIR systems, while looking into large and growing body of literature; it appears that CBIR systems techniques have been proven to be successful for information retrieval tasks. Town [27] described a system called the "Immense Picture Search" that is based on automated analysis and recognition of image content. This system provided two searching modalities which are (1) exploratory search where users could use keyword or a more complex theory and (2) similarity-based refinement where user could find the similar images based on the returned search result. The proposed system is highly suitable for medical area due to the diverse corpus of medical images. The Gray Level Neighbors Matrix (GLNM) was used to extract texture features for the retrieval of mammograms from the MIAS database. The performance of GLNM was compared to the Gabor, CDF 9/7 and Db4 wavelets, GLAM and GLCM based texture extraction methods. The Precision Recall (PR) graphs showed that the retrieval performance of the proposed method is much better than the competing methods. As for the future works, the proposed GLCM approach may be applied for texture feature extraction problems involving other databases, for example diabetic retinopathy and bloats.

Popova & Neshov [28] attempted to improve the retrieval effectiveness and accuracy of image search in huge databases. Their works recommended two sets of different feature combinations that perform well for medical image categories, IRMA. The top three features used in the research were Colour Layout, Edge Histogram and DCT Coefficients, which were all combined for a higher image re-trivial ranking than others that were based on individual features. The results returned a 14.49% improvement of the retrieval MAP (50.68% for combined features Set). In the future, the research is set to investigate new features and apply appropriate weights for the particular features used in the combination. In handling noisy image in Relevance Feedback CBIR systems, Huang *et al.*, (2013) proposed a two-step strategy. The first step was to apply a noisy elimination algorithm. The second step was to adopt Fuzzy Support Vector Machine (FSVM) to re-rank the images. Noisy-Smoothing Relevance Feedback (NS-RF) was reported to yield better retrieval performance as compared to the Support Vector Machine-Noisy-Smoothing-Relevance Feedback (SVM-NS-RF). The justification was that NS-RF can further handle the noisy images by considering different relevant images with different relevance probabilities in the image re-ranking procedure, whereas SVM-NS-RF treated the retained relevant image equally. In addition, the research noted that the effect of noisy elimination algorithm in NS-RF is sensitive to the slack factor in the proposed formula. Hence, it is proposed to conduct further investigation on the value in order to improve the noisy elimination algorithm in the future.

Burdescu et al. [30] include Cross Media Relevance Model (CMRM) and Continuous Space Relevance Model (CRM) to annotate the medical images with a segmentation algorithm based on hexagonal structure. The research resulted in an ontology created using Medical Subject Heading (MesH). Their results showed that the CRM model produced better results as compared to CMRM for image annotation and semantic-based image retrieval. Annotation of mammograms has also been explored using the mammography. For Annotation Ontology, a model was proposed for similarity calculation between breast masses based on their high, mid and low-level features. They also used the Semantic Query-enhanced Web Rule Language (SQWRL) to retrieve similar masses from annotated mammography collection in OWL. The proposed approach was tested with DEMS, which is a fully-annotated digital mammography dataset. The mid-level feature also helped to effectively rank the result set. Future work was set to include more mid-level features as well as to integrate the low-level features into reasoning phase [31].

Da Silva et al., [17] conducted three experiments using mammograms taken from the University of South Carolina and University of Sao Paulo. This technique of feature selection enhanced similarity search accuracy and lowered substantially data dimensionality, which consequently improved access methods efficiency and the CBIR system. For future work, it is recommended that a new local search into GA and the synergy between filter-based methods and the GA wrapper-based method in CBIR systems can be introduced to improve the efficiency of the proposed method. Besides those, all textual information on patients' clinical history can be incorporated into the search mechanism.

Similarly, Wei et al. [16] proposed a set of mammogram descriptors according to the BI-RADS definition. The proposed learning approach improved the classification accuracy as high as 72% and 74%. The calcification Query-By-Example (QBE) was reported to be slightly higher than the mass QBE is more effective in extracting calcification lesions as compared to mass lesions. This study resolved that in the process of feature extraction, the calcification features are able to withstand possibilities of inaccuracies, noises and misses in spot detection. Mass lesions, in feature comparisons with those of calcification, are found to be more sensitive to segmentation imprecision and extraction. This study intends to investigate long-term learning in relevance feedback for a real-time system to develop users' search information. Ramamurthy & Chandran [30], presented the texture-based image retrieval system for medical images. Gray Level Co-occurrence Matrix (GLCM) and k-means clustering algorithm were utilised in the proposed system to improve its retrieval competence. However, as the system solely used texture feature, retrieval performance was only at 50%.

To characterize the texture of breast density, MammoSys was introduced based on Two-dimensional Principal Component Analysis (2DPCA). Support Vector Machine (SVM) was used to aid the retrieval process. 2DPCA was able to extract the features of the texture of breast tissues by capturing the difference between the gray level intensities among different breast densities. This research has also produced another positive consequence, i.e. all retrieved images, sharing the same view and direction, originated from BI-RADS category of the query image. This research anticipated that there could be a combination of texture attributes with the grey-level histogram, and also with other retrieval patterns, for example, breast lesions, masses and calcifications [23].

To abridge, different researchers have conducted study on different parts of CBIR to improve the efficiency of the CBIR system as a whole. Some of researches focused on feature extraction and combine several features to improve images retrieval. In medical, researchers usually used texture and shape features to represent the images content because medical images such as breast cancer [23][13][32] and brain tumor are in grayscale except for ophthalmology, pathology and dermatology [12]. Outside of the medical area, colour features are used because it is the most recognised feature to represent the image content as compared to other features like textures and shapes [33][34]. Researchers also combine different features for a better characterisation of images content. Nonetheless, the efforts resulted in high dimensionality of feature set and hence lead to the research in feature selection. It is a common belief that selecting the best features that best describe the image content will help retrieve the most relevant ranked images [28]. There are also researchers who work on similarity measure such as new matching strategy to enhance the performance of the CBIR as well as a novel relative comparison-based similarity learning strategy to enhance CBIR performance [35][36].

Zhang [37] stated that only colour features cannot produce good results. Therefore, the author combined the colour features with texture features. In this approach, the similar images are firstly searched based on the colour features. Then, from these searched images, the searching process is performed again using texture features. For colour features, perceptually weighted histogram is generated using L*u*v colour space. Gabor filter is used to extract the texture features. To assess the proposed techniques, experiments were performed on the MPEG-7 natural image database. Validation was done by calculating the precision and recall value for retrieval result. Likewise, Singh & Hemachandran [38] proposed a CBIR system using Colour moments and Gabor texture features. To extract the colour information from the image, the image is divided into three non-overlapping segments. The first three moments of the colour distributions are extracted from each segment. HSV colour space is used for the colour feature extraction. Texture features are extracted using Gabor texture descriptor and combined with the colour feature vector. They used the Canberra distance to compute the distance between the query image and the database images. For experimental purpose, they used the Coral image gallery and the results are compared with the previous CBIR techniques. To measure the performance of the proposed techniques, precision and recall were used on the performance metrics. The authors also performed experiments using only colour or texture features but the results demonstrated that only colour or texture cannot represent the image well.

Yue *et al.,* [39] used the co-occurrence matrix for extracting the colour and texture features. They compared the performance of local colour histogram, global colour histogram and texture features for a CBIR system. For this comparison, they designed a CBIR approach using fused colour and texture features. Sakhar & Nasre [40] used the local colour histogram, global colour histogram and fuzzy colour histogram for the colour feature extraction. To extract texture, they used Tammura features and the wavelet transform. Euclidean distance was used to compute the distance between the query image and the database images. Reddy and Prasad [41] proposed a CBIR approach using colour correlogram for colour features and multi wavelet transform for texture feature extraction. RGB colour space was used to generate the colour correlogram. They separated each colour space and then apply the correlogram in 0^0, 45^0, 90^0 and 135^0. Soman et al., [42] used the discrete block wise cosine transform to extract the texture features and colour moments such as mean, skewness and deviation used to compute the colour feature vector. In this proposed approach, images are firstly searched based on texture features. Then, from these searched images, searching process was performed again using colour feature vector. Coral database was selected to perform the experiments and the results were compared with previous techniques.

4. Reducing The Semantic Gap

Semantic is one of the crucial issues in CBIR system. Researchers who worked on low level features have employed several techniques to reduce the semantic gap in CBIR. Based on the previous works, some researchers improved the efficiency of CBIR systems by means of ontology, machine learning tools as well as employing Relevance Feedback (RF) to ensure high relevant images. High-level features are semantically meaningful labels which describe the shape of a mass that can be round, oval, lobular or irregular [43]. Relevance feedback is a powerful tool and online learning to retrieve most relevant images. This strategy ask user to give some feedbacks on the results returned in the previous query round and come up with a better result based on these feedbacks. A variety of relevance feedback techniques designed to bride the semantics gap between low level visual features and high level semantic concept of each image. The general process of Relevance Feedback is as follows: First user labels a number of relevant images as positive feedback and a number of irrelevant images as negative feedback from retrieved images. Then the CBIR system then refines its retrieval procedure based on these labeled samples. These processes carried out iteratively. RF techniques are classified into two categories: that is query movement and biased subspace learning. In this biased subspace learning, all positive samples are alike and each negative sample in negative in its own way [44].

Furthermore, these descriptors form a simple vocabulary, the so-called 'object-ontology' that provides a qualitative definition of high-level query concepts. Database images can be classified into different categories by mapping such descriptors to high-level semantics (keywords) based on our knowledge[7]. Researchers could automate the images according to this ontology. In medical, ontology is developed based on some standard such as the BI-RADS standard. Similar work has been done to annotate the mammogram images using Mammography Annotation Ontology (MAO) that was developed via MAO and was also based on BI-RADS standard [44]. DeOliveira et al., [23] also used SVM for retrieval task as it is able to solve a variety of learning, classification and prediction problems. For medical applications, Liu et al., [46] presented a new image retrieval method. It utilises a convolutional neural network (CNN) that is merged with neural codes for global classification and equipped with the latest Radon barcodes for the final retrieval. The researchers investigated image search based on regions of interest (ROI) that are matched after image retrieval. Relevance Feedback (RF) is another approach to reduce semantic gap in retrieving images. RF is a supervised learning technique because user has to give feedback by labeling the returned images. In conventional text-based information retrieval systems, relevance feedback is a very reliable technique used. This technique regulates automatically existing query relevant to the preceding objects retrieved through information fed back by users to the system. This allows users to have a query which is more attuned based on their information need.

The main aim of the medical image databases therefore is to organize, search, and index the vast corpus of medical images. It is therefore crucial to have, an intelligent system that enables the capture, recognition, and comprehension of the intricate content of medical images. For these tasks to be fulfilled, content-based retrieval, by which a number of techniques used in medical images has been developed, is a promising approach. In spite of the current progress, medical content-based image retrieval has the potential to develop further and there should be more research into this area. Eventually, a competent image database, together with an intelligent retrieval system will be able to improve clinical treatment, and offer a platform for better medical research and education [47].

To enhance performance in relevance feedback scheme, Wang et al., [7] proposed a Feature Line Embedding Biased Discriminant Analysis (FLE-BDA). The margin between relevant and irrelevant samples in the local neighbourhood is maximised by reducing the gap between the query and relevant images. This subspace learning method allows a linear transformation matrix from relevant or irrelevant images that is used in dimensionality reduction. The process of retrieval comprises: 1)Keying-in a query image into the IR system. Gallery images are ranked subsequent to the calculation of the similarity values. 2) The labelling of the relevant or irrelevant images as per users'preference 3) Adopting users' feedback for a new transformation. 4) Re-ranking the gallery images to acquire the retrieval results in the next round. Two labels are ascribed to the top-ranking images according to users' preference.

Users' preference is signified by the feedback of relevant or irrelevant labels. The within class scatter is calculated from the image samples with positive labels, while the between-class scatter is calculated from those with negative labels. Based on these assigned labels, the within class and between-class weighted graphs are constructed for maximizing the margin of relevant and irrelevant samples. Then new distance between query and images are calculated. The advantages are dimensionality reduction, solve singular problem in the high dimensional space, increases generalization and robustness using Laplacian regularization. The disadvantage is computational complexity is very high due to the large-scale dataset. Early relevance feedback method has been classified into the query point movement and the reweighting method. Basically, the query point movement method enhances the estimation of the 'ideal query point' by shifting positive examples nearer to it, and negative ones afar. Fundamentally, the reweighting method is to amend the weight of each image feature element. These methods strongly assume that the target class, with an elliptical shape, significantly limits their performance. The main goal of this system is to get the best relevant and similar image as preferred by user.

5. Conclusion And Future Directions

Content based image retrieval is a technique to retrieve more relevant images. Retrieve similar images only is a standing problem in digital image processing. The performance of CBIR system is improved by introducing relevance feedback techniques in the system. Several feature modification and subspace learning based relevance feedback methods are studied. Various systems use feature modification of each image and tries to retrieve relevant images. But these systems do not suitable for high dimensional images. Several subspace learning relevance feedback methods provides more

IOP Conf. Series: Journal of Physics: Conf. Series **1019** (2018) 012044 doi:10.1088/1742-6596/1019/1/012044

relevant images compared with feature modification based methods. It also considers local information of images and aims those similar images close to but dissimilar images are far away from query image. This paper focuses on the different relevance feedback techniques in digital image processing.

Researchers have proposed several approaches to provide a linkage in the use of low level features with high-level ones to bridge the semantic gap in the medical field. Past and current researches suggested of using ontology in CBIR, applying machine learning to classify images in retrieval task and employing RF to refine the retrieved images in the first iteration. Various algorithms have been conducted by the researchers to solve this matter. However, to build a working CBIR with high-level semantic requires many efforts such as extraction salient low-level features, selection of significant features, electiveness of learning process, user friendly interface usually in case of RF and efficient similarity measure to index the images.

Acknowledgement

The authors would like to thank the Universiti Tun Hussein Onn Malaysia (UTHM), Research, Innovation, Commercialization and Consultancy Management (ORICC) office for facilitating this research activity under Post-Doctoral Fund Vote no D001.

6. References

[1] Belattar, K., Mostefai, S., & Draa, A. (2017). Intelligent Content-Based Dermoscopic Image Retrieval with Relevance Feedback for Computer-Aided Melanoma Diagnosis. *Journal of Information Technology Research (JITR)*, *10*(1), 85-108

[2] Akgül, C. B., Rubin, D. L., Napel, S., Beaulieu, C. F., Greenspan, H., & Acar, B. (2011). Content-based image retrieval in radiology: current status and future directions. *Journal of Digital Imaging*, *24*(2), 208-222.

[3] Bunte, K., Biehl, M., Jonkman, M. F., & Petkov, N. (2011). Learning effective color features for content based image retrieval in dermatology. *Pattern Recognition*, *44*(9), 1892-1902.

[4] Ribeiro, M. X., Bugatti, P. H., Traina, C., Marques, P. M., Rosa, N. A., & Traina, A. J. (2009). Supporting content-based image retrieval and computer-aided diagnosis systems with association rule-based techniques. *Data & Knowledge Engineering*, *68*(12), 1370-1382.

[5] Yildizer, E., Balci, A. M., Jarada, T. N., & Alhajj, R. (2012). Integrating wavelets with clustering and indexing for effective content-based image retrieval. *Knowledge-Based Systems*, *31*, 55-66

[6] Das, R., Thepade, S., Bhattacharya, S., & Ghosh, S. (2016). Retrieval Architecture with classified query for content based image recognition. *Applied Computational Intelligence and Soft Computing*, *2016*, 2.

[7] Wang, Y. C., Han, C. C., Hsieh, C. T., Chen, Y. N., & Fan, K. C. (2015). Biased discriminant analysis with feature line embedding for relevance feedback-based image retrieval. *IEEE Transactions on Multimedia*, *17*(12), 2245-2258.

[8] Lashari, S. A., Ibrahim, R., & Senan, N. (2015). Fuzzy Soft Set based Classification for Mammogram Images. *International Journal of Computer Information Systems and Industrial Management Applications*, *7*, 66-73.

[9] Mohanan, A., & Raju, S. (2017). A Survey on Different Relevance Feedback Techniques in Content Based Image Retrieval

[10] Santos, L. F., Dias, R. L., Ribeiro, M. X., Traina, A. J., & Traina, C. (2015, December). Combining diversity queries and visual mining to improve content-based image retrieval systems: The divi method. In *Multimedia (ISM), 2015 IEEE International Symposium on* (pp. 357-362). IEEE.

[11] de Ves, E., Benavent, X., Coma, I., & Ayala, G. (2016). A novel dynamic multi-model relevance feedback procedure for content-based image retrieval. *Neurocomputing*, *208*, 99-107.

[12] Depeursinge, A., & Miiller, H. (2011). Medical visual information retrieval based on multi-dimensional texture modeling. *Procedia Computer Science*, *7*, 127-129.

[13] Lashari, S. A., Ibrahim, R., Senan, N., Yanto, I. T. R., & Herawan, T. (2016, August). Application of Wavelet De-noising Filters in Mammogram Images Classification Using Fuzzy Soft Set. In *International Conference on Soft Computing and Data Mining* (pp. 529-537). Springer, Cham.

[14] Müller, H., Michoux, N., Bandon, D., & Geissbuhler, A. (2004). A review of content-based image retrieval systems in medical applications—clinical benefits and future directions. *International journal of medical informatics*, *73*(1), 1-23.

[15] Rui, Y., Huang, T. S., Ortega, M., & Mehrotra, S. (1998). Relevance feedback: a power tool for interactive content-based image retrieval. *IEEE Transactions on circuits and systems for video technology*, *8*(5), 644-655.

[16] Wei, C. H., Li, Y., & Huang, P. J. (2011). Mammogram retrieval through machine learning within BI-RADS standards. *Journal of biomedical informatics*, *44*(4), 607-614.

[17] Da Silva, S. F., Ribeiro, M. X., Neto, J. D. E. B., Traina-Jr, C., & Traina, A. J. (2011). Improving the ranking quality of medical image retrieval using a genetic feature selection method. *Decision Support Systems*, *51*(4), 810-820

[18] Zheng, B. (2009). Computer-aided diagnosis in mammography using content-based image retrieval approaches: current status and future perspectives. *Algorithms*, *2*(2), 828-849.

[19] Rashedi, E., Nezamabadi-Pour, H., & Saryazdi, S. (2013). A simultaneous feature adaptation and feature selection method for content-based image retrieval systems. *Knowledge-Based Systems*, *39*, 85-94

[20] Popova, A. A., & Neshov, N. N. (2013). Combining Features Evaluation Approach in Content-Based Image Search for Medical Applications.

[21] Kumar, A., Kim, J., Cai, W., Fulham, M., & Feng, D. (2013). Content-based medical image retrieval: a survey of applications to multidimensional and multimodality data. *Journal of digital imaging*, *26*(6), 1025-1039.

[22] Li, Y., Zhou, C., Geng, B., Xu, C., & Liu, H. (2013). A comprehensive study on learning to rank for content-based image retrieval. *Signal Processing*, *93*(6), 1426-1434.

[23] De Oliveira, J. E., Machado, A. M., Chavez, G. C., Lopes, A. P. B., Deserno, T. M., & Araújo, A. D. A. (2010). MammoSys: A content-based image retrieval system using breast density patterns. *Computer methods and programs in biomedicine*, *99*(3), 289-297.

[24] Marinakis, Y., Dounias, G., & Jantzen, J. (2009). Pap smear diagnosis using a hybrid intelligent scheme focusing on genetic algorithm based feature selection and nearest neighbor classification. *Computers in Biology and Medicine*, *39*(1), 69-78

[25] Sharma, H., Alekseychuk, A., Leskovsky, P., Hellwich, O., Anand, R. S., Zerbe, N., & Hufnagl, P. (2012). Determining similarity in histological images using graph-theoretic description and matching methods for content-based image retrieval in medical diagnostics. *Diagnostic pathology*, *7*(1), 134.

[26] Reddy, P. V. N., & Prasad, K. S. (2011). CONTENT BASED IMAGE RETRIEVAL USING LOCAL DERIVATIVE PATTERNS. *Journal of Theoretical & Applied Information Technology*, *28*(2).

[27] Town, C. (2013). Content-based and similarity-based querying for broad-usage medical image retrieval. *Advances in Biomedical Infrastructure*, *477*, 63-76.

[28] Popova, A. A., & Neshov, N. N. (2013). Combining Features Evaluation Approach in Content-Based Image Search for Medical Applications.

[29] Huang, W., Zhang, P., & Wan, M. (2013). A novel similarity learning method via relative comparison for content-based medical image retrieval. *Journal of digital imaging*, *26*(5), 850-865.

[30] Burdescu, D. D., Mihai, C. G., Stanescu, L., & Brezovan, M. (2013). Automatic image annotation and semantic based image retrieval for medical domain. *Neurocomputing*, *109*, 33-48.

[31] Bulu, H., Alpkocak, A., & Balci, P. (2012). Ontology-based mammography annotation and case-based retrieval of breast masses. *Expert Systems with Applications*, *39*(12), 11194-11202.

[32] Chandy, D. A., Johnson, J. S., & Selvan, S. E. (2014). Texture feature extraction using gray level statistical matrix for content-based mammogram retrieval. *Multimedia tools and applications*, *72*(2), 2011-2024

[33] Talib, A., Mahmuddin, M., Husni, H., & George, L. E. (2013). A weighted dominant color descriptor for content-based image retrieval. *Journal of Visual Communication and Image Representation*, *24*(3), 345-360.

[34] Huang, Z. C., Chan, P. P., Ng, W. W., & Yeung, D. S. (2010, July). Content-based image retrieval using color moment and gabor texture feature. In *Machine Learning and Cybernetics (ICMLC), 2010 International Conference on* (Vol. 2, pp. 719-724). IEEE

[35] Huang, W., Zhang, P., & Wan, M. (2013). A novel similarity learning method via relative comparison for content-based medical image retrieval. *Journal of digital imaging*, *26*(5), 850-865.

[36] Yildizer, E., Balci, A. M., Jarada, T. N., & Alhajj, R. (2012). Integrating wavelets with clustering and indexing for effective content-based image retrieval. *Knowledge-Based Systems*, *31*, 55-66.

[37] Zhang, D. (2004, December). Improving image retrieval performance by using both color and texture features. In *Image and Graphics (ICIG'04), Third International Conference on* (pp. 172-175). IEEE

[38] Singh, S. M., & Hemachandran, K. (2012). Content based image retrieval based on the integration of color histogram, color moment and Gabor texture. *International Journal of Computer Applications*, *59*(17).

[39] Yue, J., Li, Z., Liu, L., & Fu, Z. (2011). Content-based image retrieval using color and texture fused features. *Mathematical and Computer Modelling*, *54*(3), 1121-1127.

[40] Sakhare, S. V., & Nasre, V. G. (2011). Design of feature extraction in content based image retrieval (CBIR) using color and texture. *International Journal of Computer Science & Informatics*, *1*(II).

[41] Reddy, P. V. N., & Prasad, K. S. (2011). CONTENT BASED IMAGE RETRIEVAL USING LOCAL DERIVATIVE PATTERNS. *Journal of Theoretical & Applied Information Technology*, *28*(2).

[42] Soman, S., Ghorpade, M., Sonone, V., & Chavan, S. (2012). Content based image retrieval using advanced color and texture features. In *International Conference in Computational Intelligence (ICCIA)* (Vol. 3, No. 4).

[43] Salem, Y. B., Idoudi, R., Ettabaa, K. S., Hamrouni, K., & Solaiman, B. (2017, September). Ontology Based Possibilistic Reasoning for Breast Cancer Aided Diagnosis. In *European, Mediterranean, and Middle Eastern Conference on Information Systems* (pp. 353-366). Springer, Cham.

[44] Mohanan, A., & Raju, S. (2017). A Survey on Different Relevance Feedback Techniques in Content Based Image Retrieval.

[45] Bulu, H., Alpkocak, A., & Balci, P. (2012). Ontology-based mammography annotation and case-based retrieval of breast masses. *Expert Systems with Applications*, *39*(12), 11194-11202.

[46] Liu, X., Tizhoosh, H. R., & Kofman, J. (2016, July). Generating binary tags for fast medical image retrieval based on convolutional nets and radon transform. In *Neural Networks (IJCNN), 2016 International Joint Conference on* (pp. 2872-2878). IEEE.

[47] Arakeri, M. P., & Reddy, G. R. M. (2013). An intelligent content-based image retrieval system for clinical decision support in brain tumor diagnosis. *International Journal of Multimedia Information Retrieval*, *2*(3), 175-188.

IOP Conf. Series: Journal of Physics: Conf. Series **1019** (2018) 012045 doi:10.1088/1742-6596/1019/1/012045

MOBILE SYSTEM FOR MANAGING AND MITIGATING THE ACCOMMODATION PROBLEMS

Murtaja Ali Saare [1], Azman B Ta'a [2], Saima Anwar Lashari [3], Sari Ali Sari [4]
[1,2] School of Computing, University Utara Malaysia, Sintok 06010, Kedah, Malaysia
[3] Faculty of Computer Science and Information Technology Universiti Tun Hussein Onn Malaysia,86400 Parit Raja, Batu Pahat, Johor, Malaysia.
[4] Faculty of Information and communication Technology ,Universiti Teknikal Malaysia Melaka ,Hang Tuah jaya 76100 Durian Tunggal , Malacca, Malaysia

Mmurtaja88@gmail.com[1], azman@uum.edu.my[2], saima@uthm.edu.my[3], sanalasharisana@gmail.com[4]

Abstract. Students either local or international experience different adjustment problems while moving one place to another place, and out of those problems, finding suitable accommodation is the most challenging one. However, at Universiti Utara Malaysia (UUM), the accommodation centre for the students is utilizing the traditional method for managing and processing the issues related to the students' accommodation. Therefore, the current empirical study seeks to exploit the modern technologies, such as mobile devices to increase the communication among the students and management of resident. The present study has three core objectives i.e. identifying the requirements for managing the student accommodation, designing the mobile accommodation system and evaluating the usability of the proposed mobile accommodation system. To evaluate the usability of the system, a usability scale is used and data are collected from 40 students. The results reveal that the users accept and are satisfied with the interface functions, usability, integrity and other facilities for the application. Thus, students and accommodations staff can use the applications successfully.

1. Introduction

Student accommodation plays a significant part in a student's life. Student learning can be enhanced and made more effective when learning environment is made adequate and appropriate [1-2]. Previous studies show that accommodation plays a vital role in the physiological needs of the students, and that is considered as a condition for student's quality survival in terms of health, academic performance and learning [3]. Therefore, a favorable environment plays a significant role to maintain the academic performance of the students. In general, all students (local or international) experience different adjustment problems while moving to one place to another place. Duangpracha [4] states that leaving home carries various difficulties; out of them, finding suitable accommodation is the most challenging one. Several studies have been conducted on the accommodation and its effects on the students' performance in many western countries. However, less studies are conducted in Malaysian Universities.

In Universiti Utara Malaysia (UUM), the accommodation is provided within the campus territory. However, the system to manage and mitigate the accommodation for the students is utilized by the traditional method for managing and processing the issues related to the student's accommodation. The advancements in wireless and mobile technologies have developed many mobile-based applications in various fields, such as health, agriculture, education and entertainment.

These technologies pave the way for private and public organizations to deliver, can organize and distribute services to the public in a more well-organized and cost-effective manner [5].

Wright [6] states that the ability of online services not mainly focuses on reducing the effort and the time of the users, but also on improving client services and the services validation efficiency. Students' accommodation should be treated as an issue of great priority. An accommodation problem is considered as one of the daily stressors, affecting the students' performance [7-8]. These stressors are associated with the depressive disorders among the university students. There are several factors that are considered as the major factors that may lead towards the failure of the students in the final exam, and accommodation problem is found as one the vital factors [9-10]. At UUM, there are many residential halls that are named by different companies, and each of the residential hall has their own office, but the main authority of all the offices is under Student Accommodation Center (SAC). Therefore, the students must come and apply for the accommodation at SAC. Due to large number of students in UUM, usually there is a long queue of students. Moreover, regarding the employee, there are many branches of residential hall under Student Accommodation Center in Universiti Utara Malaysia. Each branch has different number of students depending on the level of education. Therefore, it is difficult for the employee of the student Accommodation Center to deal with the problems of the students. Furthermore, the employees are also facing the difficulties in terms of communicating with the students and determining the proper time for them to address the problem. Therefore, ICT can deal with a substantial number of procedures as suggested by Spotti, Kluzer & Ferrari [11], and it also can enhance the communication among the stakeholders. Indeed, the previous studies also assert that there is a need to utilize the modern technologies to deal with accommodation problems, especially at UUM. Olaniyi [12] affirms that there should be a convenient accommodation system for the students during study times. Therefore, keeping the basis of the facts that previous studies have identified, the present study seeks to explore the modern technologies (such as mobile devices (android or iOS platform), Web 2.0 and cloud-computing to increase the communication among the students and management of the residential halls. The result of the study relates facilitating the students' accommodation process and easing the accommodation problems.

Therefore, based on these arguments, this study strives to exploit the modern technologies to design a comprehensive system via mobile device (using android) and personal computer (using Web 2.0) to enhance the communication between the student and SAC. Some of the functions benefits the students, and they are hostel application, reporting to complaints, and requesting for services, and other functions that benefit the SAC include room allocation, student feedbacks, and student complaint list. Therefore, this project aims to develop a mobile-based application to facilitate the management of students' accommodation and mitigate the problem of students' accommodation. Thus, based on the issues highlighted, this study identifies the requirements for managing the student accommodation, and does analysis and design of the mobile accommodation system, and evaluates the usability of the proposed mobile accommodation system (MAS).

2. Related work

Colleges or universities with residential facilities play a significant role in numerable top universities in the world. It has been believed that student's educational involvement and outcomes enhance if the student lives in a university-affiliated residence. Campus residence provides a potentially effective environment that encourages openness to diversity as it provides extended opportunities for students to collaborate with peers and staff to implement programs that expose students to multicultural issues [13]. Not surprisingly, earlier studies have found that living on campus is related to increasing tolerance and openness to diversity as compared to commuting from home [14].

College students are the group of people who have more complex problems than how it is used to be over the past 10 years where the difficulties faced are on the demands of a greater academic studies. Kumaraswamy [15] states that stress, nervousness and depression among college students are causes of concern. Apart from that, there is a crucial urgency which demands for the adequate accommodation for the students which emphasizing on the implementation of programmers related to education as well as the welfare of the students in a qualitative way. It is also deemed necessary for any institution to provide the students with the best possible accommodation. Khawaja and Dempsey [16] state that rising discontent with accommodation may serve to increase the psychological strain on internationals students.

The advancements in wireless and mobile technologies have triggered a significant number of applications that had been established in various fields, such as health, agriculture, education, entertainment, among others. Furthermore, these technologies pave the way for private and public organizations to deliver, can organize and dispense services to the public in more efficient and economical manner The Internet and mobile applications are becoming so common use in our everyday life in the sense that we would not go a single day without using them[6].Thus, the ability of online services not mainly focuses on reducing the effort and the time of the users, but also improving client services and the services validation efficiency. Therefore, this study explores the development of mobile accommodation system (MAS), which utilizes the mobile computing technology for assisting the accommodation process in UUM.

3- Methodology

Methodology used in the present study comprises of five phases namely, the awareness of the problem through reading and analyzing literature, suggestion phase through the visualization of the interaction between the prototyped used, development phase using Java program, evaluation phase that consists of testing the usability and conclusion phase that summarizes all the phases. The details about the work carried out for every stage is explained below.

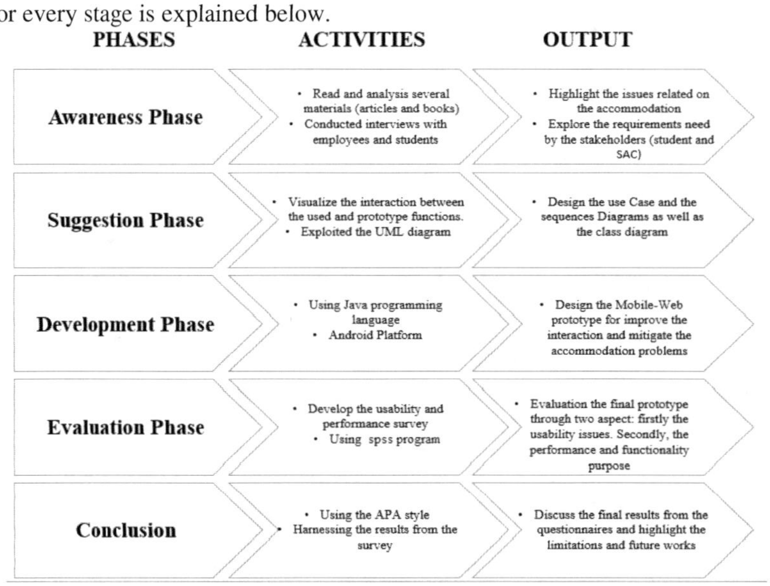

Figure 1: Proposed Methodology

3.1 Awareness Phase

Awareness phase involves requirements of the prototype as requirements identification is an important phase in the prototype. This phase also requires understanding the objectives, the scope, and the problem. Regarding this study, awareness of the current problem is highlighted by the thorough reading of the literature and from real life. As for the literature, the previous studies assert to solve the accommodation problems because the accommodation issues effect on the student performance and stress. About real world, the interviews with employees and students are conducted for this purpose, and the results indicates that there is less communication between the Student Accommodation Center (SAC) and the student. Moreover, there is lack of coordination between the staff and students to fix the problems related to accommodation issues.

3.2 Suggestion Phase

After identifying and understanding the needs of users from the first phase of design research method, this study suggests using a mobile device to build MSAS. The output of this phase is quite tentative prototype. The analysis and design of the system constitute the system prototype design using UML relationship composition. In general, the UML diagrams used are those of case diagrams, detailed sequence diagrams (for each use case) and more importantly, the class diagrams. The proposed architecture for the system can be seen in the following:

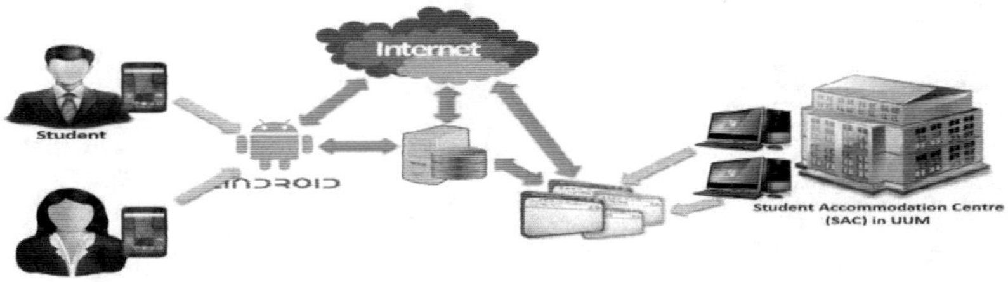

Figure 2: The architecture of the MAS (Vaishnavi and Kuechler, 2015).

3.3 Development Phase

The main tools that is used to design the Mobile System for Accommodation of students System (MSAS) is Windows 7, and used as an operating System for a PC or Proto-type for development purposes Eclipse was Used as development tools. Photoshop was used for designing interface required by the system.

3.4 System Design

This section gives detailed explanations to the design mobile system for accommodation of students (MSAS) prototype, depending on Unified Modeling Language (UML) for mobile application. The process of the application is designed in a way that helps its user to understand. UML comprises of diagrams and the flow chart of the system's architecture, and it is the standard language that helps to specify, visualize, construct and document the artifacts of the software system; the details are mentioned in the analysis and comparison section. It has been clarified in the use case diagram, activity diagram, sequence diagram and state chart diagram sequentially. The specification for every use case and the list of functional requirements and non-functional requirements are also presented.

The illustration of MSAS application requirements is one of the most important stages for the development of useful and accurate mobile application. Besides, the functional requirements are the functions or techniques used in MSAS application while the non-functional requirements of MSAS application explain a methodical and pragmatic approach to application development and the quality measurement for attributes, as example usability, performance, and reliability. To draw necessary diagrams that play a significant role in the developmental stage of mobile application. The following section of the study illustrates the design of the prototype.

3.5 Use Case Diagram

The main functional requirements in MSAS formalized through modeling methods, using case diagram is one of these methods which describe what does a certain application from an external observer's standpoint. The main goal of using use case diagram is to understand application functions and what it does; an actor is involved in doing application functions [17]. The relationship between student and use case are shown in the form of case diagram which is a three-layered architecture, and it is a tool to present the design and software [18]. The system has following main functions: login, registration, home page, booking, SAC information, complaint's report, complaint status and instant chatting as shown in the Figure 3.

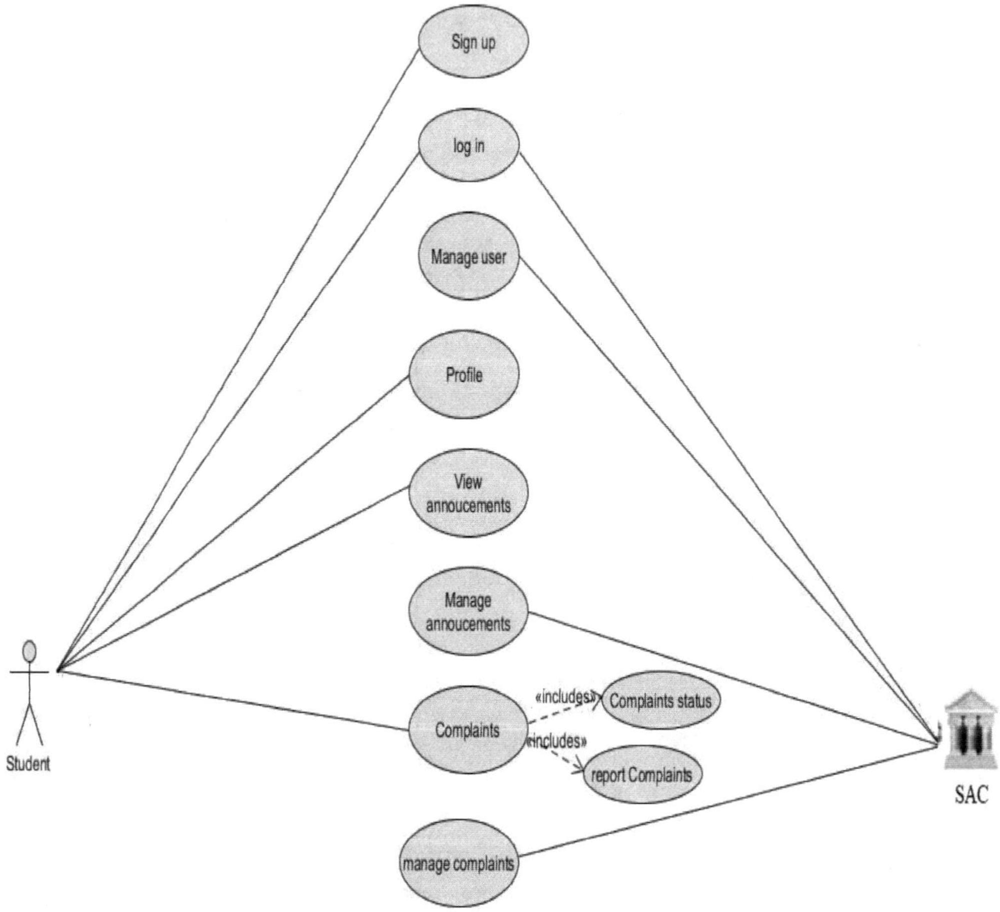

Figure 3 Use-Case Diagram

3.6 Class Diagram

ERD is a kind of graph that is used to visually present the entity in the form of set, attribute, and the relationship between entities in a database. ERD is top down approach in database design which gets started by identifying the vital data called entities, and associations between the data should be represented in the model. ERD is developed through using the technique of normalization as shown in the Figure 4.

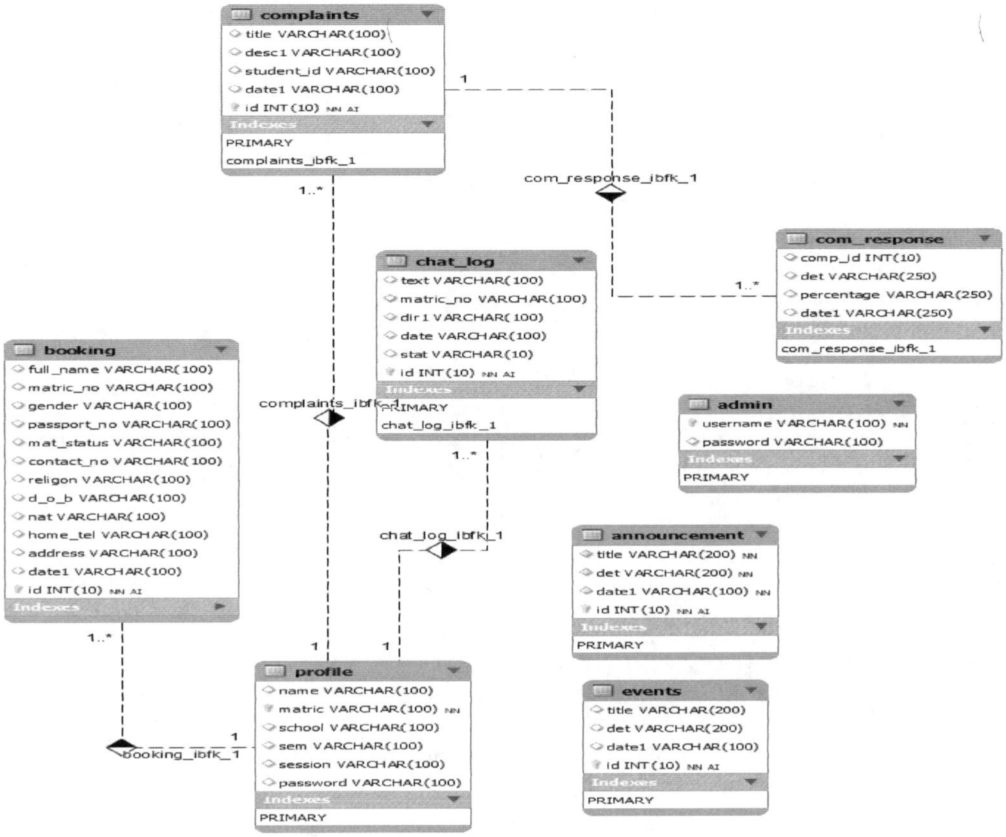

Figure 4 Conceptual Database Design

3.7 Application Interface of mobile Application and web server

The MSAS application is connected between SAC Officers, and students.

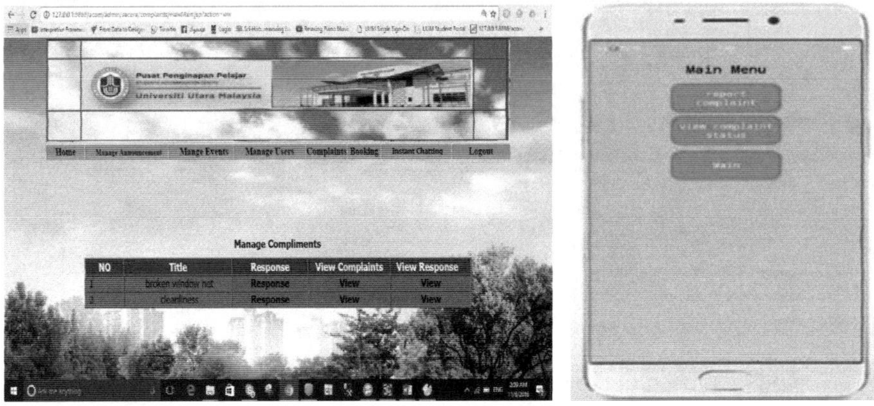

Figure 5 Application Interface of mobile Application and web server

3.7.1 Direct Connection

The mobile application sends a request to web application to get response. To ensure the privacy, the Secure Socket Layer (SSL) or another encryption type is used. There are three features of connection:

- Direct connection using standard telnet/SSL;

- User verification via simple login/password;
- UI commands converted to server commands.

3.7.2 Server-side application
The mobile application and the server-side application are directly connected. In or-der words, a software program is run on a remote server. It has a unique feature of connection i.e.: Direct connection to a server-side application.

3.7.3 Web application
The mobile application can communicate with the same web application to get in-formation from it. There are three features of connection:

- Using of a web services via specialized API;
- User verification via simple/secure login/password;
- UI commands converted to API requests.

3.7.4 Cloud Integration
Cloud Infrastructure as a Service with the organization (UUM) receives a private cloud which is pretty like the classic understanding of clouds, but that is used solely for UUM purposes, and it has better security for connecting to cloud by means of its services (API).

3.8 Evaluation Phase
Evaluation phase is the last phase of the methodology of the application. Feature test is an essential part to know if the features developed are easy to use. A usability test is one of the most significant techniques for evaluation because it needs students to use the developed application.
The evaluation process is conducted in determining the level of usefulness as well as the ability in operating the system after the development of the system by the System Usability Scale (SUS) by Bangor et al., [19]. A sample of 40 students from UUM are involved in gathering data to evaluate the perceived usability of a system which gives rich information with respect to the multilingual system.

4 Findings And Discussion
Data is collected using randomly sampling technique. Each of the respondent is given a brief description regarding the use of the MSAS and the user interface other prototype. However, the user has been given ample time to learn and explore the content of MSAS. Once they are done with the exploration, the users are given a questionnaire to evaluate the usability. Statistical package for social science (SPSS) version 23 is used to analyze the data gathered. The results of the analysis are shown in the table 1.

Table 1 Mean and standard deviation of SUS

Item number	N	Mean	Std. Deviation
I think that I would like to use MSAS frequently.	40	3.80	0.72
I found the MSAS unnecessarily complex	40	3.12	0.64
I thought the MSAS was easy to use	40	3.75	0.80
I think that I would need the support of a technical person to be able to use MSAS	40	2.82	0.87
I found that the various functions in this MSAS were well integrated	40	3.70	0.68
I thought that there was too much inconsistency in MSAS.	40	2.97	0.83
I would imagine that most people would learn to use MSAS very quickly	40	3.57	0.90
I found MSAS very difficult to use	40	3.52	0.78
I felt very confident using MSAS	40	3.37	0.70
I needed to learn a lot of things before I could get going with MSAS	40	2.70	0.88

The results of the table 1 show that all questions have high mean score which is approximately 3. Besides, the standard deviations are small, less than 1. This explains that all the statements regarding the usability of MSAS are agreed by the subjects with small biasness or influence of other factors. Reliability is addressed for the usability evaluation questionnaire. One of the most widespread reliability measures is Cronbach Alpha Coefficient (Coakes, [20]. Thus, using the Cronbach alpha reliability of the SUS is used to calculate by using SPSS 23 to determine the scale reliability which assesses the degree of internally consistent. Table 2 presents the Cronbach alpha value for System usability scale. If the measures have a Cronbach alpha of greater than 0.7, these measures satisfy the internal reliability criterion [21]. Thus, the scale is said to have a good reliability i.e. it is internally consistent.

Table 2 Reliability Statistics

Cronbach's Alpha	N of Items
792	10

Based on Table 5.8, the reliability is significant and accepted because the Cronbach alpha value is greater than 0.7 which indicates good reliability of the scales. Through the questionnaire analysis, we can observe that the user in general agree that by using mobile facilities, they make online complaint. The MSAS user has accepted and are satisfied with the application interface functions, usability, integrity and other facilities in-side the application. Based on the result, the researcher has concluded that our application has achieved the goals, and it is worth noting down. The users are satisfied by the MSAS utilization, and they claim that it would be easy to be used in the future.

5. Conclusion

In this study, MSAS is conceptualized to provide facility for students as well as the management team of SAC. The present study addresses the existing gap in the literature about the managing and mitigating the accommodation problems among students at Universiti Utara Malaysia (UUM). The findings of the present study can be applicable for administrators to formulate guidelines in terms of management for future development of student accommodation system of UUM. Hence for future development and expansion of this research, the followings are suggested. Since the scope of this research is limited, further developments should be made Thus, to cover other services of the report complaint, ii) informing users about their reserve transaction by SMS or Push mobile technology, not only be email, iii) making integrating between writing report complaint services and SAC officer services via mobile technology are needed.

References

[1]. Modebelu, M. N., & Chinyere, A. P. (2014). Environmental hazards and hostel accommodation problems: Challenges for university education in Nigeria. US-China Education Review B, Education Theory, 4 (6), 407, 413

[2]. Alias, M., Lashari, T. A., Akasah, Z. A., & Kesot, M. J. (2014). Translating theory into practice: integrating the affective and cognitive learning dimensions for effective instruction in engineering education. European Journal of Engineering Education, 39(2), 212-232.

[3]. Ezeukwu,A.(2009).Managing students housing problems on polytechnics in Anambra and Enugu states (Unpublished master's thesis, University of Nigeria)

[4]. Duangpracha, K. (2012). The accommodation challenges of international students. RMIT University.

[5]. Maake, B., & Mzee, F. A. (2014). Application of M-Gov to Provision of Education for all in Developing Nations. Computer Science and Information Technology, 2(1), 40-44.

[6]. Wright, K. B. (2005). Researching Internet-based populations: Advantages and disadvantages of online survey research, online questionnaire authoring software packages, and web survey services. Journal of Computer-Mediated Communication, 10(3).

[7]. Pat-Mbano, E., Alaka, I., & Okeoma, O. (2012). Examining the Physio, Psycho and Socio-Economic Implications of Non-Residential Policy on Imo State University Students. Canadian Social Science, 8(2), 170.

[8]. Faiz, N. S. M., Lashari, T. A., Awang, H., Mohamed, M., Sulong, M. S., & Behrend, M. (2015). Consolidating the Literature on Mutual Adjustment in the Learning and Teaching of International Students and Academic Staff. Advanced Science Letters, 21(7), 2446-2450.
[9]. Ghalawat, S., Dhingra, A., & Boora, P. (2015). Relationship of demographic variables and factors affecting stress among college. JIMS8M: The Journal of Indian Management & Strategy, 20(3), 8-14.
[10]. Demir, S., Demir, S. G., Bulut, H., & Hisar, F. (2014). Effect of mentoring program on ways of coping with stress and locus of control for nursing students. Asian nursing research, 8(4), 254-260.
[11]. Spotti, M., Kluzer, S., & Ferrari, A. (2010). ICT for L2 Acquisition by Adult Migrants: Results from a Comparative Study in the Netherlands and Sweden. Paper presented at the Proceedings of the 3rd PIXEL Conference: ICT for Language Learning. Firenze: Mulino
[12]. Olaniyi, F. O. (2014). Perceptions of continuing education programmes by in-service teachers from two universities in Africa. University of Zululand.
[13]. Kuh, G. D., Kinzie, J., Schuh, J. H., & Whitt, E. J. (2011). Student success in college: Creating conditions that matter. John Wiley & Sons.
[14]. Kuh, G. D., Kinzie, J., Schuh, J. H., & Whitt, E. J. (2011). Student success in college: Creating conditions that matter. John Wiley & Sons.
[15]. Kumaraswamy, N. (2013). Academic stress, anxiety and depression among college students-A brief review. International Review of Social Sciences and Humanities, 5(1), 135-143.
[16]. Khawaja, N. G., & Dempsey, J. (2007). Psychological Distress in International University Students: An Australian Study. Australian Journal of Guidance and Counselling, 17(01), 13-27.
[17]. Alorfi, A. S. M. (2012). Web Based Online Volunteer Trainer System (OVTRS). Universiti Utara Malaysia
[18]. Rysavy, O. & Bures, F. (2004). Formal abstract architecture for use case specifications. Proceedings 11th IEEE International Conference and Workshop on the Engineering of Computer Based systems 2004, 203-210.http/doi.org/10 1109/ECBs2004 1316700
[19]. Bangor, A., Kortum, P., & Miller, J. (2008). The system usability scale (SUS): An empirical evaluation. International Journal of Human-Computer Interaction, 24(6), 574-594.
[20]. Coakes, E. (Ed.). (2005). Encyclopedia of communities of practice in information and knowledge management. IGI Global.
[21]. Andrew, D. P., Pedersen, P. M., & McEvoy, C. D. (2011). Research methods and design in sport management. Human Kinetics.
[22]. Vaishnavi, V. K., & Kuechler, W. (2015). Design science research methods and patterns: innovating information and communication technology. Crc Press.

A Critical Review of MANET Testbed Using Mobile Robot Technology

Farkhana Muchtar*[12], Abdul Hanan Abdullah[1], Marina Md Arshad[1], Mohd Helmy Abd Wahab[3], Siti Nor Zawani Ahmmad[4], Gaddafi Abdul-Salaam[5]

[1] Faculty of Computing, Universiti Teknologi Malaysia, 81310, Skudai, Johor, Malaysia
[2] School of Computer Sciences, Universiti Sains Malaysia, 11800, Gelugor, Pulau Pinang, Malaysia
[3] Faculty of Electronic and Electrical Engineering, Universiti Tun Hussein Onn, 86400 Batu Pahat, Johor, Malaysia
[4] Instrumental and Control Engineering, University Kuala Lumpur, MITEC, Persiaran Sinaran Ilmu, Bandar Seri Alam, 81750, Johor, Malaysia
[5] Department of Computer Science, Kwame Nkrumah University of Science and Technology, Ghana

E-mail: farkhana@gmail.com, hanan@utm.my, marina@fc.utm.my, helmy@uthm.edu.my, gaddafi.ict@knust.edu.gh

Abstract. This paper is a continuation of our previous paper under the same topic, MANET testbed using mobile robot technology. In our previous paper, we studied the topic by scrutinizing all the technical aspects and presented it as a technical review. However in this paper, we study the topic and presents it as a critical review that dwells into four aspect, namely (i) purpose, accessibility and scope of testbed facilities, (ii) usability and controllability of robot mobility in testbed facilities, (iii) repeatability and reproducibility of real mobility in testbeds, and (iv) tools for MANET implementation, deployment and debugging for experiments. With the wealth of information on the topic provided in this paper, the content of this paper is expected to be a source of reference for MANET researchers who are at a crossroad when selecting the preferred mobile robot technology and approach to suit their own specific needs.

1. Introduction

Recently, MANET (Mobile Ad Hoc Network) researchers have shown increased interest in using mobile robot technology for their testbed platforms. This paper therefore extends the technical discussions of our previous paper titled *"A Technical Review on MANET Testbed Using Mobile Robot Technology"* [1] by scrutinizing it as a critical review.

2. Critical Review

2.1. Purpose, Accessibility and Scope of Testbed Facilities.

From our observations, there are three categories of objectives and purpose in robot-based MANET testbed namely; (i) MANET testbeds for specific purpose (private testbed), (ii) MANET testbeds for community usage, and (iii) MANET testbeds for public access.

In the first category, some MANET testbed facilities have been identified as those created specifically for just one particular study or experiment. Examples of some testbed facilities created for such research purposes include the CONE project [2, 3] and Roomba MADNet [4-6] with their one-off usage characteristics. Once the project was completed, no further development or improvements were made on the testbed platforms. Robot-based MANET testbeds in this category are only accessible by the owner of the testbed facilities to carry out experiments for their own research purposes.

The second testbed category was developed for various research and experimentations that could be used either within the same field or in different fields within community groups. Community members share testbed facilities that they jointly develop on a rotation basis to perform their experiments. Examples of testbed facilities used only for internal community members are MINT, Pharos (Proteus mobile node) and ARUM.

Usually, community-based testbed facilities are used in similar research areas such as in wireless networks or MANETs. MINT and ARUM testbeds are examples of testbed facilities in the same research area. However, there are also community-based testbed facilities that cater for a variety of different fields such as Pharos that uses the Proteus mobile robot for robotic research, MANETs and wireless networks.

The last category is MANET testbed facilities for public access. This category was developed as a public facility to be utilised by researchers from all over the world. In normal circumstances, public robot-based MANET testbeds can be accessed by testbed users remotely via a web based interface.

Public access testbed facilities are usually developed to serve similar research interests that are related to general networking. The type of sub-networking that can be supported are dependent on the testbed facility provider. Some public access testbed facility laboratories are available for MANET research and some of them are equipped with robot-based MANET testbed facilities. On the other hand., some public access testbed laboratory provide facilities to conduct wireless networking research that covers and includes a wider area scope such as MANET, static mesh networks, WSNs, MSNs, DTNs, Wi-Fi, WiMAX, Bluetooth, ZigBee, IoT, Cellular Networks (including 2G, 3G and 4G networks) and other user defined radio signals.

Most public MANET testbed facilities are part of federated network testbed laboratory facilities such as GENI (Global Environment for Network Innovations) Federation [7], Fed4FIRE (Federation for Future Internet Research and Experimentation) [8], CONET (Cooperating Objects Network of Excellence) Testbed Federation [9, 10] and Planet Lab [11]. Furthermore, federation members share their resources to further improve the quality and strength of each testbed laboratory and this facilitates various experiments and research from all over the world. Examples of robot-based MANET testbeds developed for public access are IoT-LAB, NITO, CONET- IT and w-iLab.t.

2.2. Usability and Controllability of Robot Mobility in Testbed Facilities

In this section, the discussion focuses on the level of usability, configurability and controllability of mobile robot movement and the activities that are involved within the robot-based MANET testbeds that are reviewed. The purpose of this discussion is to observe the types of methods, approaches and the technologies used in the MANET testbed facilities that are related to controlling and monitoring the mobile robots during experiments. Furthermore, the ease of use of the testbed interfaces when controlling and monitoring the mobile robots by the testbed users is also examined.

2.2.1. Usability. Testbed interface to manage mobile node resources can be divided into three main types, namely, (i) Command Line Interface (CLI) such as those used to control the Proteus mobile nodes in Pharos testbed and Roomba MADNet. (ii) Desktop GUI as used in the MINT and MINT-2 testbeds, and (iii) Web based GUI as used in the w-iLab.t CONET testbed.

CLI is a user interface with the lowest usability level as compared to desktop GUI and web based GUI. Although the CLI provides more flexibility and the ability for users to have more control on the testbed but it requires users to be well versed with each of the commands and their options and most users are not able to master it properly and efficiently.

Desktop GUI enables better interactions for testbed users and provides higher usability. Testbed users perform experimental configurations and are able to conduct the experiments with ease as compared to CLI. However, users need to perform some installation processes on the testbed client terminal before its desktop GUI based testbed interface can be used. Furthermore, if any changes occur when upgrades or improvements to the testbed management system are made, the desktop GUI software needs to be redeveloped and realigned to the changes on testbed management system. At the same time, users need to reinstall the software to get new updates and features. Other challenges exist, particularly when the developed desktop GUI software is not multi-platform in nature and therefore this limits the type of client platform that can be utilised when using the testbed facilities.

A testbed interfaces that uses web based GUI is apparently the most ideal. Besides ease of use, it do not have the same problem as the GUI interface testbed. Users are not required to install any software to use it but instead only a web browser within the computer terminal is required in order to access the testbed management system. Testbed users are also not required to perform any additional upgrade process when upgrades are made to the testbed management system. A web based testbed interface also enables the use of almost all the computer platforms to conduct experiments in testbed facilities, as per requirements. The user only needs compatible web browser software.

Usually, testbed facilities utilised for specific experiments or private usage, do not emphasize on the usability of the testbed system as this is not the main goal for testbeds in this category. Testbeds for CONE research and Roomba MADNet testbed facilities are examples of testbed facilities in this category as they were developed and meant for internal users only.

Testbed facilities that are meant for community usage encourage the usability of the testbed system. Usability of the testbed system allows for shorter experiment times which allows more experiments to be performed by community members using the same testbed facilities. An example of a robot-based MANET testbed in this category is the ARUM testbed. In public testbed facilities, usability stands as one of the most important factors in the design of the developed testbed system. Examples of public testbeds with high usability are CONET-IT, w-iLab.t, IoT-LAB and NITOS testbed systems.

The usability of the testbed interface is very important in public testbed facilities as it is used by large numbers of users remotely. If the usability of a public testbed is low, only a few users are able to use the testbed facilities.

2.2.2. Controllability. Controllability of mobility in robot-based MANET testbeds can be divided into three categories; which are (i) Fixed circuit-based mobility (ii) Fixed mobility model list, and (iii) User controlled mobility.

Fixed circuit-based mobility is the easiest mobility method to set up but testbed users do not have control on the mobility of the mobile nodes in order to modify them according to their needs. Fixed circuit-based mobility provides the lowest controllability of mobile nodes as it does not allow users to control the number of mobile nodes used and the timing of mobility for each mobile node. Examples of robot-based MANET testbeds that use fixed circuit-based mobility are Sensei-UU, ARUM and IoT-Lab.

Some testbed facilities provide a list of mobility models that can be used by the users such as the random waypoint mobility model, metropolitan mobility model, mobility model group and social mobility model. Mobility models used are mostly inspired by other mobility models that are used in network simulators like ns-2, ns-3, OMNeT ++, QualNet, OPNET and NETSIM. Unlike network simulators, mobility models used in robot-based MANET testbeds are used to determine the pattern of multiple mobile robots when experiments are conducted.

The highest controllability of mobility for robot-based MANET testbeds is user controlled mobility. User controlled mobility can be divided into two subcategories distinguishable as mobility in which users manually perform the setup or in the second subcategory, in which users adds other mobility model algorithms into the testbed system as required. In the first subcategory, users perform a manual setup on the mobility of mobile nodes, choosing one from the two available setup methods. The first method is by setting up multiple waypoints manually or secondly, by setting the walk path of the mobile nodes during the experiment. Examples of testbeds that allow users to set their own mobility of mobile nodes are Pharos, MiNT and w-iLab. t.

The approach used in the second sub-category on the other hand, allows users to add new mobility model algorithms as additions to the mobility models already provided in the testbed system to suit the requirements and needs of the users. An example of a testbed facility that allow users to add a mobility model into the testbed system is IoT-LAB.

2.3. Repeatability and Reproducibility of Real Mobility in Testbeds

When the level of controllability of real mobility is high, then the repeatability and reproducibility of the mobile nodes movement is high. In any scientific research that also includes MANET, the repeatability and reproducibility of the experiments are very important as the credibility, preciseness and accuracy of the experiments are determined and dependent on it. Among the reasons why mobile robot technology is selected to provide real mobility in MANET testbeds is mainly due to the fact that this method provides the most efficient method that allows the movement of the mobile nodes to be repeatable and reproducible when experiments are conducted [12]. This is because the higher the level of controllability of real mobility, the higher the repeatability and reproducibility of the movement of the mobile nodes in the experiments.

Nonetheless, not all the MANET testbeds that were reviewed in this work are repeatable and reproducible in nature. This is because, the accuracy of the mobile robot localisation that is used in the testbeds is the key factor that would determine whether or not the mobility of the mobile nodes can be repeated or reproduced. In this section, the characteristics of the repeatability and reproducibility of the mobility of the mobile nodes in the reviewed MANET testbeds are outlined.

2.3.1. Repeatability. In scientific research, observations on experiments conducted have no serious bearings and are not regarded as scientific observations if the experiments are not performed repetitively [13]. Furthermore, the repeatability of an experiment is very important to ensure that uncontrolled parameters or external factors such as wireless communication interferences and signal noise during the testbed runs can be obtained randomly through repeated tests to ensure unbiased testbed results [12].

Hence, experiments carried out in MANET testbeds must also be repeatable in their characteristics and this includes the mobility pattern of the mobile nodes. The repeatability of the movement of the mobile nodes in the MANET testbeds can be described in simple terms as the ability to repeat the movement of each mobile node in the same testbed based on position, direction of the movement and all the activities involved when moving.

To achieve repeatability in MANET testbeds is difficult and costly [14, 15]. To generate the repeatability of real mobility in testbeds is even more difficult as each mobile node requires the capability to move in the same mobility pattern, at the exact same period of time and at the same time be able to perform and obtain the same results from the previous experiments [16]. Therefore, most MANET testbed developers have used an emulation method to create a repeatable mobility mechanism in their testbed [15].

Among the reasons for the lack of repeatability in the movement of the mobile nodes movement are the lack of controllability features available in the testbed [12]. A high controllability of real mobility in the MANET testbed is achievable by developing a mobile robot localisation and monitoring system that is reliable and accurate. With a highly reliable and accurate mobile robot localisation, the movement of mobile nodes in a MANET testbed can be repeated thus reducing biasness on the experimental results.

Some of reviewed MANET testbed facilities in this work that were identified to operate with characteristics of repeatability on the mobility of mobile nodes were Sensei-UU, Kansei, and CONET, w-iLab-t, FIT IOT and NITOS laboratory testbeds. This was because all of the testbeds had mobile robot localisation and positioning systems that were reliable and accurate.

In addition, repeatable experiments can only be realized if the testbed facilities have the ability to store mobile node movements in the form of experiment configurations and descriptions. Public testbed facilities such as w-iLab-t, CONET, FIT IOT Laboratory and NITOS have high repeatability of mobility as this feature is particularly vital in public testbed facilities in order to ensure that the results obtained are regarded as scientifically valid.

The repeatability of mobility in MANET testbeds is easily implemented by using physical tracking (line tracing) to determine the movement of the mobile nodes. Control on the movements of the mobile robots is relatively easy as each mobile robot only needs to follow a black line on the floor using IR sensors and the monitoring system determines the current position of the mobile robots. However, the main disadvantage of this particular method is the limited and inflexible movement patterns that cannot be diversified.

For a virtual path method such as that used in Mobile Emulab and MINT testbeds, mobile nodes movement was found to be more flexible as it created different random waypoints for each experiment conducted and the same movement pattern could be repeated many times. This method however, required a mobile robots localisation and positioning system that was much more complex but it did allow better flexibility in providing controlled mobility of the mobile nodes.

There are also testbeds where movements of the mobile node[s] were non-repeatable as they were not equipped with a mobile robot localisation and positioning system such as CONE, MADNet Roomba serial connector and iRobotSense. Although the experiments were carried out repeatedly but the repetition used different mobile node movements as the mobility pattern was random and uncontrolled. As there was no mechanism to determine the current position and direction of each mobile node, the level of controllability of the mobile node mobility was low and results of the mobile node movements were unrepeatable.

It is believed that irrespective of whether the MANET testbed is private or public, the scope of experiment is specific or generic, the repeatability aspects of real mobility should be taken into account as this will be the determining factor that would affect the credibility of the experimental data and conclusions as whether or not, it will be accepted as scientifically valid by the research community.

2.3.2. Reproducibility. In the Oxford English Dictionary, reproducibility is defined as "the extent to which consistent results are obtained when repeatedly produced". In other words, reproducibility is the ability to completely or almost completely duplicate an experiment or study by other researchers who are conducting the experiment independently [17].

A reproducible experiment enables a research to be validated by other researchers to strengthen and verify that the findings obtained from their research are credible [18]. Any scientific hypothesis in an undertaken experiment should be reproducible to allow independent validations to be performed by other researchers as it is the core of the scientific method [18, 19]. Popper once said in his book 'The Logic of Scientific Discovery' that "a non-reproducible single occurrence poses no significance to science" [13].

MANET testbeds that are able to run reproducible experiments also require testbed facilities that can run repeatable experiments. However, repeatable testbed facilities are not enough to execute reproducible experiments. To allow the movement of mobile nodes in the testbed during an experiment to be reproducible, testbed facilities need to be able to record movements of mobile nodes with the exact similar timings and activities or processes that are involved during a specific mobile node movement.

Most of the older generation robot-based MANET testbeds (before 2010) had similar problems in con- ducting reproducible experimental testbeds in other locations. This was because robotic technology that was available at the time was either very expensive or the mobile robot localisation used depended on the local infrastructure conditions [12]. Thus, in the newer generation of robotic-based MANET testbeds (during and after 2010), the latest robotic technology was used which was cheaper and easier and hence, the experiments performed could be easily reproduced by other researchers.

Examples of robot-based MANET testbeds that can perform reproducible experiments are CONET, w- iLab.t, FIT IOT lab, Kansei testbed and NITOS. This is because all of the aforementioned testbeds use the latest easy-to-obtain robotic technology that enables real mobility for MANET experiments to be conducted with ease and better performance. Furthermore, testbed configurations provided by robot-based MANET testbeds use wider network testbed frameworks such as the GENI Framework and OMF-OML framework. Hence, testbed configurations, mobile node movements and all their activities can be easily reproduced on another testbed using the same framework.

Therefore, the development of the present robotic technology has facilitated the development of robot-based MANET testbeds that are capable of producing an environment for reproducible experiments. Additionally, whether or not the testbed developed is private, experiment-specific or is a public testbed, network testbed frameworks that are recommended for use that are widely used by others are the GENI framework and OMF-OML framework that can improve the reproducibility aspect of the experiment conducted in testbed facilities that they develop.

2.4. Tools for MANET Implementation, Deployment and Debugging for Experiments

Among the main reasons why only a few researchers have used the testbed approach to conduct their experiments is because of the fact that the development of a real implementation of any MANET solution is a highly difficult and complex process. Implementing a suggested solution of MANET using a simulation model is far much easier as researchers only need to focus on sections that are of interest to them. Developing a real implementation of a MANET suggested solution requires multiple and varied technical skills, high commitment, dedication and a long time to complete.

Mechanisms to facilitate areal implementation process of MANET is important because it can increase the amount of MANET research using testbed platforms as a tool to test the suggested solutions developed by MANET researchers. MANET implementation in real world testbeds should be made to be at least as easy as MANET implementation in network simulators. In addition, the mechanisms should also allow simulation results to be validated with testbed results in improving the quality and accuracy of the simulation model of MANET solutions.

There have been a few developed hybrid types of simulators for MANETs and WSN such as sensor sensorsim [20, 21], EM* or EMStar [22, 23] and TOSSIM [24].

Based on previous MANET testbed implementations that were reviewed, several solutions proposed by previous researchers from their testbeds were recorded. Among them are MINT that introduced a hybrid simulator that was developed. De [25] developed a modified hybrid simulator from ns-2 network simulator by replacing the simulation model of the link layer, MAC layer and physical layer of the simulator with wireless card drivers, firmware, and a real wireless channel respectively. Hence, the ns-2 based simulation model implementation could be reused as a real implementation in the testbed through the use of hybrid simulator.

In the Pharos, a Click modular router was used to facilitate MANET implementation testbed to be as simple as MANET simulations using simulation models in a network simulator [26-28]. The Click modular router is a software platform that allows the development of modular, flexible and configurable. Each component in Click performs simple router functions such as packet classification, queuing and packet scheduling. The modular design allowed Click users to focus only on the development of routing implementation specifically on the area of interest without the need of thinking about other unrelated and irrelevant components [29]. In the Pharos testbed, Click was used to validate implementation developed in an OMNeT ++ network simulator and the data obtained from the testbed was then compared with the data obtained from the OMNeT++ simulation [26-28].

MANET solution deployment and debugging in the testbed's mobile node are also very important because when conducted manually, they will reduce the autonomy and simplicity of the testbed operation. A manual deployment and debugging process of a MANET solution to each individual mobile node is not practical and efficient in any testbed operation. Therefore, a distributed deployment and distributed mechanism is required in MANET testbeds especially in mobile node facilities.

In the MANET testbeds that were reviewed, only the MINT testbed platform discussed the mechanisms for software deployment and debugging in detail. MINT developers used FIAT (Fault Injection and Analysis Tool) as the software component to enable distributed deployment and debugging of various MANET solutions which allowed experiments to be conducted in the MINT testbed with ease.

There are public access MANET testbed that provides the Click modular router in their facilities such as w-iLab.t, where its usage in testbed facilities can be found in the user's manual. However, the use of the Click modular router in public MANET testbeds has yet to be discussed critically and academically. From these observations, it was found that no research has seriously investigated on the potential use of supporting tool such as the Click modular router ns-3-click bridge and FINS framework in the development of MANET solution as part of the facilities that are available in MANET testbeds.

Table 1. Robot-Based MANET Testbed Criteria Summary

Testbed	Purpose and Accessibility	Scope	Usability	Controllability	Repeatability	Reproducibility	MANET Dev. Tools
Mobile Emulab	Public	Mobile Sensor Network	Medium	Low	High	Medium	None
MiNT-m	Community	Mobile Sensor Network	Medium	Medium	High	Low	Hybrid Simulator
MiNT-2	Community	Mobile Sensor Network	Medium	Medium	High	Low	Hybrid Simulator
Proteus	Community	Various	Low	High	Medium	Medium	Click Modular Router
w-ilab.t	Public	Wireless Network	High	High	High	High	Click Modular Router
ARUM	Community	MANET, VANET, DTN	Medium	Medium	High	Low	None
Sensei-UU	Community	Mobile Sensor Network	Medium	Medium	High	Low	None
Kansei	Public	Mobile Sensor Network	Medium	Low	High	Low	None
CONE-IT	Public	Wireless Network	High	High	High	High	None
CONE	Private	MANET, DTN	Low	Low	Low	Low	None
Roomba MADNet	Private	MANET	Low	Low	Low	Low	None
Explorebot	Community	Mobile Sensor Network	Medium	High	Low	Low	None
SCORPION	Community	MANET, DTN	Medium	Low	Low	Low	None
MOTEL	Community	Mobile Sensor Network	Medium	Medium	High	Low	None
iRobotSense	Private	Mobile Sensor Network	Low	Low	Low	Low	None

3. Conclusion

The targeted objective of this work was achieved, that is to comprehensively discuss and analyse on the topic of mobile robot usage in MANET testbeds. Thus, it is hoped that this article is able to assist potential users to easily choose their preferred mobile robot technology and approach that is suitable to his own needs.

4. Acknowledgement
The authors would like to thank the Universiti Teknologi Malaysia (UTM) for supporting this research under VOT NUMBER: Q.J130000.2528.14H81.

References
[1] Farkhana Muchtar, Abdul Hanan Abdullah, Muhammad Shafie Abd Latiff, Suhaidi Hassan, Mohd Helmy Abd Wahab, Gaddafi Abdul-Salaam 2017 A Technical Review of MANET Testbed Using Mobile Robot Technology *J. Phys.: Conf. Ser*
[2] Vingelmann P Heide J Pedersen M V Zhang Q Fitzek F H P 2014 *Computer Networks* 74 **Part** B pp 34–47
[3] Vingelmann P Pedersen M Heide J Zhang Q Fitzek F *IEEE International Conference on Communications (ICC'12)* pp. 291–296
[4] Reich J Misra V Rubenstein D Zussman G 2012 *IEEE Journal on Selected Areas in Communications* **30**(5) pp 935–950
[5] Reich J Misra V Rubenstein D 2008 *ACM SIGMOBILE Mobile Computing and Communications Review* **12**(1) pp 68–70
[6] Reich J Misra V Rubenstein D-S Zussman G Spreadable Connected Autonomic Networks (SCAN) Columbia University Academy Common
[7] Berman M Chase J S Landweber L Nakao A Ott M Raychaudhuri D Ricci R Seskar I 2014 *Computer Networks* **61** pp 5–23
[8] Turi L 2015 *Contribution to the Federation of the asynchronous SmartSantander service layer within the European Fed4fire context* Master Thesis University of Padova, Padova, Italy
[9] Dios J R M.-d., Jimenez-Gonzalez A, Bernabe A d S, Ollero A 2014 *A Remote Integrated Testbed for Cooperating Objects*, Springer Briefs in Electrical and Computer Engineering, (Switzerland: Springer International Publishing) pp 23–39
[10] Dios J R M.-d., Jimenez-Gonzalez A, Bernabe A d S, Ollero A 2014 *Electrical and Computer Engineering* (Switzerland: Springer International Publishing) pp. 59–73
[11] Kim W Roopakalu A Li K Y Pai V S 2011 *Proc. of the 2011 ACM SIGCOMM Conference on Internet Measurement Conference, IMC'11* (New York, NY, USA: ACM) pp 515–532
[12] Rensfelt O Hermans F Gunningberg P Larzon L-Å Björnemo E 2011 *The Computer Journal* **54**(12) pp 1973–1986
[13] Popper K R 2002 *The Logic of Scientific Discovery* (San Diego: Psychology Press)
[14] Kulla E Ikeda M Barolli L Xhafa F Iwashige J 2012 *Computer Science and Convergence, LNEE*, vol 114, ed J J J H Park, H-C Chao *et al* (Netherlands: Springer) pp 651–657
[15] Yoon H Kim J Ott M Rakotoarivelo T 2009 *Proc. of the 4th ACM International Workshop on Experimental Evaluation and Characterization, WINTECH '09* (New York, NY, USA: ACM) pp. 51–58
[16] Tschudin C Gunningberg P Lundgren H Nordström E 2005 *Ad Hoc Networks* **3**(2) 221–233.
[17] Casadevall A Fang F C 2010 *Infection and Immunity* **78**(12) pp 4972–4975
[18] Vitek J Kalibera T 2011 *Proc. of the 9th ACM International Conf. on Embedded Software, EMSOFT'11 (New York, NY, USA: ACM)* pp 33–38
[19] Vaidya N H Bernhard J Veeravalli V V Kumar P R and Iyer R K 2005 *Proc. of the 2005 ACM SIG- COMM Workshop on Experimental Approaches to Wireless Network Design and Analysis, E-WIND'05* (New York, NY, USA: ACM) pp. 64–69
[20] Zhou J Ji Z Bagrodia R *25th IEEE International Conf. on Computer Communications. Proceedings (INFOCOM'06)* pp. 1–13
[21] Park S Savvides A Srivastava M B 2000 *Proc of the 3rd ACM International Workshop on Modeling, Analysis and Simulation of Wireless and Mobile Systems, MSWIM '00* (ACM, New York, NY, USA: ACM) pp. 104–111

[22] Girodb L Elson J Cerpa A Stathopoulos T Ramanathan Estrin N 2004 *Proc. of the Annual Conference on USENIX Annual Technical Conf., ATEC '04* (Berkeley, CA: USENIX Association) pp 24–24
[23] Girod L Ramanathan N Elson J Stathopoulos T Lukac M Estrin D 2007 *ACM Trans. Sen Netw.* **3**(3)
[24] Levis P Lee N Welsh M Culler D 2003 *Proc. of the 1st International Conf. on Embedded Networked Sensor Systems, SenSys'03* (New York, NY, USA ACM) pp. 126–137
[25] De P 2007 *Mint: A Reconfigurable Mobile Multi-hop Wireless Network Testbed* Ph.D. Thesis State University of New York at Stony Brook, Stony Brook, NY, USA
[26] Petz A Bednarczyk A Paine N Stovall D Julien C 2010 *Technical Report* TR-UTEDGE-2010-011 University of Texas at Austin
[27] Petz A Fok C-L Julien C Walker B Ardi C 2011 *Proc. of the 3rd Extreme Conf. on Communication: The Amazon Expedition, ExtremeCom'11* (New York, NY, USA: ACM) pp 4:1–4:6
[28] Petz A 2010 The Click Convergence Layer: Putting a Modular Router Under DTN2
[29] Kohler E Morris R Chen B Jannotti J Kaashoek M F 2000 *ACM Trans. Comput. Syst.* **18**(3) pp 263–297

Mobile Ad hoc Network Testbed Using Mobile Robot Technology

Farkhana Muchtar*[12], Abdul Hanan Abdullah[1], Mohd Helmy Abd Wahab[3], Radzi Ambar[3], Hafizul Fahri Hanafi[4], Siti Nor Zawani Ahmmad[5]

[1] Faculty of Computing, Universiti Teknologi Malaysia, 81310, Skudai, Johor, Malaysia
[2] School of Computer Sciences, Universiti Sains Malaysia, 11800, Gelugor, Pulau Pinang, Malaysia
[3] Faculty of Electronic and Electrical Engineering, Universiti Tun Hussein Onn, 86400 Batu Pahat, Johor, Malaysia
[4] Faculty of Art, Computing and Creative Industry, Universiti Pendidikan Sultan Idris, 35900 Tanjung Malim, Perak, Malaysia
[5] Instrumental and Control Engineering, University Kuala Lumpur, MITEC, Persiaran Sinaran Ilmu, Bandar Seri Alam, 81750, Johor, Malaysia

E-mail: farkhana@gmail.com, hanan@utm.my, helmy@uthm.edu.my, aradzi@uthm.edu.my, hafizul@fskik.upsi.edu.my, sitinorzawani@unikl.edu.my

Abstract. MANET (Mobile Ad Hoc Network) researchers have shown increased interest in using mobile robot technology for their testbed platforms. Thus, the main motivation of this paper is to review various robot-based MANET testbeds that have been developed in previously reported research. Additionally, suggestions to heighten mobility mechanisms by using mobile robots to be more practical, easy and inexpensive are also included in this paper, as we unveils ToMRobot, a low-cost MANET robot created from an ordinary remote control car that is capable of performing a real system MANET testbed with the addition of only a few low-cost electronic components. Despite greatly reduced costs, the ToMRobot does not sacrifice any of the necessary MANET basic structures and will still be easily customisable and upgradeable through the use of open hardware technology like Cubieboard2 and Arduino, as its robot controller. This paper will also include guidelines to enable technically limited MANET researchers to design and develop the ToMRobot. It is hoped that this paper achieves its two pronged objectives namely (i) to facilitate other MANET researchers by providing them with a source of reference that eases their decision making for selecting the best and most suitable MANET mobile robots for real mobility in their MANET testbeds (ii) to provide MANET researchers with a prospect of building their own MANET robots that can be applied in their own MANET testbed in the future.

1. Introduction

The majority of existing research involving MANET evaluations have utilised simulation methods primarily. But of late, there has been an increase in awareness that the current practice of utilizing network simulators only allows for summarized assumptions in modelling the characteristics of a real system. As the researcher needs to observe the effect and influence of the MAC (Media Access Control) and the physical layer network with regards to mobility and topology changes against the links and communication quality of the MANET, it becomes clear that real mobility in MANET testbeds is vital. As such, in terms of accuracy, the data derived from this method cannot be similarly obtained by using other methods like network simulation and emulation. At present, the number of experiments conducted in MANET testbeds are significantly lesser compared to simulation based MANET experiments [1]. There are various methods of mobility implementation that are used in MANET testbeds. These methods can be categorised into two main divisions, real mobility and emulated mobility or virtual mobility as it is also known.

The main distinctive feature of emulated mobility is due to non-physical node mobility. In other words, although real implementation of MANET testbed are performed on the data-link layer and above (application layer) but node mobility is not conducted physically. Physical node movement and topology changes are carried out through the use of emulation [2]. There are several different methods of emulated mobility such as instance migration [3], on/off connection [2], RF matrix switch [4] and the use of virtual machine technology and virtual network [5].

Emulated mobility has the advantage of being repeatable and having reproducible mobility that is almost the same as the network simulator. The mobility mechanism conducted is therefore more manageable and predictable. However, emulated mobility is unable to describe the actual MAC layer and physical layer. Hence data obtained through it is less accurate than methods that use real mobility [2]. Among the wireless testbed platforms that use emulated mobility methods are the Open-Access Research Testbed for Next-Generation Wireless Networks (ORBIT) [3], Ad-hoc Protocol Evaluation (APE) testbed with emulated mobility [6], the Carnegie Mellon University Wireless Emulator (CMUWE) [7], Castadiva [8], Mobi-emu [9], Emulab [10], MOBNET [11], MobiNet [12], the Resilience Evaluation Framework for Ad Hoc Networks (REFRAHN) [13], MeshTest [4] and WISEBED [14].

Conversely, real mobility in a MANET testbed experiment is vital in order to obtain accurate and realistic results, despite the challenging nature of implementing it. Though if done correctly, real mobility implementation is able to provide extremely accurate experimental data on the impact of mobility on a MANET [15].

Previous research on MANETs have reported the use of various approaches to provide real mobility mechanisms in their MANET testbeds. Some of the methods included the use of cars [16], taxis [17], trains [18], bicycles [19], humans [20], and remote control cars [21]. Considering that the main motivation of this paper is to review various robot-based MANET testbeds that have been developed in previously reported research, other available but non-related mechanism types will not be discussed in this paper.

Since cost and other attendant concerns of acquiring robots to perform real mobility in MANET testbed experiments is a major constraint that hinders MANET researchers from using it as an evaluation tool for MANET testbeds, this paper also serves to propose and develop a low cost robot for MANET testbeds, named as ToMRobot that is built with a remote control car and low cost electronic components.

This review paper provides a two prong approach in its objectives. In other words, this review paper will allow readers to observe how mobile technology robots have been used to create well-designed MANET testbeds while at the same time providing readers with alternative solution, that is to develop their own MANET robot at a fraction of the cost yet it is still equipped with the basic structure to function well and can easily be upgraded whenever necessary.

This paper is organized as follows. In section 2, background information on former research carried out on MANET robot platforms is detailed. Section 3 describes the development of ToMRobot and finally, in section 4, the research is summarized and the future goals of this work are out lined.

2. Literature Review

This section list out and discusses earlier robots that were used in previous MANET–based research. Table 1 provides readers with a quick glimpse on the types of robots used in MANET-based research.

It has been identified from literature [22] that most of the MANET testbed platforms have used readymade robot platforms such as iRobot Roomba/Create and LEGO Mindstorms. iRobot Roomba and Create are one of the most popular robot platforms due to several factors, including the ability to carry loads up to 2kg, the presence of basic sensors for mobile robots such as bump sensors, obstacle avoidance sensors and wheel encoders, the ease of controllability via iRobot Roomba Open Interface (ROI), the large rechargeable li-ion battery (44000 I) capacity, and the self-recharging docking station as well as their competitive prices as compared to other platforms.

Historically, the use of iRobot Roomba began when it was introduced in 2002 and it became among the earliest household robots made available to the public. It functioned as an autonomous robotic vacuum cleaner while at the same time it became a popular robot choice for robotics researchers and hobbyists that used it to be the main platform for mobile robots.

iRobot Create on the other hand was introduced in 2007 and it was based on the iRobot platform due to the popularity of iRobot Roomba amongst robotics researchers and hobbyists. iRobot Create was sold at a lower price because it did not have some of the components attached to iRobot Roomba such as a vacuum cleaning component, although the component could be added separately [22].

Table 1. List of Robots Used in MANET Testbeds

Robot Name	Robot Platform	References
Traxxas Stampede RC Car Chassis	Proteus	[24-26]
iRobot Roomba	MiNT-m w-ilab.t	[23] [27-29]
iRobot Roomba/Create	Roomba MADNet	[30, 31]
iRobot Create	MiNT-2 iRobotSense NITOS	[24-26] [33] [52]
LEGO Mindstorms	CONE Sensei-UU	[36, 37] [1, 34, 35]
Acroname Garcia	Mobile Emulab Kansei	[38, 39] [40, 41]
e-puck	MOTEL	[42-44]
Pioneer 3AT	CONET-IT	[45-48]
RC car chassis	CONET-IT	[45-48]
Rogue ATV	Explorebots	[51]
Lynxmotion 4WD Rove	ARUM	[49, 50]
Turtlebot2	IoT-Lab	[53]
Wifibot	IoT-Lab	[53]

iRobot Roomba and Create were usually controlled with a robot controller (usually a combination of an embedded PC and microcontroller) through the iRobot Roomba Open Interface (ROI) protocol whereby communication was enabled via a serial port on iRobot Roomba/Create including MINT-m [23], MINT-2 [24-26], Proteus Roomba [24–26], w-ilab.t [27-29], Roomba MADNet [30, 31], SCORPION [32] and iRobotSense [33].

Apart from iRobot Roomba and Create, LEGO Mindstorms was another preferred choice as a robot platform for mobile robots in the MANET testbeds. Examples of MANET testbeds that used the LEGO Mindstorms robot platform are Sensei UU [1, 34, 35] and Cone Testbed [36, 37].

The Acroname Garcia robot platform also was a popular choice in several MANET testbeds such as the Emulab Mobile (TrueMobile) [38, 39] and Kansei Testbed [40, 41]. There have also been reports of other readymade robot platforms that were utilized, such as e-puck that was used in the MOTEL testbed [42-44], the Pioneer 3AT robot platform that was used in CONET-IT [45-48], the Lynxmotion 4WD rover robot platform used in ARUM [49, 50] and the Rogue ATV robot platform used in Explorebot [51].

Previous studies revealed that the cost for robots used in said MANET testbeds ranged from between USD $3000 to USD $6,000. Understandably, procuring the higher end price range robots were a constraint for many MANET researchers. Opportunely, the ToMRobot is able to fill this research need by providing the same benefits as the existing robots, with the difference being that the ToMRobot uses different robotic components. Propitiously, said components are easily available, cheaper and easy to install by MANET researchers themselves because the required components are common ones used by hobbyists and robot makers for making mobile robots.

3. Development of ToMRobot

This section discusses the three main parts of the robot's components used in this study: (i) the mechanical component, (ii) the electronic component and (iii) software.

3.1. Mechanical Component

The mechanical component of the robot represents the robot anatomy, where it can be compared to a human body. The main element for the mechanical component of the robot is the robot chassis because it is the main mechanical structure of the robot and acts as a base to other mechanical components.

The mechanical component of this study utilizes a modified remote control car as the robot chassis. This is the cheapest solution as the RC car body serves as the robot housing and has a DC motor, gear system to increase torque and big wheels for testbed durability. Figure 1 and Figure 2 shows the original remote control car chassis before and after modification.

Figure 1. Original Remote Control Car

Figure 2. Modified Chassis of the Remote Control Car

3.2. Electronic Component

The electronic components are elements used as inputs and outputs to the robot. Input components such as sensors, cameras, LIDAR and IMU act as the information provider to the robot in order to determine the environmental conditions and the location of the robot.

The electronic components used in this study can be divided into two parts namely: ToMRobot and ToMDuino. The components used for ToMDuino and ToMRobot will be discussed in the next section.

3.2.1. ***ToMduino Electronic Components.*** Instead of using an Arduino board, a customized version of the Arduino board, named ToMDuino (Arduino MANET Testbed), was created for this undertaking and can be seen in Figure 3 and Figure 4.

The objectives of developing ToMduino are as follows:

i. **Small.** Suitably sized for the testbed robot chassis.
ii. **Cheap.** Cheaper than the official Arduino board with several required shields. In most cases, the Arduino shield designs are not optimized for advanced Arduino project requirements, including ToMRobot.
iii. **Simplicity.** Avoiding the use of a combination of the Arduino board with several shields, which would increase the complexity of the setup.
iv. **Compatibility.** Some Arduino shields are simply not compatible or cannot be used together with other Arduino shields.

The major differences of this customized Arduino version with the official Arduino Uno R3 version are as follows:

i. Built-in dual h-bridge motor controller to control the ToMRobot DC motors.
ii. Built-in XBee socket for the ToMRobot wireless communication using the XBee protocol with a 3.3v power source board.
iii. Specific pin header for sensors, serial connection to cubieboard2 and LCD screen.
iv. No generic I/O pin header-like official.
v. Better 3.3v voltage regulator specifically for XBee.

Figure 3. ToMDuino (Actual Implementation)

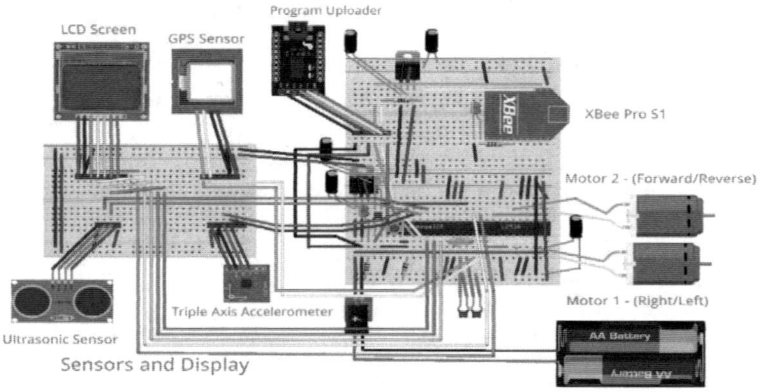

Figure 4. ToMDuino (Circuit Design)

3.2.2. ToMRobot Electronic Components. Table 2 below provide readers with a pictured view on the list of component required to build ToMRobot. Once the listed components in Table 2 have been assembled, the created robot will be similar to that in Figure 5. Figure 6 serves to illustrate how to assemble the ToMRobot.

Table 2. Description of the ToMRobot Components

Name of Component	Descriptions
Cubieboard2	To represent mobile device
USB 2.0 hub	To provide extra USB port for cubieboard2
Powerbank	To supply electric power for cubieboard2
Wi-Fi Dongle	To provide Wi-Fi connection for MANET environment testbed
XBee Pro S1	To provide remote communication during testbed
HD Webcam	To provide life HD video streaming testbed
GPS Module	To provide ToMRobot location during MANET testbed
Triple Axis Accelerometer Sensor	To provide ToMRobot location during MANET testbed
Ultrasonic Sensor	To allow ToMRobot avoiding any possible obstacle during testbed

Figure 5. ToMRobot Components for MANET Testbed (Actual Implementation)

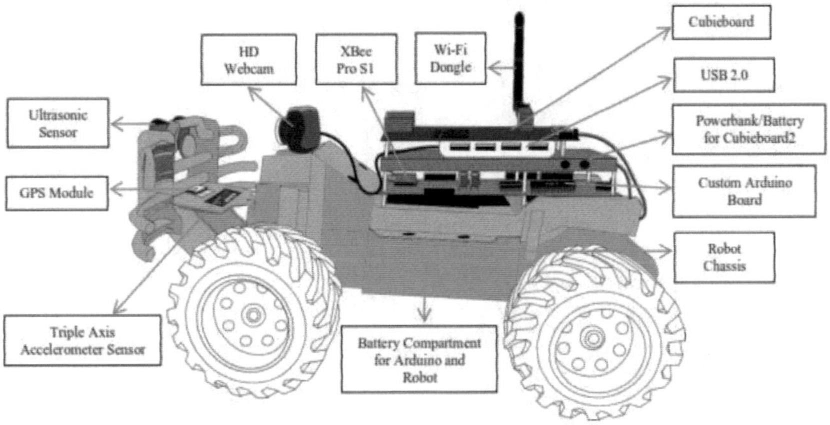

Figure 6. ToMRobot Components for MANET Testbed (Concept Design)

3.3. Software

The last component of ToMRobot is the software, which acts as the brain or control centre of the robot that is based on logic that has been set in advance by the robot builder. The software component also acts in translating the input received from the input components and using that information to make decisions before translating it into output components such as the movement of robot's wheels, legs, or arms.

There are three software that have been used in this study; Robotic Operating System (ROS), Rvis and Gazebo. ROS was used as a robot operating and monitoring system. The Rosserial protocol was used to allow the ROS to be able to communicate with the ToMRobot via XBee wireless communication. Concomitantly, by using Rosserial, it negated the need to develop another new ROS driver to control the ToMRobot, which would increase the testbed development. Through a combination of the ROS, Rosserial and Gazebo, the coordinator will give directions to the ToMRobot testbed for executing the testbed. Simultaneously, the

ToMRobot will return the current status of the ToMRobot, mobile devices (Cubieboard2) and testbed data that is captured in real time.

4. Conclusion

The two prong objectives of this paper has been met, that is (i) to be source of reference for other MANET researchers who are considering the choice of the most suitable mobile robots for real mobility in their MANET testbeds. (ii) to be source of reference for MANET researchers to build their own MANET robots that can be used in their future MANET testbed.

The approach used to develop ToMRobot has the advantage of using easily obtainable and affordable robotic components like a remote control car, which serves as the robot chassis, combined with the use of open hardware technology like Cubieboard2 and Arduino as the robot controller. These combined factors is what enables this robot to be conveniently developed by anyone who is interested to use it in their MANET testbeds. The provided guidelines in this article will aid MANET researchers who lack the technical background to build their own robot and ultimately increase the number of MANET researchers who choose testbeds as their evaluation tool.

This article also serves as a preliminary step for the development of future research projects. For future work, the ToMRobot will be used to measure energy efficiency in MANET testbeds. Two types of MANET testbeds will be designed and set up for the experiment; (i) passive mode and (ii) testbed with mobility. The remote car chassis used in this paper will also be changed to a Tamiya tank remote car chassis so that it can climb obstacles with ease and traverse crevasses without the wheels getting stuck.

The extension of this article can be referred to our paper entitled "*A Technical Review of MANET Testbed Using Mobile Robot Technology*" [54] and "*A Critical Review of MANET Testbed Using Mobile Robot Technology*" [55].

5. Acknowledgements

The authors would like to thank the Universiti Teknologi Malaysia (UTM) for supporting this research under VOT NUMBER: Q.J130000.2528.14H81.

References

[1] Rensfelt O Hermans F Gunningberg P Larzon L.-Å. and Björnemo E 2011 *The Computer Journal*, **54** (12) pp 1973–1986

[2] Yoon H Kim J Ott M, and Rakotoarivelo T 2009 *Proc. of the 4th ACM International Workshop on Experimental Evaluation and Characterization, WINTECH '09* (New York, NY, USA: ACM) pp 51–58

[3] Raychaudhuri D Seskar I M Ott M Ganu S Ramachandran K Kremo H, Siracusa R Liu H, and Singh M 2005 *Wireless Communications and Networking Conf., 2005 IEEE*, Vol. 3 pp 1664–1669.

[4] Hahn D Lee G Kim Y Walker B Beecher M and Mundur P 2010 *Proc. of the 5th ACM Workshop on Challenged Networks, CHANTS '10*, (New York, NY, USA: ACM) pp. 79–82

[5] Kim Y Taylor K Dunbar C Walker B and Mundur P 2011 *Proc. of the 3rd ACM International Workshop on MobiArch, Hot Planet '11,* (New York, NY, USA: ACM) pp 23–28

[6] Nordstrom E Gunningberg P and Lundgren H 2005 *1st International Conf. on Testbeds and Research Infrastructures for the Development of Networks and Communities*, 2005. Tridentcom'05 pp100–109

[7] Wang X Borries K Anderson E and Steenkiste P 2011 *2011 IEEE Vehicular Technology Conf. (VTC Fall)* pp 1–5.

[8] Hortelano J Cano J-C Calafate C T and Manzoni P 2009 *EURASIP J. Wireless Communication Networking* pp 47:1–47:9.

[9] Zhang Y and Li W 2001 *Proc. of the 3rd ACM International Symposium on Mobile Ad Hoc Networking & Computing (MobiHoc'02)*, (new York, NY, USA: ACM) pp 104–111

[10] Guruprasad S Ricci R and Lepreau J 2005 1st *International Conf. on Testbeds and Research Infrastructures for the Development of Networks and Communities, 2005. Tridentcom'05* pp 204–212

[11] Krop T Bredel M Hollick M and Steinmetz R 2007 *Proc. of the 2nd ACM International Workshop on Wireless Network Testbeds, Experimental Evaluation and Characterization, WinTECH '07,* (New York, NY, USA: ACM) pp 27–34

[12] P. Mahadevan, A. Rodriguez, D. Becker, and A. Vahdat, *2005 Workshop on Wireless Traffic Measurements and Modeling, WiTMeMo '05,* USENIX Association, Berkeley, CA, USA, 2005, pp 7–12.

[13] Friginal J Andrés D de Ruiz J-C and Martínez M 2015 *Computer Networks* 82 pp 114–134.

[14] Coulson G B Porter B Chatzigiannakis I Koninis C Fischer S Pfisterer D Bimschas D Braun T Hurni P Anwander M Wagenknecht G Fekete S P Kröller A and Baumgartner T 2011 *Commun. ACM* **55**(1) pp82–90

[15] Kulla E Ikeda M Barolli L Xhafa F and Iwashige J, 2012 *A Survey on MANET Testbeds and Mobility Models* (eds) Park J J J H Chao H-C Obaidat M S Kim J Computer Science and Convergence, LNEE (Heidelberg: Springer Berlin) **Vol 114** pp 651–657

[16] Ramanathan R Redi J Santivanez C Wiggins D and Polit S, 2005 *IEEE Journal on Selected Areas in Communications* **23**(3) pp 496–506

[17] Galati A Bourchas T Siby S Frey S Olivares M Mangold S, 2014 *2014 IEEE Global Humanitarian Technology Conference (GHTC)* pp 699–705

[18] Zarafshan-Araki M and Chin K-W 2010 *Computer Communications* **33**(15) pp 1850–1863

[19] Kazdaridis G and Stavropoulos D Maglogiannis V Korakis T Lalis S Tassiulas L 2014 *2014 IEEE 15th International Conf. on Mobile Data Management (MDM),* Vol. 1 pp 89–98

[20] Kulla E Oda T Ikeda M Barolli L Biberaj A and Shurdi O, 2014 *Int. J. Wire. Mob. Comput.* **7**(4) pp 318–326

[21] M. J. F. Alenazi, E. K. Çetinkaya, J. P. Rohrer, *International Telemetering Conf. Proc*

[22] De, L P Maestre J Camacho E and Alonso I 2011 *International Journal of Ad Hoc and Ubiquitous Computing* **7**(3) pp 192–201.

[23] De P Raniwala A Krishnan R Tatavarthi K Modi J Syed N A Sharma S and Chiueh T-C 2006 *Proc.of the 4th International Conf. on Mobile Systems, Applications and Services, MobiSys'06,* (New York, NY, USA: ACM) pp 124–137.

[24] Mitchell C Munishwar V Singh S Wang X Gopalan K and Abu-Ghazaleh N 2009 *Communication Systems and Networks and Workshops (COMSNETS 2009)* First International pp 1–10

[25] Munishwar V Singh S Wang X Mitchell C Gopalan K and Abu-Ghazaleh N *IEEE International Conference on Pervasive Computing and Communications (PerCom 2009)* pp. 1–6.

[26] Munishwar V Singh S Mitchell C Wang X Gopalan K Abu-Ghazaleh N 2009 *IEEE International Conf. on Pervasive Computing and Communications (PerCom 2009)* pp 1–3.

[27] Becue P Jooris B Sercu V Bouckaert S Moerman I and Demeester P *Remote Control of Robots for Setting Up Mobility Scenarios during Wireless Experiments in the IBBT w-iLab.t,* (eds) Ko- rakis T Zink M Ott M LNICST (Heidelberg: Springer Berlin) **44** pp 425–426

[28] Moerman L Korakis T Niavis, H Stavropoulos D Igoumenos I Becue P Sercu V Jooris B Neuhaus S Lyberopoulos G Theodoropoulou E and Filis K Toolkit for wireless mobility testbeds, Deliverable Report Deliverable D3.5, OpenLab (Oct. 2014)

[29] Abdelhadi A Rechia F Narayanan A Teixeira T Lent R Benhaddou D Lee H and Clancy T C, Position Estimation of Robotic Mobile Nodes in Wireless Testbed using GENI, arXiv preprint arXiv: 1511.08936.

[30] Reich J Misra V and Rubenstein D 2008 *ACM SIGMOBILE Mobile Computing and Communications Review* **12**(1) pp 68–70

[31] Reich J Misra V Rubenstein D and Zussman G 2012 *IEEE Journal on Selected Areas in Communications* **30**(5) pp 935–950

[32] Bromage S Engstrom C Koshimoto J Bromage M, Dabideen S and Hu M Menchaca-Mendez R Nguyen D Nunes B Petkov V Sampath D Taylor H Veyseh M Garcia-Luna-Aceves J JObraczka K Sadjadpour H Smith B 2009 *ACM SIGMOBILE Mobile Computing and Communications Review* **13** (1) pp 65–68

[33] Janansefat S Senturk I Akkaya and K Gloff M 2012 *37th Annual IEEE Conf. on Local Computer Networks (LCN)*

[34] Rensfelt O Hermans F Larzon L-Å and Gunningberg P, 2010 *Proc. of the 5th ACM International Workshop on Wireless Network Testbeds, Experimental Evaluation and Characterization, WiNTECH'10,* (New York, NY, USA: ACM) pp63–70

[35] Hermans F Rensfelt O Gunningberg P, Larzon L-Å and Ngai E C, *TRIDENTCOM, 2010, pp*

[36] Vingelmann P Pedersen M Heide J Zhang Q and Fitzek F 2012 *IEEE International Conf. on Communications (ICC 2102)* pp 291–296

[37] Vingelmann V Heide J Pedersen M V Zhang Q and Fitzek F H P, 2014 *Computer Networks* 74(Part B) pp 34–47

[38] Johnson D Stack T Fish R Flickinger D Ricci R and Lepreau J 2006 *The 25th Annual Joint Conf. of the IEEE Computer and Communications Societies.* IEEE Computer Society

[39] Johnson D Stack T Fish R Flickinger D Stoller L Ricci R and Lepreau J 2006 *25th IEEE International Conf. on Computer Communications (INFOCOM'*06) pp 1–12

[40] Arora A, E Ertin E Ramnath R Nesterenko M and Leal W 2006 *IEEE Internet Computing* **10**(2) pp 35–47.

[41] Ertin E Arora A Ramnath R Nesterenko M, Naik V Bapat S Kulathumani V Sridharan M Zhang H and Cao H 2006 *The 5th International Conf. on Information Processing in Sensor Networks (IPSN 2006)* pp 399–406

[42] Forster A Forster A Leidi T Garg K Puccinelli D Ducatelle F Giordano S and Gambardella L M 2010 MOTEL: Towards Flexible Mobile Wireless Sensor Network Testbeds, 2010.

[43] Förster A Förster A Garg K Puccinelli D Giordano S and Gambardella L M 2013 *Wireless Integration of Sensor networks in Hybrid architectures* 13

[44] Foerster A Foerster A Garg K Giordano S and Gambardella L M *Adhoc & Sensor Wireless Networks* **24**(3)

[45] Jiménez-González A Martínez-de Dios J R and A. Ollero A, 2010 *IEEE/RSJ International Conf. on Intelligent Robots and Systems (IROS 2010)* pp 3327–3332

[46] Jimenez-Gonzalez A Martinez-De Dios J R and Ollero A 2011 *Sensors* **11**(12) pp 11516–11543

[47] Jiménez-González A Martinez-de Dios J R and Ollero A 2013 *Robotics and Autonomous Systems* **61**(12) pp 1487–1501

[48] Dios J R M.-d Jimenez-Gonzalez A Bernabe A. d. S. and Ollero A 2014, A Remote Integrated Testbed for Cooperating Objects, Springer Briefs in Electrical and Computer Engineering, Springer International Publishing, pp 23–39.

[49] Killijian M-O Roy M and Severac G 2010 *IEEE 6th International Conference on Wireless and Mobile Computing, Networking and Communications (WiMob 2010)* pp 442–449

[50] Killijian M-O Roy M and Severac G 2012 *The ARUM Experimentation Platform: An Open Tool to Evaluate Mobile Systems Applications (eds)* RÃŒckert U Joaquin S Felix W (Heidelberg: Springer Berlin) pp. 221–234.

[51] Dahlberg T A Nasipuri A and Taylor C 2005 *Proc. of the 2005 ACM SIGCOMM Workshop on Experimental Approaches to Wireless Network Design and Analysis (E-WIND '05)* (New York, NY, USA: ACM) pp 76–81

[52] Niavis H Kazdaridis G Korakis T and Tassiulas L Enabling Sensing and Mobility on Wireless Testbeds 2012 (eds) Korakis T Zink M Ott M LNICST (Heidelberg: Springer Berlin) **44** pp 421–424

[53] Quilez R Zeeman A Mitton N and Vandaele J, Docking autonomous robots in passive docks with Infrared sensors and QR codes, 2015.

[54] Farkhana Muchtar, Abdul Hanan Abdullah, Muhammad Shafie Abd Latiff, Suhaidi Hassan, Mohd Helmy Abd Wahab, and Gaddafi Abdul-Salaam 2017 A Technical Review of MANET Testbed Using Mobile Robot Technology *J. Phys.: Conf. Ser*

[55] Farkhana Muchtar, Abdul Hanan Abdullah, Marina Md Arshad, Mohd Helmy Abd Wahab, Siti Nor Zawani Ahmmad and Gaddafi Abdul-Salaam 2017 A Critical Review of MANET Testbed Using Mobile Robot Technology *J. Phys.: Conf. Ser*

IOP Conf. Series: Journal of Physics: Conf. Series **1019** (2018) 012048 doi:10.1088/1742-6596/1019/1/012048

The Potential Use of Service-Oriented Infrastructure Framework to Enable Transparent Vertical Scalability of Cloud Computing Infrastructure

Farkhana Muchtar*[12], Abdul Hanan Abdullah[1], Mohd Helmy Abd Wahab[3] Nan Mad Sahar[3], Radzi Ambar[3], Hafizul Fahri Hanafi[4],

[1] Faculty of Computing, Universiti Teknologi Malaysia, 81310, Skudai, Johor, Malaysia
[2] School of Computer Sciences, Universiti Sains Malaysia, 11800, Gelugor, Pulau Pinang, Malaysia
[3] Faculty of Electronic and Electrical Engineering, Universiti Tun Hussein Onn, 86400 Batu Pahat, Johor, Malaysia
[4] Faculty of Art, Computing and Creative Industry, Universiti Pendidikan Sultan Idris, 35900 Tanjung Malim, Perak, Malaysia

E-mail: farkhana@gmail.com, hanan@utm.my, helmy@uthm.edu.my, nan@uthm.edu.my, hafizul@fskik.upsi.edu.my,

Abstract. Cloud computing technology has become familiar to most Internet users. Subsequently, there has been an increased growth in the use of cloud computing, including Infrastructure as a Service (IaaS). To ensure that IaaS can easily meet the growing demand, IaaS providers usually increase the capacity of their facilities in a vertical IaaS increase capability and the capacity for local IaaS amenities such as increasing the number of servers, storage and network bandwidth. However, at the same time, horizontal scalability is sometimes not enough and requires additional strategies to ensure that the large number of IaaS service requests can be met. Therefore, strategies requiring horizontal scalability are more complex than the vertical scalability strategies because they involve the interaction of more than one facility at different service centers. To reduce the complexity of the implementation of the horizontal scalability of the IaaS infrastructures, the use of a technology service oriented infrastructure is recommended to ensure that the interaction between two or more different service centers can be done more simply and easily even though it is likely to involve a wide range of communication technologies and different cloud computing management. This is because the service oriented infrastructure acts as a middle man that translates and processes interactions and protocols of different cloud computing infrastructures without the modification of the complex to ensure horizontal scalability can be run easily and smoothly. This paper presents the potential of using a service-oriented infrastructure framework to enable transparent vertical scalability of cloud computing infrastructures by adapting three projects in this research: SLA@SOI consortium, Open Cloud Computing Interface (OCCI), and OpenStack.

1. Introduction

Cloud computing is not new in the computing world. When the internet became as important as other utilities such as water and electricity, the idea of cloud computing already existed although the term 'cloud computing' still did not exist. This was largely ignored until 2007 when big players such as Amazon, Google, Sun, IBM and Microsoft started to gain interest in the idea and the term 'cloud computing' entered the mainstream [1].

IOP Conf. Series: Journal of Physics: Conf. Series **1019** (2018) 012048 doi:10.1088/1742-6596/1019/1/012048

Cloud computing is an evolution and combination of several ideas – on-demand computing, utility computing, grid computing, software as-a-service, and thin clients – where the locus of data storage and computation shifts to ever-bigger clusters of centrally managed computers [2]. In this paradigm, information is permanently stored in servers on the Internet ("Cloud") and cached temporarily on clients' hardware that includes desktops, entertainment centers, table computers, notebooks, wall computers, handhelds, sensors, monitors, etc. [3]. There are more than 20 definitions of cloud computing that seem to only focus on certain aspects of this technology [4].

Cloud computing creates a pool of computing resources (processor, memory, disk and network capacity) to allocate them on demand. In other words, this is a collection or a group of integrated and networked hardware, software and internet working infrastructure using the Internet as their communications and transport links to provide hardware, software and networking services to the end clients [5].

By merging technologies such as networking, virtualization and service-oriented architecture and wrapping it all within an Internet-based delivery model, the said platforms can provide an on-demand, always on, anywhere, anytime and anyplace, pay-per-use and as needed, highly elastic (scaled up and down in capacity and functionalities) hardware and software services to the general public, general and vertical corporate and business markets etc. [6].

Currently, cloud computing is considered an important paradigm and is the hottest topic in Information Technology (IT) [7]. Cloud computing is an omnipresent concept that is still about to reach its envisioned potential in the business domain [8, 9]. Everyone sees and uses the cloud differently. Some people offload ordinary desktop software and instead use web-based applications, others want on-demand computing with instant availability of extra processing power and extra storage. Nevertheless, the inescapable fact is that companies of all sizes have been increasingly using the cloud to increase their businesses, perhaps by enabling remote working or simply to cut costs or even to create/implement entirely new business models.

Companies can use clouds for different purposes, such as running batch jobs, hosting web applications or for backup and storage. Three main markets are associated to cloud computing:

- Infrastructure-as-a-Service (IaaS) designates the provision of IT and network resources such as processing, storage and bandwidth as well as management middleware. Examples are Amazon EC2 [10], RackSpace [11] and the new Google Compute Engine [12].
- Platform-as-a-Service (PaaS) designates programming environments and tools supported by cloud providers that can be used by consumers to build and deploy applications onto the cloud infrastructure. Examples of PaaS include Amazon Elastic Beanstalk [13, Heroku [14], Force.com [15], Google App Engine [16], and Microsoft Windows Azure [17].
- Software-as-a-Service (SaaS) designates hosted vendor applications. For example, Google Apps [18] (such as Google Docs, Google Calendar or Google Sites), Microsoft Office 365 [19] and Salesforce.com [20].

The idea of this research focuses on the development of resource management in IaaS Cloud Computing, thus allowing vertically scalable computer resources to cut across different data centers. Furthermore, integration between Data Centers is also considered and the OCCI open standard is used for the purpose of integration. The component has been developed by any open source community, thus the component does not need to be developed from scratch.

This paper is organized as follows. In section II, some related works are presented. In section III, an overview of service oriented infrastructure is described and a proposed solution is presented in section IV. Finally section V includes the conclusions and future studies.

2. Related Work

The key feature of cloud computing that allows these computational resources to be shared or outsourced transparently among a vast number of users over a scalable network of nodes is in the notion of scalability [21, 22]. In cloud computing, scalability means a way to add and remove resources to the services depending on the usage [23].

As for customers who generally do not own the infrastructure, they can merely access or rent and consume resources as a service, forego capital expenditure, and pay for what they use – the same analogy with traditional utilities like consumed electricity, or television bills that are on a subscription basis. And from the viewpoint of cloud providers: when there is a lot of traffic and users, they can scale up the service up so that it will not choke, and when there is little or no volume, they can scale the service down to a bare minimum [24, 25].

For Cloud, there are two dimension of scalability; namely, horizontal cloud scalability and vertical cloud scalability [7, 26, 27]. Horizontal cloud scalability is referred as the scalability of inter scalability. This means the scalability of the infrastructure that is not specific to one place which is related to the integration between different data centres that are located in different geographical locations [26]. The method of vertical scaling uses bigger and higher performing hardware to scale. The method of horizontal scaling relies on using more machines instead of more powerful machines. Horizontal scaling relies on commodity hardware whereas vertical scaling would require specialty hardware. Cloud computing uses horizontal scaling may be easy to implement, but it has more limitation, such as 1) there is no guarantees that sufficiently capable hardware exists or is affordable; 2) the hardware improvement is always expensive; 3) because of hardware changes involved, this approach usually involves downtime.

Contrary to horizontal scalability, vertical cloud scalability is referred as the scalability of intra infrastructure. This means that the scalability of the infrastructures inside a single data centre are related to the resources that any data centre are concerned with, such as the servers, physical storage, bandwidth, IPs, domain name, securities and others [26]. On the other hand, horizontal scaling can take full advantage of cloud computing technologies, which means 1) it can use couple normal computers or devices to implement high-performance computing, 2) it can use the capacity on demand and automatically, and 3) can upgrade without downtime.

Most cloud providers offer both scaling up and scaling out solutions, such as Amazon EC2 and Windows Azure [7]. They both support scaling up by using larger (size of) instances and scaling out by adding additional instances. We only focus on scaling out since horizontal scaling is a more cloud-native style. Vertical scaling is limited by the most powerful hardware that is available. Therefore, out of the prior works reviewed, there are only a few studies that focus on the issue of vertical cloud scalability for managing resources in cloud computing have been reported.

For this research, the idea is to look into the potential of using service-oriented infrastructure frameworks to enable transparent vertical scalability of cloud computing infrastructures by adapting three open source projects in this research: SLA@SOI consortium, Open Cloud Computing Interface (OCCI), and OpenStack.

3. Service Oriented Infrastructure

The Service Oriented Infrastructure (SOI) applies Service Oriented Architecture (SOA) to IT Infrastructure [28]. Service Oriented Infrastructure or SOI is the architecture that describes IT infrastructure in terms of service. The architecture encompasses all phases of the SOI life cycle, Design, Provisioning, Operation and Decommissioning and Management of the services. Infrastructure services typically use and/or provide a virtualized pool of shared resources (Servers, Network, Storage, Infrastructure software) that are deployed and managed in a highly automated way. An SOI provides foundational support for a Service-Oriented Application Architecture or other application architectures.

This model shows the conceptual building blocks of SOI. These conceptual building blocks must be instantiated for each type of service or other elements used in an architecture, when that architecture is implemented.

A key element of SOI as opposed to just a service-oriented design is the tight coupling of the infrastructure services to applications and consumers thus capturing the full length and breadth of the IT service chain.

Infrastructure services provide the application designer with a language that enables the designer to work with abstract concepts rather than implementation details. IT infrastructure services provide a standard response to a standard request. This definition could be extended with a classification of the bandwidth within which this statement is valid. Typically an infrastructure service provides its service based on the actual demand, but is limited to an upper level of the service.

This upper level typically is bound to the maximum capacity the service can use as provided by a resource pool supporting the service. In order to assign the correct amount of resources to deliver the service in the right quality, the service needs to monitor the actual demand as well as the actual supply. This information should be used to adjust the service.

Typically, applications will require a number of infrastructure services in order to support their application interfaces. All the required infrastructure services need to be provided from their respective pools and assigned to the support of the specific application. The application definition should specify a number of aspects:

- The required services or response of the service to a request;
- The initial capacity of the services

The specific quality constraints of the service is defined as the bandwidth in which the service is able to meet its service level requirements. Designing SOI solutions is the process of translating requirements into working solutions. With respect to the type of requirements, functional requirements and non-functional requirements can be distinguished. The functional requirements drive the capabilities within the Block Box: the (static) behavior of the IT Component on a standard request. The non-functional requirements determine the dynamics of the solution and control the provisioning of the resources for the SOI Service.

4. Propose Solution

Based on current trends, there is an increasing number of innovation and new technologies in cloud computing, such as Hybrid Cloud movement, in which many researchers try to create an environment that enables the resources of cloud service providers to be combined.

In addition, open source cloud computing is getting more active, in which there are currently more than one open source project that produces IaaS tools, including Eucalyptus, OpenNebula, and the latest being OpenStack. Moreover, most open source projects, such as KVM, libvirt, glustefs, xen and others have also grown rapidly, and the software developed have become more stable, and has many new features that meet the current demands in IaaS.

There are three open source projects that have an indirect relationship with this research. The first is a research by the SLA@SOI consortium, which focuses on the use of a Service Oriented Infrastructure Framework to attain quality SLA management from cloud service providers to cloud service consumers. The second project is the Open Cloud Computing Interface (OCCI) community. It was created to provide a specification that was agreed upon by the community as a protocol and API for all the needs of management tasks in cloud computing. The third project is OpenStack, which is an open-source software project to develop public and private clouds. OpenStack is a project based on community contribution, aiming at creating a set of open source software to help any organization to have its own cloud computing for virtual computing and storage.

The tools and technologies from all three projects will be utilized in this research to facilitate the development of a prototype. Further information on the three projects are delineated below.

4.1. ***SLA@SOI***

SLA@SOI is a European Commission funded research project committed to deliver and showcase an innovative open Service Level Agreement (SLA) Management Framework that provides holistic support for service level objectives - enabling an open, dynamic, SLA-aware market for European service providers.

The rapidly growing service-oriented economy has highlighted key challenges in IT-supported service provisioning. Service consumers are frustrated by the lack of formal, negotiable, readily enforceable SLAs. Service providers find it impractical to create personalized offerings, translate business requirements into technical manifestations, and optimize internal deployments, whilst maintaining all the individual SLAs.

SLA@SOI has addressed these challenges by providing 3 major benefits:

- Predictability & Dependability: The quality characteristics of a service could be predicted and enforced at run-time.
- Transparent SLA Management: Service level agreements (SLAs) defining the exact conditions under which services were provided/consumed could be transparently managed across the whole business and IT stack.
- Automation: The whole process of negotiating SLAs and provisioning, delivery and monitoring of services were automated, thus allowing for a highly dynamic and scalable service consumption.

SLA@SOI defined a holistic view for the management of service level agreements (SLAs), implementing an SLA management framework that could be easily integrated into a service-oriented infrastructure (SOI). The main innovative features of the project were:

- An automated e-contracting framework.
- Systematic grounding of SLAs from the business level down to the infrastructure.
- Exploitation of virtualization technologies at the infrastructure level for SLA enforcement.
- Advanced engineering methodologies for the creation of predictable and manageable services.

*4.2. **OCCI Framework***

The Open Cloud Computing Interface (OCCI) is a RESTful Protocol and API for all kinds of management tasks. OCCI was originally initiated to create a remote management API for IaaS1 model based Services, allowing for the development of interoperable tools for common tasks, including deployment, autonomic scaling and monitoring. It has since evolved into an exible API with a strong focus on interoperability, while still offering a high degree of extensibility. The current release of the Open Cloud Computing Interface is suitable to serve many other models, in addition to IaaS, including e.g. PaaS and SaaS.

OCCI began in March 2009 and was initially led by co-chairs from SUN Microsystems, RabbitMQ and Universidad Complutense de Madrid. Today, the working group has a membership of over 250 members and includes numerous individuals, and industry and academic parties.

The reasons driving the development of OCCI were:

- **Interoperability:** Allowing for different Cloud providers to work together without data schema/format translations, facade/proxying between APIs and understanding and/or dependency on multiple APIs.
- **Portability**: There was no technical/vendor lock-in and services were enabled to move between providers, hence allowing clients to easily switch between providers, based on business objectives (e.g. cost) with minimal technical cost, which enabled and fostered competition
- **Integration:** Implementation of the specifications could be easily integrated with existing middleware, third party software and other applications.
- **Innovation:** Driving modern technologies.

The Open Cloud Computing Interface was a boundary protocol and API that acted as a service front-end to a provider's internal management framework. Service consumers could be both end-users and other system instances. OCCI was suitable for both cases. The key feature was that OCCI could be used as a management API for all types of resources, while at the same time, maintaining a high level of interoperability.

*4.3. **OpenStack Cloud Computing***

OpenStack is a collection of open source technologies delivering a massively scalable cloud operating system. OpenStack is currently developing two interrelated projects: OpenStack Compute and OpenStack Object Storage. OpenStack Compute is a software that is being developed for the provision and management of large groups of virtual private servers, whereas the OpenStack Object

IOP Conf. Series: Journal of Physics: Conf. Series **1019** (2018) 012048 doi:10.1088/1742-6596/1019/1/012048

Storage is a software for the creation of redundant, scalable object storage using clusters of commodity servers to store terabytes or even petabytes of data.

The OpenStack Community Project is backed by Rackspace, NASA, Dell, Citrix, Cisco, Canonical and over 50 other organizations. OpenStack has grown to become a global software community of developers, technologists, researchers and corporations; collaborating on a standard and massively scalable open source cloud operating system. Their mission was to enable any organization to create and offer cloud computing services running on standard hardware.

OpenStack offers open source software to build public and private clouds. OpenStack is a community and a project, as well as a stack of open source software to help organizations run clouds for virtual computing or storage. OpenStack contains a collection of open source projects that are community-maintained, including OpenStack Compute (code-named Nova), OpenStack Object Storage (code-named Swift) and OpenStack Imaging Service (code-named Glance).

OpenStack provides an operating platform or toolkit for orchestrating clouds. OpenStack is more easily defined once the concepts of cloud computing become apparent, to provide scalable, elastic cloud computing for both public and private clouds, large and small. Their mission is that clouds must be simple to be implemented and massively scalable.

5. Prototype Design

As stated previously, the main component that will be developed is resource management. However, other components, such as the scheduler and management policy must be developed to support the resource management component.

At the same time, those components will not be developed from scratch, but instead will be developed through the process of additions and modifications to the source code for those components that are already in existence in the OpenStack Cloud Computing project.

OpenSTack, SLA@SOI and OCCI projects are open source projects. Hence, those components could be modified, as they also provide the source code of all the available components and are readily available. The components that are proposed in this study are as follows:

i. **Additional scheduler/resource management component**

This component enables the resources of different data centers to be combined into a single resource management. At the same time, this component is created to be integrated with management policy. The resource management component serves to monitor the resources that exist locally or at other locations that have been integrated with each other. At the same time, this component also manages the requests from cloud users, i.e. to create, use, alter and destroy virtual machines in a cloud environment. On the other hand, a scheduler is an implementer of policy management and receives information from the resource management component to determine the quantity of resources to be used in the local cloud. The scope of a scheduler is only based on local resources and resources other than the local cloud are only known by the resource management component, as integration between clouds occurs only in resource management.

ii. ***Trusted connections***

This component is created to ensure that the connection between two cloud computing data centres at different locations is secured. This is because there is the possibility of two virtual machines (usually owned by the same user) interacting with each other but are in different data centre locations.

Therefore, a secured connection is an additional requirement to ensure that the data between two (or more) virtual machines cannot be seen by unauthorized parties. In this requirement, it is recommended that OpenCA be used to generate a digital certification for the interconnection between different cloud computing data centres.

iii. Policy Management

The main role of management policy is to establish the behavior of the scheduler in determining the utilization of local resources that exist in a local cloud. In this research, management policy also serves as the determinant of resource management component behavior in implementing vertical scalability between cloud resources in different locations. In other words, besides interacting with the scheduler for the local management of cloud resources, the management policy component also interacts with the resource management component in the management of inter-cloud resources.

iv. Data Centre Integration

This is the integration that will be used between the data centres. It is proposed that the OCCI standard be used to connect different data centres.

The OCCI standard is chosen for a variety of reasons, including:

- It is an open standard. Thus, the integration between cloud computing across different cloud computing products can be implemented.
- The APIs are straightforward and simple.
- The use of RESTful Web Services instead of other complex web services approaches.
- Almost all cloud computing implementers have already started developing the interface for OCCI open standard

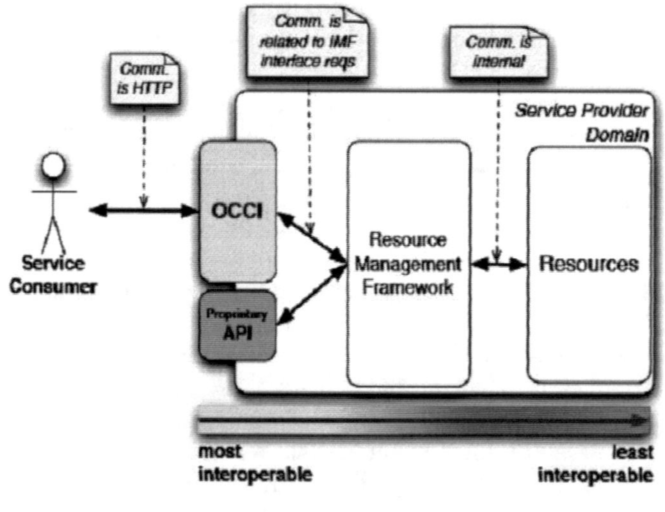

Figure 1. OCCI in the Cloud Service Provider

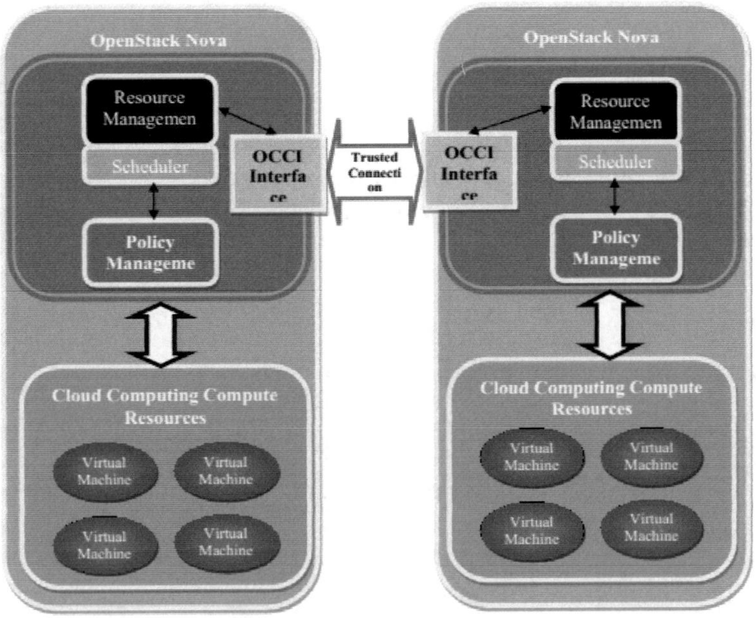

Figure 2. Prototype Design

6. Conclusion

This paper discusses the potential of using a service-oriented infrastructure framework to enable the transparent vertical scalability of cloud computing infrastructures.

The contribution of the paper can be categorized into the following: reviewing the three projects that have been proposed to be adapted in this research: SLA@SOI consortium, Open Cloud Computing Interface (OCCI), and OpenStack, and suggesting four components to be created in this research: additional scheduler/resource management component, policy management, data centre integration and trusted connections. For future works, a tool will be developed and tested in suitable case studies.

References

[1] Carolan J Gaede S Baty J Brunette G Licht A Remmell J et al. 2009 *Introduction to Cloud Computing architecture White Paper* (1st Edition)
[2] Baker S 2007, Dec 24. *Business Week*
[3] Cassarino I Geuna A Martiis 2008. *Background paper on ICT and Creative Industries CReATE* – Creating a Joint Research Agenda for Promoting ICT-Innovations in Creative Industries across, Europe EU-Project, FP7- Regions-2007-1
[4] Vaquero L M Rodero-Merino L Caceres J and Lindner M 2008 *SIGCOMM Computer Communication Review*, 39(1) pp 50-55
[5] Galen G (2008, April 7 *InfoWorld*
[6] Ganek A (2009, Apr 30). The Cloud: a leap created from combining existing technologies
[7] Seyyed Mohsen Hashemi and Amid Khatibi Bardsiri 2012 *ARPN Journal of Systems and Software* **2**(5)
[8] Leavitt N 2009 Is cloud computing really ready for prime time? *Computer* **42**(1) pp 15–20
[9] Raza M H Adenola AF Nafarieh A Robertson W 2015 *Procedia Comput. Sci.* 52 pp 1114–1119
[10] Amazon Elastic Compute Cloud (Amazon EC2). http://http://aws.amazon.com/ ec2/ [Online; accessed 13-September-2017].
[11] Rackspace. The open cloud company. http://www.rackspace.com/ [Online; accessed 13-September-2017].
[12] Google Compute Engine. http://cloud.google.com/products/compute-engine. html/ [Online; accessed 13-September-2017].

[13] AWS Elastic Beanstalk (beta). Easy to begin, Impossible to outgrow. http://aws.amazon.com/elasticbeanstalk/. [Online; accessed 13-September-2017].
[14] Heroku. Cloud application platform. http://www.heroku.com/ [Online; accessed 13-September-2017].
[15] force.com (salesforce). http://www.force.com/ [Online; accessed 13-September- 2017].
[16] Google App Engine. http://cloud.google.com/products/ [Online; accessed 13-September-2017].
[17] Microsoft Windows Azure. https://www.windowsazure.com/en-us/ [Online; accessed 13-September-2017].
[18] Google Apps for Business. http://www.google.com/intl/es/enterprise/apps/ business/products.html [Online; accessed 13-September-2017].
[19] Microsoft Office 365. http://www.microsoft.com/en-us/office365/ online-software.aspx. [Online; accessed 13-September-2017].
[20] Salesforce.com. http://www.salesforce.com/ [Online; accessed 13-September- 2012].
[21] Anderson J E Wiles F A and Young K P 2008. *Issues in Information System* **9** (No 1-2) pp 203-206
[22] Mikkilineni R and Sarathy V 2009. 1st *Workshop on Collaboration and Cloud Compu*ting *(WETICE'09)*, pp 57-62
[23] Korri T 2009. *Technical Reports in Computer Science and Engineering (TKK-CSE-B5)*, Department of Computer Science and Engineering, Helsinki University of Technology, Finlad
[24] Aymerich F M Fenu G and Surcis S 2008. In 1st *International Conference on the Applications of Digital Information and Web Technologies* pp 113–118
[25] Mc Evoy G V and Schulze B 2008. *Proc. of the 6th International workshop on Middleware for grid computing (MGC '08)* (New York, NY, USA: ACM) pp 1–6
[26] Mei L Chan L and Tse T H 2008. *Proc. of 2008 IEEE Asia-Pacific Services Computing Conference, APSCC'08* (Los Alamitos, CA, USA: IEEE Computer Society Press) pp 464–469
[27] Mei L Zhang Z and Chan W K 2009. *33rd Annual IEEE International Computer Software and Applications Conference, COMPSAC'09* (Los Alamitos, CA: IEEE Computer Society Press) pp 464-469.
[28] Yan Z 2008 *Proceedings of the 2008 IEEE Congress on Services - Part I, SERVICES '08,* (Washington, DC, USA: IEEE Computer Society) pp 99-100

IOP Conf. Series: Journal of Physics: Conf. Series **1019** (2018) 012053 doi:10.1088/1742-6596/1019/1/012053

IoT based monitoring and environment control system for indoor cultivation of oyster mushroom

M F Mohammed[1], A Azmi[1], Z Zakaria[2], M F N Tajuddin[1], Z M Isa[1] and S A Azmi[1]

[1] Centre of Excellent of Renewable Energy (CERE), School of Electrical Systems Engineering, Universiti Malaysia Perlis (UniMAP), 02600 Pauh Putra, Perlis.
[2] Faculty of Engineering Technology, Universiti Malaysia Perlis (UniMAP), 02100 Padang Besar, Perlis, Malaysia

m.fayzul@unimap.edu.my

Abstract. This paper is presenting an implement of Internet of Things (IoT) monitoring and environment control for indoor cultivation of oyster mushroom. By using IoT based remote monitoring system not only the labor cost can be reduced with cover a bigger area, but can improve the productivity of the oyster mushroom by control the most suitable environment for mush room to grow. This paper discuss about the oyster mushroom, IoT system in general, the system implementation and results that obtained from the system. In conclusion based on the developed system, authors discussed the advantages and points need to be considered when working on the system.

1. Introduction

The need of food and limitation of space or land as an agro-economic activity make urban farming technology is becoming popular and has become one of promising solution for securing food supply. Apart from that, extreme weather changes and climates affect the production of crop, thus increasing their prices and lowering the quality of the crops produced [1]. Hence, this paper present an internet of things (IoT) based monitoring and environment control for indoor cultivation oyster mushroom, which is a smart urban farming system that requires less maintenance, less manpower and saves a lot of space [2]-[3]. Furthermore, this project is dedicated to improve and enhance the conventional plantation system in general. Using IoT platform will enhance the capability of current equipment for remote monitoring purpose and at the same time log the data for analysis and references [4]-[5].

1.1. Internet of Things (IoT)

Gartner Inc. defines the IoT is the network of physical objects that contain embedded technology to communicate and sense or interact with their internal state or the external environmental. Applications of IoT in agro-industrial are monitoring, control, prediction and logistic [1],[6]. Figure 1 shows a block diagram of basic IoT system. IoT device will receive a data from the attached sensor and sent the data to the database via internet connection. Most of popular IoT connection is via WiFi module. The data sent to server must follow the API format of the platform provider, and then data will store in the database accordingly. Data can be seen or retrieved by user through the browser application whether via smart phone or computer. The browser will request data from the server and server will sent it accordingly. The process for control also using a same path, it is start from user sent command data to the server and server will store it in the database. The device will keep monitor the changes of command storage data by request it accordingly. When the command data is received by IoT device, it will be executed accordingly as programmed by system developer.

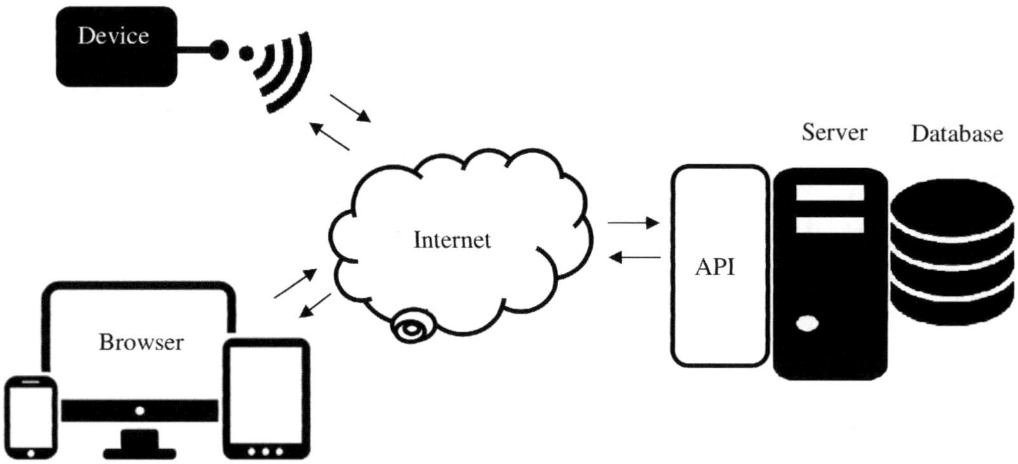

Figure 1. Basic topology of IoT system.

1.2. Indoor cultivation oyster mushroom

The indoor cultivation room was measured at 5.8 m in length, 4.57 m in wide and 2.74 m in height inside a concrete building located at Taman Pauh Indah, Arau, Perlis. The wall was made by bricks at one side and by gypsum board on the rest of the walls. The roof of the room was built up by also gypsum board under the main concrete roof. The room was installed with four rows of racks; each rack contained four iron bars distance by 15 cm from each other and 55 cm distance between the racks. Roof ventilation as 7.5 m^3/min ventilation rate (20CQT1 KDK, Malaysia), light as 10 watts for 648 lumens (E27 B22, Malaysia) and side ventilation as 2.8-3.3 m^3/min ventilation rate (P120×120×38, UK) was installed inside the room in order to facilitate fresh air and maintain proper light intensity respectively. The roof ventilations was installed with a top roof to bring air from inside to outside and the side ventilations was installed at bottom of the wall with a black filter to bring outside air to inside.

The bags were arranged horizontally through the racks by using nylon rope and made it tight by cable tie and hanged vertically. Twelve bags were arranged in one column of rope by 3 bags in together in 4 parts. The columns of rope were arranged at 25 cm distance from each other in every rack so that the air and humidity can easily flow through the spaces between the bags, preventing the temperature of the bags from increasing. Mushroom mycelia emitted heat during incubation, so the bags can be easily overheated if they touch each other [7]. The distance of the top end bag from the ceiling was 76 cm and the down end bag from the floor was 30 cm. The humidifiers as 4-5 L/hr in 2,700-2,900 rpm/min (TAY-RING TL-3600, Taiwan) were placed between the two columns of bags where the distance of humidifier from the 48 opening part of the bags (front view) was 25-30 cm, from the side view of bags was 20-25 cm and from the back view of bags was 10-15 cm. All the devices were fixed by an electrical digital timer (DG-1120, Malaysia). Instrumentations were installed at different positions inside the cultivation room to monitor temperature, humidity, light intensity and CO2 level during the experimental procedure.

The optimal value of temperature, light intensity and humidity for oyster mushroom planting is at 26-29 °C, 50-300 lux and 80-90% respectively. The temperature and humidity normally are by exhaust fan and humidifier [8]. Light intensity is provided by fluorescent lamps. The range of normal condition for oyster mushrooms indoor plantation is given in Table 1.

Table 1. Important parameters for oyster mushroom grow.

Parameters	Range	*Unit*	Controlled Equipment
Temperature	26-29	°C	Fan
Light	50-300	lux	Lamp
Humidity	80-90	% RH	humidifier

2. Hardware Implementation

The illustrated diagram for the project is as shown in Figure 2. In the project, ATMega 2560 microcontroller is used as the main control board. The high performance, low power Microchip 8-bit AVR RISC-based microcontroller combines 256KB ISP flash memory, 8KB SRAM and 86 general I/0, 32 general purpose working registers, real time counter, six flexible timer/counter with compare modes, PWM, 4USARTs byte oriented 2-wire serial interface, 16channels 10-bit A/D converter and a JTAG interface for on chip debugging. This device operates between 4.5-5.5 V.

There are four inputs that are LCD, temperature and humidity sensor (H & C), light intensity sensor (L) and real time clock (RTC). For LCD the team use Nokia 5110 LCD screen. It is an 84x48 pixel monochrome LCD display. These displays are small, only about 1.5" diagonal, but very readable and come with a blue backlight. This display is made of 84x48 individual pixels, hence the application is used for graphics and text. These displays are inexpensive, easy to use, require only a few digital I/O pins and low power consumption. Temperature and humidity sensor (DHT 22) is a low cost humidity and temperature sensor with a single wire digital interface. The sensor is calibrated and does not require extra components so measuring relative humidity and temperature is simpler. This device operates at 3.3-6 V and the range is from 0-100% RH for humidity and -40 to 80 degrees C temperature range. For light intensity measurement, BH1730 sensor is been used. This device operates at 3-5.5 V and capable to measure light intensity in between 0-500 lux. DS1302Real Time Clock (RTC) has been used to get actual time. Three pins are needed specifically for the interface (CE,I/O, SCLK) and VCC2 should be connected to +5 V. For output parts, there were 6 outputs which are LEDs, Buzzer, ventilation fan (V1 & V2), Water pump (M1), lamps (L1 & L2) and humidifier unit (H1, H2, H3 & H4). Finally, ESP8266 WiFi module has been used for push data to cloud. The controller of the system have been fabricate as shows in Figure 3.

Table 2. Main equipment.

	Model
Controller Board	Arduino Mega
Humidity & temperature sensor	DHT22
Light sensor	BH1750
Relay circuit	10 A, 250 Vac and 5V dc, active low module
Wifi module	ESP8266-01
Display	Nokia 5110 graphic LCD

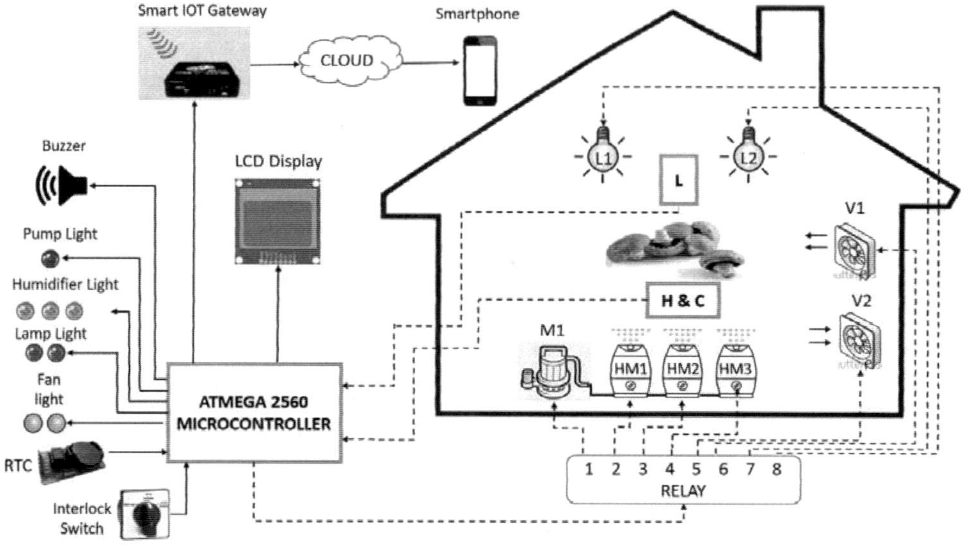

Figure 2. An Illustrated Diagram of Smart Agriculture System: Mushroom House Box

Figure 3. The controller and socket outlet for IoT Monitoring and environment control system for indoor cultivation oyster mushroom.

3. Software Implementation

In this project, Arduino Mega have been used as a controller to integrate data from sensor and sent the information to IoT platform through ESP8266-01 WiFi module. In this project, ThingSpeak have been uses as an IoT platform. The sensor reading has been recorded to the series of parameter needed by microcontroller through connection of analog or digital input of Arduino Mega, then passing this information data using serial communication to the ESP8266-01 WiFi module. This ESP8266-01 has been program to connect to available WIFI with internet connection located at the field area, receive the sensor data from Arduino Mega and send the sensor data to the specific channel to the ThingSpeak

IOP Conf. Series: Journal of Physics: Conf. Series **1019** (2018) 012053 doi:10.1088/1742-6596/1019/1/012053

IoT platform server using issuing provided HTTP API request format. The successful data submitted can be visualize at ThingSpeak IoT platform and available Android or IOS apps that linked to the same platform. Figure 4 and 5 show the algorithm data transfer by main controller and WiFi module respectively.

In order to retrieve the information on existing channel ThingSpeak server, the device such as mobile apps need to request the data on ThingSpeak server using determined HTTP API request format. The data received on mobile apps can be displayed in sensor value, graph in time domain for visualize on trend of data and can be used as a control parameter to the related device.

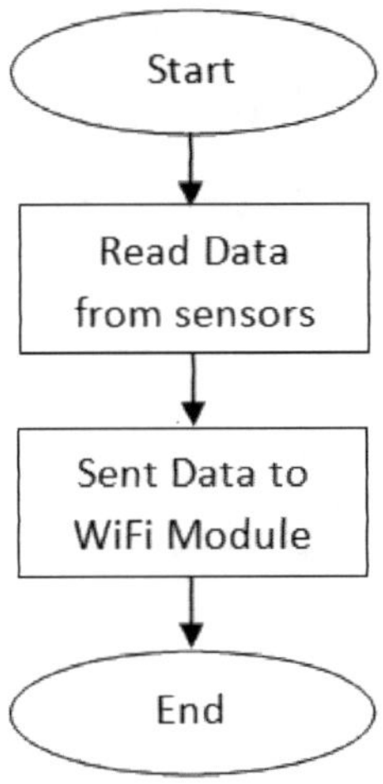

Figure 4. Algorithm for data transfer by main controller

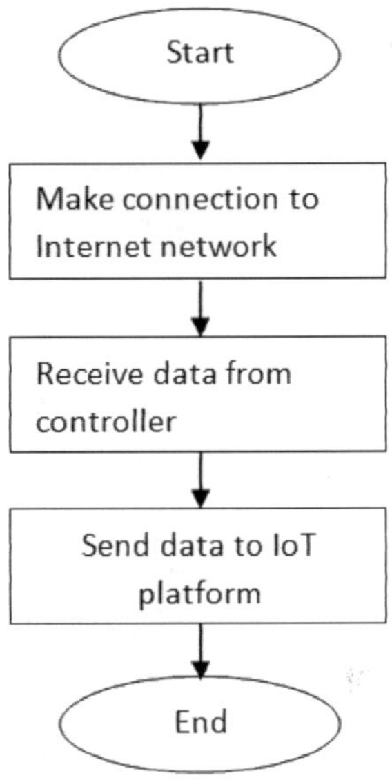

Figure 5. Algorithm for data transfer by WiFi module.

4. System monitoring Function Test

In this system, there are three parameters are monitored at a 5 minutes sampling rate; room temperature, relative humidity and light intensity. Figure 6-8 are show data with function of time that obtained from the IoT server platform. Besides using browser, the data can be duplicated and can be monitored by user via third party apps on android. This method gives flexibility to user to monitor their system. Figure 9 shows example of dashboard apps that have been used in the system.

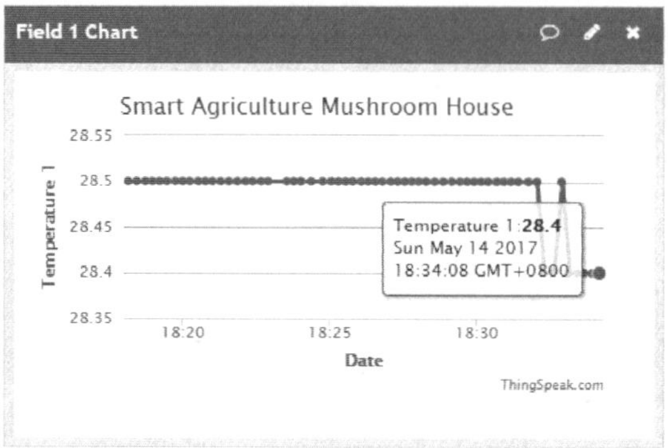

Figure 6. Room Temperature data.

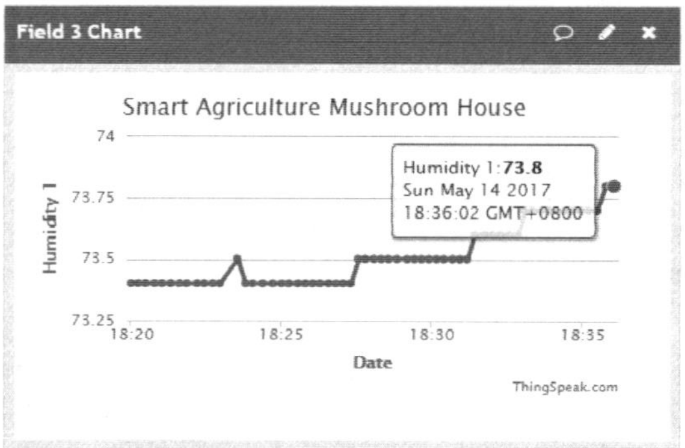

Figure 7. Room Humidity data.

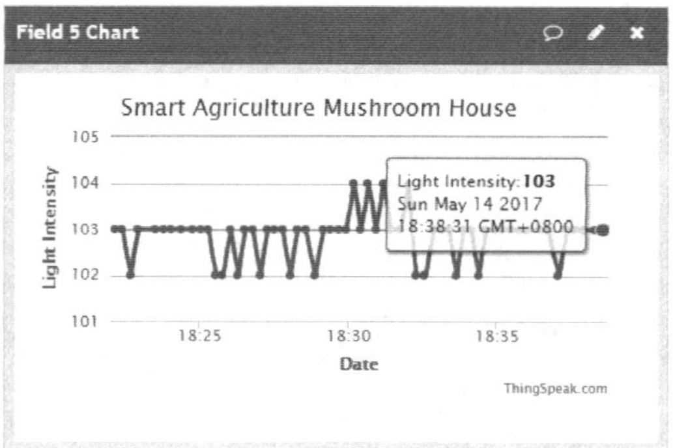

Figure 8. Room Light intensity data.

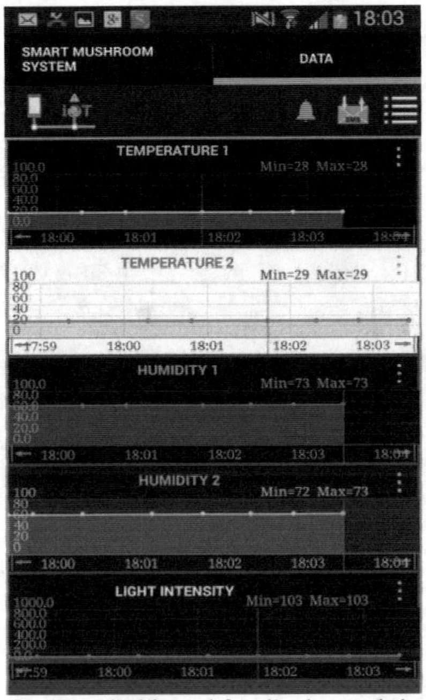

Figure 9. Apps dashboard for display real time data.

5. Conclusion

Monitoring and environment control for indoor cultivation of oyster mushroom has successful been implemented. IoT based monitoring system is promising technology to enhance the production of mushroom due to:

- Low cost in capital because of system framework is based on open source platform.
- Easy and user friendly. Data migration is easy to be done.
- Expandable. The system can be modular, easy to increase the number of data/sensor.

However, there are limitations on the system that need to be considered:

- Remote monitoring is relying on internet connection quality.
- Currently, system control via internet is not recommended because of quality of internet. However it can be done with proposed system.
- Beside the hardware and internet data cost, operation cost (mothly/yearly) of server cloud platform must be taken account. Many platforms are free only for personal or non-commercial used and have a limited access.
- A data security issue in IoT is not a confidential. If the system deals with highly sensitive data, data verification or encryption method maybe needs to apply which is required a complex algorithm at server and nodes.

References

[1] Jesús Martín Talavera, Luis Eduardo Tobón, Jairo Alejandro Gómez, María Alejandra Culman, Juan Manuel Aranda, Diana Teresa Parra, Luis Alfredo Quiroz, Adolfo Hoyos, Luis Ernesto Garreta, Review of IoT applications in agro-industrial and environmental fields, In Computers and Electronics in Agriculture, Volume 142, Part A, 2017, Pages 283-297, ISSN 0168-1699,

[2] A Mustafa Alper Akkaş, Radosveta Sokullu, An IoT-based greenhouse monitoring system with

Micaz motes, In Procedia Computer Science, Volume 113, 2017, Pages 603-608, ISSN 1877-0509

[3] Jiang Zhaohui, Xu Zhengrong . The remote monitoring of agricultural information system design and implementation of Journal of agricultural network information, 2010, (11): 40-43.

[4] Fernando Terroso-Saenz, Aurora González-Vidal, Alfonso P. Ramallo-González, Antonio F. Skarmeta, An open IoT platform for the management and analysis of energy data, In Future Generation Computer Systems, 2017.

[5] C. Cambra, S. Sendra, J. Lloret and L. Garcia, "An IoT service-oriented system for agriculture monitoring," *2017 IEEE International Conference on Communications (ICC)*, Paris, 2017, pp. 1-6.doi: 10.1109/ICC.2017.7996640

[6] K. R. S. R. Raju and G. H. K. Varma, "Knowledge Based Real Time Monitoring System for Aquaculture Using IoT," *2017 IEEE 7th International Advance Computing*

[7] Curvetto, N. R., Gonzalez Matute, R., Figlas, D., & Delmastro, S. (2004). *Oyster mushroom cultivation*. MushWorld.

[8] Md. Tariqul Islam, Zarina Zakaria, Nasrul Hamidin & Mohd Azlan Bin Mohd Ishak, "Characteristics of Indoor Mushroom Cultivation of Grey Oyster (Pleurotus pulmonarius) by Different Stages of Humidifying Treatment", World Applied Sciences Journal 34 (8): 1066-1075, 2016.

IOP Conf. Series: Journal of Physics: Conf. Series **1019** (2018) 012054 doi:10.1088/1742-6596/1019/1/012054

Studies of ENR-25/TiO_2 Composites for Electronic Materials Packaging Applications

Omar S Dahham[1], N Z Noriman[1], R Hamzah[1], T Adam[1,2], Z Shayfull[3,4], S Sudin[5], S Z Syed Idrus[6]

[1] Center of Excellence Geopolymer and Green Technology (CEGeoGTech), Faculty of Engineering Technology (FETech), Universiti Malaysia Perlis (UniMAP), Level 1 Block S2, UniCITI Alam Campus, Sungai Chucuh, Padang Besar, 02100, Perlis, Malaysia.

[2] Institute of Nano Electronic Engineering (INEE), Universiti Malaysia Perlis (UniMAP), 01000 Kangar, Perlis, Malaysia

[3] School of Manufacturing Engineering, Universiti Malaysia Perlis, Arau, 02600, Malaysia

[4] Green Design and Manufacture Research Group, Center of Excellence Geopolymer and Green Technology (CEGeoGTech),Universiti Malaysia Perlis, Kangar, 0100, Malaysia

[5] School of Computer and Communication Engineering (SCCE), Green Advanced Computing and Technology (GREAT) Research Group – CEGeoGTech, Universiti Malaysia Perlis (UniMAP)

[6] School of Human Development and Techno-communication (iKOM), Green Advanced Computing and Technology (GREAT) Research Group – CEGeoGTech, Universiti Malaysia Perlis (UniMAP)

E-mail: niknoriman@unimap.edu.my

Abstract. In this study, the synthesis and characterization of epoxidized natural rubber/ Titanium dioxide (ENR-25/TiO2) composites in two different solvent (Toluene and Chloroform) via sonication technique was conducted using Fourier transform infrared spectroscopy (FTIR) and Field Emission Scanning Electron Microscopy (FESEM). Spectroscopy results showed new peak at 1662 cm^{-1} for both ENR-25/Titania composite films, which is related to weak C-O stretching vibration. Moreover, The peak at 3390.4 cm^{-1} and 3384.3 cm^{-1} of the composites are produced by Ti-OH stretching vibration shifts. The composite with toluene solvent showed poor dispersion of TiO2, which can be seen by the agglomeration embedded in the matrix. The composite with chloroform showed same trend of composite with toluene. The poor TiO_2 dispersion in the composite can be observed by the surface as bright layers. These layers appeared because the TiO_2 particles covered the surface of purified ENR-25 matrix.

1. Introduction

Nowadays, the electronic industry increasing rapidly with smaller and better performance products to achieved advance technology. Electronic packaging of microelectronics was known as the art of enclosing, cooling, powering, chip packaging, interconnecting and protecting chips from the immediate surroundings environment as well as protecting the related immediate surroundings environment from the chips [1-6].

The electronic packaging of a the electronic system can be classified into different levels of packaging such as packaged chip, bare chip, electronic subassembly, printed circuit board assembly and electronic assembly to system or final electronic product. The aim of electronic packaging is to simplify and/or reduce the packaging levers number [7-11].

In current state, epoxy is used as a main material of the electronic material packaging to be medium for the most part in this application [12-14]. Nevertheless, epoxy contributes to the problem such as expensive and un-green material. Additionally, the using of epoxy (thermosetting matrix) also could cause other problem such as recycle difficulties, which in turn effect negatively on the environment. With increased performance requirements for smaller, more capable, and more efficient electronic systems, thermal challenges have become critical issues in electronic packaging design. Breakthroughs are required in board, advanced cooling solutions, practical design at chip and system package levels. In response to these serious needs, advanced and new materials with process developments in packaging and cooling technology are needed to provide good-controlled thermal transient behavior, high thermal transfer efficiency, lightweight, environmental compatibility and low fabrication cost.

The using of epoxidized natural rubber as a polymer matrix will provide an alternative of epoxy-based polymer usage. Furthermore, the using of titanium dioxide as moieties in ENR-25 can improve the properties of the material [15,16], which in turn increase the opportunity of this combination to be used in electronic materials packaging applications.

Therefore, the aim of this research is synthesize a new composite obtained from ENR-25 as a polymer matrix and Titanium dioxide (TiO_2) as inorganic filler for electronic packaging application. Furthermore, the structural and morphological properties of ENR-25/TiO_2 composites were investigated in this work.

2. Experimental

2.1. Materials and chemicals

The elastomer (epoxidized natural rubber with 25% mole epoxidation) was obtained from Rubber Research Institute, Malaysia (RRIM). Titanium (IV) butoxide, 98%, Ti(OCH2CH2CH3)4 is a titanium alkoxide which commonly used as a source of titanium for the synthesis of Titania or titanium dioxide (TiO2) and it was purchased from Merck, Germany. While solvents such as 2-propanol (IPA), n-hexane, chloroform and toluene were purchased from Systerm Malaysia.

2.2. Measurements and characterization technique

FTIR spectroscopy: Transmission FTIR spectra of composite films were recorded from thin KBr disc of the samples with Perkin Elmer 2000 FTIR spectrophotometer at room temperature. The samples were scanned from 4000 to 400 cm−1 with resolution of 0.4 cm−1. By using FTIR, the functional groups in a molecule can be identified.

FESEM: The morphological features of each sample were obtained using a Zeiss Supra 35 VP field emission scanning electron microscopy (FESEM). The samples were prepared with gold coating for the measurements.

2.3. Synthesis of ENR-25/ TiO_2 composites

First, 20 g of ENR-25 was dissolved in 400 mL of two different solvents (toluene and chloroform) separately and stirred for 24 h at 25 °C temperature. The rubber solution was then filtered through cotton gauze to separate the high from the low molecular weight ENR-25. The latter was precipitated

in n-hexane while stirring using a glass rod. The white precipitate stuck to the glass rod was then transferred to a Teflon dish and dried in a vacuum oven at 60 °C temperature for two days.
A 1.32 x 10-3 M (3.82 x 10-3 mg mL-1) Titania precursor solution was prepared by diluting a 40 mg (1.32 x 10-4 mol) of precursor in 10 ml IPA. Both of the ENR-25 solutions obtained from subsection 3.4.2 were used.
In a typical preparation, 10 mL of Titania precursor solution was slowly added drop wise to the 40 mL of purified ENR-25 at room temperature. Upon completion, the mixture was continuously sonication for 10 minutes to make sure it is homogenously distribute into the ENR-25 solution for both toluene and chloroform respectively. The resultant reaction mixture was then cast onto Teflon dishes and dried in vacuum oven at 60 °C for two days to obtain ENR-25/TiO_2 composites as shown in Figure 1.

(a) (b)

Figure 1. ENR-25/Titania composite film samples prepared with dissolution of ENR-25 in (a) toluene, and (b) chloroform.

3. Results and Discussion

3.1. FTIR analysis

Figure 2 presents the FTIR spectra of (a) purified ENR-25, (b) ENR-25/ TiO_2 composite in toluene, and (c) ENR-25/ TiO_2 composite in chloroform. The spectra of purified ENR-25 (Figure 2 a) indicate the presence of CH3, CH2, CH, epoxide and isoprene (C=C) [17]. The CH3 asymmetric and symmetric stretching is detected at 2963.6 and 2860 cm^{-1} respectively and CH2 asymmetric stretching at 2921 cm^{-1}.
The values mentioned are all listed in range of alkane groups, which is 2840 - 3000 cm^{-1} [18]. The CH3 at the 2963.6 cm^{-1} and CH2 at 2921 cm^{-1} are emerged as split peaks due to difference resonance for C-H stretching. Similar asymmetric and symmetric stretching of CH3 for both ENR-25/Titania composites at 2963.6 cm^{-1} and 2860 cm^{-1} were shown in Figure 2 (b) and (c) respectively. The peak at 3390.4 cm^{-1} and 3384.3 cm^{-1} were signature of hydroxyl group [19] in both ENR-25/Titania composite in dilution of (b) toluene and (c) chloroform samples respectively.
Generally, there are no significant differences among the three samples except for the peak intensity of the characteristic peaks. The broad band presented at 1662 cm^{-1} for both ENR-25/Titania composite films is related to weak C-O stretching vibration.

IOP Conf. Series: Journal of Physics: Conf. Series **1019** (2018) 012054 doi:10.1088/1742-6596/1019/1/012054

According to Xu (2015), the peaks at 875 cm^{-1}, 836 cm^{-1} and 1252 cm^{-1} belong to epoxy ring [20]. The peak at 3390.4 cm^{-1} and 3384.3 cm^{-1} are produced by Ti-OH stretching vibration shifts, which may be attributed to the effect of hydrogen bond.

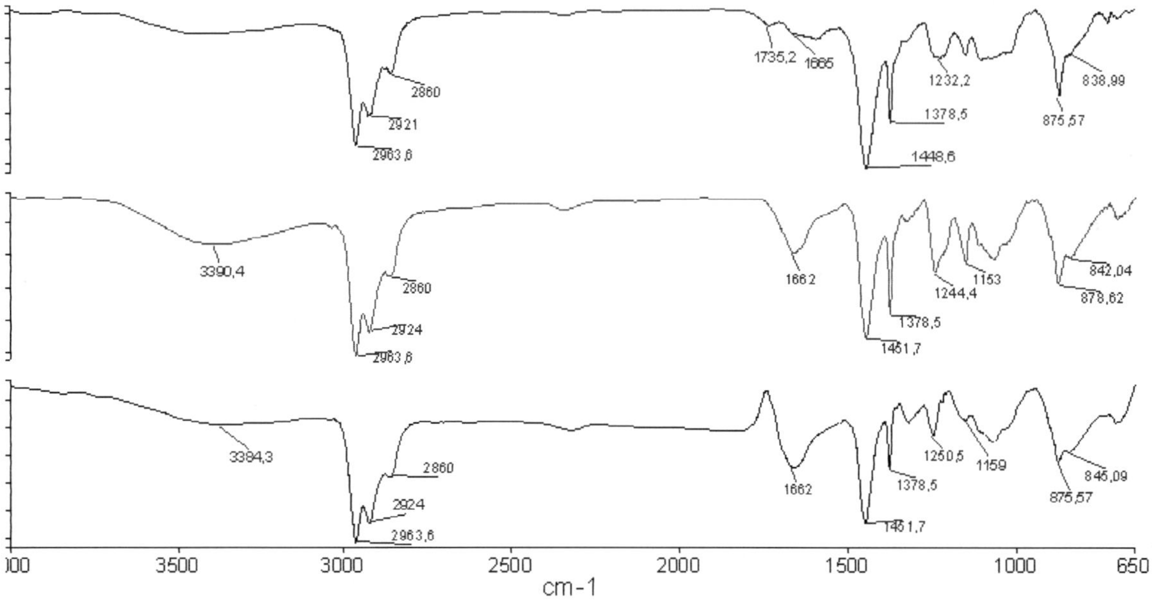

Figure 2. FTIR spectra of (a) Purified ENR-25, (b) ENR-25/TiO_2 composite in toluene and (c) ENR-25/TiO_2 composite in chloroform.

3.2. FESEM morphology

The FESEM images of purified ENR-25 and ENR-25/Titania composites with toluene and chloroform solvent are shown in figure 3 a, b and c respectively. The purified ENR-25 shows a typically smooth surface. The small amount of impurities presented on the surface of the sample could be due to the dirt of surrounding which caused by negligence of handling the sample as the ENR-25 has a sticky characteristic [21,22]. Figure 3 (b) shows the surface of ENR-25/ TiO_2 composite with toluene solvent. The non- homogenous dispersion of TiO_2 can be seen by the agglomeration embedded in the matrix. Therefore, the surface seems swelled in the most parts of the sample due to the TiO_2 agglomeration. Figure 3 (c) shows the surface of ENR-25/ TiO_2 composite with chloroform solvent. The bright phase showed the dispersion of TiO_2 particles. The TiO_2 particles covered the purified ENR-25 matrix. The non-homogenous dispersion of TiO_2 can be seen by the surface as bright layers. These layers appeared because the TiO_2 particles covered the surface of purified ENR-25 matrix [23-25].

IOP Conf. Series: Journal of Physics: Conf. Series **1019** (2018) 012054 doi:10.1088/1742-6596/1019/1/012054

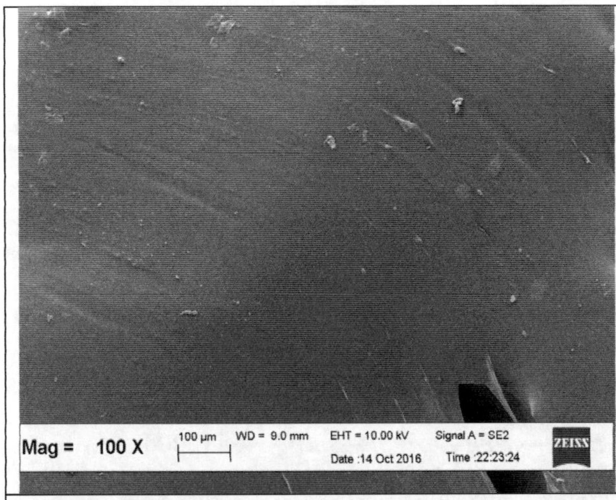

(a)

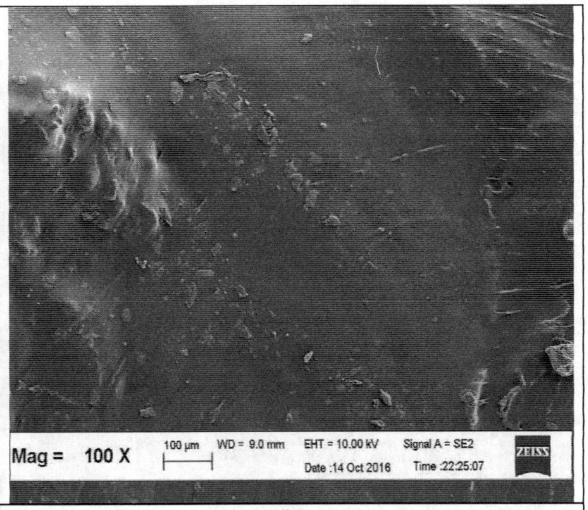

(b)

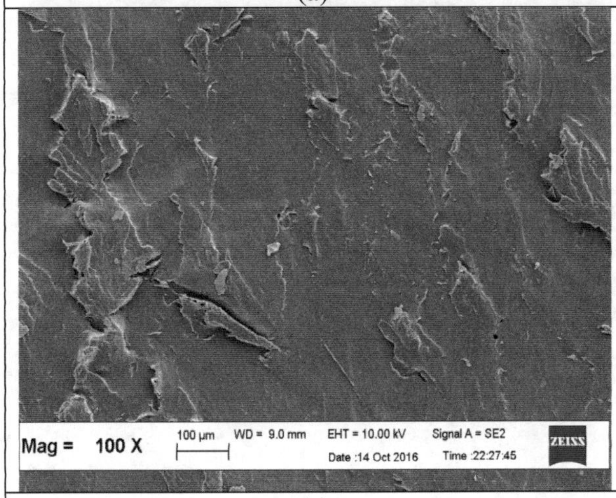

(c)

Figure 3. FESEM image of (a) purified ENR-25, (b) ENR-25/TiO_2 composite in Toluene (c) ENR-25/TiO_2 composite in Chloroform

4. Conclusion

The data obtained by FTIR measurements revealed the possible interaction of TiO_2 and ENR-25 in different dissolution solvent. The composite films were not hybrid due to the technique used to produce the ENR-25/TiO_2 composite and have been revealed by FESEM of the non- homogeneous dispersion of TiO_2 in ENR-25 matrix. The aggregation of TiO_2 nanoparticles into the ENR-25 matrix becomes severe in the purified ENR-25 dissolved in toluene compared with dissolved in chloroform.

5. Acknowledgment

The authors would like to acknowledge the Fundamental Research Grant Scheme (FRGS) Phase 1/2015 (9003-00523) under Ministry of Higher Education, Malaysia. The authors would appreciate Universiti Malaysia Perlis (UniMAP) for providing us all the facilities to make this research.

6. References

1. M. J. Yim, Y. Li, K. S. Moon, K. W. Paik and C. P. Wong, J. Adhes. Sci. Technol. **22**, 1593-1630 (2008).
2. H. Q. G. M. J. Yanping, Materials Review, **9**, 009 (2000).
3. X. Lu and G. Xu, Journal of applied polymer science, **65**, 2733-2738 (1997).
4. M. J. Yim, Y. Li, K. S. Moon, K. W. Paik and C. P. Wong, Journal of adhesion science and technology, **22**, 1593-1630 (2008).
5. X. Huang, P. Jiang and T. Tanaka, IEEE Electrical Insulation Magazine, **27** (2011).
6. S. T. Sam, M. A. Nuradibah, H. Ismail, N. Z. Noriman and S. Ragunathan, Polymer-Plastics Technology and Engineering, **53**, 631-644 (2014).
7. X. Tong, *Advanced Materials for Thermal Management of Electronic Packaging* (Springer Science+Business Media, LLC. Schaumburg, IL, 2011).
8. Y. Li, and C. P. Wong, *Materials Science and Engineering: R: Reports*, **51**, 1-35 (2006).
9. S. Rimdusit and H. Ishida, *Polymer*, **41**, 7941-7949 (2000).
10. C. Zweben, *JOM Journal of the Minerals, Metals and Materials Society*, **50**, 47-51 (1998).
11. Y. H. W. Z. W. Haishan and M. W. G. Lei, *Materials Review*, **6**, 025 (2004).
12. L. I. Lin-kai, World Plastics **9**, 007 (2005).
13. S. Rimdusit and H. Ishida, Polymer, **41**, 7941-7949 (2000).
14. H. Ardebili, E. H. Wong and M. Pecht, IEEE Transactions on Components and Packaging Technologies, **26**, 206-214 (2003).
15. O. S. Dahham, N. Z. Noriman, S. T. Sam, N. M. Al-Samarrai, Z. Shayfull, A. M. Alakrach and S. A. Elnaid, MATEC Web of Conferences ,**78**, 01075 (2016).
16. O. S. Dahham, N. Z. Noriman, S. T. Sam, N. A. S. Marwa, Z/ Shayfull and A. M. Alakrach, MATEC Web of Conferences, **78**, 01076 (2016).
17. R. Hamzah, M. A. Bakar, O. S. Dahham, N. N. Zulkepli and S. S. Dahham, J. Appl. Polym. Sci. **133** (2016), **DOI:** 10.1002/app.44123.
18. O. S. Dahham, R. Hamzah, M. A. Bakar, N. N. Zulkepli, S. S. Dahham and S. S. Ting, Polym. Test. **59**, 55-66. (2017).
19. N. N. Zulkepli and H. Ismail, Polymer-Plastics Technology and Engineering, **51**, 350-357 (2012).
20. T. Xu, Z. Jia, Y. Luo and D. Jia, Appl. Surf. Sci. **328**, 306-313, (2015).
21. N. Z. Noriman and H. Ismail, Journal of Applied Polymer Science, **123**, 779-787 (2012).
22. C. S. L., Baker, I. R., Gelling and R. Newell, *Rubber chemistry and Technology*, **58**, 67-85 (1985).
23. M. R. Kashani, S. Javadi and N. Gharavi, Smart Materials and Structures, **19**, 035019 (2010).
24. T. P. Selvin, J. Kuruvilla and T. Sabu, Materials Letters, **58**, 281-289 (2004).
25. W. E Dondero and R. E. Gorga, Journal of Polymer Science Part B: Polymer Physics, **44**, 864-878 (2006).

The Influences of Zirconium Dioxide on ENR-25/ZrO_2 Composites: FTIR and TGA Analysis

Omar S Dahham[1], R Hamzah[1], N Z Noriman[1], A M Alakrach[1], S Z Syed Idrus[2], Z Shayfull[3,4], T Adam[1,5]

[1] Center of Excellence Geopolymer and Green Technology (CEGeoGTech), Faculty of Engineering Technology (FETech), Universiti Malaysia Perlis (UniMAP), Level 1 Block S2, UniCITI Alam Campus, Sungai Chucuh, Padang Besar, 02100, Perlis, Malaysia.
[2] School of Human Development and Techno-communication (iKOM), Green Advanced Computing and Technology (GREAT) Research Group – CEGeoGTech, Universiti Malaysia Perlis (UniMAP)
[3] School of Manufacturing Engineering, Universiti Malaysia Perlis, Arau, 02600, Malaysia
[4] Green Design and Manufacture Research Group, Center of Excellence Geopolymer and Green Technology (CEGeoGTech),Universiti Malaysia Perlis, Kangar, 0100, Malaysia
[5] Institute of Nano Electronic Engineering (INEE), Universiti Malaysia Perlis (UniMAP), 01000 Kangar, Perlis, Malaysia

E-mail: rosnizahamzah@unimap.edu.my

Abstract. The characterization and thermal stability of purified epoxidized natural rubber with 25 mol% epoxidation (ENR-25) and treated epoxidized natural rubber with 25 mol% epoxidation / Zirconium dioxide (ENR-25/ZrO_2) Composites were studied using Fourier transform infrared spectroscopy (FTIR) and thermo gravimetric analysis (TGA). The ENR-25/ZrO_2 composites were treated with acetic acid and two different loading (50 and 100 mg) of ZrO_2 were used to form the composites. FTIR results show the ester (C-O) peaks at 1000- 1300 cm^{-1} due to the presence of acetic acid in the composites. The disappearance of epoxide peak at range 880-875 cm^{-1} indicated that ZrO_2 have formed hybrid bonds with opened epoxide rings. The both ENR/ZrO_2 composites showed 2 major thermal degradation steps (304 - 489°C and 493 – 551°C) for 50 mg ZrO_2 loading and (306 - 497°C 512 – 581°C) for 100 mg ZrO_2 loading indicating that the thermal stability of the composites have improved.

1. Introduction

Epoxidized natural rubber (ENR) is obtained from the chemical modification of natural rubber (NR). The ENR are characterized by having two or more epoxy units attached to NR molecule. ENR are produced by epoxidation process using peracetic acid of the NR latex [1-4]. Nowadays, the increasing interest of the people in science and technology regarding ENR makes it inevitable to understand the detail structure of ENR in term of epoxidation and physical form of rubber [5].

Even though ENR is the product from the modification process of NR, the characteristic of NR and ENR have no difference. There are a few unique characteristic of NR which including its glass transition temperature (Tg). The Tg of the NR shows that at low temperature, NR will have the glassy

IOP Conf. Series: Journal of Physics: Conf. Series **1019** (2018) 012055 doi:10.1088/1742-6596/1019/1/012055

characteristic while at high temperature; NR will have the rubbery characteristic [1,6]. In the meantime, zirconia (ZrO_2) recently has a high demand in most of the chemical based industry such as the dental industry.

It has a supportive characteristic in which it improves the mechanical properties and biocompatibility of the fixed dental prostheses (FDPs) that is being used in implantation process in dental industry [7]. Pure ZrO_2 cannot be used at the high temperature because of large changes in volume from tetragonal to monoclinic phase. This will affect the elasticity and fracture limit of the ZrO2, which will result in crack and flaws in ceramics.

At the same time, ENR-50 has epoxy characteristics, which has the characteristics of low toughness and brittleness. So, with the brittle characteristic that ZrO_2 had, the toughness and brittleness of ENR-50 can be improved. The formation of covalent bond between ENR-50 and ZrO_2 will change its chemical properties and leads to produce a new hybrid.

In this study, the characterization of the ENR-25 and ENR/ZrO_2 composites at different ZrO_2 were carried out using Fourier Transform Infrared Spectroscopy (FTIR), while their thermal stability were investigated using thermo gravimetric analysis (TGA). The experimental comprises purification of ENR-25, acid treatment of ENR-25 and the ENR-25/ZrO_2 composites synthesis.

2. Experimental

2.1. Materials

Epoxidized natural rubber, 25% mole epoxide (ENR-25) was obtained from Malaysia Rubber Research Institute (RRIM), Malaysia. Zirconium precursor, Zr(OCH2CH2CH3)4 was purchased from Fluka, USA. Acetic glacial 99.8% (CH3COOH) was purchased from Hamburg Chemical, Germany. The solvents such as N-hexane, Tetra hydro furan (THF), Toluene and 2-propanol (IPA) were purchased from System, Malaysia. All the materials and chemicals used in this study were used without any additional purification unless otherwise stated.

2.2. Composites synthesis

In this study, the synthesis of composites gone through three main stages ENR-25 purification, ENR-25 acid treatment, and ENR-25/ZrO_2 composites synthesis at different ZrO2 loading, 50 and 100 mg respectively (Figure 1).

To purify ENR-25, 20.00 g of ENR-50 was swelled and stirred in 400 mL of chloroform 25 C for 24 hours and then the solution was filtered through a cotton gauze pack to separate gel from sol content. The latter poured into n-and stirred by glass rod until white precipitate forms on the rod surface, which represented the purified ENR-25. The purified ENR-25 was transferred to Teflon dish and dried by vacuum oven at 60°C for 48 hours until the constant weight was recorded.

The acid treatment was carried out using 4 mL of glacial acetic acid added into rubber solution (200 mg Purified ENR-25 with 40 mL Toluene). The solution was refluxed and stirred at 110°C. After 3 hours, different Zr(OCH2CH2CH3)4 loading (50 and 100 mg) were added into the acid treated ENR-25 solution separately and refluxed at 60°C for 24 hours. The ENR/ZrO2 solution was poured in Teflon dishes and air-dried before dried in vacuum oven at 50°C for 24 hours.

IOP Conf. Series: Journal of Physics: Conf. Series **1019** (2018) 012055 doi:10.1088/1742-6596/1019/1/012055

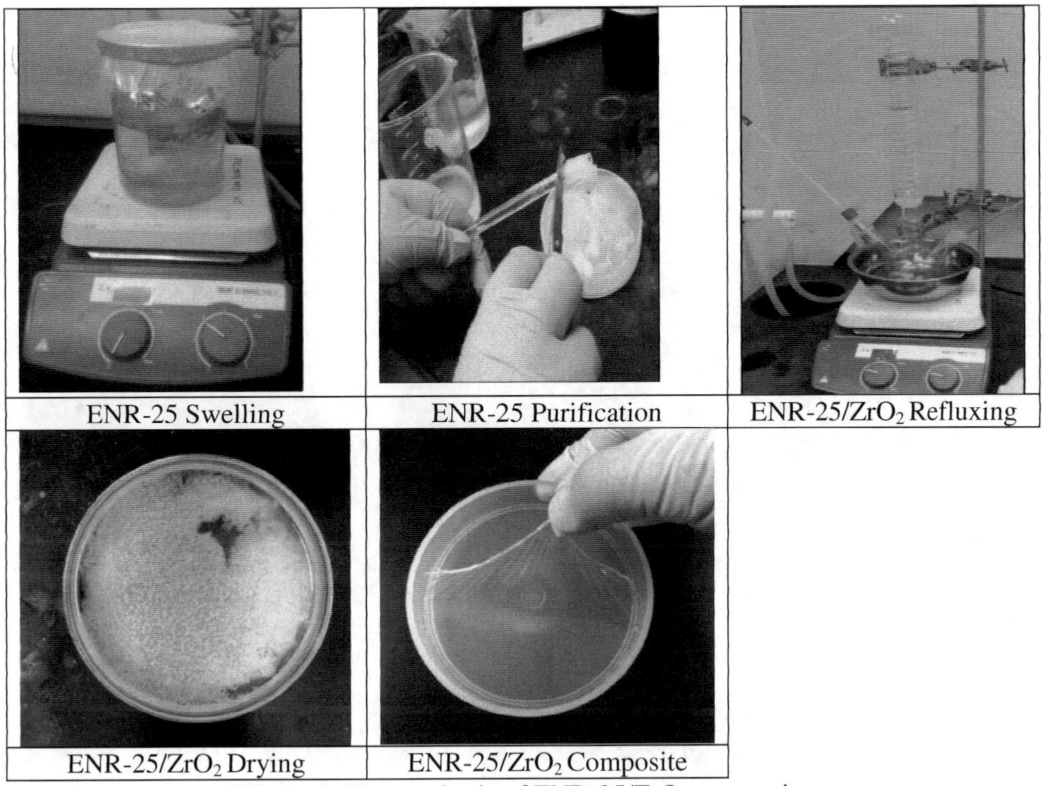

Figure 1. The synthesis of ENR-25/ZrO_2 composites

2.3. Composites characterization and analysis

FTIR analysis was carried out using *Perkin-Elmer 2000- FTIR.* About 100.00 mg of sample was placed onto zinc selenide (ZnSe) window using attenuated total reflection (ATR) method. The sample was scanned at range 4,000 cm^{-1} to 650 cm^{-1} with 16 numbers of scan.

The thermal stability of the samples was conducted using thermal gravimetric analyser (Perkin Elmer TGA-7). The samples were heated under nitrogen gas start from 30 °C to 900 °C. The heating rate was 20 °C min^{-1} in all cases.

2.4. Purified ENR-25 characterization

The FTIR spectrum of purified ENR-25 is shown in Figure 2. Generally there are five functional groups that represent the chemical structure of purified ENR-25. There are CH3, CH2, CH, epoxide and isoprene (C=C). The CH3 asymmetric at 2964 cm-1, symmetric stretching at 2860 cm^{-1} and CH2 asymmetric stretching at 2925 cm^{-1} are observed clearly.

These values are between 3100 and 2750 cm^{-1}. The CH3 at 2964 cm^{-1} and CH2 at 2925 cm^{-1} appeared as split peaks due to difference resonance for C-H stretching [1,8]. At low wavenumber, the peak at 1450 cm^{-1} corresponds to both CH3 and CH2 bending vibration and at 1378 cm^{-1} for C-H bending vibration.

The existence of the epoxide ring in the purified ENR-25 is shown by peak at 875 cm^{-1} due to half ring stretching of the epoxide while the full ring stretching is at 1234 cm-1 [9]. The C=C band of the isoprene is shown at 835 cm^{-1} [6,7] as a shoulder peak to the epoxide.

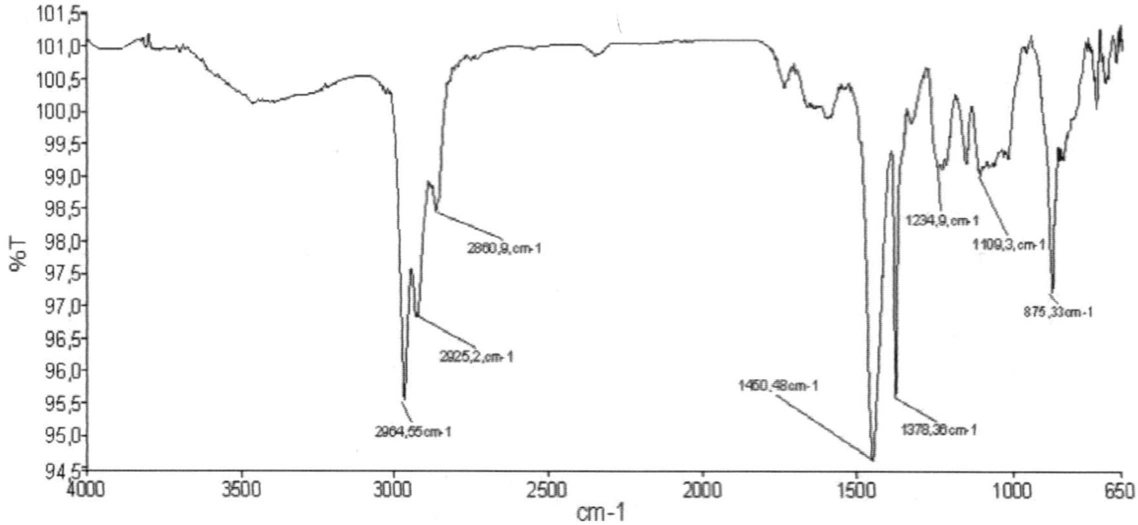

Figure 2. FTIR spectrum of the purified ENR-25

3. Results and Discussion

3.1. Fourier Transform Infrared Spectroscopy

Figure 3 shows the FTIR spectra of ENR-50/ZrO_2 composites at 50 and 100 gm of ZrO_2 loading. Generally, the composites produced few new peaks that are observable in FTIR. The peak of C=O stretching vibration at 1735.2 cm^{-1}.

The peak region at 1000- 1300 cm^{-1} is assigned to the ester (C-O) functionalities due to the presence of acetic acid in the composites. In this band, the intensity to have two or more bands is probable i.e. two adjoining band within this range.

Nevertheless, the band depletion also occurs throughout the process. The half ring stretching of the epoxide at range 880-875 cm^{-1} is disappeared as ZrO2 added. This indicated that ZrO2 has formed hybrid bonds with ring opened epoxide units. The broad peak 3400-3450 cm^{-1} indicated the existence of the OH group of that originates from precursor intermediate (Zr-OH) via its stretching vibration mode [10,11].

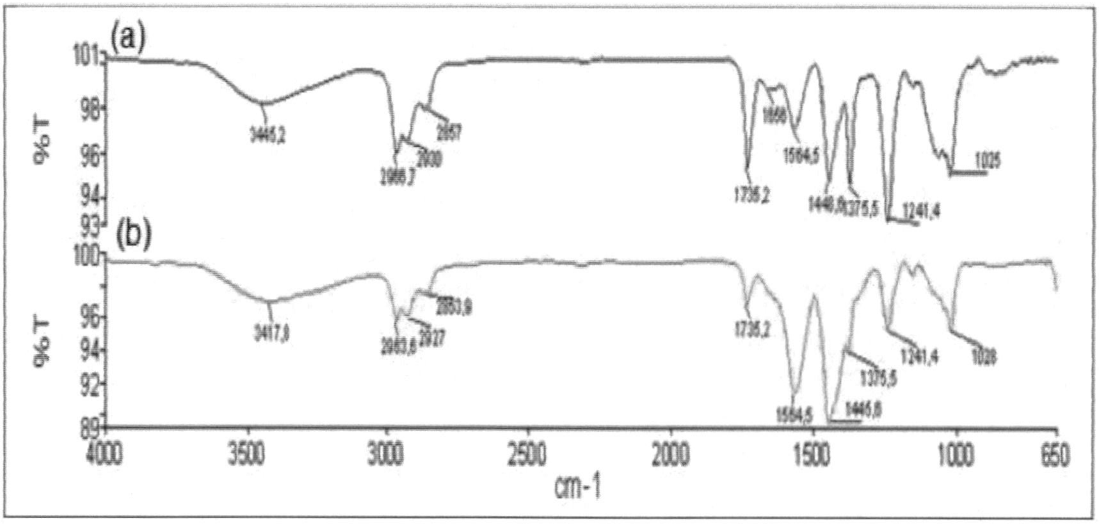

Figure 3. FTIR spectra of ENR-25/ZrO2 composite at (a) 50 mg and (b) 100 mg of ZrO_2 loading.

3.2. Thermo Gravimetric Analysis
The TG thermograms of purified ENR-25 and ENR-25/ZrO_2 composite at different ZrO_2 loading are shown in Figure 4. TG thermograms of purified ENR-25 show a single step degradation range at 382 - 466 °C [12]. While both ENR/ZrO_2 composites showed 2 major steps of thermal degradation at 304 - 489°C and 493 – 551 °C for ENR/ZrO_2 composite at 50 mg ZrO_2 loading and 306 - 497°C and 512 – 581 °C for ENR/ZrO_2 composite at 100 mg ZrO_2 loading [13,14].
The increasing of temperature over weight loss percentage of the composites, particularly with high ZrO_2 loading indicate that the thermal stability has improved as ZrO_2 loading increased [15-17]. The char residue of ENR/ZrO_2 composites indicates the high molecular mass of Zr compared with purified ENR-25.

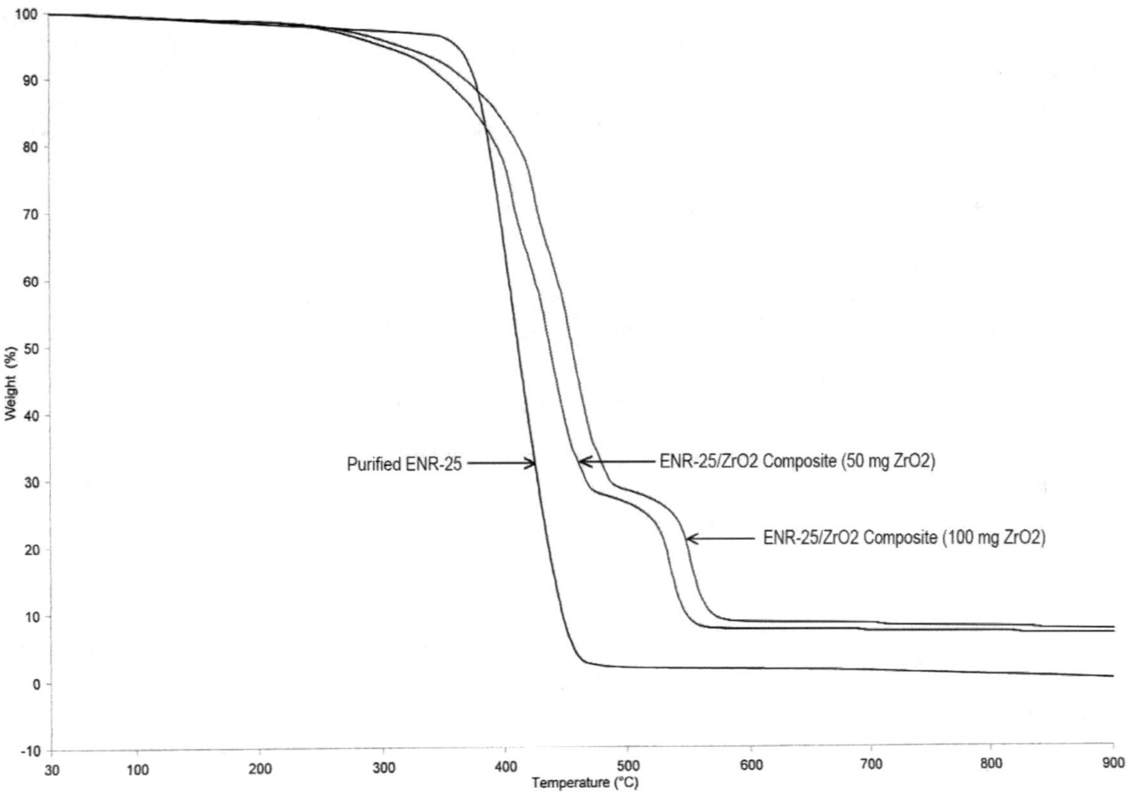

Figure 4. TG thermogram of purified ENR-25 and ENR-25/ZrO_2 composites at 50 and 100 mg ZrO_2 loading.

4. Conclusion
Based on FTIR spectra figure, the addition of ZrO_2 into ENR-25 formed bonds between Zr and epoxy unites. This type of bonding has improved the thermal stability of ENR/ZrO_2 samples as shown in TG thermogram.

5. Acknowledgement
The authors would like to acknowledge UniMAP Short Term Grant Scheme (9001 - 00521). We would also appreciate Universiti Malaysia Perlis (UniMAP) for providing us all the facilities to make this research.

IOP Conf. Series: Journal of Physics: Conf. Series **1019** (2018) 012055 doi:10.1088/1742-6596/1019/1/012055

6. References

1. R. Hamzah, M. A. Bakar, O. S. Dahham, N. N. Zulkepli and S. S. Dahham, *Journal of Applied Polymer Science* **133**, (2016).
2. O. S. Dahham, R. Hamzah, M. A. Bakar, N. N. Zulkepli, S. S. Ting, M. F. Omar, K. Muhamad, and S. S. Dahham, Polymer Testing **65**, 10-20 (2018).
3. M. Arroyo, M. A. Lopez-Manchado, J. L. Valentin and J. Carretero, Composites science and technology, **67**, 1330-1339 (2007).
4. W. Arayapranee, N. Na-Ranong and G. L. Rempel, Journal of applied polymer science, **98**, 34-41 (2005).
5. C. S. L. Baker, I. R. Gelling and R. Newell, *Rubber chemistry and Technology*, **58**, 67-85 (1985).
6. I. Khan, B. T. Poh and K. E. Lee, *Journal of Polymers and the Environment* **21**, 833-849 (2013).
7. A. Dorigato, A. Pegoretti, F. Bonddioli and M. Messori, *Composite Interfaces* **17**, 873-892 (2010).
8. W.A.K. Mahmood, M.M.R. Khan and M.H. Azarian, *J. Non-Crystalline Solids* **378**, 152-157 (2013).
9. O. S. Dahham, R. Hamzah, M. A. Bakar, N. N. Zulkepli, S. S. Dahham and S. S. Ting, Polymer Testing, **59**, 55-66. (2017).
10. M. Ochi, D. Nii, Y. Suzuki and M. Harada, *J. Mater. Sci.* **45**, 2655-2661 (2010).
11. N. N. Zulkepli and H. Ismail, Polymer-Plastics Technology and Engineering, **51**, 350-357 (2012)
12. S.D. Li, Y. Chen, J. Zhou, P.S. Li, C.S. Zhu and M.L. Lin, *J. Appl. Polym. Sci* **67**, 2207-2211 (1998).
13. S. Wang, Z. Tan, Y. Li, L. Sun and T. Zhang, Thermochimica Acta, **441**, 191-194 (2006).
14. N. Z. Noriman and H. Ismail, International Journal of Polymeric Materials and Polymeric Biomaterials, **62**, 252-259 (2013).
15. E. L. Crepaldi, G. J. de AA Soler-Illia, D. Grosso and C. Sanchez, *New Journal of Chemistry 27*, 9-13 (2003).
16. E. L. Crepaldi, G. J. de AA Soler-Illia, D. Grosso and C. Sanchez, New Journal of Chemistry, **27**, 9-13. (2003).
17. Y. Hu, S. Zhou and L. Wu, *Polymer*, **50**, 3609-3616 (2009).

The Effect of *Cyperus Odoratus* Size and Loading on the Properties of Polypropylene/*Cyperus Odoratus* (PP/CY) Composites

Omar S Dahham[1], N Z Noriman[1], R Hamzah[1], S Sudin[2], S Z Syed Idrus[2], Z Shayfull[3,4], T Adam[5]

[1] Center of Excellence Geopolymer and Green Technology (CEGeoGTech), Faculty of Engineering Technology (FETech), Universiti Malaysia Perlis (UniMAP), Level 1 Block S2, UniCITI Alam Campus, Sungai Chucuh, Padang Besar, 02100, Perlis, Malaysia.
[2] School of Human Development and Techno-communication (iKOM), Green Advanced Computing and Technology (GREAT) Research Group – CEGeoGTech, Universiti Malaysia Perlis (UniMAP)
[3] School of Manufacturing Engineering, Universiti Malaysia Perlis, Arau, 02600, Malaysia
[4] Green Design and Manufacture Research Group, Center of Excellence Geopolymer and Green Technology (CEGeoGTech),Universiti Malaysia Perlis, Kangar, 0100, Malaysia
[5] Institute of Nano Electronic Engineering (INEE), Universiti Malaysia Perlis (UniMAP), 01000 Kangar, Perlis, Malaysia

E-mail: niknoriman@unimap.edu.my

Abstract. In this research, a new bio-plastic composite was developed using *Cyperus Odoratus* (CY) as natural filler along with Polypropylene (PP) matrix. PP/CY composites were prepared using different CY size, fine size at (0-63) and coarse size at (150-250) and using different PP/CY ratio at (100/0, 95/5, 90/10 and 85/15). The extrusion process followed by injection molding process was conducted for PP/CY composites fabrication. The tensile and morphological characteristics of PP/CY samples were evaluated. The results suggest that the addition of CY content (fine size) up to 95/5 PP/CY ratio gave highest tensile strength (TS) value. The increasing of CY content, particularly CY coarse size into PP/CY composites has decreased the TS. By contrast, the high CY content, particularly CY coarse size shows highest Modulus indicating that the sample became stiffer and more rigid. The increasing of stiffness and rigidity of PP/CY composites as CY content increased has reduced the flexibility. Therefore, the elongation at break of the composites decreased. The tensile fractured surface of composites was evaluated using SEM analysis. SEM micrographs show that the low CY content, particularly CY fine size was well bonded in PP matrix in contrast to high CY content, particularly CY coarse size.

1. Introduction

The composite materials are usually used to improve the properties of products. Generally, the composite materials comprise two or more physically distinct phases and this combination produce aggregate properties, which can be different from its constituents [1,2]. Currently, the polymer composite (PC) products are extensively used in several applications due to the advantages of these products such as lightweight, high stiffness and high strength compared with the original constituent [3-5]. Moreover, PC can also enhance the chemical, thermal and physical properties along with the mechanical properties [6]. The utilization of natural fiber as reinforcement materials has attracted much attention in both industrial and academic field due to the significant advantages over synthetics fiber such as low cost, low density, non-abrasive to the equipment, non-irritation to the skin, reduced energy consumption, less health risk, recyclability, sustainability and biodegradability [7-9]. There are numerous kinds of natural fibers spread widely in the nature, particularly in tropical area such as kenaf, flax, jute, hemp, sisal, banana, kapok, henequen, coir, grass, wheat and many others. Furthermore, the waste of natural materials can also be use as natural filler to improve some properties of polymer composites [6,10]. *Cyperus Odoratus* (CY) or fragrant flats edge is a plant that can be found in most tropical and warm temperate areas such as the South America, southeastern Asia, Australia and parts of Africa. In Malaysia, this plant can be easily found in big amounts in wet areas such as lakes and abandoned paddy fields. The height of CY can extend up to 1.0 - 1.5 meters short time, which make CY a sustainable plant. Based on recent researches, natural fibers are valuable materials that are use as reinforcing agent in polymer-based materials [11,12]. It is also become a valuable alternative to synthetic fibers for various industrial applications [8].

Therefore, the aim of the current research is to evaluate the different amount and size of *Cyperus Odoratus* as natural fiber on the tensile properties of Polypropylene / *Cyperus Odoratus* Composites.

2. Experimental

2.1. Materials

Cyperus Odoratus (CY) was collected from Hutan Kampung, Jitra, Kedah, Malaysia. Polypropylene was purchased from Titan Polymer Sdn. Bhd., Malaysia.

2.2. Cyperus Odoratus preparing

First, CY leaves were dried under sunlight for 2 days. The dried CY were cut, ground and sieved into two different size, 0 - 63μm, and 150 - 250μm respectively. After sieving, ground CY were dried using vacuum oven at 60 oC for 8 hours.

2.3. Composites fabrication

The PP matrix and ground CY were mixed together in 100/0, 95/5, 90/10, 85/15 and 80/20 ration of PP/CY composites using Twin Screw Extruder according to ASTM D4101-14e. The screw speed of the Extruder was 120 rpm while the temperature-feeding zone, mixing zone, reacting zone and die zone were 175, 190, 190, and 180°C, respectively [13]. After cut it into pallets shape, the pallets were introduced through injection molding process. Standard tensile dumbbells (ASTM D638) were used in Battenfell Injection Molding Machine with temperature profile from feed to exit die zone at 150, 165, 170 and 175°C respectively.

2.4. Tensile properties

The tensile test was conducted according to ASTM D638 using universal tensile testing machine (Instron 5569). The dumbbell shaped specimens was used for this test with a crosshead speed at 50mm/min and 25 ± 2 °C. The tensile strength, elongation at break and Young's modulus were reported. The average value of the tensile properties was obtained from five specimens.

2.5. Scanning Electron Microscope

The tensile fracture surface of the specimens were analyzed using scanning electron microscope (SEM), model JSM-6460LA. The specimens were platinum coated to increase the resolution of the micrograph and to avoid the electrical charging.

3. Experimental Results and Discussion

3.1. Tensile properties

The tensile strength value of PP control and PP/CY composite at different CY loading and size are shown in Figure 1,a. From the graph below, it shows decrease in tensile strength in the overall trends of PP/CY composites for both CY fine and coarse size as the CY content increased. This was probably due to the amount of the plastic as adhesive decrease when the filler content in the composite increase. The high CY content, particularly 15wt% of CY coarse size shows the lowest reading of tensile strength at 23.94 MPa. This is attributed to the CY aggregation inside PP matrix, which in turn effected negatively on the stress transition from PP matrix to CY filler during tensile strength test. However, the low CY content, particularly 5wt% of the CY fine size shows the optimum tensile strength reading at 31.28 MPa. The relatively small particle size and high distribution of 5wt% of the CY fine size contributed to a strong interfacial adhesion between PP matrix and CY fiber, which in turn increased the strength of the composite [14,15].

Figure 1,c illustrates the tensile modulus of PP control and PP/CY composites. The tensile modulus of the composite increased slightly with the increasing of CY content.

CY filler have higher stiffness than PP matrix, which in turn contributed to the increases in the tensile modulus of the composites as shown before in Figure 1,c. The increasing of modulus led to decreasing in flexibility. Our previous researches reported that the increasing in the tensile modulus was due to the increase in the rigidity as well as the intermolecular interaction between polymer matrix and the natural filler, which in turn led to decrease in flexibility [16,17].

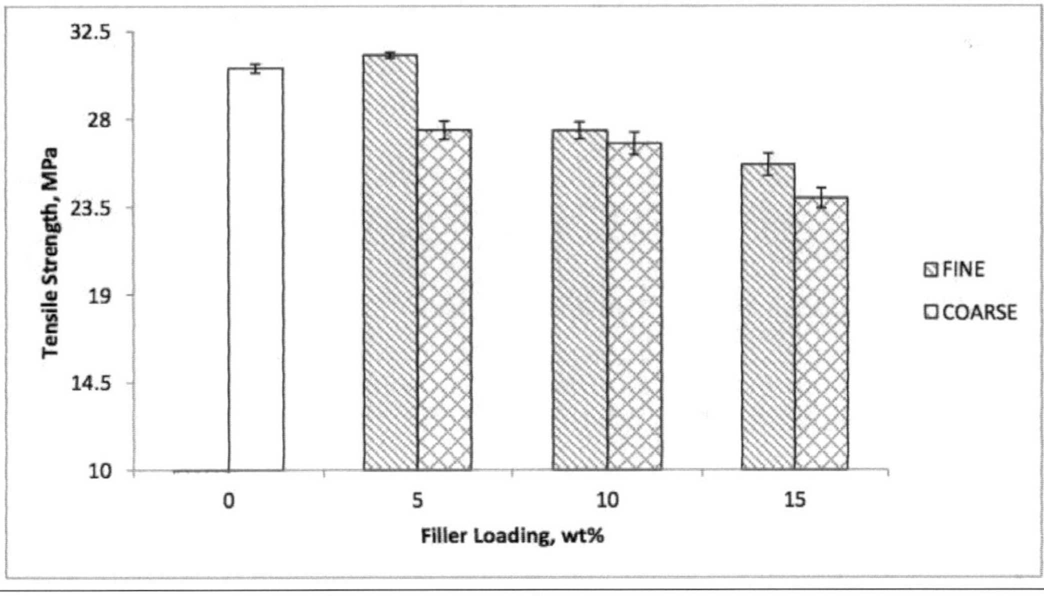

(a)

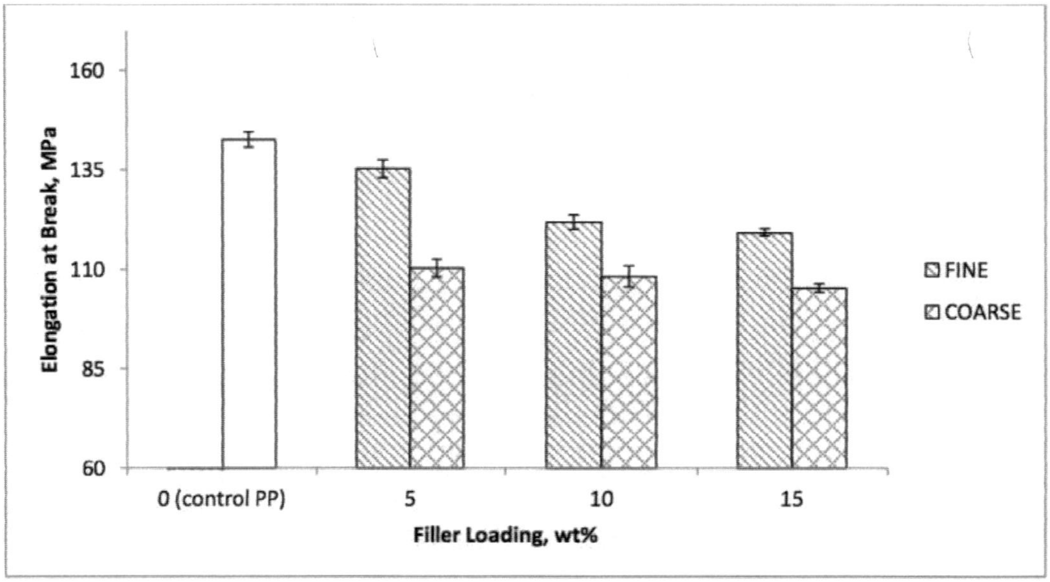

(b)

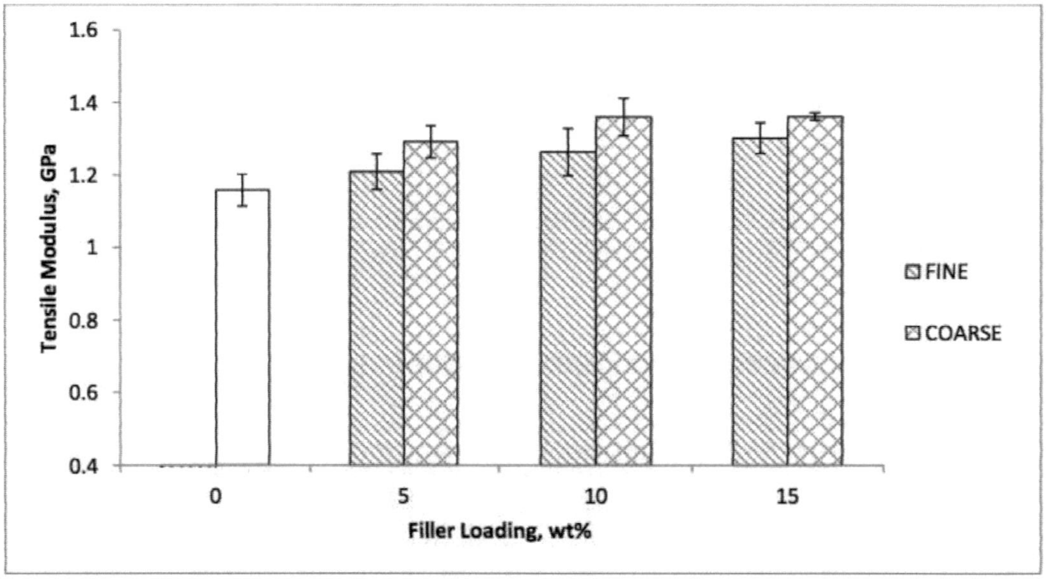

(c)

Figure 1. The effect of CY content on the (a) tensile strength, (b) elongation at break and (c) tensile modulus of PP/CY composites

3.2. Morphology

Figure 2 (a) and (b) shows the morphological comparison of the tensile fractured surface of PP/CY composites at 5wt% CY content in both CY fine and coarse size respectively. In figure 4 (a), the SEM micrograph shows better dispersion of CY, which in turn resulted a good stress concentration. Therefore, the ductility phenomenon has occurred causing increases in the tensile strength of the composite. The SEM micrograph of Figure 4 (b) shows the relatively big particle size of CY that was detached from PP matrix. The gap between CY filler and PP matrix are big compared to CY fine size. However, it still can be seen that CY was bonded on to the matrix [17,18].

Figure 4 (c) and (d) illustrate SEM micrograph of PP/CY composite at 15wt% CY content at fine and coarse sizes respectively. In both figures, it can be observed clearly the CY filler detachment from PP matrix. The poor adhesion between PP matrix and CY filler created weak interfacial regions, which can result in de-bonding and frictional pullout [19,20].

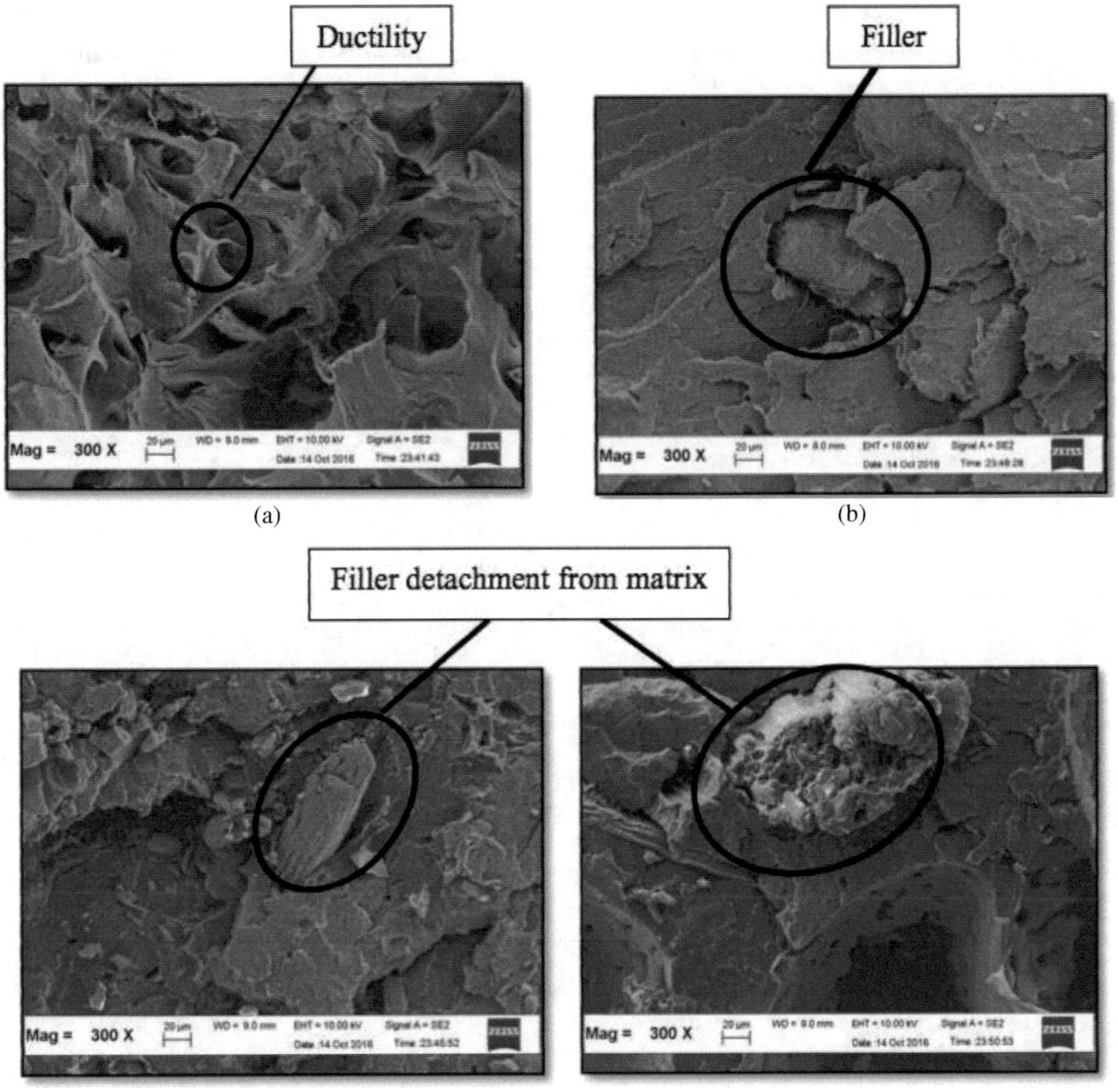

(c) (d)

Figure 2. The SEM micrographs of tensile fractured surface of PP/CY composite at (a) 5wt% CY fine size, (b) 5wt% CY coarse size, (c) 15wt% CY fine size and (d) 15wt% CY coarse size

4. Conclusion

From this research, it could be concluded that the *Cyperus Odoratus* (CY) was successfully used as natural fiber to improve polypropylene tensile properties. However, the high CY content and/or coarse CY size is not recommended due to the deterioration of tensile properties of PP/CY composite caused by CY high content and coarse size.

5. Acknowledgement

The authors would like to acknowledge the Fundamental Research Grant Scheme (FRGS) Phase 1/2015 (9003-00523) under Ministry of Higher Education, Malaysia. The authors would appreciate Universiti Malaysia Perlis (UniMAP) for providing us all the facilities to make this research.

IOP Conf. Series: Journal of Physics: Conf. Series **1019** (2018) 012056 doi:10.1088/1742-6596/1019/1/012056

6. References

1. S. Y. Fu, X. Q. Feng, B. Lauke and Y. W. Mai, Composites Part B **39**, 933-961 (2008).
2. N. Z., Noriman and H. Ismail, H. International Journal of Polymeric Materials and Polymeric Biomaterials, **62**, 252-259 (2013).
3. C. E. Bakis, L. C. Bank, V. Brown, E. Cosenza, J. F. Davalos, J. J. Lesko and T. C. Triantafillou, J. Compos. Constr. **6**, 73-87 (2002).
4. S. Ramakrishna, J. Mayer, E. Wintermantel and K. W. Leong, Compos. Sci. Technol. **61**, 1189-1224 (2001).
5. J. George, M. S. Sreekala and S.Thomas, *Polymer Engineering & Science*, **41**, 1471-1485 (2001).
6. H. Ku, H. Wang, N. Pattarachaiyakoop and M. Trada, Composites Part B **42**, 856-873 (2011).
7. K. Balaed, N. Z. Noriman, O. S. Dahham, S. T. Sam, R. Hamzah, and M. F. Omar, Int. J. Polym. Anal. Charact. **21**, 590-598 (2016).
8. F. M. Al-Oqla and S. M. Sapuan, J. Cleaner Prod. **66**, 347-354 (2014).
9. S. T. Sam, M. A. Nuradibah, H. Ismail, N. Z. Noriman S. Ragunathan, Polymer-Plastics Technology and Engineering, **53**, 631-644 (2014).
10. D. N. Saheb and J. P. Jog, Adv. Polym. Tech. ***18***, 351-363 (1999).
11. K. L., Pickering, M. A., Efendy and T. M. Le, *Composites Part A: Applied Science and Manufacturing*, **83**, 98-112 (2016).
12. S. Taj, M. A. Munawar and S. Khan, Proceedings-Pakistan Academy of Sciences **44**, 129 (2007).
13. D. Shi, J. Yang, Z. Yao, Y. Wang, H. Huang, W. Jing and G. Costa, Polymer **42**, 5549-5557 (2001).
14. O. S. Dahham, N. Z. Noriman, S. T. Sam, H. Rosniza, N. A. S. Marwa, Z. Shayfull and A. M. Alakrach, MATEC Web of Conferences **78**, 01076 (2016).
15. M. Zampaloni, F. Pourboghrat, S. A. Yankovich, B. N. Rodgers, J. Moore, L. T. Drzal and M. Misra, Composites Part A **38**, 1569-1580 (2007).
16. O. S. Dahham, N. Z. Noriman, S. T. Sam, N. M. Al-Samarrai, Z. Shayfull, A. M. Alakrach and S. A. Elnaid, MATEC Web of Conferences **78**, 01075 (2016).
17. B., Bax and J. Müssig, *Composites Science and Technology*, **68**, 1601-1607 (2008).
18. V., Fiore, T., Scalici, F., Nicoletti, G., Vitale, M., Prestipino and A. Valenza, *Composites Part B: Engineering*, **85**, 150-160 (2016).
19. M. H. Subli, M. F. Omar, I. G. Sandu, N. N. Zulkepli, M. M. A. B. Abdullah and A. V. Sandu, Mater. Plast. Buchar. **51**, 391-395. (2014).
20. Y. Lei, Q. Wu and Q. Zhang, *Composites Part A: Applied Science and Manufacturing*, **40**, 904-912 (2009).

The Influences NaOH Treatment on Polypropylene/*Cyperus Odoratus* (PP/CY) Composites: Tensile and Morphology

Omar S Dahham[1], N Z Noriman[1], R Hamzah[1], Marwa N Al-Samarrai[1], S Z Syed Idrus[2], Z Shayfull[3,4], T Adam[1,5], M Mohammed[6]

[1] Center of Excellence Geopolymer and Green Technology (CEGeoGTech), Faculty of Engineering Technology (FETech), Universiti Malaysia Perlis (UniMAP), Level 1 Block S2, UniCITI Alam Campus, Sungai Chucuh, Padang Besar, 02100, Perlis, Malaysia.
[2] School of Human Development and Techno-communication (iKOM), Green Advanced Computing and Technology (GREAT) Research Group – CEGeoGTech, Universiti Malaysia Perlis (UniMAP)
[3] School of Manufacturing Engineering, Universiti Malaysia Perlis, Arau, 02600, Malaysia
[4] Green Design and Manufacture Research Group, Center of Excellence Geopolymer and Green Technology (CEGeoGTech),Universiti Malaysia Perlis, Kangar, 0100, Malaysia
[5] Institute of Nano Electronic Engineering (INEE), Universiti Malaysia Perlis (UniMAP), 01000 Kangar, Perlis, Malaysia
[6] Center of Excellence Geopolymer and Green Technology (CEGeoGTech), School of Materials Engineering, Universiti Malaysia Perlis (UniMAP), Kompleks Pengajian Jejawi 2, 02600 Arau, Malaysia

E-mail: niknoriman@unimap.edu.my

Abstract. The bio-plastic composite has attracted the attention of people who work in academic and industrial fields due to the several advantages of this composite as compared with conventional plastic composites. However, the modification of the natural filler used in the bio-composites usually needs a modification before use. In the current work, the chemical modification on *Cyperus Odoratus* (CY) using different sodium hydroxide (NaOH) concentration at 1, 3, 6 and 9 wt.% were used to investigate the effect of this modification on the tensile properties of Polypropylene/*Cyperus Odoratus* (PP/CY) composites. The extrusion and injection molding processes were used respectively to prepare the composites. Results suggest that the presence of NaOH, particularly 3 wt.% has improved the tensile strength and the elongation at break of the composite. By contrast, the high NaOH concentration, particularly 9 wt.% has effected negatively on this properties. The tensile modulus increased linearly as NaOH concentration increased in the composites. The SEM micrograph of treated specimens at 3 wt.% NaOH shows improvement on CY surface while 9 wt.% NaOH shows deterioration on the DS surface.

1. Introduction

Nowadays, polymer Composites (PC) is used extensively in the industry to enhance the product properties. Normally, the composite materials contain two phases which are primary and secondary phase. Primary phase is the matrix while the secondary phase is reinforcement, which is the imbedded phase. The uses of natural fiber for the reinforcements have attractive the attention of both sectors in industry and academy due to the significant advantages of natural fiber as compared with synthetics fiber [1-3]. *Cyperus Odoratus* (CY) is one of the natural fibers that can be located near the paddy field. It does not give any benefits to the farmer. *Cyperus Odoratus* (CY) can be an attraction to the dangerous animal such as snake in the paddy field. Thus, it can be harmful to the farmer. This plant cans also bothering the growth of paddy plant. It will be such a waste if this plant is cut and being thrown away.

When the natural filler is use as the reinforcement of the composite, there will be some disadvantages such as lignocellulosic fiber contain high level of moisture absorption, poor surface adhesion to hydrophobic polymer, low processing temperature permissible ad many more [4-6]. The main problem is because these filler have strong hydrophilic character, which make the interfacial adhesion between this type of filler and hydrophobic polymer matrix poor.

The chemical modification is the common method used to improve the performances of natural fiber reinforced composites by linking the gap in compatibility between hydrophilic natural fibers and hydrophobic polymer matrix [4,8]. The chemical modification using alkaline is a simple and efficient treatment that can enhance the adhesion between lignocelluloses fiber and polymer matrix [8,9]. Rong et al. 2001, studied the effect of alkaline modification on the mechanical properties of sisal/epoxy composites. Results showed that the alkaline modification introduced some reactive groups and provide the fibers with higher extensibility through partial removal of lignin and hemicellulose, which in turn effected positively on the mechanical properties of the composite [10].

In our current research, we have investigated the effect of different NaOH concentration as an alkaline treatment on the tensile properties of polypropylene/*Cyperus Odoratus* (PP/CY) Composites.

2. Experimental

2.1. Materials

Polypropylene was used as a matrix in this study. It was obtained from Titan Polymer Sdn. Bhd., Malaysia. The *Cyperus Odoratus* (CY) was used as a natural fiber, which was obtained from Hutan Kampung, Jitra, Kedah, Malaysia.

2.2. Preparing of Cyperus Odoratus

The leaves of CY were initially dried under sunlight for 2 days. After drying, CY were cut, ground and sieved at 0 - 63μm particle size. The ground CY were then dried using vacuum oven at 60 °C for 8 hours. After drying, the ground CY were treated using different NaOH concentration at 1, 3, 6 and 9 wt.% respectively.

2.3. PP/CY composites fabrication

The polymer matrix (PP) and treated natural fiber (CY) were mixed together at fixed PP/CY ration (95/5) with different NaOH concentration at 1, 3, 6 and 9 wt.% respectively. Twin Screw Extruder was utilized for extrusion process according to ASTM D4101-14e. The Extruder screw speed was at 120 rpm and the temperature feeding zone, mixing zone, reacting zone and die zone were 175, 190, 190, and 180°C, respectively [11]. After that, the extrudate were molded using injection-molding process. Standard tensile dumbbells (ASTM D638) were used in Battenfell Injection Molding Machine with temperature profile from feed to exit die zone at 150, 165, 170 and 175°C respectively.

2.4. Tensile properties
According to ASTM D638, the tensile test was carried out using universal tensile testing machine (Instron 5569) and the dumbbell shaped samples was used for this test with a crosshead speed at 50mm/min and 25 ± 2 °C. The tensile strength, elongation at break and Young's modulus were reported. The average value of the tensile properties was obtained from five specimens.

2.5. Scanning electron microscope
The tensile fracture surface of the samples were examined using scanning electron microscope (SEM), model JSM-6460LA. The samples were platinum coated to enhance the micrograph resolution and also to reduce the electrical charging during analysis.

3. Results and discussion

3.1. Tensile properties
The influence of different NaOH concentration on the tensile strength, elongation at break and tensile modulus of PP/CY composites are shown in Figure 1 a, b and c respectively. It can be observed clearly that the addition of NaOH up to 3 wt.% has increased the values of the tensile strength and elongation at break (Figure 1a and b). This is due to the improvement on the fiber surface caused by NaOH solution. Mishra et al. reported that the addition of NaOH up to 5% into natural fiber/ polyester composite have improved the tensile strength while the high NaOH concentration such as 10% NaOH have effected negatively on the properties of the composites. The high concentration of NaOH increased the delignification of natural fiber, which in turn leads to damage of fiber structure [12]. Other researcher, such as Jacob et al. assessed the effects of different concentration of NaOH (0.5,1, 2, 4 and 10%) on the properties of sisal fiber-reinforced composites. Their results concluded that the maximum tensile strength value was 4% of NaOH [13]. Therefore, the values of tensile strength and elongation at break of PP/CY composite decreased steadily after 3 wt.% of NaOH concentration. However, The high concentration of NaOH, particularly 9 wt.% shows higher tensile modulus than other NaOH concentration (Figure 1c). This could be attributed to the high hydrogen bonding disruption in the network structure causing increases in surface roughness of the natural fiber. The roughness of the fiber surface improves the interaction between fiber and polymer matrix, which in turn make the composite stiffer and more rigid [14]. Thus, the tensile modulus value has increased.

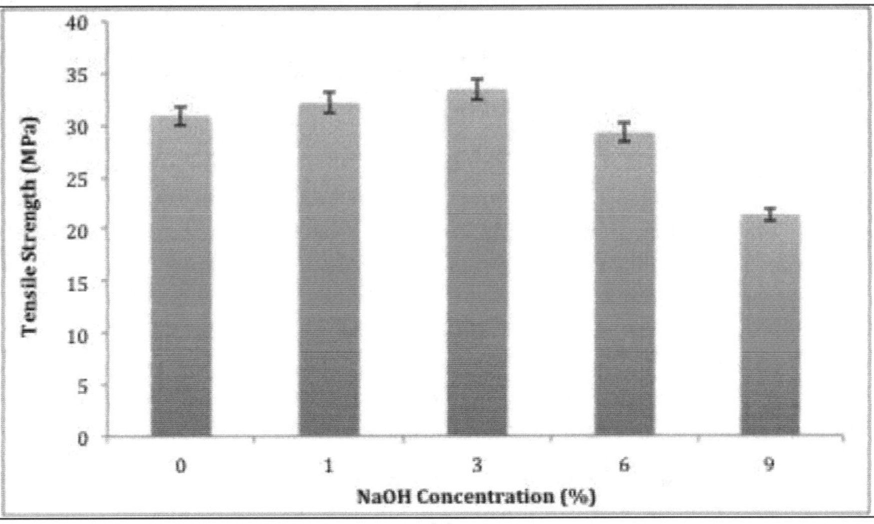

(a)

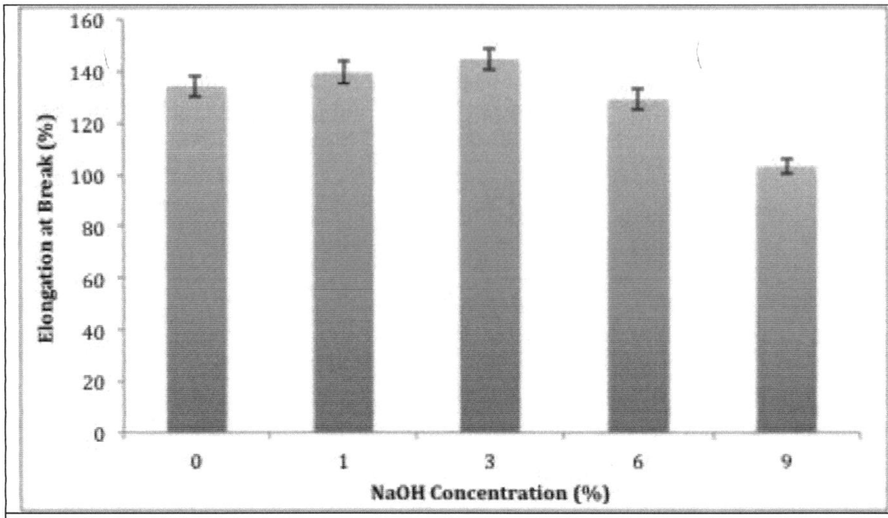

(b)

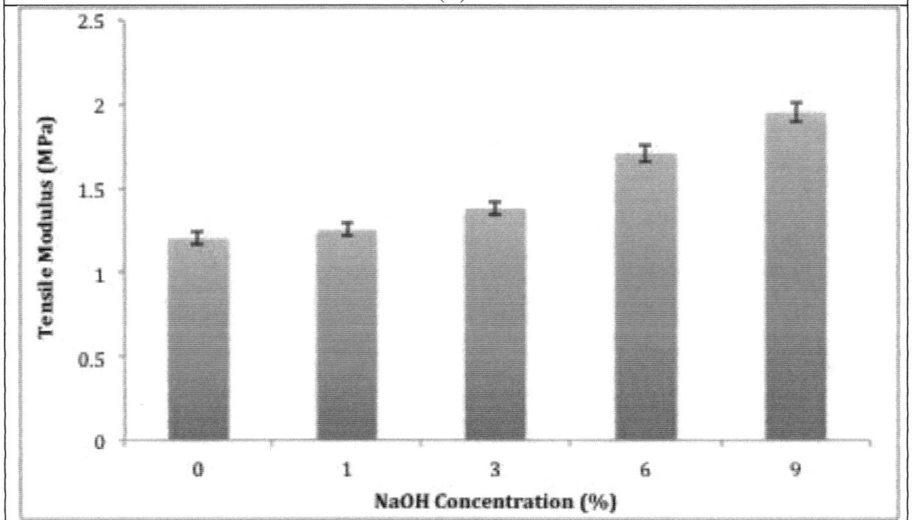

(c)

Figure 1. The effect of NaOH concentration on the (a) tensile strength, (b) elongation at break and (c) tensile modulus of PP/CY composites

3.2. Morphology

The scanning electron micrographs (Figure 2 a and b) show the tensile fractured surfaces of treated PP/CY composites at 3 wt.% and 9 wt.% NaOH concentrations respectively. The SEM micrograph of treated LLDPE/DS composite with low NaOH concentration (3 wt.%) shows relatively smoother CY surface and stronger interfacial bonding between CY fiber and PP matrix. Nevertheless, the high NaOH concentration, particularly (9 wt.%) shows rough surface (Figure 1 b) indicating that the high NaOH concentration increased the delignification of natural fiber (CY), which in turn lead to damage of fiber structure [15-17].

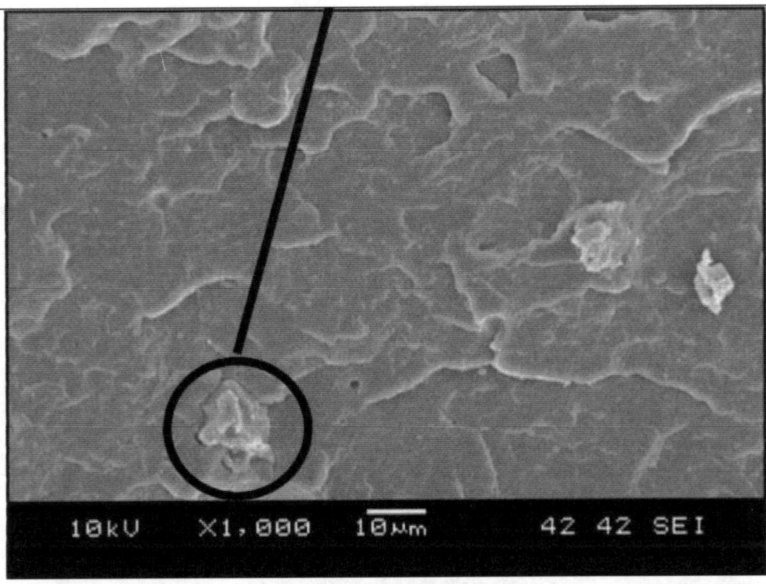

(a)

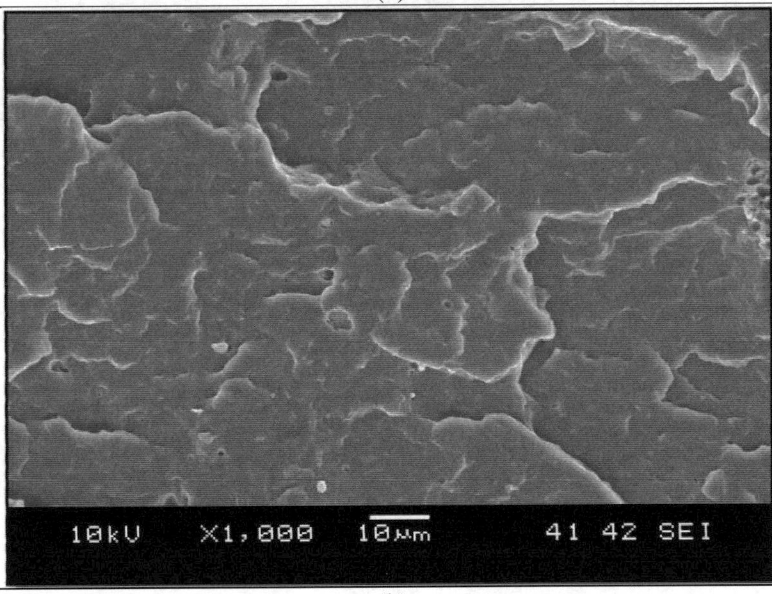

(b)

Figure 2. The SEM micrographs of tensile fractured surface of PP/CY composite at (a) 3 wt.% NaOH and (b) 9 wt.% NaOH.

4. Conclusion

The fabrication of bio-plastic composites such as Polypropylene/*Cyperus Odoratus* (PP/CY) composites can give a grate opportunity to the manufacture of green plastic products. Nevertheless, the mechanical properties of this composite were relatively poor. Consequently, the chemical modification using NaOH was necessary to improve the properties. The using of low concentration (3 wt.%) of NaOH gave better tensile properties compared with control and other NaOH concentration. The SEM micrograph of treated composite with 3 wt.% NaOH shows strong interfacial bonding between CY fiber and PP matrix, which reflected positively on the tensile properties of the composite.

5. Acknowledgement
The authors would like to acknowledge the Fundamental Research Grant Scheme (FRGS) Phase 1/2015 (9003-00523) under Ministry of Higher Education, Malaysia. The authors would appreciate Universiti Malaysia Perlis (UniMAP) for providing us all the facilities to make this research.

6. References
1. M. P. Brown and K. Austin, Appl. Phys. Letters **85**, 2503–2504 (2004). D. N. Saheb and J. P. Jog, Adv. Polym. Techn. **18**, 351-363 (1999).
2. K. Balaed, N. Z. Noriman, O. S. Dahham, S. T. Sam, R. Hamzah and M. F. Omar, Int. J. Polym. Anal. Charact. **21**, 590-598 (2016).
3. S. T., Sam, M. A., Nuradibah, H. Ismail, N. Z. Noriman and S. Ragunathan, Polymer-Plastics Technology and Engineering, **53**, 631-644 (2014)
4. V. Fiore, T. Scalici, F. Nicoletti, G. Vitale, M. Prestipino and A. Valenza, Composites Part B **85**, 150-160 (2016).
5. O. S. Dahham, N. Z. Noriman, S. T. Sam, N. M. Al-Samarrai, Z. Shayfull, A. M. Alakrach and S. A. Elnaid, MATEC Web of Conferences **78**, 01075 (2016).
6. Y. Lei, Q. Wu and Q. Zhang, *Composites Part A: Applied Science and Manufacturing*, **40**, 904-912 (2009).
7. A. Sabbagh, Composites Part B **57**, 126-135 (2014).
8. A. Orue, A. Jauregi, U. Unsuain, J. Labidi, A. Eceiza and A. Arbelaiz, Composites Part A **84**, 186-195 (2016).
9. O. S. Dahham, N. Z. Noriman, S. T. Sam, H. Ismail, S. Ragunathan, H. Rosniza and M. N. Al-Samarrai, J. Polym. Mater. **33** 647 (2016).
10. M. Z. Rong, M. Q. Zhang, Y. Liu, G. C. Yang and H. M. Zeng, Compos. Sci. Technol. **61**, 1437-1447 (2001).
11. D. Shi, J. Yang, Z. Yao, Y. Wang, H. Huang, W. Jing, and G. Costa, Polymer **42**, 5549-5557 (2001).
12. S. Mishra, A. K. Mohanty, L. T. Drzal, M. Misra, S. Parija, S. K. Nayak and S. S. Tripathy, Compos. Sci. Technol. **63**, 1377-1385 (2003).
13. M. Jacob, S. Thomas and K. T. Varughese, Compos. Sci. Technol.**64**, 955-965 (2004).
14. M. H.,Subli, , M. F. Omar, I. G. Sandu, N. N. Zulkepli, M. M. A. B. Abdullah and A. V. Sandu, Mater. Plast. Buchar. **51**, 391-395 (2014).
15. X. Li, L. G. Tabil and S. Panigrahi, J. Polym. Environ. **15**, 25-33 (2007).
16. A. Valadez-Gonzalez, J. M. Cervantes-Uc, R. J. I. P. Olayo and P. J. Herrera-Franco, *Composites Part B: Engineering*,**30(**, 309-320 (1999).
17. J. George, M. S. Sreekala and S. Thomas, *Polymer Engineering & Science*, **41**, 1471-1485 (2001).

IOP Conf. Series: Journal of Physics: Conf. Series **1019** (2018) 012058 doi:10.1088/1742-6596/1019/1/012058

The Effects of SMR L Content on the Properties of ENR-25/ RSC Compounds

Omar S Dahham[1], R Hamzah[1], N Z Noriman[1], A M Alakrach[1], S Z Syed Idrus[2], Z Shayfull[3,4], T Adam[1,5], M Mohammed[6]

[1] Center of Excellence Geopolymer and Green Technology (CEGeoGTech), Faculty of Engineering Technology (FETech), Universiti Malaysia Perlis (UniMAP), Level 1 Block S2, UniCITI Alam Campus, Sungai Chucuh, Padang Besar, 02100, Perlis, Malaysia.
[2] School of Human Development and Techno-communication (iKOM), Green Advanced Computing and Technology (GREAT) Research Group – CEGeoGTech, Universiti Malaysia Perlis (UniMAP)
[3] School of Manufacturing Engineering, Universiti Malaysia Perlis, Arau, 02600, Malaysia
[4] Green Design and Manufacture Research Group, Center of Excellence Geopolymer and Green Technology (CEGeoGTech),Universiti Malaysia Perlis, Kangar, 0100, Malaysia
[5] Institute of Nano Electronic Engineering (INEE), Universiti Malaysia Perlis (UniMAP), 01000 Kangar, Perlis, Malaysia
[6] Center of Excellence Geopolymer and Green Technology (CEGeoGTech), School of Materials Engineering, Universiti Malaysia Perlis (UniMAP), Kompleks Pengajian Jejawi 2, 02600 Arau, Malaysia

E-mail: rosnizahamzah@unimap.edu.my

Abstract. Normally, direct introduction of recycled materials into polymer matrix could cause unwanted aggregations, which could effect on the properties of the final product. However, adding third material as a compatibilizer could enhance the properties of the product. This work studied the effects of different SMR L content (2,4,6,8 and 10 Phr) as a compatibilizer on the physical, tensile properties, and cure characteristics of ENR-25/rSC compounds. The results of cure characteristics showed decreasing on scorch time (t_2) and cure time (t_{90}) while increasing on minimum torque (M_L) and maximum torque (M_H) as SMR L increased in ENR-25/rSC/SMR L compounds. The values of modulus (M100), tensile strength (Ts), hardness and crosslinking density increased gradually with increasing SMR L while elongation at break (Eb) decreased.

1. Introduction

The mixing of different polymer types together might be a valuable technique to improve the constituents with characteristics superior to those of individual components [1]. The main objective of polymer mixing is to improve the processing properties and also to enhance some properties such as (mechanical and physical properties). Furthermore, the blending of polymers could reduce the cost of the products. Nevertheless, one of the biggest flaws in polymer blending process is the phase separation, which could be due to the poor interfacial adhesion between different polymer phases [2-5].

Adding third material as a compatibilizer into an immiscible binary system could be effective solution to improve the blending of polymers. The compatibilizer could interact chemically with both the phases or could make a certain interaction with one phase and a physical interaction with the other. Consequently, the adding of suitable compatibilizer improves the physical and mechanical properties of the composites [6-9].

Several researches have studied the compatibilizer effect on the properties of polymer blends, compounds or/and composites. Teh et al. investigated the influences of epoxidized natural rubber (ENR) as a compatibilizer in melt compounded natural rubber with organ clay nanocomposites. They found that all properties such as cure characteristics, mechanical and morphological properties of the compatibilized composites are improved [10]. Other researchers, such as Angnanon et al. reported the compatibilization effect of acrylonitrile (ACN) and styrene copolymers on the natural rubber/ acrylonitrile butadiene rubber (NR/NBR) blends. The results indicated that the tensile properties and oil-swelling resistance of the blends increased as %graft copolymer increased and scanning electron microscopy (SEM) proved the improvement of interfacial adhesion between two rubber phases [11].

The effects of same compatibilizer, TOR on the physical properties, tensile properties, and cure characteristics of natural rubber / silicon catheter blends have been studied in another research [12]. Results showed that the increase in TOR content reduced the cure time and scorch time while torque improved. The physical and tensile properties of the blends improved in presence of TOR. Nevertheless, the compatibilized sample became less flexible as compared with uncompatibilized sample.

In the current research, standard malaysian rubber grade L (SMR L) was used as a compatibilizer into ENR-25/rSC blends to study the physical properties, tensile properties and cure characteristics of the compatibilized blends.

2. Experimental

2.1. Samples preparation and testing

The materials used in this study were: Natural rubber (25% mole epoxidation grade, ENR-25) as a raw material. Recycle Silicone Catheter (rSC) as filler. Standard Malaysian rubber grade L (SMR L) as a compatibilizer. N-cyclohexyl-2-benzothiazole sulfonamide (CBS) as an accelerator. Sulfur as a vulcanizing agent. Stearic acid and zinc oxide (ZnO) as organic and inorganic activators respectively. The formulation of the rubber vulcanizates is listed in Table 1.

Rubber with the additives and chemicals were mixed using Two roll mill model X-S-K/160-X-320 according to ASTM D-3184/89. Cure characteristics were studied using Rheometer (MDR 2000) according to ASTM D-2240/93. Hot-press machine was used to mold rubber compounds based on curing time at 160 ^{0}C temperature and 30000 kg pressure. To form tensile test species, Dumbbell-shaped mold was utilized. Instron machine model 5582 was utilized for tensile test based on ASTM-D/412. Durometer device model Shore A was utilized for hardness test. The vulcanized species were immersed in Toluene solvent for swelling test based on ASTM-D/3616. Crosslinking density was calculated using Flory-Rehner equation based on swelling test data [13].

Table 1. The ingredients of ENR25/rSC/SMR L compounds.

Ingredients	Phr					
	R0	R02	R04	R06	R08	R10
ENR-25	100	100	100	100	100	100
ZnO	5	5	5	5	5	5
Stearic Acid	2	2	2	2	2	2
CBS	1	1	1	1	1	1
Sulfur	2	2	2	2	2	2
rSC	10	10	10	10	10	10
SMR L	0	2	4	6	8	10

3. Results and discussion

3.1. Curing characteristics

The curing characteristics results (curing time t_{90}, scorch time t_2, maximum torque M_H and minimum torque M_L) of ENR25/rSC/SMR L compounds were summarized in Table 2. It can be seen clearly that t_{90} and t_2 became shorter as compatibilizer loading increased in the compounds. This was attributed to the unsaturated SMR L rubber, which led to increase the incorporation between ENR25 and SMR L in short time. Thus t_{90} and t_2 decreased.

M_L exhibited declining as SMR L content increased. This indicates better processability of the vulcanizates after compatibilization. This led to decrease in the viscosity of the compounds. Whereas M_H exhibited an opposite tendency, which increased as SMR L loading increased in ENR25/rSC/SMR L compounds. This could be due to the unsaturated structure of SMR L, which bonded with the ENR25 and enhanced the crosslinking density of the compounds [14-16]. This observation is also refers to the improvement of ENR25/rSC compounds incompatibility with the presence of SMR L as a compatibilizer.

Table 2 The curing characteristic of ENR25/rSC/SMR L compounds.

SMR L	t_2 (min)	t_{90} (min)	M_L (dNm)	M_H (dNm)
R0	1.12	7.1	6.9	23.9
R02	0.65	6.8	4.0	26.0
R04	0.55	6.1	3.7	27.3
R06	0.50	5.2	3.2	28.2
R08	0.45	4.5	2.5	31.5
R10	0.30	3.8	1.8	34.6

3.2. Tensile and physical properties

The effects of SMR L content as a compatibilizer on the tensile properties such as tensile strength (Ts), elongation at break (Eb) and modulus (M100) of the ENR25/rSC/SMR L compounds are shown in Figure 1 a, b, and c respectively.

It's easily observed that the Ts values increased steadily as compatibilizer loading increased (Figure 1 a). The Ts increasing was because of the high dispersion of rSC in ENR25 as SMR L added. This was led to improvement in the interfacial adhesion of the compounds by reducing the interfacial energy between phases, Thus Ts increased.

Eb percentages decreased with the existence of SMR L in ENR25/rSC/SMR L compounds (Figure 1 b). The adding of SMR L as a compatibilizer restricted the mobility and flow of rubber compounds, which in turn led to decrease in the flexibility of the compounds too [16].

M100 value improved as SMR L added in ENR25/rSC/SMR L compounds (Figure 1 c). This was due to the presence of unsaturated SMR L that made a good incorporation with ENR-25 matrix and also increased the distribution of rSC in the vulcaniziateas, which in turn developed the crosslinking density of the compounds [16-18] (Figure 1a).

Hardness values (Figure 2 b) exhibited same trend of crosslinking density (Figure 2 a). Furthermore, the hardness of the composites improved as SMR L content increased (Figure 2 b). Nah et al. reported that the improvement on hardness with existence of compatibilizer could be due to the fact that the crosslinking relative degree was higher than other compounds, which made the compatibilized rubber compounds more rigid and stiff [19].

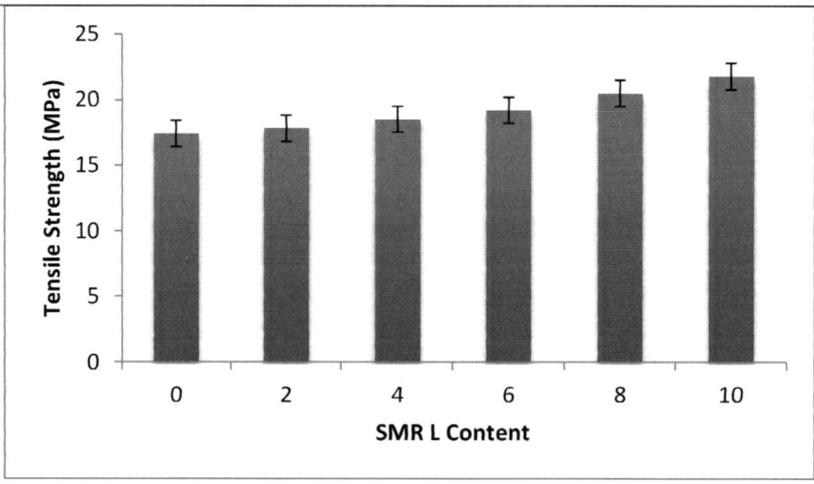

(a)

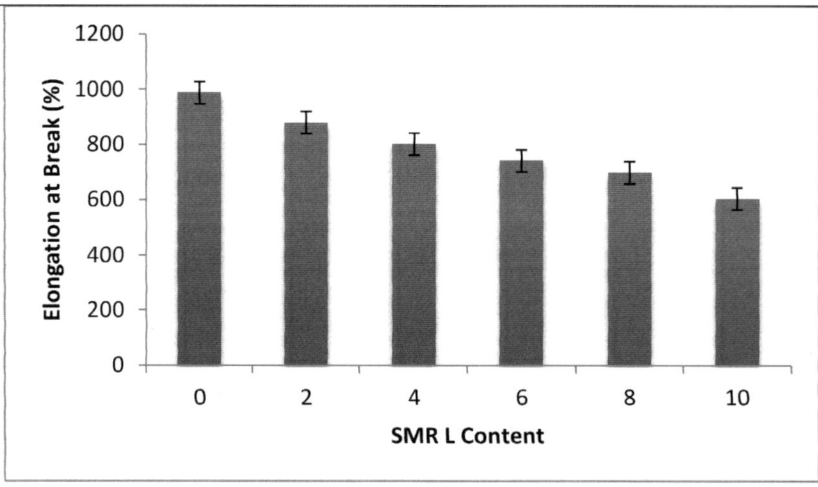

(b)

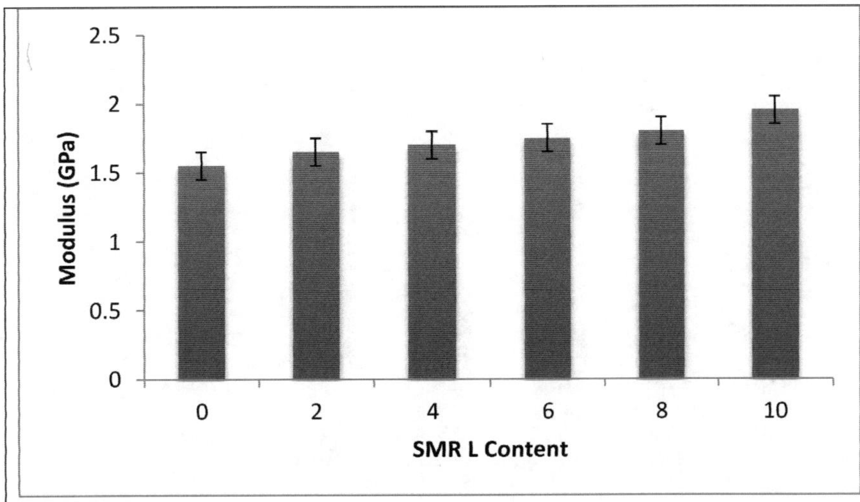

(c)

Figure 1. The variation of (a) tensile strength, (b) elongation at break, and (c) Modulus of ENR25/rSC/SMR L compounds

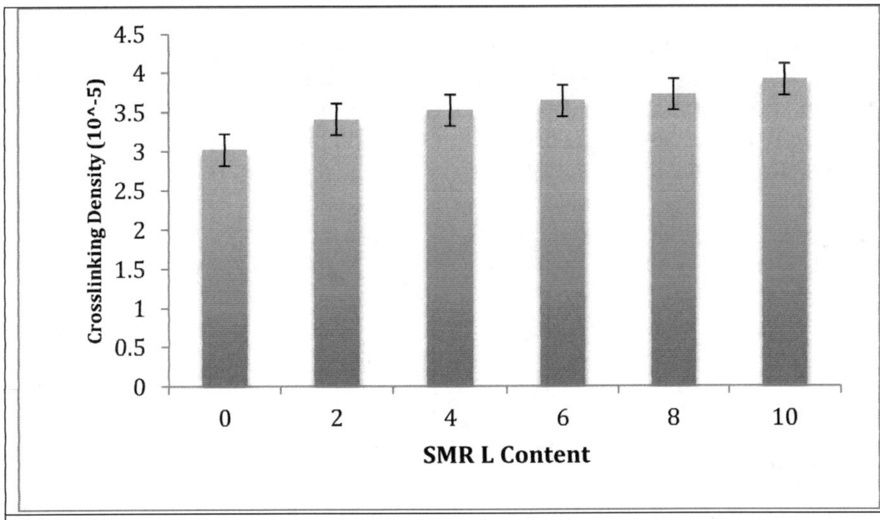

(a)

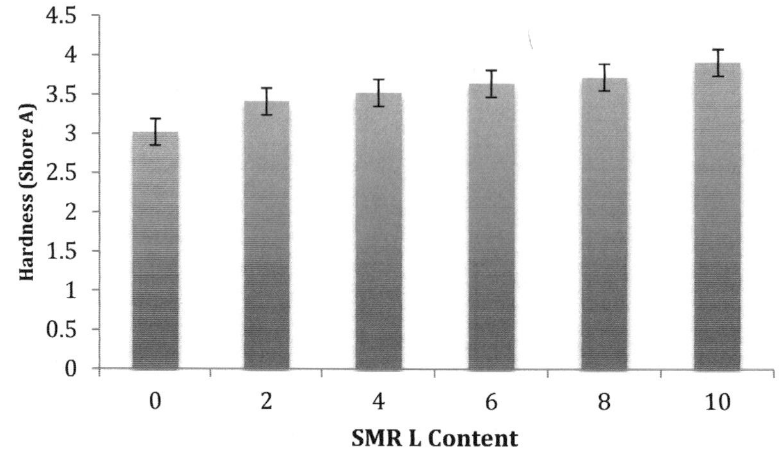

(b)

Figure 2. The value of (a) crosslinking density and (b) Hardness of ENR25/rSC/TOR compounds

4. Summary

The presence of SMR L as a compatibilizer has enhanced the overall properties of ENR25/rSC/SMR L compounds. The curing time and scorch time have decreased as of SMR L increased in the vulcanizates and also improved the toque. Besides, The physical and tensile properties of the compounds were improved as SMR L added. However, elongation at break reduced due to increase the stiffness of the vulcanizates.

5. Acknowledgments

The authors would like to acknowledge UniMAP Short Term Grant Scheme (9001 - 00521). We would also appreciate Universiti Malaysia Perlis (UniMAP) for providing us all the facilities to make this research.

6. References

1. M. N. Ismail, S. H. El-Sabbagh and A. A. Yehia, Journal of Elastomers & Plastics **31,** 255 (1999).
2. N. Z. Noriman, H. Ismail, C. T. Ratnam and A. A. Rashid, Polymer-Plastics Technology and Engineering, **49**, 228-236 (2010).
3. H. L. Snyder, P. Meakin and S. Reich, *Macromolecules*, **16**, 757-762 (1983).
4. Y. S. Lipatov, A. E. Nesterov, T. D. Ignatova and D. A. Nesterov, Polymer, **43**, 875-880 (2002).
5. J. S. Gutmann, P. Müller-Buschbaum and M. Stamm, Faraday Discussions, **112**, 285-297 (1999).
6. R. F. Bauer and E. A. Dudley, Rubber Chem. Technol. **50,** 35 (1977).
7. H. G. Ock, D. H. Kim, K. H. Ahn, S .J Lee and J. M. Maia, Eur. Polym. J. **76**, 216 (2016).
8. N. Z. Noriman, H. Ismail and A. A. Rashid, Journal of Applied Polymer Science, **126**, 56 (2012).
9. N. Z. Noriman, H. Ismail, and A. A. Rashid, Advances in Polymer Technology, **31**, 100-108 (2012).
10. P. L. Teh, Z. M. Ishak, A. S. Hashim, J. Karger-Kocsis and U. S. Ishiaku, Eur. Polym. J. **40**, 2513 (2004).
11. S. Angnanon, P. Prasassarakich and N. Hinchiranan, Polymer-Plastics Technology and Engineering **50,** 1170 (2011).
12. O. S. Dahham, N. Z. Noriman, S. T. Sam, R. Hamzah, N. A. S. Marwa, Z. Shayfull and A. M. Alakrach, *MATEC Web of Conferences* **78**, 01076 (2016).
13. P. J. Flory and J. Rehner, Journal Chemistry Physics **11,** 512 (1943).

14. S. S. Choi, Journal of applied polymer science, **75**, 1378-1384 (2000).
15. H. Ismail, S. Shuhelmy and M. R. Edyham, European Polymer Journal, **38**, 39-47 (2002).
16. H. Ismail, N. Rosnah and H. D. Rozman, Polymer, **38**, 4059-4064 (1997).
17. O. S. Dahham, N. Z. Noriman, S. T. Sam, H. Ismail, S. Ragunathan, H. Rosniza and M. N. Al-Samarrai, Journal of Polymer Materials, **33**, 647 (2016).
18. D. J. Plazek, Journal of Polymer Science Part B: Polymer Physics, **4**, 745-763 (1966).
19. C. Nah, S. C. Han, B. W. Jo, W. D. Kim and Y. W. Chang, J. Appl. Polym. Sci. **86**, 125 (2002).

The Effectiveness of Colour Constancy on Segmenting Leukemia Cells Using Unsupervised Clustering Technique

Nurul Hazwani Abd Halim[a]MohdYusoffMashor[a] Aimi Salihah Abdul Nasir[a]RoslineHassan[b]

[a]*Electronic& Biomedical Intelligent Systems (EBItS) Research Group, School of Mechatronics Engineering, University Malaysia Perlis, 02600 KampusPauh Putra, Perlis, Malaysia.*

[b]*Department of Hematology, School of Medical Science, University Science Malaysia, KubangKerian, Kelantan, Malaysia.*

[a)]Corresponding author: nurul.hazwani43@yahoo.com.

Abstract. Nowadays, colour image segmentation plays important tasks in medical field application such as brain, eyes, glaucoma and so on. This paper aims to compare the performance between different colour constancy of gray world (GW) and white patch (WP) that will be fusion with clustering algorithm to segment the nucleus of blast cells. First, both of the colour constancy algorithms are utilized on leukemia images to improve the microscopic quality of the blast cells.An unsupervised clustering, namely moving *k*-means algorithm is utilized on the various colour constancies of GWand WP for the purpose of segmentation. Both colour constancy algorithms have been analysedto identify which one can give the good segmentation performance. Since there is some noise in the segmented nucleus,the combination of median filter and region growing technique plays important tasks inreducing noise. The results show that WP is the best colour constancy in segmenting thenucleus as compared to Gray World.

1. Introduction

Leukemia is a cancer of blood which is white blood cells and its early diagnosis can prevent the rapid progression of the disease. Acute leukemia can be classified into two categories: acute lymphoblastic leukemia (ALL) and acute myelogenous leukemia (AML) [1]. ALL occurs for children of age 2-5 years old [1] while AML occurs in adults which effect in menthan women [2]. The recent development in diagnosing of the leukemia are fluorescent microscopy, polymerase chain reaction (PCR), cytochemistry, cytogenetic and molecular [3].Despite these advances, light microscopy remains the golden standard in developing countries. This is because the method is simple, rapid, and low cost. However, microscopic examination has several drawbacks, which are depending on the subjectivity in human perception, labor intensive and time-consuming [3]. Image processing method is one of the most important steps in preparing a way for a successful implementation of existing technologies for leukemia diagnosis.

In digital images, colour constancy plays important role in removing unreal colour from the image, which is caused by the acquisition condition [4]. There are several algorithms have been proposed by researchers such as Gray World [5], White Patch [6] and Iterative White Balance algorithm [6]. In general, the gray world will use the average of the pixels as their illuminant, while the maximum values in the image will be use in white patch method [6].

Many researchers worked on blood cell segmentation and have been proposed various techniques to segment the region of interest, WBC. Thresholding technique is a most common conventional algorithm and usually depend on the histogram in order to select the threshold value correctly. Green component of the RGB has been extracted and then thresholding technique has been applied for this authors in order to segment the WBC [7]. Optimal thresholding and watershed transformation are other approaches that have been introduced by them in order to segment WBC [8]. Lim proposed a double K-means clustering and watershed [9].Thus the current study proposed a combination between colour constancy (Gray World and White Patch) and moving *k*-mean clustering algorithm. The best segmentation performance was analysed either WP or Gray World algorithm that can segment nucleus successfully.

2. Methodology

The methodology for the proposed segmentation method for acute leukemia images are shown as in Figure 1.

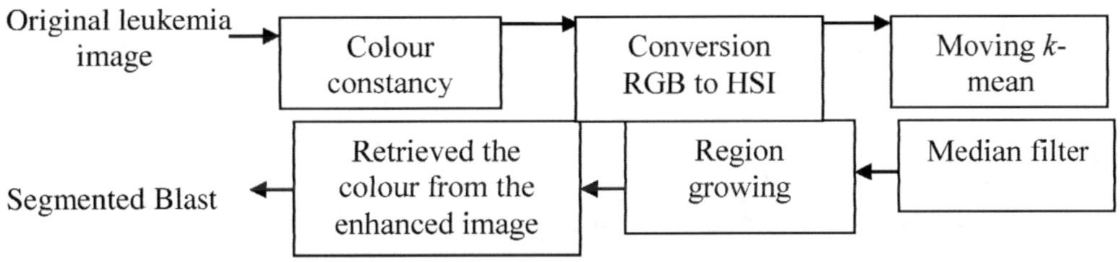

Figure 1. The proposed method for segmentation of acute leukemia.

2.1. Image Acquisition

A total of 100leukemia images that were obtained from Hospital Universiti Sains Malaysia (HUSM), were captured under 100X magnification and at a resolution of 1280×960 pixels. Samples of captured images of acute leukemia cells can be seen in Figure 2.

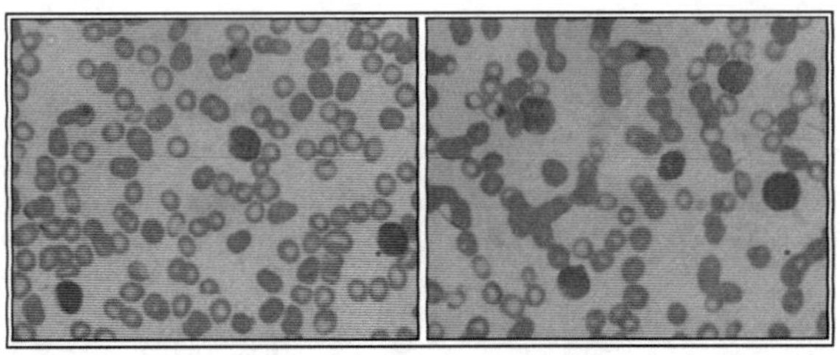

Figure 2. Sample images of acute lymphoblastic leukemia (ALL).

2.2. *Colour Constancy Algorithms*

Based on the gray world theory, the average colour of the pixels will be used in order to estimate the illuminant [6]. Assume $x \times y$ pixels are the resolution of leukemia image. Let the image $s(x, y)$ with the size of image $M \times N$ and the average colour can be calculated as below:

$$R_{avg} = \frac{1}{M \times N} \sum_{x=1}^{M} \sum_{y=1}^{N} s_R(x, y) \tag{1}$$

$$G_{avg} = \frac{1}{M \times N} \sum_{x=1}^{M} \sum_{y=1}^{N} s_G(x, y) \quad (2)$$

$$B_{avg} = \frac{1}{M \times N} \sum_{x=1}^{M} \sum_{y=1}^{N} s_B(x, y) \quad (3)$$

Where: $s_R(x, y)$, $s_G(x, y)$ and $s_B(x, y)$ are the red, green and blue channel for the image respectively.

Generally, white patch will use the determined white point as the maximum values, R_{max}, G_{max} and B_{max} based on the presumption that there is always white in the image. The formula can be calculated as below [7]:

$$R_{\max} = \max\{s_R(x, y)\} \quad (4)$$

$$G_{\max} = \max\{s_G(x, y)\} \quad (5)$$

$$B_{\max} = \max\{s_B(x, y)\} \quad (6)$$

2.3. Conversion of RGB to HSI Colour Models

Three features can be determined based on HSI colour model which ishue, saturation and intensity. The following equations were used to transform RGB to HSI colour model [10].

$$H = \begin{cases} \cos^{-1}\left\{\dfrac{1/2[(R-G)+(R-G)]}{[(R-G)^2+(R-B)(G-B)]^{1/2}}\right\} & for\ B \le G \\ 2\pi - \cos^{-1}\left\{\dfrac{1/2[(R-G)+(R-G)]}{[(R-G)^2+(R-B)(G-B)]^{1/2}}\right\} & for\ B > G \end{cases} \quad (7)$$

$$S = 1 - \frac{3}{(R+G+B)}[\min(R, G, B)] \quad (8)$$

$$I = \frac{1}{3} * (R+G+B) \quad (9)$$

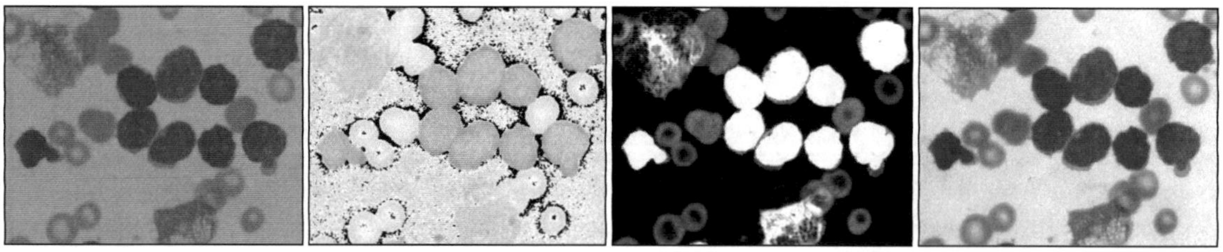

(a)Original image (b) Hue (c) Saturation (d) Intensity

Figure 3. The Transformation of RGB to HSI Colour Space

2.4. Moving k-Means Clustering Algorithm

Recently, the most commonly clustering algorithms used by researchers are *k*-means clustering and fuzzy *c*-means clustering. However, both techniques have several drawbacks such as centre redundancy, dead centre and the centre trapped in local minima which can be leading a poor result. Thus, a modified version of *k*-means clustering called moving *k*-means clustering [11] will be implementing in this research. In this study, the saturation component (from the HSI colour model) was selected to be segmented using moving *k*-means clustering as the blast cells can be seen clearly rather than other colour component as shown in Figure 3(c).

Assume a leukemia image with resolution of $x \times y$ pixels to be clustered into n_c regions. Let $s(x, y)$ as an input pixel to be clustered and c_j is $j-th$ centre (cluster) ($x = 1,2..,X$, $y = 1,2..,Y$ and $j = 1,2..n_c$). The procedures for MKM clustering are as follows [11]:

1. The centres and α_o is initialized and set $\alpha_a = \alpha_b = \alpha_o$,(where α_o is a small constant value, 0< α_o < 1/3and should be chosen to be inversely proportional to the number of centres. α_a and α_b are small constants).
2. The Euclidean distance between the pixel and all the centres is calculated. All pixels are assigned as the member of the centre with the smallest Euclidean distance.

$$c_j = \frac{1}{n_j} \sum_{y \in c_j} \sum_{x \in c_j} s(x, y) \tag{10}$$

3. The fitness of each centre is checked using the equation:

$$f(c_j) = \sum_{y \in c_j} \sum_{x \in c_j} \left(\left\| s(x, y) - c_j \right\| \right)^2 \tag{11}$$

4. Find c_s and c_l , the centre that has the smallest and the largest value of f (.).
5. If $f(c_s) < \alpha_a f(c_l)$,
   (a) The members of c_s and c_l are assigned, if $s(x,y) < c_l$, where $x, y \in c_l$, and leave the rest of the members to c_l .
   (b) The positions of c_s and c_l are recalculated based on:

$$c_s = \frac{1}{n_s} \sum_{y \in c_s} \sum_{x \in c_s} s(x, y) \tag{12}$$

$$c_l = \frac{1}{n_l} \sum_{y \in c_l} \sum_{x \in c_l} s(x, y) \tag{13}$$

   Note: c_s will give up its members before step 5(a) and, n_s and n_l in (7) and (8) are the number of the new members of c_s and c_l respectively, after the reassigning process in step 5a.
6. α_a is updated according to $\alpha_a = \alpha_a - \alpha_a / n_c$ and steps 4 and 5 are repeated until $f(c_s) \geq \alpha_a f(c_l)$
7. All pixels are reassigned to the nearest centre and the centre positions are recalculated using (5).
8. α_a and α_b are updated based on $\alpha_a = \alpha_o$ and $\alpha_b = \alpha_b - \alpha_b / n_c$ respectively, and steps 3 to 7 are repeated until $f(c_s) \geq \alpha_b f(c_l)$.

Results & Discussions

Comparisons between the gray world and white patch algorithms have been performed to validate the best colour constancy in segmenting the nucleus of blast cells. Input images of each colour constancy have been utilized for MKM clustering. Figure 4 (a), (b) and (c) are examples of acute leukemia images that will be conduct in this paper. Both of the original image and after enhancement image have been analysed based morphological features of blast cells. The results after implementing the gray world and white patch can be seen in Figure 5 and Figure 6 (a), (b) and (c), respectively. By using white patch algorithm, the contrast of cytoplasm and blast cells were significantly increased rather than the gray world. The output of the gray world image was tended to brighter and appear likely to original images.

IOP Conf. Series: Journal of Physics: Conf. Series **1019** (2018) 012059 doi:10.1088/1742-6596/1019/1/012059

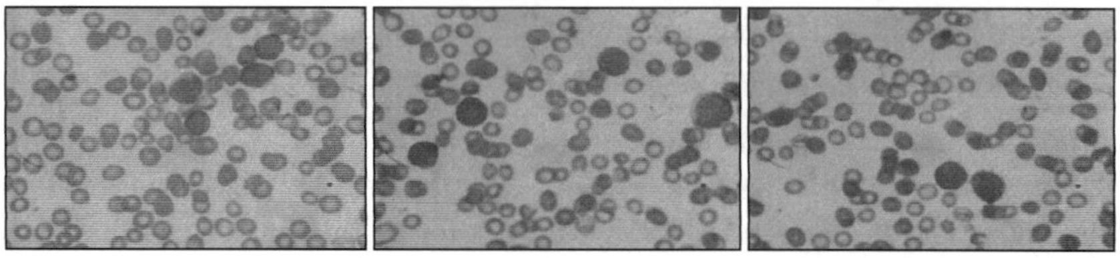

(a)ALL1 (b) ALL2 (c) ALL3

Figure 4. Original image for acute leukemia

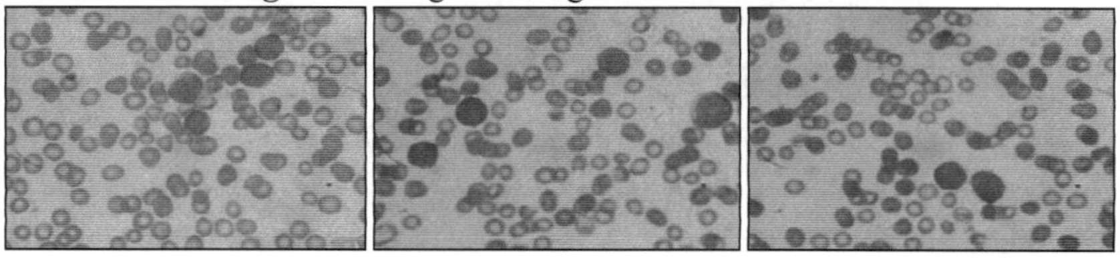

(a)ALL1 (b) ALL2 (c) ALL3

Figure 5. Result after applying gray world.

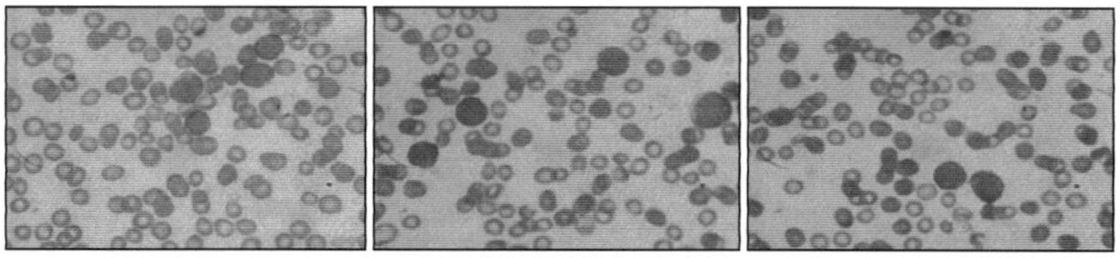

(a)ALL1 (b) ALL2 (c) ALL3

Figure 6. Result after applying white patch.

A saturation component from HSI colour model was used as input for MKM clustering. Figure 7 and Figure 8 are shown the saturation colour component was extracted from GW and WP images, respectively. Based on the resultant results, the blast cells appear as the brightest areas when saturation component was used on both gray world and white patch but RBC and background areas can be seen more clearly using white patch algorithm.

(a)ALL1 (b) ALL2 (c) ALL3

Figure 7. Saturation component that has been extracted from gray world

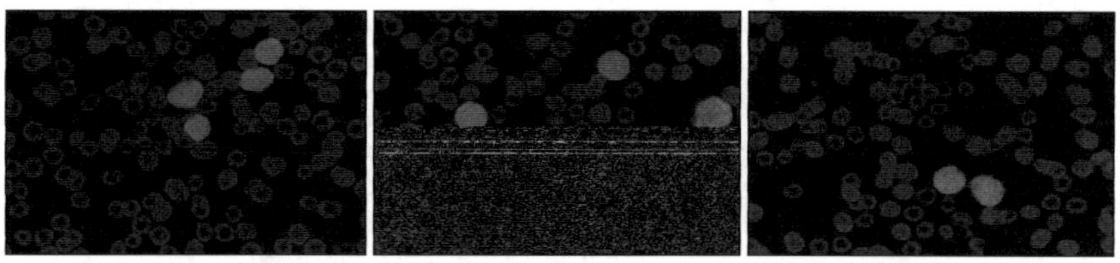

(a)ALL1 (b) ALL2 (c) ALL3

Figure 8. Saturation component that has been extracted from white patch

In Figure 9 and Figure 10 are shown the MKM segmentation results using saturation component for the gray world and white patch, respectively. From the images, MKM will be cluster into three categories such as blast cell, RBCs and background.

IOP Conf. Series: Journal of Physics: Conf. Series **1019** (2018) 012059 doi:10.1088/1742-6596/1019/1/012059

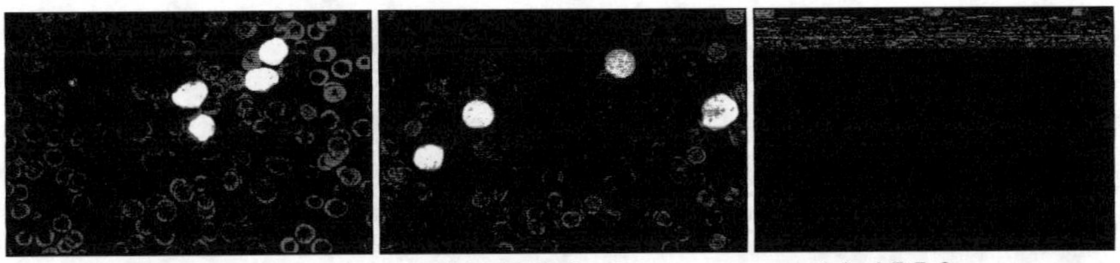

(a)ALL1 (b) ALL2 (c) ALL3

Figure 9. Results after applying moving *k*-means clustering on saturation component images for gray world

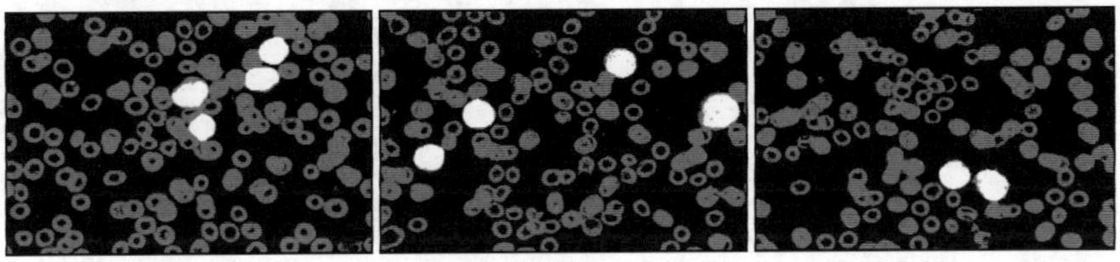

(a)ALL1 (b) ALL2 (c) ALL3

Figure 10. Results after applying moving *k*-means clustering on saturation component images for white patch

Figure 11 and Figure 12 indicated the retrieved images based on the gray world and white patch, respectively. Our proposed technique is not working very well on ALL2 for gray world compared to white patch due to having the similarities the pixels' value between blast cells and also RBCs.

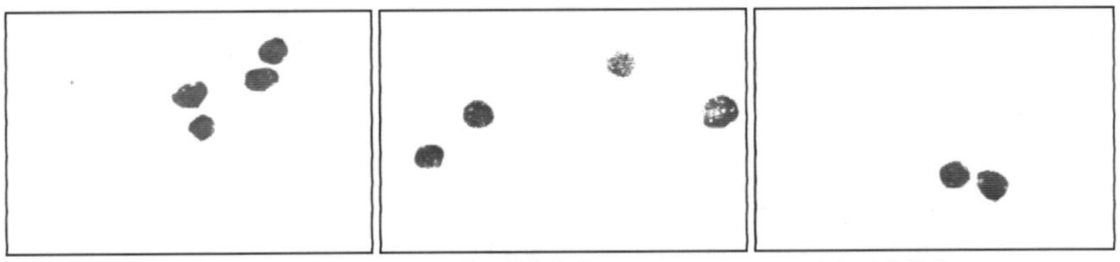

(a)ALL1 (b) ALL2 (c) ALL3

Figure 11. Results after moving *k*-means clustering image has been retrieved based on gray world image.

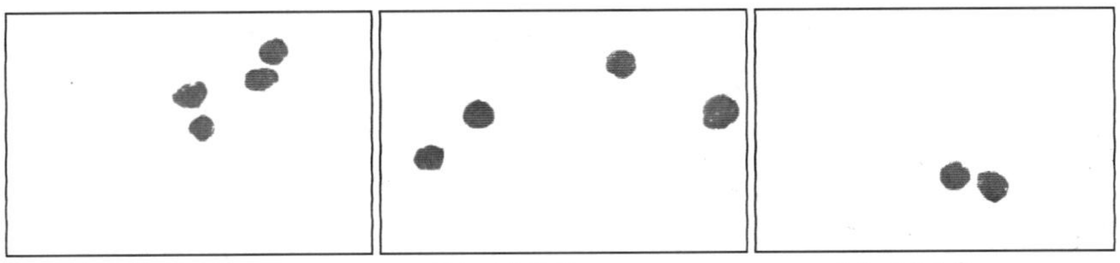

(a)ALL1 (b) ALL2 (c) ALL3

Figure 12. Results after moving *k*-means clustering image has been retrieved based on white patch image.

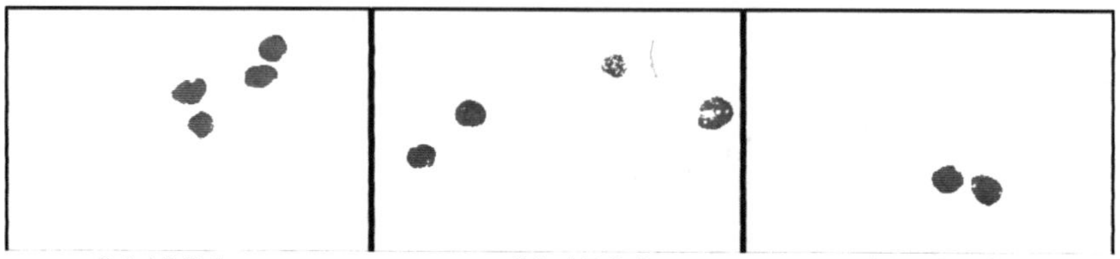

(a)ALL1 (b) ALL2 (c) ALL3

Figure 13. Results after implementing median filter and region growing on gray world image.

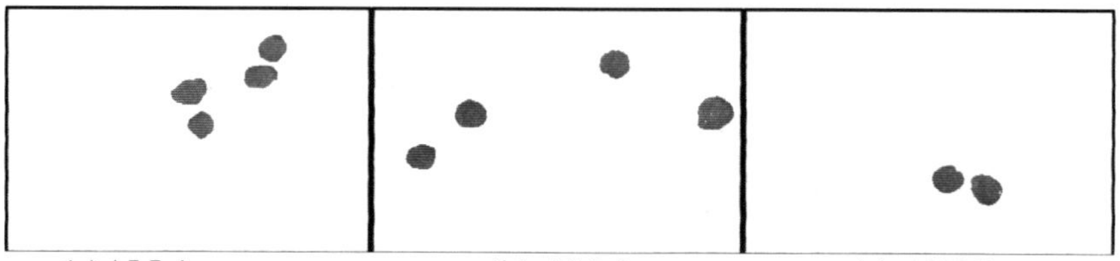

(a)ALL1 (b) ALL2 (c) ALL3

Figure 14. Results after implementing median filter and region growing based on white patch image.

There are some small areas of RBCs that could not be eliminated using the clustering process since it has similar pixel values with blast cells. To overcome this problem, the fusion of the 7 × 7 pixels' median filter with region growing were processed for smoothing the image, and to get rid of any large unwanted regions from the image, respectively. Non-blast cells were eliminated from the images since it has less than 1000 pixels. The results after implementing the region growing technique and median filter are shown in Figure 13 and Figure 14. The segmented nucleus has been successful obtained by using WP since it still maintains the features of the nucleus rather than Gray World algorithm.

3. Conclusion

Segmentation using various colour constancy algorithms and MKM clustering has been presented. In this study, comparison between GW and WP has been making to analyse the significant of it in segmenting the acute leukemia images. It is shown that WP algorithm has proven a better outcome in segmenting nucleus from acute leukemia images compared to GW algorithm.

Acknowledgments
The authors gratefully acknowledge to the team members and University Science Malaysia (USM).

4. References

[1] S. Rajendran, "Image Acquisition and Retrieval Systems for Leukaemia Cells" Master's thesis, Department of Computer Science, UM, 2007.

[2] J.N. Jameson, L.K. Dennis, T.R. Harrison, E. Braunwald, A.S. Fauci, S.L. Hauser, D.L. Longo, "Harrison's principles of internal medicine", New York: McGraw-Hill Medical Publishing Division, 2005.

[3] F. Thung, and I.S. Suwardi, "Blood parasite identitication using feature based recognition," in International Conference on Electrical Engineering and Informatics (ICEEJ), 2011, pp.1-4.

[4] R.A.A.Raof,M.Y. Mashor ,R.B.Ahmad, S.S.Md.Noor , "Image Enhancement Using White Balance Technique in Ziehl-Neelsen Sputum Slide Images",Proc. of International Postgraduate Conference on Engineering(IPCE 2010),16-17 October 2010.

[5] G. Buchsbaum," A special processor model for object color perception ", in Journal of Franklin Institute, Vol 310,pp.1-26, 1980.

[6] V.Cardei, B.Funt, and K.Barndad, "White point estimation for uncalibration images," in Proceeding of the Seventh Color Imaging Conference (CIC'99),pp 97-100,1999.

[7] M.Mohamed, and B.Far, "An enhanced threshold based technique for white blood cells nuclei automatic segmentation," in e-Health Networking, Applications and Services (Healthcom), 2012 IEEE 14th International Conference on, vol., no., pp.202-207, 10-13 Oct. 2012.

[8] M. Mohamed, B.Far and C.Naugler.,"Chronic lymphocytic leukemia cell segmentation from microscopic blood images using watershed algorithm and optimal thresholding,"in Electrical and Computer Engineering (CCECE), 2013 26th Annual IEEE Canadian Conference on, Regina, SK, 2013, pp.1-5.

[9] L.HueyNee, M.Y. Mashor,N.Z. Supardi,;R. Hassan,,"Color and morphological based techniques on white blood cells segmentation,"in Biomedical Engineering (ICoBE), 2015 2nd International Conference on ,vol., no., pp.1-5, 30-31 March 2015.

[10] R. C. Gonzalez and R. E. Woods, Digital Image Processing, 3rd ed.,Prentice Hall, 2007.

[11] M.Y. Mashor,International Journal of the Computer 8, (2), pp. 50–65(2000).

IOP Conf. Series: Journal of Physics: Conf. Series **1019** (2018) 012060 doi:10.1088/1742-6596/1019/1/012060

Effects of Date Seeds Size and Loading on Properties of Linear Low-Density Polyethylene/Date Seeds Powder Composites

Abduati Alnaid[1], N Z Noriman[1], Omar S Dahham[1], R Hamzah[1], Marwa N Al-Samarrai[1], S Z Syed Idrus[2], Z Shayfull[3,4]

[1] Center of Excellence Geopolymer and Green Technology (CEGeoGTech), Faculty of Engineering Technology (FETech), Universiti Malaysia Perlis (UniMAP), Level 1 Block S2, UniCITI Alam Campus, Sungai Chucuh, Padang Besar, 02100, Perlis, Malaysia.
[2] School of Human Development and Techno-communication (iKOM), Green Advanced Computing and Technology (GREAT) Research Group – CEGeoGTech, Universiti Malaysia Perlis (UniMAP)
[3] School of Manufacturing Engineering, Universiti Malaysia Perlis, Arau, 02600, Malaysia
[4] Green Design and Manufacture Research Group, Center of Excellence Geopolymer and Green Technology (CEGeoGTech),Universiti Malaysia Perlis, Kangar, 0100, Malaysia

E-mail: niknoriman@unimap.edu.my

Abstract. In this work, the influences of different date seeds (DS) size, fine size (125 - 250 μm), medium size (250 - 500 μm) and coarse size (500, 1000 μm) and different DS content (5, 10, 15, 20 wt%) on the tensile properties of linear low density polyethylene / date seeds (LLDPE/DS) composites were evaluated. The extrusion and injection molding process were used respectively to form LLDPE/DS composites. Results show that the addition of DS up to 10 wt%, particularly DS fine size has increased the tensile strength (TS) of LLDPE/DS composites. However, high DS content, particularly DS coarse size decreased the TS of the composites. The percentage of elongation at break (Eb) decreased while the modulus increased gradually as DS content increased in LLDPE/DS composites, which made the composite stiffer and less flexible. The SEM shows clearly the good distribution of DS fine size and the agglomeration of DS coarse size.

1. Introduction

Generally, the history using cellulose fiber with plastic to form bio composite started in 1908, later extending to melamine and urea and reaching commodity status with glass fiber / plastics composites [1,2]. In the last decade, the bio plastic composites have been utilized in several applications [3-9]. Biopolymer composites, started to improve novel polymer-based composites that one day could revolutionize the automotive industry. Aims to obtain high-performance biopolymer composites that might reduce the car structures weight. The lightweight of the biopolymer composite will subsequently increase the efficiency of device products such as airplanes, cars, etc.

Polyethylene (PE) is the cheapest and highest production share of any thermoplastic type, representing about 29% of the plastic production in the world [10]. PE is produced through the polymerization of ethylene (C2H4) that obtains macromolecules containing of a repeating monomer unit (CH2-CH2). Linear low-density polyethylene (LLDPE) is the most affordable variant of PE that has poor mechanical properties and high ductility at room temperature among all semi-crystalline polymers [10]. Gu and Raj et al studied the mechanical properties of LLDPE / aspen fibers bio-composites [11]. In a recent research, Mijiyawa et al. studied the thermal degradation of LLDPE/birch but did not analyze the resulting mechanical damage [12].
Date seeds (DS) are by-products of some date processing industries. Generally, there is no exact use of DS but sometimes they are used as an organic fertilizer of the soil and/or as livestock feed [13]. Normally, DS comprise approximately 78% carbohydrates, 5% moisture, 1.5% ash, 10% oil and 5.5% protein [14]. The DS powder after grinding might be used as natural filler in thermoplastics and DS oil content could be a natural lubricant and improves the mixing and molding process. Earliest study on using DS flour as natural filler for production showed that, in comparison with flax fiber and pistachio shell flour, DS filler had higher melt flow index compared to flax fiber and shell flour, [15]. With high disposal of DS particularly in Middle East countries with no specific use for them, it seems reasonable idea to study the DS powder as natural filler for thermoplastic-based composite.
Therefore, the main objective of this work is to investigate the effects of different date seeds powder size and loading on the tensile properties of linear low density polyethylene / date seeds powder (LLDPE/DS) composites.

2. Experimental

2.1. Materials
LLDPE matrix with density 0.920 g cm^{-3} and melt flow index 1.00 g 10 min^{-1} was used in this work. It was purchased from PT. Lotte Chemical Titan Nusantara, Banten, Indonesia. The date seeds were purchased from Syarikat Abdul Ghaffar sdn. Bhd., Penang, Malaysia.

2.2. Preparation of LLDPE/DS composites
First, DS were ground and sieved using Crasher Model RT34 and lab sieves respectively. After grinding and sieving, DS powder was dried using vacuum oven at 60°C for 24 h. Dried DS along with LLDPE pellets was mixed and introduced into a twin-screw extruder feeder model JSW-100 tonne with 60 rpm screw speed The extruder barrel temperature was 160°C while the die temperature was 165°C.
The Amount of the feed was fixed at 1,000 g/h. The extrudate was cooled in a water bath to room temperature and pelletized using pelletizer. Pellets were then vacuum dried at 60°C for 8 hours under. This process was carried out for four different DS loading (5, 10, 15 and 20 wt. %) and three different DS sizes, fine size (125 - 250 μm), medium size (250 - 500 μm) and coarse size (500 - 1000 μm).

2.3. Injection molding
According to (ASTM D638), standard tensile dumbbells were prepared using the Battenfell Injection Moulding Machine with a temperature profile of 160, 165, 170 and 175°C from feed to exit die zone respectively.

2.4. Tensile properties
According to ASTM D638, the tensile test for LLDPE/DS composites was carried out using universal tensile testing machine, Instron Machine model 5569. The test on dumbbell shaped specimens of 4mm thickness was performed at a cross-head shape of 50mm/min at 23 ± 2°C. Tensile strength, elongation at break and Young's modulus were reported and five specimens were used and take the average values of the results.

2.5. Scanning electron microscopy
The Tensile fracture surface of LLDPE/DS composites was analyzed using SEM (JSM-6400, JEOL). All samples were gold-coated to prevent the electron charging effect during test.

3. Results and Discussion

3.1. Tensile properties
The tensile strength (Ts), elongation at break (Eb), Young's modulus variations of LLDPE /DS composites at different DS size and loading are shown in Figure 1 a, b and c respectively.
It can be clearly observed that the addition of DS up to 10 wt%, particularly DS fine size has increased the tensile strength (TS) of LLDPE/DS composites. This could be due to the synergism effect, which means the interaction between DS filler and LLDPE matrix has exceeded their maximum properties [16, 17]. The increasing of DS content into LLDPE has effected negatively on the tensile strength (TS) of the composites. The TS reduction after 10 wt% of DS, particularly for DS coarser size was due to aggregation phenomena of DS in LLDPE/DS composites, which in turn increased the difficulty of the stress transmission from LLDPE matrix to DS filler [18].
The values of the elongation at break (Eb) have decreased as DS content increased in LLDPE/DS composites causing reduce the chains mobility inside the composite, which in turn decreased the flexibility of the composite. This made the composite harder and stiffer [18]. In conjunction with Eb, the value of modulus increased linearly with increasing of DS content in LLDPE/DS composites. This might be due to the increasing of the adhesion bonding between DS, particularly DS fine size and LLDPE matrix which made the composites stiffer [19].
In general, the fine size of DS showed better tensile properties than DS coarse size at same loading in the composite. This was due to the good dispersion of DS fine size in LLDPE matrix that improved matrix/fiber bonding. This matter explained clearly in SEM micrographs of tensile fracture surface of the composites.

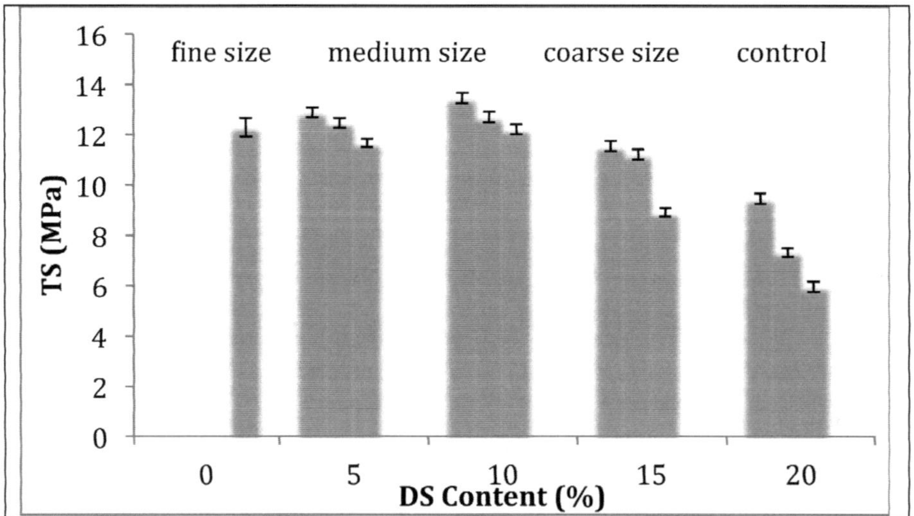

(a)

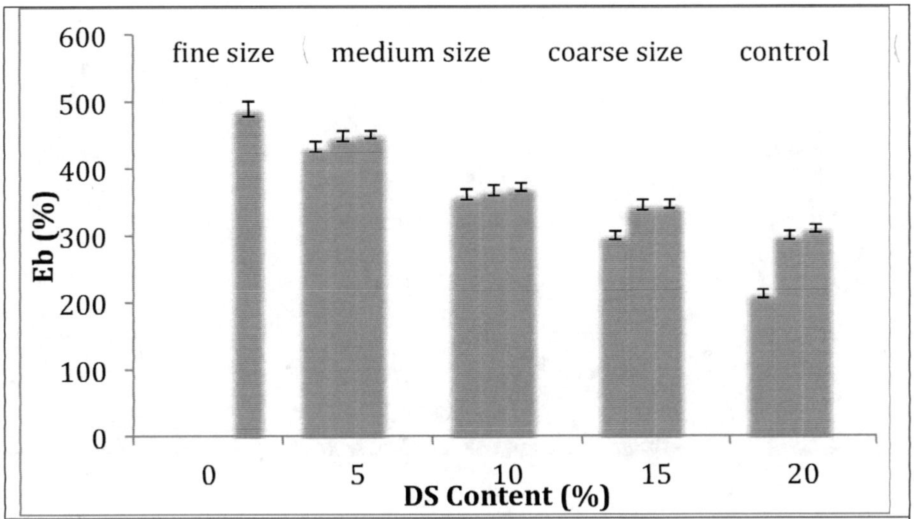

(b)

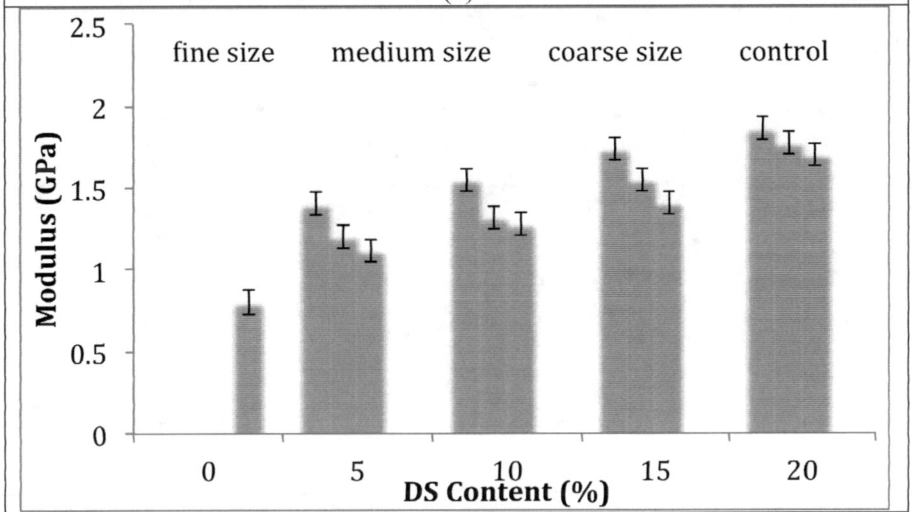

(c)

Figure 1. The values of (a) tensile strength, (b) elongation at break and (c) and modulus of LLDPE/DS composites at different DS size and loading.

3.2. Scanning electron microscopy

The scanning electron microscopy (SEM) micrographs of tensile fracture surface of LLDPE/DS composite at 10 wt.% DS fine size, 10 wt.% DS coarse size, 20 wt.% DS fine size and 20 wt.% DS coarse size are shown in figure 2 a, b, c and d respectively. Results show that the low DS content (10 wt.% DS), particularly DS fine size have numerous tearing lines on the tensile fracture surface indicating that the sample need more energy to break the matrix (Figure 2 a). This phenomenon attributed to the high surface area of the date seeds that strongly bonded with LLDPE matrix [17]. By contrast, the high DS content (20 wt.% DS), particularly DS coarse size show DS agglomeration and rough surface causing breakage and pull out of DS particles, which in turn reduce the stress transition from LLDPE matrix to DS filler. Thus, the tensile strength value of high DS content has decreased [20,21].

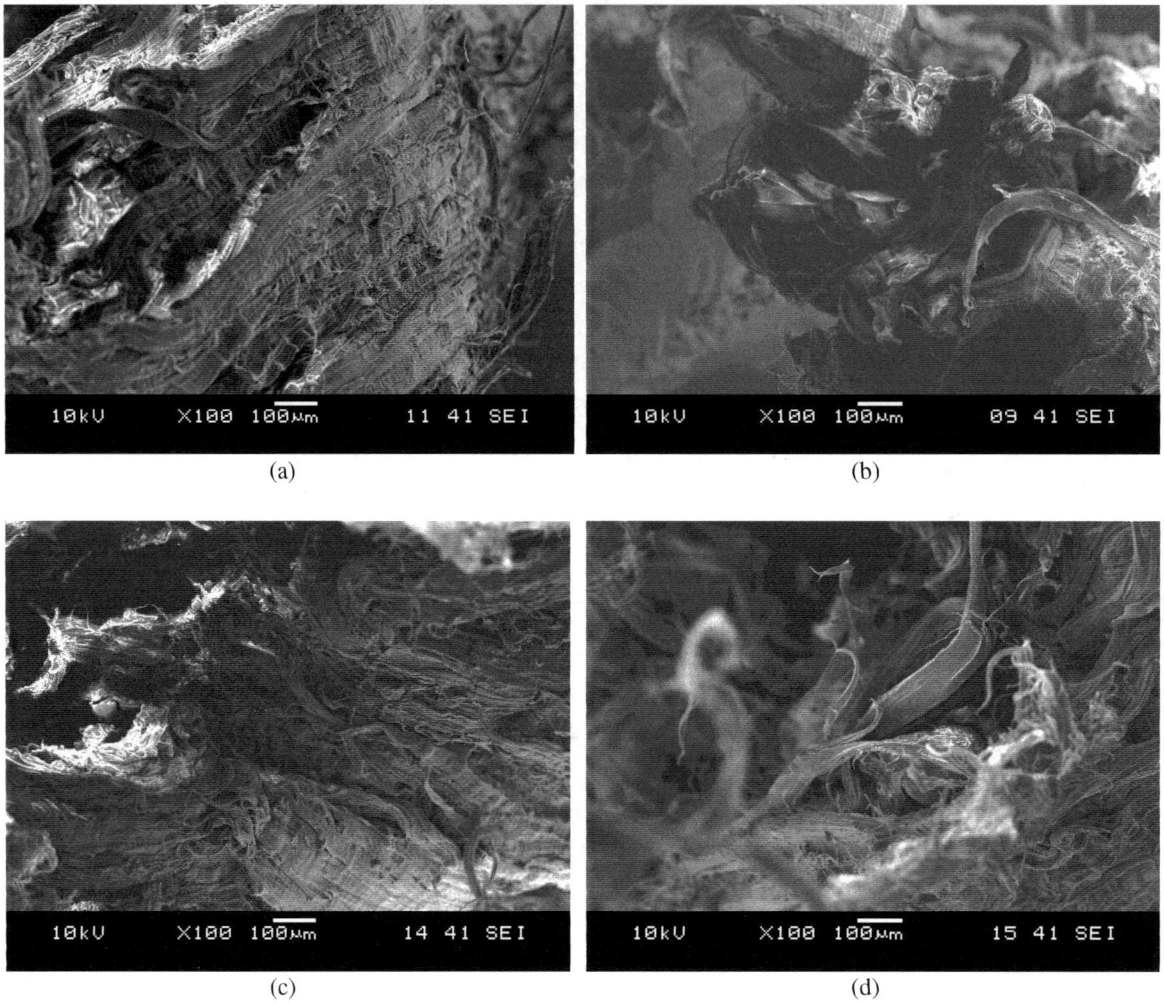

Figure 2. The SEM micrographs of LLDPE/DS composite at (a) 10 wt.% DS fine size, (b) 10 wt.% DS coarse size, (c) 20 wt.% DS fine size and (d) 20 wt.% DS coarse size

4. Conclusion

According to the results of this work, it can be concluded that the fine size of date seeds, particularly at DS 10 wt% exhibited better properties than DS coarse size and other DS content in LLDPE/DS composites due to the high DS distribution in LLDP matrix. However, the high DS content reduced the flexibility and made the composites harder and stiffer.

5. Acknowledgement

The authors would like to acknowledge the Fundamental Research Grant Scheme (FRGS) Phase 1/2015 (9003-00523) under Ministry of Higher Education, Malaysia. The authors would appreciate Universiti Malaysia Perlis (UniMAP) for providing us all the facilities to make this research.

6. References

1. C. Bakis, L. C. Bank, V. Brown, E. Cosenza, J. F. Davalos, J. J. Lesko and T. C. Triantafillou, J. Compos. Constr. **6**, 73-87 (2002).
2. K. Balaed, N. Z. Noriman, O. S. Dahham, S. T. Sam, R. Hamzah and M. F. Omar, Int. J. Polym. Anal. Charact. **21**, 590-598 (2016).
3. V. K. Thakur, M. K. Thakur and R. K. Gupta, Int. J. Polym. Anal. Charact. **19**, 256-271 (2014).

4. J. Holbery and D. Houston, Jom. **58**, 80-86 (2006).
5. D. N. Saheb and J. P. Jog, *Advances in polymer technology*, **18**, 351-363 (1999).
6. M. Pervaiz and M. M. Sain, *Macromolecular materials and engineering*, **288**, 553-557 (2003).
7. K. Goda, M. S. Sreekala, A. Gomes, T. Kaji and J. Ohgi, *Composites Part A: Applied science and manufacturing*, **37**, 2213-2220 (2006).
8. M. A. Dweib, B. Hu, A., O'donnell, H. W. Shenton and R. P. Wool, *Composite structures*, **63**, 147-157 (2004).
9. S. T. Sam, M. A. Nuradibah, H. Ismail, N. Z. Noriman and S. Ragunathan, Polymer-Plastics Technology and Engineering, **53**, 631-644. (2014).
10. O. I. Nkwachukwu, C.H. Chima, A.O. Ikenna and L. Albert, Int. J. Ind. Chem. **4**, 1–13 (2013).
11. R. Gu, B. V. Kokta, D. Michalkova, B. Dimzoski, I. Fortelny, M. Slouf and Z. Krulis, J. Reinf. Plast. Compos. **29**, 3566–86 (2010).
12. F. Mijiyawa, D. Koffi, B. Kokta and F. Erchiqui, 29th international conference of the polymer processing society (PPS-29) Nuremberg, Germany, (2013).
13. F. Banat, S. Al-Asheh and A. Al-Makhadmeh, Chemical Engineering Technology **27**, 80–86 (2004).
14. S. Besbesa, C. Bleckerb, C. Deroanneb, N. Drirac and H. Attiaa, Food Chem. **84**, 577–584 (2004).
15. A. Ghazanfari, S. Panigrahi and L. Tabil, CSAE/SCGR Paper No. 05-079 CSAE/ SCGR Secretary, PO Box 316, Mansonville, QC, Canada. (2005).
16. O. S. Dahham, N. Z. Noriman, S. T. Sam, N. M. Al-Samarrai, Z. Shayfull, A. M. Alakrach and S. A. Elnaid, MATEC Web of Conferences **78**, 01075 (2016).
17. M. H. Subli, , M. F. Omar, I. G. Sandu, N. N. Zulkepli, M. M. A. B. Abdullah and A. V. Sandu, Mater. Plast. Buchar. **51**, 391-395 (2014).
18. H. Ismail and R. M. Jaffri, Polym. Test. **18**, 381-388 (1999).
19. O. S. Dahham, N. Z. Noriman, S. T. Sam, H. Ismail, S. Ragunathan, H. Rosniza and M. N. Al-Samarrai, J. Polym. Mater. **33** 647 (2016).
20. M. S., Huda, L. T. Drzal, M. Misra and A. K. Mohanty, *Journal of Applied Polymer Science*, **102**, 4856-4869 (2006).
21. A. Arbelaiz, B. Fernández, G. Cantero, R. Llano-Ponte, A. Valea and I. Mondragon, *Composites Part A: Applied Science and Manufacturing*, **36**, 1637-1644 (2005).

IOP Conf. Series: Journal of Physics: Conf. Series **1019** (2018) 012061 doi:10.1088/1742-6596/1019/1/012061

Effects of Sodium Hydroxide Treatment on LLDPE/DS Composites: Tensile Properties and Morphology

Abduati Alnaid[1], N Z Noriman[1], Omar S Dahham[1], R Hamzah[1], S Sudin[2], S Z Syed Idrus[3], Z Shayfull[4,5], M Mohammed[6]

[1] Center of Excellence Geopolymer and Green Technology (CEGeoGTech), Faculty of Engineering Technology (FETech), Universiti Malaysia Perlis (UniMAP), Level 1 Block S2, UniCITI Alam Campus, Sungai Chucuh, Padang Besar, 02100, Perlis, Malaysia.
[2] School of Computer and Communication Engineering (SCCE), Green Advanced Computing and Technology (GREAT) Research Group – CEGeoGTech, Universiti Malaysia Perlis (UniMAP)
[3] School of Human Development and Techno-communication (iKOM), Green Advanced Computing and Technology (GREAT) Research Group – CEGeoGTech, Universiti Malaysia Perlis (UniMAP)
[4] School of Manufacturing Engineering, Universiti Malaysia Perlis, Arau, 02600, Malaysia
[5] Green Design and Manufacture Research Group, Center of Excellence Geopolymer and Green Technology (CEGeoGTech),Universiti Malaysia Perlis, Kangar, 0100, Malaysia
[6] Center of Excellence Geopolymer and Green Technology (CEGeoGTech), School of Materials Engineering, Universiti Malaysia Perlis (UniMAP), Kompleks Pengajian Jejawi 2, 02600 Arau, Malaysia

E-mail: niknoriman@unimap.edu.my

Abstract. In this research, the effect of different sodium hydroxide (NaOH) concentration (3, 6 and 9 wt.%) on the tensile properties of linear low density polyethylene / date seeds (LLDPE/DS) composites were investigated. The size of DS particles was measured at (125 - 250 μm) and the ratio of LLDPE over DS was fixed at 90/10 wt.%. The samples were prepared using extrusion process (screw speed at 60 rpm and barrel and die temperature at 160°C and 165°C respectively) followed by injection moulding process (from feed to exit die zone temperature at 160, 165, 170 and 175°C respectively). Results suggest that the addition of NaOH particularly 3 wt% has improved the tensile strength. Additionally, the elongation at break has improved as well. However, the high concentration of NaOH, particularly 9 wt.% effects negatively on these properties. In contrast, the Modulus shows different trend, which increased as NaOH concentration increased in the composites. The SEM section shows the improvement on DS surface as low NaOH concentration used (3 wt.%) and deterioration on the DS surface as high NaOH concentration used (9 wt.%).

1. Introduction

Generally, polymer composites consist of filler such as fiber as reinforcing materials and polymer as matrix. There are several types of plastic polymers in the literature that were used to form polymer composites, such as low, linear low and high density polyethylene, polyether ether ketone, polypropylene and other [1,2]. Traditional fiber, such as glass fiber, silica, carbon, metal oxide, and many others were used extensively as reinforcing components in polymer composites [3-6]. However, the using of natural fiber as reinforcing materials in plastic composites has increased rapidly in the last decade due to the advantages of natural fiber over traditional fiber. These advantages include low cost, low density, recyclability and biodegradability, which in turn have attracted the attention of researchers and engineers and make natural fibers potential replacement for traditional fibers in several applications such as automobile, construction, plastics and packaging [7-9].

However, the low interfacial characteristics between natural fiber and polymer matrix could effect negatively on the properties, particularly the mechanical properties of the polymer composites due to the hydrophilic behavior of natural fibers. Therefore, the chemical modification could be the suitable technique to improve the interface of natural fibers. Several researches were conducted to evaluate the effect of different type of chemical modification of natural fiber on the properties of polymer composites [10,11]. Chemical modification using alkaline is one of the most treatment methods used in thermoplastic, thermoset and elastomer. The alkaline modification improves the properties of polymer itself [12]. Additionally, alkaline modification increases the surface roughness of the fiber due to the disruption of hydrogen bonding in the network structure. Moreover, this type of treatment can remove a part of wax, lignin and oils covering the external surface of the fiber cell wall, depolymerizes cellulose and exposes the short length crystallites [13]. Middle East countries are well known for palm trees and largest producers of date fruit in the world [14]. The nature of Middle East people is consumes date fruit daily and date pits are usually discarded as waste with no use or value. However, this type of waste material, date pits comprise significant constituents such as minerals (rich in potassium), oils (bout 10%) and fibers (up to 46.4%) that may be used for specific purposes. There are limited researches in the literature that studied the potential of date pits as fillers for thermoplastics. Ghazanfari et al. have evaluated the effect of date pits flour with HDPE. Their results showed that date pits have decreased the melt flow index (MFI), and improved the thermal conductivity of the HDPE/date pits composites [15].

In our present research, we have investigated the effect of sodium hydroxide (NaOH) as alkali treatment on the tensile properties of linear low density polyethylene / date seeds powder (LLDPE/DS) composites.

2. Experimental

2.1. Materials

Polymer matrix, LLDPE (density 0.920 g cm^{-3}, melt flow index 1.00 g 10 min^{-1}) was purchased from PT. Lotte Chemical Titan Nusantara, Banten, Indonesia. The date seeds were obtained from Syarikat Abdul Ghaffar sdn. Bhd., Penang, Malaysia. The Sodium Hydroxide (NaOH) was purchased from Anchor Chemical Co. (M) LTD.

2.2. Fabrication and treatment of the composites

DS were milled using Crasher (RT34) and then sieved using lab sieves to obtain (125 - 250 μm) DS size. After sieving, chemical treatment was applied to sieved DS using sodium hydroxide (NaOH). The DS were treated in 3%, 6% and 9% of NaOH for 24 hours separately. Treated DS powder was dried using vacuum oven at 60°C for 24 hours. Dried DS was mixed with LLDPE and introduced into feeder of twin-screw extruder (JSW-100 tonne) with ratio 90/10 wt.% of LLDPE/DS. The screw speed was 60 rpm and the extruder barrel and die temperature were 160°C and 165°C respectively. The feed amount of the mixture was fixed at 1,000 g/h. The extrudate was cooled in a water bath to room temperature and pelletized using pelletizer. Pellets were then dried using vacuum oven at 60°C for 8 hours.

2.3. *Injection molding*
The standard tensile dumbbells were prepared using the Battenfell Injection Moulding Machine according to ASTM D638. Different temperature profile was used in this process, 160, 165, 170 and 175°C from feed to exit die zone respectively.

2.4. *Tensile properties*
The tensile test was conducted for LLDPE/DS composites using Instron Machine (model 5569) according to ASTM D638. The test on dumbbell shaped specimens of 4mm thickness was carried out at a cross-head shape of 50mm/min at 23 ± 2°C. Results such as tensile strength, elongation at break and Young's modulus were recorded. Five samples were used for each batch and take the average values of the results.

2.5. *Scanning electron microscopy*
The Tensile fracture surface of treated LLDPE/DS composites was examined using SEM device model JSM-6400, JEOL. All specimens were gold-coated to avoid the effect of electron charging during analysis.

3. Results and Discussion

3.1. *Tensile properties*
The effect of different NaOH concentration on the tensile strength, elongation at break and young modulus properties of LLDPE/DS composites are shown in Figure 1 a, b and c respectively. The 3 wt.% of NaOH treatment shows highest tensile strength and elongation at break (Figure 1a and b). This is due to the improvement on the fiber surface in presence of NaOH. Jacob et al. evaluated the influence of different NaOH concentration (0.5,1, 2, 4 and 10%) on the sisal fiber-reinforced composites. Their results concluded that the maximum tensile strength value was 4% of NaOH [16]. Other researcher, such as Mishra et al. found that the 5% of NaOH treated natural fiber-reinforced polyester composite had better tensile strength than 10% NaOH treated composites. This is because at high NaOH concentration, excess delignification of natural fiber occurs which in turn lead to damage of fiber structure [17]. Thus, the tensile strength of LLDPE/DS composite decreased gradually after 3 wt.% of NaOH concentration. However, The high NaOH concentration such as 9 wt.% shows higher modulus than other NaOH concentration. This could be due to the high disruption of hydrogen bonding in the network structure causing increase in surface roughness of the fiber. The increasing of fiber surface roughness can increase the interaction between fiber and polymer matrix, which in turn make composite stiffer and more rigid. Therefore, modulus value increased.

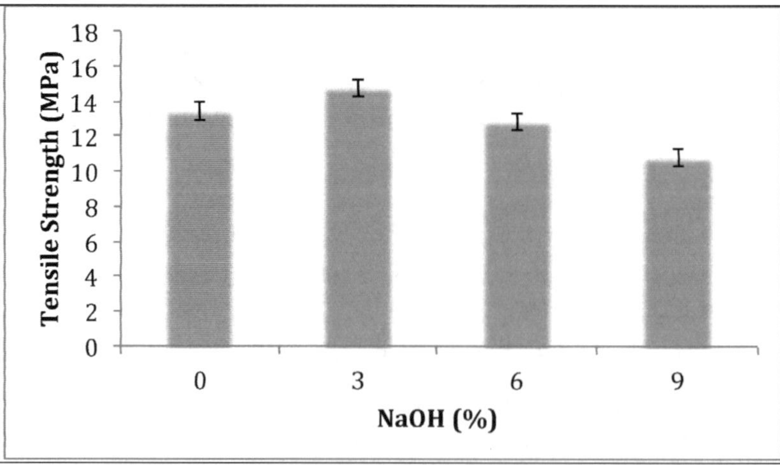

(a)

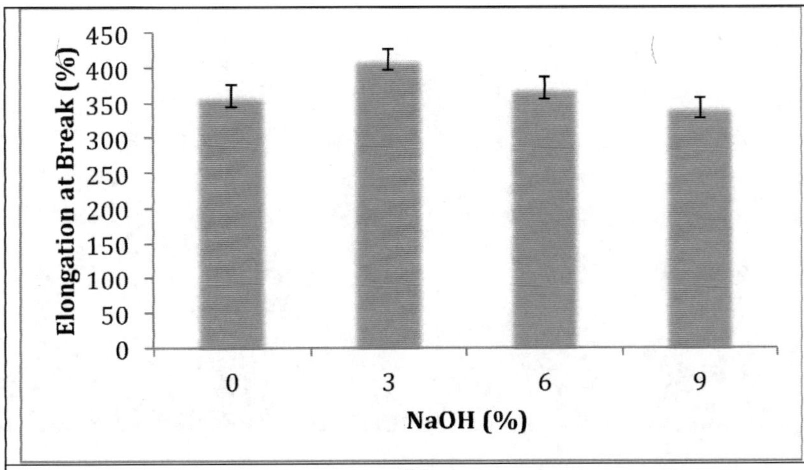

(b)

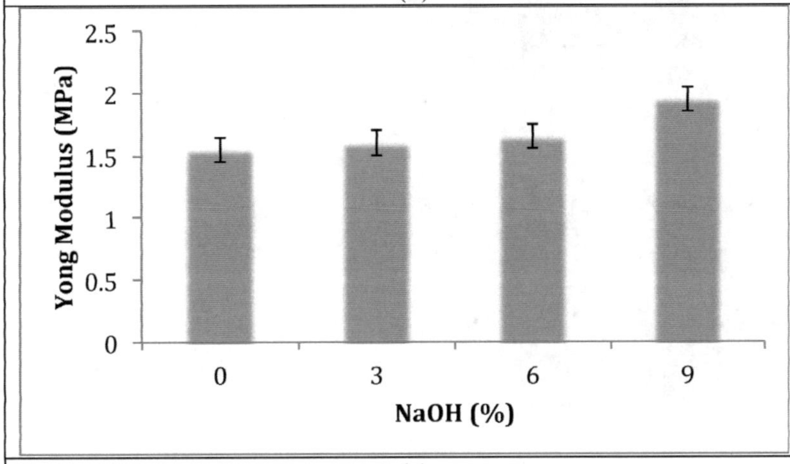

(c)

Figure 1. The effect of NaOH on the (a) tensile strength, (b) elongation at break and (c) Modulus of LLDPE/DS composites.

3.2. Scanning electron microscopy

The Scanning electron micrographs of the tensile fractured surfaces of treated LLDPE/DS composites at 3 wt.%, 6wt.% and 9 wt.% are taken to evaluate the fracture behavior and the interaction between date seeds filler and LLDPE matrix (Figure 2 a, b, an c respectively). The SEM micrograph of treated LLDPE/DS composite with low NaOH concentration (3 wt.%) shows relatively smooth DS surface and strong interfacial bonding between DS and LLDPE matrix [18]. However, the high NaOH concentration, particularly (9 wt.%) shows rough surface (Figure 1 c) indicating that the high NaOH concentration, excess delignification of natural fiber occurs which in turn lead to damage of fiber structure [19,20].

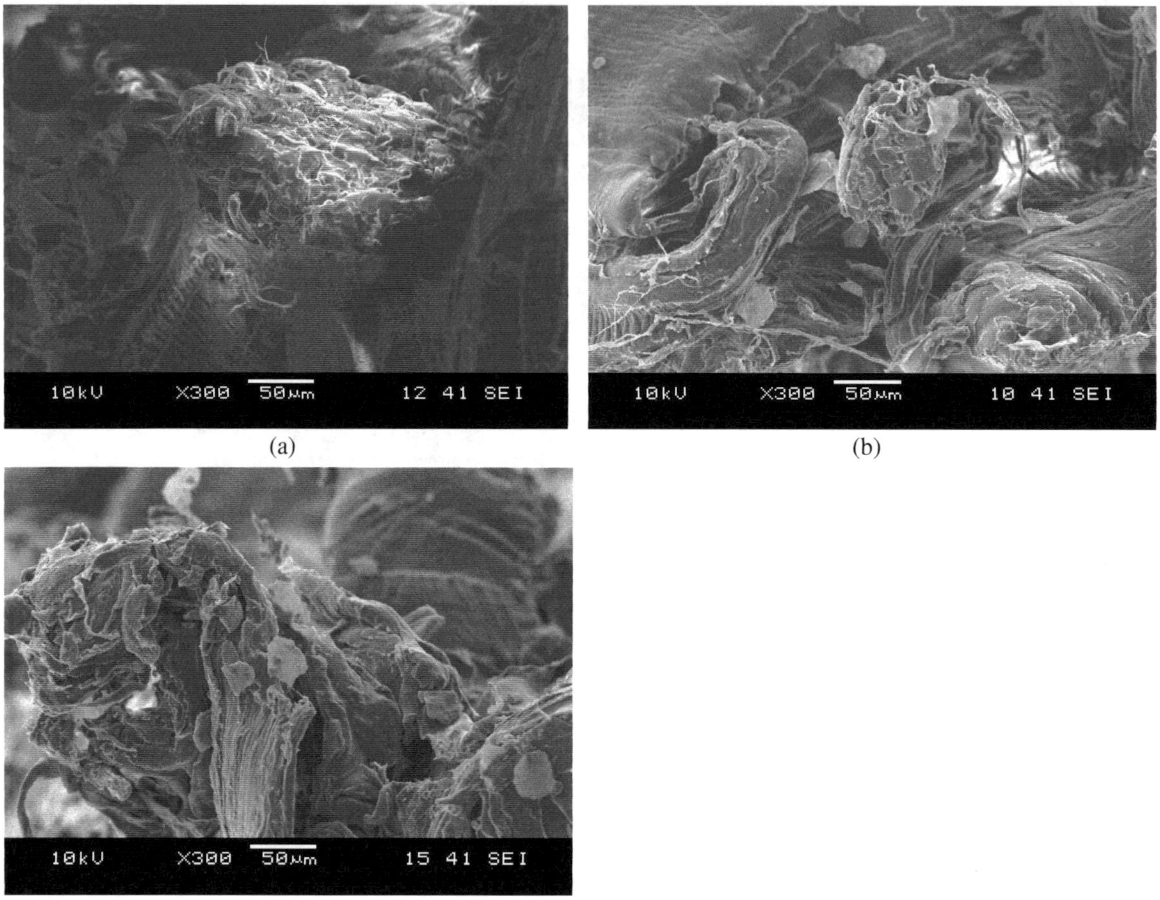

(c)

Figure 2. The tensile fracture surface of treated LLDPE/DS composite at (a) 3 wt.% NaOH, (b) 6 wt.% NaOH and (c) 9 wt.% NaOH.

4. Conclusion

In conclusion, fabricating polymer (low linear density polyethylene) with natural fiber such as Date seeds could give a new opportunity to the manufacture of green plastic products. However, the mechanical properties of this composite were relatively poor. Therefore, the chemical modification using NaOH was necessary to improve the properties. The using of low concentration (3 wt.%) of NaOH gave better tensile properties compared with control and other NaOH concentration.

5. Acknowledgement

The authors would like to acknowledge the Fundamental Research Grant Scheme (FRGS) Phase 1/2015 (9003-00523) under Ministry of Higher Education, Malaysia. The authors would appreciate Universiti Malaysia Perlis (UniMAP) for providing us all the facilities to make this research. We would also acknowledge UniMAP Short Term Grant Scheme (9001 - 00521).

6. References

1. V. K. Thakur, M. K. Thakur and R. K. Gupta, Int. J. Polym. Anal. Charact. **19**, 256-271 (2014).
2. M. H. Subli, , M. F. Omar, I. G. Sandu, N. N. Zulkepli, M. M. A. B. Abdullah and A. V. Sandu, Mater. Plast. Buchar. **51**, 391-395 (2014).
3. M. M. Thwe and K. Liao, Composites Part A **33**, 43-52 (2002).

4. I. Balberg, *Carbon*, **40**, 139-143 (2002).
5. R. Zhu, C. H. Chung, K. C. Cha, W. Yang, Y. B. Zheng, H. Zhou, T. B. Song, C. C. Chen, P. S. Weiss, G. Li and Y. Yang, *ACS nano*, 5, 9877-9882 (2011).
6. N. Z., Noriman and H. Ismail, H. International Journal of Polymeric Materials and Polymeric Biomaterials, **62**, 252-259 (2013).
7. K. Balaed, N. Z. Noriman, O. S. Dahham, S. T. Sam, R. Hamzah and M. F. Omar, Int. J. Polym. Anal. Charact. **21**, 590-598 (2016).
8. A. K. Mohanty, A. Wibowo, M. Misra and L. T. Drzal, Composites Part A **35**, 363-370 (2004).
9. O. S. Dahham, N. Z. Noriman, S. T. Sam, N. M. Al-Samarrai, Z. Shayfull, A. M. Alakrach and S. A. Elnaid, MATEC Web of Conferences **78**, 01075 (2016).
10. X. Li, L. G. Tabil and S. Panigrahi, J. Polym. Environ. **15**, 25-33 (2007).
11. O. S. Dahham, N. Z. Noriman, S. T. Sam, H. Ismail, S. Ragunathan, H. Rosniza and M. N. Al-Samarrai, J. Polym. Mater. **33** 647 (2016).
12. J. Gassan and A. K. Bledzki, *Composites Science and Technology*, **59**, 1303-1309 (1999).
13. A. K. Mohanty, M. Misra and L. T. Drzal, Composite Interfaces **8**, 313-343 (2001).
14. W. Al-Shahib and R. J. Marshall, International journal of food sciences and nutrition, **54**, 247-259 (2003).
15. A. Ghazanfari, S. Emami, S. Panigrahi and L. G. Tabil, Journal of Composites Materials, **42**, 77–89 (2008).
16. M. Jacob, S. Thomas and K. T. Varughese, Compos. Sci. Technol. **64**, 955-965 (2004).
17. A. Gomes, T. Matsuo, K. Goda, J. Ohgi, *Composites Part A: Applied Science and Manufacturing*, **38**, 1811-1820 (2007).
18. S. Mishra, A. K. Mohanty, L. T. Drzal, M. Misra, S. Parija, S. K. Nayak and S. S. Tripathy, Compos. Sci. Technol. **63**, 1377-1385 (2003).
19. D. Bachtiar, S. M. Sapuan and M. M. Hamdan, *Materials & Design*, **29**, 1285-1290 (2008).
20. M. N. Izani, M. T. Paridah, U. M. K. Anwar, M. M. Nor and P. S. H'ng, *Composites Part B: Engineering*, **45**, 1251-1257 (2013).

IOP Conf. Series: Journal of Physics: Conf. Series **1019** (2018) 012062 doi:10.1088/1742-6596/1019/1/012062

New Slow Release Fertilizer from ENR-50/RH/Urea Composites: Effect of Sodium Chloride Concentration

Marwa N Al-Samarrai, R Hamzah[1], S T Sam[2], N Z Noriman[1], Omar S Dahham[1], S Z Syed Idrus[3], S Sudin[4]

[1] Center of Excellence Geopolymer and Green Technology (CEGeoGTech), Faculty of Engineering Technology (FETech), Universiti Malaysia Perlis (UniMAP), Level 1 Block S2, UniCITI Alam Campus, Sungai Chucuh, Padang Besar, 02100, Perlis, Malaysia.
[2] School of Bioprocess Engineering, Universiti Malaysia Perlis (UniMAP), Kompleks Pengajian Jejawi 3, 02600 Arau, Perlis, Malaysia.
[3] School of Human Development and Techno-communication (iKOM), Green Advanced Computing and Technology (GREAT) Research Group – CEGeoGTech, Universiti Malaysia Perlis (UniMAP)
[4] School of Computer and Communication Engineering (SCCE), Green Advanced Computing and Technology (GREAT) Research Group – CEGeoGTech, Universiti Malaysia Perlis (UniMAP)

E-mail: rosnizahamzah@unimap.edu.my

Abstract. A green and renewable slow release material (SRM) was developed using combination of urea, epoxidized natural rubber (ENR-50), NaCl and rice husk (RH). The RH was used as a support to store urea within its fibrous structure after chemical modification with 7% sodium hydroxide (NaOH). ENR-50/NaCl was used as a coating material for RH/urea beads. Two different content of salt (5% and 10% NaCl) was used to study the effect of pore formation on the composites. FTIR spectroscopy reveals that urea was successfully absorbed into RH and the recorded peaks were overlapped with the main components of RH such as cellulose, hemicellulose and lignin. UV-Vis reveals that at 10% salt loading releases urea higher than 5% salt loading. This is due to the formation of many pores in this ENR-50 composite.

1. Introduction

Natural rubber is commercially accessible as latex from tropical tree *Havea brasiliences.* Epoxidized natural rubber-50 (ENR-50) contains 50% epoxidation mole of natural rubber. The physical properties of ENR-50 are higher mechanical strength such as tensile, elongation, hardness and compression set. It also has higher density, higher viscosity and lower refractive index. ENR-50 has higher oil and solvent resistivity. It is more resistant to non-polar solvent or hydrocarbon and reduces its resistivity in polar solvent. It also has lower air and oxygen permeability across the ENR membrane because of the simple diffusion of oxygen than air into the ENR membrane. The glass transition temperature (T_g) of ENR-50 is at -18 ^{o}C. The physical and chemical properties of ENR, for example heat and swelling resistance, can change as per the epoxide content. ENR likewise has higher polarity than virgin NR due to the presence of epoxide groups in its structure [1,2].

Rice husk (RH) are major agricultural wastes from paddy rice. Raw RH commonly made up of 35 % cellulose, 25 % hemicellulose, 20 % lignin, 17 % ash from a total of 94 % silica and 3 % moisture by weight [3,4]. The composition of RH is 23 % of the total paddy weight. The physical properties of RH are biodegradable, cheap, low density, abundant and easy to obtain [5]. Rice straw is an organic material, which commonly used and it contains many mineral nutrients. There are 0.5–1.5% N, 0.2–1.0% P, 0.8–1.0% K, and Si [6]. RS contains easily decomposable C and has a relatively wide C/N ratio compared with RS compost. RS is an economical soil revision to enhance fertilizer recovery efficiencies [7].
The fertilizers are applied to the soil in terms of slow release for market efficiency, economic cultivation and biomass production increment. The coated RH and RS are considered as a suitable candidate for development of slow release material in agriculture. This slow release coating technology supplies consistent nitrogen to the plants and reduce loss and contamination impacts. The coating technology intended for slow release of nutritional content of fertilizer to the plants. It additionally synchronizes the discharge rate with the nutritious request of the plants [8].

2. Methodology

2.1. Purification of ENR-50
About 20g of ENR-50 was swelled in 400 mL of chloroform and stirred for 24 hours at room temperature. The solution undergoes filtration process through a pack of cotton gauze to separate the high molecular weight from the low molecular weight of ENR-50 [9]. The latter precipitated in *n*-hexane while stirring using a glass rod. The white precipitate stuck to the glass rod transferred to a Petri dish and dried in a vacuum oven at 50 °C for two days. The purified sample was recorded every day until a constant reading is attained. The sample was then characterized by FTIR technique.

2.2. Cellulose pretreatment
The RH fibers were washed, cleaned, shredded and dried in a convection oven at temperature of 60 °C. The dried RH fibers were grinded using lab grinder with a 250 μm screen mesh to achieve a uniform particle size. The cellulose pretreatment involved the process of soaking the mesh RH in 7 wt% sodium hydroxide (NaOH) for 24 hours. The 7 wt% of NaOH prepared by dissolve 7 g of NaOH solution in 100 mL of distilled water. The cellulose filtered using centrifugal method to separate the cellulose/NaOH mixture. After that, the pH value of the sample checked by using pH paper. If the pH value is higher, the sample washed using distilled water until neutral to remove all the NaOH solution. Then, the sample dried on a white paper under sunlight. The sample dried in a vacuum oven at 50 °C for 48 hours. After that, the sample analyzed using FTIR technique [10].

2.3. Preparation of RH/urea beads
The solvent prepared by dissolving 60 g of urea fertilizer in 500 mL of distilled water to produce 12 wt% urea. The prepared sample from cellulose pretreatment was dissolved in 12 wt% urea solution for 24 hours at room temperature. Then, the sample was filtered using centrifugal method to separate the RH/urea composite and solution mixture. Then, the sample was dried on a white paper under sunlight. The sample was further dried in a vacuum oven at 50 °C for 48 hours. After that, the RH/urea beads were characterized by FTIR technique [10].

2.4. Preparation of ENR-50/salt composites
About 10 mg of sodium chloride was dissolved in 10 mL of tetrahydrofuran. Another sample was prepared using 20 mg of sodium chloride dissolved in 10 mL of tetrahydrofuran. About 200 mg of purified ENR-50 was dissolved in 40 mL of chloroform. The different sodium chloride solutions and ENR-50 solution were dissolved together to form ENR-50/5% salt composite and ENR-50/10% salt composite respectively. The solution was then stirred for 24 hours at room temperature. This solution does not need to undergo drying process.

2.5. Preparation of RH/urea beads coated ENR-50/salt composite
About 20 mg of RH/urea beads was dissolved into 50 mL of ENR-50/5% salt composite and 50 mL of ENR-50/10% salt composite separately (0.08 wt% ENR-50). Then, the mixtures were stirred for 24 hours at room temperature until homogenous solution occurred. The mixtures was dried in fume cupboard at 50 °C for 24 hours.

2.6. Formation of pores on RH/urea beads coated ENR-50/salt composite
The mixtures was dissolved in 200 mL distilled water. Then, the mixtures was filtered using filter paper and funnel. After that, the samples were dried using vacuum drier. The RH/urea beads coated ENR-50/5% salt composite and RH/urea beads coated ENR-50/10% salt composite were characterized using FTIR technique.

2.7. Release study
The slow release study of the beads was conducted in distilled water; the following experiments were carried out. About 5 g of dry samples of 5% salt loading and 10% salt loading were added to the conical flask containing 200 mL of distilled water. Then the conical flask put into incubators with the temperature 25 °C. Each day, 2 mL solution sampled for urea contents, and additional of 2 mL distilled water was added to the flask to maintain a constant amount of solvent. This experiment was carried out for 30 days. Then, it was characterized using UV-vis spectrometer [11].

2.8. Kinetic study
A standard urea concentration was prepared by using 100 ml aqueous solution having concentration between 0.2 and 1.0 mg/ml. 20 mg of urea was dissolved in 100 ml of distilled water to obtain 0.2 mg/ml. UV-vis spectrometer was used to record the UV-visible absorbance spectrum of the prepared solution. The spectrum with absorbance on the y-axis (dependent variable) and concentration in mg/ml on the x-axis (independent variable) was plotted as the calibration curve. Microsoft Excel was used to draw the equation of the graph (Eq. 1.). The gradient of the graph was used as molar absorptivity for Beer-Lambert equation. Then, the concentration was calculated to plot a graph of concentration versus time.

$$A = \epsilon C \ell$$
$$C = \frac{A}{I \, x \, \varepsilon} \qquad (1)$$

3. Results and Discussion

3.1. Characterization of purified ENR-50 and ENR-50/salt composite
The FTIR spectrum of purified ENR-50 is shown in Figure 1(a) and compared with FTIR spectra of ENR-50/salt at both 5% and 10% salt loadings (Figure 1(b,c)) . Generally, 5% and 10% salt loading showed approximately similar FTIR spectra. Both showed broad OH absorption band at range 3433-3420 cm^{-1}. For both salt loading, CH_3 peaks occurred at range 2966-2963 cm^{-1} and 2857-2850 cm^{-1}. The CH_2 peaks at 2924-2921 cm^{-1} and 1448 cm^{-1}. The C-H peak remains similar like purified ENR-50 at 1378 cm^{-1}. The present of the epoxide group at 878-875 cm^{-1} and C=C at 838 cm^{-1} indicated that the chemical structure of purified ENR-50 didn't affect by the salt loading. The salt only dispersed amongst the polymeric chains of purified ENR-50 and not causing the ring opening reaction of the epoxide group [12].

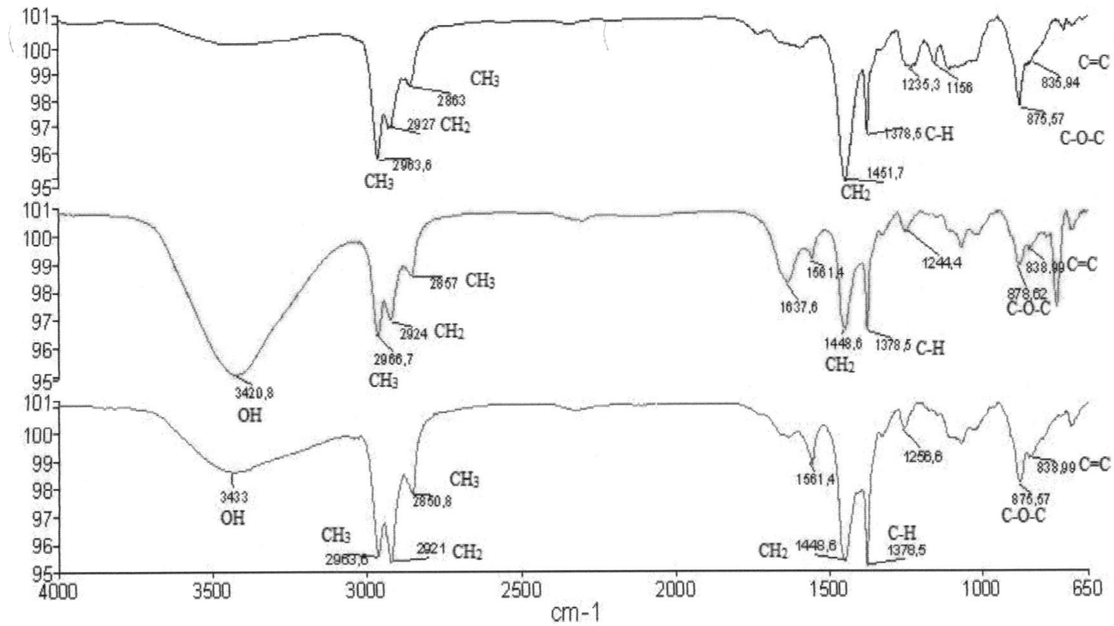

Figure 1: FTIR spectra of (a) purified ENR-50, (b) ENR-50/5% salt composite and (c) ENR-50/10% salt composite

3.2. Characterization of RHRS/urea beads

The FTIR spectrum of pretreated RH is shown in Fig. 2(a) and compared with FTIR spectra of RH/urea beads in Fig. 2(b). Generally, the peaks of RH/urea beads are similar to pretreated RH. The most characteristics peaks of urea at 1667 and 1629 cm^{-1} are assigned to carbonyl (C=O) and N-H bending vibration respectively. In the region 3200-3500 cm^{-1}, there are also peaks, which may be attributed to urea compounds, like 3425 cm^{-1} assigned to symmetric stretching vibration of NH_2. Also peak at 2917 cm^{-1} can be assigned to C-H stretching vibration of absorbed water. However, this region is overlapped with RH spectrum. It shows that no chemical interaction involves between RH and urea but it just involves physical interaction. Therefore, more evident and reliable data related to increasing amount of urea in extruded sample are in region between 1629-1667 cm^{-1} [13].

3.3. Characterization of RH/urea beads coated with ENR-50/Salt composite

The FTIR spectra of both RH/urea beads with ENR-50/5% and ENR-50/10% salt composite are shown in Fig. 2(c) and 2(d) respectively. Generally, both of the spectra are almost similar based on its characteristics peaks. The peaks observed at 3357 cm^{-1} and 3434 cm^{-1} are assigned to OH of groups [14,15]. This OH groups was due to the hydroscopic nature of salt that exist in the composites. At high loading of salt, the OH peaks are more broad and intense as compare to low loading of salt. The peaks at 2917 cm^{-1} and 1464 cm^{-1} are attributed to the C-H stretching vibrations. The peaks at ranges 2846-2850 cm^{-1} are assigned to CH_3 symmetry stretching. In addition, the peaks at 1031-1063 cm^{-1} could be suggested as C-O stretching vibration. The peak at 1645 cm^{-1} at 10% salt loading which is not detected earlier in 5% salt loading is due to the C=C aromatic which is dominant in this composite than 5% salt. At high salt loading, aromatic group of lignin plays a crucial role to more space for urea to be absorbed into cellulose fibers. The peak at 1319 cm^{-1} was due to C-O vibration. The C-O-C asymmetric stretching is shown at 1118-1163 cm^{-1}. In addition, the peak at 895 cm^{-1} is assigned to C-H deformation vibration [12, 16-19].

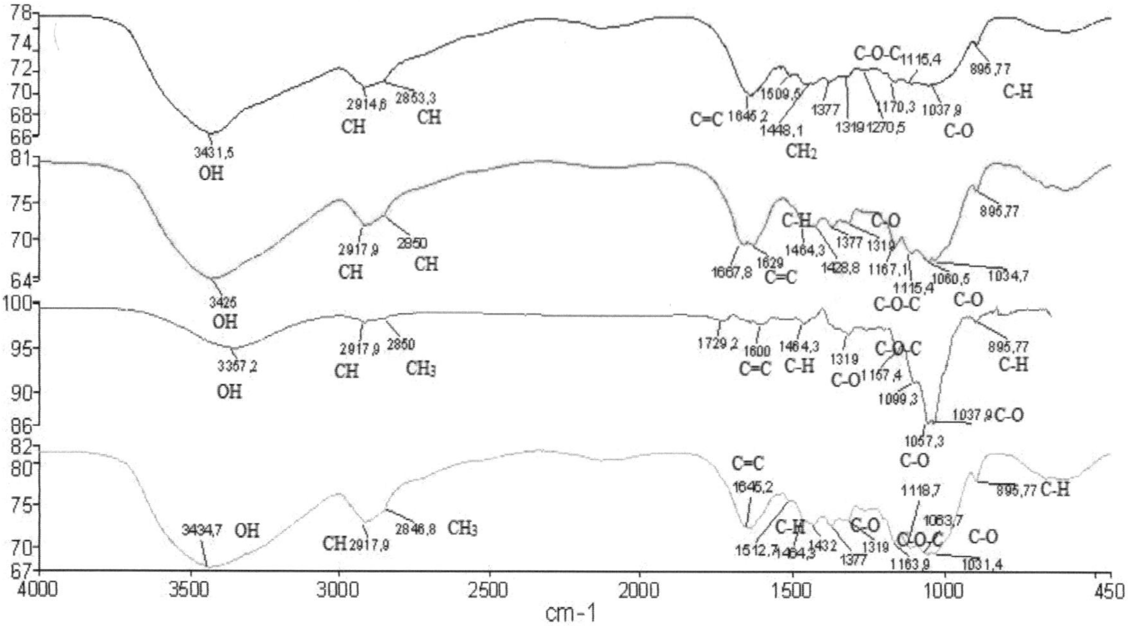

Figure. 2: FTIR spectra of (a) pretreated RH, (b)RH/urea beads, (c)RH/urea beads coated with ENR-50/5% salt composite and (d)RH/urea beads coated with ENR-50/10% salt composite

3.4. Release study

The UV-Vis spectra of slow release study of the RH/urea beads coated with ENR-50/salt at 5% and 10% salt loadings are shown in Fig. 3. Generally, the absorbance value for 10% salt loading is higher than 5% salt loading. This indicated that the composites at10% salt loading released more urea than the composites at 5% salt loading. At 10% salt loading represented high amount of salt of as compared to 5% salt loading. Thus more pores are available to release urea at 10% salt loading as compared to 5% salt loading.

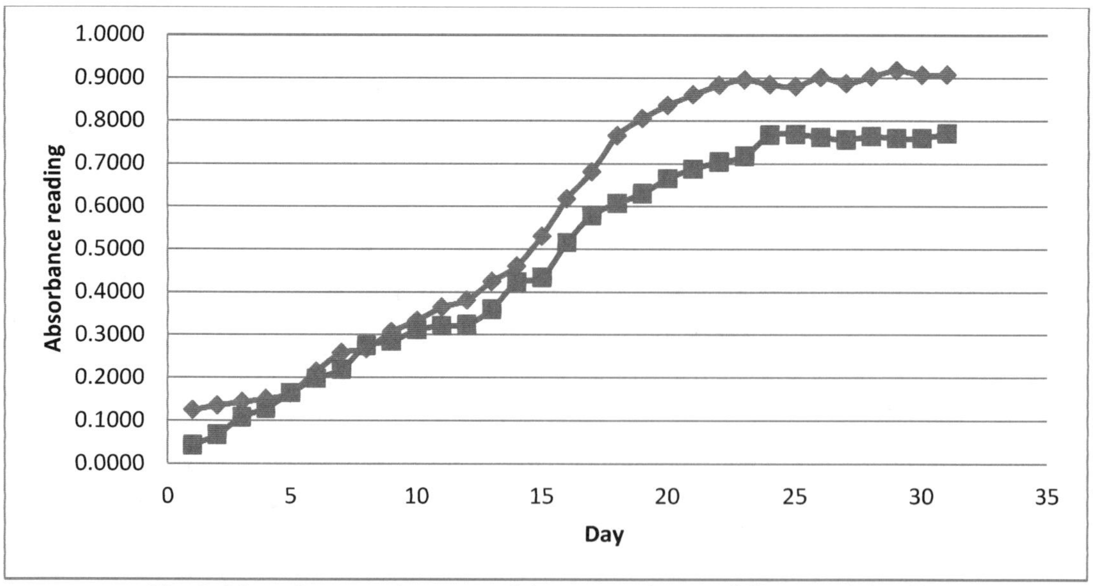

Fig. 3: The release study curve of the composites at different salt loading

3.5. Kinetic study

The slow release mechanism can be divided into three stages. First stage is a lag period in which water penetrated the ENR-50 coating without urea release. Then, a constant release period followed when urea dissolved and flowed through the coating which can be said as burst effect. Finally, there came a stage of decline until the release of urea ultimately ended [20].

4. Conclusion

In conclusion, FTIR spectroscopy showed that the urea was successfully absorbed into RH and the recorded peaks were overlapped with the main components of RH i.e. cellulose, hemicellulose and lignin. UV-Vis showed that the composite at 10% salt loading releasesed urea higher than composite at 5% salt loading. This was due to number of pores formation in this composite.

5. Acknowledgement

The authors would like to acknowledge UniMAP Short Term Grant Scheme (9001 - 00521). We would also appreciate Universiti Malaysia Perlis (UniMAP) for providing us all the facilities to make this research.

6. References

1. R. Hamzah, M. A. Bakar, O. S. Dahham, N. N. Zulkepli and S. S. Dahham, J. Appl. Polym. Sci., **133** (2016).
2. O. S. Dahham, R. Hamzah, M. A. Bakar, N. N. Zulkepli, S. S. Ting, M. F. Omar, K. Muhamad, and S. S. Dahham, Polymer Testing **65**, 10-20 (2018).
3. A. Kumar, B. Sengupta, D. Dasgupta, T. Mandal, and S. Datta, *Reviews in Environmental Science and Bio/Technology* 15, 47-65 (2016).
4. M. A. Fuad, Z. Ismail, Z. M. Ishak and A. M. Omar, European Polymer Journal, **31**, 885-893 (1995)
5. M. Suzeren Jamil, I. Ahmad and I. Abdullah, *Journal of Polymer Research* **13**, 315-321 (2006).
6. A. Polthanee, V. Tre-loges and K. Promsena, *Paddy and Water Environment* **6**, 237-241 (2008).
7. P. J. A. Van Asten, P. M. Van Bodegom, L. M. Mulder and M. J. Kropff, *Nutrient Cycling in Agroecosystems* **72**, 255-266 (2005).
8. M. Y. Naz, and S. A. Sulaiman, *Journal of Controlled Release* **225,** 109-120 (2016).
9. O. S. Dahham, R. Hamzah, M. A. Bakar, N. N. Zulkepli, S. S. Dahham and S. S. Ting, Polym. Test. **59**, 55-66 (2017).
10. J. W. Dormanns, J. Schuermann, J. Müssig, B. J. Duchemin and M. P. Staiger, *Composites Part A: Applied Science and Manufacturing* **82**, 130-140 (2016).
11. G. Z. Zhao, Y. Q. Liu, Y. Tian, Y. Y. Sun and Y. Cao, *Journal of Polymer Research* **17**, 119-125 (2010).
12. W. L. Tan and M. A. Bakar, *Journal of Alloys and Compounds* **561**, 40-47 (2013).
13. P. Rychter, M. Kot, K. Bajer, D. Rogacz, A. Siskova, and J. Kapusniak, *Carbohydrate polymers* **137**, 127-138 (2016).
14. N. N. Zulkepli and H. Ismail, Polymer-Plastics Technology and Engineering, **51**, 350-357 (2012).
15. N. Z. Noriman, H. Ismail and A. A. Rashid, A. A. Journal of Applied Polymer Science, **126** (2012).
16. K. Jantanasakulwong, N. Leksawasdi, P. Seesuriyachan, S. Wongsuriyasak, C. Techapun, and T. Ougizawa, *European Polymer Journal* **84**, 292-299 (2016).
17. T. Xu, Z. Jia, Y. Luo, D. Jia and Z. Peng, *Applied Surface Science* **328**, 306-313 (2015).
18. W. A. K. Mahmood and M. H. Azarian, *Current Applied Physics* **15**, 599-607 (2015).
19. W. A. K. Mahmood, M. M. R. Khan and M. H. Azarian, *Journal of Non-Crystalline Solids* 378**,** 152-157 (2013).
20. L. Wu, M. Liu and R. Liang, *Bioresource Technology* **99**, 547-554 (2008).

IOP Conf. Series: Journal of Physics: Conf. Series **1019** (2018) 012063 doi:10.1088/1742-6596/1019/1/012063

Slow Release Material from Epoxidized Natural Rubber and Rice Husk Composites for Agriculture Applications

Marwa N Al-Samarrai[1], R Hamzah[1], S T Sam[2], N Z Noriman[1], Omar S Dahham[1], S Z Syed Idrus[3], T Adam[1,4]

[1] Center of Excellence Geopolymer and Green Technology (CEGeoGTech), Faculty of Engineering Technology (FETech), Universiti Malaysia Perlis (UniMAP), Level 1 Block S2, UniCITI Alam Campus, Sungai Chucuh, Padang Besar, 02100, Perlis, Malaysia.
[2] School of Bioprocess Engineering, Universiti Malaysia Perlis (UniMAP), Kompleks Pengajian Jejawi 3, 02600 Arau, Perlis, Malaysia.
[3] School of Human Development and Techno-communication (iKOM), Green Advanced Computing and Technology (GREAT) Research Group – CEGeoGTech, Universiti Malaysia Perlis (UniMAP)
[4] Institute of Nano Electronic Engineering (INEE), Universiti Malaysia Perlis (UniMAP), 01000 Kangar, Perlis, Malaysia
E-mail: rosnizahamzah@unimap.edu.my

Abstract. This synthesis and characterization of slow release material from epoxidized natural rubber (ENR) and rice husks (RH) composites was conducted. The alkali modification of rice husks was carried using different NaOH concentration. ENR-50 and RH composites were analyzed by FTIR. Furthermore, the release behavior of the composites was invistigated using UV-visible spectroscopy. From the purified ENR-50, it can be characterized the peak epoxide group identified at 875 cm^{-1} and for the isoprene at 835 cm^{-1}. For RH/urea beads, the CH_2 peak of cellulose was obtained at 1457 cm^{-1}. However, the CH_2 is overlapped with cellulose at 1741 cm^{-1} and 1726 cm^{-1} that resulted for the partial opening of epoxide ring from carbonyl peak for hemi-cellulose and lignin, there are C=C stretching from aromatic group can be obtained at 1512 cm^{-1} and 1595 cm^{-1}. The release of the composite at 7% NaOH concentration was higher than the composite at 1% NaOH concentration.

1. Introduction

Natural rubber (NR) is a Malaysia's commodity that obtained from latex tree *Havea Brasiliences*. NR that contain 99% of cis-1,4-polyisoprene is an elastomeric material and contributes to high degree of crystallization as well as high mechanical properties for natural rubber. Epoxidized natural rubber (ENR) is a modified NR while epoxidized natural rubber-50 (ENR-50) is a modified natural rubber with 50 mole percentage of epoxidation and 50 mole percent of isoprene. There are various fields that used the ENR such as polymer blends, polymer modifications, polymer composites and advanced green materials [1,2]. Both NR and ENR are considered as green polymer and green materials due to non-toxic and non-hazardous properties [3].

Rice husk (RH) are one of the major by-product, which would always be removed during the rice-milling process [4]. The RH contains cellulose, hemicelluloses and lignin [5]. The exterior structure of rice husk is the silica coated with a thick cuticle and surface hairs. It provides with other special husk to form physical interaction matrix [6]. The chemical composition of rice husk is cellulose (40-50%),

lignin (25-30%), ash (15-20%) and moisture (8-15%). The common organic fibers mostly contains the same composition like a rice husk [7]. Rice husk is suitable as filler in polymer composites due to the low cost, low density, high specific strength and modulus, and recyclability [8].
Slow release is a method that active ingredient is release as a function of time [9]. The release of the active ingredient may be controlled by microbial release or water and it is due to chemical properties of materials. Fertilizer is the important thing in the food security that might be as a input material for the sustainable development of the crop production [10]. The slow release fertilizer (SRF) is a fertilizer that been coated with another materials to release it active ingredient slowly to the soil for a period time. This method is suitable in agriculture and it consistently releases the nitrogen to the plants. It also improves the soil quality, property control and the germination rate [11].
Ultraviolet and visible spectroscopy is a technique to infrared spectroscopy and it will be used to analyze solutions of organic compound and complex metal ions. UV-Vis spectroscopy is used to determine the concentration of the solution. According to the Beer-Lambert law, the relationship of the absorbance of a solution with the concentration of the absorbing species in the solution and the path length is directly proportional. This law also is the principle behind absorbance spectroscopy [12]. The law may be written in terms of the absorption A for a single wavelength which is equation 2.

$$A = abc \qquad (2.0)$$

Where a is the molar absorptivity of the compound or molecule solution ($M^{-1}cm^{-1}$), b is the path length of the cuvette or sample holder (usually 1 cm) and c is the concentration of the solution (M) [4].

2. Experimental Part

2.1. Materials and chemicals

All chemicals were obtained commercially and used without further purification unless otherwise mentioned as tabulated in Table 3.0. n-hexane were from Systerm, Malaysia. Epoxidized natural rubber of 50% epoxidation level (ENR-50) was purchased from the Rubber Research Institute, Malaysia. Rice Husk (RH) were from Merbuk MDF Sdn. Bhd., Kedah, Malaysia. Urea fertilizer was purchased from PETRONAS Fertilizer (Kedah) Sdn. Bhd., Kedah, Malaysia. Tetrahydrofuran (THF) from BASF Petronas Chemicals Sdn. Bhd., Selangor, Malaysia. Isopropanol (IPA) was obtained from Primechem Malaysia Sdn. Bhd., Johor, Malaysia. Chloroform was obtained from Octagon Chemicals Sdn. Bhd., Selangor, Malaysia. Sodium chloride (NaCl) was purchased from BIS Chemicals Sdn. Bhd., Selangor, Malaysia.

2.2. Purification of ENR-50

About 20.00 g of ENR-50 was swelled in 400 mL of chloroform and it stirred for 24 hours at room temperature [13]. The cotton gauze was used to separate the high molecular from the low molecular weight ENR-50. Next, the precipitate was obtained by stirring the low molecular weight ENR-50 in n-hexane by using glass rod. The white precipitate that form on the surface of glass rod was removed onto the Teflon Petri dish and dried in a vacuum oven at 50 °C for two days. The weight of purified ENR-50 was recorded daily until the constant weight of ENR-50 achieved. The dried sample was characterized using FTIR analysis.

2.3. Pretreatment of Cellulose

The rice husk was washed with tap water to remove excess of soil and dirt and finally with distilled water. Then, it was dried in a convection oven at temperature of 60 °C to remove excess moisture for 2 hours. The dried rice husk was stranded and through the grinder with 500 μm mesh screen to achieve a uniform particle size. The rice husk was soaked in the 7 wt% of sodium hydroxide (NaOH) solution for 24 hours and abbreviated as cellulose pre-treatment process. The 7 wt% of NaOH solution was prepared by dissolving 7.0 g of NaOH in 100 mL of distilled water. The filter funnel and filter paper were used to filter the mixture and to separate the cellulose from its NaOH mixture. Then, the pH value of the supernatant was measured using pH paper. The sample was washed repeatly using distilled water until neutral was attained. Then, the sample was air dried under sunlight and later dried in a vacuum oven for 48 hours at 50 °C. This steps was repeated by soaking the rice husk in the 1 wt% of sodium hydroxide (NaOH) solution. The FTIR was used to analyze the sample [14].

2.4. Pretreatment of Cellulose /urea beads

A 60 g of urea fertilizer was dissolved in 500 mL of distilled water to prepare a thick mixture of 12 wt% of urea fertilizer solution [9]. The sample obtained from treated of rice husk was soaked in 12 wt% urea solution for 24 hours at room temperature. The sample was filtered to separate the cellulose/urea solution mixture by using filter funnel. Then, the sample was air dried under sunlight and later in a vacuum oven for 48 hours at 50 °C. The FTIR analysis was used to characterize the sample

2.5. Pretreatment of Cellulose/Urea Beads Coated with ENR-50/Salt Composite

The solution mixture was mixed with 20.0 mg of cellulose/urea beads. The homogenous solution was formed after stirring the mixture for 24 hours at room temperature. The mixture was air dried for 48 hours and then further dried using vacuum oven at 50 °C for 24 hours.

2.6. Release study

The distilled water was used to conduct the slow release of the SRM in this experiment. About 200 mL of distilled water was used to soak the composite obtained from the preparation of beads. About 5.0 g of dry sample were added into the conical flask that containing 200 mL of distilled water and it being placed in an incubator at 25 °C. About 5 mL of solution was taken daily to investigate the content of urea in the solution. Thus, 5 mL of distilled water was added to maintain the constant amount of solvent. The experiment has been carried out for 30 days and the UV-Vis spectrophotometer was used to characterize the samples.

2.7. Kinetic study

A standard urea fertilizer concentration was prepared for kinetic study reason. 100 mL aqueous solution was prepared having concentration between 0.2 mg/mL and 1.0 mg/mL. 20 mg of urea fertilizer dissolved in 100 mL of distilled water to obtain 0.2 mg/mL. UV-Vis spectrophotometer was used to record the UV-Visible absorbance spectrum of the prepared solution. The spectrum with absorbance on the y-axis (dependent variable) and concentration in mg/mL on the x-axis (independent variable) was plotted as the calibration curve. The gradient of the graph was used as molar absorptivity for Beer-Lambert equation. Then, the concentration was calculated to plot a graph of concentration versus time.

3. Results and Discussion

3.1. FTIR of ENR-50 and ENR-50/Salt Composite

Figure 1(a) shows the FTIR spectrum of purified ENR-50. Generally, there are five types of peaks presented the CH_3, CH_2, CH, epoxide, and isoprene (C=C). The CH_3 peaks are detected at 2863 cm^{-1} and 2963 cm^{-1} while CH_2 at 2927 cm^{-1} and 1451 cm^{-1} [13]. The CH peak at 1378 cm^{-1}. The evidence of epoxide group at 875 cm^{-1} is the C-O epoxide half ring stretching [7]. The stability of isoprene group is shown by C=C peak at 838 cm^{-1}.

The FTIR spectrum of ENR-50/salt composite is shown in Figure 1(b). The OH peaks is broad as compared to purified ENR-50. This is due to the hydroscopic nature of NaCl. It is well known that most of the salt such as NaCl, $MgCl_2$ etc is easily trapped moisture contain its ionic structure. This explains the incorporation of moisture to ENR-50 matrix in the presence of salt. Thus, OH peak of ENR-50/salt composite is detected at 3427 cm^{-1}. The CH_3 peaks at 2963 cm^{-1} and 2853 cm^{-1}. The CH_2 peaks at 2921 cm^{-1} and 1448 cm^{-1}. The CH peak at 1378 cm^{-1} [15]. The epoxide groups are remained at 875 cm^{-1} and isoprene at 835 cm^{-1}. The rest of the peaks almost similar to the purified ENR-50 as and this suggest that the loading of salt does not change the chemical structure of purified ENR-50.

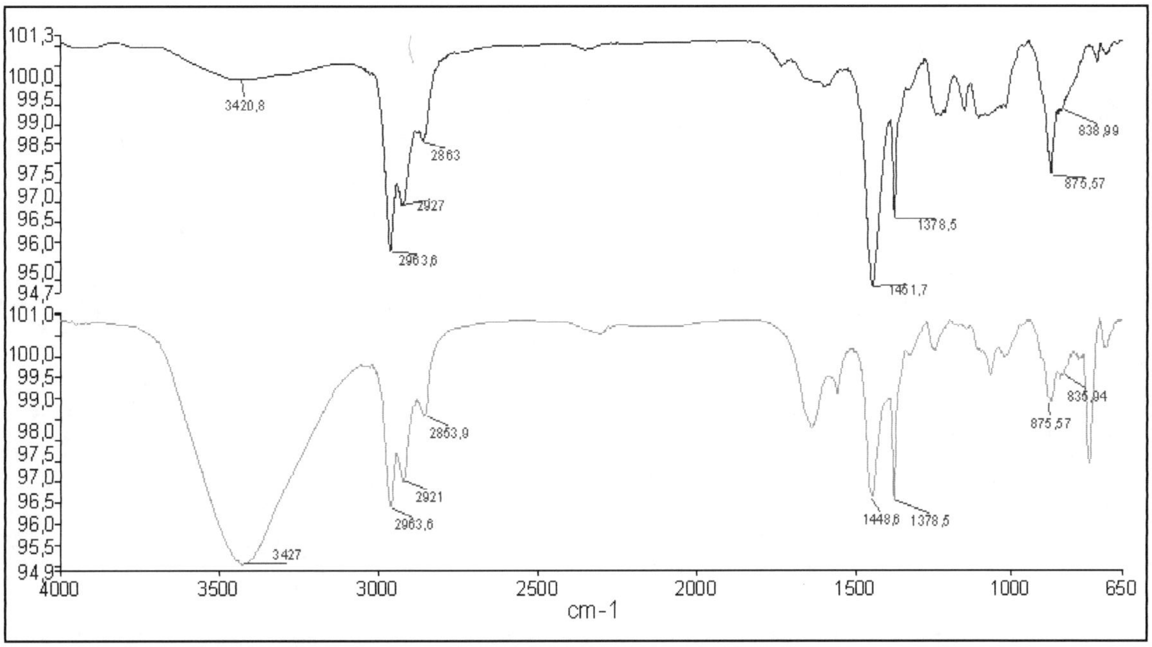

Figure 1. FTIR spectra of purified ENR-50 and ENR-50/salt composite

3.2. FTIR of pretreated RH and RH/urea beads

Figure 2 shows the FTIR spectra of treated RH with (a) 1% NaOH, (b) RH/urea beads with pretreated 1% NaOH, (c) 7% NaOH and (d) RH/urea beads with pretreated 7% NaOH. In general, RH contains cellulose, hemi-cellulose and lignin. The pretreated of RH with different concentration of NaOH is carried out to expand the volume of fibrous cellulose structure in RH. The expanded volume is necessary to occupy more urea into it. Thus, it is acceptance for RH to contain cellulose, hemi-cellulose and lignin after the pretreatment process.

Typically, there are several groups of peaks attributed to the structure of cellulose, hemi-cellulose and lignin. The OH stretching of cellulose is located at 3353 cm^{-1} in Figure 2(a) [15], The CH_2 peak of cellulose was obtained at 1457 cm^{-1} due to the C-H bending as shown in Figure 2(b). The broad and strong peak in Figure 2 were obtained between 1153 cm^{-1} to 1159 cm^{-1} due to the C-O stretching vibration from tertiary alcohol [16]. For the hemicellulose, The CH_2 is overlapped with cellulose at 1741 cm^{-1} and 1726 cm^{-1} (Figure 2(a,c)) that can be seen as a result of partial opening of epoxide ring from carbonyl peak [17]. This peak also shows the carbonyl stretching vibration which is C=O stretching of carboxylic group [18]. For lignin, there are C=C stretching from aromatic group can be obtained at 1512 cm^{-1} and 1595 cm^{-1} (Figure 2(c,d)) due to the removal of lignin after alkali treatment [8].

The spectrum of rice husk after treatment with 1% NaOH and 7% NaOH shows the differences peak from 1058 cm^{-1} to 1043 cm^{-1} in Figure 2(a,c). The physical interaction between epoxy groups and cellulose groups might be occurred. After mesh urea, the new peaks in Figure 2(b,d) was observed at 3460 cm^{-1} and 3442 cm^{-1} for RH/urea beads with pretreated 1% NaOH and RH/urea beads with pretreated 7% NaOH. This is due to the hydroscopic nature of NaCl to trap moisture. Thus, most of the urea peaks overlapped with RH such as CH_3, CH_2 and CH types.

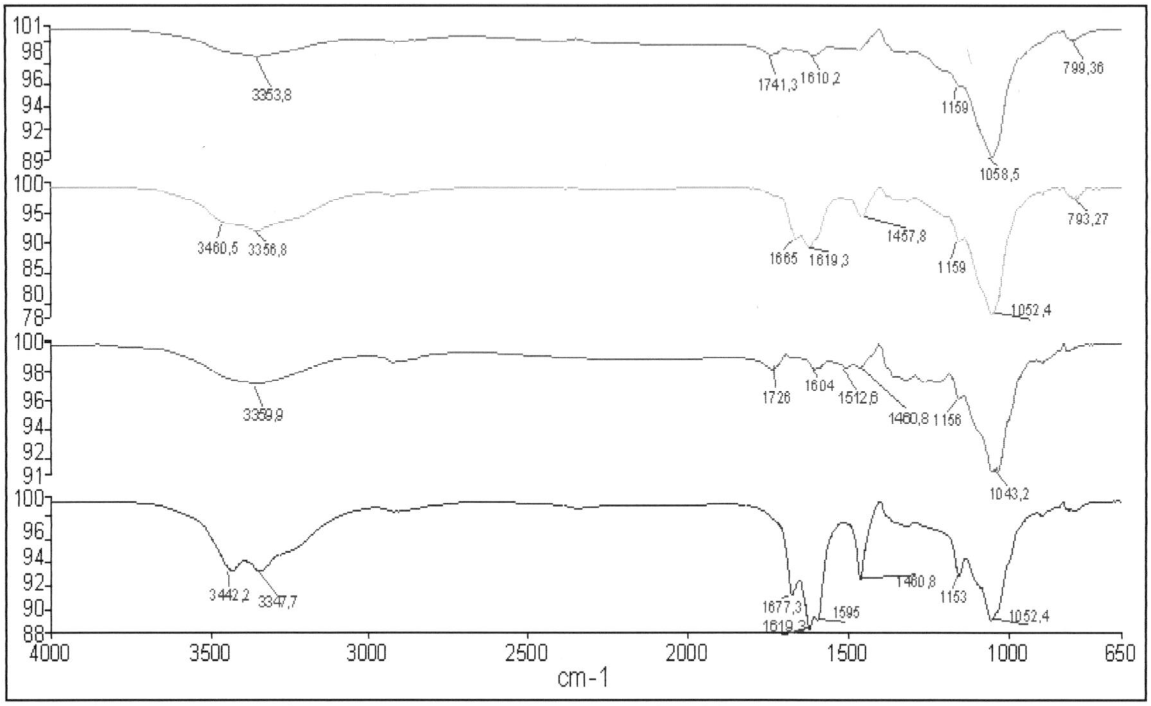

Figure 2. FTIR spectra of (a) RH 1% NaOH, (b) RH/urea beads 1% NaOH, (c) RH 7% NaOH and (d) RH/urea beads 7%

3.3. Evaluation of release study

The Figure 3 shows the release curve of absorbance versus time from RH/urea beads 1% NaOH coated with ENR-50/NaCl composite and RH/urea beads 7% NaOH coated with ENR-50/NaCl composite for 30 days.

The trending for both graph shows the release of urea fertilizer was gradually increased. This figure shows the amount of urea release from the RH/urea beads for 1% NaOH is higher than RH/urea beads for 7% NaOH. This is due to the formation of pores on the surface of RH/urea beads coated with ENR-50/NaCl composite that more available to release the urea fertilizer.

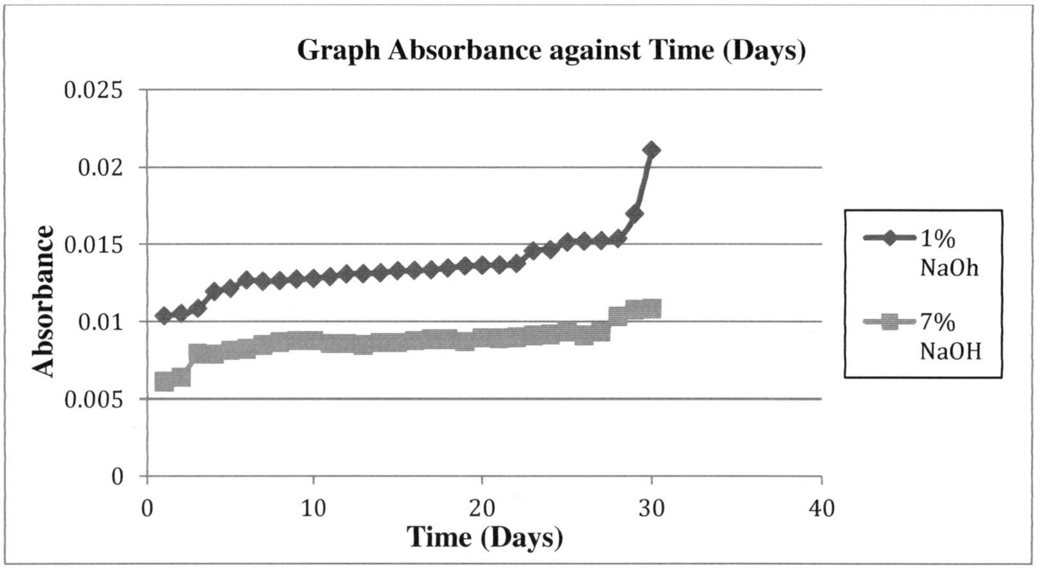

Figure 3. The release study of treated composites at different NaOH concentration.

3.4. Kinetic study of release rate

The standard curve was obtained from the absorbance values for different concentration of urea fertilizer. The standard urea concentrations were chosen between 0.2 to 1.0 mg/mL and the value of a is obtained from gradient of the standard curve. Then, its values substituted into Beer-Lambert equation to find the concentration.

Figure 4 shows the amount of urea fertilizer released at RH/urea beads 1% NaOH coated with ENR-50/NaCl composite, about 49.14%, 62.99% and 100% (wt%) of urea fertilizer are released at day 1, 15 and 30 of water incubation respectively. At RH/urea beads 7% NaOH coated with ENR-50/NaCl composite, the urea fertilizer released 56.26%, 79.73% and 100% (wt%) at day 1, 15 and 30 respectively.

The release study was divided into three stage which are the water penetrated the ENR-50 coating at the first stage. For the second stage, the urea released is constant by following the period of time and its flowed through the coating. Then, the urea fertilizer released at a stage of deadline that finally ended.

The slow release mechanism involves the absorption of water by the coating material which cause it to swell and will transform into hydrogel. Thus, ENR-50 is cross-linked hydrophilic polymer which has an ability to absorb water and cannot discharged easily even under extended pressure. The core urea fertilizer then dissolves in the hydrogel water and diffuses slowly through a grid of the swollen hydrogel from within the hydrogel to water in the soil via mass transfer of water. Water was diffused through the coating on the surface then slowly penetrated the RH/urea beads layer to dissolve the urea fertilizer. The urea fertilizer will flow out gradually between hydrogel free water and soil moisture.

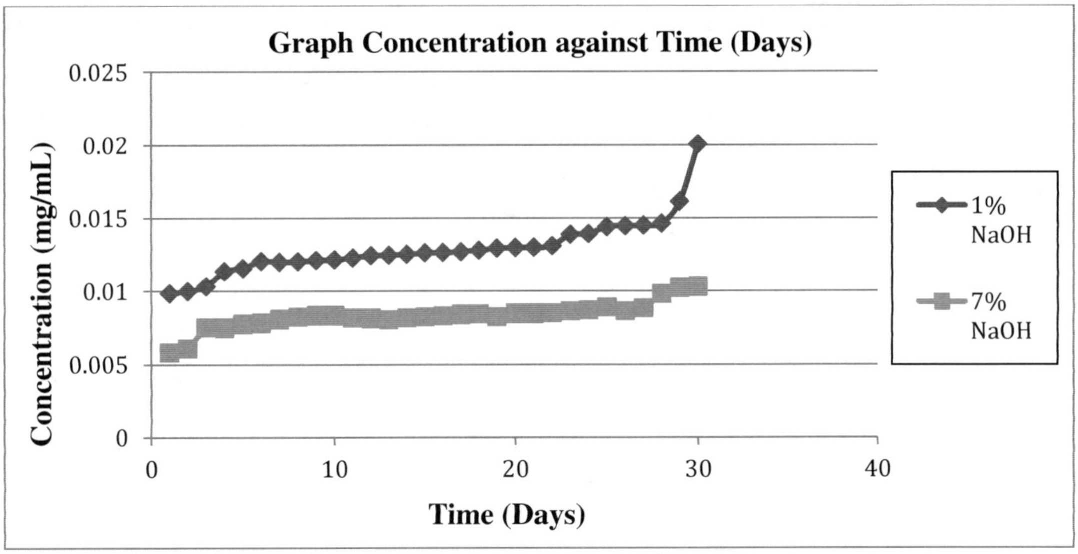

Figure 4. The release study curve from the graph concentration against time

4. Summary

In conclusion, the peak of epoxide group in the composite was identified at 875 cm^{-1} and for the isoprene at 835 cm^{-1}. The CH_2 peak of cellulose in the composite was obtained at 1457 cm^{-1} due to the C-H bending. However, the CH_2 is overlapped with cellulose at 1741 cm^{-1} and 1726 cm^{-1}. The release of urea from treated composite at 7% NaOH was higher than treated composite at 1% NaOH.

5. Acknowledgement

The authors would like to acknowledge UniMAP Short Term Grant Scheme (9001 - 00521). We would also appreciate Universiti Malaysia Perlis (UniMAP) for providing us all the facilities to make this research.

IOP Conf. Series: Journal of Physics: Conf. Series **1019** (2018) 012063 doi:10.1088/1742-6596/1019/1/012063

6. References

1. R. Hamzah, M. A. Bakar, O. S. Dahham, N. N. Zulkepli and S. S. Dahham, J. Appl. Polym. Sci., **133** (2016).
2. O. S. Dahham, R. Hamzah, M. A. Bakar, N. N. Zulkepli, S. S. Ting, M. F. Omar, K. Muhamad, and S. S. Dahham, Polymer Testing **65**, 10-20 (2018).
3. I. S. A. Rahim and M. A. Ahmad ***9***, 64–69 (2015).
4. O. S. Dahham, N. Z. Noriman, S. T. Sam, N. M. Al-Samarrai, Z. Shayfull, A. M. Alakrach and S. A. Elnaid, MATEC Web of Conferences **78**, 01075 (2016).
5. B. Ndazi, J. V. Tesha, S. Karlsson and E. T. N. Bisanda, Journal of Materials Science **41**, 6978–6983 (2006).
6. S. Turmanova, S. Genieva and L. Vlaev, International Journal of Chemistry **4**, 62–89 (2012).
7. G, Liu, L. Zotarelli, Y. Li, D. Dinkins and Q. Wang, *Controlled-Release Fertilizers Slow-Release Fertilizers Controlled-Release Fertilizers*, 1–7. (2014).
8. R. Syafri, I. Ahmad and I. Abdullah, Sains Malaysiana **40**, 749–756 (2011).
9. L. C. Medina, J. B. Sartain and T. A. Obreza, Hort Technology **19**, 13–15 (2009).
10. L. Xie, M. Liu, B. Ni, X. Zhang and Y. Wang, Chem. Eng. J. **167**, 342–348 (2011). (
11. B. Azeem, K. Kushaari, Z. B. Man, A. Basit and T. H. Thanh, J. Controlled Release **181**, 11–21 (2014).
12. R. Gandhimathi, S. Vijayaraj and M. P. Jyothirmaie, International Journal of Pharmaceutical Research & Analysis **2**, 72–78 (2012).
13. O. S. Dahham, R. Hamzah, M. A. Bakar, N. N. Zulkepli, S. S. Dahham and S. S. Ting, Polym. Test. **59**, 55-66 (2017).
14. N. A. Samad, R. Othaman and I. Abdullah, International Journal of Materials Engineering Innovation **5**, 61-69 (2014).
15. N. Z. Noriman, H. Ismail and A. A. Rashid, Journal of Applied Polymer Science, **126**, 56 (2012).
16. R. T. Conley, "Qualitative Analysis" in *Infrared Spectroscopy*, 2nd edited by Boston (Allyn and Bacon Inc., 1972), pp. 112-140
17. W. A. K. Mahmood, M. M. R. Khan and M. H. Azarian, J. of Non-Cryst. Solids **378**, 152-157 (2013).
18. M. Taherdanak and H. Zilouei, *Fuel* **115**, 714-719 (2014).

IOP Conf. Series: Journal of Physics: Conf. Series **1019** (2018) 012064 doi:10.1088/1742-6596/1019/1/012064

The Use of Augmented Reality Technology for Primary School Education in Perlis, Malaysia

Izwan Nurli Mat Bistaman[1], Syed Zulkarnain Syed Idrus[2], Salleh Abd Rashid[1]

[1] School of Human Development and Techno-communication (iKOM), Universiti Malaysia Perlis (UniMAP)
[2] School of Human Development and Techno-communication (iKOM), Green Advanced Computing and Technology (GREAT) Research Group – CEGeoGTech, Universiti Malaysia Perlis (UniMAP)

izwanz_80@yahoo.com, syzul@unimap.edu.my, salleh@unimap.edu.my

Abstract. The advanced of the technology evolving currently enable the use of innovative learning tools for various fields especially in education. Different technologies have been implemented in the educational sector to enhance the method of teaching and learning. Augmented Reality (AR) is a technology with vast potential and great pedagogical that offers new methods for the purpose of education. AR with computer generated system allows user to see the real world environment with virtual objects superimposed or composite in real world. Although AR will enhance the educational outcome, the domain specific, pedagogical and psychological aspects have to be considered for real use in class. The strategies such as collaborative learning were considered when designing and engaging an AR learning environment. Collaboration is to work together from one to other party where they can communicate each other to learn and get some mutual benefits. This research gives a brief insight into the potential and challenge of Augmented Reality in education to enhance the method of teaching and learning from conventional to the technological by using the AR technology.

Keyword: Augmented Reality; Collaborative Learning; Learning Experience; Education.

1. Introduction

Education in Malaysia is overseen by the Ministry of Education where each state and federal territory has an education department to co-ordinate educational matters. Currently, the process of teaching and learning more to the traditional with verbal educational method and it is not necessarily effective [1]. The past studies have shown that student who received verbal method is very difficult to understand and to solve the problem compare than the explanation in verbal and visual method [2]. In fact, many of educational institution still use the ineffective method to deliver their lectures which will cause the lack of student interest. To measure the effectiveness of learning, retention (remember presented information) and transfer (able to use information to solve the problem) are the factors that can help students understand what they have learned. Thus, educators have to find the appropriate method in order to deliver the teaching to the student and to give better approaches to improve their learning experiences.

Primary school education in other word, the place for student development and playing, activities, experience and social interaction practiced. In Malaysia, primary education begins at age seven and lasts for six years which is also known as Standard 1 to 6. The primary school divided into two levels, year 1 to 3 are classified as Level One while years 4 to 6 are considered as Level Two. Generally, this place is to prepare student before enrolling to secondary school.

Learning experience is very important part in the process of teaching and learning for primary school education because of student in this age need to get more attention in order to increase their motivation and satisfaction in classroom [3]. The one way method of teaching and learning must be change to be more sense as such as sight, hear, sound and touch. The past researcher believed that the use of Augmented Reality technology as a promising tool to enhanced the student motivation and interest, then support the teaching and learning process in educational context.

The AR technology is a computer generated system that allowed user to see the objects in the real world environment. Refer to [4] Games can be a new approach to implement for children in term of learning compare than lectures. Regarding to the past studies, games can be a part of teaching programming to preschool children. Educators must be consider to bring this emerging technology to support and enhance learning by integrated with the use of computers, multimedia materials, internet, simulations game and immersive technology such as 3D virtual world and Augmented Reality [5]. The creative strategies should be implemented as such as organize sets of activities to support learning and at the same time to meet pedagogical practice [6].

Awareness of the critical process to acquiring, storing and applying the information towards student nowadays, the emergent technologies should be utilize for educational context to deliver the course content so that fit to the human cognition [7]. The evolving of technology recently, especially in audio and visual become a new approach to the educational field to implement and gives possible impact on teaching and learning process as well as to offer new environment for primary school education. Nevertheless, augmented reality technology has their potential compared then other technology as such as multimedia, game and online learning. Although AR has provided new opportunities to enhance the education system, it depends to the educators to identify the new technologies to ensure the improvement in education [8].

The subjects tested in Primary School are Malay Comprehension, Malay Writing, English Comprehension, English Writing, Science and Mathematics. Regarding to the Wikipedia, teaching is the imparting of knowledge by a teacher or other knowledgeable person. Teaching method comprises the principles and methods used for instruction. An application is a method that used or designed to help student to perform specific task. The teaching application is the method that the way how the teacher used to the student by using the materials that can support the teaching. The material is including documents, graphic, software, audio and visual that performed to the classroom. To ensure the process of teaching and learning running smoothly, collaboration among teacher and children must be perform as well. Previously, teaching and learning in today's classroom become complex and daunting [9].

The advanced of the technology enable the use of innovative learning tools for education such as Augmented Reality. Augmented Reality (AR) is a variation of Virtual Reality. AR allows the users to see the real world, with virtual objects superimposed upon or composited with the real world. AR for interactive learning in education is the greater context of immersive virtual learning environment. Through the use of advanced technology an empty space is turned into a very rich educational experience. AR commonly capable to overlay computer graphics onto the real world and allow users to see the real world at the same time as a virtual imagery attached to real location and objects [10].

Virtual Reality (VR) is an experiential medium, which is combines with advances in image processing, tracking mechanisms and intuitive human computer communication. To develop the educational application using this approach, domain specific, pedagogical and psychological aspects have to be considered for real use in classrooms [10]. Learning in classroom for children by adapting to the games is becoming effectiveness to the process of learning in term of make collaborative task among teacher and student. Collaboration is working together between one party and another party to seek mutual benefit. Collaborative task in education allow teacher and children to communicate each other and learn through play and exploration. Children also enjoy an enhanced experience by use the multiple input devices [11].

The Augmented Reality (AR) for interactive learning in education is a new approached in Malaysia. This research paper concentrated on the use of Augmented Reality for primary school education. The use of these applications, particularly in education has great potential to enhance teaching and learning and at the same time to create a collaborative task among teacher and student.

Currently, teaching and learning in Malaysia start from primarily to high school, where the teachers have total control of their student. The problem statement of this study are more to the process of teaching and learning in primary school, where the conventional methods used is not effective for children such as by using chalk to write, book to story and image to explain. The appropriate equipment is needed in response to instructor activity without the need for manual control [9]. Using AR paradigms for teaching student is becoming increasingly popular because student are moving towards a new level of inter-action with technologies and there is a need to student to educational contents through the use of attractive technologies. According to Jim Scrivener, the teacher's main role is to "help learning to happen" which includes "involving" students in what is going on "by enabling them to work at their own speed, by not giving long explanations, by encouraging them to participate, talk, interact, do thing, etc. [12].

A creative and comprehensive developmentally appropriate primary curriculum designed to build skills and facilitate option learning experiences, while enhancing self-concept and preparing student for future educational success. Nowadays, teaching and learning methods in primary school more to the one way process. The elements of 5 sense methods such as sight, hear and touch will be applied in term of teaching and learning. Students will be excited and able to adapt themselves to the class environment with fun and interactive games. It needs to meet the minimum quality in term of minimum inputs (facilities, teaching materials, qualified teachers and community support), processes (effective leadership, monitoring and evaluation accountability, community participation, effective teaching and learning and student assessment) and outputs such as high student learning, social and individual returns [13]. The curriculum will be divided into modules from low to high levels. So that, to enhanced traditional teaching and learning methods for primary school education in classroom for future using Augmented Reality tools is needed.

The rapid expansion of the education system will cause the improving quality of education by employing interactive teaching and learning process with the limited resources, which is also one of the cost effective ways of improving quality of education [14]. The emerging of AR in education system enable to reduce the previous problem and make it enhanced from traditional to the new technology using AR application and to create collaborative task among teacher and student. Augmented Reality (AR) is the ideal solution for all educational application and potential to explore. The improvements of the technology nowadays make me aware and try to enhance the educational fields with the effective ways. Based on the studies, AR has potential to engage, stimulate, and motivate student to explore class materials from different angles.

Using this application, teacher can help student to teach subjects where they could not feasibly gain real-world first-hand experience and to enhance collaboration between teacher and student. By using the technological approach, hopefully it can help to foster student creativity and imagination to the subject. Besides that, this method also can help student to take control of their learning at their own pace and on their own path. The adaptation of this application will change the process of teaching and learning where it can create an authentic learning environment suitable to various learning styles.

The research is focusing on the teaching and learning in education for Primary School Education. The research will be focuses on the best way to develop application system using AR in term of teaching and learning for Primary School education. Besides that, to enhance the conventional system to the technological approached in Augmented Reality environment with the evolving technology and to ensure that the process of learning become fun and effective.

Basically, lessons at primary school based on subjects that provided by the Ministry Of Education. For this study, Syllabus will be divided into modules and sub-modules for Science subject. The research will involve surveys, questionnaires and detailed studies involving framework and collaboration with the selected Primary Schools in Perlis.

IOP Conf. Series: Journal of Physics: Conf. Series **1019** (2018) 012064 doi:10.1088/1742-6596/1019/1/012064

Currently, the teaching method is based on the syllabus provided to the teacher where the lectures is more to explanation by using white board, books, still photo and other materials. The lacking of the previous method is where student only hear and try to understand the lesson. Unfortunately, level of the maturity among them is different where some of student can easily to understand, and some of them are difficult to understand. By using Augmented Reality tools, teaching and learning process is become fun where student can play game by using marker provided such as alphabet, then when the marker is show to the camera, the monitor will display that alphabet and including audio and video which is can make interactive learning and create learning experience.

2. Augmented Reality

Augmented Reality (AR) is a live, direct or indirect, view of a physical, real world environment whose elements are augmented by computer-generated sensory input such as sound, video, graphics or GPS data. AR interfaces offer seamless interaction between the real and virtual worlds, a tangible interface metaphor and a mean for transitioning between real and virtual worlds [15]. Augmented Reality (AR) is a variation of Virtual Reality (VR). AR allows the users to see the real world, with virtual objects superimposed upon or composited with the real world.

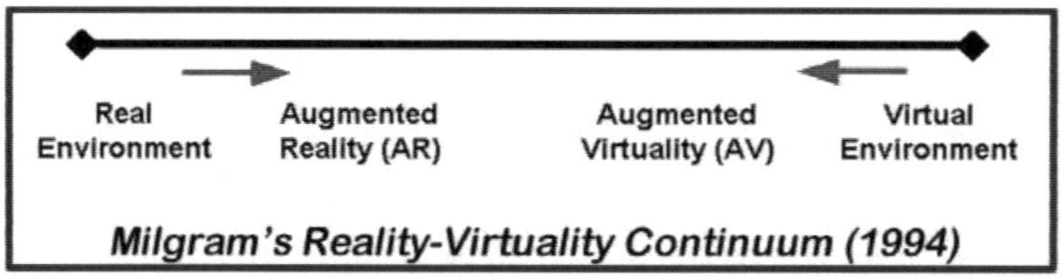

Figure 1. Milgram's Reality-Virtuality Continuum (1994)

Augmented Reality (AR) can be defined as a system of tools that allows a person to view one or more virtual objects in real world environment. AR technology allows for viewing things in real environment that impossible to others see the real world as an imagery attached to real location and objects. In other word, this application enables person to interact with the real world. AR is extension of Virtual Reality (VR) which is known as Mixed Reality (MR). AR is a kind of application and technology where by using computer generates image or object and mixed them into real environment. This technology has already been applied in various fields such as engineering, industrial design, military, medical science, education and others.

Currently, the term AR is defined differently by researchers. The article entitled Augmented Reality in Education by Mark Billinghurst, has defined AR as a situation where user can see virtual imaginary or objects in the real world by computer generated where the user view the world through a handheld or head mounted display (HMD) either see through or overlays graphics in a real time [16]. Furthermore, AR can categorize by three properties which is a combination of real and virtual objects, interactive in real time and 3D registration of the virtual and real objects [17]. However, AR is a type of application and technology through computer and generated the image and mixes them into real environment [18]. Figure 2 show the diagram of Augmented Reality system.

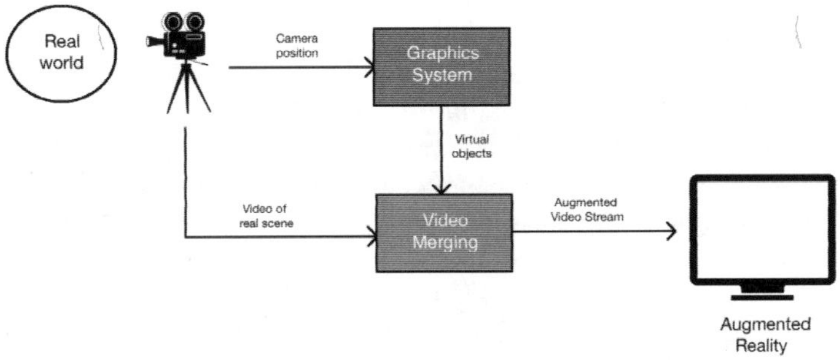

Figure 2. Augmented Reality System

Moreover, AR is quiet similar with Virtual Reality, because AR is a variation or extension of VR [19]. The AR will strength the visualization by image or object generated blend into the real environment [20]. The user can't see the real world around him in VR, while AR technology allowed the user to see the real world environment with the virtual image overlaid.

There are two taxonomies of AR application in term of development. Marker based AR and Markerless AR. Marker based AR need the marker registration the position of a virtual object to be displayed into the user perception of the real world. AR may operate with devices, software, sensor, camera and display device. Markerless AR operates by incorporating global positioning system (GPS) that identifies coordinates or locations.

3. Studies

The technological advances enable the use of innovative learning tools for education. To ensure the implementation of this technology in education more effectively, various factors need to be considered, particularly traditional teaching methods and technological methods using AR. Refers to the previous study entitled Traditional vs. Modern Teaching Methods : Advantages and Disadvantages of Each, the traditional method of teaching is deeply teacher-centered, where the methods puts the responsibility totally to the teacher and it is believed that, if students are present in the lesson and listen the explanation and examples, they successfully understand. According to this study, the traditional method is not the best mode of delivery for the purpose of communication [21]. Compared with the traditional method, modern methodology is much more student centered, where based on the research made, 12.5% of the teacher prefer traditional method, 25% of them demanded to use both, and 62.5% believe that modern technology is more effectively.

To achieve more effective in term of teaching and learning, the appropriate strategy should be implemented and requires the use of appropriate methodologies and pedagogies to meet the demands of the current generation of students, new technologies and changing the educational environments [14]. The advancement of technology, particularly in education has resulted in the teaching and learning process has been blended with a variety of methods and strategies. Refers to the past studies, Blended Learning (BL) or known as hybrid learning describes a learning environment that either the combination of teaching methods, delivery methods, media formats or a mixture of all these. This new dimension expected that to provide student with more control over learning. Students at this level are prefer something that would appeal to them, such as an interactive game and containing multimedia elements.

The introduction of graphics-oriented computer has great changes in education, particularly in animations. The use of multimedia elements in education such as using animated images will attract the children in term of helping them understand and remember the content. The object oriented animation features have better understanding compare with the complex explanation and it can help children learn faster and easier [22]. At the same time, this element will foster the teaching and learning process and increase the attention rates. Besides, a good educational technique in augmented reality learning system for kindergarten through the graphical images such as 2D, 3D and animation would make both of teacher and children to collaborate each other.

This research give a brief insight into the potential and challenges of using collaboration Augmented Reality (AR) in education within the greater context of immersive virtual learning environments. Nowadays, the conventional process of learning in Malaysia based on the MyGovernment.com the official website, the subject of education is truly vital to the establishment of a developed and knowledgeable society. In term of AR system development, the appropriate equipment or devices need to be considered and to ensure the system implementation successful. The element of devices which is consist of input device such as computer or laptop and webcam, monitor or screen as an output and AR tools software to generate objects to the real environment when the marker is shown to the webcam. Marker is very important part to design and the complexity of pattern should be reduced because of each marker is unique and it would facilitate the webcam to captures and identify the marker [18]. Users can operate the marker whether turn and move around as long as the printed surface must face to the webcam. The object would disappear when users occlude marker of the AR pattern with hand.

According to the enhance the traditional teaching and learning process methods in education, AR technology has matured to the point where it can be applied to a much wider range of application domains, and education is an area where this technology could be especially valuable. The educational experience offered by Augmented Reality is different for several reasons including to support of seamless interaction between real and virtual environments, to use of a tangible interface metaphor for object manipulation and the ability to transition smoothly between reality and virtuality [16]. The implementation of this technological method to enhanced the process of teaching and learning method, collaborative task should be consider in order to perform AR system.

To enhance the collaborative task in education using AR applications, one of the most important purposes of an educational environment is to promote social interaction among users located in the same physical space [10]. To become a successful of learner, communication and collaboration are essential part to ensure the involving of students in real world task through the dialogue and interaction among them, curriculum objectives will come alive where learning group in classroom can accomplish meaningful learning and learning experience to children [23]. Working with collaborative peer-work groups is more effective because student try to reach a common goal, sharing tools and activities. These elements show the positive children interdependence when they try to use marker together and learn something among them.

The visualization technique is interesting method that enable users especially children not only to observe how objects behave and interact, but also provide children with visual that they may capture, and preserve the essence of physical phenomena with effectively compare than do verbal descriptions [24]. Furthermore, to perform the AR in education specifically for Primary School, there are several things should be provided to ensure the process of teaching and learning through the AR application running effectively. Student in Primary school should be exposed to the AR application where teaching methods are more to the fun games to interact them. In response to the frustration with this situation, we need augmented the classroom technology with various sensors and computer processing [25].

The process of teaching and learning should be enhanced parallel with technology evolving to ensure children in future exposed to the varieties of aspect especially in augmented reality fields. Learning should be continued to ensure that our community is rich with useful knowledge, particularly in the field of AR and being able to create something new. Regarding to the past researcher, they believe many advantages when integrating AR technologies toward teaching and learning process. The AR technology enable supports seamless interaction between real and virtual environments and allows the use of a tangible interface metaphor for object manipulation [26]. Hence, AR technology will provide instructors with high students' understanding in the classroom by augmenting physical devices with virtual annotations and illustrations [27].

IOP Conf. Series: Journal of Physics: Conf. Series **1019** (2018) 012064 doi:10.1088/1742-6596/1019/1/012064

4. AR Collaboration

Based on the studies in [11], there are two different experiments were performed using the AR games as a prototype for answering two research hypotheses. First, the role of immediate feedback and size of the display on the levels of collaboration and the effects on learning and motivation levels is made. The first step was to make sure the prototypes were well tested. For this, they used the University's Laboratory, under different lighting conditions. The augmented reality would not function in totally dark environment because it is based on the camera recognition of the black and white printed marker.

The main procedure was divided into two issues and two different experiments aimed at studying those two issues. The issue are the effects of the game had on learning and the effects different design variables had on collaboration levels. This research also investigated the effect on student's motivation. This was also due to the importance that promoting successful collaborations has on the Primary School environment. The first experiment is about an initial school aimed at knowing if the game really aids student in novel knowledge acquisition. Each class went through three different phases. A pre-test phase, where students answer random set of questions about the subject being taught. It consisted of playing the game in groups of four children. Next, a post test was performed and the measure of different in results, which will be described in the next section. The second experiment was performed to assess the collaboration levels and to investigate the influence of the following variables on those levels of collaboration which are the display size and the immediacy of the feedback. During both experiments, and whenever the game play started, a screen of recording software was automatically launched and kept running in background. Therefore, the children were being digitally videotaped while they played the game.

The participants of the experiment aimed at assessing the effects on learning were different from the participants and school of the second experiment, aimed at investigating effects on collaboration levels. The participants of the first experiment were twenty-two children aged from 5 to 6 years old who interacted with the system in a collaboration setting using projector as display device. The participants of the second experiment were thirty-six different children aged from 3 to 5 years old, who played the game in groups of four or five.

Finally, based on the results, motivation levels were high because children never gave up the game. Even when feedback showed them they were wrong, nobody quit the game until reaching the solution. The children's overall reaction to the system was very positive in both experiments and most importantly, they conclude that the system didn't make the learning process go wrong. The results suggest that immediate feedback may play a role in increasing the number of collaboration behaviours and interactions among kindergarten children.

The research from [28] was designed Alien Contact! Simulation. The AR game developed for middle and high school to teach math, language arts, and scientific literacy skills. Basically, the AR game concept based to the Alien invasion where the student require to work in group four, carrying four roles which is a chemist, a cryptologist, a computer hacker and an FBI agent. Student had to solve the problem given by collaborating with their team and sharing information with them. This simulation game show the positive impact especially in collaborative task among them and increase their high level of student engagements.

SMART (System of Augmented Reality for Teaching) is a collaborative AR system designed for the purpose of teaching 2nd grade level student about different modes of transportation and types of animals. This study included a devises such as AR marker, laptop, software, web camera and projector. To achieve the target of study, they conducted with the 54 students (aged 7-8) from three local primary schools. For each school, the participants were divided into two groups. The First group, for control group which used a traditional teaching and learning method. Second group used the SMART system as a learning tool. The result shows the positive impact on students' motivation, learning experience and collaboration among student.

Based on the previous reviews, most of the researchers believed the AR technology will give potential for educational benefit and positive impact for student in order to deliver the teaching and learning process. The advantage of using AR technology to generate objects or virtual information in the real world environment is the unique technic for this technology. The AR provides a combination of real and virtual environments to the system that is unachievable for other technologies.

Besides, the AR technology enable the use of 3D object registration of the virtual and real objects in an AR allows users to view the learning content in 3D perspectives. Regarding to [11], the AR technology can help students who have difficulty visualizing complex learning concept and the AR can assist student to understand and learn new concepts with new environment that cannot be viewed in real world.

5. Conclusion

The augmented reality system has very positive impact on the whole class collaboration. Motivation, enjoyment and curiousity are important ingredients for any kind of educational game. Augmented reality technology and tangible interfaces are well accepted by today's Primary School Education children and by their teachers as well. Interactions among teacher and children with AR system will enhance the motivation and collaboration.

By utilizing proper instructional strategies, AR may provide students with numerous benefit and lead to the effective learning experiences. The AR technology also give the opportunities to create educational experience that are more engaging and attractive.

References

[1] Mayer R E 2003 *The Promise of Multimedia Learning: Using the Same Instructional Design Method Across Different Media, Learning and Instruction* vol.13, no.2, pp. 125-139

[2] Mayer R E 1997 *Multimedia learning: Are We Asking the Right Questions? Educational Psychologist* vol.32, pp.1-19

[3] Perez-Lopez D and Contero M 2013 *Delivering Educational Multimedia Contents Through An Augmented Reality Application: A Case Study On Its Impact On Knowledge Acquisition and Retention* (Tojet) vol.12, no.4

[4] Gibson J P 2003 *A Noughts and Crosses Java Applet to Teach Programming to Primary School Children. In Proc. of the 2nd Int. Conf. on Principles and Practice of Programming in java*, PPPJ, vol.42. Computer Science Press (New York: NY) pp.85-88.

[5] Diserio A, Ibanez M B and Kloos C D 2012 *Impact of An Augmented Reality System on Students Motivation for A Visual Art Course* Computers & Education pp.1-11

[6] Kreijns K, Acker F V, Vermeulen M & Buuren H V 2013 *What Stimulate Teachers To Integrate ICT in Their Pedagogical Practice? The Use of Digital Learning Materials in Education* Computer in Human Behavior, vol.29, pp.217-225

[7] Dror I 2008 *Technology Enhanced Learning: The Good, The Bad and The Ugly* Pragmati cognition vol.2, no.2, pp. 215-223

[8] Martin S, Diaz G, Sancristobal E, Gil R, Castro M & Peiri J 2011 *New Echnology Trends in Education: Seven Years of Forecasts and Convergence* Computer & Education, 57, 1893-1906. http://dx.doi.org/10.1016/j.compedu.2011.04.003

[9] Cooperstock J R 2001 *The Classroom of the Future: Enhancing Education through Augmented Reality Centre for Intelligent Machines* (McGill University)

[10] Kaufmann H 2011 *Journals Collaborative Augmented Reality in Education* Institute of Software Technology and Interactive Systems (Vienna University of Technology: Favoritenstrasse) 9-11/188.

[11] Campos P, Pessanha S, Jorge J 2011 *Fostering Collaboration in Kindergarten Through an Augmented Reality Game The Int. Journal of Virtual Reality* 2011 10(3):33-39.

[12] Scrivener J 2005 *Learning Teaching* (Oxford:Macmillan)

[13] Lockheed M E and Verspour R 1991 *Improving primary Education in developing countries* (London: Macmillan)

[14] Serbessa D D 2006 *Tension between Traditional and Modern Teaching-Learning Approaches in Ethiopian Primary Schools* (CICE Hiroshima University) *Journal of Int. Cooperation in Education* Vol.9, No.1, (2006) pp.123-140

[15] Billinghurst M, Diinser A 2011 *Evaluating Reality Systems. Handbook of Augmented Reality*. pp. 289-307.

[16] Billinghurst M 2002 *Augmented Reality in Education* New Horizons for Learning.

[17] Kaufmann H 2003 *Collaborative Augmented Reality in education in Proceeding of Imagina 2003 Conference* (Monte Carlo: Monaco) pp.1-4

[18] Hsieh M C, Lee J S 2008 *AR Marker Capacity Increasing for Kindergarten English Learning Proc. of the Int. MultiConf. of Engineers and Computer Scientists 2008* Vol. 1 IMECS 2008, 19-21 March, 2008, (Hong Kong)

[19] Azuma R T 1997 *A Survey of Augmented Reality* Presence-Teleoperators and Virtual Environments vol.6 no.4 pp 355-385

[20] Hsieh M C and Koong Lin H C 2011 *A Conceptual Study for Augmented Reality E-Learning System Based in Usability Evaluation* CISME Vol.1 No.8, pp 5-7

[21] Boumova B V 2008 *Traditional vs Modern Teaching Methods: Advantages and Disadvantages of Each* Department of English and American Studies. English Language and Literature. (Masaryk university: Faculty of Arts)

[22] Chee H M, Onn C W, Sun Y G 2009 *Preliminary Study For The Usage of Multimedia in Education.* Symp. on Progress in Information & Communication Technology 2009 (Faculty of Information and Communication Technology, Universiti Tunku Abdul Rahman: Malaysia)

[23] Tinzmann M B, Jones B F, Fennimore T F, Bakker J, Fine C and Pierce J 1990 *New Learning and Thingking Curricula Require Collaboration* Retrieved from American Institute for Research website http://www.ncrel.org/sdrs/areas/rpl_esys/collab.htm

[24] Akpinar Y and Bal V 2006 *Teachers' Collaborative Task Authoring to Help Students Learn a Science Unit* Educational Technology & Society, 9 (2), 84-95.

[25] Shelton B E 2002 *Augmented Reality and Education, current Projects and the Potential for Classroom Learning by New Horizzons for Learning* Vol. 9 No. 1.

[26] Singhal S, Bagga S, Goyal P & Saxena V 2012 *Augmented Chemistry: Interactive Education System Inter. Journal of Computer Applications* http://dx.doi.org/10.5120/7700-1041

[27] Coffin C, Bostandjiev S, Ford J & Hollerer T 2008 *Enhancing Classroom and Distance Learning Through Augmented Reality*

[28] Dunleavy M, Dede C & Mitchell R 2009 *Affordances And Limitations of Immersive Participatory Augmented Reality Simulations for Teaching and Learning* Journal of Science Education and Technology 18 (1), 7-22, (Springer Netherlands)

IOP Conf. Series: Journal of Physics: Conf. Series **1019** (2018) 012065 doi:10.1088/1742-6596/1019/1/012065

Chemical and Hydrophobic Properties of PLA/HNTs-ZrO_2 Bionanocomposites

A M Alakrach[1], N Z Noriman[1], Omar S Dahham[1], R Hamzah[1], Mohammed A Alsaadi[2], Z Shayfull[3,4], S Z Syed Idrus[5]

[1]Center of Excellence Geopolymer and Green Technology (CEGeoGTech), Faculty of Engineering Technology (FETech), Universiti Malaysia Perlis (UniMAP), Level 1 Block S2, UniCITI Alam Campus, Sungai Chucuh, Padang Besar, 02100, Perlis, Malaysia
[2]Nanotechnology & Catalysis Research Centre (NANOCAT), University of Malaya, 50603 Kuala Lumpur, Malaysia
[3]School of Manufacturing Engineering, Universiti Malaysia Perlis, Arau, 02600, Malaysia
[4]Green Design and Manufacture Research Group, Center of Excellence Geopolymer and Green Technology (CEGeoGTech),Universiti Malaysia Perlis, Kangar, 0100, Malaysia
[5]School of Human Development and Techno-communication (iKOM), Green Advanced Computing and Technology (GREAT) Research Group – CEGeoGTech, Universiti Malaysia Perlis (UniMAP)

Email: niknoriman@unimap.edu.my

Abstract. The utilization of biopolymers in the applications that meet demands in society for sustainability and environmental safety has been limited. That is owing to the poor mechanical and thermal properties of biopolymers. The goal of this study is to improve the limitations of the biopolymers properties. polylactic acid (PLA) as one of the most important biopolymer was selected to investigate the effect of incorporating of natural halloysites nanotubes HNTs with different concentrations (0–8 wt%) on PLA chemical and hydrophobic properties. In the second series, Zirconium dioxide (ZrO_2) nanoparticles were added to the PLA/HNTs film with the optimum mechanical properties in the first series. the chemical structure and surface wettability were investigated for both of PLA/HNTs and PLA/HNTs-ZrO_2 bionanocomposites. The addition of ZrO_2 resulted in appearance of OH groups in the FTIR graph. In addition, the increasing of ZrO_2 loading led to increase the hydrophobicity of the PLA nanocomposites.

1. Introduction
Recently, the environmental effect of non-degradable plastic packaging wastes is raising general universal concern. These concerns caused an increased demand for biodegradable packaging materials from renewable sources (biopolymers) as an alternative to artificial plastic packaging materials. In addition to the packaging applications, biopolymers can be applied to a wide range of applications such as agriculture, electronic, outdoor furnishings, automotive industry, and construction. One such polymer is PLA, has been extensively investigated over the last several decades as a biocompatible, biodegradable, thermoplastic polyester and derived from renewable resources [1, 2]. However, PLA, like other biopolymers, exhibits some weaknesses in some properties like low impact strength, poor gas barrier, low heat resistance, and slow crystallization rate [3, 4].

The incorporation of micro or nano-fillers into PLA is a promising method to terminate the weak properties of PLA. Different kinds of nanoparticles have been investigated to improve PLA properties such as CNT, Organoclay, Kaolinite, graphite, etc [5]. Owing to its multi-purpose features, naturally occurring tubular halloysite clay (HNTs) has attracted many interests of the researchers. These nanomaterials can be utilized as nano-filler for biopolymers. The development of thermoplastics matrices with HNTs has been investigated extremely by different research groups. HNTs can used to enhance the characteristics of polymers such as mechanical strength, thermal stability, crystalline behavior etc [6].

Unluckily, the applications of halloysite nanotubes (HNTs) in the natural situation is currently limited because of its highly hydrophilic internal and external surfaces, HNTs tend to agglomerate into clusters during the processing of nanocomposites. the interfacial adhesion of HNTs and PLA is not ideal because of the existing of hydroxyl groups only on the sides of unmodified HNTs [7, 8].

Owing to its good mechanical and biocompatibility properties, ZrO_2 is one of the most important metal oxides, which utilized extensively in different applications such as various implant devices and esthetic devices. ZrO_2 demonstrates variety of advantages like corrosion resistance, mechanical strength and fracture toughness. The enhanced properties of ZrO_2 endorsed to high and nano crystallinity of ZrO_2 which resulted to better biocompatibility and bioactivity nature [9, 10]. Recently, ZrO_2 nanoparticles have attracted the researchers in many attractive scientific and technological aspects in different research areas such as like gas sensors, solid fuel cells, high durability coating, catalytic agents, etc. This variety of applications according to the good mechanical and electrical properties, great dielectric constant, and wide band gap of ZrO_2 [11].

In this work, halloysite nanotube (HNTs) and ZrO_2 nanoparticles were used as hybrid fillers to enhance the properties of polylactic acid PLA. The hydrophilic properties and wettability of PLA/HNTs and PLA/HNTs-ZrO_2 were studied by contact angle measurements and the structure of the nanocomposites investigated by FTIR.

2. Experimental

2.1. Materials

Raw materials that used in this study are listed in Table 1. All raw materials are listed along with their functions and respective suppliers.

Table 1. List of raw materials, their functions and suppliers.

Raw materials	**Function**	**Supplier**
Polylactic Acid (PLA)	Biodegradable polymer used as a matrix.	Nature Works, LLC (MN).
Halloysites Nanotubes (HNT)	Enhancing the mechanical and thermal properties of polymer	Sigma-Aldrich chemicals company
Zirconia	For enhancing PLA properties.	Merck, India Pvt. Ltd.
Chloroform solutions	solvent	EM Science (NJ)
Distilled water	Wash the beaker	Labratory Unicity Alam

2.2. Preparation of ZrO_2 Nanoparticles

Zirconium isopropoxide (ZrTIP), acetic acid glacial, and deionized water were used to synthesize ZrO_2 via the sol-gel method. Initially, a ZrTIP (2.8 ml; 97%) solution was mixed in concentrated/pure acid acetic acid glacial (6 ml; >90%). After that, ZrTIP was titrated slowly into the acetic acid glacial solution. Next, DI water (36 ml) was slowly added under vigorous stirring at an ambient temperature and mixing/rate of 350 - 500 rpm for 5 hours. Once a clear solution was obtained (sol – transparent solution), the mixture was stirred for another hour at 80 °C. To keep the reaction temperature constant, the beaker was covered with an aluminum foil and a thermometer was placed near the surface of the mixture. The solution was subsequently converted to a gel mixture at a reaction temperature of 60 °C. The continuation of the reaction will eventually result in a white suspension, which was subsequently dried overnight (12 hours at 80 °C) until it is amorphous (to obtained smallest crystal size structure from drying process), resulting in dried ZrO_2 nanoparticles (yellowish color). Next, the ZrO_2 nanoparticles was manually grounded using an agate mortar. ZrO_2 powder (0.70 g) was calcined for 5 hours at 500 °C, and directly stored in a controlled humidity and temperature conditions. This method was utilized by Amir et al. (2015) [12].

2.3. Preparation of PLA/HNTs-ZrO_2 Nanocomposites Films

PLA/HNTs-ZrO_2 films were prepared using the method of solution casting with some modifications to the method utilized by De silva *et al.* [13]. PLA/HNTs solutions were prepared by dissolving 3 g of PLA pellets in 60 ml pure chloroform. Then, the solution was stirred vigorously for 4 hours. The 2 wt%, 4 wt%, 6 wt% and 8 wt% of HNTs were added to the PLA/chloroform solution and stirred for 24 hours. After that the solutions were sonicated for 30 minutes. The mixture was then poured into petri dishes and left for 24 hours to dry and evaporate. The obtained PLA/HNTs nanocomposites were named as PLA/HNT0, PLA/HNT2, PLA/HNT4, PLA/HNT6 and PLA/HNT8. Once the mixture had evaporated, the plastic film was removed and collected from the petri dish.

For the second series, the 1 and 3 wt% of zirconia (ZrO_2) concentration were prepared by dissolving them in the solution of optimum value of PLA/HNTs formulation. The solution was stirred for 24 hours to allow the dispersion and then sonicated it for 30 minutes. The mixture was then poured into petri dishes and left to evaporate. When the mixture had evaporated, the plastic film was removed and collected from the petri dish.

Table 2. Formulation of PLA/HNTs-ZrO_2 nanocomposites films.

Sample code	(PLA) (wt%)	HNTs (wt%)	ZrO_2 (wt%)
Neat PLA	100	0	0
PLA/HNTs 2	98	2	0
PLA/HNTs 4	96	4	0
PLA/HNTs 6	94	6	0
PLA/HNTs-ZrO_2 -1	95	4	1
PLA/HNTs-ZrO_2 -3	93	4	3

2.4. Fourier Transform Infrared Spectrometer (FTIR)

Chemical properties for series 1 and series 2 were done by using fourier transform infrared spectroscopy (FTIR). This FTIR analysis was conducted to identify the chemical interaction occurring inside the PLA/HNTs and PLA/HNTs-ZrO_2 films and the chemical composition uniformly of the samples. All samples were examined using the Thermo Scientific IS10.

2.5. Surface Wettability

Contact angle (θ) measurements of the PLA/HNTs and PLA/HNTs-ZrO_2 films were carried out using a goniometer (Rame - Hart model 250 - Standard Goniometer- Tensiometer equipment) under room temperature. The medium was water and the droplet size was 4 μL. A total of 10 measurements at different places on the films were taken for each HNTs and ZrO_2 concentration.

3. Results and discussions

3.1. Fourier Transform Infrared Spectroscopy (FTIR)

First, ATR-FTIR analyses were carried out for the neat PLA sample. Figure 1 shows that IR spectrum of PLA has mainly five regions; the C–C stretching at 871 cm^{-1}, the C–O stretching peaks at 1265, 1180, 1129, 1090 1045 cm^{-1}, the C–H deformation peaks at 1360, 1370, 1384 1454 cm^{-1}, C=O ester carbonyl groups at 1755 and 1760 cm^{-1}, and the C–H stretching at 2995, 2944 and 2985 cm^{-1} [14].

FTIR spectra of PLA/HNTs films revealed that some HNTs functional groups contributed to the nanocomposite film including O–H stretching of inner-surface hydroxyl groups 3695 cm^{-1}, O–H stretching of inner hydroxyl group 3621 cm^{-1} and O–H deformation of inner hydroxyl groups 911 cm^{-1}

[15-17]. In addition, the Si–O (in-plane stretching) at 1006 cm^{-1} of HNTs appeared as a shoulder to the vibration band at 1043 cm^{-1} of PLA spectra for the nanocomposite film. FTIR spectra peaks of few functional groups of HNTs such as perpendicular Si–O stretching at 1118 cm^{-1} and 750 cm^{-1} and symmetric stretching of Si–O at 799 cm^{-1} are not visible in the full range spectra of nanocomposite film due to their lower intensities compared to the other main peaks.

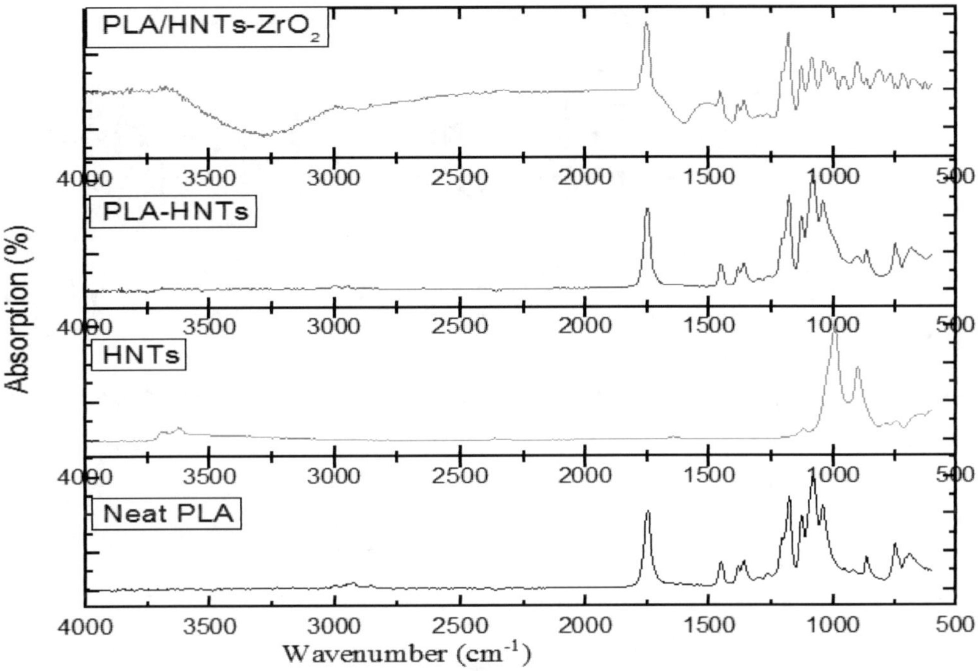

Figure 1. FTIR spectra of neat PLA, pure HNTs, PLA/HNTs and PLA/HNTs-ZrO_2 nanocomposites.

The addition of ZrO_2 nanoparticles to PLA/HNTs nanocomposites film, led to appearance of the vibrations of the O-H bond in the range of 3136–3857 cm-1. this is due to sorption of H_2O molecules on ZrO_2 surface. The peaks which appear at 1764, 1634 and 1385 cm-1 are due to the bending vibration of Zr–OH groups [18].

3.2. Surface Wettability

Figure 2 demonstrates the plot of contact angle versus HNTs loadings in the PLA/HNTs nanocomposites films. The measured θ value of neat PLA films 67.27 was close to the other reported y values of PLA films prepared by casting solution method. De Silva et al. revealed that the only significant difference of y values is between 2 and 7 wt % and it indicates that PLA/HNTs films became more hydrophobic at high HNTs loading. When adding 2 and 4 w/w % of HNTs into PLA matrix, the polarity of the composites increased due to the presence of OH groups of HNTs. Consequently, surface energy of the composite films increased and therefore θ values tend to decrease slightly at these two concentrations compared to neat PLA film. When HNTs loading increased to 8 wt%, the contact angle values slightly increase. Owing to increase HNTs loading to high values, the surface roughness increases, the material's surface energy reduces and becomes hydrophobic [19].

Figure 3 shows the plot of contact angle of PLA/HNTs-ZrO_2 nanocomposites versus ZrO_2 loading. The results revealed that the contact angle decreased with the increasing of ZrO_2 loading. This result might be explained by the polarity of ZrO_2. Which with the increasing of ZrO_2 the film became more hydrophilic, that caused to decrease the contact angle values.

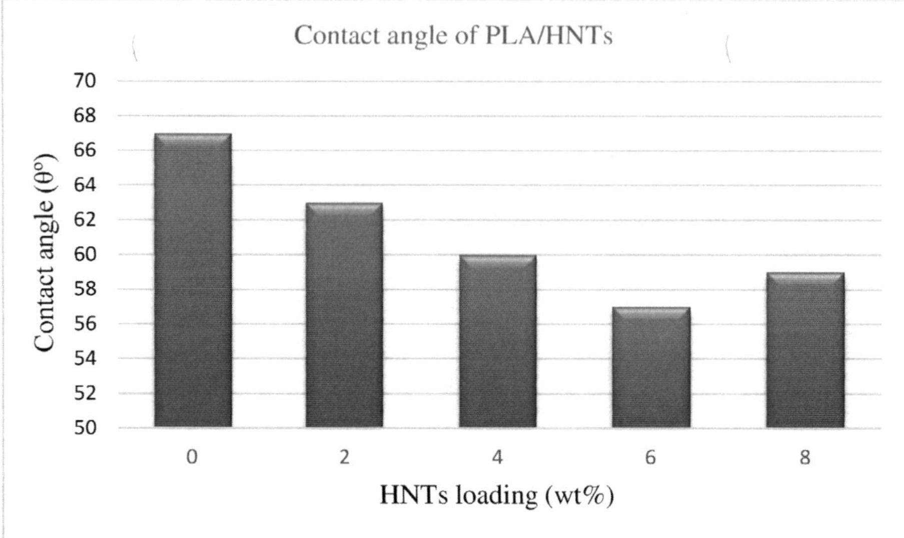

Figure 2. Plot of contact angle (θ) versus HNTs loading of PLA/HNTs nanocomposites films.

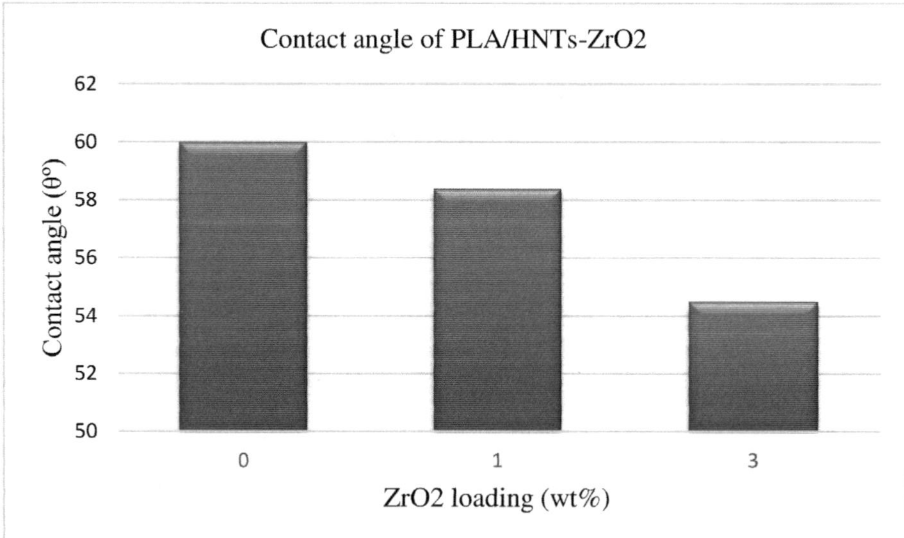

Figure 3. Plot of contact angle (θ) versus ZrO_2 loading of PLA/HNTs nanocomposites films.

Acknowledgments
The authors would like to acknowledge the Fundamental Research Grant Scheme (FRGS) Phase 1/2015 (9003-00523) under Ministry of Higher Education, Malaysia. The authors would appreciate Universiti Malaysia Perlis (UniMAP) for providing us all the facilities to make this research.

References

[1] Khalf A and Ward A 2010 Mater. Des. 31 2414-2421
[2] S. T. Sam, M. A. Nuradibah, H. Ismail, N. Z. Noriman and S. Ragunathan, Polymer-Plastics Technology and Engineering, **53**, 631-644 (2014).
[3] Zubir NHM, Sam ST, Zulkepli NN and Omar MF 2017 Polymer Bulletin. 3 1-16
[4] Obuchi S and Ogawa S 2010 Poly (lactic acid): Synthesis, Structures, Properties, Processing,

and Applications.(Hoboken, NJ, USA) 457-467
[5] Gonçalves C, Gonçalves IC, Magalhães FD and Pinto AM. 2017 Polymers. 9 269
[6] Tham WL, Poh BT, Ishak ZM and Chow WS 2014 J. Therm. Anal. Calorim. 118 1639-1647
[7] Papoulis D, Komarneni S, Nikolopoulou A, Tsolis-Katagas P, Panagiotaras D, Kacandes HG, Zhang P, Yin S, Sato T and Katsuki H 2010 Appl. Clay. Sci. 50 118-124
[8] Yuan P, Southon PD, Liu Z, Green ME, Hook JM, Antill SJ and Kepert CJ 2008 J. Phys. Chem. C 112 15742-15751
[9] Pielichowska K, Bieliński D, Dworak M, Kilian E, Macherzyńska B and Błażewicz S. 2017 Int. J. Polymer. Sci. 2017
[10] Pradhaban G, Kaliaraj GS and Vishwakarma V 2014 Prog. Biomat. Sci. 3 123-130
[11] Maskell WC 2000 Solid State Ionics. 134 43-50
[12] Amir I, Nur M, Julkapli NM, Bagheri S and Yousefi AT 2015 REV INORG CHEM. 35 151-178
[13] Viswanathan V, Laha T, Balani K and Seal S. 2006 Mater. Sci. Eng., R 54 121-285
[14] Sarkar D, Mohapatra D, Ray S, Bhattacharyya S, Adak S and Mitra N 2007 Ceram. Int. 33 1275- 1282
[15] N. N. Zulkepli and H. Ismail, Polymer-Plastics Technology and Engineering, **51**, 350-357 (2012)
[16] O. S. Dahham, R. Hamzah, M. A. Bakar, N. N. Zulkepli, S. S. Ting, M. F. Omar, K. Muhamad and S. S. Dahham, Polymer Testing, **65**, 10-20 (2018)
[17] O. S. Dahham, R. Hamzah, M. A. Bakar, N. N. Zulkepli, S. S. Dahham and S. S. Ting, Polym. Test. **59**, 55-66. (2017)
[18] Singh AK and Nakate UT. 2014 Sci. World J. 21 2014 [15] Oliaei E and Kaffashi B 2016 Polymer 8 104-114
[19] Oliaei E and Kaffashi B 2016 Polymer 8 104-114

IOP Conf. Series: Journal of Physics: Conf. Series **1019** (2018) 012066 doi:10.1088/1742-6596/1019/1/012066

The Effects of Tensile Properties of PLA/HNTs-ZrO_2 Bionanocomposites

A M Alakrach[1], N Z Noriman[1], Omar S Dahham[1], R Hamzah[1], Mohammed A Alsaadi[2], Z Shayfull[3,4], S Z Syed Idrus[5]

[1]Center of Excellence Geopolymer and Green Technology (CEGeoGTech), Faculty of Engineering Technology (FETech), Universiti Malaysia Perlis (UniMAP), Level 1 Block S2, UniCITI Alam Campus, Sungai Chucuh, Padang Besar, 02100, Perlis, Malaysia
[2]Nanotechnology & Catalysis Research Centre (NANOCAT), University of Malaya, 50603 Kuala Lumpur, Malaysia
[3]School of Manufacturing Engineering, Universiti Malaysia Perlis, Arau, 02600, Malaysia
[4]Green Design and Manufacture Research Group, Center of Excellence Geopolymer and Green Technology (CEGeoGTech),Universiti Malaysia Perlis, Kangar, 0100, Malaysia
[5]School of Human Development and Techno-communication (iKOM), Green Advanced Computing and Technology (GREAT) Research Group – CEGeoGTech, Universiti Malaysia Perlis (UniMAP)

Email: niknoriman@unimap.edu.my

Abstract. The applications of biopolymers that are requested in the society for sustainability and environmental safety have many limitations. That is can be attributed to the limited mechanical and thermal properties of biopolymers. The aim of this study is to diminish the limitations of the biopolymers properties. Polylactic acid (PLA) as one of the famous biopolymers was selected to investigate the effect of incorporating of hybrid fillers, natural halloysites nanotubes HNTs and Zirconium dioxide (ZrO_2) nanoparticles. In the first stage, HNTs was added to PLA matrix in different loading (0, 2, 4, 6 and 8 wt%). In the next stage ZrO_2 was incorporated to PLA/HNTs in two loadings (1 and 3 wt%). The mechanical properties (tensile strength, elongation at break and Young's modulus) were investigated for both of PLA/HNTs and PLA/HNTs-ZrO_2 bionanocomposites. PLA films incorporated with 4 wt% loading of HNTs resulted in optimum mechanical properties due to the homogenous distribution of halloysite nanoclay, which increased the tensile strength and elongation at break by 70% and 27% respectively. Unfortunately, the addition of ZrO_2 reduced the mechanical properties of PLA/HNTs-ZrO_2.

1. Introduction

Polylactic acid (PLA) is one of the promising biodegradable polymer material, owing to its unique properties such as high modulus, high strength, good biocompatibility and compatibility, and high optical transparency. Consequently, PLA is undoubtedly one of the good candidates for the replacement of petroleum based polymer in a lot of applications. Recently, various industrial applications started using PLA such as food packaging, automotive parts, and medical devices.

However, PLA exhibits low thermal stability and poor resistance properties, Thus, the conception of a bionanocomposite has been initiated as one of the solutions to improve the properties of the biopolymers [1-3]. Bio-nanocomposites usually are produced by the combination between a biopolymer and a nanofiller. The good distribution of nanofiller into the biopolymer matrix plays a vital role in the improvement of the biopolymer properties, such as physico-chemical mechanical, thermal, and barrier properties [4]. Numerous kinds of nanofiller have been incorporated to prepare PLA bio-nanocomposites, these includes organoclay [5], CNT [6], and graphitic nanofiller [7], etc.

Natural Halloysite nanotubes (HNTs) with the chemical structure (Al2Si2O5(OH)4.2H2O) is a 1:1 layered silicate clay mineral where each tube consists of an outer tetrahedral (Si–O) sheet and an outer octahedral (Al–OH) sheet. Recently, Halloysite nanotubes (HNTs) have engaged the researchers' attention as a new category of nanofiller for reinforcing the mechanical, thermal, crystallization, fire and other polymers properties such as PP, PS, PA6, PA12 and epoxy resins, etc. [8, 9]. HNTs are compatible and durable in the high levels of shear mixing. In addition, they are available in huge amounts, as a result, halloysites are much more affordable than CNTs and graphene [10]. On the other hand, HNTs tend to be agglomerated during the preparation of polymer/HNTs nanocomposites, that is because of the negative charge on the outer surface and the large surface energy of HNTs [11].

Surface modifications to HNTs may give the chance to expand the basal spacing of HNTs by the intercalation of inorganic and organic additives either on the outer surface or in their interlayers, which may help in the producing of homogeneous mixture of HNTs with polymers during the preparation. Furthermore, precipitating of suitable nanoparticles on the HNTs surface also provide good wetting and bonding of HNTs with polymers [12].

Zirconium dioxide or Zirconia (ZrO_2) material has unique characteristics such as excellent thermal stability, high refractive index and high hardness. The nanoparticles zirconia has been utilized successfully nanocomposite industry with high refractive index, high hardness, and improved scratch resistance. ZrO_2 nanopowder was employed to decrease the thermal expansion coefficient of PES to be able for using in the connector of optical fiber. However, most of the studies that utilized ZrO_2 nanoparticles in the fabrication of polymer/ZrO_2 nanocomposites focused on the thermal stabilities and infrequently on the surface mechanical properties of the polymer/ZrO_2 nanocomposites [13].

In this study, ZrO_2 nanoparticles was prepared from zirconium isopropoxide (ZrIP) by sol-gel method, and incorporated into PLA/HNTs mixture to prepare a homogeneous PLA/HNTs-ZrO_2 nanocomposites. Moreover, PLA/HNTs-ZrO_2 thin films were fabricated using solution casting at different PLA/HNTs-ZrO_2 compositions.

2. Experimental

2.1. Preparation of ZrO_2 Nanoparticles

Zirconium isopropoxide (ZrTIP), acetic acid glacial, and deionized water were used to synthesize ZrO_2 via the sol-gel method. Initially, a ZrTIP (2.8 ml; 97%) solution was mixed in concentrated/pure acid acetic acid glacial (6ml; >90%). After that, ZrTIP was titrated slowly into the acetic acid glacial solution. Next, DI water (36 ml) was slowly added under vigorous stirring at an ambient temperature and mixing/rate of 350 - 500 rpm for 5 hours. Once a clear solution was obtained (sol – transparent solution), the mixture was stirred for another hour at 80 °C. To keep the reaction temperature constant, the beaker was covered with an aluminum foil and a thermometer was placed near the surface of the mixture. The solution was subsequently converted to a gel mixture at a reaction temperature of 60 °C. The continuation of the reaction will eventually result in a white suspension, which was subsequently dried overnight (12 hours at 80 °C) until it is amorphous (to obtained smallest crystal size structure from drying process), resulting in dried ZrO_2 nanoparticles (yellowish color). Next, the ZrO_2 nanoparticles was manually grounded using an agate mortar. ZrO_2 powder (0.70 g) was calcined for 5 hours at 500 °C, and directly stored in a controlled humidity and temperature conditions. This method was utilized by Amir et al. (2015) [14].

2.2. Preparation of PLA/HNTs-ZrO_2 Nanocomposites Films

PLA/HNTs-ZrO_2 films were prepared using the method of solution casting with some modifications to the method utilized by De Silva *et al.* [15]. PLA/HNTs solutions were prepared by dissolving 3 g of PLA pellets in 60 ml pure chloroform. Then, the solution was stirred vigorously for 4 hours. The 2 wt%, 4 wt%, 6 wt% and 8wt% of HNTs were added to the PLA/chloroform solution and stirred for 24 hours. After that the solutions were sonicated for 30 minutes. The mixture was then poured into petri dishes and left for 24 hours to dry and evaporate. The obtained PLA/HNTs nanocomposites were named as PLA/HNT0, PLA/HNT2, PLA/HNT4, PLA/HNT6 and PLA/HNT8. Once the mixture had evaporated, the plastic film was removed and collected from the petri dish. For series 2, the 1 and 3 wt% of zirconia (ZrO_2) loading were prepared by dissolving them in the solution of optimum value of PLA/HNTs formulation. The solution was stirred for 24 hours to allow the dispersion and then sonicated it for 30 minutes. The mixture was then poured into petri dishes and left to evaporate. When the mixture had evaporated, the plastic film was removed and collected from the petri dish.

Table 1. Formulation of PLA/HNTs-ZrO_2 nanocomposites films.

Sample code	PLA (wt%)	HNTs (wt%)	ZrO_2 (wt%)
Neat PLA	100	0	0
PLA/HNTs2	98	2	0
PLA/HNTs4	96	4	0
PLA/HNTs6	94	6	0
PLA/HNTs8	92	8	0
PLA/HNTs-ZrO_2 1	95	4	1
PLA/HNTs-ZrO_2 3	93	4	3

2.3. Tensile Testing

Tensile strength (TS), modulus of elasticity (E) and elongation at break (ε) of PLA/HNTs (series 1) and PLA/HNTs-ZrO_2) (series 2) were measure with TA TX Plus Texture Analyzer (Stable Micro System, UK) based on the ASTM D 882-02 standard. Strip of films (10.0 x 1.0 cm) were clamped between the manual grips (without sharp edge). Five strips of each HNTs loading and another five strips of each ZrO_2 loading was used.

3. Results and discussions

3.1. Tensile Properties of PLA/HNTs

Figure 1 shows the TS of PLA films at different loadings of HNTs nanoclay. Tensile strength measures the highest stress value that a film can resist being stretched before cracking. Figure 1 exhibits that TS increased with the increasing of HNTs up 0 to 4 wt%, and thus decreased slightly from 4 to 8 wt%. the optimum value of TS was 37 MPa at HNTs loading of 4 wt%, which the tensile strength increased by 43% compared to the neat PLA. However, the TS values of the PLA/HNTs bio-nanocomposite films were still high compared to the neat PLA films. The enhancement in tensile strength of PLA/HNTs bio-nanocomposites was attributed to the reinforcement provided by the good dispersion of halloysites nanotube. Better dispersed nanoclay may promote higher tensile strength due to an increase contact surface area and interaction between HNTs and the PLA matrix [16].

Similarly, to the tensile strength, Fig. 2 exhibits that the elongation at break (EAB) was increased up to 4 wt% of HNTs loading, therefore, the values of elongation at break started decreasing. This results can be explained as a Small quantity of HNTs could work as plasticizer of PLA, which dispersed the impact energy during the break. In addition, the reduction in toughness values of the PLA/HNTs nanocomposites with relatively high HNTs loadings should be attributed to the weakened interfacial interactions at high loadings of nanofillers [17].

The young's modulus (YM) or modulus of elasticity is the slope of the linear section of the stress-strain curve during the film undertakes the elastic deformation. The aim of measuring the YM is to determine the resistance deformation, which can be utilized to reflect the stiffness and strength of the film [18]. Figure 3 illustrates a similar trend with tensile strength(TS) which YM increased with the increasing of HNTs loading up to 4 wt%, thus decreased with increasing loading of HNTs from 4 to 8 wt%.

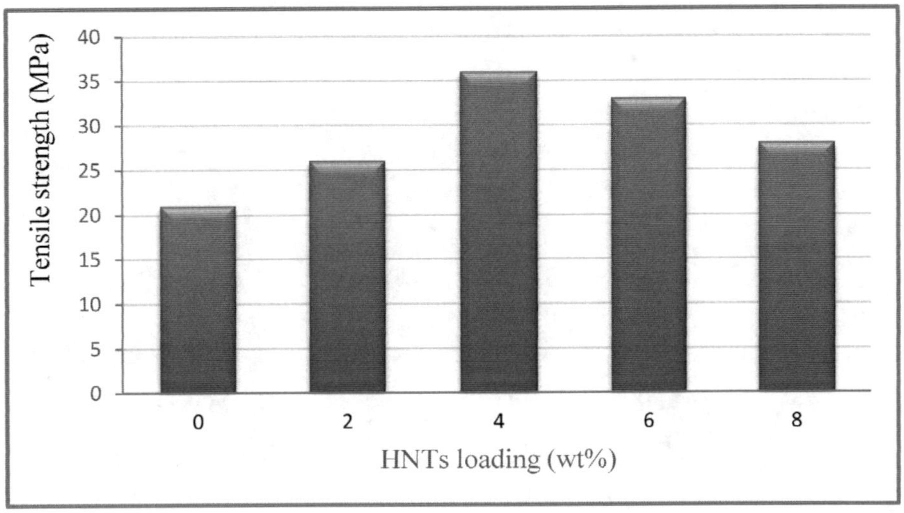

Figure 1. Tensile strength of PLA bionanocomposite films with different HNTs loadings.

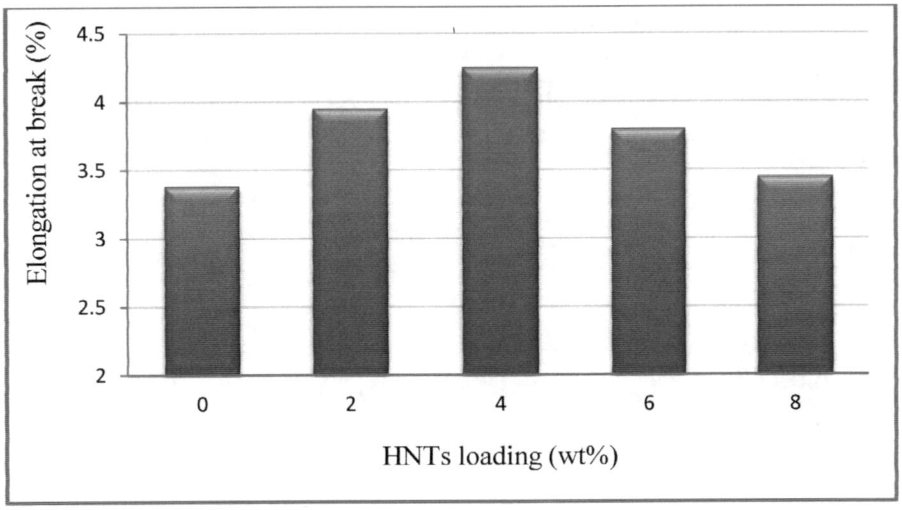

Figure 2. Elongation at break of PLA bionanocomposite films with different HNTs loadings.

3.2. Mechanical Properties of PLA/HNTs-ZrO_2

The effect of the addition of ZrO_2 on the mechanical properties (tensile strength, elongation at break and young's modulus) were shown in table 2.

The tensile strength (TS) of optimum 4 wt% PLA /HNTs film is the highest compared to 1 and 3 wt% of ZrO_2 loading film. In addition, with the increasing of ZrO_2 loading, the values of TS decreased. This reducing in TS values could be attributed due to the poor interfacial interaction between ZrO_2 and PLA.

The results exhibit a decreasing in the elongation at break with the increasing of ZrO_2 loadings, as shown in Table 2. Which the lowest value of elongation at break was measured at 3 wt% HNTs loading. The decreasing of elongation at break of PLA/HNTs-ZrO_2 with the increasing of zirconia loading can be explained by the increasing of the film rigidity.

Finally, the young's modulus of PLA/HNTs-ZrO_2 nanocomposites values also decreased with the increasing of ZrO_2 loadings. This could probably be attributed to the effect of stiffening of the ZrO_2 filler inside the PLA/HNTs film.

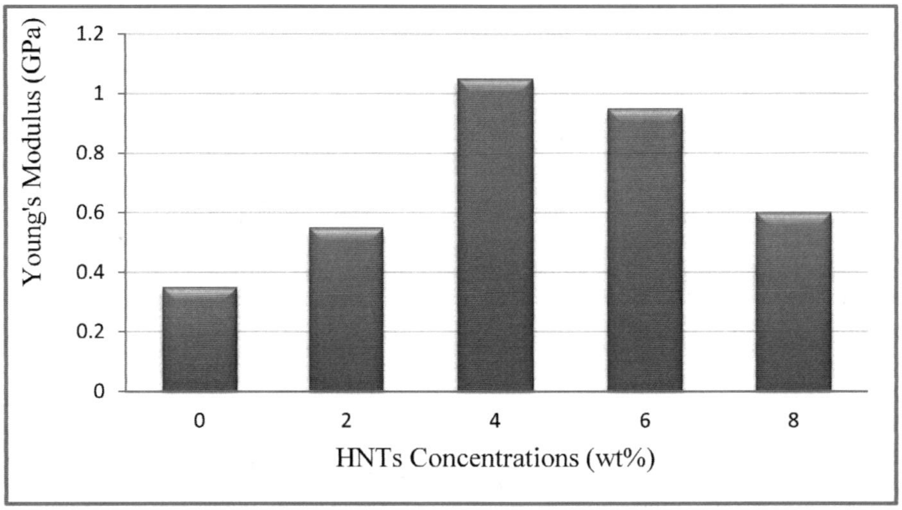

Figure 3. Young's modulus of PLA bionanocomposite films with different HNTs loadings.

Table 2. Mechanical properties of PLA/HNTs-ZrO_2

Sample	Tensile Strength MPa	Elongation at Break %	Young's Modulus GPa
PLA/HNTs	36	4.3	1.10
PLA/HNTs-ZrO_2-1	28	3.6	0.82
PLA/HNTs-ZrO_2-3	24	1.5	0.75

Acknowledgments

The authors would like to acknowledge the Fundamental Research Grant Scheme (FRGS) Phase 1/2015 (9003-00523) under Ministry of Higher Education, Malaysia. The authors would appreciate Universiti Malaysia Perlis (UniMAP) for providing us all the facilities to make this research.

References

[1] Tham WL, Chow WS, Poh BT and Mohd Ishak ZA 2016 *J.Compos. Mater.* **50** 3925-3934
[2] Yaacab ND, Ismail H and Ting SS 2016 *Procedia Chem.* **19** 757-762
[3] S. T., Sam, M. A., Nuradibah, H. Ismail, N. Z. Noriman and S. Ragunathan, Polymer-Plastics Technology and Engineering, **53**, 631-644 (2014).
[4] N. Z., Noriman and H. Ismail, H. International Journal of Polymeric Materials and Polymeric Biomaterials, **62**, 252-259 (2013).
[5] Shabanian M, Hajibeygi M, Hedayati K, Khaleghi M and Khonakdar HA 2016 *Mater. Des.* **110** 811-820
[6] Desa M, Zaidi MS, Hassan A, Arsad A, Arjmandi R and Mohammad NB 2016 *J. Appl. Polym. Sci.* **133**
[7] Yu WW, Zhang J, Wu JR, Wang XZ and Deng YH 2017 *J. Appl.Polym. Sci.* **134**
[8] Murariu M, Dechief AL, Peeterbroeck S, Bonnaud L and Dubois P 2012 *J. Polym. Environ.* **20**

932-943
[9] Pasbakhsh P, Churchman GJ and Keeling JL 2013 *Appl. Clay. Sci.* **74** 47-57
[10] Guo J, Qiao J and Zhang X 2016 *J. Appl. Polym. Sci.* **133**
[11] Deng S, Zhang J, Ye L and Wu J 2008 *Polym. J.* **49** 5119-5127
[12] Deng S, Zhang J and Ye L 2009 *Compos. Sci. Technol.* **69** 2497-2505
[13] Hu Y, Zhou S and Wu L 2009 *Polym. J.* **50** 3609-3616
[14] Amir I, Nur M, Julkapli NM, Bagheri S and Yousefi AT 2015 *REV. INORG. CHEM.* **35** 151-178
[15] De Silva RT, Pasbakhsh P, Goh KL, Chai SP and Chen J 2014 *J. Compos. Mater.* **48** 3705-3717
[16] Osman AF, Alakrach A, Kalo H, Dahham OS, Al Bakri Abdullah MM 2015 *AMM.* **754** 24-28
[17] Liu M, Zhang Y and Zhou C 2013 *Appl. Clay. Sci.* **75** 52-59
[18] Dahham OS, Noriman NZ, Sam ST, Marwa NA, Shayfull Z and Alakrach AM 2016 *MATEC Web of Conferences* **78** 01076

IOP Conf. Series: Journal of Physics: Conf. Series **1019** (2018) 012067 doi:10.1088/1742-6596/1019/1/012067

Recycled Natural Rubber Latex Gloves Filled Chloroprene Rubber: Effects of Compatibilizers

A M Alakrach[1], R Hamzah[1], N Z Noriman[1], Omar S Dahham[1], Z Shayfull[2,3], S Z Syed Idrus[4] , S Sudin[5]

[1]Center of Excellence Geopolymer and Green Technology (CEGeoGTech), Faculty of Engineering Technology (FETech), Universiti Malaysia Perlis (UniMAP), Level 1 Block S2, UniCITI Alam Campus, Sungai Chucuh, Padang Besar, 02100, Perlis, Malaysia
[2]School of Manufacturing Engineering, Universiti Malaysia Perlis, Arau, 02600, Malaysia
[3]Green Design and Manufacture Research Group, Center of Excellence Geopolymer and Green Technology (CEGeoGTech),Universiti Malaysia Perlis, Kangar, 0100, Malaysia
[4]School of Human Development and Techno-communication (iKOM), Green Advanced Computing and Technology (GREAT) Research Group – CEGeoGTech, Universiti Malaysia Perlis (UniMAP)
[5]School of Computer and Communication Engineering (SCCE), Green Advanced Computing and Technology (GREAT) Research Group – CEGeoGTech, Universiti Malaysia Perlis (UniMAP)

Email: rosnizahamzah@unimap.edu.my

Abstract. In this study, the effects of the compatibilizers; trans-polyoctylene (TOR) and epoxidized natural rubber (ENR-50), on the cure characteristics and tensile properties of recycled natural rubber latex gloves (rNR-G) filled chloroprene rubber (CR) were investigated. Blends consisting of rNR-G in a weight ratio 20 phr in CR and The compatibilizers TOR and ENR-50 at different concentrations, i.e. 2, 4, 6, 8 and 10 phr, were prepared using a two roll mill. Incorporation of TOR resulted in a reducing of the cure characteristics, meanwhile ENR-50 increased them. On the other hand, both of TOR and ENR-50 were improved tensile properties. These effects are attributed to the enhanced interfacial adhesion between rNR-G and phases through the compatibilizing effect induced by TOR and ENR-50, and the good dispersion of the rNR-G phase in the CR matrix. The results confirmed that TOR and ENR-50 could potentially be used as compatibilizers for the recycled natural rubber latex gloves filled chloroprene rubber.

1. Introduction

Synthetic rubber has been extensively utilized for many products in various environments. That is due to its unique characteristics such as flexible elasticity, energy absorption, and fluid resistance [1, 2]. However, the life cycle of rubber and its performance remarkably damages over time owing to several physical and chemical factors, such as temperature, humidity, ozone and radiation [3, 4]. For example, chloroprene rubber (CR) is one of the common rubbers, this rubber maintains its performance over time, and is used for high voltage electric wire covers and ignition line covers. Unfortunately, CR as other types of rubber becomes stiff and fragile after long-term utilizing under high-temperature conditions [3, 5].

At present, the natural rubber glove (NRG) becomes one of the most important necessities in the health demands. After its use, NRG is generally convert to waste rubbers which is difficult to biodegrade because polyisoprene chains of NR are crosslinked either by sulphur or peroxide system [6, 7]. The problem of NRG waste started in growing as a direct result of diseases diffusion and population growth. In addition to NRG waste from hospital and other toxic hazardous waste, more than 11% of NRG products considered as a rejected or defect products during the processing of latex rubber. All of the huge waste rubber quantities might cause serious environmental problem and great health risk to public [8-10]. To solve this problem, many research investigated the reuse of NRG waste in several applications such as mat and large tire [11].

Recently, the rNR-G started in use which blended with virgin synthetic rubber and NR [6, 8]. In this present work, the rNR-G waste was blended with CR by using TOR and ENR-50 as compatibilizers. As reported in the previous work concerning CR/ rNR-G, it was expected that the 20 phr addition of rNR-G resulted the optimum properties of CR/ rNR-G blend. As a result, this ratio was depended in this work to study the effect of different compatibilizers with different ratios.

2. Experimental

2.1. Materials

Recycled natural rubber latex gloves (rNR-G) waste was obtained from a local company, Top Gloves Co. Sdn Bhd, Malaysia. Chloroprene rubber (CR), trans polyoctylene (TOR) and epoxidized natural rubber (ENR-50) were purchased from RRIM Guthrie Group Sdn. Bhd. The compounding chemicals such as zinc oxide (activator), stearic acid (activator), sulfur (vulcanizing agent) and N-cyclohexyl-2-benzothiazole sulfonamide (CBS) (accelerator) were supplied by Anchor Chemical Co. (M) Ltd.

2.2. Formulation and testing of CR/ rNR-G compounds

rNR-G waste was cut and masticated by using two-roll mill (X(S)K-160x320) at room temperature for 30 minutes to break the crosslinking of the latex glove, to be able for grinding. After the grinding process, the rNR-G were sieved to obtained filler with particulate size of 300 – 600 μm (fine size), and the rest was considered as coarse size fillers. After preparation the gloves waste in suitable sizes, the compounding was done as shown in table 1.

Table 1. The formulation of rNR-G filled CR with compatibilizer

Ingredients	Compounds (phr)				
Sample	1	2	3	4	5
CR	100	100	100	100	100
Sulphur	2	2	2	2	2
CBS	1	1	1	1	1
ZnO	5	5	5	5	5
Stearic acid	2	2	2	2	2
rNR-G	20 (Based on results in series 1)				
Compatibilizers	2	4	6	8	10

2.3. Characterizations and testing of CR/ rNR-G compounds

After preparing all the specimens, the cure characteristics were done by using the Rheometer Monsanto Model MDR 2000. By the cure test, the minimum torque (ML), maximum torque (MH), cure time (t90), and scorch time (t2) were studied. In addition, the tensile test was studied according to the ASTM D412 method, which was done by using the universal testing machine Instron 5582. Tensile strength, elongation at break and the modulus at 100% elongation were investigated.

3. Results and discussion

Based on the results in the first series, 20 phr of fine rNR-G filler was chosen as the optimum loading and filler size to be implemented in this series. This series studies the effect of the type and concentration of compatibilizers on the 20 phr fine rNR-G filled CR. both of TOR and ENR-50 were utilized as compatibilizers with various concentrations (0, 2, 4, 6, 8 and 10 phr).

3.1. Cure Characteristics

Figure 1 shows the results of scorch time (t_2) for CR/rNR-G with different loading of compatibilizers (TOR and ENR-50). It can be released that there is an increasing trend of scorch time as the loading of ENR-50 increased. On the other hand, the scorch time of CR decreased as the levels of TOR within CR increased. This results can be attributed to the low molecular weight of TOR, that was led to lowering the viscosity of CR/rNR-G, therefore the processing of CR became easier, thus resulting in faster scorch time as more TOR was incorporated into the compound [12]. On contrast, ENR-50 possesses a high molecular weight, as a result, the incorporation of ENR-50 raised the viscosity of CR/rNR-G which increased the scorch time as more ENR-50 was added.

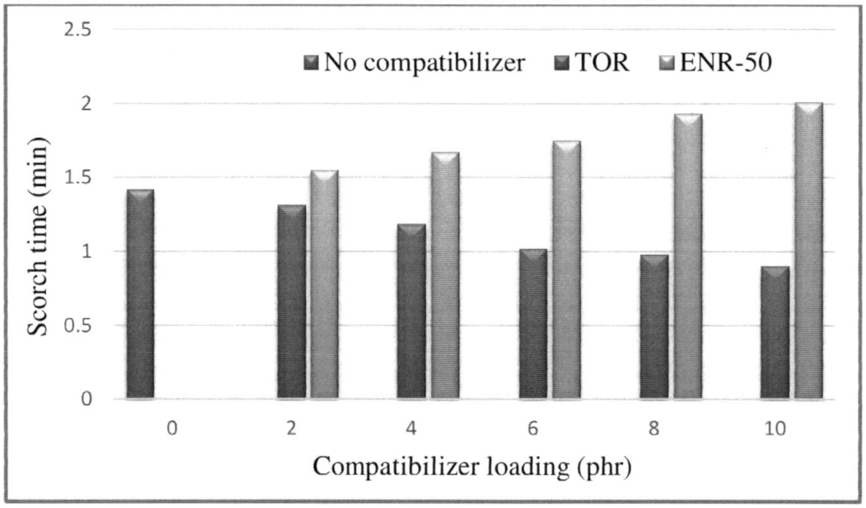

Figure 1. The scorch time of CR/rNR-G compound with compatibilizers.

From the results in figure 2, the cure time t_{90} without compatibilizers was higher compared to CR/rNR-G with compatibilizers. The increasing of TOR concentration led to decrease the cure time to steadily from 2 phr loading to 6 phr. However, after 6 phr loading, the cure time did not exhibit any increasing. Indicating that at 6 phr loading of TOR, further loading will have no influence on the cure time of CR/rNR-G. On the other hand, the increasing of ENR-50 loading yielded an increasing in the cure time.

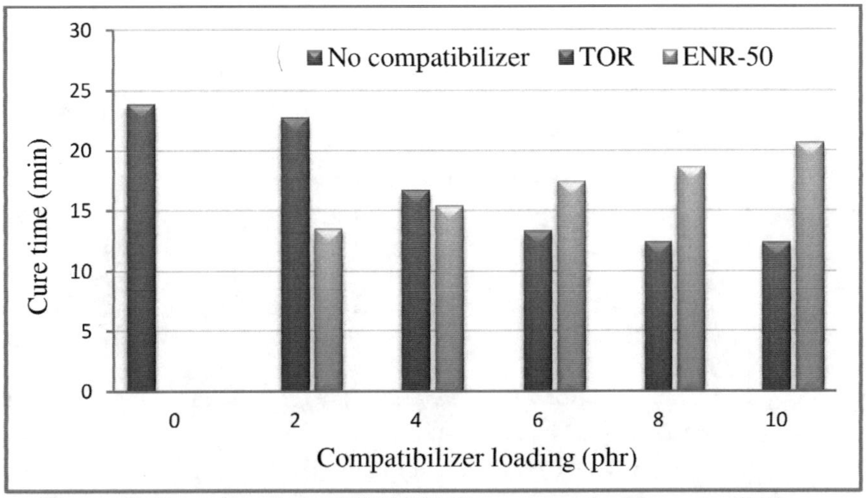

Figure 2. Cure time (t_{90}) of CR/rNR-G compound with compatibilizers.

Table 2. The cure characteristic of CR/rNR-G compound with compatibilizers.

Compatibilizers (phr)		M_L (dNm)	M_H (dNm)
0	-	7.79	50.17
2	TOR	7.23	46.29
	ENR-50	4.50	25.08
4	TOR	6.47	41.36
	ENR-50	5.89	34.29
6	TOR	5.66	36.56
	ENR-50	6.51	43.50
8	TOR	4.88	31.76
	ENR-50	7.30	47.60
10	TOR	3.42	30.77
	ENR-50	8.10	48.98

Table 1 demonstrates the results of minimum torque (M_L) and maximum torque (M_H), the results showed an opposite effect for adding TOR and ENR-50, which with the increasing of TOR loading, the ML and MH values were decreased, that was owing to the processing aid effect of TOR, which it helps to reduce the amount of torque generated during the processing stage. Meanwhile, the adding of ENR-50 increased them.

Figure 3 shows the cure rate index (CRI) of CR/ rNR-G with different loading of compatibilizer TOR and ENR-50. It can be seen that the cure rate of the rubber matrix increased with higher loading of TOR. This phenomenon was due to the characteristic of TOR itself, which possesses smaller particle size, low surface area, chemically active surface, porous surface, high moisture content and uneven shapes, which lead to maximization of interaction area between filled rubber and compatibilizer [13]. In addition, the results showed that the decrease of cure rate as higher loading of ENR-50 was incorporated. This is due to ENR-50 is a polymer with high molecular weight. Thus the incorporation of ENR-50 will increase the viscosity of the filled CR, and then lowering the CRI.

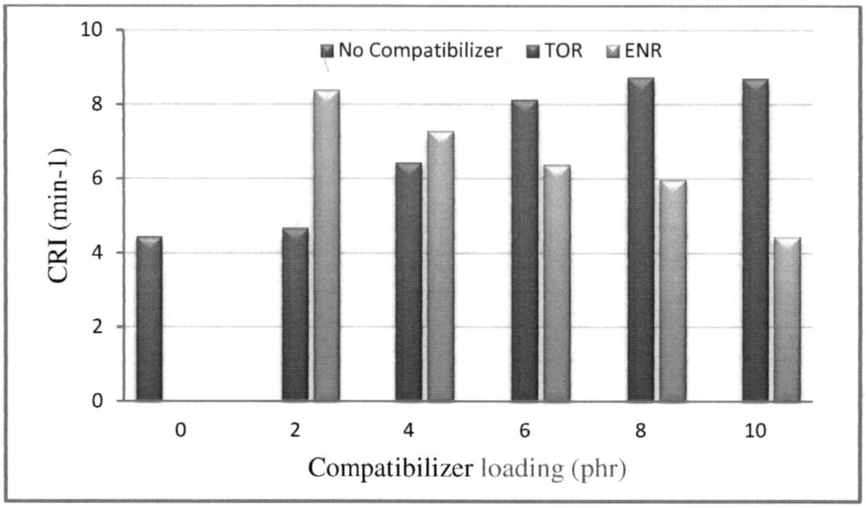

Figure 3. The cure rate index (CRI) of CR/rNR-G compound with compatibilizers.

3.2. Tensile Properties

Table 3 shows the result of tensile strength, Elongation at break and tensile Modulus at 100% elongation of CR/ rNR-G with different loading of compatibilizer TOR and ENR-50. It can be noticed that there is an increasing in the tensile strength values with the increasing of compatibilizers loading. But the incorporation of ENR-50 showed higher tensile strength of the CR compared to TOR. Both of TOR and ENR-50 helped to improve the distribution of the rNR-G filler within the CR. With the better distribution, the interaction between rubber-filler would improve [14, 15]. Thus, the tensile strength improved with the incorporation of compatibilizers. The lower tensile strength of TOR was probably attributed to the weak stress transfer from the matrix to filler although there was still some interfacial adhesion. On the other hand, ENR-50 enhanced the interfacial bonding between the CR and rNR-G filler, thus the interfacial layer can transfer the stress and the bonding after the deformation occurred. Therefore, the incorporation of ENR-50 would increase the tensile strength more than TOR.

The trend showed that the elongation at break increased as the loading of compatibilizers increased. From the results, ENR-50 showed better improvement compared to TOR. That is because ENR-50 possesses a high molecular weight comparing to TOR. As a result, ENR-50 has a good ductility and toughness which can load large deformation before experiencing failure. Therefore, it can be deduced that the incorporation of ENR-50 improves the ductility and toughness of the filled CR, then it improved the elongation at break more than TOR. Finally, both of caused an increasing in the modulus values. However, ENR-50 granted the filled CR higher modulus compared to TOR. This probably due to ENR-50 giving better dispersion of rNR-G filler in the matrix CR. Therefore, a promoting the rubber-filler interaction was occurred, which improved the modulus of the filled CR [16].

Table 3. The tensile characteristic of CR/rNR-G compounds with compatibilizers.

Compatibilizers (phr)		Tensile Strength (MPa)	Elongation at break (%)	Modulus at 100% (MPa)
0	-	1.92	175.23	209.86
2	TOR	2.15	176.33	210.87
	ENR-50	2.38	178.66	211.61
4	TOR	2.34	178.94	211.23
	ENR-50	2.51	180.16	213.54
6	TOR	2.47	180.46	213.89
	ENR-50	2.73	182.64	214.44

8	TOR	2.61	182.87	214.37
	ENR-50	2.79	184.23	216.86
10	TOR	2.89	184.32	215.46
	ENR-50	2.93	186.78	218.42

4. Summary

According to the results obtained, it can be concluded that by taking the optimum loading of fine rNR-G filler which was 20 phr, TOR and ENR-50 effect CR differently. In the cure characteristics, TOR improved the processing time by lowering the scorch time and cure time. Not only that, the filled CR also became easier to process, as TOR also lowers the minimum and maximum torque. On the other hand, ENR-50 increased the processing time and made the processibility more difficult. This is especially obvious as the CRI of TOR increased and ENR-50 decreased as both respective loadings increased. However, both compatibilizers exhibited improvement for the tensile test, which they increased tensile strength, elongation at break and tensile modulus at 100% elongation.

Acknowledgments

The authors would like to acknowledge UniMAP Short Term Grant Scheme (9001 - 00521). We would also appreciate Universiti Malaysia Perlis (UniMAP) for providing us all the facilities to make this research.

References

[1] N. Z. Noriman, H. Ismail and A. A. Rashid, Journal of Applied Polymer Science, **126** (2012)
[2] N. N. Zulkepli and H. Ismail, Polymer-Plastics Technology and Engineering, **51**, 350-357 (2012)
[3] Kuwahara H, Sudo S, Iijima M and Ohya S 2010 *Polym. Degrad. Stab.* **95** 12
[4] N. Z. Noriman, H. Ismail, C. T. Ratnam and A. A. Rashid, Polymer-Plastics Technology and Engineering, **49**, 228-236 (2010)
[5] Rodas CO, Zaïri F, Naït-Abdelaziz M and Charrier P 2015 *Int J Solids Struct* **58**
[6] Riyajan SA, Intharit I, Tangboriboonrat P 2012 *Ind. Crop. Prod.* **36** 1
[7] Hall WJ, Zakaria N and Williams PT 2009 *Waste Manage.* **29** 2
[8] H. Ismail, T. B. Khoon, N. Hayeemasae and S. Husseinsyah, *BioResources*, **10**, 1495-1505 (2015)
[9] R. R. A. M., Mato and G. R. Kassenga, Resources, conservation and recycling, **21**, 1-16 (1997)
[10] B. Hagos, I. O. Kibwage, M. Mwongera, J. N. Muthotho, I. M. Githiga and G. G. Mukindia, East African Medical Journal, **74**, 224-226 (1997)
[11] Rajan V, Dierkes W, Joseph R and Noordermeer J 2006 *Prog. Polym. Sci.* **31** 9
[12] O. S. Dahham, N. Z. Noriman, S. T. Sam, N. M. Al-Samarrai, Z. Shayfull and A. M. Alakrach, MATEC Web of Conferences **78**, 01076 (2016)
[13] H. Ismail, H. C. Leong, Polymer Testing, 20(5), 509-516 (2001)
[14] N. Z. Noriman, H. Ismail and A. A. Rashid, Advances in Polymer Technology, **31**, 100-108 (2012).
[15] B. G. Scares, A. S. Sirqueira, M. G. Oliveira and M. S. Almeida, *Macromolecular Symposia* **189**, **1**, 45-58 (2002)
[16] C. Nah, *Polymer international*, **51**, 245-252 (2002)

The Effect of Epoxidized Natural Rubber as A Compatibilizer on the Properties of Standard Malaysian Rubber/Ethylene Propylene Diene Monomer Blends

A M Alakrach[1], R Hamzah[1], N Z Noriman[1], Omar S Dahham[1], Z Shayfull[2,3], S Z Syed Idrus[4], S Sudin[5]

[1]Center of Excellence Geopolymer and Green Technology (CEGeoGTech), Faculty of Engineering Technology (FETech), Universiti Malaysia Perlis (UniMAP), Level 1 Block S2, UniCITI Alam Campus, Sungai Chucuh, Padang Besar, 02100, Perlis, Malaysia
[2]School of Manufacturing Engineering, Universiti Malaysia Perlis, Arau, 02600, Malaysia
[3]Green Design and Manufacture Research Group, Center of Excellence Geopolymer and Green Technology (CEGeoGTech),Universiti Malaysia Perlis, Kangar, 0100, Malaysia
[4]School of Human Development and Techno-communication (iKOM), Green Advanced Computing and Technology (GREAT) Research Group – CEGeoGTech, Universiti Malaysia Perlis (UniMAP)
[5]School of Computer and Communication Engineering (SCCE), Green Advanced Computing and Technology (GREAT) Research Group – CEGeoGTech, Universiti Malaysia Perlis (UniMAP)

Email: rosnizahamzah@unimap.edu.my

Abstract. In this study, the effects of epoxidized natural rubber (ENR-50) as compatibilizer on the properties of Standard Malaysian Rubber/Ethylene Propylene Diene Monomer (SMR L/EPDM) blend were investigated. Standard Malaysian Rubber/Ethylene Propylene Diene Monomer/ Epoxidized Natural Rubber (SMR L/EPDM/ ENR-50) blends were prepared by two-roll mill with five different compositions (i.e., 80/20/0, 80/20/1, 80/20/3, 80/20/5, 80/20/7 and 80/20/10), which the (SMR L/EPDM) content fixed at 80/20 which is the optimum ratio for (SMR L/EPDM) depending on our previous study. Cure characteristics, mechanical properties studies were performed to determine the compatibility of SMR L/EPDM blends in the presence of ENR-50 as a compatibilizer. The optimum cure characteristics showed by SMR L/EPDM blends with the presence of ENR-50, which lower scorch time t2 and cure time t90 and increase the cure rate index CRI. In addition, SMR L/EPDM blends with the presence of ENR-50 have improved the physical properties such as swelling test, crosslink density and hardness test.

1. Introduction

Polymer blend is conducted to achieve desired requirements that cannot be obtained by the current available individual polymer. In addition, blending of rubbers introduces a superiority to the physical properties of the final vulcanized product [1, 2]. Unfortunately, the blending of two rubbers sometimes leads to immiscibility, which depends on the individual properties of each rubber. As a result, the desired properties will not be gained without a third component. The compatibilizer is one of the solutions of this problem, which is able to mediate an attractive interaction between the phases [3].

The interfacial adhesion can be improved either interact chemically with both phases or will have a specific interaction with one phase and a physical interaction with other. The compatibilizer may form an interface between the immiscible blend components so that imposed stresses can be transferred between the phases via covalent bonds along the copolymer backbone. However, it is important to understand the blend morphology because it has a strong effect on the properties of polymer blends [4].

Epoxidized natural rubber (ENR-50) is manufactured from natural rubber by chemical modification method, by changing some of the double bonds with epoxide parts and contains 50 moll% of the epoxide group. ENR-50 is a polar rubber and possesses properties similar to some of synthetic elastomers. It has reported that with the increasing of the ENR's polarity, the compatibility of ENR-50 will increase. That is because of the presence of the epoxide groups in ENR, which introduces polarity and make it more attractive for rubber-rubber or rubber-plastic blending industry [5].

According to the polarity, ENR-50 is miscible with the polar rubbers such as ethylene propylene diene monomer (EPDM), thereby offering significant properties as well as high strength. In this work, the effect of ENR-50 as a compatibilizer on the cure characteristics and physical properties of standard Malaysian rubber/ethylene propylene diene monomer (SMR L/EPDM) blends were carried out.

2. Experimental

2.1. Materials

Standard Malaysian natural rubber grade L (SMR L) and ethylene propylene diene monomer (EPDM) rubber were supplied by Rubber Research Institute Malaysia (RRIM). Epoxidized natural rubber (ENR-50) that used as a compatibilizer was purchased also from Rubber Research Institute Malaysia (RRIM). The other chemicals that used in the compounding process such as zinc oxide and steric acid (S. A.) as activators, N-cyclohexyl-2-benzothiazole sulphonamide (CBS) as accelerator, butylated hydroxytoluene (BHT) as antioxidant and sulphur as vulcanizing agent were supplied by Anchor Chemical Co. (M) Ltd.

2.2. Formulation and testing of CR/ rNR-G compounds

The SMR L/EPDM/ ENR-50 blends were prepared by two roll mill. The different parts per hundred resin (phr) of ENR-50 were used i.e., 0, 1, 3, 5, 7 and 10, whereas the weight percent of SMR L and EPDM was maintained at 80 and 20, respectively, as the optimum ratio was selected from series (based on 1st series). The (80/20/1, 80/20/3, 80/20/5, 80/20/7 and 80/20/10) ratios were compared to the control ratio (80/20/0).

2.3. Characterizations and testing of CR/ rNR-G compounds

After compounding process, rubber compounds were taken out from freezer and defrozen at room temperature for 30 minutes for the purpose of curing characteristic test. curing characteristic test was done by using Rheometer Model HT-M2000. Rheometer was heated until achieve the desired curing temperature which is 160 °C. The test was run for 20 minutes. 4.0 g of uncured compound rubber was weighed and placed on the heated rotor. and the heated top die cavity was immediately brought down on to the lower dies thus filling the cavity. The machine plots a graph of torque verses time. Curing characteristic such as minimum torque M_L maximum torque M_H. cure time (t_{90}) and scorch time (t_2) Were obtained.

2.4. Hardness Test

Hardness of the samples was measured according to ASTM D-2240-81 using a Shore A type durometer which employed a calibrated spring to provide the indenting force. Since the hardness reading decreased with time after establishment firm contact between the indentor and the sample, the recording immediately after establishment to firm contact were taken. The measured values of hardness were obtained at three different points distributed over test piece. Three test pieces were used and their average value was determined.

2.5. Swelling Test and Crosslink Density

A cured test piece of above dimension was weighed using an electronic balance before swollen in toluene. After that, all parts of test piece were fully immersed in toluene. After 48 hours, test piece was taken out, wiped, weighed again rapidly. Swelling bottle that contains test sample and toluene need to be closed neatly because of rapid evaporation of toluene. Precautions need to be taken into account during handling toluene because this solvent is hazardous. Swelling percentage was calculated by the equation 1.

$$\textit{Swelling percentage} = (M_2\text{-}M_1) \times 100 / M_1 \qquad (1)$$

Which M_1 is the initial mass of the tested piece (g) and M_2 is the mass of it after swollen in toluene.

3. Results and discussion

3.1. Cure Characteristics

The effect of ENR-50 as a compatibilizer on the scorch time (t_2), cure time (t_{90}), Minimum Torque (M_L) and Maximum Torque (M_H) of SMR L/EPDM blends is shown in Table 1, figure 1 and 2.

Table 1, cure characteristics of (SMR)/(EPDM) with different ENR-50 loading.

ENR-50 (phr)	Minimum torque M_L (dNm)	Maximum torque M_H (dNm)
Control	3.05	14.40
1	2.65	11.95
3	3.22	13.70
5	3.60	15.85
7	3.27	16.90
10	2.95	15.65

From the results that illustrated in figure 1, it can be seen that with the increasing of ENR-50 content, t_2 of the rubber compounding also increased till reached the ratio 5 phr. After that decreased gradually from 7 phr until 10 phr. Generally, the addition of ENR-50 led to increase crosslink density and shorter scorch time because the epoxy ring in ENR-50 activates the transformation of adjacent double bonds in the rubber molecules into free radicals. These radicals can quickly react and form additional crosslink with sulfur. But, the increasing of ENR-50 concentration caused a dilution of the curing agent content and an increase in the viscosity, that was resulted in a stoichiometric imbalance and led to a change in the reaction mechanisms. In addition, the increasing of ENR-50 content caused an increasing in the molecular weight of the epoxy during curing. This led to the formation of a second rubbery phase, all these reasons were a result to increase the scorch time with the increment of ENR-50 ratio [6].

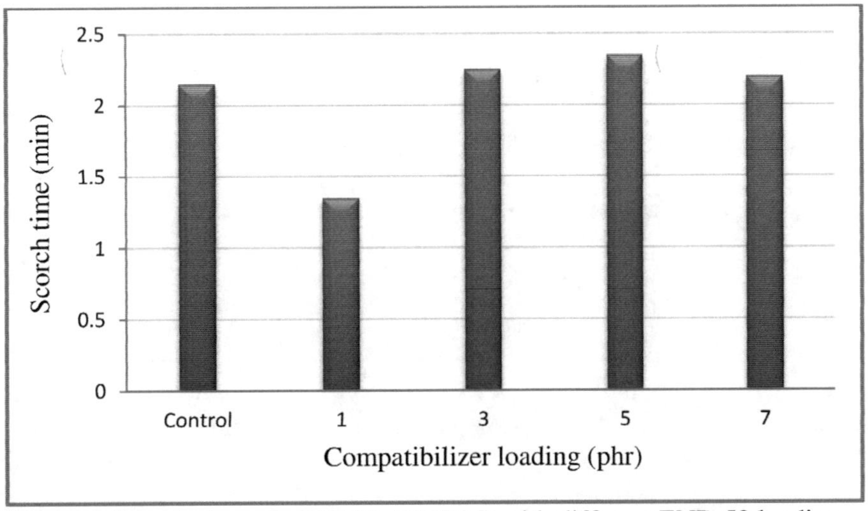

Figure 1. Scorch time of of (SMR)/(EPDM) with different ENR-50 loading

Figure 2 illustrates the effect of ENR-50 as a compatibilizer on the cure time (t_{90} of SMR L/EPDM blends. The results showed that the optimum t_{90} value of the blends were found by adding of ENR 50 at 1 phr but after this ratio there is no significant change with the increasing of ENR 50 content. The shorter cure time shown by SMR L/EPDM with addition 1 phr ENR 50 due the presence of ENR 50 which catalyzed the crosslink reaction [3].

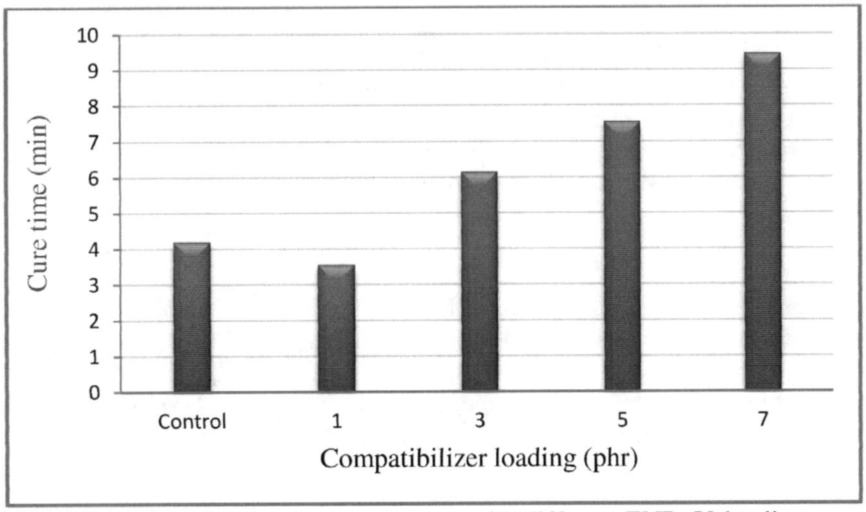

Figure 2. cure time of (SMR)/(EPDM) with different ENR-50 loading.

In addition, Figure 3 shows a cure rate index (CRI) of SMR L/EPDM filled ENR 50. It has denoted that the cure rate index (CRI) decreased until 7 phr and increased gradually up to 10 phr of ENR 50 loading. The decreased in cure rate index with increasing ENR-50 content might be due to the lower viscosity. In blends, the lower viscosity component tends to form a continues phase which it will become the more ingredient govern the curing process [7]. The SMR L/EPDM with addition at 1 phr of ENR 50 shows the optimum cure rate index. It can be seen that at this loading, the rubber matrix is filled with 1 phr of ENR 50 content tend to crosslink efficiently.

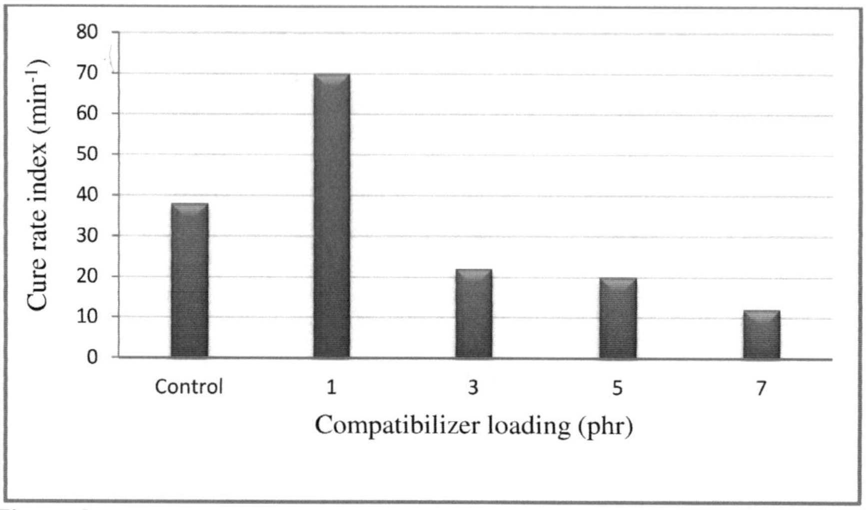

Figure 3. cure rate index of (SMR)/(EPDM) with different ENR-50 loading.

Minimum torque is related to the viscosity after addition of ENR-50. It can be seen that the trend increased gradually from 1 until 5 phr loading and it started decreasing gradually to 10 phr loading. At 1 phr of ENR 50 showed the lowest minimum torque compared to the control and different loading of ENR 50, which it can be selected as the optimum minimum torque value in this series. This result was attributed to the effect of the adjacent double bond of the oxirane group of ENR 50, which provides an acceleration in the vulcanization reaction [8].

Furthermore, the result showed an increasing in the maximum torque significantly with the increasing of ENR 50 loading. Similarly, the maximum torque values decreased from 7 to 10 phr. Maximum torque values were determine a measurement of the stock modulus, which increases the interaction between the rubber matrix and the compatibilizer. Moreover, as can be seen, the lowest value of maximum torque showed by 1 phr of ENR 50 addition and it also selected as the optimum maximum torque in this series. This can be explained by the poor interaction between rubber matrix and ENR 50 [3].

3.2. Hardness Test

Table 2. shows the effect of hardness test on SMR L/EPDM blends with different loading ENR 50. As can be seen that, the hardness values increased with the increasing of different loading of ENR 50 until reached 5 phr and started to decrease gradually until 10 phr. The addition of different loading of ENR 50 showed values higher than the control value. However, SMR LIEPDM blend at 1 phr was selected as an optimum value of hardness compared to other values. This indicate that, the flexibility and elasticity of the rubber chain were less when ENR 50 at 1phr loading incorporated into rubber matrix, which also in more rigid rubber vulcanizate and increased the hardness. This observation might be attributed to the addition of ENR 50 on SMR LIEPDM. Which SMR L is non-polar rubber whereas EPDM and ENR 50 are polar rubber. It is generally observed, the mechanical response of the blend is related closely with compatibility and a synergistic effect is often obtained with miscible or partially miscible blends [9].

3.3. Swelling percentages

Swelling resistance is an indicator of the degree of crosslinking because the degree of swelling is proportioned inversely to the amount of crosslink. The effect of swelling percentage on SMR L/EPDM blends with different loading of ENR 50 is illustrated in table 2. As observed swelling percentage increased while the crosslink density decreased. SMR L/EPDM blend with 1 phr loading showed the highest value compared to other different loading. The decrement of crosslink lead to the increase in molecular weight between two adjacent crosslink. Thus, the diffusion of toluene into vulcanizate was expected to be in its maximum value, which resulted in high percentage of swelling [10, 11].

Table 2, cure characteristics of (SMR)/(EPDM) with different ENR-50 loading.

ENR-50 content (phr)	Swelling percentages (%)	Hardness test (Shore A)
Control	275	42.8
1	245	44.3
3	238	46.0
5	242	46.8
7	235	45.9
10	237	45.4

4. Summary

The addition of different contents of ENR-50 into the optimum blend ratio from series 1 made a significant effect on the cure characteristics and physical properties. The optimum blend showed by SMR L/EPDM blends with addition 1 phr of ENR 50. The ratio of 1 phr of ENR 50 showed better improvement towards the rubber compound such as curing characteristics; lowered both of t_{90} and t_2, higher cure rate index (CRI). Meanwhile it also showed the improvement in physical properties such as swelling percentage, crosslink density and hardness test compared to other contents of ENR 50.

Acknowledgments
The authors would like to acknowledge UniMAP Short Term Grant Scheme (9001 - 00521). We would also appreciate Universiti Malaysia Perlis (UniMAP) for providing us all the facilities to make this research.

References

[1] Jovanović S, Samaržija-Jovanović S, Marković G, Jovanović V, Adamović T and Marinović-Cincović M 2016 *Composites Part B* **98**
[2] N. Z. Noriman and H. Ismail, International Journal of Polymeric Materials and Polymeric Biomaterials, **62**, 252-259 (2013)
[3] N. Z. Noriman, H. Ismail and A. A. Rashid, Journal of Applied Polymer Science, **126**, (2012)
[4] O. S. Dahham, N. Z. Noriman, S. T. Sam, N. M. Al-Samarrai, Z. Shayfull and A. M. Alakrach, MATEC Web of Conferences **78**, 01076 (2016)
[5] Ismail H and Ooi ZX 2010 *Polym. Plast. Technol. Eng* **49** 7
[6] Tanjung FA, Hassan A and Hasan M 2015 *J. Appl. Polym. Sci.* **132** 29
[7] Lee YK, Jeong YT, Kim KC, Jeong HM and Kim BK 1991 *Polym. Eng. Sci.* **31** 13
[8] Teh P, Ishak ZM, Hashim A, Karger-Kocsis J and Ishiaku U 2004 *Eur. Polym. J.* **40** 11
[9] Ismail H and Leong H 2001 *Polym. Test.* **20** 5
[10] Ismail H and Hairunezam H 2001 *Eur. Polym. J.* **37**1
[11] G. B. McKenna, K. M. Flynn and Y. Chen, *Polymer*, **31**, 1937-1945 (1990)

The ICT-Induced on Behavioral of Lecturer and Society Change in Libya Universities

Abdul Ghafar Mansor M .M . Mohamad[1], Syed Zulkarnain Syed Idrus[1,2], Amani Ali Elmetwely Ali Ibrahim[1]

[1]School of Human Development and Techno-communication (iKOM), Universiti Malaysia Perlis (UniMAP)
[2]School of Human Development and Techno-communication (iKOM), Green Advanced Computing and Technology (GREAT) Research Group – CEGeoGTech, Universiti Malaysia Perlis (UniMAP)

Corresponding author: abdulghafarmmm@gmail.com

Abstract

Information communication and technology has changed the lifestyle of people and influences their behavior towards their daily activities. ICT has become a significant part of people, community or society by influence behavior people towards their attitude, and it has several of activities that can change individual activities pertaining their behavior to interact with the system. However, previous studies have been investigated ICT in a different context in Libya but there is lack behind to explore what motivate, encourage and induced lecturer to use ICT for teaching and learning in Libya university. This study intends to investigate the ICT induced on behavioral of lecturer and society change of Libya University towards the use of ICT. This paper is conceptual by employed technology acceptance model (TAM) and theory of planned behavior (TPB) and attitude, perceived ease of use, perceived usefulness, behavioral intention, and social influence were used to explain the use of ICT. Therefore, this paper found that when lecturers have an adequate knowledge about ICT may influence them to use. The future study can focus on empirical evidence to support the lecturer behavior towards the use of ICT.

Keywords: attitude, perceived ease of use, perceived usefulness, behavioral intention, social influence, ICT.

IOP Conf. Series: Journal of Physics: Conf. Series **1019** (2018) 012069 doi:10.1088/1742-6596/1019/1/012069

1. Introduction

Information communication and technology has changed the lifestyle of people and influences their behavior towards their daily activities [8]. ICT has become a significant part of people, community or society by influence behavior people towards their attitude, and it has several of activities that can change individual activities regarding their behavior to interact with the system [28]. Therefore, higher education institution around the world has been invested and curiously adopts ICT to their institution, which had a major change to their university context of teaching as well as the organization entirely.

The ICT has induced on societal change, which has led to information knowledge that people have access to information through the use of technology. However, inducing of ICT occurred through social influence as major influence between ICT and society to accept the use of ICT [27]. ICT has changed the method of teaching and learning in many universities, and they perceived the potential of ICT either from student or lecturer. Although, Several studies have been indicated that there are various factors that can motivate or influence the use of ICT in the context of the university including ease of use, usefulness, enjoyment, effort expectancy, and others were categorized as induced to use ICT [2][3][5][6][23][27]. However, ICT can change the attitude and they people interact with each other's when they perceive the usefulness and benefit of using ICT. Some studies highlight that to understand the benefit and changing the behavior of the individual to use ICT through the social influence such as friend or family influences them to use ICT.

Moreover, ICT induce has been studies in a different context in higher education institution research such as ICT performance in education, adoption, intention to use, acceptance and implementation of ICT, which had been studied for the last two decades [15], but little attention has been focused on behavior of lecturer in university and how ICT change their behavior. Though previous studies have been explored the influence and impact of ICT among the university in Libya, there is the big portion that most of the research are lack behind from previous research. Researchers neglect and ignore the motivation and encouragement of lecturer to use ICT and they are behavioral after adopting ICT by changing the method and process of delivering the lecture to the student. This study intends to contribute to the body of knowledge by investigating induce of lecturer behavior towards the use of ICT in Libya university.

This study employed TAM model and TPB theory change to investigate the induced of ICT among the lecturer in Libya Universities. This paper uses the conceptual technique by adopting the variable of perceived ease of use, usefulness, attitude behavioral intention and social influence toward the use of ICT in Libya universities.

2. Background of Libya education

Libya is located in North Africa with a population of 6,404,219. Libya got independent in 1951 from Italy, which they are practicing presidential system of government and the official language is Arabic. However, Libya is among of oil countries in the world with sound economic [24].

Accordingly, for the past few decades, the literacy of Libya is between 50 to 60 percent but now every individual has a right to education. The government provides free education from basic education to postgraduate study for every Libyan citizen, government compulsory for every citizen to hold at least school leaving certificate and more than 90 percent in Libya can read and write [12][15]. Additionally, Libyan government provides the opportunity for any citizen who prefers to study outside the Libya either to study degree or postgraduate study is fully supported by the government.

Accordingly, the education system of Libya is classified into two public and private, which both are provide or operate basic education or primary education, secondary school, polytechnic, vocational, technical school, university, and postgraduate. In addition, children between 6 years of age to 15 years to attend primary and secondary school for free, and it is optional for citizen either to continue or further their study into the degree, vocational, technical school, diploma or postgraduate.

The language of an instructor is Arabic and English as a medium of teaching and learning. Currently, the English language is taken as a course in entire faculties in all universities in Libya [15]. According to [15], highlighted that some faculty in Libya University are using the English language to teach their studies such as medicine and engineering.

Furthermore, in 2005, Libyan government partnership with United Nations Educational, Scientific and Cultural Organization and other private company to install LAN and WAN in public university to have access to the internet. Although, number of students enrolled in to universities increased every year, which has led some universities to implement the use of ICT as new method of deliver lecturer to the student such as E-learning installation of projector inside the lecturer room, using power point, video/ audio and any others application of ICT [12]. Therefore, implementation of ICT in Libya University is not fully adopting in some university and there is required to investigate behavior of users of ICT.

3. Literature review

Study has been investigating the impact ICT to influence behavioral and societal change in different context of research such as [28] studied the impact of ICT on travel behavior through the accessibility of the website and the finding ICT change the way access the travel behavior when they interact on website than face to face. Moreover, most of the developed and developing countries based on their different political and social backgrounds, follow different paradigms in educational innovations and reforms concerning information and communication technologies (ICT), from the perspective of education [17]. Using ICT in most of the university is helping lecturer to manage and engage lecturer and student in a meaningful communication and learning activities [3]. Similarly, ICT helped lecturer to manage the course, research material and student assignment faster than traditional assessment. Additionally, [13] stated that provide common ICT tools such as the internet, projector and video will motivate lecturer to use ICT for teaching in the lecture room. Hence, teaching student with modern ICT tools will provide easiness to the lecturer as well as clearly understand to the student such video clip relating to the lessen [26]. [9] revealed that ICT has changed the way of receive knowledge, communication between lecturer and student, as well as level of access information relating to research methodology is improve the education system. Hence, lecturer realized that utilized a certain technology can be induced their behavior and help to use ICT for teaching designs for better respond to today's students' abilities, interests and learning styles.

Furthermore, [21] stated that ICT has been changed the learning process and educational environment in term of teaching and learning from traditional perspective to the use of ICT system. ICT has changed numerous of learning process and method of delivering lecture to the student, which has been motive and influence both lecturer and student as a new method of communication. However, [6] argued that teacher has influenced to use some tools which will improve the effective use of ICT in teaching and learning. Therefore, this has been proved that using ICT can change the educational society from the traditional process of teaching to use of ICT. Many universities have perceived the potential of change the method of delivering lecture to student has to increase the performance of lecturer as well as student engagement. In the context of Libya university, ICT can persuade lecturer to be active in the class, time save and enhance easiness of communication between lecturer and the student, may increase or bring a strong relationship to the educational environment.

4. Technology acceptance model

[10] developed technology acceptance model to investigate the degree of acceptance of new system or technology. The model proposed how the organization would accept or reject new technology is easy to use or difficult to operate or interact. The model developed two variables perceived ease of use and usefulness to measure acceptance of the technology. This TAM has shown the relationship between the individual beliefs and intention to use or accept new technology. In the past, the technology acceptance model sought to explain why people accepted or rejected a particular technology ([18]; [10]. However, the researcher has been expanded and added an extension to the TAM model in a different context of research such as learning management system (LMS) [3]. Use of ICT in sports education [19], mobile learning [18], e-learning [24].

Previously, TAM has gained popularity on acceptance of technology ICT and system either the user will accept or reject to use a particular system [9]. Information technology model is one of the influential models that use to determine the use of technology or system. Although, TAM has several limitations which are called for the extension and for more explanation such as time factor, human control environmental or socal change and user behavior is limited measure and explain from TAM model [22]. Although, the previous study has been used TAM to investigate acceptance of ICT in education such as learning management system (LMS) and the result from the study revealed that TAM is significance to explain or investigate acceptance of LMS [3].

Therefore, TAM model has been employed in this paper to find out ICT-induced on the behavior of lecturer towards the use of ICT and society change of university in Libya. Although, [10] suggested that perceived usefulness is the degree to which a person believes that using a particular system would enhance his or her job performance [4]. From the outset, the researchers intend how ICT motivate and encourage lecturer and change their behavior toward teaching and learning to perceive both the usefulness and would probably indicate their ease of use of ICT tools in their classroom environments.

5. Theory of planned behavior

The theory of planned behavior (TPB) is the extension of the theory of reasoned action in 1980. TPB developed to measure the behavior of human to engage in specific behavior. The theory is intended to explain the ability and intention of individual [1]. TPB adding an additional variable of perceived behavioral control. The TPB consisted of three factors, attitude towards the behavior, subjective norms, and perceived behavioral control that predict behavior and intention. TPB has predicted behaviors. Also, all the variables of the TPB have a significant effect on intention and behavior. Therefore, the theory measures users' behavioral intention to adopt and use a given ICT. Since measuring actual system use is difficult, behavioral intention is normally used as a reliable predictor of actual use [10]. Though, the previous study has been employed this theory to explain the different context of research in the field of ICT. In this paper, TPB adopts to measure how ICT can change the behavior of lecturer to use a specific ICT. [20] examined the TPB theory on technological innovation, the findings reveal that TPB can measure the behavior of user to use technology.

6. Behavioral intention

According to TAM, behavioral intention defines the actual use of a given IS system and therefore determines technology acceptance. According to [27] highlighted that there are various factors influence users to use ICT, which has enumerated of individual differences, social influences, beliefs, attitudes and situational influences as factors that encourage individual intention to use technology and encourage them to accept it. [11] revealed that individual behavior can influence and change other people's intention to perform a certain behavior, which it will engage their behavioral intention to use ICT. [3] found that lecturer has the intention to use ICT but their fairness of use. Meanwhile, fairness o the lecturer is affected their intention and courage to use ICT in the classroom may have negated influence to use ICT.

6.1 Perceived ease of use

[10] explained that the perceived ease of use is "the degree to which a person believes that using a particular system would be free of effort" [10]. According to [10], the intention to use technology can be influenced when the user has seen the technology as ease to use. In this case, this paper employed ease of use to investigate how easiness of ICT motivates lecturers to use ICT for teaching and learning student in the lecture room. Although potential users could believe that a given technology is useful, they may, at the same time be against (for some reason) its use in their lecturer. They may perceive that there aren't sufficient performance benefits for using mobile learning technologies [25]. [3] stated that lecturer can have the courage to use ICT when the perceived that a particular system is ease to use will motivate or influence to use by lecture. [2] claimed that using ICT tools in the classroom is not the major problem but how to motivated lecturer to seen ICT tools as ease to use in order to bring the strong relationship between lecturer and using the ICT tools in the classroom. [7] argued that to influence lecturer to use IC, there is required to understand what can motivate and encourage lecturer to use new ICT innovation. Similarly, when a lecturer in Libya universities induced that use a certain ICT tool in ease to operate will motivate and change their behavior to use.

6.2 Perceived usefulness

Perceived usefulness is another major determinant in the adoption of technology. It is the tendency of users to believe that the new technology will help them perform their job better [10]. [3] claimed that lecturer will use ICT when there is awareness of benefit of use and usefulness of the system that It's free from harm or difficult to operate will change the behavior of lecturer to use ICT for teaching and learning. [23] found that instructor understands the level to which they perceive ICT tools can enhance their efficacy in the performance of a task. [14] argued that lecturers in the university have a significant factor to influence and perceived usefulness of ICT when the university provides training and workshop on a particular ICT tool. Therefore, the usefulness of ICT can enhance teaching and learning performance of lecturer in Libya universities.

6.3 Attitude

[1] define attitude as "an individual's positive or negative feelings (evaluative affect) about performing the target behavior" (p. 216). .hence, literature revealed that lecturer attitude towards the use of ICT can influence their behavioral intention to use. Attitude towards anything or practice to utilize ICT in class tend to open up professionals to the different accessible choices, along these lines making the teaching practices possible. Therefore, distinctive research indicates that the successful reconciliation of ICT into the classroom relies on the attitude of the instructors [17].

[6] studied the influence of ICT toward the teacher attitude to use ICT for teaching and learning process, the findings reveal that ICT has high reliability to influence the attitude of the teacher to use ICT. [17] argued that empirical evidence revealed that instructors always have a positive attitude towards the use of ICT. Therefore, positive attitude of lecturer will help them to see the input and give the impact of ICT knowledge to the students. Similarly, when the Libya lecturer in universities is encouraged to use ICT, they can transfer the knowledge to the student and society as well.

6.4 Social influence

Social influence is defined as "the degree to which an individual perceives that important others believe he or she should use the new system" [29]. However, Social influence is the extent to which people are influenced and motivated by either through friend or family to use a specific technology or system. Furthermore, social influence is determined by changing or influencing people in the society and change the behavioral intention of the individual to use ICT [5]. According to [16], academic staff in the higher institution can influence change other lecturers to use ICT and benefit them in term of teaching and learning process. [11] claimed that there different factors that encourage and persuade lecturer to use ICT when they perceived social influence from their co-lecturers will change their behavioral intention to use ICT.

7. Conclusion

This paper is conceptual research that focuses on lecturer in Libya University. Therefore, this paper concludes that ICT-induced is lacking or limited among the lecturer to understand the potential of using ICT. The objective of this study is to investigate factors motivate and encourage lecturer behavior to use ICT as well as how this influence change society in Libya university. TAM model and TPB theory have employed to persuade the behavioral intention of lecturer and how ICT encourage lecturer to use. The lecturer can use ICT for teaching and learning when there is training or social influence from friend or co-lecturer to motivate that using a particular system or ICT tools are easy to use and the usefulness will benefit them. Hence, each university has been implementing the use of ICT as a new method of teaching and learning process such as how to operate or communicate with the student can be another way to motivate lecturer and it can change them from traditional method to new modern ICT tools for teaching and learning. Moreover, this paper recommends that each university in Libya should provide training to their lecturers on how to operate any ICT tools relating to teaching student inside lecture room such as a projector, power point, video clips and others. Though there is limited influence or motivation among the lecturer to use ICT, university management can to those that have knowledge about how to use or operate ICT to teach or explain to them when they are in need of help. The university also can provide help center for any lecturer who needs some help regarding how to operate the ICT to teach student inside the lecture room.

Additionally, future research can look for deep investigate either qualitative on quantitate method to provide good evidence to support the ICT induce among lecture in Libya universities.

References

[1] Ajzen, I., & Fishbein, M. (1980). Understanding attitudes and predicting social behaviour. *MIS quarterly*

[2] Al Shemmari, J. M. (2015). *Information and communication technology acceptance model for empirically testing primary science teachers' use of ICT in Kuwait* (Doctoral dissertation, University of Southern Queensland).

[3] Alharbi, S., & Drew, S. (2014). Using the technology acceptance model in understanding academics' behavioural intention to use learning management systems. *International Journal of Advanced Computer Science and Applications*, *5*(1), 143-155.

[4] Arrigo, M., Kukulska-Hulme, A., Arnedillo-Sánchez, I. and Kismihok, G., 2013. Meta-analyses from a collaborative project in mobile lifelong learning. *British Educational Research Journal*, Vol. 39, No 2, pp.222-247.

[5] Attuquayefio, S. N., & Addo, H. (2014). Using the UTAUT model to analyze students' ICT adoption. *International Journal of Education and Development using Information and Communication Technology*, *10*(3), 75.

[6] Aydin, M. K., Semerci, A., & Gürol, M. (2016). Teachers' Attitude towards ICT Use in Secondary Schools: A Scale Development Study. *International Association for Development of the Information Society*.

[7] Barak, M. (2014). Closing the gap between attitudes and perceptions about ICTenhanced learning among pre-service STEM teachers. *Journal of Science*

[8] Braccini, A. M. (2013). Does ICT influence organizational behaviour? An investigation of digital natives leadership potential. In *Organizational Change and Information Systems* (pp. 11-19). Springer Berlin Heidelberg.

[9] Camilleri, M. A., & Camilleri, A. C. (2017). The Technology Acceptance of Mobile Applications in Education. *In Sánchez, I.A. & Isaias, P. (Eds) 13th International Conference on Mobile Learning (Budapest, April 10th). Proceedings, pp., International Association for Development of the Information Society.*

[10] Davis, F. D. (1989). Perceived usefulness, perceived ease of use, and user acceptance of information technology. *MIS quarterly*, 319-340.

[11] Durodolu, O. O. (2016). Technology acceptance model as a predictor of using information system'to acquire information literacy skills. *Library Philosophy and Practice*. *Education and Technology*, *23*(1), 1-14

[12] ElZoghbi, N. B., & Khashkhush, A. S. (2013, January). Overcoming the barriers to implement electronic learning in Higher education (HE). In *Proceedings of the International Conference on e-Learning, e-Business, Enterprise Information Systems, and e-Government (EEE)* 333

[13] Hergüner, G. 2012. Views of Students in the Department of Recreation and Sport Management on Distance Education. TOJET: *The Turkish Online Journal of Educational Technology*, 11 (4)

[14] Ibrahim, J. S., & Shahzad. A., (2015). The Adoption of Blended Learning in Higher Education Institution in Nigeria. International Journal of Management and Social Sciences Research (IJMSSR), 4(10).

[15] Khalid, K. A. A. (2017). *Learning and Teaching in English: A Case Study of Higher Education in Libya* (Doctoral dissertation, Liverpool John Moores University).

[16] Kocaleva, M., Stojanovic, I., & Zdravev, Z. (2015). Model of e-learning acceptance and use for teaching staff in Higher Education Institutions. *International Journal of Modern Education and Computer Science*, *7*(4), 23.

[17] Krause, M., Pietzner, V., Dori, Y. J., & Eilks, I. (2017). Differences and Developments in Attitudes and Self-Efficacy of Prospective Chemistry Teachers Concerning the Use of ICT in Education. *Eurasia Journal of Mathematics, Science and Technology Education*, *13*(8), 4405-4417.

[18] Mac Callum, K., & Jeffrey, L. (2014). Comparing the role of ICT literacy and anxiety in the adoption of mobile learning. *Computers in Human Behavior*, *39*, 8-19.

[19] Ntshakala, T. T. (2016). *Using the technology adoption model for the modelling of teachers' technology awareness factors* (Doctoral dissertation).

[20] Otieno, O. C., Liyala, S., Odongo, B. C., & Abeka, S. O. (2016). Theory of reasoned action as an underpinning to technological innovation adoption studies. World Journal of Computer Application and Technology 4(1), 1-7.

[21] Oye, N. D., Iahad, N. A., & Rabin, Z. A. (2011). A model of ICT acceptance and use for teachers in higher education institutions. *International Journal of Computer Science & Communication Networks*, *1*(1), 22-40.

[22] Oye, N. D., Noorminshah, A., & Rahim, N. A. (2011). Examining the effect of technology acceptance model on ict usage in nigerian tertiary institutions. *Journal of Emerging Trends in Computing and Information Sciences*, *2*(10), 533-545.

[23] Prieto, J. C. S., Migueláñez, S. O., & García-Peñalvo, F. J. (2015, November). Behavioral intention of use of mobile technologies among pre-service teachers: Implementation of a technology adoption model based on TAM with the constructs of compatibility and resistance to change. In *Computers in Education (SIIE), 2015 International Symposium on* (pp. 120-125). IEEE.

[24] Rhema, A., & Miliszewska, I. (2014). Analysis of student attitudes towards e-learning: The case of engineering students in Libya. *Issues in Informing Science and Information Technology*, *11*, 169-190.

[25] Sampson, D. G., Isaias, P., Ifenthaler, D. and Spector, M. (Eds.). 2012. *Ubiquitous and mobile learning in the digital age*. Springer Science & Business Media, New York, USA.

[26] Shelly, G.B. and Vermaat, M.E. 2012. *Discovering computers' fundamentals: your interactive guide to the digital world*. United State of America: Course Technology, Cengage Learning.

[27] Teo, T. (2013). A comparison of non-nested models in explaining teachers' intention to use technology. *British Journal of Educational Technology*, *44*(3).

[28] Van Wee, G. P., Chorus, C. G., & Geurs, K. T. (2013). ICT, travel behavior and accessibility: A review of literature, a conceptual model and a research agenda. *Journal of Transport and Land Use*, *6*(3), 1-16.

[29] Venkatesh, V., Thong, J. Y., & Xu, X. (2012). Consumer acceptance and use of information technology: extending the unified theory of acceptance and use of technology. *MIS quarterly*

IOP Conf. Series: Journal of Physics: Conf. Series **1019** (2018) 012070 doi:10.1088/1742-6596/1019/1/012070
*** Formatting Issue - Best Available Paper/Graphic ***

Acceptance Factor of Mobile Learning Application for Adult Learners in Life Long Learning Education

Azizah binti Safie[1], Muhammad Rafie Mohd. Arshad[2] and Norhaliza binti Idris[3]

[1, 3] Department of Information Technology & Communication, Polytechnic of Balik Pulau, Pulau Pinang, Malaysia
[2] School of Computer Science, Universiti Sains Malaysia, Pulau Pinang, Malaysia

E-mail: azizah@pbu.edu.my

Abstract. Mobile technology has introduced a new technology that can be used as teaching aids to enrich the teaching and learning process and it have become increasingly a part of education for those who study or teach at the higher learning institution and school levels. The usage mobile technology in learning which also known as mobile learning (m-learning) is one type of learning model allowing learners to obtain learning materials anywhere and anytime using mobile technologies as learning aided tool. Academics and practitioners alike are showing interest in examining its ability to support online learning. However, earlier studies have highlighted the challenges in promoting m-learning adoption. The existing m-learning literature has mainly focused on technology-related factors to examine m-learning adoption. However, few studies have investigated this topic from the users' motivation perspective and even fewer from an adult learner perspective. Therefore, this study employs the uses and delight theory to provide a better understanding of what motivates m-learning adoption in adult learners. Futher research is needed to examine the determination factors in Malaysia specifically with adequate samples. As suggestion, m-learning needs to be build in such a way to ensure the adoption of usage is higher for adult learner.

KEYWORDS: Mobile learning; M-learning adoption; Lifelong learning, Learning applications, Determination factors

1. Introduction

Previously, communication through telephone provided the ability to talk to someone from a distance. However, the capability of the telephone has expanded to include other features. Nowadays, most of the mobile devices combine the features of traditional telephone, text messaging, a diary or planner, wireless internet connection and personal computer capabilities.

1.1 Evolution of Mobile Learning (M-Learning)

Currently mobile technology has introduced new technology that can be used as learning aids to enrich the teaching and learning process. Mobile devices or smartphone have become increasingly a part of learning material for higher learning institution and also in school levels. The usage of mobile technology in learning which also known as mobile learning (m-learning) is one type of learning model that allowing learners to gain learning materials anywhere and anytime using mobile technologies as aided learning tool. M-learning is a kind of learning model that allows learners to obtain learning materials anywhere and anytime using mobile technologies and the Internet.

IOP Conf. Series: Journal of Physics: Conf. Series **1019** (2018) 012070 doi:10.1088/1742-6596/1019/1/012070

M-learning is an innovation that leads the technology of mobile phone usage in education. This learning type allows learning process to be continuous anytime regardless of geographical location [1], [2]. M-learning presents to be a new education medium for helping people to acquire knowledge and skill in a ubiquitous manner. M-learning is not only providing the educational content but also provide the good adaptation to adequate visualization on the small screen of mobile devices. Furthermore, m-learning is flexible and gives freedom feelings to students [3], which allows mobile learners to have access to a wide variety of richly diversified learning resources [4]. M-learning is being the digital support of adaptive, investigative, communicative, collaborative, and productive learning activities in remote locations, proposes a wide variety of environments in which the learners can operate.It is a collection of methods that people learn or stay connected with their learning environments including their classmates, instructors, and instructional resources while going portable. In generally m-learning is considered to increase the performance of learners by making learning accessible and eliminates geographic boundaries and provides collaborating learning environment.

However, this new method is still overwhelmed with some questions among researchers, educators and learners because the learning focus is being distracted by the technology itself [5]. Some of the main discussion argues the ability of m-learning in achieving one of the main objective in learning process [6], whereby learning should be able to provide a real understanding of what is learned [7].

1.2 Lifelong Learning

Famous words seek knowledge from the cradle until one dies describing the nature of life-long learning. It pictures the various types of education that one has to go through in one's life: formal education, non-formal education and informal education. Formal education involves structured learning with a planned curriculum, proper classroom and qualified teachers; non-formal education involves being outside formal settings such as on a field trip to a museum; and informal education occurs out of spontaneous situations.

The term of lifelong learning has been interpreted as the training of a workforce capable of adapting to a rapidly changing world [8], [9], [10]. Those researchers mention that the basic idea of lifelong learning is deliberate learning that can and should occur throughout each person's life. The concept of lifelong learning as life-wide learning that joints its successful endeavor on the potential of distance education via computer-mediated communication [11]. Lifelong learning is often used in relation to adult education or continuing education. Therefore, all forms of education that one receives after one has completed one's formal education can be classified as part of the process of lifelong learning [10].

The aim of lifelong learning is to improve knowledge, skills and competence with a personal, civic, social or employment-related perspective [12]. The unique characteristic of lifelong learning is the fact that it centres around individuals, such as the learners themselves. In Life long learning, the main of the learning practice is the learner: a person with certain physical characteristics who assumes a number of social roles and who has a number of characteristics that relate to how they practise learning. For the reason that of the lifelong learning it centres on the individual, the use of technology in delivering becomes essential part. The use of technology such as m-learning can provide a flexible learning framework which is often preferred by adult learners [10], [11].

2. Literature Review

First, this literature review will look at the previous studies related to the m-learning in term of the trends of researchs, disciplines and courses which often used m-learning. Second, this literature review will look at the adult learners and their characteristics.

2.1 Previous Research of M-learning

Study related to the m-learning has intensified by various parties whether the academia for education purposes or industrial sector in order to expand, improve or streamline the products and services related to m-learning. Result of the studies conducted, the definition of m-learning is becoming clear whereas in the early 2000s. Most of the researchers said m-learning is an e-learning that use mobile computational devices such as smartphone, android phone, or else from digital cell phone [13]. This

***** Formatting Issue - Best Available Paper/Graphic *****

definition means that the evolution of m-learning is related to the tools, devices, or equipment to be used while in terms of learning it is still the same concept as e-learning. M-learning is not only learning through mobile phones or learning over a wireless connection but it is the evolution of e-learning, which completes the missing component of an e-learning solution that most suits for those mobile users in education institutes. However, in general most researchers define m-learning as a learning process that utilize mobile technology in order to be connected to the information anytime and anywhere. Previous researchs of m-learning fall into two broad research directions: evaluating the effectiveness of m-learning, and designing m-learning systems. Figure 1, shows the review of trends from m-learning studies that show the distribution of m-learning studies by research purpose [14].

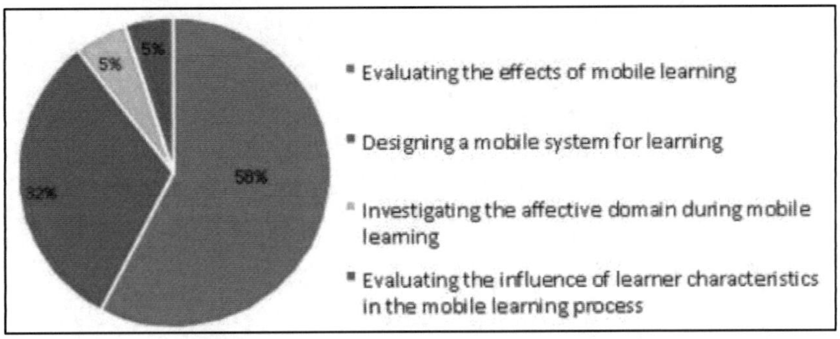

Figure 1. Distribution of mobile learning studies by research purpose

The research purposes of most mobile learning studies center on effectiveness, followed by mobile learning system design. Based of the following research [14], the highest topics of the researchs in m-learning is evaluating the effects of mobile learning, 58%, followed by designing a mobile system for learning, 32%. Whereas, the topic of investigating the affective domain during mobile learning and evaluating the influence of learner characteristics in the mobile learning process was only 10% from the distributions. While some of the studies found trends in mobile learning research identified into seven major findings (refer Table 1) that it is particularly significant that most studies of m-learning focused on effectiveness and identify this as a significant new finding [15].

Table 1. Trends in m-learning research

No.	Trends in M-Learning
1	The research purpose of mobile learning studies focused on effectiveness, followed by mobile learning system design.
2	Surveys and experimental methods were the preferred research methods, regardless of whether the research purpose focused on evaluation or design.
3	Research outcomes in mobile learning were significantly positive.
4	Mobile phones and PDAs were the most commonly used device for mobile learning, but these may be replaced in the future by new emerging technologies.
5	Mobile learning is most prevalent at higher education institutions, followed by elementary schools.
6	Mobile learning most frequently supports students in the professions and applied sciences, followed by the humanities and formal sciences.
7	The most highly cited articles fell into categories of mobile learning system design followed by effectiveness.

Figure 4 shows that m-learning in educational contexts most frequently focus on use in supporting professional subjects and applied sciences (29%), followed by the humanities (20%), and formal sciences (16%). Studies showed mobile learning was often used in computer and language courses [14], [15]. The study found that mobile learning is also widely used in courses related to environmental studies, forestry and health sciences, but considerably less so in other courses such as statistics or law.

IOP Conf. Series: Journal of Physics: Conf. Series **1019** (2018) 012070 doi:10.1088/1742-6596/1019/1/012070
***** Formatting Issue - Best Available Paper/Graphic *****

Figure 2. Distribution of disciplines and courses

2.2 Adult Learners and Their Characteristics

The word 'adult' can refer to 'a stage in a life cycle'. In other words, biologically, a person is first a child and then a youth before reaching the adult stage. The term 'adult' can also refer to the social role given to a person who plays a fully independent or responsible role in society. Malcolm Knowles (1913-1997) was an American educator who theorized adult education. He brought the term andragogy into use as a term associated with adult education. According Malcolm Knowles, andragogy is the art and science of adult learning, thus andragogy refers to any form of adult learning [16]. Knowles has proposed a more systematic approach to defining the term "adult", in which understanding what an adult is means considering both social and psychological aspects [16], [17]. Generally, to Knowles, a person is an adult if he or she is capable of performing the social roles typically assigned by his or her culture, such as parent, spouse, worker, or responsible citizen. Psychologically, one is considered an adult when he or she is capable of making decisions and taking full responsibility of the outcomes of those decisions. In other words, adult learners essentially responsible for his or her own life. Dr. Knowles gave adult learners 5 characteristics [16], [17], [18]:

(1) Self-concept: As a person matures his self concept moves from one of being a dependent personality toward one of being a self-directed human being
(2) Experience: As a person matures he accumulates a growing reservoir of experience that becomes an increasing resource for learning.
(3) Readiness to learn. As a person matures his readiness to learn becomes oriented increasingly to the developmental tasks of his social roles.
(4) Orientation to learning. As a person matures his time perspective changes from one of postponed application of knowledge to immediacy of application, and accordingly his orientation toward learning shifts from one of subject-centeredness to one of problem centredness.
(5) Motivation to learn: As a person matures the motivation to learn is internal.

3. Research Model and Finding

In this study, four (4) research papers which have same topic on adoption of M-learning for adult learners have been reviewed. The paper are; **1.** Adult learners' intention to adopt mobile learning: A motivational perspective by Kamarul, Tan and Rashid (2015), **2.** Toward a learner-centered system for adult learning by Hermans, Kalz and Koper (2013), **3.** Research on Mobile Learning in Adult Education by Liu and et al. (2013) and **4.** Adult Learners and Mobile Learning: A Case Study by Faizah (2012). From the listed research paper above, we observed some of the issues and finding from several perspectives views.

3.1 Finding Issues

There are several finding issues are highlighted from the surveys as shown in Table 2.

Table 2. Research finding

Research purposes	Study	Participants	Methods	Finding
Adult learners' intention to adopt mobile learning: A motivational perspective.	Kamarul, Tan and Rashid (2015)	Adult University students	Web Survey	This study suggested that adult learners' intention to adopt m-learning is influenced by cognitive, affective and social needs through attitude.

IOP Conf. Series: Journal of Physics: Conf. Series **1019** (2018) 012070 doi:10.1088/1742-6596/1019/1/012070
***** Formatting Issue - Best Available Paper/Graphic *****

Toward a learner-centered system for adult learning	Hermans, Kalz and Koper (2013)	Adult University students	Experimental Research	A model of an e-learning system that integrates the concepts of Virtual Learning Environment, Personal Learning Environments, and Social Network Sites for the adult learner. It puts the learner in control of his/her learning supported by a personal context (personalized workspace) for managing the learning process [20].
Research on Mobile Learning in Adult Education	Liu and et al. (2013)	-	Literature Research	The literature on mobile learning with adult learners has provided preliminary evidence that can inform issues with use of mobile devices to enhance teaching, learning, and training [21].
Adult Learners and Mobile Learning: A Case Study by Faizah	Faizah (2012)	Lecturers	Case Study (Thematic Analysis)	M-Learning has great potential, and the participants have all claimed to be implementing it. However, the quality of their implementation was heavily impacted by their understanding of M-Learning and the needs of adult learners [18].

***** Formatting Issue - Best Available Paper/Graphic *****

Figure 3. Adult learners' intention to adopt m-learning.

4. Conclusion

The important reason for this study is there have been no extensive studies that have been done on m-learning acceptance for adult learners in Malaysia. Therefore, this study endeavors to address these issues. It also is to contribute the body of knowledge in the fields of m-learning and technology acceptance in lifelong education in Malaysia by providing information needed to the academicians and mobile application developer in Malaysia. It is important and necessary to promote academically the new technology to users in Malaysia. It also will help Malaysia academicians to understand the users' view towards the utilization of m-learning as well as to understand the variety of factors that can influence their intention and acceptance. With information on how to use m-learning by identifying the factors as realized by the learners, it is expected that the finding from this research will be applicable and helpful. Hopefully this research will give a contribution, suggestions and guideline to the application developer in the industry to produce the most suitable framework features of mobile application.

References

[1] Brown, K., Campbell, S., C., and Ling, R. (2011). Mobile Phones Bridging the Digital Divide for Teens in the US: Future Internet.

[2] Tina Lim, Mansor Fadzil, and Norziati Mansor (2011). Mobile Learning via SMS at Open University Malaysia: Equitable, Effective, and Sustainable. Mobile Learning via SMS at OUM. The International Review of Research in Open and Distance Education Learning, Vol 12, No 2 (Feb 2011).Open University Malaysia, Malaysia

[3] Low, L. & O'Connell, M. (2006), 'Learner-centric design of digital mobile learning'. Paper presented at Learning on the Move, Brisbane, Australia. Retrieved 25 October 2006 from https://olt.qut.edu.au/udf/OLT2006/gen/static/papers/Low_OLT2006_paper.pdf

[4] Yi, C. C., Liao, W. P., Huang, C. F. & Hwang, I. H. (2009). Acceptance of mobile learning: a respecification and validation of information system success. In Proceeding of World Academy of Science, Engineering and Technology

IOP Conf. Series: Journal of Physics: Conf. Series **1019** (2018) 012070 doi:10.1088/1742-6596/1019/1/012070
***** Formatting Issue - Best Available Paper/Graphic *****

[5] Göth, C., Frohberg, D., andSchwabe, G. (2006). The Focus Problem in Mobile Learning : IEE International Workshop.

[6] Baggio, B., and Beldarrain, Y. (2011). Anonymity and Learning in Digitally Mediated Communications - Authenticity and Trust in Cyber Education : IGI Global

[7] Huitt, W. (2011).Bloom et al.'s Taxonomy of the Cognitive Domain.Educational Psychology Interactive. Valdosta, GA: Valdosta State University. Retrieved on 11 January 2012 : http://www.edpsycinteractive.org/topics/cogsys/bloom.html [pdf]

[8] Sharples, M. (2010).A Theory of Learning for the Mobile Age. : Nottingham University

[9] Knapper, Christopher K., and Cropley, Arthur, J. 2000. Lifelong Learning in Higher Education. 3rd ed. London: Kogan Page. 238 pp

[10] Nordin, Norazah & Embi, Mohamed & Yunus, Melor. (2010). Mobile Learning Framework for Lifelong Learning. Procedia - Social and Behavioral Sciences. 7. 130-138. 10.1016/j.sbspro.2010.10.019.

[11] Rozhan Idrus, R., & Hanafi Atan (2007). Lifelong and lifewide distances education in Abdul Razak, N., Kamarul Kabilan, M. and Silong, A. D. (eds) Online Lifelong Learning in Malaysia: research and practice (pp. 1– 8). Serdang: Universiti Putra Malaysia Press.

[12] Hamidah Yamat, Ahmad Ismail & Aliza Alias. (2007). English Language Skills for Lifelong Learning: A Question of Acquisition or Learning? in Proceedings of International Conference on Lifelong Learning, Bangi: Faculty of Education, University Kebangsaan Malaysia Press.

[13] Baggio, B., and Beldarrain, Y. (2011). Anonymity and Learning in Digitally Mediated Communications - Authenticity and Trust in Cyber Education : IGI Global

[14] Hwang, G. J., & Tsai, C. C. (2011). Research trend in mobile and ubiquitous learning: a review of publications in selected journal from 2001 to 2010. British Journal of Education Technology, 42(4), E65–E70.

[15] Wu, Wen-Hsiung & Wu, Yen-Chun & Chen, Chun-Yu & Kao, Hao-Yun & Lin, Che-Hung & Huang, Sih-Han. (2012). Review of trends from mobile learning studies: A meta-analysis. Computers & Education. 59. 817–827. 10.1016/j.compedu.2012.03.016.

[16] Kearsley, G. (2010). Andragogy (M.Knowles). The theory Into practice database. Retrieved from http://tip.psychology.org

[17] Faizah, AM. & Hazadiah, MD. (2009). Catering for adult learners through the transformation of Higher Education. Bulletin of Higher Education Research, 14.

[18] Faizah, AM. (2004). Adult learners' characteristics and their academic reading strategies: a case study. Unpublished Doctoral Thesis. Bangi, Malaysia: Universiti Kebangsaan Malaysia.

[19] Kamarul, FH. , Tan, FB. , and Rashid, A (2015). Adult learners' intention to adopt mobile learning: A motivational perspective. British Journal of Educational Technology, Vol 46 No 2 2015.

[20] Henry Hermans, Marco Kalz, Rob Koper, (2013) "Toward a learner-centered system for adult learning", Campus-Wide Information Systems, Vol. 31 Issue: 1, pp.2-13, https://doi.org/10.1108/CWIS-07-2013-0029

[21] Liu, M., Geurtz, R, Karam, A., Navarrete, C. and Scordino, R. (2013), Research on Mobile Learning in Adult Education. In W. Kinuthia & S. Marshall (Eds.) On the Move: Mobile Learning for Development, Charlotte, NC: Information Age Publishing.

Histogram equalization smoothing for determining threshold accuracy on ancient document image binarization

Mahendar Dwipayana[1,a)], Fitri Arnia[2], Zuhar Musliyana [1,b)]

[1]*Department of Information System, Faculty of Computer Science, Ubudiyah Indonesia University, Jalan Alue Naga, Desa Tibang, Banda Aceh*
[2]*Magister Electrical Engineering, Syiah Kuala University, Darussalam Banda Aceh*

[a)] *mahendar@uui.ac.id,* [b)] *zuhar@uui.ac.id*

Abstract. Ancient documents are inheritance that must be preserved. The documents contain historical, scientific, social, religious information, etc. Converting ancient documents into digital image formats is one of ways to preserve the inheritance and can be stored into a computer. However, images of ancientdocuments have many blemishes caused by age, moisture, flood, etc. Therefore, special techniques are needed for those images to be restored and can improve the legibility of the ancient documents' images. In this study, the image restoration process uses separation of background and foreground/text on histogram equalization such as research conducted by Fitri Arnia in 2008. Through histogram equalizationimages can be seen the distribution of pixels from the intensity of black color "0" to white "1". The distribution of pixels on histogram equalization describes the curves of foreground/text and curves of background. Among the histogram curves, the determination of thresholdvalues can be done so as to clarify the foreground/text and background areas on images of ancient documents. The lowest point between the two curves is the lowest pixel (local minima) which is used as the threshold value. However, the selection of such threshold values in some cases is very difficult to determine because there are still many fluctuations in the curve at the lowest curve. Therefore, this study proposesa histogram smoothing method in the ancient documents' images to minimize curvature fluctuations and to determine more accurate threshold values. In this research, average filtering method is used for smoothing the histogram image. This filter successfully refines the histogram and makes the image of the restoration or binary image display the value of the ancient document image readability increases.

Keywords: HistogramEqualization, Smoothing Histogram, Average Filtering, Thresholding

1. Introduction

Many of ancient documents found so far are in very bad condition due to their age, humid storage and so on. In those documents there are many disturbances that make the document difficult to read. Therefore, it is necessary to restore the information contained in ancient documents by converting it first into digital format / digitalization so that reconstruction can be done.

In this study the process of restoration of ancient documents using background and foregraound separationtechniques such as in existing researches and by using histogram equalization [1]. Histogram equalization is useful in fulfilling pixel gradation level and adding color contrast between background and foreground/text. In histogram, the lowest curve or threshold value is obtained which is the reference point for separation between background and foreground.

Histogramequalization method on that study [1] has successfully eliminated fox and noise. This method is not like method used in the research of Otsu [2] that automatically divides the gray level image. The method Otsu used was a large threshold value so that the pixels obtained accumulate on the black color causing some text to be affected. In contrast to *wafa Bousella*, et al., they used the maximum likelihoodand k-means clustering -based estimation methods [3] and iterations with recursive algorithms in separating the background and foreground/text [4].

In this study, the process of restoration of ancient documents' images using four smoothing histogram methods. Those are mean filtering, median filtering, wiener filtering, and cubic spline which are smoothing methods in histogram equalization to facilitate the determination of threshold values. Of the four smoothing methods will be obtained a different binary image on each method. The difference in the results of this binary image will be measured using the recall and precision parameters. These parameters are useful to determine the ability of each smoothing method in restoring ancient document images.

2. Methodology

Subjectin this research is the field of science of digital image processingby using processing method for image quality enhancement. The objects of this study are images of ancient documents derived from Acehs inscribedin Arabic Malay. The steps of test performed in this study can be seen in Figure 1 as follows.

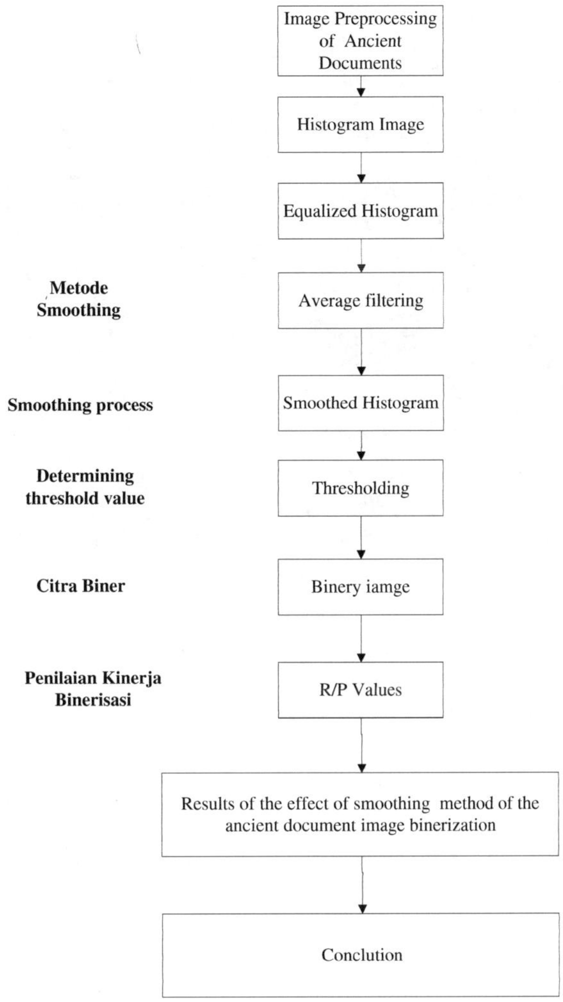

Figure 1. Flow of Implementation Method

Image Preprocessing

Pre-precessing is the processing of image data for further analysis. Pre-processing includes mostly is by changing the colored image (RGB) into a grayscale image. Grayscaling is served to simplify an image model to make the image easier to be processed

In general, to generate grayscale image the following formula is written:

$$S = \frac{r + g + b}{3} \quad (1)$$

Where S is the grayscale image by searching for the mean of each layer of r (red), g (green), and b (blue). Below is achange image from RGB image to grayscale image.

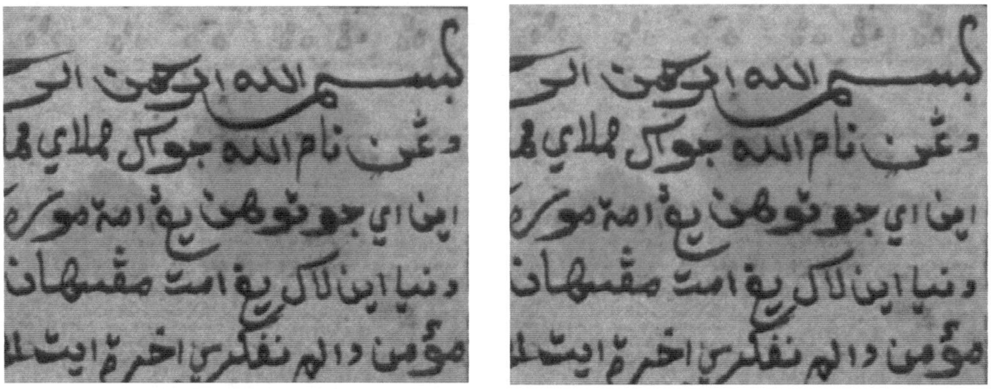

a. Colour Image(RGB) b. Grayscale Image

Figure 2. Changes of RGB image to grayscale

Histogram Equalization and Normalization

A histogram in a digital image is a graph that represents the color distribution of a digital image showing the intensity of pixel values of an image. The mapping of pixel values in the histogram is as follows [1]:

$$h(n_k) = n_k \tag{2}$$

Where $n_k$is the axis denoting the pixel value (k = 1-255) and $h(n_k)$ is the ordinate representing the number of pixels for each pixel.

In this study the histogram in an image should be normalized before the histogram equalization process. The benefits of histogram normalization is to see statistics of an image divided by the total number of pixels in the image. Normalization of histogram can be defined as below [1]:

$$p(n_k) = \frac{h(n_k)}{n} = \frac{n_k}{n} \tag{3}$$

Histogram equalization is an image enhancement technique by manipulating each image pixel in which the spread of the original image histogram is not evenly distributed because the pixel distributiondoes not keep the entire level of gradation available on histogram [1]. This process results in pixel values evenly distributed in the interval (0-255).

Histogram is equalized mathematically and can be performed with the following equation [1]:

$$T(n_k) = \sum_{j=1}^{k} p(n_k) \tag{4}$$

Where $p(n_k)$ is the ordinate that states the number of pixels for each pixel. While $T(n_k)$ is the location where the $n_k$intensity value will be mapped.

IOP Conf. Series: Journal of Physics: Conf. Series **1019** (2018) 012071 doi:10.1088/1742-6596/1019/1/012071

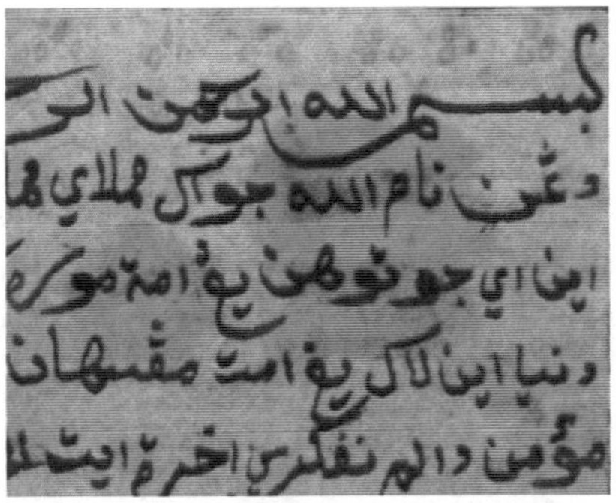

a.

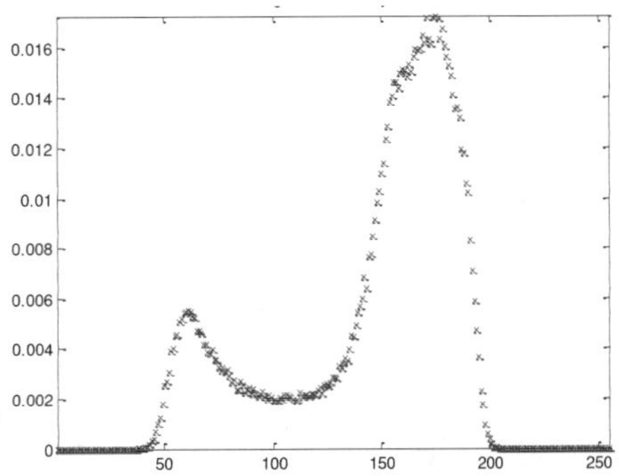

b.

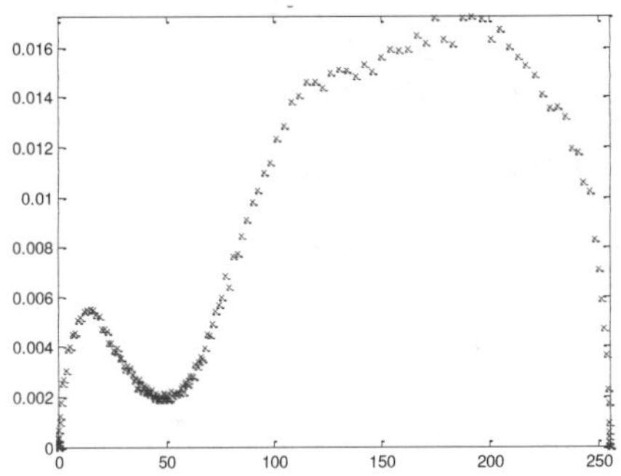

c.

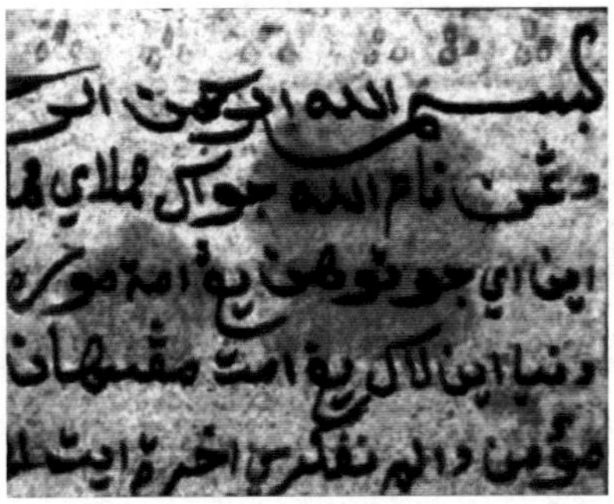

d.

Figure 3. (a) Document image, (b) Image histogram, (c) Equalized histogram, (d) Imagery of histogram equalization

Average filtering

Average filtering is filter which issearching foraverage values of data set [5]. The formula of calculating Average filtering is as follows:

$$\overline{X} = \frac{1}{n}\sum_{i=1}^{n} x_i \tag{5}$$

Where $\overline{X}$ is the average, n is the number of data, $x_i$is i value and i is the initial value.

Thresholding dan Binerisasi

Thresholding or determining the threshold value is the process of separating pixels according to the degree of gray they have. [7] The threshold value of the histogram represents the object and the background. The provisions in determining the threshold values are as follows.

$$g(x,y) = \begin{cases} 1 & \text{if } f(x,y) > T \\ 0 & \text{if } f(x,y) \leq T \end{cases} \tag{6}$$

Where $g\ (x,\ y)$ is the image segmentation between object and backround, T is threshold value, and $f(x,\ y)$ is the image dimension. If $f(x,y) > T$ then it is called background, if $f(x,y) \leq T$ then it is called object or foreground.

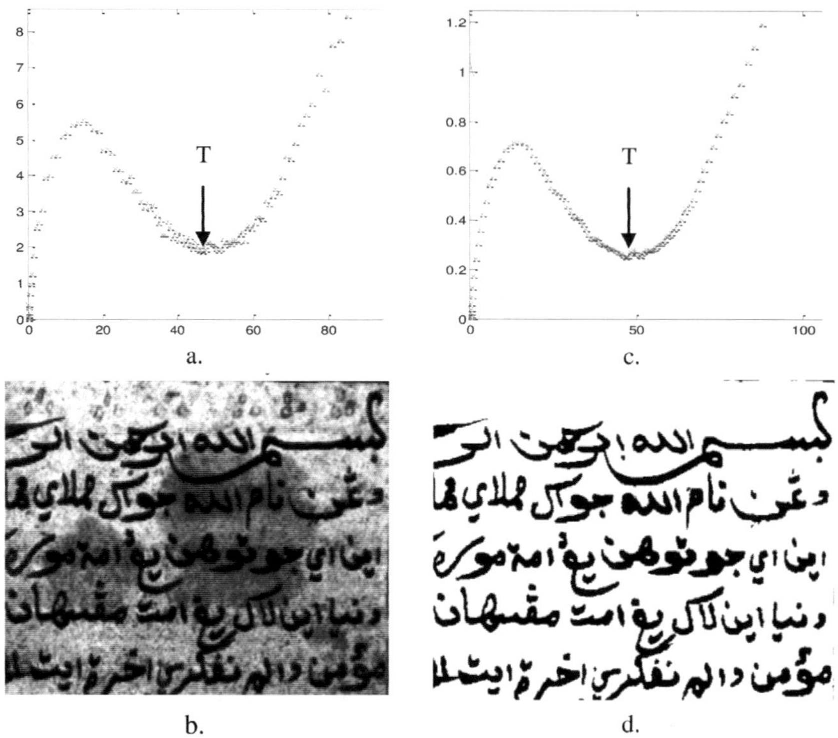

Figure 6. (a) equalization histogram (b) the image of equalization histogram, (c) the smoothed histogram, (d) binarization result

3. Materials

The imagesof ancient documents used were ancient documentsinscribed with Arabic that had been digitized and had been pre-processed i.e changed it in grayscale to facilitate the binarization process. Images used were in the format of ".tif" which ahve dimensions of 1320x2000 pixels per image. The images of ancient documents used in this study were 10 images with characteristics of 5 low-noise images and 5 high-noise images. Software testing used in this study was MATLAB application.

3.1 Methods

The method used to test readability value using Recall and Precision.Recall parameters is the size of the number of relevant documents retrieved from document set at the time the query is applied. While precision is a measure of the accuracy or relevance of query results. A application of method research is by counting all characters of readable texts and unreadable texts. With these parameters,the precentage of readability value before and after the proposed method appliedwill be obtained.

Recall and precision in this study were used to evaluate retrival of text characters against applied methods by measuring the number of relevant and irrelevant characters. Recall and precision are typically rated in percentages of 1 to 100%. High recall value means little false negative and high precision value means little false positive. Recall and precision can be calculated by the following equation.

$$Recall = \frac{NCD}{GT} \tag{7}$$

$$Precision = \frac{NCD}{TR} \tag{8}$$

Where NCD is the correct number of characters detected in the binarization result document. GT is the total number of characters contained in the original document. And TR is the number of characters detected in the binarization result document including the correct and damaged characters.

The GT (Ground-truth) of the document image is searched manually by counting the number of characters read and the damaged characters in the original document image. The detection of NCD and TR by following GT (Ground-truth) [13].

4. RESULTS

Ancient document images produce a histogram which is a representation of the appeared color intensity. An equalized histogram has unfavorable image curve that causes the determination of the threshold value to be very difficult. Histogram smoothing method is required so thresholding can be done. In this research method used in smoothing curve at local minima curve is Average filtering. Image histogram before and after average filtering performed can be seen in figure 7.

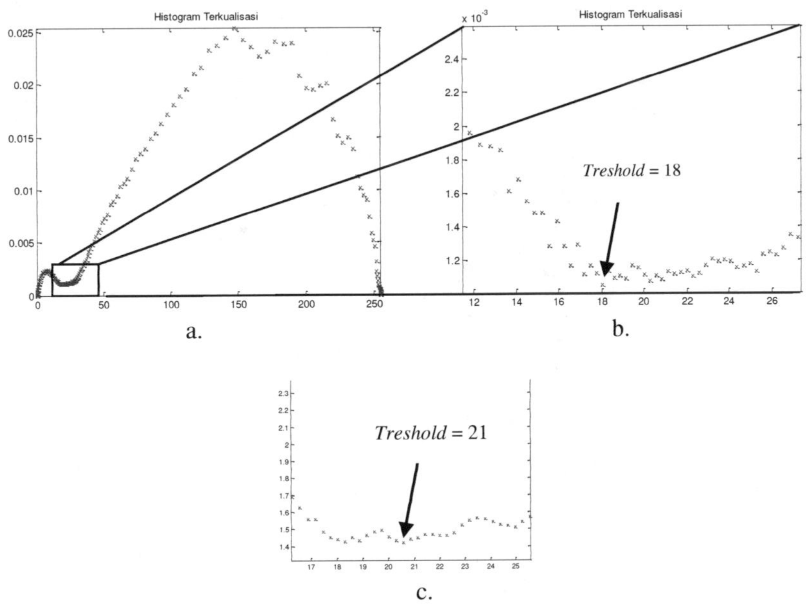

Figure 7. (a) Equalized histogram, (b) Enlarged Histogram, (c) H. Average filtering

From the above smoothing histogram, we get different threshold values before and after filtering process. To find out the results of threshold values generated by some ancient document images, it can be seen in table 1 and table 2. Table 1 shows some images that have low noise with difference before and after smoothing process performed. While table 2 shows images that has high noise qualification and show the difference before and after the smoothing process performed.

Table 1. The Results of Threshold ValuesOf Some Low-Noise Images

File	Histeq	Average

2l.tif	18	21
2r.tif	14	14
3l.tif	28	29
3r.tif	26	25
4l.tif	26	26

Table 2. Results of Threshold Values From Some High-Noise Images

File	Histeq	Average
1l.tif	26	24
20l.tif	30	29
23l.tif	14	15
23r.tif	18	18
24l.tif	13	13

From Table 1 and Table 2 above there are several variations of threshold values of each method. If we look carefully at the result of determination of threshold values in each method is not much different from the result of the threshold values in the histogram. But from the side of the binarization results it becomes important because one number on the different threshold value will affect the number of successfully restored characters.From the binarization results there are differences that on**figure 8.a** noises contained in the image are less than in image of **figure 8.b**

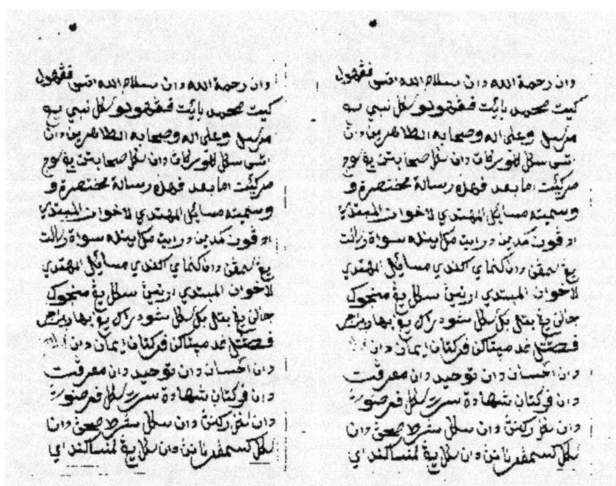

a. b.

Figure 8. Binarization image difference (a) "21" threshold value and (b) "18" threshold value

To find out the binarization performance or readability values of documents in the ancient documents' images above, this research uses Recall and Precision parameters to find out howmuch the successful percentage is in the restoration of characters contained in those images. Here are some Recall and Precision results.

Table 3. table of recall and precision of low-noise images

Image Name	Recall and Precision
	Average Filtering
2l.tif	98.81%
2r.tif	97.78%
3l.tif	99.08%
3r.tif	99.52%
4l.tif	99.28%

Table 4. table of recall and precision of high-noise images

Image name	Average Filtering	
	Recall	Precision
1l.tif	94.50%	96.30%
20l.tif	97.51%	98.73%
23l.tif	90.93%	97.34%
23r.tif	97.75%	99.54%
24l.tif	68.73%	98.31%

5. Discussion

The determination of threshold values on histogram has constraints on unsmoothed histogram curves. Using the smoothing histogram method in this study has proven very helpful. Average Filtering successfully refines the histogram and clarifies curves so that the lowest points (local minima) look more clearly. The result of binarization on each method is known its difference after Recall and Precision calculations. Recall and Precision count how many characters are successfully restored and the damaged characters are compared to the characters in original images.

6. Conclusion

From the table of determination of threshold values, the different valuescan be seen. The determination values of threshold values before and after the smoothing histogram process look very much different. However, in binarization results the ancient documents'imagesresult in slightly different readability values. This depends on the level of noise or condition of the ancient document images.

REFERENCES

[1] F. Arnia dan K. Munadi, "Metode Restorasi Citra Manuskrip Kuno Berbasis Histogram Terekualisasi", Seminar Nasional Teknologi Informasi, hal. A12, 59-63, 2008.

[2] R. C. Gonzalez dan R. E. Woods, *Digital Image Processing,* 2nd Ed. Practice Hall, 2002.

[3] R. Munir. "Pengolahan Citra Digital dengan Pendekatan Algoritma" Bandung, Penerbit Informatika, 2004.

[4] J. Utama, "Akuisisi Citra Digital Menggunakan Pemrograman Matlab", Jurnal Majalan Ilmiah UNICOM, Vol. 9, No.1, 2011.

[5] B. Yuwono, "Image Smoothing Menggunakan Mean Filtering, Media Filtering, Modus Filtering dan Gaussian Filtering", Jurnal UPN Veteran Yokyakarta, Vol 7, No.1, 2010.

[6] R. S. Lasijo, "Fitting Kurva dengan Menggunakan Spline Kubik", INTEGRAL, vol. 6, no. 2, 2010.

[7] E. Kavallieratou dan H. Antonopoulou, "Cleaning and Enhancing Historical Image", LNCS3708, pp. 681-688, 2005

IOP Conf. Series: Journal of Physics: Conf. Series **1019** (2018) 012072 doi:10.1088/1742-6596/1019/1/012072

The Enrichment Methods Viewpoint Oriented Requirements Definition (VORD) With The Capability Model Integration (CMMI) And Proto Personas Methods For Needs Analysis

Muhammad Bayu Wibawa[1,a)], I Made Wiryana[2,b)]

[1]*Department of Informatics Engineering, Faculty of Computer Science, Ubudiyah Indonesia University, Jalan Alue Naga, Desa Tibang, Banda Aceh*
[2] *Gunadarma University, Jakarta*

[a)]mbayuw@uui.ac.id, [b)]mwiryana@staff.gunadarma.ac.id

Abstract. Software needs begins with a statement of the need for the parties involved in the system. Needs a description of the statements coming from the client, user, or stakeholder. Some of the activities that must be done to get a system needs so that produce a specification document that be the benchmark for the developers in the development of a system, including: requirements elicitation, requirements analysis and negotiation, documentation requirements, requirements validation. One method that can be applied to requirements engineering is a Viewpoint Oriented Requirement Definition (VORD) method. VORD is one method used to define the need to approach the viewpoint system users. The results of the analysis of this VORD method produce the required specification with the help of a tool, that is VORD tool, the VORD method is still too general, so need some method to support the VORD methods, including: proto personas methods and a CMMI approach. This method is very helpful for independent software developers to build a better system, and the result can be a documentation for developers and consumers.

Keyword : *Requirement Engineering, VORD Methode, Proto Personas, CMMI, Requirement Specification*

1. Introduction

Software development begins with a statement about the need for the parties involved in the system. Description needs statement from the client, users or stakeholders that determine the features needed in a system. The need not only to describe the needs of users, but also standards organizations, government and industry. It is often overlooked by the developers in terms of attention to the quality of a system to be built. Some of the developers are still less attention to documents related to the system to be built, and this will result in a system needs to be not optimal.

Needs a system can be analyzed by several methods, one can use methods Viewpoint Oriented Requirement Definition (Vord). Vord is one method to analyze the needs of the system by using viewpoint. A method angle approach was developed to assist in the specification of the interaction of the system. Vord focused on the external entities that interact with the system [5,12,13]. Vord methode still too general, that takes an approach that can help Vord method approach that aims to increase the maturity of the organization by providing guidance on improving the process of developing a product and service. This approach is called with the Capability Maturity Model Integration (CMMI) for Development. CMMI consists of practical things which refers to the maintenance and development activity [2,4,]. Vord is part method rather than the process of software engineering, or collectively, the Requirement Engineering (RE). Requirement Engineering is the leading phase of software engineering processes, in which software requirements from users and customers is collected, understood and defined. The results of this phase of the engineering requirements documented in the requirement specification. Requirements Specification contains a mutual agreement about the problems that want to be solved between developers and customers, and is the starting point towards the next process design software (Sommerville, 2003)[1,6,9,13,16]. To get data about problems and solutions desired by consumers, then need a tool that can map out the problems and the solution, the tool i.e. Proto Personas. Proto Personas are an a method for obtaining and ensuring harmony between the user and the analysis to the needs of the system. In making an analysis, first make a sketch on an analytical paper, then the assumptions of people put in a related paper that has been . From these assumptions will be studied and analyzed [8]. The results of the analysis can be described using Use Case diagrams. A picture of the Use Case Diagram reflects the functional system requirements that must be met in accordance with the mutual agreement between the developers and consumers [8]

The research focuses on enriching the Vord method is used as a method to analyze the needs of the system which was seen from the external entity. The most important thing from this research is how to get the data and next do the mapping of the data obtained. The results of the mapping, then becomes a documentation about the needs of the system that is built or called the Requirement Specification.

2. Methodology

In this study there are several stages to the needs of the system, including:

1. The use of methods, Proto Personas are to support of the VORD methods, particularly to obtain information about the needs of the users on the system. The Proto Personas format can bee seen in Figure 1.

1. Sketch and Name	2. Behavioral demographic information
3. Pain point and Name	4. Potential Solutions

Figure 1. Proto Personas Format

2. Use case, helping to represent the interaction between actors in the system to the needs of the function of any type of actor.

3. The method of VORD, this method to analyze the needs of the users of the system with how to determine the viewpoint and then explain the viewpoint needs into the table service and point of view. VORD method of requirement analysis can bee seen in Figure 2.

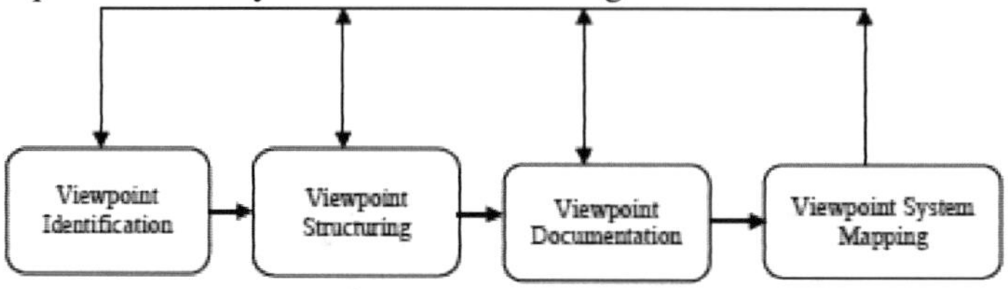

Figure 2. VORD Method of Requirement Analysis.

4. CMMI, is an approach that practices presents detailed activity that is already standard, especially in software development.
   Stages of research can be seen in Figure 3.

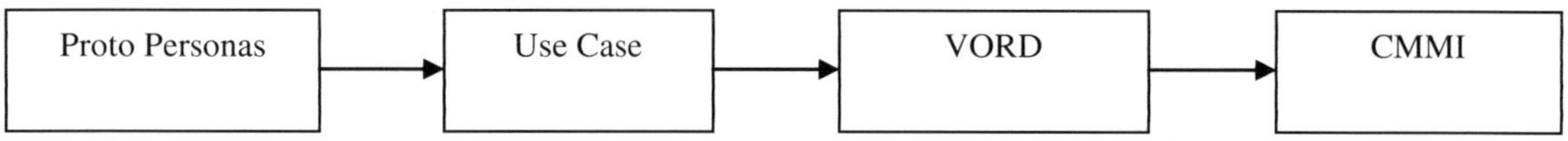

Figure 3. Stages Of Research Methods

3. Materials

Use the tools in the form of software or hardware on this research aim to ease in applying the methods used in this study, some of the tools used in this research are: Google Form, the use of the facilities of google form for applying the method, the results of personas proto google form this will be data that contains about consumer problems and solutions desired by consumers. Mysql, use Mysql in this research as a database. ASP.NET, basically VORD has had its own data processing as a tool that has been mapped from proto personas, but researchers more made their own software to make it easier in terms of modifying if there is a change. The output of this tool generates the VORD documentation system requirements of the projects examined. The purpose of the Diagram, the use of this diagram is to illustrate the needs of system functions and describe the flow of system usage by the user. Diagrams are used, among others, use case diagrams and sequence diagram.

4. Results

The results of the needs analysis system on this research in the show in the form of pictures, among them: viewpoint template and viewpoint service templates, use case diagrams. This result was obtained on the basis of the methods applicable to the method of assisted with some of the VORD and other support methods:

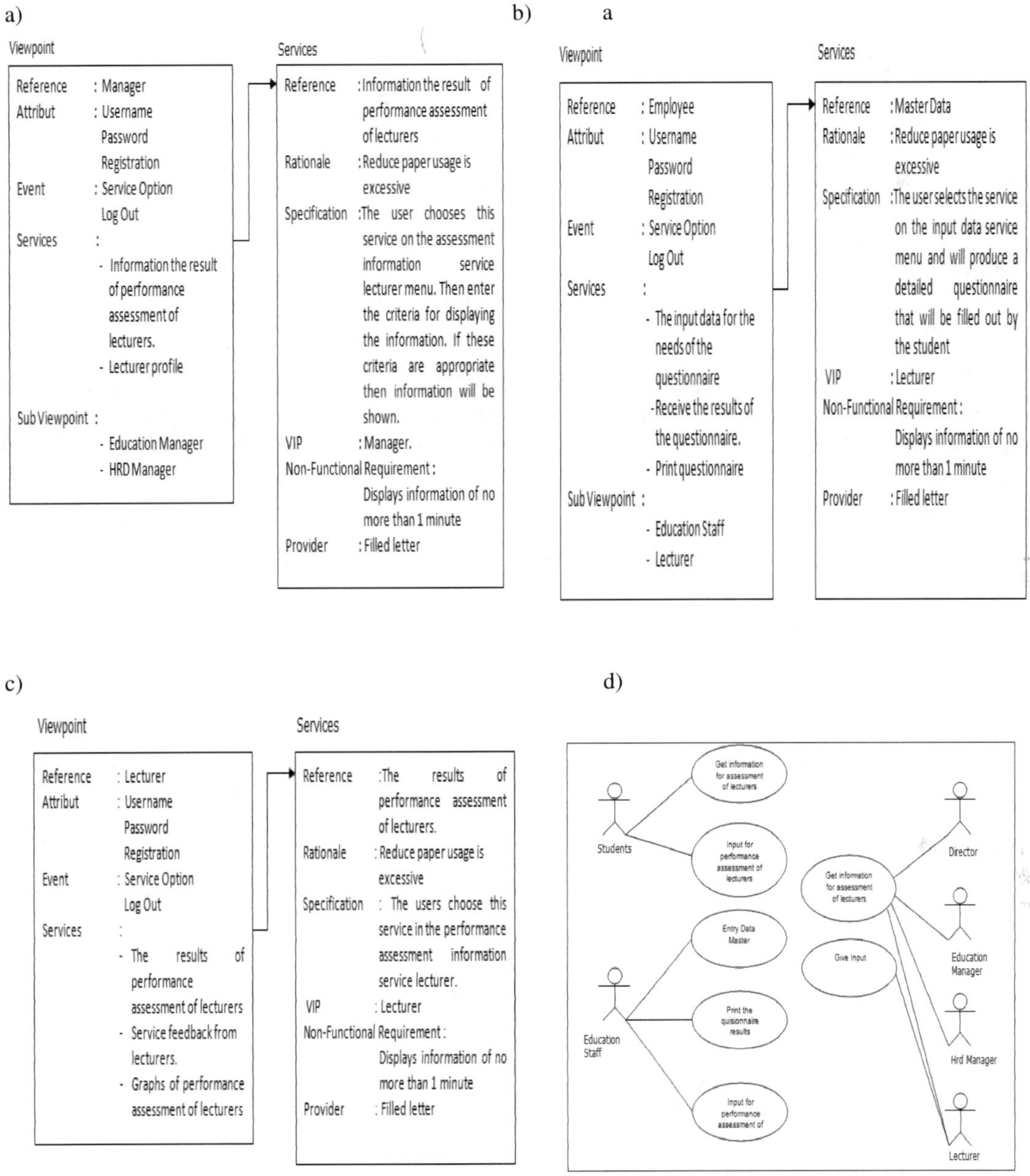
a)
Viewpoint
Reference : Manager
Attribut : Username
Password
Registration
Event : Service Option
Log Out
Services :
- Information the result of performance assessment of lecturers.
- Lecturer profile
Sub Viewpoint :
- Education Manager
- HRD Manager
Services
Reference : Information the result of performance assessment of lecturers
Rationale : Reduce paper usage is excessive
Specification : The user chooses this service on the assessment information service lecturer menu. Then enter the criteria for displaying the information. If these criteria are appropriate then information will be shown.
VIP : Manager.
Non-Functional Requirement : Displays information of no more than 1 minute
Provider : Filled letter
b) a
Viewpoint
Reference : Employee
Attribut : Username
Password
Registration
Event : Service Option
Log Out
Services :
- The input data for the needs of the questionnaire
- Receive the results of the questionnaire.
- Print questionnaire
Sub Viewpoint :
- Education Staff
- Lecturer
Services
Reference : Master Data
Rationale : Reduce paper usage is excessive
Specification : The user selects the service on the input data service menu and will produce a detailed questionnaire that will be filled out by the student
VIP : Lecturer
Non-Functional Requirement : Displays information of no more than 1 minute
Provider : Filled letter
c)
Viewpoint
Reference : Lecturer
Attribut : Username
Password
Registration
Event : Service Option
Log Out
Services :
- The results of performance assessment of lecturers
- Service feedback from lecturers.
- Graphs of performance assessment of lecturers
Services
Reference : The results of performance assessment of lecturers.
Rationale : Reduce paper usage is excessive
Specification : The users choose this service in the performance assessment information service lecturer.
VIP : Lecturer
Non-Functional Requirement : Displays information of no more than 1 minute
Provider : Filled letter
d)
Students
Get information for assessment of lecturers
Input for performance assessment of lecturers
Entry Data Master
Print the quisionnaire results
Input for performance assessment of
Education Staff
Get information for assessment of lecturers
Give Input
Director
Education Manager
Hrd Manager
Lecturer

IOP Conf. Series: Journal of Physics: Conf. Series **1019** (2018) 012072 doi:10.1088/1742-6596/1019/1/012072

5. Discussion

Requirements Engineering (RE) is the leading phase of software engineering processes, in which software requirements (needs) of users and customers is collected, understood and defined. The results of this phase of the engineering requirements documented in the requirement specification. Requirements Specification contains a mutual agreement about the problems that want to be solved between developers and customers. This process is very important in the development of the system, because it can know the problem of consumer solutions and desired by consumers. To generate a good documentation also need tools that can help map out the results of consumer needs the power to fine tune and VORD methods with CMMI approach, proto personas and use case diagrams

6. Conclusion

VORD is a method of identifying the needs of the system software using the system user point of view, while CMMI is a process improvement method that consists of detailed practices thus providing elements of a process that is more effective. CMMI is a model approach that is able to assess the ability and maturity in developing software systems. VORD and CMMI can be mapped into the CMMI framework because the VORD in identifying system requirements still are too general. In the VORD is not yet there are steps detailed work identifying needs. It is therefore necessary the CMMI framework can assist in identifying the need clearly. Mapping the VORD into CMMI is done by combining each stage in the VORD into existing practices in CMMI process area, particularly for its development. The results of this mapping in the form requirement specified that the results obtained from the implementation method into the VORD and VORD tool.

References

[1] Andriyani, Yanti. 2011. Software Requirement Engineering Specification Pattern Pada Aplikasi Sistem Informasi Registrasi Akademik. Riau. Journal Generic, Vol. 6, No. 2, Juli 2001, pp. 44-51, ISSN : 1907-4093

[2] Alegria Hurtado Ariel Julio, Bastarrica, Cecilia, Maria. 2019. Implementing CMMI using a Combination of Agile Methods. Colombia. CleiElectronic Journal, VO. 9, No. 1, JUNE 2006

[3] Ahamad Fias Mr, Khozium Osama Mohamed, Dr. Ubiquitous Computing and Communication Journal, Vol. 3, No.1

[4] CMMI Product Team. 2006. CMMI for Development,Version 1.2. Carnegie Mellon University.

[5] Chakraborty Abhijit., et al. 2012. The Role of Requirement Engineering in Software Development Life Cycle. 2012. Bangladeh. Journal of Emerging Trends on Computing and Information Sciences, Vol. 3, No. 5, May 2012.

[6] Firesmith, G. Donal. 2002. Requirement Engineering. USA. Journal of Object Technology, Vol. 1, No. 4, September-October 2002.

[7] Firesmith, G. Donal. 2006. Requirement Engineering Tasks. USA. Journal of Object Technology, Vol. 5, No. 8, November-December 2006.

[8] Gothelf Jeff. 2013. Lean UX Applying Lear Principles to Improve User Experience. Cambridge. O'Reilly Media, Inc.

[9] Hull Elizabeth, Jackson Ken , and Dick, Jeremy. 204. Requirements Engineering Third Edition. London. Springer.

[10] Salem, M Amed. 2010. Requirement Analysis Through Viewpoint Oriented Requirement Model (VORD). Sacramento. (IJASCA) International Journal of Advanced Computer Science and Applications, Vol. 1, No. 5, November 2010.

[11] Sommnerville Ian. 2003. Software Engineering. Jakarta. PT. Gelora Aksara Pratama.

[12] TOGAF.2009. TOGAF Version 9.1.US Department.

[13] Wiegers Karl, Beaty Joy. 2013. Software Requirements Third Edition, Karl Wiegers and Seilevel.

[14] Young. R. Ralp. 2014. The Requirements Engineering Handbook. Boston. Artech House, INC.

Improvement of Data Exchange Security on HTTP using Client-side Encryption

Zuhar Musliyana[1,a], Mahendar Dwipayana[1,b], Ayu Helinda[1,c] and Zahrul Maizi[2, d]

[1]Department of Information System, Faculty of Computer Science, Ubudiyah Indonesia University, Jalan Alue Naga, Desa Tibang, Banda Aceh, Indonesia.
[2]Department of Informatics Engineering, Faculty of Computer Science, Ubudiyah Indonesia University, Jalan Alue Naga, Desa Tibang, Banda Aceh, Indonesia.

[a]zuhar@uui.ac.id, [b]mahendar@uui.ac.id, [c]ayu@uui.ac.id, [d]zahrulmaizi@uui.ac.id

Abstract. Hypertext Transfer Protocol (HTTP) is a standard data exchange protocol over the web. Currently, HTTP is still widely used to handle communication between client and server. However the related researches show that data exchange via HTTP is still vulnerable because the data sent is in the form of plaintext. To overcome the problem, the available solutions are using Hypertext Transfer Protocol Secure (HTTPS) and Secure Hypertext Transfer Protocol (SHTTP). However, some results of related researches show HTTPS and SHTTP still have some weaknesses. Based on these problems, this study proposes a method of securing HTTP using Client-side Encryption. In this research, the encryption algorithm used is Advanced Encryption Standard (AES) dynamic key. Encryption method was performed by encrypting the data on the side of the web browser before it was sent, then the decryption process was done on the web server side. The testings of this research were performed in case of login authentication. The results show the proposed method has successfully secured data exchange on HTTP.

1. Introduction

Data exchange security is one of the main issues in various web-based applications. In the data communication side, Hypertext Transfer Protocol (HTTP) is still widely used as one of the standard protocols that governs data exchange via web. Whereas the use of this protocol can cause security problems because the data exchange process is not encrypted [1]. The application of encryption process is one of the most commonly used ways in securing submitted data. This has been widely applied on Content Management System (CMS) WordPress, Joomla, Drupal which use cryptographic hash function of Message-Digest Algorithm 5 (MD5). But this encryption process works on the side of web server so that data sent from web browser is still in the form of plaintext.

Currently, it is known that there are several methods that can be used to improve data exchange security on HTTP by using Hypertext Transfer Protocol Secure (HTTPS) [2]. However, related researches show this solution is still vulnerable to SSL Stripping and Man-In-The-Middle (MITM) attacks either through the Poisoning Address Resolution Protocol (ARP) or through Fake Certificate

Attack or Phishing Attack [1], [3], [4]. Another solution is to use Secure Hypertext Transfer Protocol (SHTTP) by encrypting the application layer using the combination of Rivest Shamir Adleman (RSA) and MD5 algorithms [5]. However SHTTP is still rarely used and researches on this method have not been reviewed widely.

Based on the problems presented above, this paper proposes the improvement of security during data exchange process through HTTP using client-side encryption method by performing data encryption process at application level on the web browser side. In this method, there are some cryptographic algorithms that can be used, including Data Encryption Standard (DES), Advanced Encryption Standard (AES), Triple Data Encryption Standard (3DES) and Rivest Shamir Adleman (RSA). This research uses AES dynamic key 128 bits algorithm [6]. The selection is based on the consideration that AES has balance between security and flexibility in various softwares and hardwares [7]. In addition, AES published by the National Institute of Standards and Technology (NIST) as a substitute for DES is also one of the symmetric-key algorithms used as standard algorithms until today [8]. Testings of this research were done in case of login authentication using HTTP. The research results show that the proposed method can secure the data exchange on login authentication process of HTTP. The next section of this paper describes the research methods used, materials, methods and results.

2. Methodology

In this study, the proposed method to secure data exchange on HTTP is using Client Side Encryption. Client Side Encryption is one of methods used to improve the security of data stored on cloud server [9]. The use of this method was utilized to perform data encryption which can be on web browser side before data is sent to the web server via HTTP like Figure 1.

Figure 1. Client side encryption

As discussed in the introduction, this study uses AES dynamic key as Client-side Encryption algorithm. AES dynamic key is the development of AES algorithm that generates dynamic encryption keys based on the time value. The dynamic key consists of 3 main components i.e. time value, salt value and user password with length of 128 bits respectively. The generation of 128 AES dynamic keys can be shown in equation 1 [6].

$$K = K \oplus K2 \quad (1)$$

Where
K = AES Key (128 bit)
K1 = Time value when User logs in $\oplus$ Salt Value
K2 = Password User $\oplus$ Salt value

Security is an important factor in an algorithm. In standard AES algorithm, to be able to break AES key will take 1.02 x 1018 years [10]. When viewed from the key usage period, this dynamic key generator concept has a better security value because it can minimize the use of static key in long-term

IOP Conf. Series: Journal of Physics: Conf. Series **1019** (2018) 012073 doi:10.1088/1742-6596/1019/1/012073

period. This happens because the key that is used will change each time the encryption process. This change is caused by an ever-changing mix of time values. This is also in line with NIST's recommendation as an agency that publishes AES on the possibility of security concerns over the long-term use of AES static keys [11].

The implementation of Client-side Encryption on HTTP using the AES dynamic key algorithm can be seen in Figure 2.

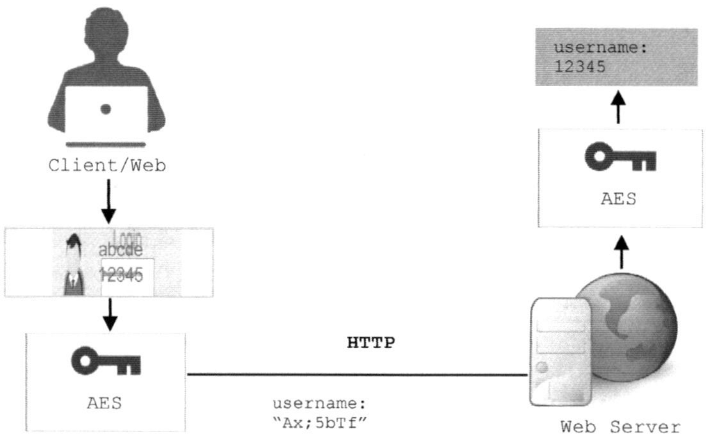

Figure 2. Client side encryption method

Login authentication process like Figure 2 can be explained as: client / web browser sends plaintext data in the form of username and password to web server. Before sent, the data is encrypted first using Client Side Encryption method so that the data through HTTP will be in the form of ciphertext. Furthermore, on the web server the ciphertext data is decrypted into plain text.

2.1 Material
In this study, observation of data exchange on HTTP was analyzed using wireshark software. XAMPP is a web server software used in the simulation process. On the client side the implementation of AES dynamyc key algorithm was written using client-side programming (javascript). While on the server side, PHP programming language was used. In addition, this study also tested the response time of the proposed method. The number of responded time samples observed is 25 samples.

2.2 Methods
As described in the previous section, this section will describe the Client-side Encryption proposed to secure the data exchange on HTTP. Before this method is applied, first of all it is necessary to understand the standard scheme of data exchange process on HTTP as in Figure 3 [12].

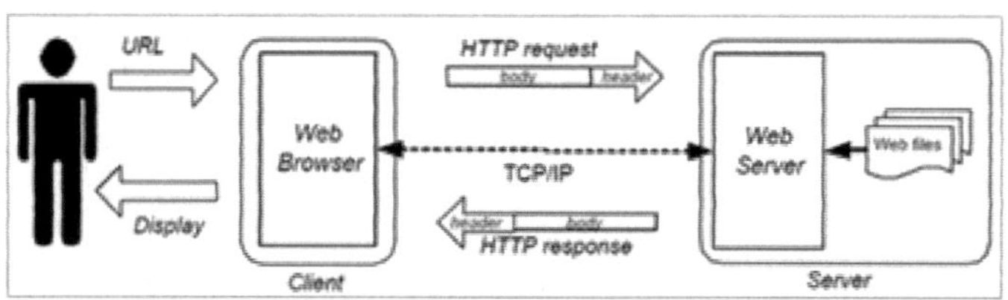

Figure 3. Scheme of data exchange process on HTTP

IOP Conf. Series: Journal of Physics: Conf. Series **1019** (2018) 012073 doi:10.1088/1742-6596/1019/1/012073
***** Formatting Issue - Best Available Paper/Graphic *****

Scheme in Figure 3 is the flow of data exchange process on HTTP. In the scheme, the problem occurs is data on HTTP request namely header and body are sent in the form of plaintext. To solve the problem, HTTP request data will be encrypted before it is sent as described in Figure 2. While pseudocode of AES dynamic key encryption and decryption process proposed in this research can be seen in Figure 4 and Figure 5.

AES Dynamic Key Encryption

```
Input: U (Username), P (Password), K (Key), Salt, WM (Time_Start)
WM = get time when the user logged in {key length 128 bit}
Salt = value of salt (static/dynamic) {Key length 128 bit}
K1 = Salt ⊕ WM
K2 = Salt ⊕ P
K = K1 ⊕ K2
Output: CT (Chipertext)
    (Nr, w) ← Key expansion(K) {Nr: Total round, w: Array byte key round}
    CT = P
    AddRoundKey(CT, w[0...3])
    for i=1 → Nr do
        SubBytes(CT)
        ShiftRows(CT)
        if i ≠ Nr then
            MixColumns(CT)
        end if
        AddRoundKey(CT, w[(i*4)...(i*4+3)])
    end for
```

Figure 4.

AES Dynamic Key Decryption

```
Input: CT(Chipertext), K(Key), Salt, WM, WP(Time of Process), TW(Time tolerance), BW(Time limit), use_key(UK)
WP = get the time value when the authentication process
TW = 30 (second)  / optional
BW = WM+TW
UK = 0
Output: P (Plaintext:Username dan Password)
    while (WM ≤ BW) do
        if WM = WP
            K1 = Salt ⊕WP
            K2 = Salt ⊕P
            K = K1 ⊕K2 {Key length 128 bit}
            if use_key ≤ 0
                (Nr, w) ← KeyExpansion(K) {Nr: Total round, w: Array byte key round}
                P = CT
                AddRoundKey(P, w[Nr * 4...Nr * 4 – 3])
                for  i=1 → Nr do
                    InvSubBytes(P)
                    InvShiftRows(P)
                    AddRoundKey(P, w[(Nr – i) * 4...((Nr – i) * 4) + 3])
                    if i ≠ Nr then
                        InvMixColumns(P)
                    end if
                end for
                use_key++
            end if
        end if
        WM = WM+1
    end while
```

Pseudocode of AES dynamic key encryption

IOP Conf. Series: Journal of Physics: Conf. Series **1019** (2018) 012073 doi:10.1088/1742-6596/1019/1/012073
***** Formatting Issue - Best Available Paper/Graphic *****

Figure 5. Pseudocode of AES dynamic key decryption

Figure 4 describes the generation of dynamic key stage process in AES encryption algorithm. The Salt value, Password (P), and the start time value (WM) i.e. the time value when user login is the initial input data to be used in the key generator process with the length of 128 bit, respectively. The first stage of key generation is to perform an Exclusive Or (XOR) operation between Salt and WM values to generate K1. The same operation is also performed between the Password value and the salt value to generate K2. The last step is to perform XOR operation between K1 and K2. These values will then be used as AES (K) encryption key with length of 128 bit. While P is the plaintext to be encrypted. The next stage is performing 4 main processes of AES encryption which are AddRoundKey, SubByte, ShiftRows, and MixColomns to produce ciphertext (CT). After going through encryption process, then ciphertext data will be sent from web browser to web server through HTTP.

On the web server side, data of ciphertext and K were received as inputs in the AES description process as shown in Figure 5. In the decryption process, looping time was performed until it satisfied BW value condition to obtain a match between WM and WP values. When a match was found, the salt value would be subjected to XOR operation with WP value to generate K1 and password with salt values to generate K2. Furthermore, the value of K was obtained by performing XOR operation between K1 and K2. This value would be used as a decryption key with a length of 128 bits. The next process was the inverse of the encryption process by using inverse transformation of all basic transformations in the encryption process i.e. InvSubBytes, InvShiftRows, InvMixColumns. While the AddRoundKeys transformation is self-inverse on condition of using the same key.

3. Results

This section will explain the implementation of Client Side Encryption on HTTP. The observations were performed before and after the application of AES encryption on HTTP communication path. In the initial test, data transmission test through the login form was performed as shown in Figure 6. Account user information used is username = tes and password = tes13579.

***** Formatting Issue - Best Available Paper/Graphic *****

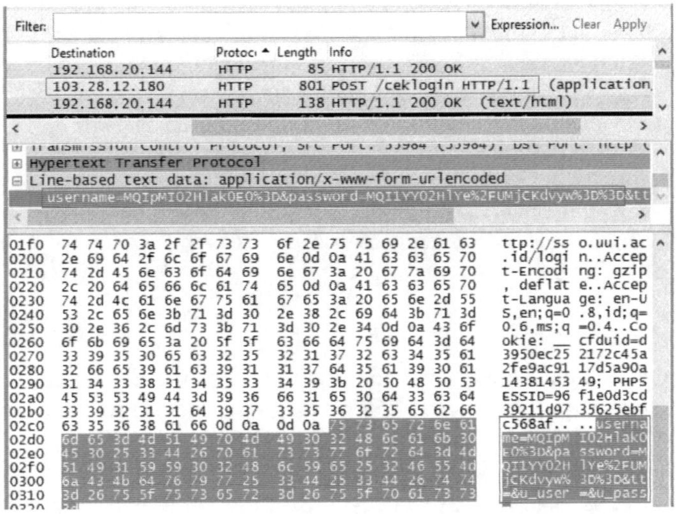

Figure 6. The experiment of HTTP login authentication

After user has sent data like Figure 6, then data packet was analysed on HTTP. This data packet was analysed using Wireshark software. The results of the analysis can be shown in Figure 7.

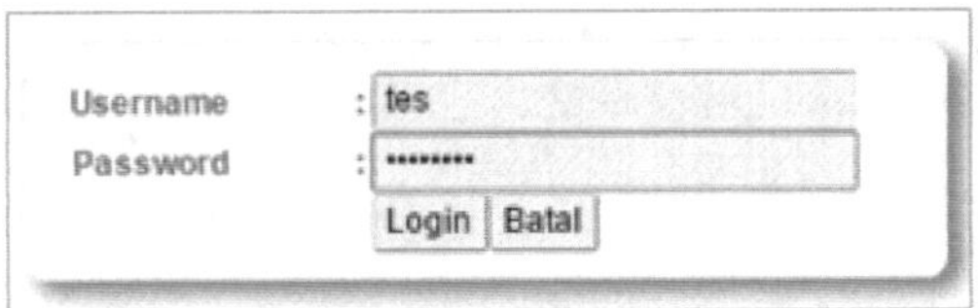

Figure 7. HTTP capture request before using Client-side encryption

Figure 7 shows the data of account login user sent via the HTTP is still in the form of plaintext so it is very risky if this information known by others. Figure 8 below shows the analysis results of data packet after AES encryption is applied on the web browser side. After using Client-side Encryption, data of account user is successfully on the web browser side a sent via HTTP protocol is form of ciphertext.

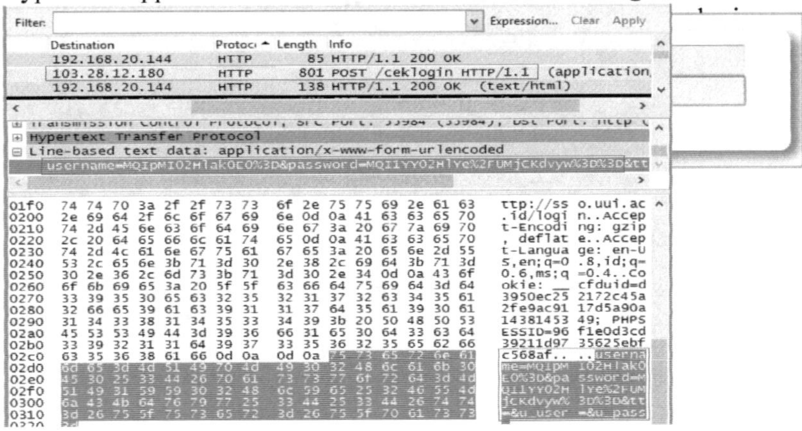

IOP Conf. Series: Journal of Physics: Conf. Series **1019** (2018) 012073 doi:10.1088/1742-6596/1019/1/012073
***** Formatting Issue - Best Available Paper/Graphic *****

Figure 8. HTTP capture request after using Client-side Encryption

The next observation was performed to know the response time of the proposed method. This observation used 25 test samples. The response time value obtained in each test can be seen in Figure 9.

Figure 9. Response time of HTTP

4. Discussion

Based on the observed data, Client Side Encryption method with AES Dynamic Key encryption can be applied to solve data security problem on HTTP request by encrypting data on web browser side before data is sent. However, the application of the method has an impact on increasing the response time value generated with an average of 0.0054s. Meanwhile, the initial response time value before this method applied is 0.0025s. The increase of response time value is due to the computational time required for data encryption process using AES algorithm.

5. Conclusion

The results of this study show that Client-side Encryption method with AES encryption dynamic key can improve data exchange security on HTTP because it is able to encrypt data at application level in web browser side before data is sent to web server.

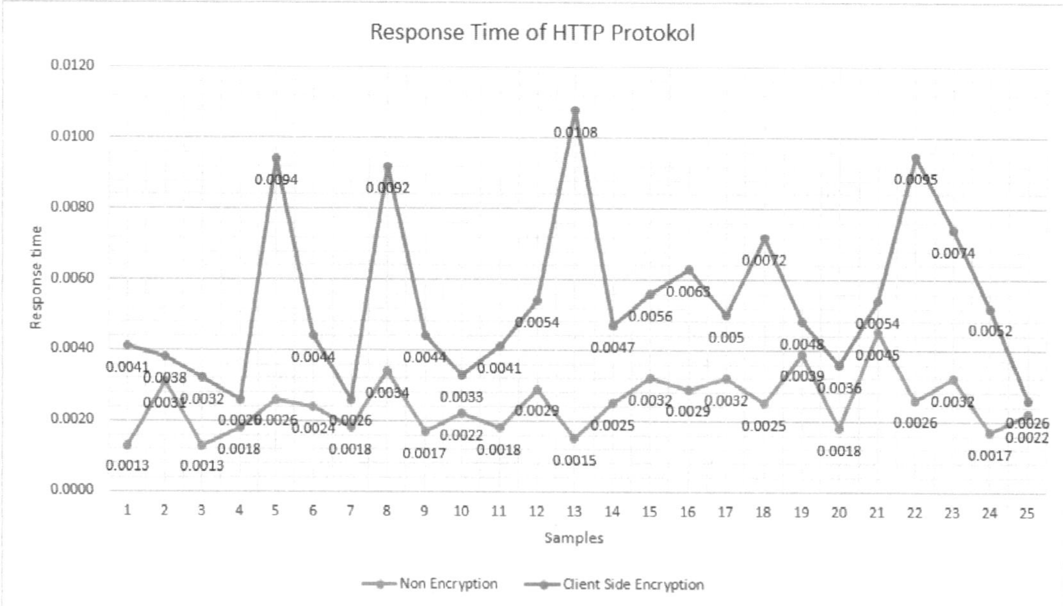

References

IOP Conf. Series: Journal of Physics: Conf. Series **1019** (2018) 012073 doi:10.1088/1742-6596/1019/1/012073
***** Formatting Issue - Best Available Paper/Graphic *****

[1] I. Homoliak, D. Ovsonka, M. Gregr and P. Hanacek, "NBA of obfuscated network vulnerabilities' exploitation hidden into HTTPS traffic," The 9th International Conference for Internet Technology and Secured Transactions (ICITST-2014), London, 2014, pp. 310-317.
[2] W. M. Shbair, T. Cholez, J. Francois and I. Chrisment, "A multi-level framework to identify HTTPS services," NOMS 2016 - 2016 IEEE/IFIP Network Operations and Management Symposium, Istanbul, 2016, pp. 240-248.
[3] S. Stricot-Tarboton, S. Chaisiri and R. K. L. Ko, "Taxonomy of Man-in-the-Middle Attacks on HTTPS," 2016 IEEE Trustcom/BigDataSE/ISPA, Tianjin, 2016, pp. 527-534.
[4] Puangpronpitag, S.; Sriwiboon, N., "Simple and Lightweight HTTPS Enforcement to Protect against SSL Striping Attack," International Conference of Computational Intelligence, Communication Systems and Networks (CICSyN), 26 Jul. 2012
[5] Adam S., (1999, Mei.). An Overview of SHTTP. National Academy of Sciences Cryptography. USA. [Online]. Available: http://www.homeport.org/~adam/shttp.html
[6] Z. Musliyana, T.Y. Arif, R. Munadi, "Security Enhancement of Advanced Encryption Standard (AES) using Time-Based Dynamic Key Generation," ARPN Journal of Engineering and Applied Sciences, Vol. 10, No. 18, Oct. 2015
[7] B. Bilgin, B. Gierlichs, S. Nikova, V. Nikov and V. Rijmen, "Trade-Offs for Threshold Implementations Illustrated on AES," in IEEE Transactions on Computer-Aided Design of Integrated Circuits and Systems, vol. 34, no. 7, pp. 1188-1200, July 2015.
[8] M. Bedoui, H. Mestiri, B. Bouallegue and M. Machhout, "A reliable fault detection scheme for the AES hardware implementation," 2016 International Symposium on Signal, Image, Video and Communications (ISIVC), Tunis, Tunisia, 2016, pp. 47-52.
[9] S. Kumar Da, Md. A. Hossain, Md. Ar. Sardar, R. K. Biswas, "Performance Analysis of Client Side Encryption Tools," International Journal of Computer Science Issues, Vol. 2, 16 Sept. 2014.
[10] M. Arora. (2012, Juli.). How secure is AES against brute force attacks?. Freescale Semiconductor,USA.[Online].Available:http://www.eetimes.com/document.asp?doc_id=1279619.
[11] NIST. (2012, Juli.). Recommendation for Key Management. National Institute of Standard and Technology(NIST),USA.[Online].Available:http://csrc.nist.gov/groups/ST/toolkit/key_management.html.
[12] NTU. (2012, May.). HTTP (HyperText Transfer Protocol), USA. [Online]. Available: https://www.ntu.edu.sg/home/ehchua/programming/webprogramming/HTTP_Basics.html

IOP Conf. Series: Journal of Physics: Conf. Series **1019** (2018) 012074 doi:10.1088/1742-6596/1019/1/012074

ARgot: Text-Based Detection Systems In Real Time Using Augmented Reality For Media Translator Aceh-Indonesia With Android-Based Smartphones

Zalfie Ardian[1,a], P. Insap Santoso[2,b], and Bimo Sunarfri Hantono[3,c]

[1]Department of Information System, Faculty of Cumputer Science, Ubudiyah Indonesia University, Jalan Alue Naga, Desa Tibang, Banda Aceh, Indonesia.
[2]Department of Electrical and Information Technology, Faculty of Engineering, Gadjah Mada University.
[3]Department of Electrical and Information Technology, Faculty of Engineering, Gadjah Mada University.

[a]zalfie.ardian@uui.ac.id, [b]insap@mti.ugm.ac.id, [c]bimo@te.ugm.ac.id

Abstract. Language is a media of communication which aims to understand each ethnic and certain nation. The understanding of language allows the increase of awareness and closeness value among certain individuals. Nevertheless, considerable number of Languages appears to be the causes of difficulty and misunderstanding on the meaning of a certain word. This research suggests one of the solutions in solving such problem. Hence it eases the process of understanding the meaning of the words of a language. Especially "bahasa Aceh". This application called ARgot. ARgot is an application of translation of Aceh – Indonesia's Language. This application is able to work as real time application without having a need of an input media such as a text which has to type through using a good quality of camera and having characteristic of autofocus in camera. This research emphasizes on the process of detection and tracer text which available within the society. This research is also intended to see how the technology Augmented Reality is able to be applied in markerless text and have different font of variety. The result of the research is a software for a android smartphone, as media translator in a real time, Hence it is able to be used almost in every kind of smartphone. Based on the trial and result of respondent, this application is able to attract the interest of user and it is also able to be applied in a real life.

1. Introduction

Language is an important element in the life of mankind as a communication tool to interact each other. That is why the language becomes one of the crucial factors in the life of society in the world. Indonesia is a diverse country that has a multi-ethnic group, race, religion, and language diverse region, then the language is a tool to unite the nation. Indonesia, which has population of hundreds of lives, has recorded more than 750 local languages, making Indonesia the country with the largest regional languages in the world. However, there are approximately 160 languages which are endangered due to the lack of depth of the local language.

There are two definitions of language. First definition declared that language is as a means of communication among society in the form of symbols of the sounds produced by the human vocal organs. Secondly, language is a communication system that uses vocal symbols (speech sounds) that are arbitrary [1]. Another definition, language is a form and not a state or an arbitrary sound symbol system, or also a system of many systems, a system of an order or an order of the systems [2]. Written text is one of the most common methods for conveying information in our daily life. However, when the text is found in a foreign language for some individuals, such information cannot be delivered [3]. So, to mitigate this problem, it requires an assisting device that acts as translators, ranging from a simple dictionary to electronic devices that simplify the translation process.

The application of AR technology in the text of the application of Augmented Reality technology that is rarely used in text-based markers. AR research largely focused on the use of marker, to display objects, and interactions using additional markers. There is still little use of AR with Markerless methods, especially the use of AR-based text, meaning that the image is detected to be converted into text form. Moreover, the use of AR by using the smartphone will add value to the higher mobility.

2. Methodology

Augmented Reality (AR) is a concept that combines real-world objects and objects generated virtual world of a computer system by adding information on the real object so that the boundary between the two became very thin. This system is closer to the real environment, because the reality is preferred in this system. With the help of AR technology, a real environment around us will be able to interact with the digital form (virtual). Information about the object and the environment around us can be added to the AR system then displays the information on a layer of real world in real-time so it seems that information be interactive and real. AR concept was first introduced by Thomas Caudell in 1990 when he worked at Boeing [13].

There are differences between the marker and Markerless methods, methods of using the marker is to use a box-shaped black and white image which allows applications to track and detect orientation and to adjust positions based on 3D objects. Meanwhile, according to Madden [16], Markerless is AR which is used to track objects that exist in the real world without a special marker. To perform object tracking, the system relies on the natural Markerless AR-feature and not a fiduciary marker. Moreover, no additional markers that must be made in advance which will be added to the environment. Another advantage is the possibility to issue Markerless information and characteristics of the environment for future use. However, the ability of tracking and registration techniques become more complex objects [17]

2.1 Materials

Vuforia uses Computer Vision technology to identify and track the planar images (Target Image) and simple 3D objects such as boxes, in real-time. Image registration capability allows developers to adjust the position and orientation of a virtual object, such as 3D models and other media, in connection with the real world image when it is viewed through a mobile device camera. The virtual object and then track the position and orientation of the image in real-time so that a user's perspective on objects in accordance with the perspective of the target image, then the virtual object can be displayed in the real world. Vuforia SDK supports various types of targets including 2D and 3D images Markerless, multi 3D targets, and forms a frame marker. Additional features of the SDK includes a local collision detection using virtual button, you can choose a target image in real-time, and the ability for computing at runtime. Vuforia provides Application Programming Interfaces (API) in C ++, Java, and Objective-C.

2.2 Methods

There have been a lot of mobile applications for the translators which are available, but most are still using text input method in terms of its use. Augmented Reality technology offers something new, how

IOP Conf. Series: Journal of Physics: Conf. Series **1019** (2018) 012074 doi:10.1088/1742-6596/1019/1/012074

objects are detected in real time, this is what's behind what if as text object and do real-time detection and generate output translator translation resulting media-based Augmented Reality. Vuforia makes the detection and tracking of text elements by doing scanning of camera view called Region of Interest (ROI). This technology uses two different ROI, one for detection and one for tracking.

In terms of performance, the detection of the word too much in one tracking will lead to a longer detection time. When the text detection is enabled, Vuforia tries to match words that are found in the ROI with the words in a set of reference database called wordlist. A word database that contains words that are used to process the translator. As shown in Figure 1 below

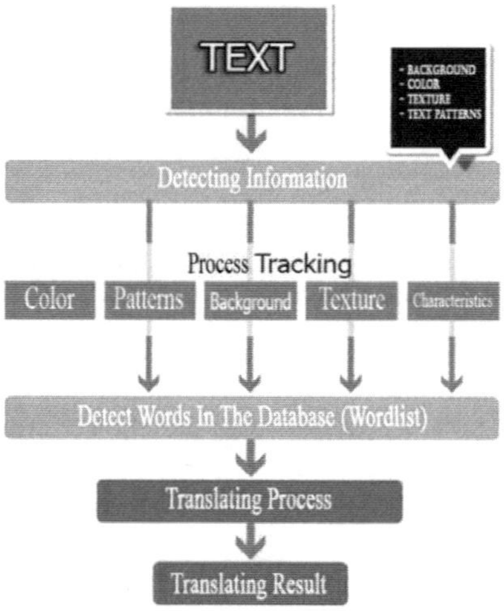

Figure 1. Text detection process.

The test aims to determine the minimum requirement in recognizing the text as the target image. The test will use two types of smartphones with different capabilities and specifications. The test is performed with the determination of a distance of 10 cm as shown in Figure 2.

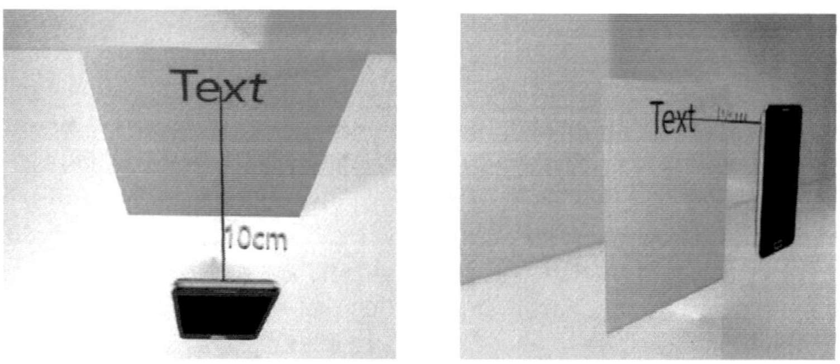

Figure 2. Distance for detection.

IOP Conf. Series: Journal of Physics: Conf. Series **1019** (2018) 012074 doi:10.1088/1742-6596/1019/1/012074

3. Results

Tests were carried out to look at the system's ability to detect the text, but this has not translated initially testing, this test is to look at the success rate of the system when it detects text. the testing process as shown in Figure 3.

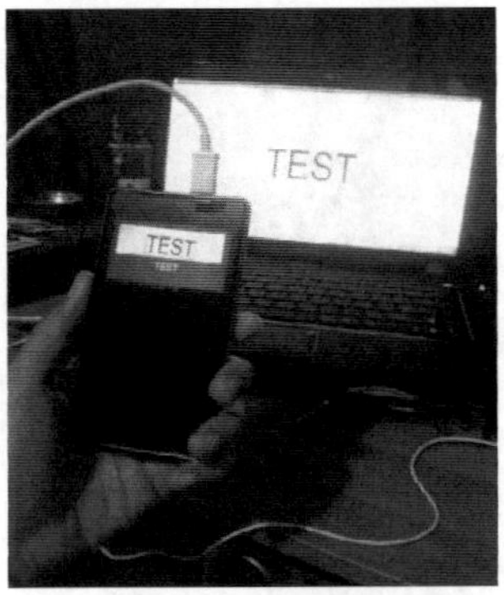

Figure 3. Test of text detection.

Based on the tests performed, it can be concluded that the level of difficulty of the system if it detects the text with the four different conditions are like those described above. Then the result can be seen in Table 1.

Table 1. Testing outcomes.

No	Test	Testing Outcome		
		Figure 1	Figure 2	Figure 3
1	Tests with different fonts	3/5 (60%)	5/5 (100%)	5/5 (100%)
2	Test with a different background and color	5/5 (100%)	5/5 (100%)	5/5 (100%)
3	Tests with different thickness levels	5/5 (100%)	5/5 (100%)	3/5 (60%)
4	Tests with different spacing between letters	5/5 (100%)	5/5 (100%)	0/5 (0%)

Based on the analysis performed on the test, the results can be concluded that the application ARgot does not have the problem of the text that is around as long as the text still has a pattern stardard to a letter and this application also does not have a problem to the colors on the condition that the text color too much contrast with background so that the system is not difficult in terms of text detection. The test results summarized in the Table 2.

Table 2. Test result.

Test	Detection	
	Andromax U	Andromax U 2
P 1	Succeded	Succeded
P 2	Succeded	Succeded
P 3	Succeded	Succeded
P 4	Succeded	Succeded
p1 + p3	Succeded	Succeded
p1 + p2 + p3	Succeded	Succeded
p1 + p3 + p4	Succeded	Succeded

Information :
P1 = Tests with different fonts
P2 = Test with a different background color and
P3 = Tests with different thickness levels
P4 = Tests with different spacing between letters

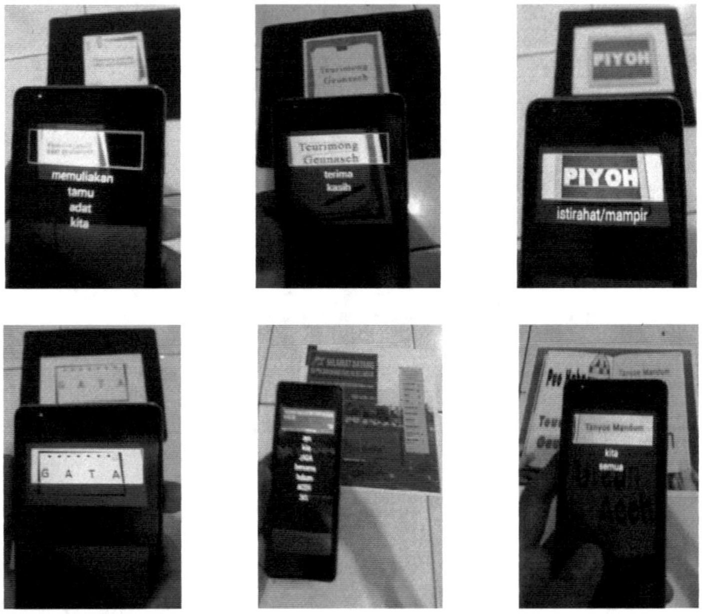

4. Discussion

Assets are the main components that will be used in displaying the AR. In the ARgot manufacture, it consists of several parts, they are asset Vuforia Word List (VWL) which serves as a database of text, and the text as an image asset targets. Vuforial Word List is a database of the vuforia package that contains the texts to be detected. In this application, use the file named "bahasa_aceh_utf8" containing words in Acehnese language, which is the word that will appear later to detect the text that will be translated. As a side note, if the word to be detected is not contained in this database, the word is not able to recognize that the process will fail. System applications are built on the Android-based smartphone consists of several major parts, such as import of assets, the implementation of a splash screen functions, initialization, rendering, and deinisialisation. In displaying text, text that serves as a target image to be imported first into the source code. This is because the program still capture text as an image before it is converted to text, the conversion process uses OCR (Optical Character Recognition), so that the letters in the image one by one match each database that is then displayed on the screen.

ARgot takes input text data based on words found on Vuforia wordlist (VWL), a database named Bahasa_aceh_utf8. After drawing terdetesi as text, then the system looks for the text in a database that will be displayed. Before displaying the text, first have to initialize the camera and render settings. First choose which camera to use, because there are smartphone has two cameras, then initialize the camera, turn on the camera and start tracking function. After initializing the camera, rendering initialization can be done. Initialization rendering aims to generate textures using OpenGL, make arrangements to be displayed using OpenGL and OpenGL settings for text display. After displaying text and completed applications are used, so that it is not stored in the buffer should be done deinisialisasi smartphone cameras to triangulate. This test uses the Android-based smartphone with a wide range of specifications. This test uses four different types of smartphone specifications. Smartphones used has a significant hardware differences, in terms of performance processor and RAM capabilities so that it can be variations to test the system ARgot. Thus, it can be seen the influence of processor, RAM, GPU, the performance of the system ARgot.

5. Conclusion

From the results of the design, implementation, testing and the elaboration can be concluded as follows:
1. Applications ARgot is a translator for the English media applications Aceh are designed using the Java programming language that can be operated on Android devices, to display using Vuforia Augmented Reality SDK and rendering using OpenGL and OCR technology as an auxiliary reading of the image as text.
2. By applying the Augmented Reality uses a technique Markerless on ARgot, does not restrict users to detect certain text or images. So that users are free to detect any text as recorded on the database. With a variety of tests performed on the performance of ARgot ARgot has proven to work well and can be known limits of ARgot.
3. Augmented Reality can be innovation bari text in android-based translator media thus enhancing the quality of software for education, especially for studying Indonesian culture.
4. From a survey of 30 responders were obtained by distributing questionnaires obtained more than 80% expressed an interest and 85% of respondents stated ARgot really helpful if applied as a media translator. However the detection process that sometimes imperfect can create confusion for users.

6. References

[1] S. Keraf, Kadhipta, 1st ed. Jakarta: Balai Pustaka, 1991
[2] W. Mackey, Analisis Bahasa. Surabaya: Usaha Nasional, 1986
[3] V. Fragoso, S. Gauglitz, S. Zamora, J. Kleban, and M. Turk TranslatAR: A mobile augmented reality translator 2011 497–502
[4] M. Petter, V. Fragoso, M. Turk, and C. Baur, "Automatic text detection for mobile augmented reality translation," 2011, pp. 48–55
[5] B. Parhizkar, K. Oteng, O. Ndaba, A. Lashkari, and Z. Gebril, "Ubiquitous Mobile Real Time Visual Translator Using Augmented Reality for Bahasa Language," IACSIT Press, vol. 3, Apr. 2013
[6] M. Khan, M. Zaman, and T. Islam, "Application Of Augmented Reality : Mobile Camera Based Bangla Text Detection And Translation," Brac University, 2012
[7] V. F. Bauset, J. M. Orduna, and P. Morillo, "Performance Characterization on Mobile Phones for Collaborative Augmented Reality (CAR) Applications," in 2011 IEEE/ACM 15th International Symposium on Distributed Simulation and Real Time Applications (DS-RT), 2011, pp. 52–53
[8] C.-L. Lai and C.-L. Wang, "Mobile Edutainment with Interactive Augmented Reality Using Adaptive Marker Tracking," in 2012 IEEE 18th International Conference on Parallel and Distributed Systems (ICPADS), 2012, pp. 124–131
[9] Wildan, Tata Bahasa Aceh untuk Madrasah Dasar dan Madrasah Lanjutan, vol. III. Banda Aceh: Global Educational Consultant Institute, 2002
[10] SIL, Bahasa-Bahasa di Indonesia (Language of Indonesia). Jakarta: SIL Internasional Cabang Indonesia, 2006
[11] M. Toha, Laporan Kekerabatan dan Pemetaan Bahasa-Bahasa di Propinsi Nanggroe Aceh Darussalam. Banda Aceh: Balai Bahasa, 2008

[12] B. Parhizkar, Z. M. Gebril, W. K. Obeidy, M. N. A. Ngan, S. A. Chowdhury, and A. H. Lashkari, "Android mobile augmented reality application based on different learning theories for primary school children," presented at the 2012 International Conference on Multimedia Computing and Systems (ICMCS), 2012, pp. 404–408

[13] F. Miller, A. Vandome, and J. McBrewster, "Augmented Reality," VDM Publ. House, 2009.

[14] J. Kim and H. Jun, "Implementation of image processing and augmented reality programs for smart mobile device," in 2011 6th International Forum on Strategic Technology (IFOST), 2011, vol. 2, pp. 1070–1073

[15] O. Bimber and Raskar, "Spatial Augmented Reality," K Peters, 2005

[16] L. Madden, Augmented Reality Browsers for Smartphones: Programming for JUNAIO, LAYAR, and WIKITUDE, First Edition., vol. 1. Wiley Publishing Inc., 2012

[17] "Markerless Augmented Reality." [Online]. Available: http://www.arlab.com/blog/markerless-augmented-reality/. [Accessed: 23-Apr-2013]

[18] R. Azuma, "A Survey of Augmented Reality," Eleoperators Virtual Environ., vol. 6, pp. 355–385, Aug. 1997

[19] J. Palenchar, "Survey: Samsung, Android Still On Top In U.S. Cellphone Market," TWICE This Week Consum. Electron., vol. 27, no. 25, pp. 56–56, Dec. 2012

[20] "App Components | Android Developers." [Online]. Available: http://developer.android.com/guide/components/index.html. [Accessed: 10-Dec-2013].

[21] "Auggie Awards," Augmented World Expo. [Online]. Available: http://augmentedworldexpo.com/auggies/. [Accessed: 29-Nov-2013]

[22] "Vuforia SDK Architecture | Vuforia Developer Portal." [Online]. Available: https://developer.vuforia.com/resources/dev-guide/vuforia-ar-architecture. [Accessed: 25-Apr-2013]

[23] Hanafi, "Sistem Aplikasi Mobile Advertising Berbasis Grafis dan Animasi (studi: PT. XL AXIATA TBK)," Thesis, Universitas Gadjah Mada, Yogyakarta, 2012

[24] Fred D Davis, "Technology Acceptance Model for Empirically Testing New End-User Information Systems: Theory and Results," Massachussetts Inst. Technol. Boston, 1986

IOP Conf. Series: Journal of Physics: Conf. Series **1019** (2018) 012075 doi:10.1088/1742-6596/1019/1/012075

Simulating a Carbon Trading Advantage from the Municipal Solid Waste Management: The Role of Waste-to Carbon in Malaysia

Iwan Budhiarta[1,a], Chamhuri Siwar[2], and Hassan Basri[3]

[1] Department of Management, Faculty of Economics, Ubudiyah Indonesia University, Jalan Alue Naga, Desa Tibang, Banda Aceh, Indonesia
[2] Institute of Environment and Development, National University of Malaysia, UKM - Bangi, 43600, Malaysia
[3] Faculty of Engineering & Built Environment, National University of Malaysia, UKM - Bangi, 43600, Malaysia

[a]iwan.budhiarta@live.com

Abstract. Management of municipal solid waste in Malaysia has proved to be efficiently, effectively, and adequate. This study has used existing literature and empirical evidences for analyzing the various issues related to carbon emission trading and solid waste management. A measurement of carbon potential value will be conduct after the municipal solid waste generation data was collected completely, and review the advantage of carbon emission trading from the municipal solid waste management in Malaysia, evaluate the benefit of carbon emission trading as an alternative investment tools and basic framework required to implement them. The article suggests that how an appropriate carbon emission trading system may give more benefit and economical value, and sustainable way of developing the future income generation in environmental sector in Malaysia. In other hand, the changing legislative environment is likely to make current practices highly non-optimal and increase pressures for a change of waste management strategy. The article will become more valuable to all those who have some interests in environmental economic and alternative investment sectors and would like to reduce the global warming and climate change effect and develop the potential value related to carbon emission trading in developing countries.

1. Introduction

In this recent years, the issues of municipal solid waste generation has taken a great attention, especially in an emerging countries where population is still rising substantially and there were limited landfills to accommodate daily waste from communities [1]. Solid waste presents unique and complex problems partly because it is both a source of pollution and secondary resources with waste management systems aiming to achieve a balance between environmental, technical, economic and regulatory factors [2]. In the other hand, there are many economic potential values from solid waste generation that can be taken as a community income generation.

Prior to the mid-1990s, waste management (WM) issues tended to revolve around public health and safety issues [18], and the optimization of landfill gas (LFG) capture and utilization [7]. The subject of its impact on greenhouse gases (GHG) has only been addressed recently, making it a relatively young study. This reflects an increasing acknowledgement of wastes contribution to GHG which is reflected in recent development in GHG mitigation policies internationally [19]. More importantly, there has been a movement to divert waste from landfills in order to reduce the negative environmental impact of landfills such as leachate contamination, GHG emissions and space limitation [16]. Composting has thus been widely acknowledged as an alternative to landfills.

In Malaysia, waste management and energy generation are undergoing significant changes, in reaction to new climate-driven statutory regulations. Climate change is also a serious issue, and Malaysia is embarking on a number of voluntary actions to reduce the emissions of greenhouse gases (GHGs) that can intensify climate change. Among the efforts to slow the potential for climate change are measures to reduce emissions of carbon dioxide (CO_2), methane (CH_4) and other non-carbon-dioxide GHGs, and promote long-term storage of carbon in forests and soil. Management options for MSW provide many opportunities to affect these processes, directly or indirectly.

Following a century of exponential growth in industrialisation together with population growth over time, the threat is reaching crisis point. The government at some point must confront the climate change problem by setting limits on CO_2 emissions. Reducing and converting CO_2 is a cost-competitive and safe way to achieve large-scale reductions in free-air emissions. Carbon emission trading is relatively new and booming issue in financial management area which aim is bringing down an air pollution by providing economic incentives to those who can achieve a significant reduction of carbon emission. The mechanism behind carbon emission trading is simple, yet ingenious. There is a central authority which prescribes a limit or a cap to the amount of air pollutant which can be emitted. Since the total of such caps is fixed, the total pollution or emission cannot exceed this level.

The flexibility of the hedge fund structure is the central reason these investment vehicles are so popular. The flexibility allows funds to specialize on certain investment sectors or strategies, and they can move quickly to the hot areas of the markets. One of the hot areas of the markets is the movement toward green investments and strategies. With this increased focus on energy and related areas, Carbon Hedge Funds have become a popular investment vehicle to focus on "carbon" as a new asset class. This article will overview the carbon hedge fund strategies.

2. Generation Rate of Municipal Solid Waste in Malaysia

Economic growth in Malaysia has brought prosperity; it has started to impose costs of industrial pollution and degradation of the urban environment [5]. The official estimated KL's population in 2007 is 1.604 million according to statistic department 2007. In KL alone, the estimated solid waste generation was 4000 tons per day in the year 2000 [8]. Kuala Lumpur alone in 1991 spent roughly about RM 25.2 million cost of managing solid waste; the comparative figures of the damage costs of haphazard open damping landfill practice were RM 178.30 per ton [5]. With the increase in municipal waste generation from 5.6 million tons in 1997 up to 8.0 million tons in 2000, there is an urgent need for a better managed disposal option [3]. Table 1 shows solid waste composition of selected locations in peninsular Malaysia and Figure 1 shows the corresponding graphical representation.

Table 1. Solid waste composition of selected locations in Peninsular Malaysia (In Percentage) [21].

Waste composition	Kuala Lumpur	Shah Alam	Petaling Jaya
Garbage	45.7	47.8	36.5
Plastic	9.0	14.0	16.4
Bottles/glass	3.9	4.3	3.1
Paper/cardboard	29.9	20.6	27.0
Metals	5.1	6.9	3.9
Fabric	2.1	2.4	3.1
Miscellaneous	4.3	4.0	10.0

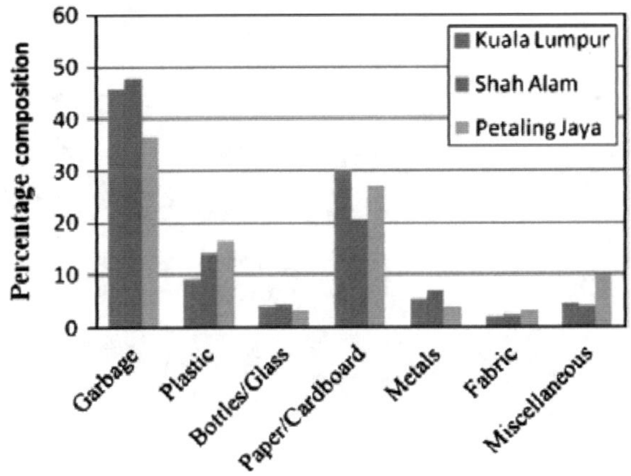

Figure 1. Solid waste composition of selected locations in Peninsular Malaysia.

Table 2 shows the recyclable components and their percentage share and recycling rate in tons per year. The graphical representation of percentage recyclables is shown in Fig. 2. Unfortunately, the attention paid by the authority towards this direction is not sufficient enough to tackle this issue. The Agenda 21 of the UNCED 1992, defined environmentally sound technologies protect the environment; moreover, recycling most of the wastes and handle residual wastes in more acceptable manner. Though recycling activity in Malaysia is rising up, the recycling industry still needs to be enhanced. Since 1993 a major effort of recycling was lunched by the Ministry of Housing and Local Government but unfortunately limited recycling activities taken place [1]. In Kuala Lumpur, by 2005 the recycling of the waste generated planned to reach 16% and 22% by 2020 [13].

IOP Conf. Series: Journal of Physics: Conf. Series **1019** (2018) 012075 doi:10.1088/1742-6596/1019/1/012075

Table 2. Recyclable components and their percentage share.

Recyclable SW components	Mass (kg/Cap./year)	Percentage	Recycling rate (tons/year)
Food waste	387.63	56.80	155041.11
Mix paper	35.59	16.50	14235.00
Mix plastic	61.87	15.30	24747.00
Textile	07.12	01.30	2847.00
Rubber and leather	03.29	00.60	1314.00
Wood	02.19	00.40	876.00
Other combustible	00.00	00.00	00.00
Yard	25.73	04.70	10293.00
Fine	03.83	00.70	1533.00
Glass	06.57	01.20	2628.00
Ferrous	13.14	02.40	5256.00
Aluminum	00.55	00.10	219.00

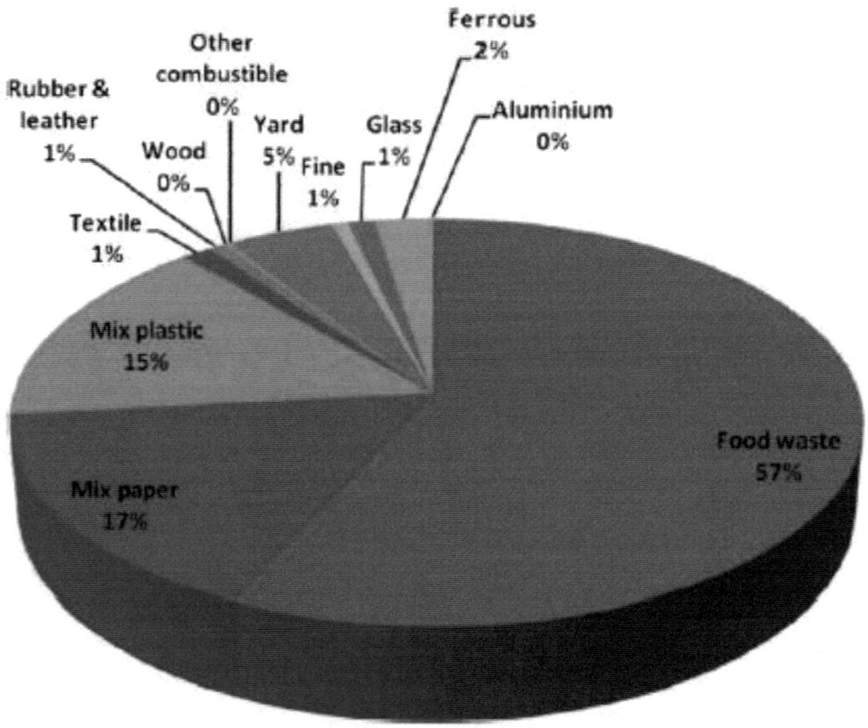

Fig. 2. MSW recyclables of Kuala Lumpur.

3. Waste in the emissions trading scheme

Emissions trading will help reduce emissions, encourage and support global action on climate change, and help put Malaysia on a path to sustainability. This factsheet explains how emissions trading will affect the waste sector.

4. Greenhouse gas emissions from waste

The greenhouse gases carbon dioxide, methane and nitrous oxide are emitted into the atmosphere from waste treatment and disposal. These emissions come from solid waste disposed in landfills, commercial and domestic wastewater treatment, and waste incineration. A distinction is drawn for sources of carbon dioxide in the waste sector; where the emissions come from disposing or treating organic matter, these emissions are considered part of the natural organic cycle and are not counted. Only carbon dioxide emissions from burning fossil fuel origin materials such as plastic are included in the emissions trading scheme.

The waste sector was responsible for Malaysia's greenhouse gas emissions in a local, regional and global scope. It is the only sector that has reduced its greenhouse gas emissions below 1990 levels, in its case by 26 percent. This reduction has come from improved landfill management systems, the installation and operation of landfill gas recovery technologies, as well as by reducing organic waste disposed at landfills through recycling and composting.

5. Potential impacts of the emissions trading scheme on the waste sector

The cost of emission units is expected to be passed on to customers of landfills (the people and organizations depositing waste) through increased prices for waste disposal. There is no exact way of generalizing a cost increase to all users of landfills, as emissions from individual landfills are noticeably affected by landfill gas collection systems and the incentives and infrastructure for recycling and organic materials diversion. Users of landfills have markedly different items in their waste, which all have different potentials for landfill gas generation. Any price increases will also be affected by competition from other landfills and the management policies and priorities of the landfill owner – usually a local authority.

6. Methodology

6.1 Carbon Valuation Tool

The emissions trading scheme will include methane emissions from landfills that deposit solid waste. Such waste must be partially household waste. Methane emissions occur as a result of the biodegradation of organic matter contained in landfills. The emissions trading scheme will also include carbon dioxide, methane and nitrous oxide emissions from any future solid waste incineration plant that combusts household waste. If electricity is generated from waste incineration, then any emissions are counted in the industrial processes sector, which has a different entry date to the waste sector. The emissions associated with treating domestic, commercial and residential wastewater (including septic tanks) will not be covered by the emissions trading scheme. Emissions of these gases are extremely difficult to measure accurately at individual sites.

In estimating the level of provision to be made through the Waste Development Framework for Municipal Solid waste the following assumptions have been made that future waste arisings will be dependent upon the amount of waste produced per person, the average number of people living in each dwelling and the increase in number of dwellings.

From the recent market price at Europe Emission Exchange, standarded price that will be used for any metric ton carbon emission is about EUR10 - EUR15.

IOP Conf. Series: Journal of Physics: Conf. Series **1019** (2018) 012075 doi:10.1088/1742-6596/1019/1/012075

Table 3. Waste Per Ton Estimates of GHG Emissions for Alternative Management Scenarios.

Material	GHG Emissions per Ton of Material Source Reduced (MTCE)	GHG Emissions per Ton of Material Recycled (MTCE)	GHG Emissions per Ton of Material Landfilled (MTCE)	GHG Emissions per Ton of Material Combusted (MTCE)	GHG Emissions per Ton of Material Composted (MTCE)
Aluminum Cans	(2.24)	(3.70)	0.01	0.02	-
Steel Cans	(0.87)	(0.49)	0.01	(0.42)	-
Copper Wire	(2.00)	(1.34)	0.01	0.01	-
Glass	(0.16)	(0.08)	0.01	0.01	-
HDPE	(0.49)	(0.38)	0.01	0.25	-
LDPE	(0.62)	(0.46)	0.01	0.25	-
PET	(0.57)	(0.42)	0.01	0.30	-
Magazines	(2.36)	(0.84)	(0.08)	(0.13)	-
Newspaper	(1.33)	(0.76)	(0.24)	(0.20)	-
Office Paper	(2.18)	(0.78)	0.53	(0.17)	-
Phonebooks	(1.72)	(0.72)	(0.24)	(0.20)	-
Textbooks	(2.50)	(0.85)	0.53	(0.17)	-
Medium Density Fiberboard	(0.60)	(0.67)	(0.13)	(0.21)	-
Food Scraps	-	-	0.20	(0.05)	(0.05)
Yard Trimmings	-	-	(0.06)	(0.06)	(0.05)
Grass	-	-	(0.00)	(0.06)	(0.05)
Leaves	-	-	(0.05)	(0.06)	(0.05)
Mixed Paper, Broad	-	(0.96)	0.09	(0.18)	-
Mixed Paper, Resid.	-	(0.96)	0.07	(0.18)	-
Mixed Paper, Office	-	(0.93)	0.13	(0.16)	-
Mixed Metals	-	(1.43)	0.01	(0.29)	-
Mixed Plastics	-	(0.41)	0.01	0.27	-
Mixed Recyclables	-	(0.79)	0.04	(0.17)	-
Mixed Organics	-	-	0.06	(0.05)	(0.05)
Mixed MSW	-	-	0.12	(0.03)	-
Carpet	(1.09)	(1.96)	0.01	0.11	-
Personal Computers	(15.13)	(0.62)	0.01	(0.05)	-
Clay Bricks	(0.08)	-	0.01	-	-
Concrete	-	(0.00)	0.01	-	-
Fly Ash	-	(0.24)	0.01	-	-
Tires	(1.09)	(0.50)	0.01	0.05	-

IOP Conf. Series: Journal of Physics: Conf. Series **1019** (2018) 012075 doi:10.1088/1742-6596/1019/1/012075

6.2 Carbon Hedge Fund Mechanism

The European Commission launched the European Climate Change Program (ECCP) in June 2000 with the objective to identify, develop and implement the essential elements of an EU strategy to implement the Kyoto Protocol. All 25 EU countries simultaneously ratified the Kyoto Protocol on 31 May 2002. The European Union Emission Trading Scheme (EU ETS) is a significant part of the ECCP and currently constitutes the largest emissions trading scheme in the world.

Actual trading in EU ETS emission allowances began January 1st, 2005. By the end of the same year, almost 400 million tonnes of carbon equivalent had been traded, representing a turnover in excess of EUR 7 billion. The impact of the release of sensitive information regarding the ETS net position in carbon emission allowances can be dramatic, as was illustrated in April 2006. First phase EU ETS carbon, in the form of the allowance expiring in December 2007, written Dec-07, had reached EUR 30 per tonne at their high in April 2006. Prices subsequently plumetted to below EUR 10 per tonne in a few days beginning May 2006 after EU figures on actual 2005 emission levels suggested emission caps to selected industries had been too generous to have a significant impact on emission practice. Emission caps for the second phase (2008-2012) are currently under review because of this apparent generosity of NAP levels in the first phase.

Risk management and hedging is a useful tool to reduce market place liability. Hedging is a strategy designed to minimize exposure to an unwanted business risk, while still allowing the business to profit from an investment activity. There could be hedgers on both sides of the market and that hedging is essentially a risk reduction technique more than anything, which allows informed traders and commodity dealers to profit from their intuitive knowledge of future changes in the difference between futures and spot prices. Many examples exist to demonstrate the mitigation of risk to an institution or financial portfolio. New products are constantly created and available on both over-the counter and exchange traded markets. It would be wise to consult with a qualified Commodity Trading Advisor or broker to discuss the analysis for an on-going risk management solution or a onetime only hedge.

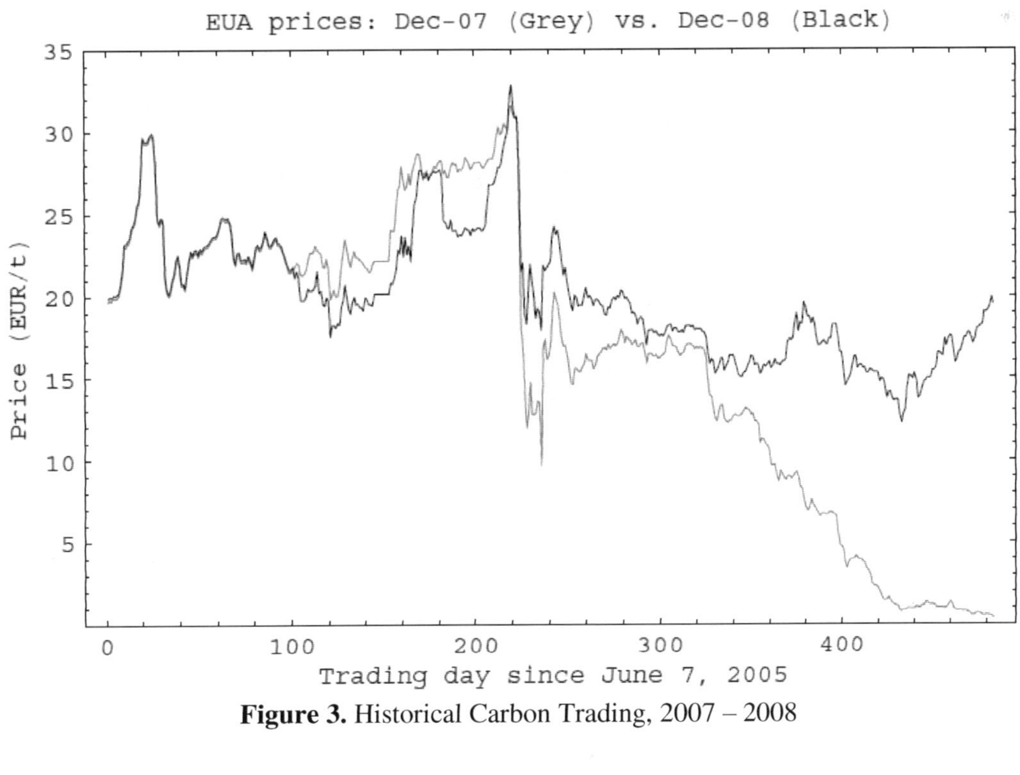

Figure 3. Historical Carbon Trading, 2007 – 2008

There is no doubt that developing nations are especially vulnerable, and even the currency tends to be tied to the price of those particular commodity items until it manages to be a fully developed nation. For example, one could see the nominally "fiat" money of Cuba as being tied to sugar prices, since the lack of hard currency paying for sugar means less foreign goods per peso in Cuba itself. In effect, Cuba needs a hedge against a drop in sugar prices, if it wishes to maintain a stable quality of life for its citizens. Whether you are a large producer or a whole set of consumers, you need to hedge your foreign currency exposures and the respective exchange futures commodities.

One has to understand that the primary goal of commodity price risk management is to protect the economic value of your business from the negative impact of commodity price fluctuations, at the lowest possible cost. Because commodity price volatility also provides opportunity for gains, a secondary goal is to strike a balance between risk and return. Risk management provides the ability to accurately budget on cash flow receipts.

It is also fair to explain that some form of risk taking is inherent to any business activity. Some risks are considered to be "natural" to specific businesses, such as the risk of oil prices increasing or decreasing is natural to oil drilling and refining firms. Other forms of risk are not wanted, but cannot be avoided without hedging. Not all hedges are financial instruments: a producer that exports to another country, for example, may hedge its currency risk when selling by linking its expenses to the desired currency. Banks and other financial institutions use hedging to control their asset-liability mismatches, such as the maturity matches between long, fixed-rate loans and short term (implicitly variable rate) deposits.

For example, rising fuel prices can cut into your profits and make budgeting difficult. That's where we need to set up this mechanism, if your business purchases oil, you can certainly budget with greater certainty by effectively managing the risks associated with fluctuating fuel prices.

6.3 Basic Carbon Hedging Strategy

This explanation should be kept as simple as possible. First, it can be used an example of frozen concentrated soya juice ("SJ"), as a commodity example and need to know a few things about this commodity before start this simulation. Here's the specifications for orange juice futures contracts.
Contract Size: 10,000 pounds (4536 kg) of soya juice
Point Value: $1.50 per point
Minimum price move: 5 points or 5/100 ths of a cent = $7.50 per contract

The assumption is taken about a soya juice producer first (SJ farmer). This person has to sell his soya juice in six months later. The problem is that any price drop in the soya juice market would have a negative effect on what the farmer can get for his crop once it's harvested.

The farmer can get around a large part of that risk by establishing a basic short commodity hedging strategy in the soya juice futures market. This may give the farmer some price protection, sort of like an insurance policy against large price fluctuations. Furthermore, the current price for soya juice in the cash market on February 1st is 90 cents per pound (fictional). The SJ farmer feels that's a fair price to cover his costs and make a profit. The farmer also knows that he will have about 10,000 pounds of SJ to bring to the market at harvest in six months. What the farmer does is sell his crop now using the futures market to protect that 90 cent sale price in the future. The farmer goes into the futures market and sells 1 contract (10,000 pounds of SJ) at the current market price of $1 per pound. Now let's fast forward 1 month into the future and see how this protects his profit margins.

On August 1st the futures price of SJ has dropped to 70 cents per pound and the cash or current price for SJ drops to 65 cents per pound because there looks to be a bumper crop of SJ this year. This situation doesn't look good as the SJ producer needed to get 90 cents a pound to cover his costs and make a profit. The farmer will be getting $3750 less for his SJ crop. The decimal point has been omitted and the calculation looks like this: 9000 - 6500 = 2500 X 1.50 = $3750 loss per contract.

Does the money loss from the crop? What about the SJ contract from the farmer sold in the futures market a few month ago? The farmer sold 1 contract at $1 per pound. If the farmer plan to buy that contract back right now he would only have to pay 70 cents a pound. The farmer has a profit of $4500 for the futures contract. The decimal point has been omitted and the calculation looks like this 10000 - 7000 = 3000 X 1.50 = $4500 profit per contract.

Furthermore, what the hedge has done to partially protect the SJ farmer's price risk is being analyzed. The $3750 cash loss is offset by the $4500 profit in the futures market, leaving the farmer with a theoretical profit on the hedging strategy of $750. This is not a bad deal.

Commodity Futures Fact: A commodity hedging strategy does not remove all price risks. In fact, there are costs associated with trading in commodity futures markets that must be factored into any hedging strategy. Those costs include the commissions paid on the futures trades and the costs associated with placing money in the futures account to cover initial margin requirements (good faith deposits) and maintenance margin calls (additional deposits to cover adverse price variations).

The market is continuing to another month into the future. On July 1st another report shows that the first report overestimated the SJ supply and the price has risen to $1.20 a pound and the cash price of SJ has gone up to $1.05 because of the simple economics of supply and demand. What will happen to the farmer's cash? The farmer can now get $2250 more for his SJ. The calculation looks like this: 10500 - 9000 = 1500 X $1.50 = $2250 more profit.

The profit must be analyzed first. The farmer shouldn't run out and buy a new BMW yet. It will cost him $1.20 per pound to buy back the futures contract he sold at $1. That gives him a loss of $3000 for his futures hedge. The calculation looks like this: 10000 - 12000 = 2000 X $1.50 = $3000 loss.

The result about how the commodity futures hedge has limited his potential profit margin. The $2250 gain on the cash price of the SJ crop is offset by the $3000 loss he currently has on his commodity futures hedge. The net result of liquidating the hedge right now would be a loss of $750.

This two examples show the importance of maintaining the hedge (regardless of price fluctuations) until the crop is ready for delivery. The cash price and the futures price will converge and become almost equal at the expiration month of the futures contract except for costs such as carrying charges (also known as "the basis"). By liquidating the futures contract and breaking the protection of the hedge before expiration, the farmer then becomes at risk-to-price fluctuations. The farmer also loses money on the costs associated with the futures portion of the hedge itself.

Commodity Futures Fact: In a liquid market, the number of speculators (people looking to profit from price fluctuations) far outnumbers the number of hedgers (those protecting themselves against price risks), but the hedgers generally carry much larger open positions.

6.4 Implementation of Waste-to-Carbon Simulation

This followed examples of hedging or mitigation of risk by using exchange traded derivatives. Contract size 1,000 Certified Emission Reductions (1 Lot). The Price Quotation is in points with two decimal places, equivalent to a value of EUR 0.01 per CER. The Minimum Price Change is EUR 0.01 per CER, equivalent to a value of EUR 10 per contract.

A waste manager which was collaborated with a hedge fund manager has an energy-related investment portfolio closely resembling the EUREX (European Energy Exchange). The Investment (hedge fund) Manager believes the global and regional economy is worsening with deteriorating expected returns.

The next two to three weeks are reports of quarterly US crude oil inventories. Until the report exposes, he is concerned of the results from a short term crude oil market correction. Without the privilege of foresight, he is unsure of the magnitude the earnings figures will produce. He now has an exposure to market risk.

The investment manager thinks of his options. The greatest risk is to do nothing, if the market falls as expected, he risks giving up all recent gains. If he sells his portfolio early, he also risks being wrong and missing further rally's. Then he realizes a hedge is the best option to mitigate his short term risk. He begins by calling his CTA (Commodity Trading Advisor) and after consultation places an order to sell short the CO^2 Derivative Index on the EUREX. Now his result is when the market falls as expected, he will off-set any losses in the portfolio with gains from the Index hedge and his portfolio continues upward, he will continue making profits. A month later the investment manager again calls his CTA and closes the hedge by buying back the equivalent number of contracts on the EUREX. Regardless of the resulting market events, the investment manager was protected during the period of short term volatility. There was no risk to their investment portfolio.

Simulation on Data - 2007 :

Total sum of Kuala Lumpur population	= (+/-) 1,600,000
Organic waste generation volume predicted per person per day	= +/- 0.5 kilogram
Total organic waste generation volume of Kuala Lumpur per month	= 30 days x 0.5 kilogram
x 1,600,000	= 24,000,000 kilogram
	= 24,000 tonnes-per-month

➔ Kuala Lumpur may generate organic waste for **<u>24,000 tonnes / month</u>**

Density

Asumption of organic waste composition	= 70 percent
	= 70% x 24,000 tonnes = 16,800 tonnes
1 tonnes of organic waste x (21-CH4)	= 1 CER
16,800 tonnes of organic waste x (21)	= 352,800 CER

Simple Equation for Carbon Emission Trading (CET) from Recycled-Reduced Organic Waste :

July 1st

1 CER = EUR 20 (as assumption)
CET = 352,800 CER x EUR 20 = EUR 7,056,000
Converted to Malaysian Ringgit ➔ EUR (1 EUR = MYR 5.05) (exchange rate at July 1st, 2009)
= EUR 7,056,000 x MYR 5.05
= **RM 35,632,800** / month of July

Aug 1st

1 CER = EUR 10 (as assumption)
CET = 352,800 CER x EUR 10 = EUR 3,528,000
Converted to Malaysian Ringgit ➔ EUR (1 EUR = MYR 5.05) (exchange rate at Aug 1st, 2009)
= EUR 3,528,000 x MYR 5.05 = **RM 17,816,400** / month of August

On August 1st the cash or current price for Carbon drops from EUR 20 to EUR 10 per 1 CER and the futures price of Carbon has dropped from EUR 23 to EUR 11 per 1 CER. Because of the carbon price is falling in August, this situation doesn't look good as it needs to get EUR 10 per 1 CER in cash to cover his costs and make a profit. The monthly return is decreased almost a half from the one month. The decimal point has been omitted and the calculation looks like: EUR 20,000 – EUR 10,000 = EUR 10,000 loss per contract.

The appropriate strategy should be implemented is short (or sell) in the carbon futures price at the EUREX market. The waste manager goes into the futures market and sells 1 contract (1,000 CER) at the current market price of EUR 23 per pound. Now lets fast forward 1 month into the future and see how this protects his profit margins. The calculation will be EUR 23,000 – EUR 11,000 = EUR 12,000 profit per contract.

At last, what the hedge has done to partially protect the waste manager's price risk. The EUR 10,000 cash loss is offset by the EUR 12,000 profit in the futures market, leaving the waste manager with a theoretical profit on the hedging strategy of EUR 2,000. This is not a bad deal.

Further Simulation

Scenario 1: Price falls in the cash and futures market
At this point, a waste manager receive a nett-return, which is the profit return from the futures market is offset with the loss from the cash market.

Cash Market	Futures Market	Base Profit/Loss
10 July: EUR 10/ CER	Sell at EUR 13/ CER	
10 Aug: EUR 8/ CER	Buy at EUR 11/ CER	
(-) EUR 2/ CER	(+) EUR 2/ CER	nett

Scenario 2: Price up in the cash and futures market
When this situation happened, a waste manager will still receive a nett-return, which is the profit return from the futures market is offset with the loss from the cash market.

Cash Market	Futures Market	Base Profit/Loss
10 July: EUR 10/ CER	Sell at EUR 11/ CER	
10 Aug: EUR 13/ CER	Buy at EUR 14/ CER	
(+) EUR 3/ CER	(-) EUR 3/ CER	nett

Scenario 3: Price up in the cash and Price falls in the futures market
At the other market situation, a waste manager will receive double profit-return, which is the profit return from the futures market is accumulated with the profit from the cash market.

Cash Market	Futures Market	Base Profit/Loss
10 July: EUR 10/ CER	Sell at EUR 11/ CER	
10 Aug: EUR 13/ CER	Buy at EUR 8/ CER	
(+) EUR 3/ CER	(+) EUR 3/ CER	(+) EUR 6/CER

Scenario 4: Price falls in the cash and Price up in the futures market
At this last point, a waste manager will receive double loss, which is the loss from the futures market is accumulated with the loss from the cash market.

Cash Market	Futures Market	Base Profit/Loss
10 July: EUR 10/ CER	Sell at EUR 11/ CER	
10 Aug: EUR 8/ CER	Buy at EUR 14/ CER	
(-) EUR 3/ CER	(-) EUR 3/ CER	(-) EUR 6/CER

5. Conclusion

From the scenarios above, first and second scenarios may give a waste manager fully price protection from hedging activity. The third scenario can be used as a profit generator by waste manager. The last fourth scenario cannot support waste manager to gain some profit return from the CER. The investment manager may also limit the risk potential by applying the first, second and third scenario of carbon emission trading in the futures market.

Since the setting is incomplete by nature there exists an interval for arbitrage-free prices and we have chosen one price based on a local risk minimizing criteria. One can come up with explicit formulas for pricing and hedging under the assumption that the market's net position is common knowledge among the market participants. Under the more realistic setting where the market does not observe the net position directly the initial design of the regulatory structure for the financial instruments associated with the global carbon market will play a significant role in determining how the instruments will trade and the type of information that is available to regulators, market participants, and the general public.

Existing regulatory models for financial markets, including the GHG cap-and-trade systems currently in operation, provide useful lessons for the design of a market-based approach to limit GHG emissions. Building upon these lessons, policymakers can implement an appropriate regulatory structure for allowance-based financial instruments at the outset, ensuring an efficient and effective approach to reducing the nation's GHG emissions.

References

[1] Agamuthu, P., 2001. Solid Waste: Principles and Management. Institute of Biological Sciences, University of Malaya.
[2] Emery, A., Davies, A., Griffiths, A., Williams, K., 2007. Environmental and economic modelling: a case study of municipal solid waste management strategy in Wales. Resource, Conservation and Recycling 49, 244–263.
[3] Fauziah, S.H., Agamuthu, P., 2003. A Comparative Study on Selected Landfills in Selango, Kuala Lumpur Municipal Solid Waste Management. Institute of Biological Sciences, Faculty of Science, University Malaya.
[4] Hadi, A.S., Johari, M.Y., 1996. The Malaysian Urban System. Paper Presented at Comparative Urban Studies Meeting, Asian–Pacific Center, Fukuoka, Japan, and Mimeo.
[5] Hassan, M.N., Zakaria Z., Rahaman, R.A., 1999. Managing Costs of Urban Pollution in Malaysia: The Case of Solid Waste. Paper presented in MPPJ Seminar Petaling Jaja, Malaysia.
[6] Hassan M.N., Chong T.L., Rahman, M.M., Salleh, M.N., Zakariah, Z., Awang, M., Yunus, M.N., 2000. Solid Waste Management – What's the Malaysian Position, Seminar on Waste to Energy 2000, Universiti Putra Malaysia.
[7] Komilis Dimitris P. , Robert K. Ham., 1999a . The effect of lignin and sugars to the aerobic decomposition of solid wastes. Department of Environmental Studies, University of the Aegean. Athens, Greece.
[8] Murad, M.W., Siwar, C., 2006. Waste management and recycling practices of the urban poor: a case study in Kuala Lumpur city, Malaysia. Online Journal of Waste Management Research, ISWA, UK.
[9] Nasir, A.A., 2007. Institutionalizing Solid Waste Management in Malaysia: Department of National Solid Waste Management. Ministry of Housing and Local Government Malaysia, Power Point Presentation, December 6.
[10] Nasir, M.H., 2004. Lecture Notes. Faculty of Environmental Studies, UPM.
[11] Norbu, T., Visanathan, C., Basnayake, B., 2005. Pretreatment of municipal solid waste prior to landfilling. Waste Management 25, 997–1003

[12] Omar, M.N., 1999. Characterization and Management of Solid Waste in Klang Valley, Master Thesis, Faculty of Engineering, University of Malaya, Kuala Lumpur Management; Paper Presented in Seminar on Managing Solid Waste Held in Peatling Jaya, Selangor on March 27.

[13] Pillay, M.S., 1999. Municipal Solid Waste Management System: A Situational Appraisal and Possible Solution: Paper Presented in Seminar on ''Local Communication and the Environment" Organized by EPSM, 24–25th October, 1998 Shah's Village Hotel.

[14] Rigamonti, L., Grosso, M., Giugliano, M., 2009. Life cycle assessment for optimizing the level of separated collection in integrated MSW management systems. Waste Management 29 (2), 934–944.

[15] Siraj, M., 2006. Waste Reduction: No Longer an Option but a Necessity: in Bernam. Kuala Lumpur.

[16] Slater, R., Gilbert, J., Frederickson, J. (2001) Review of Composting Practices in the UK. In proceedings: Profit from Waste. Institute of Mechanical Engineers, London.

[17] Tchobanoglous, G., Theisen, H., Vigil, S., 1993. Integrated Solid Waste Management-Engineering Principles and Management Issues. McGraw-Hill, New York.

[18] Tchobanoglous and PR O'Leary, Landfilling, in Handbook of Solid Waste Management, F. Kreith (ed.). McGraw-Hill. Inc., New York. 1994.

[19] Themelis, Nickolas J. and Ulloa, Priscilla A, (2007). Methane generation in landiflls, Renewable Energy, 32:1243-1257.

[20] Troschinetz, A.M., Mihelcic, J.R., in press. Sustainable recycling of municipal solid waste in developing countries. Waste Management. <http://dx.doi.org/10.1016/j.wasman.2008.06.004>.

[21] Wahid, A.G., Hassan, M.N., Muda, A., 1996. Domestic and Commercial waste: Present and Future Trends. CAP-SAM National Conference on the State of the Malaysian Environment, Penang: RECDAM.

The Factors Affecting on E-procurement Usage: The Moderating Role of Power

Luay Daoud[1,] Marhaiza Ibrahim[1]

[1]Tunku Puteri Intan Safinaz School of Accountancy, College of Business, Universiti Utara Malaysia, 06010 Sintok, Kedah, Malaysia.

**E-mail:* Luaydaoud1@gmail.com

Abstract. The paper evaluates features facilitating or impeding e-procurement utilization among firms in Jordan. E-procurement B2B is an information technology that links a business relationship between two parties. Power is the key determinant in adopting and using such technology. Regrettably, the role of power is almost disregarded in e-procurement utilization literature. By using the integrated template of the Technology-Organization-Environment (TOE) framework and Diffusion of Innovations (DOI) theory with power as a moderator, this paper helps to illustrate e-procurement usage. This argument could enhance e-procurement usage comprehension and help to rectify variability and inconsistency of findings in the literature.

Keywords: E-procurement usage, Diffusion of Innovations (DOI) theory, Technology–Organization–Environment (TOE) Framework, Power, Jordan.

1. Introduction

The usage of Information System (IS) among Jordanian firms is comparably low in spite of the immense of Information Technology (IT) being used in their routine business operation when compared with other emerging economies worldwide (Lutfi, Idris, & Mohamad, 2016). Other than limiting the capacity of the firms to completely leverage, the restricted use of IT in routine business operation also hinder the firms to refine their competitiveness in the market (Global Information Technology (GIT) Report, 2012). In addition, due to the low level of IT related solution deployment, Jordanian firms' capacity is obstructed from sustaining their competitiveness and refining their business productivity (Lutfi, Idris, & Mohamad, 2017).

To improve the firms' performances, they must consistently enhance their interior procedures and operations. The purchasing process which involves the incursion of enormous administrative expenses between buyers and suppliers is one of the cost-intensive operational chains. To overcome such predicament and minimize the business cost, many firms have adopted e-procurement to simplify the complication of the procurement process. (Zunk, Marchner, Uitz, Lerch, & Schiele, 2014).

This study examines key elements that facilitate or inhibit the firms to e-procurement usage to improve our comprehension of the current e-procurement usage among Jordanian firms. With the aid of Technology-Organization-Environment framework (TOE), and the role of power as a moderator between TOE factors and e-procurement usage.

Power exercise is one of the most important characteristics of B2B relationship that can affect the usage decision (Alsaad, Mohamad, & Ismail, 2014). In initiating B2B relationships, power is a norm. However, the role of power exercise has been overlooked in e-procurement as not many studies have considered its effect on e-procurement usage (Alsaad et al., 2014; Ke, Liu, Wei, Gu, & Chen, 2009). Previous studies largely investigated elements that facilitate or inhibit e-procurement usage. Hameed and Counsell (2012) argued that the inconsistency in the literature findings gives a narrow vision as to how these factors influence the usage decision.

This study provides some contributions to two groups of people namely researchers and practitioners. For researchers, this study has integrated Diffusion of Innovation (DOI) theory and TOE framework with the role of power as a moderator helps the researchers to comprehend the precursors to the e-procurement usage. In addition, it is proven in the study that the suggested framework in term of e-procurement practices is applicable and suitable for the usage of firms in developing countries. On the other hand, for the practitioners, this study emphasizes on significant factors that advocate greater use of e-procurement, and the results serve both academicians and practitioners especially in the developing countries.

2. Previous Studies on E-Procurement

The traditional ways of handling the purchasing process are much ineffective and inefficient due to the excessive complications when compared to the use of e-procurement. As a result of e-procurement usage in handling the purchasing process, firms can save enormous resources in the context of cost and time (Zunk et al., 2014). E-procurement usage enables the organization to access wider market and streamlining purchasing processes (Daoud & Ibrahim, 2017). DespiteAccording to Presutti (2003), e-procurement is about the efficient purchasing process management by removing paper-based logs and reports and administering the procurement via internet-based communication systems.

Generally, there are two adoption stages namely pre-adoption stage and post-adoption stage (Lin, 2014; Hameed & Counsell, 2012). The post-adoption stage emphasizes on understanding how to put an innovation into use. Such stage involves innovation design and implementation of operations to promote its widespread and quick receipt of the innovation (Lin, 2013; Ramdani & Kawalek, 2007; Rogers, 2003). Sadly, researchers paid less consideration towards issues like pre-adoption of e-procurement which serves as the first step in innovation dissemination rather than innovation usage (Abduljalil & Zainuddin, 2015; Alamin, Yeoh, Warren, & Salzman, 2015). Hence, this study focus on post-adoption stage (e-procurement usage) among firms in Jordan.

Although there are many studies being conducted by the researchers worldwide on the factors that effect on e-procurement, a few studies in the middle east especially in Jordan. This paper serves to eliminate the research gap by investigating the factors that affect the usage of e-procurement among Jordanian firms.

3. Theoretical Framework

There are most popular theories applicable in illustrating the e-procurement usage which is the DOI Theory and TOE framework (Sila, 2013; Zhu, Kraemer, & Xu, 2006). These theories focus on the function of technological, organizational, and environmental determinants and how do these determinants impact the e-procurement usage. Thus, in this study, the two significant theories are used to illustrate the conceptual research model.

Rogers (1995) proposed that the DOI theory assesses an innovation based on five innovation characteristics known as innovation attributes, namely, relative advantage, compatibility, complexity, observability, and trialability. However, previous studies show that three out of five attributes namely relative advantage, compatibility, and complexity, are always related to the usage of technology (Tornatzky & Klein 1982; Wu, Zsidisin, & Ross, 2007). Therefore, these three key attributes are used in this paper.

In contrast, TOE model addresses three contextual factors, comprising of technological, organizational, and environmental factors that can affect the e-procurement innovation usage among the firms. In brief, TOE is the main framework that motivated almost all the prior studies in explaining the antecedents of e-procurement usage (Hassan, Tretiakov, & Whiddett, 2017).

4. Conceptual framework and hypotheses

This paper helps to evaluate the factors that facilitating or inhibiting e-procurement usage among Jordanian firms. Figure 1 exhibits a research framework, identifying the association between the independent variables and the dependent variable with the role of power as a moderator between TOE factors and e-procurement usage. The three main attributes that influence on e-procurement usage are examined from the contexts of technology, organization, and environment. The hypotheses of the study play a major role in completing the research model.

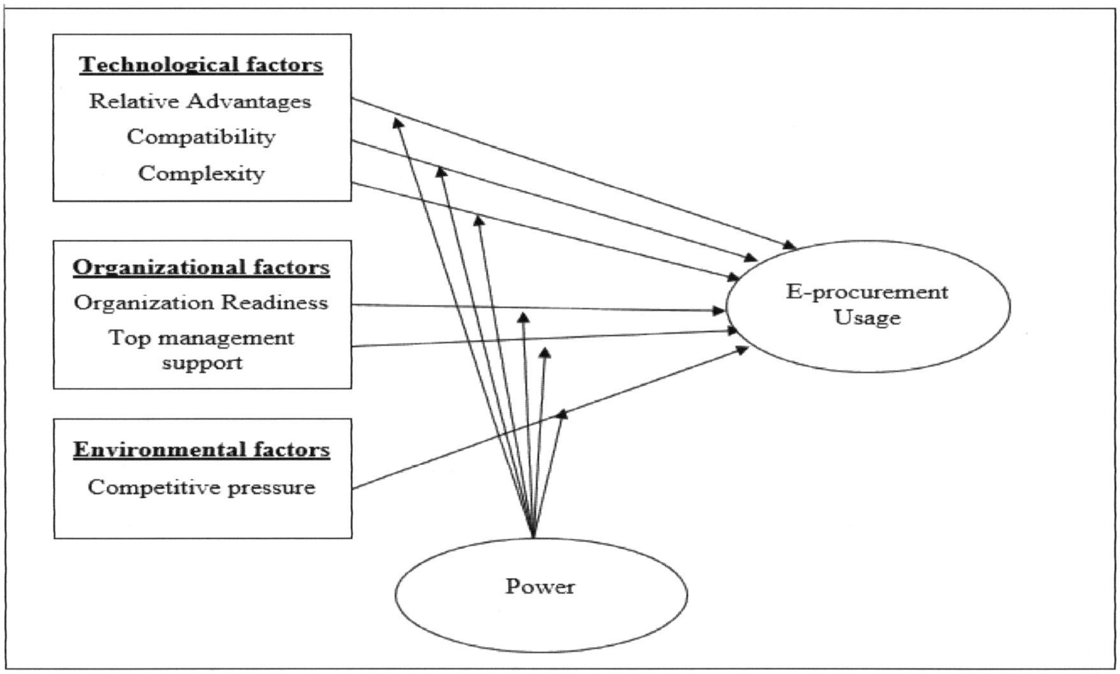

Figure 1: Proposed framework

4.1 Technological Factors

The perceived relative advantage is defined as the degree to which an innovation is recognized as more sophisticated than its substitute (Rogers, 2003). Many empirical studies found significant effects of the relative advantage on the use of IT/IS innovation (Ali, Rahman, & Ismail, 2012; Zhu et al., 2006). On the contrary, a few studies discovered the insignificant role of relative advantage to explain use of IT-related innovations (Lutfi et al., 2017; Yoon & George, 2013).

Relative advantage is considered the most important factor in IS usage (Garg, & Choeu, 2015; Rahayu, & Day, 2015). It is argued that organizations that do not perceive the benefits of e-procurement usage to be significant would consider the implementation of the system to be unnecessary (Hassan, Tretiakov, & Whiddett, 2011). Thus, this study hypothesizes the following:

H1: Relative advantage has a positive influence on e-procurement usage.

According to Moore and Benbasat (1991), compatibility is explained as "the degree to which an innovation is perceived as being consistent with the existing values, needs, and past experiences of potential adopters" (p. 195). Compatibility is one of the greatest driving force in the use of technology in a firm when compared with other innovation attributes (Grandon & Pearson, 2004; Zhu et al., 2006). This implies that the higher the innovation's compatibility, the higher the degree of adoption and usage. Hence, this paper hypothesizes the following:

H2: Compatibility has a positive influence on e-procurement usage

Rogers (2003) describes Complexity as "The degree to which an innovation is perceived as difficult to understand and to use" (p. 257). In addition to that, Al-Hudhaif and Alkubeyyer (2011) believe that the increasing innovation's complexity will decline the degree of IT adoption and usage. Hence, this paper hypothesizes the following:

H3: Complexity has a negative influence on e-procurement usage.

4.2 Organizational Factors

Organizational factors refer to the features and attributes of a firm that affect the adoption and usage of technology (Maryeni, Govindaraju, Prihartono, & Sudirman, 2014). The capacity of a firm in acquiring new technology is defined as the organizational readiness. Prior studies which are focusing on the role of organizational readiness in technology innovation usage predominantly acknowledge its significant impact on the usage of technology (Lutfi et al., 2017; Zhu & Kraemer, 2005). Hence, this implies that the increasing organizational readiness will increase the likelihood of a firm to use e-procurement. This is represented by the following hypothesis:

H4: Organizational readiness has a positive influence on e-procurement usage.

Top management support is defined as the level of the effort and resource support for innovation given by the top management (Premkumar, 2003). In forecasting organizational acceptance of technological innovations, top management support is determined as one of the best variables (Jeyaraj, Rottman, & Lacity., 2006). Prior study has proved that the top management support generally provides a positive indicator for the adoption of organizational technology innovations (Al-Qirim, 2007; Alsaad, Mohamad, & Ismail, 2017; Ifinedo, 2011). According to Ifinedo (2011), the organization inclines to ignore the adoption of new technology innovation especially when the top management gives little or no support and this will eventually generate adverse outcomes. Hence, this paper hypothesizes the following:

H5: Top management support has a positive influence on e-procurement usage.

4.3 Environmental factors

Environmental factors can be defined as exterior elements that are out of the firm's top management control (Jeyaraj et al., 2006). According to Hassan et al. (2011), competition pressure is the degree to which the effort of a firm in use e-procurement innovation is affected by any other firms or business allies in the market. The intense competition pressure will push organizations to provide quicker responses to customers' needs, reduce lead times and provide better customization (Zhu & Kraemer, 2005). Prior studies have concluded that the competitive pressure is significant influencing on IT/IS usage (Alsaad et al., 2017; Lutfi et al., 2016; Teo, Lin, & Lai, 2009). Lutfi et al. (2016) also believed that competition pressure serves as one of the prominent factors for technology adoption and usage. Hence, this paper hypothesizes the following:

H6: Competition pressure has a positive influence on e-procurement usage.

4.4 The Moderating Effect of Power

Power is defined as the capacity of an organization to influence another organization to behave in a certain pattern (Ratnasingham, 2000). The firm which has power over another firm owns the capability to influence that firm's actions and decisions. According to Alsaad et al. (2014), the authorized power from business partners is anticipated to be one of the most prominent determinants of IT/IS adoption and usage.
To enhance firm's competency, dominant actors can invest in technology innovation, where the advantage from such IT cannot be exploited without the alliance of its business partners (Alsaad et al., 2014). If such situation arises, the dominant actors will manipulate their power ability to affect organizations that rely on it to invest in comparable technologies (Al-Hakim, Abdullah, & Ng, 2012; Chan, Chong, & Zhou, 2012; Chong, Chan, Goh, & Tiwari, 2013; Hertwig, 2012).
Characteristics of innovation may play an insignificant role in innovation usage decision if the firm recognizes hostile power. On the contrary, the firm will assess the innovation attributes in making innovation usage decision when the firm recognizes a non-coercive power (Alsaad et al., 2014). Thus, innovation attributes will play a key function in influencing the decision of innovation usage in the firm. Hence, this paper hypothesizes the following:

H7(A): Power moderates the relationship between the Relative Advantages and e-procurement usage positively.

H7(B): Power moderates the relationship between the compatibility and e-procurement usage positively.

H7(C): Power moderates the relationship between the Complexity and e-procurement usage positively.

H7(D): Power moderates the relationship between the organization readiness and e-procurement usage positively.

H7(E):Power moderates the relationship between the top management support and e-procurement usage positively.

H7(F): Power moderates the relationship between the competition pressure and e-procurement usage positively.

5. Conclusion

This paper examines the significance of power in e-procurement usage decision and discusses the usage of various influence strategy may have an impact on the role of other factors on e-procurement usage. Top management can exploit the results of this study to have a better comprehension on the determinants that facilitate the e-procurement usage. Moreover, the results of this paper allow the top management to handle the impacts of these factors in a more effective way. One of the significant contributions of this paper to the accounting and IS field is the investigation of the association between technological advancement and e-procurement usage. This paper also serves as an imperative base for the ultimate evolution of successful IT strategies.

6. References

[1] Abduljalil, K., & Zainuddin, Y. (2015). Intrinsic and extrinsic motivation as attitude factors towards adoption of accounting information system (AIS) in Libyan SMEs. *International Journal of Academic Research in Accounting, Finance and Management Sciences*, **5(1)**, 161-170.

[2] Alamin, A., Yeoh, W., Warren, M., & Salzman, S. (2015). An empirical study of factors influencing accounting information systems adoption. In *Proceedings of the Twenty-Third European Conference on Information Systems* (pp. 1-11). ECIS.

[3] Al-Hakim, L., Abdullah, N. A. H. N., & Ng, E. (2012). The effect of inter-organization trust and dependency on e-procurement adoption: a case of Malaysian manufacturers. *Journal of Electronic Commerce in Organizations (JECO)*, ***10*(2)**, 40-60.

[4] Al-Hudhaif, S. A., & Alkubeyyer, A. (2011). E-commerce adoption factors in Saudi Arabia. *International Journal of Business and Management*, ***6*(9)**, 122-133

[5] ALIa, A. Z. W. A. D. I., Rahman, M. S. A., & Ismail, W. N. S. W. (2012). Predicting continuance intention to use accounting information systems among SMEs in Terengganu, Malaysia. *International Journal of Economics and Management*, ***6*(2)**, 295-320.

[6] Al-Qirim, N. (2007). The adoption and diffusion of e-commerce in developing countries: The case of an NGO in Jordan. *Information Technology for Development*, ***13*(2)**, 107-131.

[7] Alsaad, A. K., Mohamad, R., & Ismail, N. A. (2014). The moderating role of power exercise in B2B E-commerce adoption decision. *Procedia-Social and Behavioral Sciences*, ***130***, 515-523.

[8] Alsaad, A., Mohamad, R., & Ismail, N. A. (2017). The moderating role of trust in business to business electronic commerce (B2B EC) adoption. *Computers in Human Behavior*, ***68*,** 157–169.

[9] ASE. (2017). listed companies, https://www.ase.com.jo/

[10] Basu, V., Hartono, E., Lederer, A. L., & Sethi, V. (2002). The impact of organizational commitment, senior management involvement, and team involvement on strategic information systems planning. *Information & Management*, ***39*(6)**, 513-524.

[11] Chan, F. T., Chong, A. Y. L., & Zhou, L. (2012). An empirical investigation of factors affecting e-collaboration diffusion in SMEs. *International Journal of Production Economics*, ***138*(2)**, 329-344.

[12] Chong, A. Y. L., Chan, F. T., Goh, M., & Tiwari, M. K. (2013). Do interorganisational relationships and knowledge-management practices enhance collaborative commerce adoption? *International Journal of Production Research*, ***51*(7)**, 2006-2018.

[13] Daoud, L., & Ibrahim, M. (2017). Akademia Baru. *Journal of Advanced Research in Business and Management Studies*, *9*(1), 36-44.

[14] Garg, A., & Choeu, T. (2015). The Adoption of Electronic Commerce by Small and Medium Enterprises in Pretoria East. *The Electronic Journal of Information Systems in Developing Countries*, ***68***, 1-23.

[15] Global Information Technology Report, (GITR). (2012). Available from: http://www3.weforum.org/docs/Global_IT_Report_2012.

[16] Grandon, E. E., & Pearson, J. M. (2004). Electronic commerce adoption: an empirical study of small and medium US businesses. *Information & management*, ***42*(1)**, 197-216.

[17] Hameed, M. A., Counsell, S., & Swift, S. (2012). A meta-analysis of relationships between organizational characteristics and IT innovation adoption in organizations. *Information & management*, ***49*(5)**, 218-232.

[18] Hassan, H., Tretiakov, A., & Whiddett, D. (2011). Extent of e-procurement usage: An empirical study of small and medium sized New Zealand manufacturing businesses. In *Proceedings of the American Conference on Information Systems (AMCIS)*.

[19] Hassan, H., Tretiakov, A., & Whiddett, D. (2017). Factors affecting the breadth and depth of e-procurement use in small and medium enterprises. *Journal of Organizational Computing and Electronic Commerce*, (just-accepted).

[20] Hertwig, M. (2012). Institutional effects in the adoption of e-business-technology: evidence from the German automotive supplier industry. *Information and Organization*, **22(4)**, 252-272.

[21] Ifinedo, P. (2011). An empirical analysis of factors influencing Internet/E-business technologies adoption by Smes in Canada. *International Journal of Information Technology & Decision Making*, ***10*(4)**, 731–766.

[22] Jeyaraj, A., Rottman, J. W., & Lacity, M. C. (2006). A review of the predictors, linkages, and biases in IT innovation adoption research. *Journal of Information Technology*, ***21*(1)**, 1–23.

[23] Ke, W., Liu, H., Wei, K. K., Gu, J., & Chen, H. (2009). How do mediated and non-mediated power affect electronic supply chain management system adoption? The mediating effects of trust and institutional pressures. *Decision Support Systems*, ***46*(4)**, 839-851.

[24] Lin, H. F. (2013). Examining the factors influencing knowledge management system adoption and continuance intention. *Knowledge Management Research & Practice*, ***11*(4)**, 389-404.

[25] Lin, H. F. (2014). Understanding the determinants of electronic supply chain management system adoption: Using the technology–organization–environment framework. *Technological Forecasting and Social Change*, ***86***, 80-92.

[26] Lutfi, A. A., Idris, K., & Mohamad, R. (2016). The influence of technological, organizational and environmental factors on accounting information system usage among Jordanian small and medium-sized enterprises. *International Journal of Economics and Financial Issues*, **6(s7)**, 240–248.

[27] Lutfi, A. A., Idris, K., & Mohamad, R. (2017). AIS usage factors and impact among Jordanian SMEs: The moderating effect of environmental uncertainty, *Akademia Baru Journal of Advanced Research in Business*, **6(1)**, 24–38.

[28] Maryeni, Y. Y., Govindaraju, R., Prihartono, B., & Sudirman, I. (2014). E-commerce adoption by Indonesian SMEs. *Australian Journal of Basic and Applied Sciences*, ***8*(14)**, 45-9.

[29] Moore, G., & Benbasat, I. (1991). Development of an instrument to measure the perceptions of adopting an information technology innovation. *Information Systems Research*, ***2*(3)**, 193-222.

[30] Premkumar, G. (2003). A meta-analysis of research on information technology implementation in small business. *Journal of Organizational Computing and Electronic Commerce*, ***13*(2)**, 91-121.

[31] Presutti, W. D. (2003). Supply management and e-procurement: creating value added in the supply chain. *Industrial marketing management*, *32*(3), 219-226.

[32] Rahayu, R., & Day, J. (2015). Determinant factors of e-commerce adoption by SMEs in developing country: evidence from Indonesia. *Procedia-Social and Behavioral Sciences*, ***195***, 142-150.

[33] Ramdani, B., & Kawalek, P. (2007). SMEs & IS innovations adoption: a review & assessment of previous research. *Academia. Revista Latinoamericana de Administración*, **(39)**, 47-70.

[34] Ratnasingam, P. (2000). The influence of power on trading partner trust in electronic commerce. *Internet Research*, ***10*(1)**, 56-63.

[35] Rogers, E. (1995). Diffusion of innovations, (4th ed). New York: Free Press.

[36] Rogers, E. M. (2003). Diffusion of innovations (5th ed.). New York: Free Press.

[37] Sila, I. (2013). Factors affecting the adoption of B2B e-commerce technologies. *Electronic commerce research*, ***13*(2)**, 199-236.

[38] Teo, T. S., Lin, S., & Lai, K. H. (2009). Adopters and non-adopters of e-procurement in Singapore: An empirical study. *Omega*, ***37*(5)**, 972-987.

[39] Tornatzky, L. G., & Klein, K. J. (1982). Innovation characteristics and innovation adoption-implementation: A meta-analysis of findings. *IEEE Transactions on engineering management*, **(1)**, 28-45.

[40] Wu, F., Zsidisin, G., & Ross, A. (2007). Antecedents and outcomes of e-procurement adoption: an integrative model. *IEEE Transactions on Engineering Management*, ***54*(3)**, 576-587.

[41] Yoon, T. E., & George, J. F. (2013). Why aren't organizations adopting virtual worlds? *Computers in Human Behavior*, ***29*(3)**, 772-790.

[42] Zhu, K., & Kraemer, K. L. (2005). Post-adoption variations in usage and value of e-business by organizations: Cross-country evidence from the retail industry. *Information Systems Research*, ***16*(1)**, 61–84.

[43] Zhu, K., Kraemer, K. L., & Xu, S. (2006). The process of innovation assimilation by firms in different countries: a technology diffusion perspective on e-business. *Management science*, **52(10)**, 1557-1576.

[44] Zunk, B., Marchner, M., Uitz, I., Lerch, C., & Schiele, H. (2014). The role of E-procurement in the Austrian construction industry: Adoption rate, benefits and barriers. *International journal of industrial engineering and management*, **5(1)**, 13-21.

IOP Conf. Series: Journal of Physics: Conf. Series **1019** (2018) 012077 doi:10.1088/1742-6596/1019/1/012077

The influence of Human Resource Management Information System (HRMIS) Application towards Employees Efficiency and Satisfaction

Ahmad Suffian Mohd Zahari[1], Mohd Amir Harun[2], Siti Fatimah Mardiah Hamzah[3] and Suzila Mat Salleh[4]

E-mail: ahmadsuf@tganu.uitm.edu.my

[1]Faculty of Business & Management
University Teknologi MARA (Cawangan Terengganu), 23000 Dungun, Terengganu, Malaysia

Abstract. The transformation of human resource information management activities from traditional practice to urban computerization method in some way lead to dissatisfaction amongst some government employees. However, this transformation is necessary in E-Government era as its allow employees to accomplish human resource function in more effective and efficient manner. In response to this emerging trend, government has introduced HRMIS application for government organization. HRMIS application is one of the integrated e-government services that provide single interface for government employees to perform human resource function and act as a central database for public sector. In relation to this, a study is conducted to investigate the influence of Human Resources Management Information Systems (HRMIS) towards employees' efficiency and satisfaction after the implementation of the system. The study involves 191 employees from different departments at Terengganu Police Contingent. A simple random sampling was used to collect the data and were analysed using reliability analysis, descriptive frequency analysis, and Pearson's correlation analysis. When these variables were tested simultaneously, the result shows that HRMIS application, user characteristics and organization support have positive influence on Terengganu Police Contingent employees' satisfaction. The study also provides valuable insight to various stakeholders particularly to the policy makers, and organizational members about the importance of HRMIS enforcement in the organization.

1. Introduction

The implementation of information technology systems has become a necessity in a rapidly changing times. In line with the development of technology in the Public Service, a system has been introduce to public sector agencies namely Human Resource Management Information System (HRMIS). According to [1], HRMIS is one of the Malaysia e-government flagships, develop to standardize human resource functions appropriately. Main objective of the application is to enable a competent management resource through modernization of technology-based for a better and conducive working environment.

2. Literature Review

HRMIS is a system that provides database for Human Resource Management (HRM) information and acts as a business intelligence for top management for decision making. This web-based application is accessable through the internet at any location, however it is subjected to higher level confirmation. The system operates using architecture that provides a flexible and secure infrastructure. Operate as a subsystem of MIS, HRMIS also useful to organize, measure and assessing current human resources operations with regard to production (outcomes). Besides, the implementation is aims to enhance the effectiveness and efficiency of the human resource operation. Listed objectives as follows:

- To achieve effective staffing and right-sizing of civil service through better availability of human resource management information;
- To automate human resource management operational processes;
- To improve paperless human resource management capabilities among agencies such as electronic distribution of human resource policy, manuals, and circulars; and
- To provide an open and flexible system which will fulfil and improve the information needs of operational and managerial processes at a different level of agencies.

Furthermore, HRMIS application function as a tool to ease human resource process in the organization. To evaluate its efficiency, it is important to review the application and measure the level of effectiveness in relation to satisfaction among employees (users) in public sector. The application is becoming gradually complex in data-intensive for its use especially in execution of all operations, other than decision making and planning. This system has helped human resources managers and professionals to take more rapid, accurate and effective feedbacks and decisions [2]. With the use of web-based technology, human resource functionality is easy, fast and up-to-date with efficient management transaction data between employees and top managers. Thus, HRMIS is a growing domain trend in human resource management. It has been announced as HR innovation revolution [3] and is expected to be fully utilized in all HRM government departments for more efficient service [4].

[5] highlighted human resource challenges to be more strategic, flexible, cost-effective and user-friendly with the implementation of the existing HRMIS system. It is an indication that the use of information technology has the potential to reduce administrative costs, increase production, rapid response, improve decision makers and improve the effectiveness of employee-friendly services at the same time. Cost reduction, quality service and healthy work culture are three key choices that drive human resource efficiency through the use of the latest information technology [6]. With the use of the HRMIS, employees will be able to cupdate information at any time, referring to and printing salary slips that have been uploaded by human resource executives and to know the amount of leave remaining for vacation planning and other uses. Besides, with the implementation of HRMIS, HRM becomes easier and enables management to focus more on organizational objectives and more important activities and strategic planning plans.

Human resource executives has found that technology and information as important inputs to help them make decisions and implement processes that can bring effectiveness, as well as time and cost savings to the organization. Internationally, some organizations have successfully used web-based technology to assist human resource functions.

Overall, HRMIS implementation in the organization will benefit the enhancement of user satisfaction, which is the user can check the status of each transaction made through the HRMIS system. The entry of basic personnel information and service profiles should not be repeated repeatedly when dealing through existing systems; reducing manual and overlapping activities; reducing the use of large file space; paper usage and wastage can be eliminated and easy and user-friendly system. In addition, the HRM function in the HRMIS system has taken into account the elements of improving processes and workflows so that human resource managers can play a more strategic role and processing time especially for operational functions to be faster in speed through the use of the latest technologies.

Currently the process of applying for and checking leave, leave approval, cancellation of leave, submitting employee overtime (OT) claims, updating personal records including employee arrival records is still being done manually. This process is subject to all employees. Workers as the most important asset of an organization, play a role in improving organizational performance. Hence, the use of technology is especially important in today's IT era which uses information at the fingertip easily and quickly. However, it can not take over the role of employees without skilled management.

IOP Conf. Series: Journal of Physics: Conf. Series **1019** (2018) 012077 doi:10.1088/1742-6596/1019/1/012077

Other assets such as finance, buildings, tools, technology and so forth are only valuable if employees use their policies and expertise in management.

However, there are still weaknesses in this HRMIS system and with the feedback from the end user, the employee itself provides a solution space for the system, and it is also up to the employees whether to continue using and accepting them [7]. Implementation of the new system does not automatically bring the constant benefit to the organization but the level of its use is a requirement [8]. Looking at implementation of HRMIS, non-mandatory use of typical features is often an option by employees who are disturbed in their use via web-based technology and not as a habit through manual methods. Organizations can also see the real attitude towards the implementation of HRMIS and the new level of system use and the factors that cause the use of the HRMIS system.

According to [9], the relationships towards information system success derive on the importance of six HRMIS application criteria namely information quality; system quality; service quality; system use/usage intentions; user satisfaction and net system benefits.

a. HRMIS application

i. Information quality;
Refers to quality information where it related to user satisfaction to HRMIS system and also users' intentions (user characteristics) toward using the system.

ii. System quality;
System quality is related towards organizational support, where continuous support (in e.g. of ergonomics, response, technology driven and etc.) will enhance the HRMIS usage. Such limited user within IT knowledge tends to users' characteristics (jaded, technostress, phobia etc.).

iii. Service quality;
Service quality directly impacts usage intentions and user satisfaction with the system. Continuous support (in e.g. of training, documentation, helpdesk support etc.) will enhance the HRMIS usage.

iv. System use/usage intentions;
HRMIS system used is posited to influence user satisfaction within the information system, where it also influences usage intentions.

v. Employees satisfaction
Employees satisfaction directly influences the usage of the HRMIS system where there is a positive relationship between system usage and user satisfaction.

b. Organizational Support

[9] study shows that there is proof for a positive relationship between perceived organizational support and affective organization commitment. Organizational Support Theory (OST) rectifies that perceived organizational support increase affective organizational commitment (e.g. employee's welfare) and thus helps the organizations to reach its goals [10]. [11] define person characteristics such as; age, gender, and IT experience performed high impact on user's satisfaction (behaviour) towards IT usage. In this context, end-user satisfaction (behaviour) influenced by their feelings (situation) and by organizational (IS) support (training, discussion, support).

c. Employees Satisfaction

According to [12], satisfaction referring to someone feelings or attitude based on various factors upon given situation. The constant change of attitude due to technology constraint can lead employees to disobey or mistreated the importance of using HRMIS on gathering information and data resource for daily operation tasks. Upon technological rapid changes, senior employees tend to violate HRMIS system due to various factors such as bad experience on managing the system [13]. Earlier studies have discovered that there are strong relationships between employee's satisfaction and intended /actual use of information system [14]. Employee's satisfaction refers to the successful relations between the information system and its user which is the employees [3]. To summarize the literature review; documentation, user interface, timeliness, accuracy, relevance, and ease of use of the HRMIS system replicate the employees' satisfaction within the system. Individual impact and organizational impact also play an important role towards HRMIS longevity usage.

To summarize, the ongoing employment of HRMIS application system is to standardize its usage as a central database for public sector management resources towards efficiency and effectiveness. The transformation of rather a traditional way to a more urban computerization system managing and handling information could lead to dissatisfaction on several government employees. Measuring employee's satisfaction and technology acceptance is a must for identifying the acceptance and satisfaction of Terengganu Police Contingent Headquarters employees on managing HRMIS system to compute information on human resource. Therefore the objective of this study is to identify the relationship between HRMIs application and organization support on employees satisfaction at Terengganu Police Contingent.

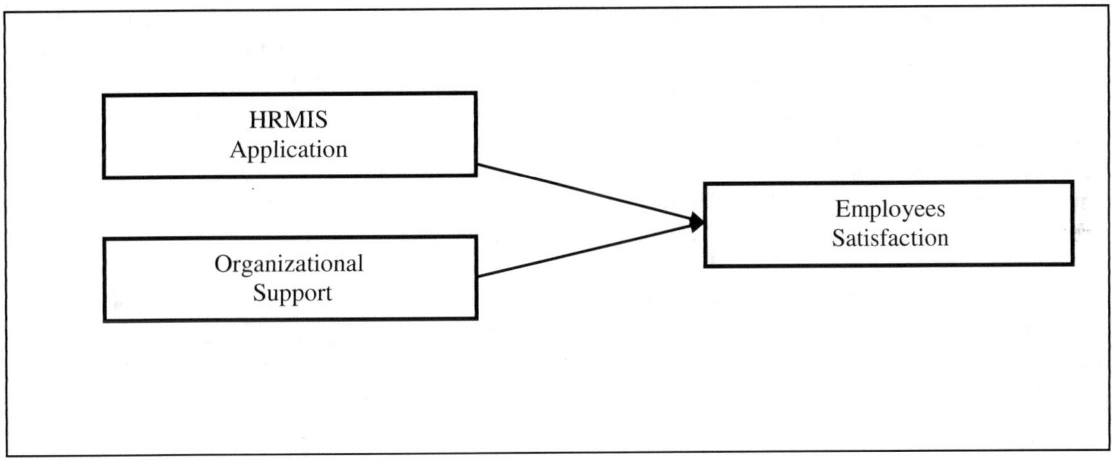

Figure 1 Proposed Conceptual Framework

Hypothesis 1

H_1: HRMIS application is positively related towards employees' satisfaction;

Hypothesis 2

H_2: Organizational support is positively related towards employees' satisfaction.;

3. Research Methodology

The population of this study consists of the government servant that works in Terengganu Police Contingent, Terengganu. Since the population of Terengganu Police Contingent, Terengganu is 380; the sample size selected was 191 respondents.

A set of questionnaires which was adopted from previous studies and formulated was designed and distributed face-to-face to the employees. The questionnaire comprises of 4 sections which contains demographic information, organizational commitment and employee motivation and sharing of knowledge. The respondents were asked to rate organizational commitment and work motivation using a five-point scale, ranging from "1-strongly dissatisfactory" to "5-strongly satisfactory".

Researchers are using correlations to seek significant relationship between two variables. Correlation research defines the variables that have mutual interaction between them. This is to examine the relationship between those 2 variables. The sampling technique that will be used is simple random technique. This is because by applying this technique, each member within the populations will have an equal chance of being selected as a respondent. For the case of Terengganu Police

Contingent employees, this sampling technique will give an ease of assembling the sample. It is a fair and constant technique in order to select a sample from the population where each member is given equal opportunities to be selected. All data and information gathered from the questionnaire then will be evaluated and processed through Statistical Package for Social Science (SPSS) system.

4. Results

4.1 Reliability Analysis

Coefficient meanwhile is used to access the reliability. Cronbach's Alpha value is the compliance value depending on the correlation between the questions or expressions. This value indicates reliability levels of the questions in scales. If Cronbach's Alpha value is over 0.70 and even more, the scale is considered reliable. Based on the table 1 above, Cronbach's Alpha showed that overall positively correlated to each other and is internally consistent. Overall findings using Cronbach's Alpha tools in quest on the reliability test (ALPHA) method shows that all IV and DV are important and positively interrelated to each other's in this research

4.2 Pearson Correlations

The correlation then will measure the relationship between two (2) sets of data. The range will be somewhere in between -1 and 1. To test the Pearson Correlation, the researcher will use every significant IV and pair it to the respective DV to resolve any correlation between them. After the questionnaires been keyed in through SPSS system, the Pearson Correlation tools had been used to analyse the IV and DV to prove their relationship towards hypothesis made by the proposal research before. Based on the table above, Cronbach's Alpha showed that overall positively correlated to each other and is internally consistent. Overall findings using Cronbach's Alpha tools in quest on the reliability test (ALPHA) method shows that all IV and DV are important and positively interrelated to each other's in this research.

Table 1: Cronbach's Alpha Result

Variable	Cronbach Alpha	No of Items	Strength of Association
HRMIS application	0.932	5	Excellent
Organizational support	0.949	6	Excellent
Users satisfaction	0.933	5	Excellent

Table 2: Pearson Correlation

		Mean HRMIS Application	Mean Organizational Support	Mean Employees Satisfaction
Mean HRMIS Application	Pearson Correlation	1	.807**	.767**
	Sig. (2-tailed)		.000	.000
	N	97	97	97
Mean Organizational Support	Pearson Correlation	.807**	1	.806**
	Sig. (2-tailed)	.000		.000
	N	97	97	97
Mean Employee's Satisfaction	Pearson Correlation	.767**	.806**	1
	Sig. (2-tailed)	.000	.000	
	N	97	97	97

From table 2, the Pearson Correlation range for the relationship between HRMIS Application (IV) and employee's satisfaction (DV) stands at .767. The relationship between these IV and DV according to Pearson Correlation stands at (0.6 and 0.8) (strong correlation) where (r= 0.767). This show that there is a significant positive correlation between HRMIS application and employee's satisfaction and the strength of the correlation is strong. Based on table 2, the Pearson Correlation ranges for the respective IV (organizational support) and the DV (employees' satisfaction) stands for .806. This range between 0.8 and 1.0 range and it indicates a very strong correlation between both IV and DV (r=.806). The Pearson Correlation findings proves out that hypothesis 2 shows a positive correlation between organizational support and employee's satisfaction and the strength of the correlation is very strong.

5. Conclusions and Suggestions

Overall result had proved that all the hypotheses had been answered well. There are two independent variables highlighted (HRMIS application, and Organizational support) where questions designed to determine whether there are positive relationship with the dependent variable (Employees satisfaction). This study is important to rectify the corrective action to ensure that HRMIS application is perform efficiently and effectively in Terengganu Police Contingent. The result of this study hopefully can be uses as a guideline for the enhancement of future implementation on HRMIS application throughout government organization and can further applies in other agencies. Application of the HRMIS system can also save cost, and time especially when the paperless concepts is introduced and fully applied.

Studies may be conducted involve more respondents to see the extent of HRMIS application acceptance by employees in the future. In addition, the application can be a benchmark to other e-government projects. As such, it will help to reduce the failure percentage of e-government projects particularly HRMIS application. [15] supported that Information System also influence by self-readiness of employees, awareness and openness towards technology. To sum up, three suggestions recommended to assist the management of the Terengganu Police Contingent to improve the efficiency of the HRMIS application in the future: -

1. Upgrade IT facilities to support the usage of HRMIS application such as enhancing integration of the main server capabilities.
2. Emphasis on the full utilization of HRMIS application among the employees.
3. Provide hands on training on HRMIS application usage to employees.

References

[1] Yusoff M, and Radzi M, 2002, E-Government Opportunities and Challenges Malaysia's Experience Retrieved on August 16 2017 from http:// egovaspac.apdip.net/topics/country/my/radzi-yusoff

[2] Bingol D, 2010, Hum Res Manage Ankara Beta

[3] Delone W H and Mclean E R, 2003, The delone and mclean model of information systems success A ten-year update *J.Manage. Info. Sys.* 19 4 9–30

[4] Adamson L and Zampetti R 2001 Web-based manager self-service adding value to the work In A J Walker Ed Web-based human resources 24-35 New York McGraw Hill

[5] Snell S, Stueber D and Lepak D, 2002, Virtual HR departments getting out of the middle In R L Heneman and D B Greenberger Eds Human resource management in virtual organizations 81-101 Greenwich CT Information Age Publishing

[6] Yeung A Brockbank W and Ulrich D, 1994, Lower cost higher value human resource function in transformation *Hum. Res. Plan.* 17 3 1-15

[7] Venkatesh V Morris M G Davis G B and Davis F D, 2003, User acceptance of information technology Toward a unified view *MIS Quar.* 27 3 425-478

[8] Devaraj S Kohli R, 2003, Performance impacts of information technology Is actual usage the missing link *Management Sc.* 49 3 273-289

[9] Rhoades L and Eisenberger R, 2002, Perceived organizational support A review of the literature *J. App.Psy.*87 698-714.

[10] Rhoades L Eisenberger R and Armeli S, 2001, Affective commitment to the organization The contribution of perceived organizational support *J. App. Psy.* 86 825-836

[11] Elie-Dit-Cosaque C Pallud J and Kalika M, 2012, The influence of individual contextual and social factors on perceived behavioral control of information technology A field theory approach *J. Manage. Info. Sys.* 28 3 201-234

[12] Wixom B H and Todd P, 2005, A theoretical integration of user satisfaction and technology acceptance *Info.Sys. Res.* 16 1 85-102

[13] Bailey J E and Pearson S W, 1983, Development of tool for measuring and analyzing computer user satisfaction *Manage. Sc.* 29 5 530-545

[14] Athanassopoulos A Gounaris S and Stathakopoulos V, 2001, Behavioral responses to customer satisfaction An empirical study *J. Mar.* 35 5/6 687-707

[15] Mazlan Mohd Sappri, Ahmad Suhaimi Baharudin and Sangeetha Raman, 2016, The Moderating Effect of User Involvement and Self Readiness and Factors that Influence Information System Net Benefits Among Malaysian Public Sector Employees. An International Journal of Applied Engineering Research. Volume 11, No 18, Page 9659-9670.

Exploring Health Communication in Social Media: A Comparative Review

Suzila Mat Salleh[1], Noor Hafiza Mohammed[2], Nurul Ulfa Abdul Aziz[3], Siti Fatimah Mardiah Hamzah[4], Ahmad Suffian Mohd Zahari[5].

Faculty of Business and Management
Universiti Teknologi MARA, Terengganu.

suzila@tganu.uitm.edu.my

Abstract. With the advent of internet, a new platform of Information Communication and Technology in health is becoming a popular trend to satisfy the craving of health consciousness people seeking health related information. Researchers published health related papers and articles while patients, their friends and families share the experiences through social media such as facebook, twitters, blogs. People can seek readily available health related information effortlessly by the convenience of their fingers tip. Therefore, the purpose of this paper is to review health communication for the past four years (2013-2016) and to identify the channels of communication used of health communication in social media. Eight papers were selected in this study related to health communication and promotion. The papers were reviewed based on their elements, methods and approaches used by researchers in their study. There are also several differences outcome in these papers. It can be concluded that social media play a main role in disseminating information and improve knowledge to the masses on health communication as people find information on health through the easiest internet channel.

1. Introduction

As Information Communication and Technology (ICT) revolutionize information distribution in each and every day of our lives, equivalently many new health technologies frontiers are discovered and becomes a breakthrough. Without doubt, healthcare has become one of the most important elements in our society. Advance in new researches, acquired knowledge and breakthrough in illness treatments, curing and prevention give way for the opportunity to the development of healthcare. According to [1], the implementation of ICT for health communication is vital for cost-effective, appropriate and individually tailored health care.

[2] reported that based on the US Harris Survey, there were 93% respondents believe online information on health, 85% health information ease to read and good quality of health information is 82%. This shows that people will use online social communities or social media as a platform to improve their knowledge on health and dissemination information on health. Furthermore, Facebook, Twitter and YouTube are some of the elements in social media that have created a new era in health communication [3].

2. Related Literature

Previous studies focus on health communication and the use of ICT in implementation of healthcare among respondents.

Therefore, this paper will analyze eight articles associated to health communication on social media for the past four years (2013-2016). The related research was selected based on the communication channel used in health communication.

2.1. New Technology

The use of mobile communication technologies in daily use for health communication is different between communities and depends on cultural characteristics and demographic. The survey on the use of mobile technologies and social media has been done to support provider-patient communication in an urban Latino community. From 130 respondents, [4] found that facebook/myspace was the most frequently used of social media by parents and youth but not for health communication. Furthermore, parents were more interested in receiving email and text messages on health.

While in China, rabies is a serious health problem whereby health education on rabies is needed. The purpose of the study by [5] are to compare and measure the effect on knowledge of rabies. A total of 350 randomly selected villages were allocated in using Short Messaging Service through handphone (SMS) and rabies health information sessions (IS). The results showed that Short Messaging Service is one of the medium in promoting public health to respondents.

2.2. Health Communication

[6] done research to improve the knowledge of health communication focus on the implications for a planned ICT supported interactive health channel. A qualitative research with multi-methods was done using content analysis. From the data analysis, two themes emerged was communicating health and environment for health communication. The results showed that health communication is an essential element of health promotion practice.

2.3. Tweeting as Health Communication

[7] analyzed the use of Twitter in disseminating information related to health, building relationships among participants and encourage them to improve their health. The study involved American Heart Asssociation, American Cancer Society, and American Diabetes Association. The researcher has identified 1,583 tweets in examining with the organisations' use of twitter by using content analysis. [7] argument relies on Twitter's interactive features, message functions and topics related. Furthermore, the researher points out that majority of the tweets were related to organization-related topics and personal health-related tweets represented a small portion of the sample.

2.4. Social Media

Social media is one of the important tools in disseminating information on health. During analysing the accounts, the researcher found that 66.7% had Facebook accounts, 80.4% twitter accounts while 50.9% You-Tube accounts. However, the researcher emphasis on Facebook pages of 34 State Health Departments that was done over a 200 day period using NVivo10 to analyse through manual coding. [8] noted that content provided on Facebook and health conditions affect their populations. However other research done by [9] analyzed message features predictive of user engagement on diabetes related Facebook pages. The results showed that specific features predicted different forms of engagement.

2.5. Online blogs

The aim of [10] paper is to analyzed the use of online doctor blogs or patients would increase the patient to seek out treatment. There were 278 respondents involved in this survey who discussed information on health with doctors through blogs and using structural equation model analysis. The findings showed that there were a positive effect between health communication, situated learning and health knowledge involvement.

3. Findings

Based on the above discussion, Table 1 described the summaries of health communication and social media. Papers that have been reviewed discussed more on social media as a channel in disseminating information on health.

Table 1: Characteristics of the studies

No	Author (Year)	Title (Journal)	Principle	User Context (Number of Respondents)	Communication Channel (Strategy)
1	Amina Jama Mahmud et al (2013)	Health communication in primary health care - A case study of ICT development for health promotion.	• Health communication • Health promotion • E-health • Information Communication Technology • Primary health care	A total of 251 respondents involved by using participatory observation in informal interviews.	• Interpersonal • Group • ICT mediated health communication
2	Lee, Y. C et al (2014)	The effects of situated learning and health knowledge involvement on health communications.	• Situated learning • Health knowledge involvement • Health communication	There were 278 adults participated in online communication using doctor blogs discussed on medical information with doctors.	• Online platform of blogs
3	Smaldone, A. et al (2015)	Adolescent and parent use of new technologies for health communication: A study in an urban Latino community.	• Health communication • Social media technology	There were 142 youths and parents seek information on health by using technology.	• Twitter, Facebook/Myspace, Instant Messaging • Videoconference (Skype/OOvoo)
4	Rus, H.M et al (2016)	Health communication in social media: Message features predicting user engagement on diabetes-related Facebook pages.	• Social media • Facebook • Health communication	Content analysis of 500 Facebook posts.	• Facebook
5	Jha, A. et al (2016)	The use of social media by state health departments in the US: Analyzing health communication through Facebook.	• Social media • Health communication • State health department • Facebook	There were 2,597 Facebook posts during 200 day study period were coded.	• Facebook
6	Wu, H. et al (2016)	Community-based interventions to enhance knowledge, protective attitudes and	• Health education • Intervention study	Selected villagers of 350 respondents were randomly allocated into three	• Short message service via mobile phone (SMS)

		behaviors towards canine rabies: Results from a health communication intervention study in Guangxi, China.		intervention (SMS, IS and SMS+IS)	• Rabies education information session (IS) • Combination of the two (SMS+IS)
7	Park, H. et al (2016)	Tweeting as Health Communication: Health Organizations' Use of Twitter for Health Promotion and Public Engagement	• Health Organization • Social media	A total of 16 twitter accounts is being associated with the organizations using content analysis.	• Twitter
8	Kreslake, J.M. et al (2016)	Developing effective communication materials on the health effects of climate change for vulnerable groups: A mixed method study.	• Climate change • Global warming • Health impacts • Health disparities • Health literacy • Health communication • Chronic disease management	There were 122 respondents were recruited from online community bulletin boards in selected metropolitan areas using non probability purposive sampling design.	• Online survey assessed within-subject changes in knowledge, beliefs and behavioral intentions

4. Conclusions

Noted that the above table summarizes health communication papers reviewed and relevant social media they employed. Thus, this paper focuses on sharing and exchanging health related knowledge using Information Communication and Technology as a communication channel. Most of the research done in this paper, use social media such as Facebook, Twitter, Skype, Instant Messaging, Myspace and Short Message Service (SMS) via mobile phone as a platform.

In conclusion, with daily improvement of technology, promoting health information through social media is becoming a new trend as people naturally seek information through the easiest, effortlessly and cheapest mean of getting information.

References

[1] Ratzan S.C (2010) Moving from IEC to IHC— the time is now. Journal of Health Communication: International Perspectives, 15(7):691–694.

[2] Chen SF, Hu WC, Ho CJ: Analysis of Taiwan's quality health information websites from the perspective of health literacy. Taiwan Health Care J 2011, 10:2–6.

[3] Moorhead, S. A., Hazlett, D. E., Harrison, L., Carroll, J. K., Irwin, A., & Hoving, C. (2013). A new dimension of health care: Systematic review of the uses, benefits, and limitations of social media for health communication. Journal of Medical Internet Research, 15(4), e85. doi:10.2196/jmir.1933.

[4] Smaldone, A., Stockwell, M.S., Osborne, J. C., Cortes, Y., Bekele, E., & Green, N.S. (2015). Adolescent and parent use of new technologies for health communication: A study in an urban Latino community. Journal of Public Health Research 2015; volume 4:376

[5] Wu, H., Chen, J., Zou, L., Zheng, L., Zhang, W., Meng., Z, Magalhaes, R.J.S., Wang, Y., Kang, J., and Sun, X. (2016) Community-based interventions to enhance knowledge, protective attitudes and behaviors towards canine rabies: results from a health communication intervention study in Guangxi, China. BMC Infectious Diseases.16:701.DOI 10.1186/s12879-016-2037-6

[6] Amina Jama Mahmud, Olander1, E., Eriksén, S. and Haglund, B.J.A. (2013) Health communication in primary health care - A case study of ICT development for health promotion. BMC Medical Informatics and Decision Making 2013, 13:17. http://www.biomedcentral.com/1472-6947/13/17

[7] Park, H., Reber, B.H., and Chon M.G. (2016). Tweeting as Health Communication: Health Organizations' Use of Twitter for Health Promotion and Public Engagement. Journal of Health Communication, 21:188–198, 2016. DOI: 10.1080/10810730.2015.1058435

[8] Jha, A., Lin, L., and Savoia, E. (2016) The Use of Social Media by State Health Departments in the US: Analyzing Health Communication Through Facebook. Journal Community Health 41:174–179. DOI 10.1007/s10900-015-0083-4

[9] Rus, H.M., & Cameron, L.D. (2016) Health Communication in Social Media: Message Features Predicting User Engagement on Diabetes-Related Facebook Pages. 50:678–689. DOI 10.1007/s12160-016-9793-9

[10] Kreslake, J.M., Price, K.M. and Sarfaty, M. (2016). Developing effective communication materials on the health effects of climate change for vulnerable groups: a mixed methods study. BMC Public Health.16:946. DOI 10.1186/s12889-016-3546-3

IOP Conf. Series: Journal of Physics: Conf. Series **1019** (2018) 012079 doi:10.1088/1742-6596/1019/1/012079

Design of an Arduino-based home fire alarm system with GSM module

N N Mahzan, N I M Enzai, N M Zin and K S S K M Noh

Faculty of Electrical Engineering, Universiti Teknologi MARA (UiTM) Terengganu, 23000 Dungun, Terengganu, Malaysia
najwa.mahzan@uitm.edu.my, nurid333@tganu.uitm.edu.my, adzianie@uitm.edu.my, kusyahidah@uitm.edu.my

Abstract. This paper describes the design of a home fire alarm with Arduino-based system by means of GSM Module. The project purposely is for house safety where the main point is to avoid the fire accidents occurred to the residents and the properties inside the house as well. It utilizes Arduino Uno board in conjunction with ATmega328 chip. The main controller used is certainly the ATmega328 which controls the home fire alert subjected to the temperature sensor. An LM35 temperature sensor is used to detect the heat from the fire. An alert message will be sent to the user via short message service (SMS) via GSM module. When the system detects the temperature of 40^{0}C or more, it will immediately display an alert notification on LCD display and simultaneously sending an SMS alert to the users upon the high raise temperature in the house. Results from the test are documented and discussed. Through this system, it can help users to improve their safety standards by having immediate response in preventing accidents. This will eventually allow the users to protect their lives and the properties as well from the disaster.

1. Introduction

Amongst the occurred disasters happened in the resident area, fires have been known as the dangerous tragedy that could cause destruction, property and life losses [1]–[4]. In many disasters, fires have become recurrent, destructive and most influential disasters if compared to others hazards. With the rapid development of urban construction, the occurrence probability of the great fire and other special disaster also increased year by year [5]–[7]. Fires in the early detection and early warning are two important ways to extinguish the fire promptly and avoid great casualties and property loss. Therefore, the requirement of placing intelligent fire alarm system is important within buildings especially in the buildings where contain many people inside or valuable belongings.

According to the statistic stated by Fire and Rescue Department of Malaysia (JBPM), in 2016 only, almost 5500 of fire cases reported are concerned with the households followed by transportations, electrical appliances and leaking gasses [8]. Fires in the households are often triggered by many common factors investigated which are from cooking equipment, smoking in the house, electrical appliances, candles, curious children, faulty wiring and many more. If the fire occurs when the residents are in the house, the possibility to extinguish the fire is a bit high. It is because the residents themselves can take immediate precaution from the fire to be spread all over by using fire extinguisher or call the fireman instantly. The main concern of this project is when the residents are not at home or are not aware of the existence of the fire in the house. Having said that, the home fire alert is purposely designed to alert the house residents whenever any possibilities for having fire disaster prompted in their house.

For this project, the development of home fire alert is built based on Arduino board as the main controller board that interacts with GSM module which works in the communication part. The interaction is for the user to know the current situation in the house. This system works totally on wireless network communication as GSM module is performed by sending an SMS to the user. The microcontroller inside the Arduino board is used as the mastermind of the circuit where it controls the circuit flows and execute all the decision as well.

The GSM Module is responsible for the communication part of the circuit. It takes information from the Arduino on where to send information and what information needs to be sent. It uses a GSM SIM card for communication purposes. It is basically just a modem which uses serial communication to interface with and needs Hayes compatible AT commands for communicating with the Arduino. The alert message and the phone number of the recipient are given by the user through the project codes. As soon as fire is detected (temperature will hit certain temperature limit) an SMS will be sent to the recipient's phone number from the SIM card inserted into the module for giving information to the user upon fire detection in the house.

2. Methodology

Basically, the design and development of this project are divided into two main parts which are hardware architecture and software details. In the hardware architecture, the design of the circuit was constructed and the prototype of the project was built. While in the software development, the whole complete prototype was operated via programming codes.

2.1 Hardware architecture

Since Arduino is the main board, microcontroller on it which is ATmega328 is used as the main controller to manage the circuit accordingly. It is a well-known open source microcontroller-based kit for creating digital devices and interactive tool that can interact with LEDs, LCD display, switches, buttons, motors, speakers and many more. The Arduino system offers a set of analog and digital pins that can be integrated to many other boards and circuits which absolutely have different functions in a design. Arduino board provides a USB serial communication interfaces for loading the codes from computer. To do the codes, Arduino has prepared its own software called integrated development environment (IDE) which completely supports C and C++ programming languages. Figure 1 shows the Arduino UNO board that is used throughout the project.

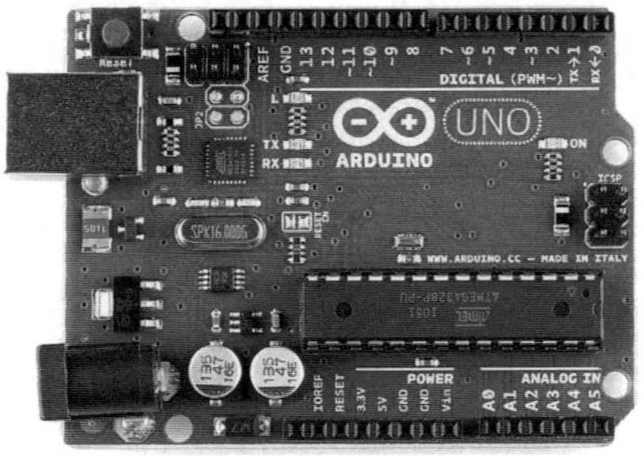

Figure 1. The Arduino UNO board which uses microcontroller ATmega328 onboard.

For GSM module, GSM SIM900A type is selected to carry the task in communication part. It can operate on Dual-Band 900-1800 MHz and designed solely for outside Europe and USA usage. It has an established performance, industrial grade interface standard plus embedded TCP/IP protocol which makes it to be presentable and suitable for the electronics project [9]. Since it consumes small of power in its operation, thus it is said able to communicate with any low power consumption microcontroller. It can be interfaced by using many interfaces which some are I2C, SPI interface, PWM, antenna pad, two serial interface and so forth. Figure 2 shows the GSM SIM900A device before connecting to the Arduino board.

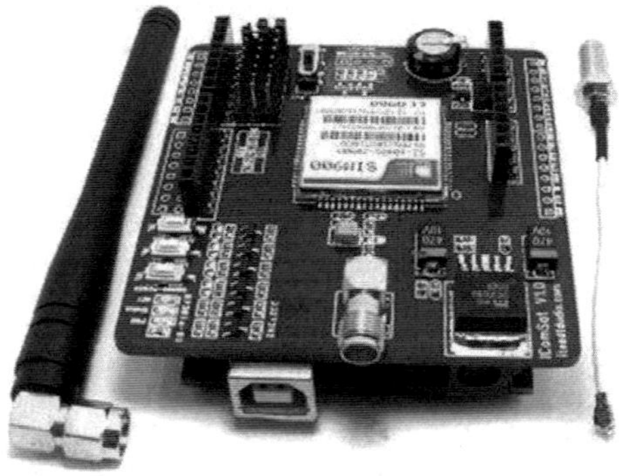

Figure 2. The GSM SIM900A module to communicate with Arduino board.

Other than the two main components mentioned, another component which is essential in this project is the temperature/heat sensor. LM35 is chosen for measuring the temperature since it is an analog and linear temperature sensor which has a linear relationship between output voltage and temperature changes (°C). Besides, the LM35 does not require external calibration to deliver the accuracy of ±(1/4)°C at room temperature or ±(8/4) °C at the temperature range of -55°C to +150°C [10]. Due to its low-input impedance, linear output, and accurate inherent calibration of LM35, it does make the sensor's readout and control circuitry interfacing become easy [10], [11]. The device is usually used with single power supply or with plus or minus supplies. It can receive an input range of -2V up to 35V and result the output voltage range of -1V to 6V. Figure 3 displays the LM35 sensor which is used to measure the temperature for this project.

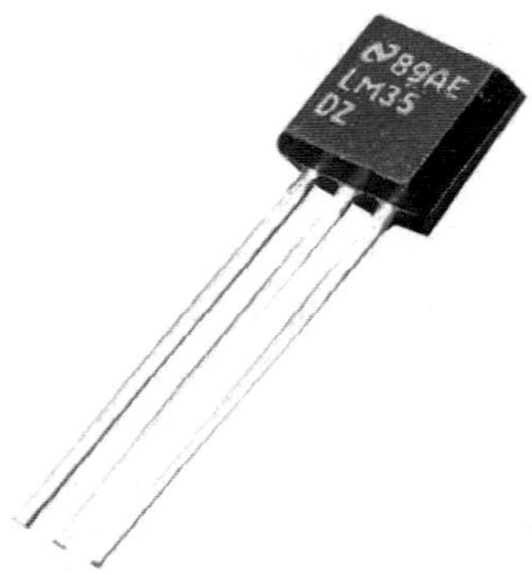

Figure 3. LM35 sensor by Texas Instrument.

Figure 4 presents the block diagram for the home fire alert while figure 5 displays the hardware architecture of the system. The hardware design entails of two main components which are the connections between Arduino UNO and the GSM SIM900A and Arduino UNO with the temperature sensor, LM35. When a fire has broken out in the house, LM35 will trigger the heat. Upon reaching the temperature of 40°C, it will directly send signal to the Arduino informing the high temperature. The increased temperature will make the Arduino to alert the user about the situation through GSM module. An SMS will be sent promptly to the user to let the user know the existence of the fire in the house. At the same time, existence of the fire will be notified as well on the LCD display.

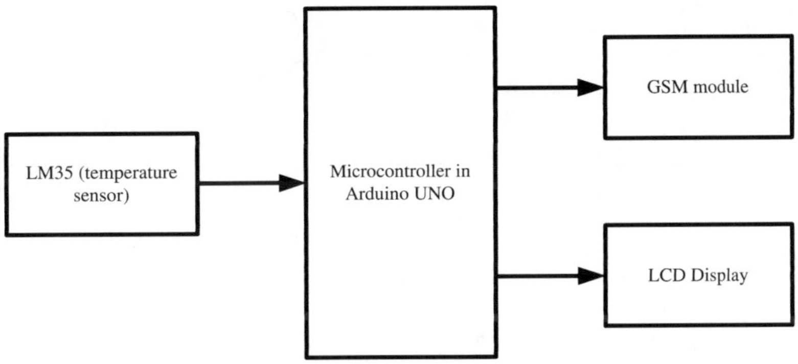

Figure 4. Block diagram of the home fire alert system.

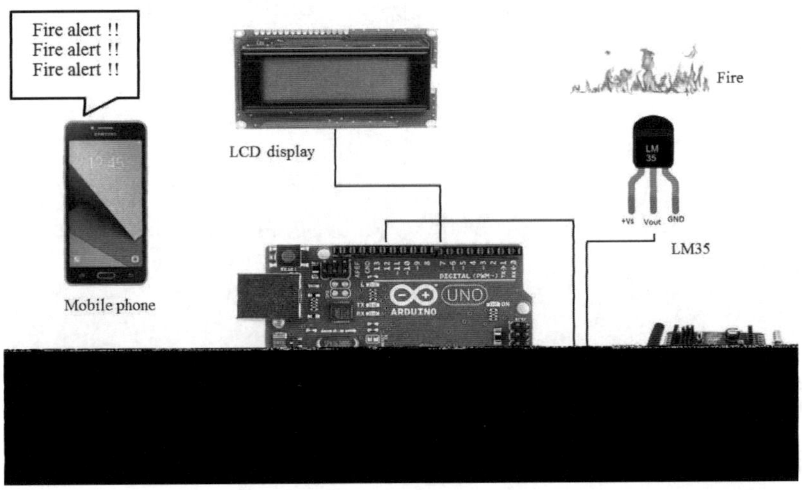

Figure 5. Hardware architecture of the home fire alert.

2.2 Software development

The software of the project is based on the flow chart in figure 6. When the system begins, LM35 will always sense the surrounding temperature. Whenever a fire is broken out, even a small little fire, a temperature rise is occurred. When that happens, LM35 can detect the temperature value instantly. At the time when the temperature reaches 40°C or above, microcontroller on the Arduino UNO board will notify

IOP Conf. Series: Journal of Physics: Conf. Series **1019** (2018) 012079 doi:10.1088/1742-6596/1019/1/012079

GSM module to send an alert message to the user. Value of temperature limit that can be triggered by LM35 can be changed in the code upon request by the user. The limit is not constrained to any value since the LM35 sensitively senses any surrounding heat with regards to the temperature range it can count (-55°C to +150°C). During hot weather in Malaysia, the temperature can reach up to almost 38°C. Thus, the limit temperature to be detected; 40°C is definitely agreeable in accordance to the Malaysia's weather. Figure 6 and 7 illustrate the flow chart of the home fire alert and the pseudo codes it represents for the project respectively. By referring to both figures, the complete program can be constructed later in Arduino IDE software.

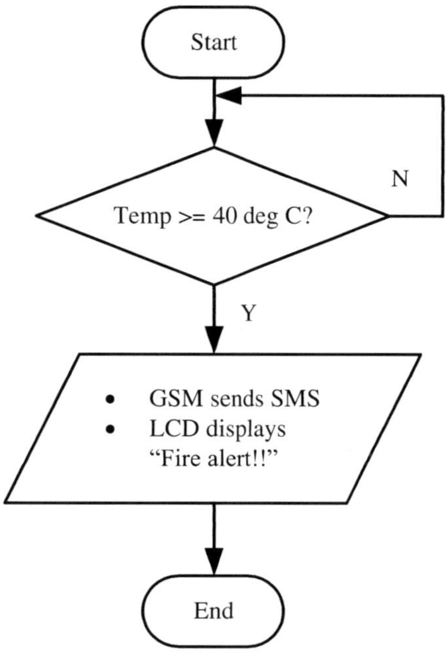

Figure 6. Flowchart of the home fire alert system.

```
START
  IF TEMPERATURE >= 40 DEG
     GSM ACTIVATED
     SEND SMS TO USER
    DISPLAY "FIRE ALERT!" ON LCD
    DISPLAY
  ELSE
     CHECK TEMPERATURE AGAIN
```

Figure 7. Pseudo code for the project based on the flow chart.

2.3 Schematic diagram

The overall project schematic diagram is shown in figure 8. While figure 9 illustrates the connection schematic diagram between Arduino UNO board with LM35 and GSM module. Since the Proteus© software does not have the GSM module component, it is replaced with the input connector available which the connector pins will be attached directly to the Arduino UNO board. For the particular part of detecting fire from LM35, the code is written and portrayed in figure 10. In the code, measured temperature from surrounding is basically identified in voltage. It is because LM35 is an analog sensor which values voltage unit. The voltage unit received will then be converted into knowable standard SI unit; degree Celcius in the subroutine called 'CheckTemp()'.

IOP Conf. Series: Journal of Physics: Conf. Series **1019** (2018) 012079 doi:10.1088/1742-6596/1019/1/012079

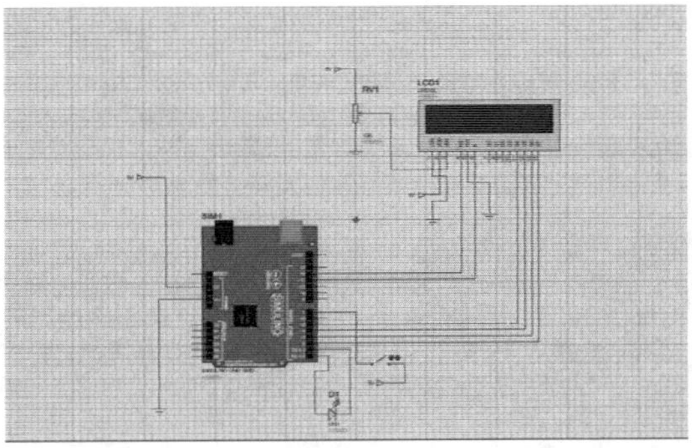

Figure 8. Project schematic diagram.

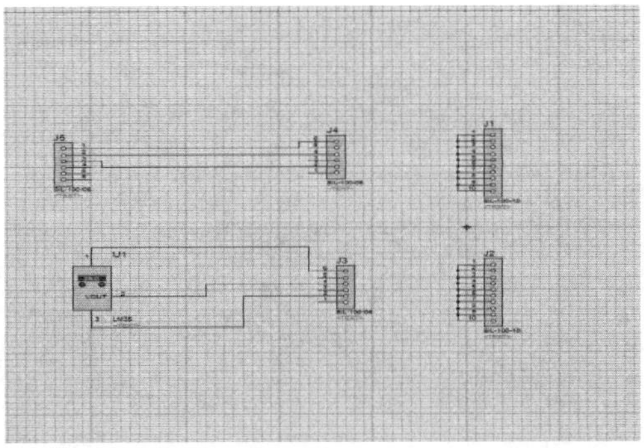

Figure 9. Schematic diagram for connection between Arduino UNO board with GSM module and LM35.

```
void CheckFire()
{
lcd.setCursor(0,0);
lcd.print("Fire Scan - ON");
Temp_alert_val=CheckTemp();
if(Temp_alert_val>165)

{
 Fire_Set=1;
lcd.setCursor(0,1);
lcd.print("Fire Alert! SMS Sent!");
while(sms_count < 3) //Number of SMS Alerts to be sent
{
SendTextMessage(); // Function to send AT Commands to
GSM module
}

}
}
```

Figure 10. The program code for measuring temperature.

```
float CheckTemp()
{
temp_read=analogRead(sensor); // reads the sensor output
(Vout of LM35)
temp_read=temp_read*5;   // converts the sensor reading to
temperature
temp_read=temp_read/10; // adds the decimal point
return temp_read; // returns temperature value in degree
celsius
}
```

Figure 11. The program code for checking temperature and converting it to degree Celcius.

3. Result and discussion

Few tests were done to observe the system's performance. The tests were completed by applying heat near to the LM35. Figure 12 shows the condition which the LM35 is in ready mode (ready to detect fire but detect none). Figure 13 displays when the LM35 detected fire and alert message will appear on the LCD display. While figure 14 exhibits the condition after the fire has been put out. Figure 15 reveals the SMS received by the user when fire alert is notified by the system.

Figure 12. When fire is still not detected but LM35 is in ready mode.

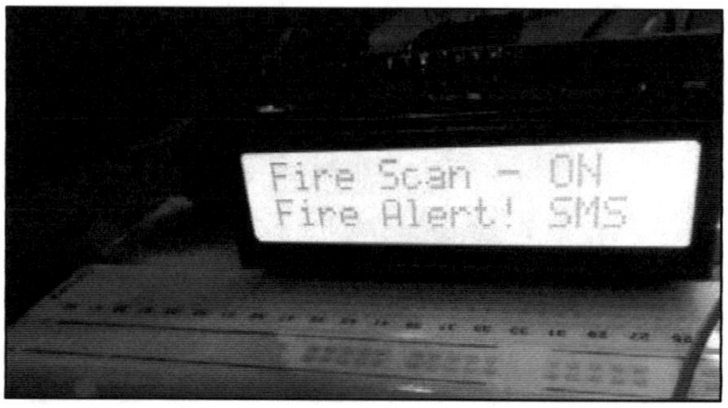

Figure 13. When fire is detected.

Figure 14. When the fire has been put out.

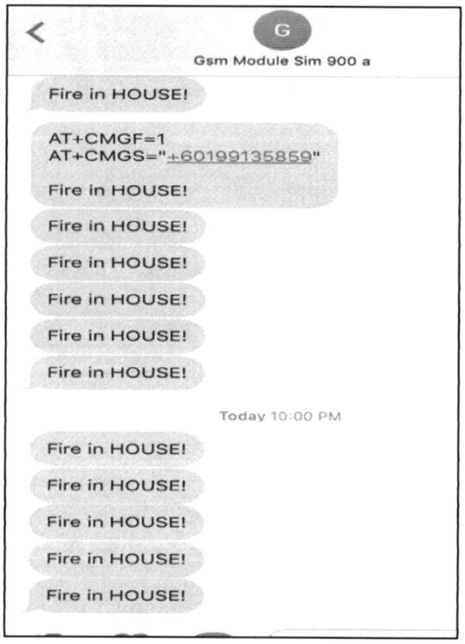

Figure 15. SMS received by the user to notify the fire existence.

Table 1 shows the analysis of the output temperature during the test measurements. When fire is noticed, LCD display only shows the readiness of LM35 sensor to detect fire. As soon as LM35 detects fire, LCD display will notify user on the display screen "Fire alert!" which is synchronized with the SMS sent by GSM module to the user alerting the fire existence. After fire has been stopped, LCD display will show "Fire shut! Safe now" on the screen.

Table 1. The analysis of the output temperature on LCD display and GSM module.

Components involved	Before detecting fire	When detecting fire	After detecting fire
LCD Display	"Fire scan – on"	"Fire alert!"	"Fire shut! Safe now"
GSM	-	Fire in HOUSE!	-

4. Conclusion

This project has been made in order to help building owner to overcome the problem which is fire spreading whenever the owner is not in the building. The unpredictable situation or critical situation always occurs in the building or resident areas without the residents' notice. Based on the results obtained, the home alert system is doable and functional to the residents to protect their houses. In fact the system built is cheap in value compared to other existing alarm system in the market and easy to apply to the houses. The ability to detect heat or high temperature is undeniable because of the use of LM35 in the system. This device can be applied in varied areas due to its flexibility and simplicity in handling; for instance in houses, hostels, hotel industries, factories, vehicle industries and many more areas which are related to the crowd, people or beneficial things. Users can simply apply the device in their interested area to protect the area from the existence of fire. Whenever the temperature reaches the limit (40°C), the device will instantly alert the users by sending a message via GSM. This will make the users become aware of the dangerous situation and can easily prevent it from happening by quick prevention (use fire extinguisher, call firemen etc.).

Acknowledgements

We would like to thank all technicians in the FKE lab UiTM Cawangan Terengganu Kampus Dungun for their assistance throughout this project. The help and comments for the project have greatly improved the manuscript.

References

[1] L Chun-yuan, "Design of Intelligent Fire Alarm System Based on GSM Network," no. Iceoe, pp. 393–396, 2011.
[2] M Faris, M Fuzi, A F Ibrahim, M. H. Ismail, N. Syakira, and A. Halim, "HOME FADS : A Dedicated Fire Alert Detection System Using ZigBee Wireless Network," pp. 53–58, 2014.
[3] S Suresh, "Home Based Fire Monitoring And Warning System," 2016.
[4] L I U Fei, Z Zhe, Y A O Hao-wei, and L Dong, "Application of Aspirating Smoke Detectors at the Fire Earliest Stage," *Procedia Eng.*, vol. **52**, pp. 671–675, 2013.
[5] H Elbehiery, "Developed Intelligent Fire alarm system," *Jounal Am. Sci.*, vol. **2**, no. August, pp. 1016–1025, 2012.
[6] K Sen, J Sarkar, S Saha, A Roy, D Dey, and S Baitalik, "Automated Fire Detection and Controlling System," *Int. Adv. Res. J. Sci. Eng. Technol.*, vol. **2**, no. 5, pp. 34–37, 2015.
[7] H Mori, "Configuration-Free Propagation System for Early Fire Alerts," 2016.
[8] J B dan P Malaysia, "Statistik Kebakaran Mengikut Jenis Kebakaran 2016," 2016. [Online]. Available: http://www.data.gov.my/data/ms_MY/dataset/jbpm-statistik-kebakaran-mengikut-jenis-kebakaran-2016. [Accessed: 20-Jul-2017].
[9] Z Jifei, "Intelligent power failure alarm based on ATmega128 and SIM900A," *Knowledge Guide*, 2014.
[10] Texas Instruments, "LM35 Precision Centigrade Temperature Sensors," 2016.
[11] P Y Mulge, "Remote Temperature Monitoring Using LM35 sensor and Intimate Android user via C2DM Service," vol. **2**, no. June, pp. 32–36, 2013.

IOP Conf. Series: Journal of Physics: Conf. Series **1019** (2018) 012080 doi:10.1088/1742-6596/1019/1/012080

Asmaul Husna Learning through Gamifications and Adaptation of Signalling Principle

Arifah Fasha Rosmani, Mohd Hafidz Zakaria

Universiti Teknologi MARA, Cawangan Perlis, Kampus Arau, 02600 Arau, Perlis.
arifah840@perlis.uitm.edu.my

Abstract. Asmaul husna is the 99 names of attributes which is owned by Allah SWT. All names in asmaul husna must be translated with the word "Maha", which shows the perfection in divine attribute. Unfortunately, in this modern age, the lack of knowledge about asmaul husna especially its meaning is the most common problem among muslim society. By learning asmaul husna via traditional methods such as text books and white board, muslims are getting less motivated and have no desire to learn more about asmaul husna. Due to the emerging of gamification, the traditional methods are less effective, and it has several limitations in terms of equipment for the learning process. Learning asmaul husna through gamification could increase motivation, offers challenges and trigger excitement to play, while applying multimedia principles could ease in the learning process by reducing the cognitive load and excess memory. The aim of this project is to learn on asmaul husna attributes and meanings, identify suitable gamification elements and multimedia principles to be adapted into the design and development of Asmaul husna Learning Game for students. This game consists of gamification elements in learning such as points, rewards, countdown timer, and discovery and multimedia learning principles which is the signalling principle. Heuristic evaluation has been initiated and from the test results, this game is highly recommended for students.

1. Introduction

The term Human Computer Interaction (HCI) has been founded and widespread since the 1980s and hence, has elevated since then [1]. Until today, various researchers have been trying to investigate the most appropriate HCI guidelines and principles to assist the relationship between human and machine. These principles and guidelines have been utilized in many computer applications to facilitate both party.

For this project, a multimedia learning principle and gamifications elements has been adopted as a medium to ease the use. There are 12 principles erected by Mayer after a thorough research and lab works that can be adapted into computer applications towards assisting the learning process and reducing cognitive load [2]. However, we only concentrate on one principle in this research which is the signalling principle as shown in Table 1 [3].

Table 1. Multimedia Principles – Signaling Principle

Multimedia Principles	Description
Signalling Principle	People learn better when signals/signs/cues that highlight the organization of the essential material are added.

Besides the use of multimedia principles, the gamifications elements are also vital to be incorporated to increase motivation, offers challenges and trigger excitement for the user in the learning process. Gamification also act as a tool for raising engagement in e-learning platforms [4], [5]. [6] has introduced various types of game elements that can be used in gamification, but we have chosen only four as listed below in Table 2.

Table 2. Game elements

Game Elements	Description
Points	Points are values awarded for an action or a combination of actions. Use to encourage people to do things by collecting them.
Countdown	Players are only given a certain amount of time to do something.
Discovery	Players are encouraged to navigate through their learning environment and uncover pocket of knowledge.
Rewards	Players get rewards for reaching specific objectives.

2. Background of research

Asmaul husna is basically the 99 names of Allah, the names which are attributes of Allah that can be found in Holy Quran. The problem related to Asmaul husna is when nowadays muslims are unable to memorize and understand the 99 names. This is due to the lack of knowledge in the meaning and benefits of asmaul husna. Even though, they have learned or memorized the names, they failed to apply it. For example, after "*solat*" muslims are encouraged to perform *dhikr* but many of them ignored this practise. Usually, muslims can read the names of Allah as they have learned and sing it since childhood, but the essence of the meaning was not embedded during the learning process. Hence, these names are later disregarded. Furthermore, gamification technique is yet to be implemented extensively in asmaul husna learning process and learning in old-fashioned ways reduced their desire to learn this essential information.

If this issue continues, students will have no intention to learn more about Islam or to be closer to their Creator via His beautiful names. This will then lead to neglect of faith and astray. Besides that, they cannot apply asmaul husna and its meanings, in their supplications (*doa*), as well as unable to feel and see the greatness of names of Allah SWT. To prevent these issues, we reckon to develop a game that will assist them so that they can understand the meaning and value of asmaul husna.

Thus, this project aims to develop a game for students, to give them better understanding about asmaul husna so that they can apply in their daily life such as by comprehending the meaning of asmaul husna and apply it in their supplications or use it as *dhikr* [7].

3. Methods

Six methodology phases were involved in this research, which were planning and analysis, design, development, testing, refinement and documentation.

IOP Conf. Series: Journal of Physics: Conf. Series **1019** (2018) 012080 doi:10.1088/1742-6596/1019/1/012080

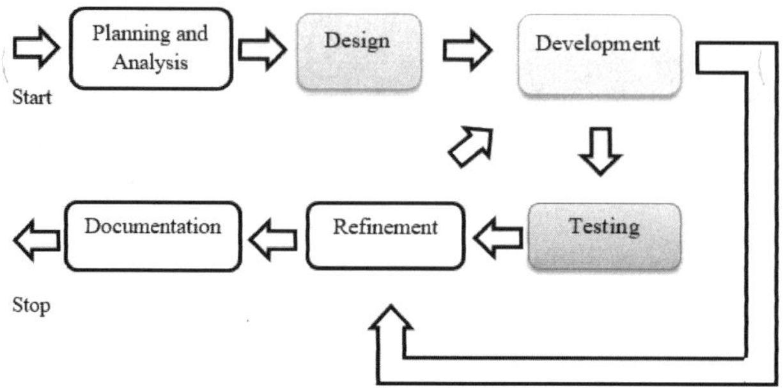

Figure 1. Methodology

3.1. Planning and Analysis
In the planning and analysis phase, the planning of completion of each phase were discussed and the project objectives and expectation were clearly stated. Analysis was done by using information gathering and data collection method. For information gathering, the information needed were about asmaul husna, multimedia principles, gamification and gamification elements. This information was obtained through books, articles, journals, and other relevant resources, as it is important to make sure that the information included in the game content are precise.

3.2. Design
Interface design is very important to attract end user's attraction. It should be properly created as it may give an impact to the end users' understanding about the game content. This is because there is quite a high possibility that almost 80 percent of all gamification apps failed due to poor design [8].

3.2.1. Storyboard
Storyboard is the best way to illustrate the idea before the real design and could ease the construction process. It is the sequence of drawing that used as a guideline before developing the real game interface and usually done through illustrations or sketching. There were a lot of aspects need to be considered in this phase, for example gamification elements, use of colour, and multimedia elements and principle. The interface appearance need to be design based on recognition, visual appearance, consistency and well organization.

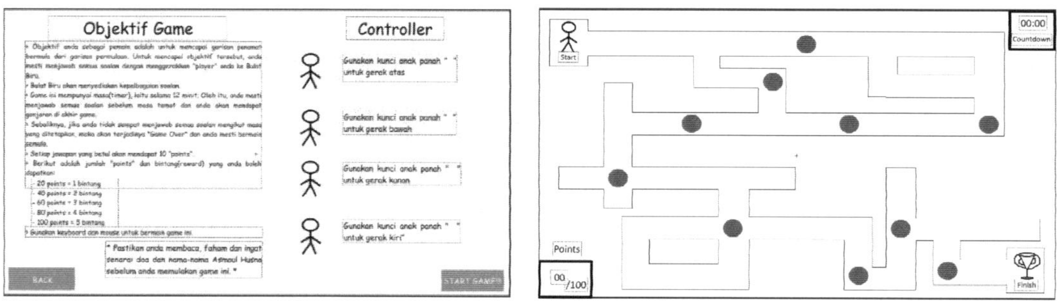

Figure 2. Storyboard

3.3. Development
The game was produced during this phase involving hardware and software such as laptop, Adobe Photoshop CS6 and Adobe Animate CC 2017. It is essential to identify the suitable hardware and software to avoid problems such as, compatible issue.

IOP Conf. Series: Journal of Physics: Conf. Series **1019** (2018) 012080 doi:10.1088/1742-6596/1019/1/012080

3.4. Test
Several tests have been conducted throughout the project, but only heuristic test that will be discussed in this paper. These testing were important to gain the users feedback on the functionality and usability of the app. The participants for these tests were UiTM Perlis students and experts from Computer Science fields, this test is performed to find any flaws in the application.

3.5. Refinement
After the testing process was completed, the game was amended to resolve any problems or errors occurred based on the test results.

3.6. Documentation
The final phase was documentation process. In this phase, all data, information and activities that are related were documented.

4. Game Architecture
The game was based on the architecture as below which consists of 20 asmaul husna names related to students, gamification elements (points, rewards, countdown, discovery) and multimedia learning principle (signalling principle). It consists of three main pages which are supplications page, game instruction and controller page and game play page.

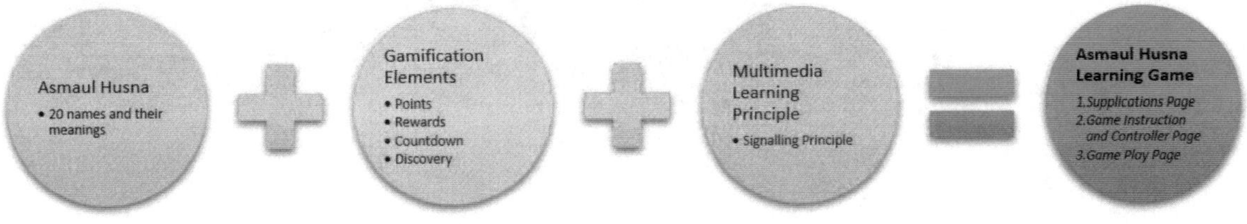

Figure 3. Game Architecture

4.1. Supplications Page
Supplications (*doa*) pages of the game application. Students must read, understand and memorize these types of supplications, names of Allah and their meanings along with the supplications that are provided. Each supplication is highlighted with black font as a cue so that students can understand the keyword from these supplications. Students can proceed to other supplications by pressing the "Seterusnya" button and go back to previous supplications page by pressing the "Kembali" button.

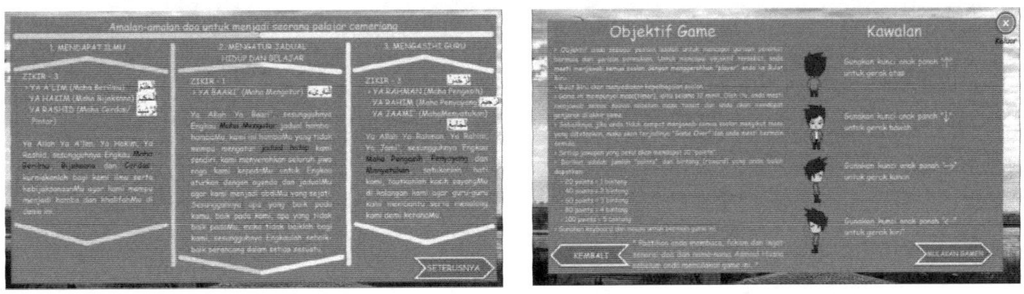

Figure 4. Supplication Page (black font is used as a cue for keyword) and Game instruction page.

4.2. Game Instruction and Controller Page
The purpose of this game instruction and controller page is to remind to the students about the rules and the controller that is used. The main objective of this game is to reach the finish line from the start line. To accomplish that, students need to answer all questions by moving their "player" to the Blue

Circle. Blue Circle is a signal that will provide a variety of questions. This game also contains countdown timer, which is 12 minutes. Therefore, students must answer all questions before the time runs out and they will receive their reward at the end of the game. Otherwise, if students are unable to finish the game according to the time given, then the game is over, and they need to restart the game. Students will get 10 points if they answer one question correctly, otherwise none. Here is the sum of points and stars (rewards) that students can obtain, 20 points - 1 star, 40 points - 2 stars, 60 points - 3 stars, 80 points - 4 stars and 100 points - 5 stars. There is also a reminder that students need to prepare their mind by reading, understanding and memorizing the lists of supplications and names of asmaul husna before they start the game. If they are ready, they can play the game by pressing the "Start Game!!!" button. If not, then they can press the "Back" button to read the lists of supplications and names of asmaul husna again. For the controller, students need to use keyboard and mouse. For the keyboard, they need to press "UP arrow" key to move up, "DOWN arrow" key to move down, "RIGHT arrow" key to move right and "LEFT arrow" key to move left. For the mouse, they only need to press and choose the correct answer from the question.

Figure 5. Game Play (blue circle as a signal for questions) and Question Page (coloured button as a sign for answer's options)

4.3. Game Play Page
After students press the "Mulakan Game!!!" button, the maze game will appear. However, there is a reminder that works as a signal written as "Sila klik pada halaman game ini dengan mouse anda supaya "player" anda boleh berjalan". The reason why there is a reminder is because after students press the "Mulakan Game!!!" button and immediately press the keyboard arrow button, the "player" cannot be moved. So, to avoid confusion, the reminder is placed at the top of the game interface. When students move their "player" to the blue circle, it will display a question. Each blue circle provides different question. If students can answer the question correctly, points will be added and a "Betul!" sign will appear. Otherwise, points will not be given, and the correct answer will pop-up. This game is also equipped with countdown timer which spare some duration to think before answering the question. There is also a variety of buildings and structures such as mosque, police station, stall food, school, an apartment and tennis court to attract students while playing the game. The number of stars will be given based on students' points and it is placed at the right side of the game interface. Lastly, when the "player" arrived at the finish line, they will win the game and received trophy as a winning sign.

5. Test, Result and Discussion
Heuristic test was conducted to three lecturers from Computer Science department in UiTM Cawangan Perlis. This test is one of the types of usability evaluation and was performed for evaluating the user interface design or to identify any usability problem in this Asmaul husna Learning Game. In heuristic evaluation, an expert will review, or inspect a product using specific guidelines (Barnum, 2011). Through this evaluation, the developer will obtain a feedback from the users early in the designing process. For the heuristic evaluation, certain criteria were evaluated such as content, interface design, and general feedback. The scoring level is between 1 (strongly disagree) – 5 (strongly agree). From the

Heuristic Evaluation, it can be concluded that both lectures were agreed with the content provided in this game application.

Table 3: Heuristic test results

Criteria (Content)	Mean	Criteria (Interface design)	Mean	Criteria (General Feedback)	Mean
1. The words used are easy to understand.	4.0	1. Color used are eye-catching for students.	4.0	1. Interesting	5.0
2. The buttons in the game application are neatly arranged.	4.5	2. The background image is suitable for this game application.	4.5	2. User friendly	5.0
3. The list of supplications based on Asmaul Husna are easy to read, memorize and understand.	4.5	3. The font is in the appropriate size.	4.0	3. Enjoyable	5.0
4. All questions that are provided are different and require time to think before answering the question.	4.5	4. Sentences and words used in this game application is understandable.	4.0	4. Valuable	5.0
5. The time that is given is enough to answer all questions.	4.5	5. The reminder is displayed clearly.	4.0		
6. The sum of points that are awarded are reasonable based on the correct answer as well as the number of stars.	4.0	6. Graphics in the game application is interesting.	4.0		
7. Instructions are clearly and easily understood.	4.5	7. The design of the "player" is good.	4.5		

Therefore, based on the heuristic evaluation results, it clearly defines that the use of multimedia principles which is signalling principle and gamifications elements which are points, rewards, countdown timer, and discovery are significant in the design and development of this application. The relationship between the heuristic results towards the principles and elements are as listed below:

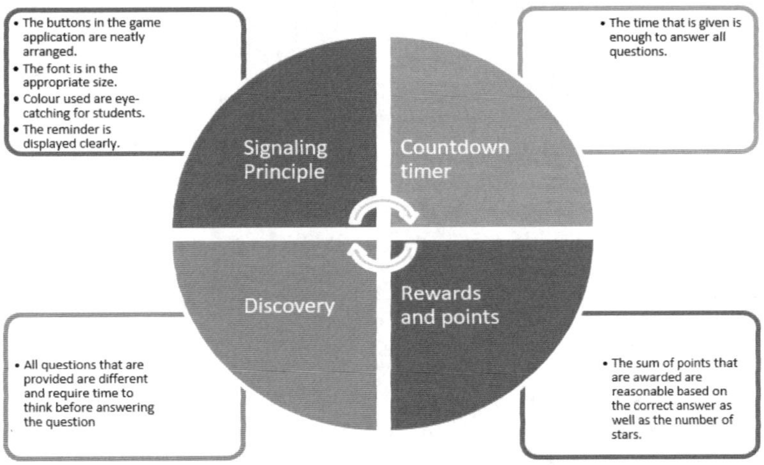

Figure 6. Relationship between heuristic results towards the signalling principle and gamification elements.

Furthermore, the criteria of interesting, user-friendly, enjoyable and valuable have contributed in increasing motivation, offering challenges and triggering excitements to the users.

6. Conclusion

Asmaul husna learning game is a learning tool that has been developed for students to concurrently play and learn asmaul husna. This game was developed using Adobe Animate CC 2017 and Adobe Photoshop CS6. Several multimedia elements were used to attract students' attention for example audio, video, animation and text. Based on the result from Heuristic test, this game had achieved its objective which is to design and developed asmaul husna learning game for students incorporating multimedia principles; signalling principle and gamifications elements; points, rewards, countdown timer, and discovery. Based on the observation that has been carried out, it can be concluded that this game has several advantages such as it can help students to learn asmaul husna supplications. Besides, students can have fun playing game while learning. In addition, students can use these supplications after praying or whenever they want. For future, it is better if a database could be implemented in this game to store score points and could be enhanced by adding voice narration that can pronounce arabic word to help students understand and pronounce it clearly.

References

[1] J. Grudin, "Three faces of human-computer interaction," *IEEE Ann. Hist. Comput.*, vol. 27, no. 4, pp. 46–62, 2005.
[2] R. E. Mayer, *Multimedia learning*, 2nd Editio. CAMBRIDGE UNIVERSITY PRESS, 2009.
[3] R. E. Mayer, *The Cambridge Handbook of Multimedia Learning (2nd Edition)*. 2014.
[4] C. I. Muntean, "Raising engagement in e-learning through gamification," *6th Int. Conf. Virtual Learn. ICVL 2011*, no. 1, pp. 323–329, 2011.
[5] D. Strmecki, A. Bernik, and D. Radosevic, "Gamification in e-learning: Introducing gamified design elements into e-learning systems," *J. Comput. Sci.*, vol. 11, no. 12, pp. 1108–1117, 2015.
[6] Bunchball, "Gamificaiton 101: An Introduction to Game Dynamics," pp. 1–15, 2012.
[7] W. M. Wan Mohd, *Penyucian Hati Dengan 99 Asmaul Husna*. SMK Publications & Distributors, 2015.
[8] Gartner, "Newsroom Gartner Says by 2014 , 80 Percent of Current Gamified Applications Will Fail to Meet Business Objectives Primarily Due to Poor Design," 2012. [Online]. Available: https://www.gartner.com/newsroom/id/2251015.

IOP Conf. Series: Journal of Physics: Conf. Series **1019** (2018) 012081 doi:10.1088/1742-6596/1019/1/012081

An integration between information systems engineering and software engineering theories towards engineering a novel framework of web-based systems success for institutions based on students' perceptions

Ibrahim Eskandar Ibrahim Fadhel[1], Syed Zulkarnain Syed Idrus[2], Amani Ali Elmetwaly Ali Ibrahim[3] and [4]Mazni Omar

[1]Department of Computer and Information Systems in National Institute of Administrative Science (NIAS), Aden, Yemen & School of Human Development and Techno-communication, Universiti Malaysia Perlis, Perlis, Malaysia.

[2]School of Human Development and Techno-communication (iKOM), Green Advanced Computing and Technology (GREAT) Research Group – CEGeoGTech, Universiti Malaysia Perlis (UniMAP), Perlis, Malaysia.

[3]School of Human Development and Techno-communication, Universiti Malaysia Perlis, Perlis, Malaysia.

[4]College of Science and Arts, School of Computing Universiti Utara Malaysia. Kedah, Malaysia.

ibrahim.eskandar@yahoo.com

Abstract. Large numbers of dissatisfying systems users, organizations and higher rate of systems failure is still a problem. In targeting this failure, measure of systems success categorized as a solution to increase the rate of success in future systems initiatives. Success measure of systems is an important part of software's engineering and information systems engineering to improve systems performance. However, studies showed a lack of research done on systems success evaluation in the universities of world, Arab region and Yemeni as special. Systems especially the web-based faces high rate of failure in Arab region, it was observed that many funded system projects failed, it was also, noted that in Yemen systems were completed and implemented without success measuring. Data analysis and interpretation for the web based information systems (WIS) success framework related to higher education can be taken as a latest phenomenon for the incorporation of theories of TAM, DM 2003 and software engineering ISO 25010. Moreover, this is also responsive and effective for assessing computer anxiety and management support aspects to explain satisfaction, loyalty and net impact. The present study is enthusiastic to outline the need for highlighting student perspective where exists dearth of research in this regard and dire need for engineering framework that works as a tool for organizations to success measure of their systems. Though, prior scholarly work has outlined significant role and importance of designing quality university WIS framework from the aspect of WIS developers not from the side of users. Henceforth, the present study aims to strive for highlighting the perception of users (students) regarding to universities WIS.

1. Introduction

The importance and vitality of measure of any software and/or any information system can be tracked in several disciplines [1]. Notably, software success measures and information system assessment has become more critical, consequently resulting in incurring challenges regarding its measurement [2]. Accordingly, it has also become important to underline challenges and demands related to systems that deplete the effectiveness and success of the measurement [2]. Therefore, assessment and test of the success of any particular information systems or software whilst keeping in view the context that it is used in has become more important these days [2]. With regards to this, scholars in the area are encountering issues whilst examining the success of a system. For this, it is important to understand that taking customer on the central position to examine the responsiveness and effectivity of a system is important [2]. For this, culture is playing a robust role as success prospects vary from one culture to another. This is due to the fact that every culture comes with certain norms, values and codes of conduct which influence the perceptions and the way systems are used [3].

The country Yemen ranks amongst the poorest Arab economies whereby, it has landed at 160th place in the human development index (HDI) [4]. Though, the economy is striving via accelerating its educational development over the past decade [5]. Sadly, this proliferation has also resulted in different issues such as wastage of resources, duplication of materials, poor systems and increasing numbers of dissatisfied users and organizations [1,5,6,7]. Pertaining to Arab economies, there is a dearth of research on success of systems whereby, the critical review of the literature has highlighted underlined no single study pertaining to Yemen in this regard [1,7,8]. The current global economy has witnessed rapid advancements in the domains of information technology (IT) and information systems. Academic institutions such as universities have been reported as amongst the earliest developers of WIS for the purpose of attracting target audience [9,55].

In fact, least developed countries suffered from low-level economies that in turn effect all aspects of life in the countries negatively. So, scientific research and higher education hardily affected which lead to don't pay any importance for the research process at all [1,7,8]. Markedly, majority of the available web assessment and engineered frameworks are not robust in assessing the quality of the systems. Henceforth, it is essentially important to understand the quality of the systems in the context of Yemen and develop a success quality measurement framework from the perspective of users respectively [10,11,12]. It was pointed out that success measure is essentially important to be examined from the perspective of users to outline the actual worth of the systems [13].

2. Problem Statement

Notably, there is a dearth of integrative research on variables that could significantly affect success [2,11,38]. In line with inconsistency and mixed results [2,39,40] there are considerable gaps in the knowledge pertaining to what actually causes IS success [2]. With a limited engineering approach for building quality WIS framework, the existing systems are not capable of meeting the needs of users under the higher education domain wherein, they are also failing to sufficiently measure systems success since being domain-specific [10,11,12,16, 18]. Till date, little is known about the success of the website systems [41] which is similar to Arab countries whereby, empirical research pertaining to IS success is largely absent [8]. High rate of systems fails, which lead to business risks and financial losses [2,7,8,33,34,35,42]. There is a cognitive gap between users' expectations from the systems and what the users perceives [21]. Investigating variables of quality of systems based on the context is highly needed since there is no common definition for the quality of information [10,11,43].

It is still a challenge to engineer a framework that could measure a service quality in the domain of higher education [26]. Due to lack of studies, little is known about how loyalty towards non-commerce systems can influence. Up to date research relationships between these two concepts i-e satisfaction and loyalty are unclear [3,8,44]. Likewise, measuring systems` success and quality in Mukalla's universities is not yet mature this is also in line with high rate of systems dissatisfying users and organizations [1,6,7,8]. As comprehensively discussed, failure of systems is still a major problem for organizations, in targeting this failure, measuring of systems success emerges as a solution to increase the rate of success in future systems initiatives [1,7,8,34,37 42]. Success or failure of any system is one of the most prominent streams in the research. Therein, understanding pertaining to why some systems fulfill their expectations and other fail in doing is a complex arena for research scholars in the area [34]. In parallel, the quality of the systems has also remained one of the major issues that requires a framework for better understanding [16].

Henceforth, it is important to measure and understand the success and quality of university`s web-based system in the context of Hadhramaut universities to forward a comprehensive framework that could provide a robust guideline to such kinds of systems from the perspectives of the users. This study seeks to developing a universities web based systems novel framework for the higher education domain in the country by measuring the success of Hadhramaut universities web-systems and determine the quality factors towards achieving user satisfaction.

3. Literature Review

Assessing the success of systems is critical to outline how a particular system is performing and to what length, it is capable of identifying any issues; resolving user problems, and forwarding suggestions to meet the user requirements [1,2,10,11,13,14,15, 16,17,18]. Principally, satisfaction of the student and quality of the systems should be the most important element and accordingly, WIS should be developed to satisfy students through meeting their expectations [19]. Today, researchers have agreed to the fact that users' satisfaction is a more accessible measure than the rest and the success of WIS is completely dependent on the users' satisfaction. Importantly, users' satisfaction can be adaptable in specific contexts (9,19,20 ,21,54]. Earlier frameworks sadly have failed to identify prominent factors and sub-factors that can potentially enhance satisfaction of users and help eliminate aspects like computer anxiety [22]. Therein, quality characteristics should also be included as core components for systems success assessment by the future studies to help refine the measurement items in a new context [23]. It has been confirmed that net benefit must be defined within the context of system under study and within the frame of reference of those who evaluate the system benefits [24].

Organizations can outline the effectiveness and success of their web information systems through highlighting satisfaction and benefits of all such factors that play valuable role in the determination of success [2,19,21,24]. Organizations henceforth need a well-structured framework and advanced tools to assist the examination of complex web-based systems and their quality against defined standards [11,23,25,26]. Similarly, it has been asserted scholarly work in the academics is expanding the researches horizon in different fields [2]. The development and formulation of robust system quality assessment framework is thus, highly important [16,18]. The users' satisfaction antecedents such as quality components have been highlighted in several comprehensive research studies. Therein, the literature has guided that systems` quality is principally one of the urgent issues with limited empirical attention. Moreover, there is limited association between different metrics and frameworks as well [11,16,21,27,28]. Importantly, notable authors have underlined that computer anxiety is important for consideration to develop a balanced system satisfaction framework [29].

Accordingly, management support plays a vital role in systems` success, affecting the satisfaction and should therefore be deployed in comprehensive studies in future [1,30]. Furthermore, culture also plays a critical role in this regard hence, scholars need to examine this construct to forward more generalizable results in this regard [31]. It is important to note that people as users switch from one system to another mainly due to experiencing dissatisfaction. Likewise, users also switch to alternative system when they have doubts regarding the quality of the systems and the information that it provides [32]. Bundles of studies concerning to system implementation have surfaced outlining businesses incurring massive financial losses due to system failures [33]. Therein, numerous organizations can also be tracked providing statistical records pertaining to these losses. Taking an example, 27,000 students at the university of Massachusetts, Stanford, and Indiana university ended up facing severe consequences due to the failure of Information system of their respective universities [34].

It has been reported that practitioner surveys consistently suggest that more than 50 percent of systems have not achieved their goals [2]. Accordingly, 56 percent of the 5400 mega information technology projects have failed over the past decades. Regrettably, the assessment of such systems is entirely based on the stakeholders which is a tricky and expensive entity [2]. Shane Hastie (2015), based on CHAOS`s report (2015), has asserted that there is still working to be done to achieve considerable successful outcomes from systems and the percentage whereby, the success of projects is still very low

[35,36]. It should be mentioned that the failure of systems is a big concern to the organizations, business world, marketplace and stakeholders [8,37]. As per the statistical records from international data corporation, 43 billion US dollars have been spent on information systems [8]. This conclusively indicates that there is much interest in practice studies in a valid means for benchmarking these systems in order to track how the ICT investments are resulting to explain the benefits to the organization, and to better plan for future systems and ICT investments [8]. This is consonant with the issue in the Yemen as its` higher education system is aimed to enhance and improve the management of the institution and facilitate users especially the students [8]. Universities in the country are generally not satisfied with the information systems whereby, they are also experiencing challenges pertaining to justifying these massive investments and expenses [8].

4. Methodology

Survey is an appropriate approach to study complex phenomenon of technology implementation in an organizational setting [45,46] to assess individual perceptions, attitudes and behaviours [47]. Moreover, surveys serve as a means of measuring the research variables for explaining the relationship of each research variable in a study [45,48]. Research could be quantitative, qualitative or a combination of both. The best method depends on the research objective and purpose a particular research is aiming to address [1,49,50].

In the present research, the quantitative approach has been selected whereby, the data will be collected through using the questionnaire technique. In parallel, the instruments have been pretested via users. Also, study, instruments and frameworks are further validated by one specialized expert and five PhD specialized academic lecturers from different universities in Yemen and Malaysia. This study will use the census method to collect the data from 2017 respondents (bachelor students of level four) from all four universities located in Mukalla.

5. Pilot Study

A total of 40 students participated in the pilot study. The test of reliability consistency is Cronbach's coefficient alpha which is used for multipoint scale items [51,52]. The higher amount of coefficient indicates the better measures. Ideally, the Cronbach's coefficient alpha should be greater than 0.70 [53]. Table 1 indicates the results of reliability for pilot study.

Table 1. Pilot Study.

Construct	*No. Items*	Cronbach's alpha
Information Quality	6	0.770
System Quality	6	0.897
Services Quality	6	0.862
Efficiency	6	0.850
Functionality	6	0.907
Reliability	6	0.877
Usability	6	0.894
Security	5	0.794
Management Support	4	0.832
Computer Anxiety	4	0.842
Perceived Ease of use	4	0.709
Perceived Usefulness	4	0.763
Student Satisfaction	7	0.864
Loyalty	7	0.817
Net Benefit	5	0.849

6. The Instrument

The instrument consists of 12 independent variables (IV), which hypothesized as will significantly affect the first dependent variable (DV) students' satisfactions, then the dependent variable students' satisfactions will work as independent variable that hypothesized as will significantly affect the second dependent variables loyalty and benefit.

First IV Perceived Information Quality (PINFQ) will measuring the accuracy, content and understandability using: The information outputs of my university web system (including on-screen and printed outputs) are Complete. The information outputs of my university web system (including on-screen and printed outputs) are concise. The information outputs of my university web system (including on-screen and printed outputs) are easy to understand. It is easy to find what I'm looking for when using my university web system. My university web system provides the precise information I need.

Second IV Perceived System Quality (PSYSQ) will measuring the adaptability and sophistication using: It is easy for me to become skilful by using my university web system. In general, I find my university web system is easy to use. My university web system applied modern technology. My university web system is well integrated. My university web system has a short time lag between input and output of data as example (registration process). My university web system has a short response time for on-line enquiry.

Third IV Perceived Service Quality (PSERVQ) will be measuring the empathy of tech personal support and integrity using: Service support staff for my university web system are prompt in responding to my queries. Service support staff of my university web system understands my needs. Service support staff responds in a cooperative manner. The information received from my university web system is adequate. It is easy for me to get the desired information while interacting with my university web system.

Forth IV Perceived Efficiency (PEFF) will measuring the time behaviour and accessibility using: It is possible to switch between pages of my university web system within reasonable time. It is possible to find in my university web system what I want in a reasonable time. My university web system enables me to get on to it quickly. My university web system does not use advertises or unwanted plug-ins. I can access my university web system from my favourite browser. It is easy to get and browse any part on my university web system.

Fifth IV Perceived Functionality (PFUN) will measuring the navigation and search using: It is easy to go to the home page while I'm browsing any other page in my university web system. I'm able to move from one page to another page without getting lost while browsing my university web system. While navigating my university web system pages, I can immediately tell where I'm browsing in the system. While using my university web system, I can easily navigate backwards through previously visited pages. My university web system provides varied search options (e.g. By faculty, courses, etc..). Search hints are provided when wrong search keywords are used.

Sixth IV Perceived Reliability (PREL) will measuring maturity, fault tolerance, recoverability, availability and reliability using: My university web system never stops unexpectedly. When there is a problem in some part or parts in my university web system I still can browse and perform some of process. In case of interruption of fault, my university web system recovers properly. My university web system allows information to be readily accessible to me. In general, my university web system is available 24/7. I believe that my university web system is reliable.

Seventh IV Perceived Usability (PUSA) will measuring the user interface aesthetics and protection from users' error using: The interface design of my university web system is attractive. All interface elements are well combined and harmonious in my university web system. My university web system interfaces are pleasant. My university web system protects me from making errors when interring data. My university web system errors messages clearly indicate to me how to correct the problem. In my university web system, it is easy to recover from the error quickly.

Eighth IV Perceived Security (PSEC) will measuring the security privacy and trust using: I believe my university web system is secure. Overall, I trust my university web system. My university web system has adequate security features that make you feel secure while using. I believe that the information

offered by my university on the university web system is sincere and honest. The output information of my university web system is secure.
Ninth IV Perceived Usefulness (PUSE) will be measured using: Using my university web system improves my performance. Using my university web system is useful. Using my university web system would enhance my effectiveness. My university web system is efficient.
Tenth IV Perceived Ease of Use (PEOU) will be measured using: I find my university web system flexible to interact with. My interactions with my university web system during doing online process were clear and understandable. My university web system is convenient for me. My university web system is laid out in a modern and fashionable.
Eleventh IV Perceived Management Support (PMSU) will be measured using: Management support discusses problems regarding the university web system and provides all necessary resources to improve it. Management support encourages using the university web system. Management support frequently mentions the various problems, matters related to university web system. Management support much interested in university web system usage rate.
Twelfth IV Computer Anxiety (COAN) will be measured using: Using my university web system let me feel bad. Using my university web system let me feel anxiety. Using my university web system let me feel uncomfortable. Using my university web system let me feel nervous.
First DV Students' Satisfaction (STSA) will be measured using: My university web system is of high quality. My university web system has met my expectations. My interaction with my university web system is very satisfying. Overall, I am satisfied by using my university web system. Overall, I'm happy with my university web system. My university web system serves my need as well as a student. My university web system is enjoyable.
Second DV Benefit (BENE) will be measured using: My university web system helps me to retrieve my information easier and quickly. My university web system saves my time. Overall, I obtained benefits from using my university web system. My university web system is an important and valuable aid to me. My university web system has a large, positive impact on me as a user.
Third DV loyalty (LOYA) will be measured using: I will be using more of my university web system in the future. I will recommend my university web system to others. I will say positive things about my university web system to others. I like using my university web system. I use my university web system frequently. I tend to use my university web system. I encourage to use my university web system.

7. Conclusion

High rate of system fails, limited of engineered frameworks that can help organization to know about the quality and benefits of their systems towards the users' satisfactions and loyalty, lack of theoretical grounding, lack of data collection and empirical data. In the Arab word knowledges about systems quality and success are lack in line with almost nonexistence studies in Yemen with the high numbers of systems dissatisfying users and organizations. There is a hardness in engineered a framework that measures quality factors in the higher education domain. Together with inconsistent results in management support, computer anxiety and quality factors towards users' satisfaction and users' satisfaction towards benefit of the systems. In parallel, relationship of users' satisfaction towards loyalty of users to the systems is ambiguous. Researches are invited to investigate these issues. It's highly recommended to engineer a framework that make integration between the theories of software engineering and information systems such as ISO 20510 and DM 2003. It's of most important to conduct a comprehensive study that develop a novel framework, can solve all or parts of the issues mentioned above. This research aims to contribute the body of knowledge by targeting these issues. PLS and SPSS tools are going to be used because research will develop a framework and predict results. By rely on the literature and consults of the experts PLS and SPSS are the suited tools. Pilot tested showed excellent result. So, researcher can be performing the data collection since the instrument are well developed and validated. After collecting the data will spreading the final result in the next article.

References

[1] I. E. I. Fadhel, "An evaluation of information system success based on students' perspective: The case of Hadramount University." Universiti Utara Malaysia, 2015.
[2] W. H. DeLone and E. R. McLean, "Information systems success measurement," Found. Trends® Inf. Syst., vol. 2, no. 1, pp. 1–116, 2016.
[3] H. Mohammadi, "Investigating users' perspectives on e-learning: an integration of TAM and IS success model," Comput. Human Behav., vol. 45, pp. 359–374, 2015.
[4] U. Nations, "Briefing note for countries on the 2015 human development report Yemen," 2015.
[5] H. Rakels, "The future of ICT in Yemeni higher education," Yemen, 2013.
[6] AbdulMonsif, "Status of computer and administrative academic research in Yemen," Hadramout University, Mukalla, 2015.
[7] F. Baheshwan, "Evaluation of information systems in Yemen based on user's' Perspective?," Hadramout University, Mukalla, 2016.
[8] D. A. Khred, "systems development in our country," 2017.
[9] L. Hasan and E. Abuelrub, "Assessing the quality of web Sites," INFOCOMP J. Comput. Sci., vol. 7, no. 4, pp. 11–20, 2008.
[10] T. Mebrate, "A framework for evaluating academic website's quality from students' perspective." TU Delft, Delft University of Technology, 2010.
[11] E. W. Mwangi, "Information quality assessment framework: case of the national safety net program single registry system." Strathmore University, 2016.
[12] K. K. Singh and P. Kumar, "A model for website quality evaluation-a practical approach," Int. J. Res. Eng. Technol., vol. 2, no. 3, pp. 61–68, 2014.
[13] R. D. Galliers and D. E. Leidner, Strategic information management: challenges and strategies in managing information systems. Routledge, 2014.
[14] P. E. D. Irani Zahir, "The propagation of technology management taxonomies for evaluating investments in information systems," J. Manag. Inf. Syst., vol. 17, no. 3, pp. 161–177, 2000.
[15] D. D. J. Suwawi, E. Darwiyanto, and M. Rochmani, "Evaluation of academic website using ISO/IEC 9126," in Information and Communication Technology (ICoICT), 2015 3rd International Conference on, 2015, pp. 222–227.
[16] Sugiyanto, R. Siti, and Sarwosri, "The improvement of software quality model for academic websites based on multi-perspective approach," J. Theor. Appl. Inf. Technol., vol. 86, no. 3, 2016.
[17] N. Urbach and B. Müller, "The updated DeLone and McLean model of information systems success," in Information systems theory, Springer, 2012, pp. 1–18.
[18] D. I. Zahran, H. A. Al-Nuaim, M. J. Rutter, and D. Benyon, "A comparative approach to web evaluation and website evaluation methods," Int. J. Public Inf. Syst., vol. 10, no. 1, 2014.
[19] M. J. Lee, C. Huh, and M. F. Jones, "Investigating quality dimensions of hospitality higher education: From Students' Perspective," J. Hosp. Tour. Educ., vol. 28, no. 2, pp. 95–106, 2016.
[20] S. Bairamzadeh and A. Bolhari, "Investigating factors affecting students' satisfaction of university websites," in Computer Science and Information Technology (ICCSIT), 2010 3rd IEEE International Conference on, 2010, vol. 1, pp. 469–473.
[21] R. Vaezi, A. Mills, W. Chin, and H. Zafar, "User Satisfaction research in information systems: Historical roots and approaches," Commun. Assoc. Inf. Syst., vol. 38, no. 27, pp. 501–532, 2016.
[22] M. L. Cheok and S. L. Wong, "Predictors of e-learning satisfaction in teaching and learning for school teachers: A literature review," Int. J. Instr., vol. 8, no. 1, pp. 75–90, 2015.
[23] N. Forsgren, A. Durcikova, P. F. Clay, and X. Wang, "The Integrated user satisfaction model: Assessing information quality and system quality as second-order constructs in system administration," Commun. Assoc. Inf. Syst., vol. 38, no. 1, p. 39, 2016.
[24] H. H. Alshibly, "Investigating decision support system (DSS) success: a partial least squares structural equation modeling approach," J. Bus. Stud. Q., vol. 6, no. 4, p. 56, 2015.

[25] A. I. ELdesouky, H. Arafat, and H. Ramzey, "Toward complex academic Web-Sites Quality evaluation method (QEM) framework: quality requirements phase definition and specification," Comput. Syst. Eng. Dep., 2008.
[26] V. Teeroovengadum, T. J. Kamalanabhan, and A. K. Seebaluck, "Measuring service quality in higher education: Development of a hierarchical model (HESQUAL)," Qual. Assur. Educ., vol. 24, no. 2, pp. 244–258, 2016.
[27] S. Laumer, "Information quality dimensions: two exploratory case studies with enterprise content management system users," 2016.
[28] S. Petter, W. DeLone, and E. R. McLean, "The past, present, and future of' IS success,'" J. Assoc. Inf. Syst., vol. 13, no. 5, p. 341, 2012.
[29] B. H. Malik, C. Shuqin, A. G. Mastoi, N. Gul, and H. Gul, "Evaluating citizen e-satisfaction from e-government services: A Case of Pakistan," Eur. Sci. J., vol. 12, no. 5, 2016.
[30] Wang and J. Song, "The relation of perceived benefits and organizational supports to user satisfaction with building information model (BIM)," Comput. Human Behav., vol. 68, pp. 493–500, 2017.
[31] B. Berger, A. Geimer, and T. Hess, "Will They Stay or Will They Go? An examination of the factors influencing user loyalty towards news websites," in Proceedings of the 50th Hawaii International Conference on System Sciences, 2017.
[32] K. Rouibah, P. B. Lowry, and L. Almutairi, "Dimensions of business-to-consumer (B2C) systems success in Kuwait: Testing a modified DeLone and McLean IS success model in an e-commerce context," in Web-Based Services: Concepts, Methodologies, Tools, and Applications, IGI Global, 2016, pp. 1223–1255.
[33] C. Maier, S. Laumer, A. Eckhardt, and T. Weitzel, "Analyzing the impact of HRIS implementations on HR personnel's job satisfaction and turnover intention," J. Strateg. Inf. Syst., vol. 22, no. 3, pp. 193–207, 2013.
[34] Y. K. Dwivedi et al., "Research on information systems failures and successes: Status update and future directions," Inf. Syst. Front., vol. 17, no. 1, pp. 143–157, 2015.
[35] S. W. Shane Hastie, "Standish Group 2015 Chaos Report - Q&A with Jennifer Lynch," 2015. [Online]. Available: https://www.infoq.com/articles/standish-chaos-2015. [Accessed: 12-Dec-2016].
[36] Chaos, "The standish group report CHAOS," 2014.
[37] R. Moh'd Al-adaileh, "An evaluation of information systems success: A user perspective-The case of Jordan telecom group," Eur. J. Sci. Res., vol. 37, no. 2, pp. 226–239, 2009.
[38] S. Petter, W. DeLone, and E. R. McLean, "Information systems success: The quest for the independent variables," J. Manag. Inf. Syst., vol. 29, no. 4, pp. 7–62, 2013.
[39] S. Mardiana, J. H. Tjakraatmadja, and A. Aprianingsih, "Validating the conceptual model for predicting intention to use as part of information system success model: The case of an indonesian government agency," Procedia Comput. Sci., vol. 72, pp. 353–360, 2015.
[40] K. C. Snead Jr, S. R. Magal, L. F. Christensen, and A. A. Ndede-Amadi, "Attribution theory: a theoretical framework for understanding information systems success," Syst. Pract. Action Res., vol. 28, no. 3, pp. 273–288, 2014.
[41] D. Stefanovic, U. Marjanovic, M. Delić, D. Culibrk, and B. Lalic, "Assessing the success of e-government systems: An employee perspective," Inf. Manag., vol. 53, no. 6, pp. 717–726, 2016.
[42] M. Bloch, S. Blumberg, and J. Laartz, "Delivering large-scale IT projects on time, on budget, and on value," Harv. Bus. Rev., 2012.
[43] A. L. McNab and D. A. Ladd, "Information quality: the importance of context and trade-offs," in System Sciences (HICSS), 2014 47th Hawaii International Conference on, 2014, pp. 3525–3532.
[44] R. Mosahab, O. Mahamad, and T. Ramayah, "Service quality, customer satisfaction and loyalty: A test of mediation," Int. Bus. Res., vol. 3, no. 4, p. 72, 2010.
[45] B. P. See, "Human cognitive perspective of extended use of information systems." Universiti Utara Malaysia, 2012.

[46] A. Majchrzak, R. E. Rice, A. Malhotra, N. King, and S. Ba, "Technology adaptation: The case of a computer-supported inter-organizational virtual team," 2000.
[47] H. F. Weisberg, J. A. Krosnick, and B. D. Bowen, An introduction to survey research and data analysis. Scott, Foresman & Co, 1989.
[48] R. W. Zmud and L. E. Apple, "Measuring technology incorporation/infusion," J. Prod. Innov. Manag., vol. 9, no. 2, pp. 148–155, 1992.
[49] M. D. Myers and D. Avison, "Qualitative research in information systems," Manag. Inf. Syst. Q., vol. 21, pp. 241–242, 1997.
[50] R. K. Yin, Case study research: Design and methods. Sage publications, 2013.
[51] A. Davarpanah and N. Mohamed, "Human resource information systems (HRIS) success factors in a public higher education institution context," in Research and Innovation in Information Systems (ICRIIS), 2013 International Conference on, 2013, pp. 79–84.
[52] L. J. Cronbach, "Response sets and test validity," Educ. Psychol. Meas., vol. 6, no. 4, pp. 475–494, 1946.
[53] J. Pallant, SPSS survival manual. McGraw-Hill Education (UK), 2013.
[54] S. M. Almahamid, A. F. Tweiqat, and M. S. Almanaseer, "University website quality characteristics and success: lecturers' perspective," Int. J. Bus. Inf. Syst., vol. 22, no. 1, pp. 41–61, 2016.
[55] L. Hasan, "The website of the university of Jordan: Usability evaluation," Int. Arab J. e-Technology, vol. 3, no. 4, pp. 258–269, 2014.

IOP Conf. Series: Journal of Physics: Conf. Series **1019** (2018) 012082 doi:10.1088/1742-6596/1019/1/012082

Investigation on Suitability of Pulse Duration Parameter for HeLa Cells Line Proliferation Properties towards Anti-Cancer Application

Azra Hazwanie Azizulkarim[1], Muhammad Mahadi Abdul Jamil[1*], Mohamad Nazib Adon[1], Muhammad Shukri Ahmad[1], Johan Mohamad Sharif[2], Md Asri Ngadi[2], and Radzi Ambar[1,3]

[1]Biomedical Modelling and Simulation (BIOMEMS) Research Group, Faculty of Electric and Electronic Engineering, Universiti Tun Hussein Onn Malaysia, 86400 Batu Pahat, Johor, Malaysia
[2]Department of Computing Science, Faculty of Computing, Universiti Teknologi of Malaysia, 81310 Skudai, Johor, Malaysia
[3]Embedded Computing System (EmbCoS) Research Group, Department of Computer Engineering, Faculty of Electrical and Electronic Engineering, Universiti Tun Hussein Onn Malaysia, 86400 Batu Pahat, Johor, Malaysia

**Email:* mahadi@uthm.edu.my

Abstract. In order to bring up a cells or tissue with a gene or DNA, Electroporation parameter such as pulse duration, field strength and number of pulse plays the most important aspect. Thus, the aim of this paper is to optimize the pulse duration parameter for HeLa cells line. This is very important because nowadays, research on EP factors is still in limited mode. Moreover, suitable and applicable pulse duration parameter would ensure successful treatment with least possible side effect. In this paper, HeLa cells or known as cervical cancer cells was used to induce into a single pulse electric field. Therefore, to achieve the aim of this study, field strength of 500V/cm was preferred and the pulse duration was varied in the range of 100μs, 500μs, 1ms, 2ms and 5ms to obtain lowest proliferation rate and inhibit growth of HeLa cells. The result obtained shows that HeLa cell induced to 500V/cm with pulse duration of 100μs, induced the lowest proliferation rate (45%) as compared to initial reading density (control) after 48 hours of culture. Thus, further research based on this work may pave the way for an anti-cancer application through the study of the effect of pulse duration parameter on HeLa cells.

1. Introduction

Electroporation (EP) is a process of the biophysical effect on cells induced by an external electrical field [1]. It is a molecular biology technique in order to create pores through a cell membrane, increase rapidly the permeability of the cell membrane, allowing chemicals, drugs, or DNA to be introduced into the cell [2]. In modern molecular medicine and biotechnology field, tissue electroporation is performed by using special electrodes placed in the target area of the body.

Moreover, in an experimental setup, EP is done by insertion of a biological sample between two electrodes, by that produce an electric field between the electrodes either in-vitro or in-vivo EP [5]. Cell EP by in-vitro is used mainly for transfection by DNA introduction, but many other interventions are possible, including microbial killing. On the other side, in-vivo electroporation of tissues enhances molecular transport through tissues and into their constitutive cells whilst ex-vivo electroporation provides manipulation of cells that are reintroduced into the body [5].

By applying electrical pulses across cells can have a variety of outcomes; from no effect to Reversible EP or to Irreversible EP, Reversible EP will allow transient molecular transport through the pores and after a few minutes, cellular membrane will reseal and cell functions are restored. In contrast, Irreversible EP offers an ability to destroy unwanted tissue in the body and could be practiced as in the absence of chemo drug to destroy the cancerous cells. There are many benefits of EP technique, such as, this technique does not change and modify the biological structure and function of the target cells, and it is safer, high ability and efficiency and less immunologic [6-11].

Furthermore, the efficient and capability of EP is achieved by controlling the pulse duration, one of the important parameter in EP technique which have the biggest clout on the outcome of the experiment [1-5]. In addition, pulse duration need to be controlled for a particular cell type, cell orientation, cell size and cell density. Hence, in this study, the efficiency of EP was determined by measuring the percentage of proliferation rate of electroporated HeLa cells.

2. Materials and Methods

2.1 Cell Culture

In this research, HeLa cells line or known as cervical cancer cells were used in this experiment. The HeLa cells line were cultured in an incubator with 5% CO2. Then, the HeLa cells line were cultured in the medium RPMI-1640 enhanced with 10% Fetal Bovine Serum (FBS) and 1% antibiotic Penicillin-streptomycin from Gibco, USA. The humidified incubator is set to 37°C for cell growth. To perform cells subculture, the confluence of HeLa cells must reach 80 – 90 % confluent and then the HeLa cells were sub-cultured every 3 to 5 days shown in Figure 1.

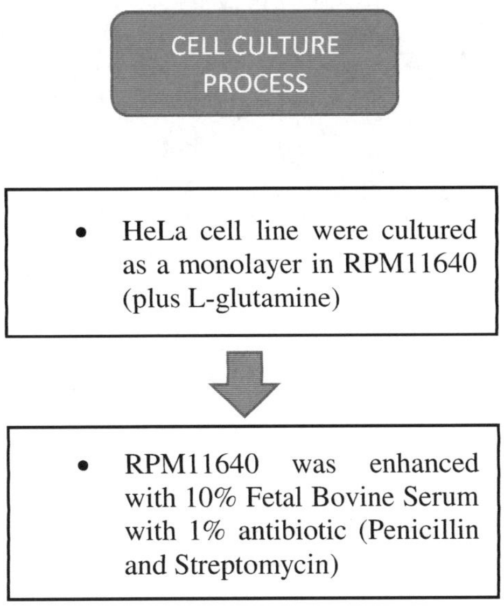

- The cultured cells were incubated in a humidified incubator containing 5% Co2 at 37°C.

- The cells were sub-cultured every 3-5 days whenever it reached 80% to 90% confluent.

Figure 1. Flow chart of the cell culture process.

After HeLa cells reached 80 – 90 % confluent, the old medium was first aspirated and discarded. Then, 3ml of Phosphate Buffer Saline (PBS) was added for washing the HeLa cells and later the PBS was removed from the old culture flask [12-13]. Once the PBS is removed, 2ml of Trypsin Express solution was introduced into the 25cm^2 old culture flask and left for 10 – 15 minutes in the incubator for cell detachment [12-13]. This is because Trypsin Express solution works very well in a warm condition. After incubation process, the HeLa cells were monitored under a high resolution microscope to make sure the cells are circular in shape and fully detached from the culture flask surface after trypsinization process in Figure 2.

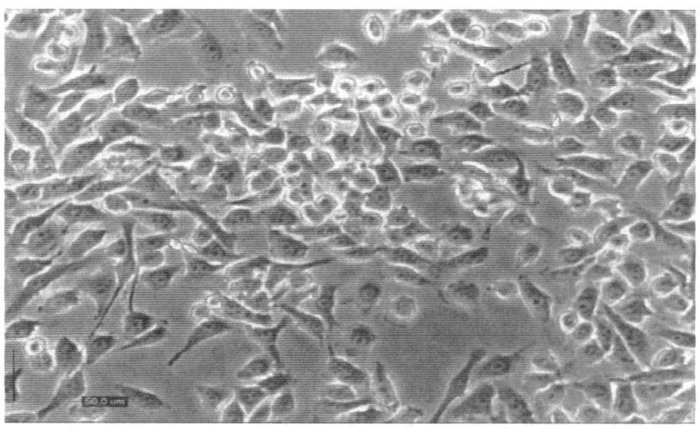

(a) The cells before detachment process

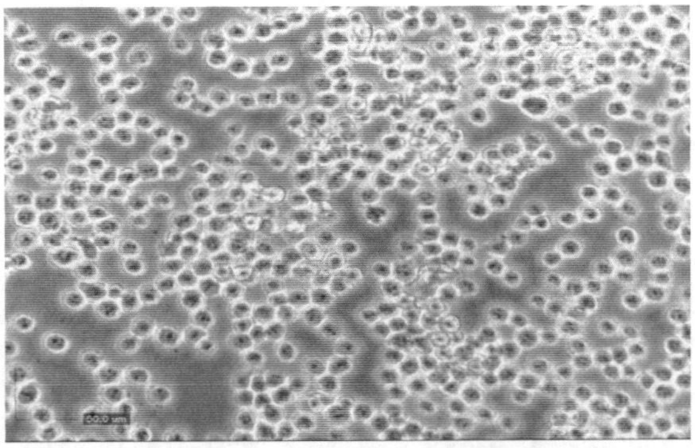

(b) After trypsinization, HeLa cells appeared rounded and fully detached from the flask surface

Figure 2. Comparison between the HeLa cells conditions.

2.2 *Electroporation (Ep)*

Figure 3 shows an ECM830 device from BTX Harvard Apparatus that was used for electroporating the HeLa cells line in suspension. Moreover, ECM830 has two modes of operation, where the first mode is a low voltage mode between pulse length 10ms until 999ms (1ms resolution), whereas high voltage mode ranging from 10μs to 600μs (1μs resolution). In addition, EP process involves several parameters and the parameters to measure the effectiveness of EP effect are the pulse duration, electric field strength and the number of pulse.

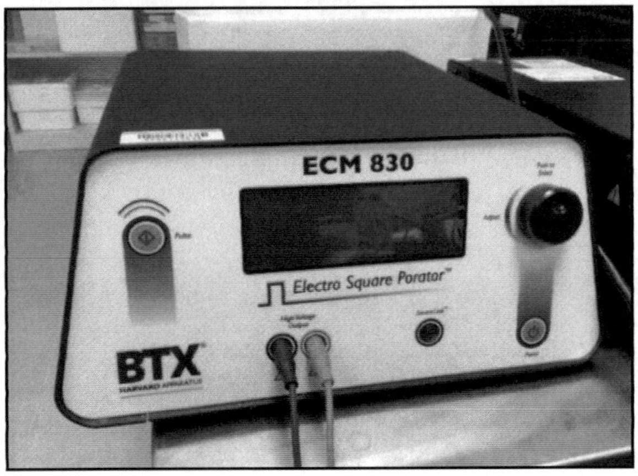

Figure 3. ECM830 device

2.3 *Measurement Of Cell Proliferation*

After 48 hours in culture mode, HeLa cells were detached and counted by using Hemocytometer method. Figure 4 shows the Hemocytometer method that was conducted to monitor the cell's proliferation growth rate. It is a device used to count the number of cells using Trypan Blue solution.

Based on the method, after 48 hours the cell densities were compared with the initial plating density (control) to observe the effect of EP on HeLa cells. Furthermore, experiments were performed in triplicate for consistency, and the mean of HeLa cells proliferation's rate was calculated for analysis purpose.

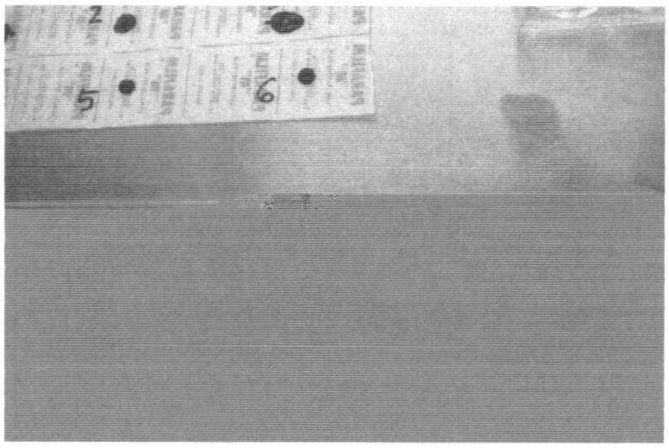

a) Mix together (HeLa cells) with Trypan Blue solution.

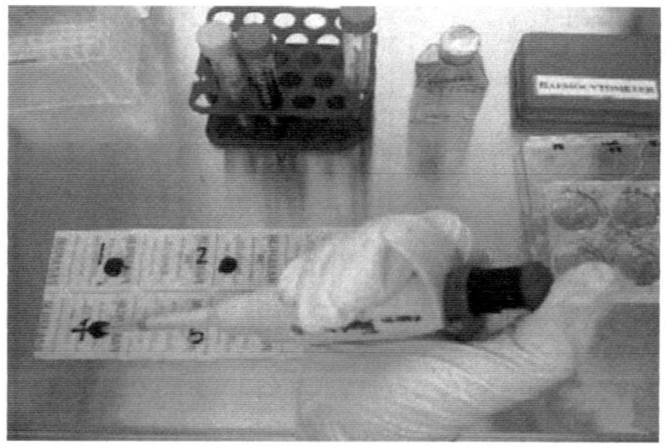

b) With a glass cover-slip in place, Cells mixed with Trypan Blue Solution (1:1) transferred into the chamber.

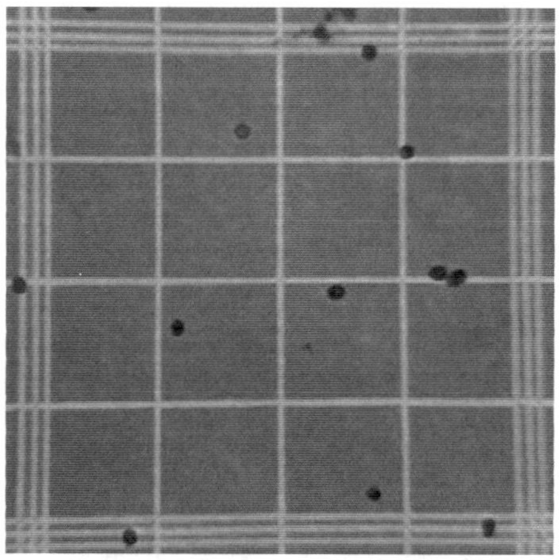

c) Process of counting the number of cells (blue staining shows-viable cells, non-viable cells remain opaque)

Figure 4. Hemocytometer method to count cells number

3. Experimental results

In this experiment, HeLa cells were exposed with different pulse duration parameter to monitor cells growth proliferation rate. EP technique involves several important parameters and the most effective EP result support by the length of the field applied known as pulse duration. The pulse duration parameter is controlled by the commercial EP device.

After harvesting the cells with Trypsin Express solution and neutralizing with an equal volume of growth medium, 0.8 ml of cells suspension was then poured into a 4mm electrode gap cuvette. An electric field with an intensity of 500V/cm was used in electroporating the cells and different pulse duration ranging between 100µs-5ms was utilized.

Subsequently, the cells suspension after exposed with EP technique were then seeded into a 6-wells plates containing 1.2ml of complete growth medium and were incubated at 37°C and 5% of CO2. Similarly, for control, same method was repeated from same in initial flask but without exposed with EP technique. All the flasks are cultured under the same condition and were monitored for 48 hours.

Moreover, after 48 hours in culture condition, the cells was detached and counted using Hemocytometer method for each treatment to study the EP effect on HeLa cell's proliferation rate. The results are shown in Table 1, and quantitatively revealed the dependence of cell's proliferation rate on pulse duration parameter.

Furthermore, Figure 5 show images of HeLa cells exposed by various pulse duration of 100µs, 500 µs, 1ms, 2ms, and 5ms with pulse amplitude (electric field) of 500V/cm after 48 hours in cultured as compared to initial plating number of cells. It was found that the cell's proliferation rate started to reduce by 40% at 100µs for 500V/cm as compared to initial number of seeding cells (control group) after 48 hours in culture conditions.

Table 1. HeLa Cells Proliferation Rate for Different Pulse Duration Parameter after 48 Hours

Different pulse duration (S)	`Proliferation rate of HeLa cells after 48 Hours (%)
CONTROL	55.00
100µs	10.00
500µs	22.50
1ms	22.50
2ms	12.50
5ms	12.50

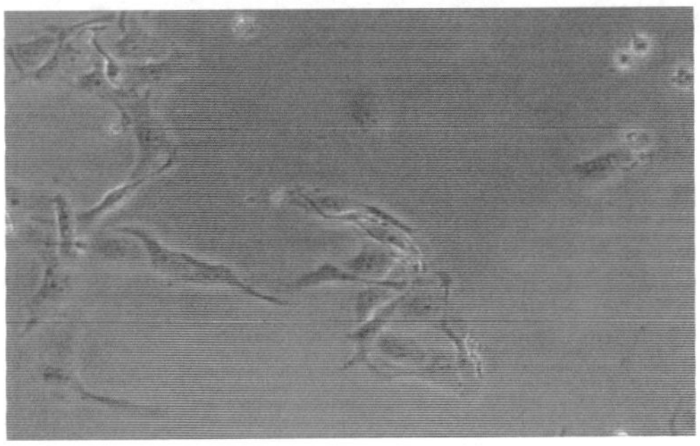

(a) Pulse duration parameter: control

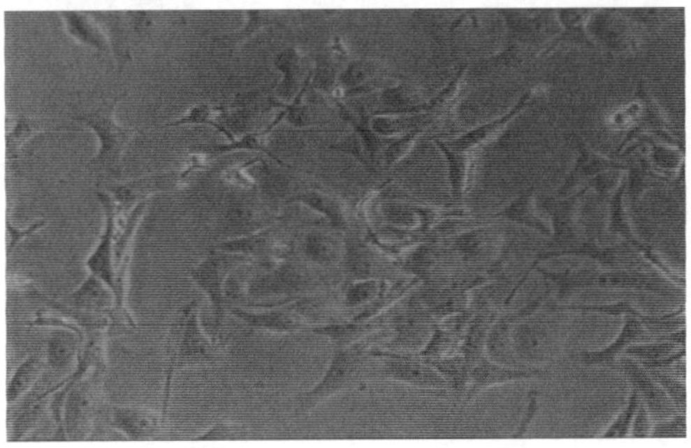

(b) Pulse duration parameter: 100µs

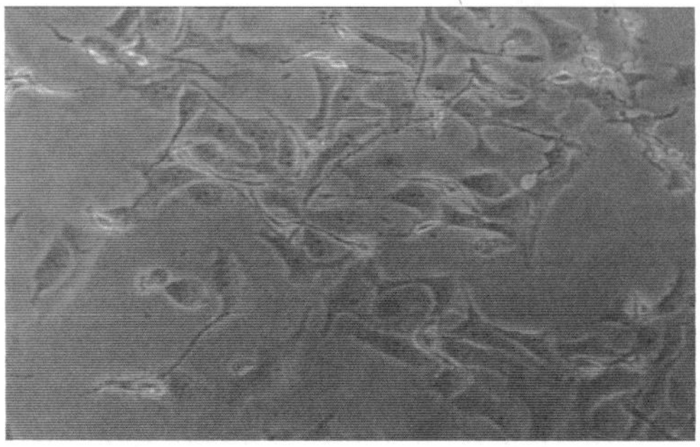

(c) Pulse duration parameter: 500μs

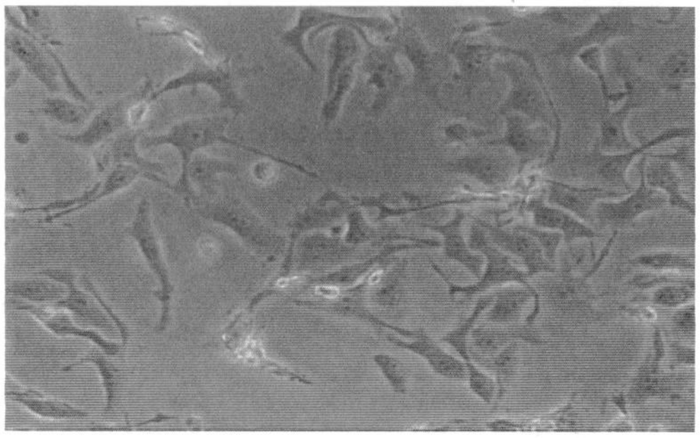

(d) Pulse duration parameter: 1ms

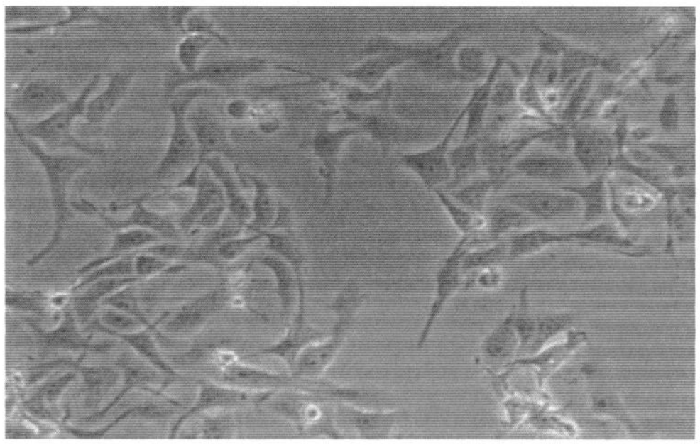

(e) Pulse duration parameter: 2ms

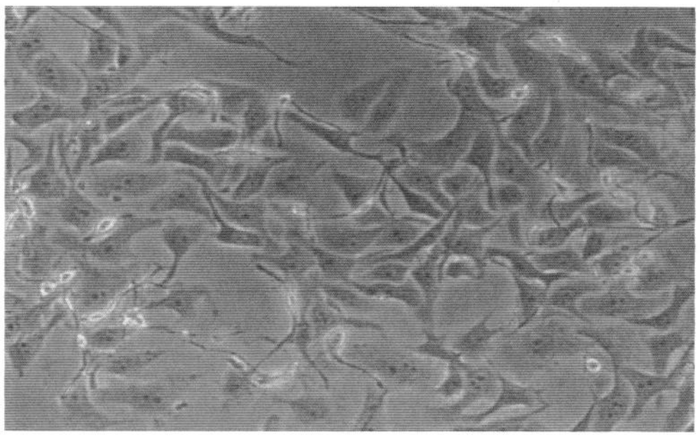

(f) Pulse duration parameter: 5ms

Figure 5. HeLa Cells Growth Rate images after 48 hours.

In addition, Figure 6 shows that the lowest proliferation rate of HeLa cells was achieved at a pulse duration of 100μs at 500V/cm. Besides, this short and fast pulse duration inhibit the proliferation rate of HeLa cells and it revealed the 100μs at 500V/cm is 45% lower in cell's proliferation rate than the control group. The results of the experiment showed that HeLa cells proliferation is considerably affected when treated with short pulse duration. At this stage, qualitative observation towards HeLa cells at this pulse duration showed a change in color of media from red to orange due to the electrolysis process at short pulse duration. Moreover the color changes occurred because of pH of media from neutral to acidic. Therefore, this changes in pH of the media might contributed to the higher number of cell death at shortest pulse duration.

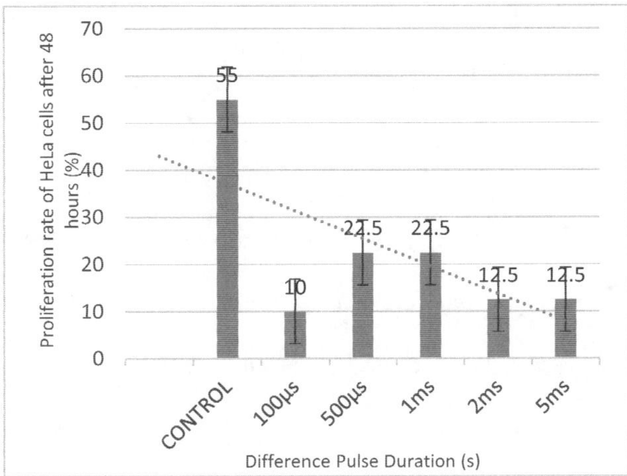

Figure 6. HeLa cells proliferation rate for different pulse duration parameter after 48 hours.

The result demonstrated that short pulse duration works the best to inhibit the HeLa cells proliferation rate and suitable for anti-cancer cell's application. This is because shortest pulse duration could result to electrolysis of the media and reduce the proliferation rate of HeLa cells. The experiments were performed in triplicate for consistency, and the mean or average proliferation factor was calculated.

4. Conclusion

In this study, the fundamental ideas and methods of EP were highlighted and the suitability of parameter of EP method is performed with shortest pulse duration to inhibit the HeLa cells proliferation rate. To achieve this objective, the culture and sub-culture technique of HeLa cells is done practically. Experimental data shows that HeLa cells exposed to 100μs with 500V/cm in single pulse shows the best anti-proliferation compared to other parameters. Later, the study will be continued by combining both EP method and porcupine bezoar (PB) extract that has the potential in becoming chemopreventive agents and requires further study and analysis. Finally, the outcome of this experiment might open the door in finding an alternative method for anti-cancer treatment.

5. Acknowledgement

The authors would like to thank Ministry of Education Malaysia for financial support of this work through ORICC, Universiti Tun Hussein Onn Malaysia (UTHM) for funding this research via University Contract Grant Vote U551 and Vote U775.

References

[1] R. Sundrarajan, et al. (2011). *Effective Use of Electrical Pulses on Cancer Cells to Control Proliferation*. IEEE.
[2] Somphop Rodamporn (2011).*Optimal Parameters of Electroporation for Gene and Tissue*, The 2011 Biomedical Engineering International Conference (BMEiCON-2011).

[3] Rafael V. Davalos, et al. (2004).*Electrical Impedance Tomography for Imaging Tissue Electroporation*,IEEE Transaction on biomedical engineering, **Vol. 51**, no. 5, May 2004.
[4] James C. Weaver, et al. (2000).*Electroporation of Cells and Tissues*, IEEE Transaction on Plasma Science, **Vol. 28**, No. 1, February 2000.
[5] Antoni Ivorra and Boris Rubinsky (2006) .*Impedance Analyzer for in vivo Electroporation Studies*, Proceedings of the 28th IEEE, EMBS Annual International Conference, New York City, USA, Aug 30-Sept 3, 2006.
[6] R. Sundararajan, et al. (2011). *Effect of Irreversible Electroporation on Cancer Cells*. Electrical Insulation and Dielectric Phenomena (CEIDP), 2011 Annual Report Conference. IEEE. 16-19 Oct. 2011. 164 – 167.
[7] Charpentier KP, et al. (2011). *Irreversible Electroporation of the liver and liver hilum in swine*. HPB 2011; 13:168–173.
[8] Phillips MA, et al. (2012).*Irreversible Electroporation on the small intestine*.BrJCancer2012; 106:490–495.
[9] Jvosé A, et al. (2012).*Irreversible Electroporation shows efficacy against pancreatic carcinoma without systemic toxicity in mouse models*. Cancer Lett 2012 ;317:16–23.
[10] Schoellnast H, et al. (2013).*The delayed effects of irreversible Electroporation ablation on nerves*. Eur Radiol2013; 23:375–380.
[11] Guo Y, et al. (2010*). Irreversible Electroporation therapy in the liver: longitudinal efficacy studies in a rat model of hepatocellular carcinoma* .Cancer Res 2010 ;70:1555–1563.
[12] Mamman H, Abdul Jamil M. *Investigation of electroporation effect on HT29 cell lines adhesion properties*.2015 2nd International Conference on Biomedical Engineering (ICOBE). IEEE; 2015 Mar 30-31. p.1-5
[13] Mamman H, Sadiq A,Adon M, Abdul Jamil M. *Study of electroporation effect on HT29 cell migration properties.* 2015 IEEE International Conference on Control System, Computing, and Engineering (ICCSCE 2015). IEEE; 2015 Nov 27-29. p. 342-346
[14] Ridzwan BH, Rosli R, Nihayah M & Asma A (2003), *Preliminary Study On The Anatomy, Growth and Haemotology Parameters of Landak Borneo (Thecurus Crassispinas)*.Proc.25th Malaysian Soc.Animan Production Annual Conference.1-3 August,Melaka.
[15] Onik G, Mikus P, Rubinsky B. *Irreversible electroporation : implications for prostate ablation. Technol Cancer Treat* 2007;6:295–300.
[16] Jvosé A, Sobrevals L , Ivorra A, Fillat C. *Irreversible electroporation shows efficacy against pancreatic carcinoma without systemic toxicity in mouse models.* Cancer Lett 2012 ;317:16–23.
[17] Deodhar A, Monette S, Single GW, et al. *Renal tissue ablation with irreversible electroporation: preliminary results in a porcine model.* Urology2011; 77:754–760.
[18] Li W, Fan Q, Ji Z ,Qiu X, Li Z.*The effects of irreversible electroporation (IRE) on nerves*.PloSOne2011;6:e18831.
[19] Guo Y, Zhang Y, Klein R, et al. Irreversible electroporation therapy in the liver: longitudinal efficacy studies in a rat model of hepatocellular carcinoma .Cancer Res 2010 ;70:1555–1563.
[20] James C. Weaver, *"Electroporation of Cells and Tissues"*, IEEE Transaction on Plasma Science, **Vol. 28**, No. 1, February 2000.

IOP Conf. Series: Journal of Physics: Conf. Series **1019** (2018) 012083 doi:10.1088/1742-6596/1019/1/012083

Monowalk: A Travelling Companion

Siti Norashikin Taufik[1], Aida Mustapha[1], Noor Azah Samsudin[1], Mohd Helmy Abd Wahab[2]

[1] Faculty of Computer Science and Information Technology, Universiti Tun Hussein Onn Malaysia, 86400 Parit Raja, Batu Pahat, Johor, Malaysia

[2] Faculty of Electrical Engineering, Universiti Tun Hussein Onn Malaysia, 86400 Parit Raja, Batu Pahat, Johor, Malaysia

E-mail: aidam@uthm.edu.my

Abstract. In this era of modernization, there are various public transportations that facilitate the movement of users, among which is the Monorail service. However, beyond the route information such as the train schedule and the ticket fare provided by the Monorail service, users do not have access to other sources of information relating to local attraction, nearby places, good food and reasonable accommodation along theirroute. The MonoWalk Travelling Companion is proposed specifically for commuters or tourists who use the Monorail services. Aside from helping them to plan their journeys better and more systematic, the application can also become a medium for advertising local businesses and attractions.

1. Introduction

Among important elements in establishing a monorail system includes ticket pricing, operation cost, maintenance implications, service quality and passenger's comfortability [2]. Kuala Lumpur (KL) Monorail is an inter-city monorail that conveniently connects the northern and central of Kuala Lumpur, Malaysia. Its route includes the trade centre, shopping complexes, and an array of hotel chains in Kuala Lumpur. KL Monorail is operated by KL Monorail System Sdn. Bhd., a wholly owned subsidiary KL Infrastructure Group Bhd. The operator provides a comprehensive information related to their train and bus services such as the routes, operating hours, train frequency and ticket fares. The web-based information portal is also convenient and informative.

Nonetheless, the web-based information portal by the Monorail operator is not without drawbacks. Firstly, users need the Internet access to view the website. Secondly, all the information is only related to services by the Monorail. Beyond the route information such as the train schedule and the ticket fare [2], urban commuters have to find others sources of information relating to local attraction, nearby places, good food and reasonable accommodation.

Research has shown that the establishment of public transportation such as the monorail system influences businesses in the neighboring surroundings [3]. The benefits of this scenario is two-sided. First, it provides opportunity for market growth or new businesses around the area. For example, a residential area within the proximity to the stations gives new potential commercial area. Second, it provides travellers or commuters more options for the nearby attractions. The main goal of this project is to develop a travelling application that helps users to plan their travel routes and to find such nearby attractions for shopping, eateries and accommodation. This is part of a larger effort to ensure travel convenience and safety among the urban commuters [4].

The remainder of this paper is organized as follows. Section 2 presents reviews on similar applications to the proposed MonoWalk. Section 3 presents the methodology for developing the MonoWalk. Section 4 presents the prototype development and evaluation, and finally Section 6 concludes the research.

2. Related Work

There are a number or monorail information and travelling applications available on both website and mobile. This application can help users who are traveling to search different places of interest. This section reviews applications that are available today that help user to access travelling information easier and simpler only using a smart phone. This review is necessary to show areas where the MonoWalk Travelling Companion application would be of advantage.

2.1. Trivago.com

Trivago [5] is a site for traveling that focusing on hotels. This website is operating online hosted at `http://www.trivago.com/`, which serves as its official website and provides information about hotel and the comparison prices with each hotel that it suggest to user. The site compares prices for over 730,700 hotels from more than 200 booking sites, such as Expedia, Hotels.com and Priceline.com. Trivago provides some of the activity such as booking hotel online, comparison with each hotel that is suggested and shows the location for each Hotel.

2.2. Booking.com

Booking.com [6] is an online booking website started as a small start-up in Enschede in 1996, based in Amsterdam, Netherlands and since 2005 owned and operated by United States-based Priceline. Booking.com offers online accommodation booking. It claims to have over 700,000 properties globally under contract and that it deals with more than 900,000 room nights reservations per day. In 2013, it accounted for more than two thirds of Priceline's revenue. Booking.com is available in more than 41 languages. This application is provide suggestion for user to choose which want and where the Hotel that he or she want to go. This is also provide a free cancellation and make changes booking hotel. This website need user to open the website for booking the hotel.

2.3. Hipmunk.com

Hipmunk [7], stylized as hipmunk, is a consumer-oriented online travel company headquartered in San Francisco, California. It was co-founded by Adam Goldstein and Reddit co-founder Steve Huffman in 2010. The company focuses on the visual organization of flight search results. It received increased media attention when Google announced its flight search service in September 2011. This website provided suggestion for users to choose any activity, for example transportation, hotels and related packages. This application allow user to open the website for search the activity that user needed in this website.

Table 1 shows the comparison for existing application with the MonoWalk Companion Travelling application. The comparisons are made in terms of the system requirement such as the log in capability, new registration for user, calendar at system, good information for user, and place suggestions for user especially traveler user.

In the proposed application, registration and account log-in are not necessary as the application resides on mobile phone and all data will be stored at local directory once installed.

Table 1. Comparison of Existing Applications

System Requirement	Trivago	Booking	Hipmunk	MonoWalk
Log-in	Yes	Yes	Yes	No
New Registration	Yes	Yes	Yes	No
Calendar-based	Yes	Yes	Yes	Yes
Information	Yes	Yes	Yes	Yes
Suggest Place	Yes	Yes	Yes	Yes
Responses	Yes	Yes	Yes	Yes

3. Methodology

Methodology represents the means, techniques and procedures used in analyzing the requirements and designing a system to be developed. The methodology is developed to improve the management and control of the system. In a system development, precision and design in the development process are important to avoid any mistake. Figure 1 shows the steps in the prototyping methodology [8], which are the planning, analysis, design and implementation.

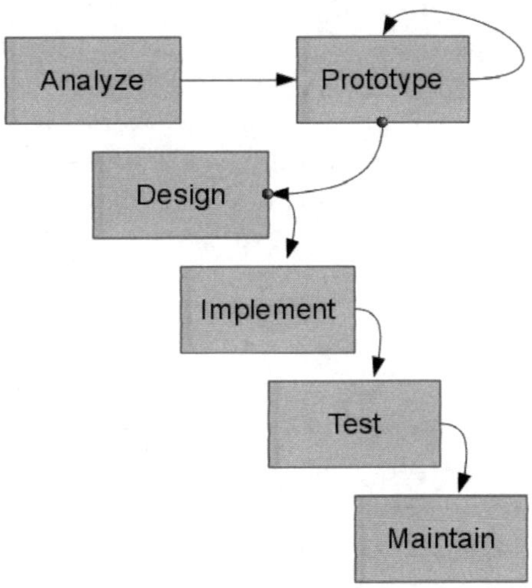

Figure 1. Prototyping Model

As shown in the figure, after implementation, if there is any problem on the system, prototyping methodology can revert back to implementation and design stage to perform corrections. Then, if a system has been completed with no problem, the process can proceed to implementation.

3.1. Planning phase

In the planning phase, the process is to identify the areas in the proposed system. The project activities must be designed to meet the master project timeline using a Gantt chart. In this phase, several key points were identified; background to design the course without using the application, application objectives to be developed to address the problems were identified, the functions and scope of potential applications users. In addition, the application also identifies the importance for depicting the inherent value in the system applications that will be developed from the perspective of MonoWalk application and user applications. Moreover, the application focuses on the target users.

3.2. Analysis Phase

Any information concerning the application of the system developed is collected and analyzed in more detail during the analysis phase. The application development strategy also study carefully so that it fits well with the objectives of development have application. All the hardware and software requirements are specified in more detail so that adequate and functioning properly. In addition, the information requirements for the application system was also identified that in accordance with user requirements. It also includes processes for identifying the programming language used, modules for user and system requirements in the implementation processes MonoWalk Travelling Companion application.

In addition, in this phase, comparison with existing application such Trivago.com, Booking.com, and Hipmunk.com was carried out as shown in Section II. In the process of application development, developers need hardware and software to generate the necessary input and output of the result of application development. The application is developed using Android Studio with SQLite database.

3.3. Design Phase

The design phase describes in detail the design of the UML diagram, database and interface application being developed. Four types of UML diagrams have been specified, which are the Use Case Diagram, Sequence Diagram and Activity Diagram. The modules are also developed according to what has been analyzed. In general, the overall design of the system MonoWalk Travelling Companion application is developed based on the results of a detailed analysis of system requirements and user application during the analysis phase. It involves processes such as designing the user interface, the basic structures of the application, the describe output, and the program identification codes to the system interface applications.

Use case diagrams are used to explain the functions available in the system that will be developed. In the use case diagram, it involves three important entity of actors, use cases, and relationships. Figure 2 shows the use case diagram for the MonoWalk Travelling Companion application. Actor involved is the user. User can make a choice based on the options that are given in the application. Actor represents the monorail passengers who receives information on travel distance and cost to a destination he or she wishes to choose. The following describes activities in the sequence diagrams based on the use case diagram in Figure 2:

- Upon entering, the user can see the first pages that shows the main pages that shows main current location, calculate and destination. Customers must be able to select current location, calculate or destination if she or he wants.
- Current location shows the main monorail line. Customers must be select Current location, before they can select their next location.
- After the customers have select the current location, they can choose the next location they want to go. There are eleven destination that user can select, which are KL Central, Tun Sambanthan, Maharajalela, Hang Tuah, Imbi, Bukit Bintang, Raja Chulan, Bukit Nanas, Medan Tuanku, Chow Kit, and finally Titiwangsa.
- After user chooses their next location, user will be directed to the list of activities that shows Attraction Places, Eateries and Accommodations. Customers must be select one of list activity for the next page.
- After the passenger has chosen the activity, they can choose a list of alternatives, such as the list of hotels, shopping malls and eatery places. After user choose the alternative, she or he can see the information details about the places that their want to go.

- In the calculate page, user can total all their transportation prices at the attraction place, eateries and accommodation. From that, users can generate their own budgeting based on the total price.
- Finally, in the destination page, user can insert the name of places that she or he wants to go at the attraction place, eateries and accommodation. Then the name of place was saved at the database. From that, admin can generate the manual report which one the place gets more attraction from tourist.

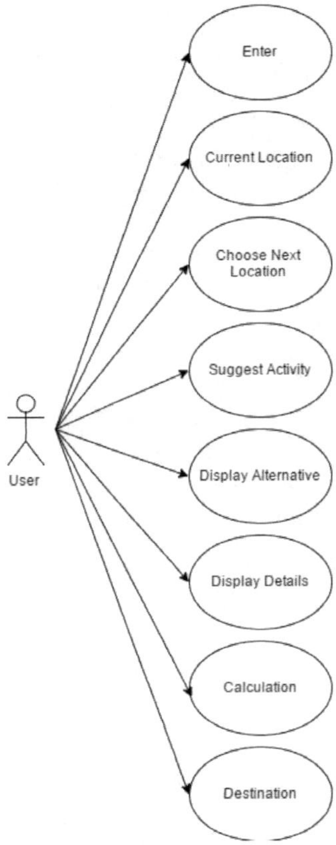

Figure 2. Use Case Diagram

Figure 3 shows the activity diagram for MonoWalk Travelling Companion application. The activity helps to build the application by determining the process flow. At the same time, it can help analyst to identify problems if the application is developed different from the activity diagram.

3.4. Prototyping and implementation phase

In this phase, all the programming codes are implemented based on the design in previous phase, resulting a a prototype called the MonoWalk Travelling Companion. However, at this stage, if there is a need or vulnerability, the analysis and design phases will be repeated until the MonoWalk Travelling Companion application system is working well and accepted by users. The database designed in the previous phase will be developed together with the information system interface that has been identified. The process is also repeated for the phases that otherwise would occur after prototype system MonoWalk Travelling Companion application is tested by developers and users.

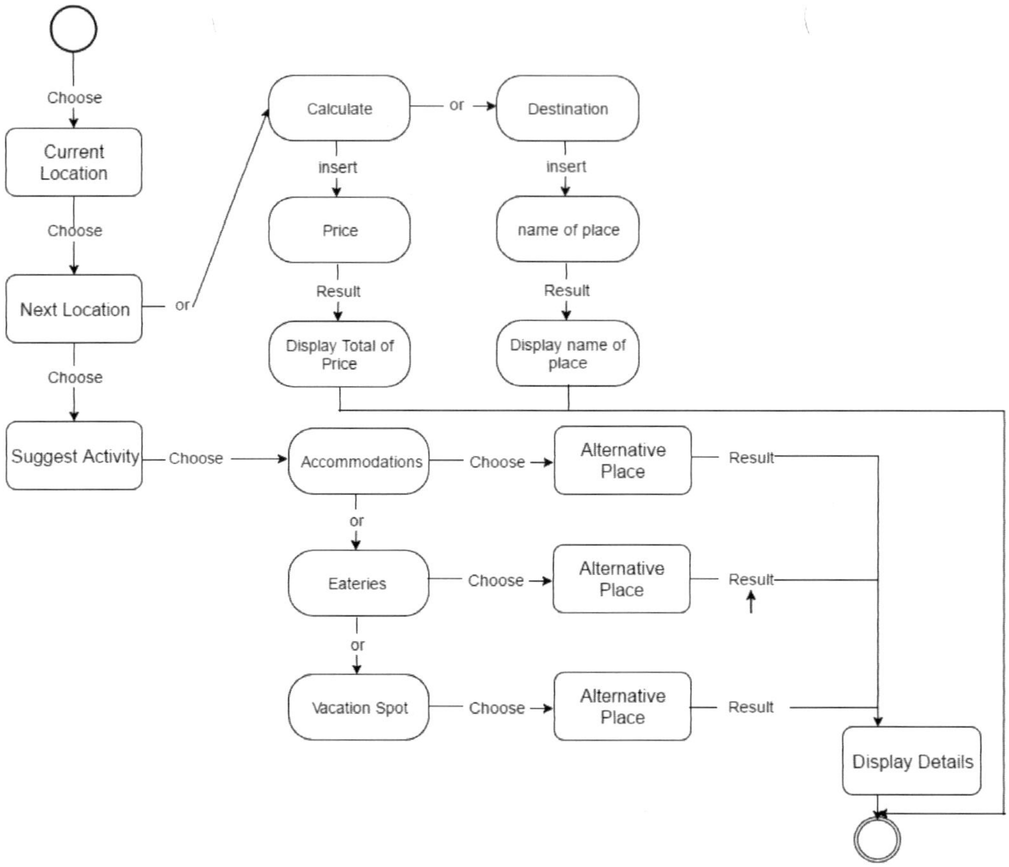

Figure 3. Activity Diagram

3.5. Testing and implementation phase

In this phase, the prototype of MonoWalk Travelling Companion application will be fully implemented and tested with real users. The modules in the application system MonoWalk Travelling Companion will be tested to run well in this phase. There is a user-acceptance testing to ensure they meet the needs from use. In addition, this testing is very important to ensure the system does not have error and the resulting application is in accordance to the initial requirement planning. The interface will also be tested to ensure all components are functioning.

4. Implementation

This section presents the details of the application interface as well as the activities during the testing phase. Figure 4 shows the page upon entering the application. The user are not required to login, because the application is open suitable to any user. In the second page, the user can choose their current location. The current location is designate for main platform that is KL Central. Furthermore, user can calculate their budgeting. For the destination user need to insert there places she or he want to go for generate the MonoWalk report. Next, Figure 5 shows the Monorail platform for the next location that user can go. The side figure shows the alternate interface that user can choose such as the attraction place, eateries and accommodation.

Figure 6 shows the display details if user chooses one particular alternative as the attraction place. This page is about the price of Monorail and the distance from the main platform to next platform. In this page, user also can choose their attraction place that she or he wants to go. The

IOP Conf. Series: Journal of Physics: Conf. Series **1019** (2018) 012083 doi:10.1088/1742-6596/1019/1/012083

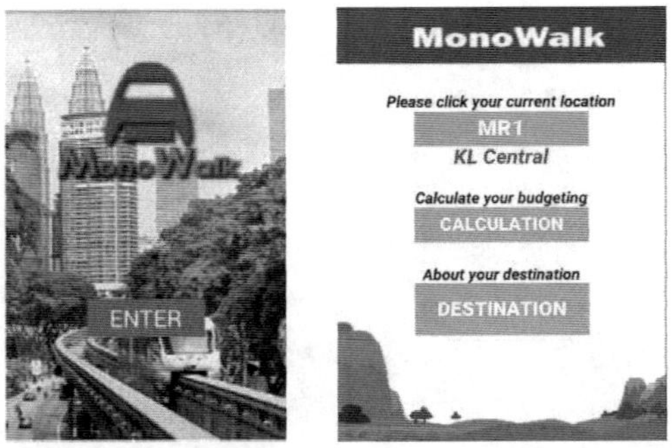

Figure 4. Main pages for MonoWalk

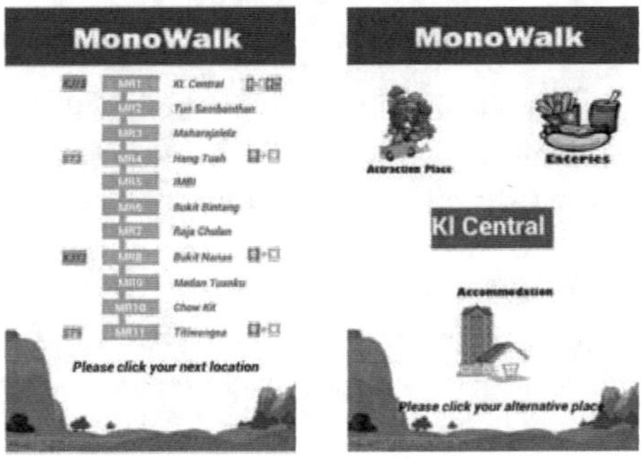

Figure 5. The Monorail Platform and Alternate View

side figure shows the page that displays price of transportation from the next location monorail platform that user chooses after clicking at the suggestion place that application provide. From the display, user can make a decision about their budgeting and what kind transportation that can she or he must use. The flow of process is also same with other alternatives such as the eateries and accommodation.

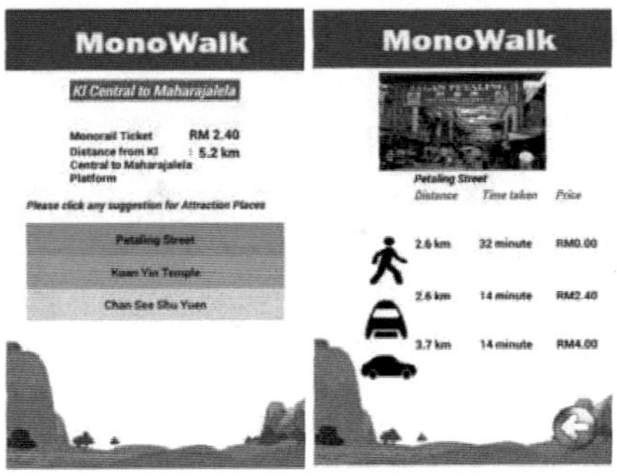

Figure 6. Display Details

Figure 7 shows the calculate page. From this page, user can insert their price of transportation that she or he want use to go the place. After insert the price, user can see the total price at the bottom application. From the total price, user can make their own budgeting. Next is the destination page, where the user inserts places she or he wants to go to generate the summary. After the user inserted the place, the data will be saved into the SQLite database.

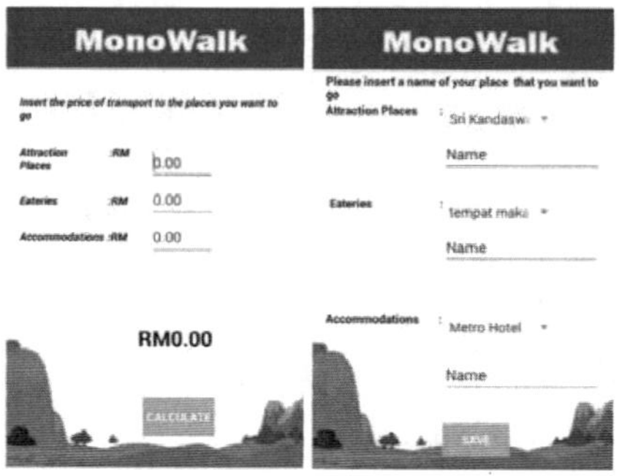

Figure 7. Calculation and Destination Page

5. Testing

Testing is an activity that will be conducted after the development process is complete. The main purpose of this process is to make sure the application that has been developed function according to its specifications. Testing is important to identify any error or unwanted functionality failures in any of the process involves in the application that has been develop. This is crucial for the application to handle all kinds of unexpected input by the use.

Table 2 shows the results after testing the button MR1 (KL Central). The testing is necessary to make sure the button was functionality and can proceed to the next process.

Table 2. Functionality Testing for Button MR1: Main Current Location

Functionality	Input	Expected Result	Real Result	Comment
Button MR1 (KL Central) to the next page		Display the list of Button Next Location	Display the list of Button Next Location	

Table 3 shows the result after testing a button Next Location to the next page. The testing to make sure the button was functionality and can proceed to the next process.

Table 3. Functionality Testing for Button Next Location

Functionality	Input	Expected Result	Real Result	Comment
Button Next Location to the next page		Display the suggestion activity	Display the suggestion activity	

Table 4 shows the real result after testing a button Attraction Places to the next page. The testing to make sure the button was functionality and can proceed to the next process.

Table 4. Functionality Testing for Button Attraction Place

Functionality	Input	Expected Result	Real Result	Comment
Button Attraction Page to the next page		Display list of alternative places	Display list of alternative places	

Table 5 shows the result after testing a button Eateries to the next page. The testing to make sure the button was functionality and can proceed to the next process.

Table 5. Functionality Testing for Button Eateries

Functionality	Input	Expected Result	Real Result	Comment
Button Eateries to the next page		Display list of alternative places	Display list of alternative places	

Table 6 shows the result after testing a button Accommodation to the next page. The testing to make sure the button was functionality and can proceed to the next process.

Table 6. Functionality Testing for Button Accommodation

Functionality	Input	Expected Result	Real Result	Comment
Button Accommodation to the next page		Display list of alternative places	Display list of alternative places	

Table 7 shows the result after testing a button Alternative to the next page. The testing to make sure the button was functionality and can proceed to the next process.

Table 7. Functionality Testing for Button Alternative Places

Functionality	Input	Expected Result	Real Result	Comment
Button Alternative Places to the next page		Display list of transportation price	Display list of transportation price	

Table 8 shows the result after testing for Calculation. The testing is to make sure the calculation of total price is correct.

Table 8. Functionality Testing for Calculation

Functionality	Input	Expected Result	Real Result	Comment
Calculating transportation price (Place + Eateries + Accommodation)	Attraction Places = 1.20, Eateries = 1.20, Accommodations = 1.15	The Total price is 3.45	The Total price is 3.45	

Finally, Table 9 shows the real result after testing a Destination. The testing to make sure the name for each place was save into database after user insert the data.

Table 9. Functionality Testing for Destination

Functionality	Input	Expected Result	Real Result	Comment
The name of place will be saved in database	Tun Sambanthan	The name place of Tun Sambanthan will be saved at database	The name place of Tun Sambanthan will be saved at database	

6. Conclusions

In conclusion, the MonoWalk Travelling Companion application is able to help users especially traveler from another countries to search places of interest along the Monorail route. This application also requires no Internet connection, hence beneficial for travelers from different countries aside from the urban commuters.

Acknowledgments

This project is sponsored by Universiti Tun Hussein Onn Malaysia and partially supported by Research Gates IT Solution Sdn. Bhd.

References

[1] Nordin NH, Masirin, MIM, Ghazali, MI, Azis MI 2016 Passenger Rail Service Comfortability in Kuala Lumpur Urban Transit System. *MATEC Web of Conferences.* Vol. 47. EDP Sciences

[2] Goh MWH 2013 *Mobile-based route planner for Klang Valley rail transportation.* Doctoral dissertation, Universiti Tun Abdul Razak, Malaysia

[3] Gunawan B 2016 *A case study on the business models of convenience store in the surrounding areas of MRT stations in Taipei.* Master's Thesis, National Taiwan University of Science and Technology

[4] Sham R, Hussein, MZSM, Ismail HN, Latif AA, Mohamed S, Othman R, Yatim NHM 2013 An improved travel safety for urban commuters using an iTracks system. In *2013 IEEE 3rd International Conference on System Engineering and Technology (ICSET)*, 365–368

[5] Trivago.: Trivago. Retrived on October 26, 2017, from `http://www.trivago.com.my/` (2011)

[6] Booking.: Booking. Retrived on October 26, 2017, from `http://www.booking.com/` (2005)

[7] Hipmunk.: Hipmunk. Retrived on October 26 , 2017 from `https://www.hipmunk.com/about` (2010)

[8] Dennis A, Wixom BH, Roth RM Systems Analysis and Design, 3rd Edition. John Wiley & Sons, Inc.

IOP Conf. Series: Journal of Physics: Conf. Series **1019** (2018) 012084 doi:10.1088/1742-6596/1019/1/012084

NFC-based Data Retrieval Device

Mohd Helmy Abd Wahab[1], Nur Farezan Mohamed Suhaimi[1], Mohamad Farhan Mohamad Mohsin[2], Aida Mustapha[3], Noor Azah Samsudin[3], Radzi Ambar[1]

[1] Faculty of Electrical Engineering, Universiti Tun Hussein Onn Malaysia, 86400 Parit Raja, Batu Pahat, Johor, Malaysia

[2] College of Arts and Sciences, Universiti Utara Malaysia, 06010 UUM Sintok, Kedah, Malaysia

[3] Faculty of Computer Science and Information Technology, Universiti Tun Hussein Onn Malaysia, 86400 Parit Raja, Batu Pahat, Johor, Malaysia

E-mail: helmy@uthm.edu.my

Abstract. This paper describes the design and development of data retrieval system using near field communication (NFC) protocol to read and transfer data from the device storage panel located at the recycle bin. In existing systems, data are manually collected from the storage device using the SD card and sent for upload to regional workstation of the data center, which is located at the central server. The device automatically establishes the NFC connection with the recycle bin panel and once the connection is established, data will automatically be transferred to the device and the current data storage in the recycle bin will be erased. Next, the collected data will be uploaded to the server through regional workstation.

1. Introduction

Wireless data transfer is the transmission of data between two or more points that are not connected by any network cable. The wireless system is also considered as the system for "world without borders" given the advanced technology in Information Technology (IT). Examples of wireless communication systems include radio frequencies, bluetooth device, infrared, and Wireless fidelity (Wi-Fi). Wireless technology is a critical factor in establishing the real potential of information-supported educational activities that occur anytime and almost anywhere [1].

In line with the current campaign by the Malaysian government to support green environment, wireless technology has also been applied in managing recyclable wastes through a reward and redemption system. Many efforts have been done pertaining waste management system, among which is the smart recycle bin [2]. The implementation of the recycle bin is based on reward points for any recyclable wastes thrown in the particular bin. At this stage, the prototype uses Radio-Frequency Identification (RFID) in managing the reward point system to encourage the community to support the campaign of green technology. The point is calculated based on the weight measured by the device panel and the information is stored into the panel itself as well as the RFID card for further usage. Information stored in the panels then need to be collected by the authorized agency to enable end user to check the accumulated points.

Nonetheless, at present the current prototype does not have a mechanism or device to read the data from the device panel for collection by the authorized waste collector. It is impractical to require a waste collector to operate the device manually as it involves technical and requires knowledge in IT and many waste authority collectors are not technology literate people. Therefore, there is a need for a wireless data transfer device for data collection prior uploading to the web-based redemption system at the regional workstation. The aim of this project is to utilize the Near Field Communication (NFC) to read the data from the smart recycle bin by approaching the device to the recycle bin. In it hoped that data collection can be performed in parallel with waste collection time.

According to work presented by [3], NFC is a short range radio frequency technology that exchanges data using a wireless system such as phone and sensor. This method could speed up the process of data collection and reduce missing data if it is being performed manually. It also reduces the hardware failure due to its minimum data interaction requirement. Figure 1 illustrates the Near Field Communication (NFC) system.

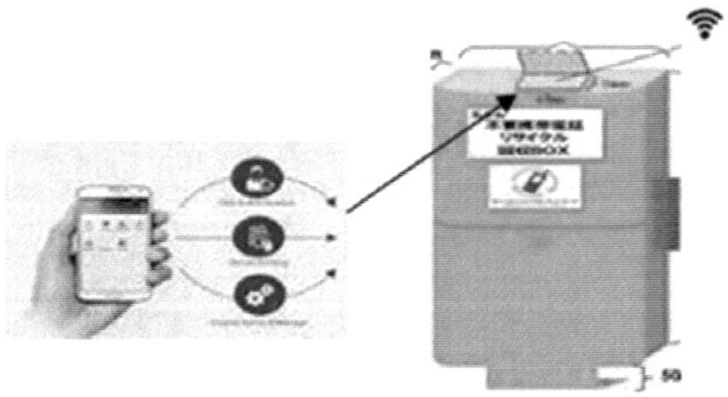

Figure 1. Near Field Communication

As illustrated in Figure 1, the NFC component will be attached to the existing device (e.g: smart recycle bin). At the moment, the smart recycle bin store information regarding user transactions and point information into the recycle bin system via the SD card as a storage. At this stage, data is being transferred manually operated by authorized people from the smart recycle bin to the data centre at the regional workstation. This often leads to missing data, failure of hardware, and high training cost to train the waste collector to handle the device. With NFC, data transfer will be easier to handle with very minimal human intervention. Since the device will be embedded with the program to automatically read the data and to establish the connection with the smart recycle, the data will automatically be transferred to the device once the connection is set and the current data storage in the recycle bin panel will automatically be erased. The collected information then will be uploaded to the web-based redemption system, which is located at the regional workstation.

The remainder of this paper is organized as follows. Section 2 presents reviews on related works on NFC-based data transfer. Section 3 presents the hardware design and embedded programming required for developing the NFC-based data transfer. Section 4 presents the evaluation results, and finally Section 5 concludes the research.

2. Related Work

Near Near Field Communication (NFC) is a short, high frequency wireless communication technology which has emerged from the convergence of contactless identification just like RFID and network technologies such as Bluetooth and Wi-Fi. NFC devices can exchange information on respective capabilities, swap the record, and initiate a long-term communication [4]. NFC is able to act as both a reader and a tag. By using a standard smartphone, sharing of information is possible. The data will be transferred into the smartphone by tapping the two devices together.

However, not every smartphone has an NFC technology. The use of popular contactless payment, for example credits cards and debit cards can also support NFC [5]. The work presented by [6] has developed a new NFC and smart phone-based toll collection system. This system proposed a new billing system and payment method, which were based on NFC technology with the support of OBU multi-technology and the smartphone. With this system, the bill could just appear directly on the smart phone. NFC technologies are also used in the medical field for promoting efficient healthcare management. NFC has been used with the assistance of Android-based mobile phone to check whether a patient is given a correct prescription from the hospital [7]. Each medicine also has an NFC tag that acts as a monitoring tool by the doctor using his or her mobile phone.

NFC and RFID are two closely related wireless communication technologies that are used globally for a vast number of applications such as access control, asset tracking and contactless payments. Table 1 shows the differences between RFID and NFC. NFC is now available in the majority of mobile phones and this is the most important difference between NFC and RFID.

Table 1. Comparison between NFC and RFID

	NFC	RFID
Operating Frequency	13.56Mhz	13.56Mhz
Communication	Two-way	One-way
Standards	ISO 14443	ISO 14443, 15693, 18000
Scan Distance	Up to 10cm	Up to 1m
Scan tag simultaneously	No	Yes

At present, there are many technologies that have been used in wireless data transfer. The use of technology in wireless data transfer is tabulated in Table 2.

Table 2. Previous Research Technology on Data Transfer

System/Technology	Description	Year
Android smartphone (Bluetooth, NFC) [9]	A system that communicated via bluetooth from a smart phone for data collection and sent it simultaneously to the central server for data storage.	2012
Arduino Microcontroller, Wifi Transceiver [10]	Sending data by using UDP (User datagram protocol) Arduino: using SPI to URT chip on Wifily Board	2013
Arduino controller, xbee module [12]	The monitoring system for an AGV that needed to read the data to understand the operational status and travelling routes of the vehicles as well as providing the fault alarm and current status of the materials.	2012

As shown in Table 2, RFID enables a one-way wireless communication, typically between an unpowered RFID tag and a powered RFID reader. RFID tags can be scanned at distances of up to 100 meters without a direct line of sight to the reader. RFID is used globally for asset tracking in warehousing, airport baggage handling, livestock identification and much more. RFID operates at a range of radio frequencies each with their own set standards and protocols. Meanwhile, NFC operates at 13.56 MHz and is an extension of High Frequency (HF) RFID standards. NFC therefore shares many physical properties with RFID such as one way

IOP Conf. Series: Journal of Physics: Conf. Series **1019** (2018) 012084 doi:10.1088/1742-6596/1019/1/012084

communication and the ability to communicate without a direct line of sight. NFC is capable of two way communication and can therefore be used for more complex interactions such as card emulation and peer-to-peer (P2P) sharing. However, NFC is limited to communication at close proximity, typically 5cm or less. Only a single NFC tag can be scanned at one time [8].

3. Materials and Methods

This paper proposed for the use of NFC to capture information stored on a tag attached to the Smart Recycle Bin [2]. The device automatically establishes the NFC connection with the recycle bin panel and once the connection is established, data will automatically be transferred to the device and the current data storage in the recycle bin will be erased. Next, the collected data will be uploaded to the server through regional workstation. In order to achieve the objective, two core important components are the hardware design as well as the embedded programs.

3.1. Hardware Design

Figure 2 depicts the block diagram of data transfer from the Smart Recycle Bin to the regional workstation using NFC technology. Data retrieval device using NFC is graphically represented in Figure 3, where the device is an integrated of Breakout board PN532 and Arduino Uno. Meanwhile, the overall system flow is shown in Figure 4.

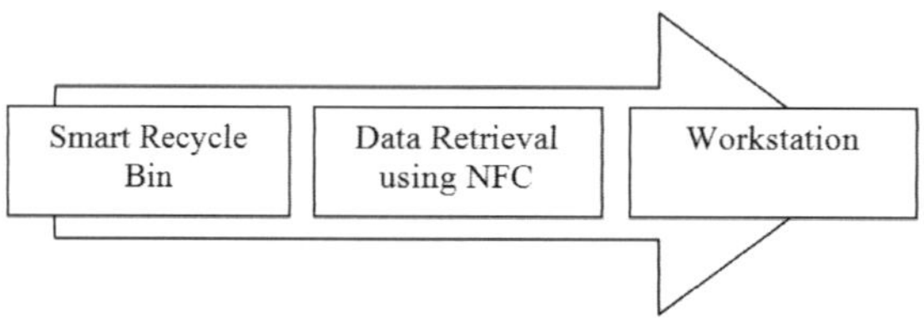

Figure 2. Block Diagram of Data Transfer

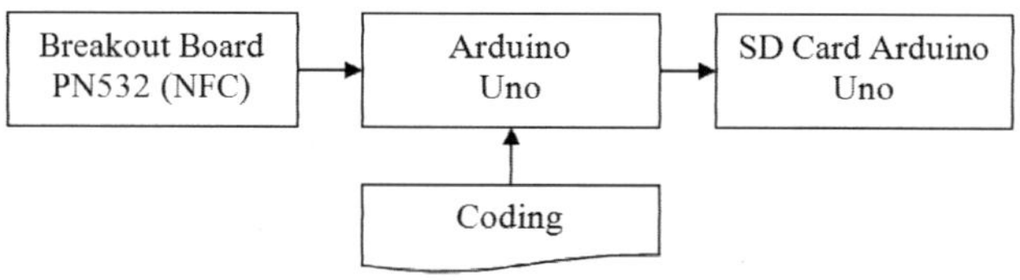

Figure 3. Block Diagram of Data Retrieval System

Figure 5 (left) shows the Arduino Uno, which is a type of kit used as the device. It is a platform with an open-source electronic prototyping that is flexible and easy to use on software and hardware components [12]. The language supported by this microcontroller is the Arduino programming language, which is a high level C language that is easier to understand and write as compared to other assembly programming languages.

The PN532 as shown in Figure 5 (right) was the platform to perform the wireless data transfer integrated with Arduino Uno as the system controller of this project. PN532 was a highly integrated transmission module for contactless communication at 13.56MHz that included micro controller functionality based on an 80C51 core. The transmission module utilised an outstanding modulation and demodulation concepts that were completely integrated for different kinds of passive contactless communication methods and protocols at 13.56MHZ.

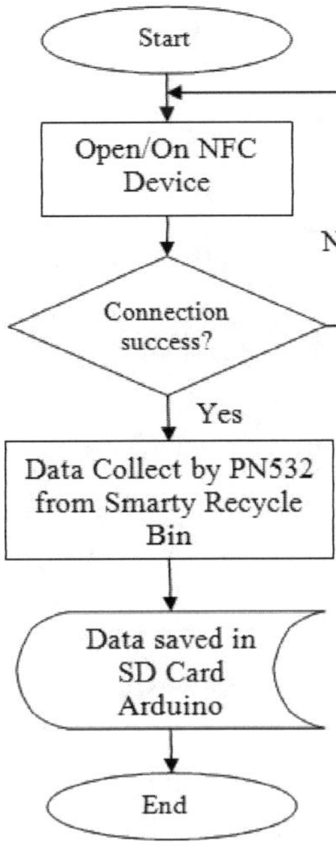

Figure 4. Overall System Flow

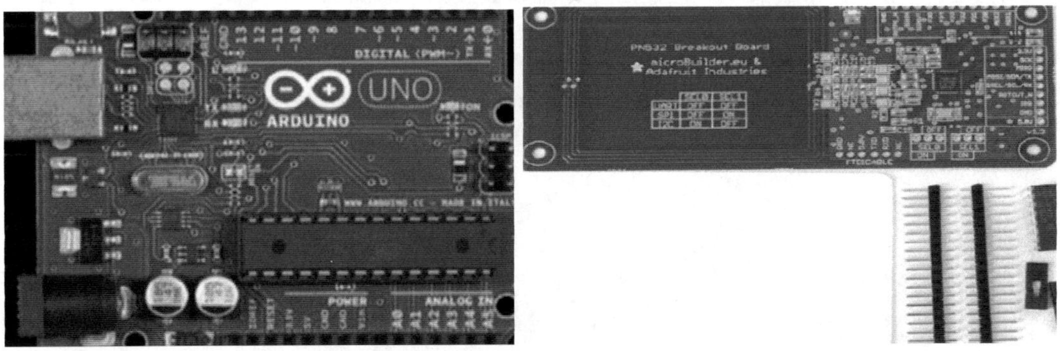

Figure 5. Arduino Uno Controller (left) and Breakout Board PN532 (right)

The PN532 supported four different operating modes, which were Reader/Writer mode supporting ISO 14443A, ISO 14443B in Reader/Writer mode only, Card interface mode supporting ISO 14443A and NFCIP-1 mode. The white paper on PN532 is available at `http://www.nxp.com/documents/short_data_sheet/PN532_C1_SDS.pdf`.

3.2. Embedded Programs
The device was programmed using the C programming language for embedded system. Several libraries needed to be correctly configured such as the following.

- Library configuration: The PN532 and Arduino needed to be set up using Arduino IDE as shown in Figure 6. The AdafruitPN532 library had the ability to read MiFare cards, including the hard-coded ID numbers, as well as to authenticate and read/write EEPROM chunks. It could work with both the breakout and shield using either an SPI or I2C connection [13].

- Arduino IDE: Arduino IDE (Figure 6) is an easy-to-use development kit for beginners, yet flexible enough for advanced users. It runs on multi-platform such as Mac, Windows, and Linux. In this project, Arduino IDE was used to code the data collected with Breakout board PN532.

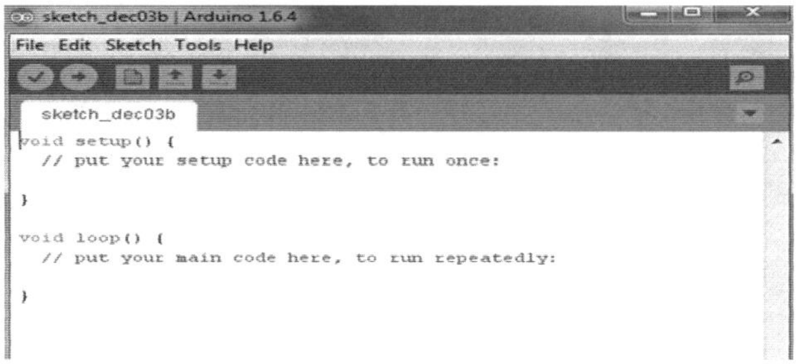

Figure 6. Arduino IDE

4. Results

The testing results on the use of NFC-based data retrieval device for the Smart Recycle Bin is shown in Figure 7. Two types of analysis were carried out to evaluate the performance of the prototype from the perspective of the hardware design.

4.1. Device sensitivity analysis

In this analysis, sensitivity analysis was done in three different angles as shown in Figure 7. This was to indicate that whether the device could be detected on each angle or vice versa.

Figure 7. Prototype of the Data Retieval Device

Table 3 indicates that the device sensitivity detection had been done based on the angle as illustrated in Figure 8. From the figure, note that angles D and E were more sensitive than angles A and C, and angle B was indicated as not sensitive at all.

4.2. Sensitivity based on Distance

The results in Table 3 shows the angle of sensitivity, its strength of sensitivity can also be represented by distance. Table 4 indicates sensitivity based on distance.

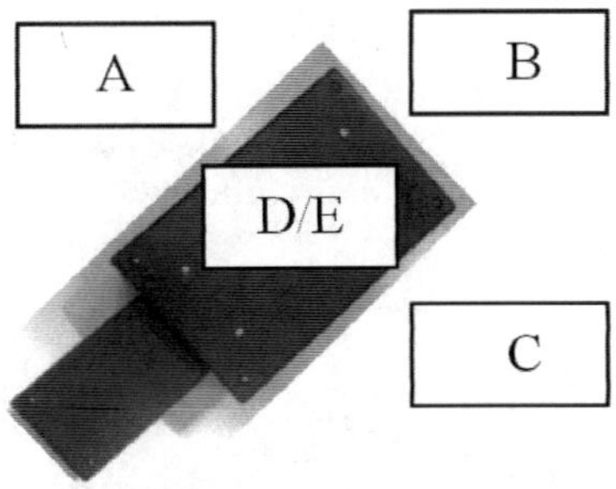

Figure 8. Data Retrieval Device (prototype) with Angle Labeled

Table 3. Sensitivity Analysis

Angle	Sensitivity
A	Less Sensitive
B	Not Sensitive
C	Less Sensitive
D/E	More Sensitive

Table 4. Distance Sensitivity

Distance (cm)	The sensitivity	Time Taken	Angle
2	Sensitive	2.08	A
4	Sensitive	2.28	A
6	Less Sensitive	2.40	A
8	Not sensitive	-	A
2	Not sensitive	-	B
4	Not Sensitive	-	B
6	Not Sensitive	-	B
8	Not sensitive	-	B
2	Sensitive	2.05	C
4	Sensitive	2.27	C
6	Less Sensitive	2.42	C
8	Not sensitive	-	C
2	Very sensitive	2.08	D
4	Sensitive	2.28	D
6	Less Sensitive	2.40	D
8	Not sensitive	-	D
2	Very sensitive	2.05	E
4	Sensitive	2.29	E
6	Less Sensitive	2.45	E
8	Not sensitive	-	E

5. Conclusions

This paper described the development of a data retrieval system using the NFC protocol, which consisted of two major parts; hardware design and embedded programming. The hardware part consisted of the use of PN532 and the software used in this project was Arduino IDE. The device was developed to overcome the problems of the current existing system which is manually done to handle the data for the smart recycle bin. By applying the device, the smart recycle bin would require minimum human operation which could reduce the possibility of recycle bin panel malfunction.

Acknowledgments
This project is sponsored by Universiti Tun Hussein Onn Malaysia and partially supported by Research Gates IT Solution Sdn. Bhd.

References

[1] Mathews, JB 2015 Why are Wireless services Important to State and Education Leaders? Atlanta: Southern Regional Education Board
[2] Wahab MHA., Tomari MR, Kadir AA, Jabbar MH 2014 Smart Recycle Bin a Conceptual Approach of Smart Waste Management with Integrated Web Based System. In *Proceeding of International Conference on it Convergence and Security 2014*. October, 28-30 2014, Beijing, China
[3] Dudwadkar A, Gore A, Nachnani T, Sabhnani H 2013 Near Field Communication in Mobile Phone. *International Journal of Engineering and Advanced Technology*, **13** 1 309–313
[4] Ahson SA, Ilyas M 2012 *Near Field Communication Handbook*. USA: CRC Press
[5] Gopichand G, Chaitanya TK, Kumar RR 2013 Near Field Communication and Its Application in Various Fields. *International Journal of Engineering Trends and Technology*, **4** 4 1305–1309
[6] Dias J, Matos, NJ, Oliveira, ASR 2014 The charge Collector System (A new NFC and smartphone-based Toll Collection Sytem). *Procedia Technology*, Volume **17**
[7] Sethia D, Gupta D, Mittal T, Arora U, Saran H 2014 NFC Based Secure Mobile Healthcare System. In *Proceedings of the Sixth International Conference on Communication Systems and Networks (COMSNETS)*. 6-10 Jan 2014, Bangolore, India. pp.1–6
[8] Wan R 2006 Introduction to RFID Technology. *IEEE Pervasive Computing*, Jan-March 2006, pp. 25–33
[9] Yi WJ, Jia W, Saniie J 2012 Mobile Sensor Data Collector using Android Smartphone. In *Proceedings of 2012 IEEE 55th International Midwest Symposium on Circuits and Systems (MWSCAS)*, 5-8 Aug. 2012, Idaho, USA. pp.956–959
[10] Torresen J, Hafting Y, Nymoen K 2013 A New Wi-Fi based Platform for Wireless Sensor Data Collection. In *Proceedings of International Conference on New Interfaces for Musical Expression 2013*, 27 -30 May 2013, Seoul, Korea, pp.337–340
[11] Ghrairi Z, Hribernik KA, Hans C 2012 Intelligent Wireless Communication devices for Efficient data Transfer and Machine control. In *International Conference on Communication, Computing and Control Application (CCCA)*, 6-8 Dec 2012. University of Bremen, Germany
[12] Murphy TE 2012 *Introduction to Arduino Microcontrolle*. Hands-on Research in Complex Systems. 1729 June 2012. Jiaotong University, Shanghai, China
[13] Setiawan BP 2016 *Pengenalan ARDUINO*. URL: `http://mkpraktis.blogspot.my/2013/09/pengenalan-arduino.html`

IOP Conf. Series: Journal of Physics: Conf. Series **1019** (2018) 012085 doi:10.1088/1742-6596/1019/1/012085

Performance Analysis Of Needleman-Wunsch Algorithm (Global) And Smith-Waterman Algorithm (Local) In Reducing Search Space And Time For Dna Sequence Alignment

FN Muhamad[1,a], RB Ahmad[2,b], SM Asi[1,c] , MN Murad[3,d]

[1]School of Computer & Communication Engineering, Universiti Malaysia Perlis, Perlis.

[2]Faculty of Informatics and Computing, Universiti Sultan Zainal Abidin, Terengganu

[3]School of Manufacturing Engineering, Universiti Malaysia Perlis, Perlis.

E-Mail: [a]fatimah_noni@yahoo.com, [b]badli@ unisza.edu.my, [c]salina@unimap.edu.my, [d]mnasir@unimap.edu.my

Abstract. Generally, sequence alignment is the process of comparing two sequences to identify similarities and differences between them, then, a typical approach to solve this problem is to find a good and plausible alignment between the two sequences. The data representation in this work is DNA sequence. This work intends to analyze large sequences as well as reducing the search space and time complexity without compromising the accuracy and efficiency. This is by evaluating the performance of Needleman-Wunsch algorithm (global) and Smith-Waterman algorithm (local) based on the Dynamic Programming algorithm. Dynamic Programming algorithm is guaranteed to find optimal alignment by exploring all possible alignments and choosing the best through the scoring and traceback techniques, which is NP-hard to optimize. Implementation of parallel technique using OpenMP on Needleman-Wunsch algorithm and Smith-Waterman algorithm to identify the strengths and weaknesses for both algorithms. By using C Programming, Needle and Smith programs are developed based on the algorithms (respectively). The analysis concluded that the scoring and traceback techniques used in Needle and Smith are able to align an optimal alignment and improved the performance in searching similarity as well as reduced gaps and mismatch. OpenMP directives able to parallelize the codes and execute it faster, with four cores it can get an execution time of around 60% reduced.

Keywords. sequence comparison, sequence alignment, global alignment, local alignment, similarity, optimal alignment, scoring, traceback, match, mismatch, gaps, parallel.

1. Introduction

The data representation in this project is DNA sequence, Deoxyribonucleic Acid is a sequence of string code in which almost all genetic information is encoded, a means of storing biological data. Specifically, the information is encoded using four key chemicals, adenine, thymine, guanine and cytosine (abbreviated as A, T, G and C) [1]. An example of a genome sequence is shown in Figure 1. The DNA sequences of hundreds of organisms have been decoded and stored in databases. This biological sequence data can be obtained from variety of public and private databases.

```
CCTTCATCTAGGAGTTGAGAAGGGTAGATAAGATTCTTGGATACTAGGTATTTAAGAACTTT
CTCAGATGAAAGGAAGCTGGGAACAAAGTAAGAAAGAATACCTTTTAGGATTCACAAAATT
ATGAGAAGTCAGCCACATACGGTAGGTCAGCTTTTTAATGTATTTGCTCCTTTTCTTATTCAT
```

Figure 1. Sub-sequences of DNA for Western Gorilla in FASTA format.

With the growing amount of data, it became impractical to analyze DNA sequences manually, so faster algorithms and tools are needed. Basically, the fundamental procedure of analyzing sequence content is sequence comparison. Sequence comparison is regarded as one of the most fundamental problems of computational biology, which is usually solved with a technique known as sequence alignment. Sequence alignment can be defined as the problem of finding which parts of the sequences are similar and which parts are different. So, a measure of how similar they are is also desirable, then, a typical approach to solve this problem is to find a good and plausible alignment between the two sequences. Finding the optimal alignment is a way to determine the similarity.

Basically, a good alignment has minimum gaps and mismatches inserted into the sequences, but at the same time maximize the number of positions in the aligned strings that match. This problem can then be solved by applying the Dynamic Programming Algorithm that is one of the major strategies for designing algorithms in sequence alignment. The problem of finding optimal alignment can be solved efficiently by determine the best scoring alignment and perform the traceback technique will result a good optimal alignment. To compare two sequences, a scoring system is needed in calculating scores for match, mismatch and gaps. The optimization problem in pairwise sequence alignment is finding the highest scoring (global or local) alignment for a pair of sequences which is optimization problems whose decision versions are NP-complete are called NP-hard. Many of the optimization problems that arise in sequence alignment are NP-hard [2].

Therefore, because of the complexity of the algorithm, there is a need for a methodology that could reduce the computation time while delivering accurate results. These algorithms require quadratic time for each comparison of two sequences [3]. There are many algorithms written that use the approach of Dynamic Programming. However, the global alignment was first introduced in Needleman-Wunsch algorithm based on Dynamic Programming for biological sequences comparison [4]. Later, the improvement of Needleman-Wunsch algorithm proposed Smith-Waterman algorithm to find the best local alignment [5]. The Needleman-Wunsch algorithm and Smith-Waterman algorithm are developed to solve the sequence alignment problems that have computational complexities.

Since the nature of Dynamic Programming requires to store and compute an *mxn* matrix, many researchers spending most of the effort to reduce calculation time. Parallel computing has developed rapidly for scalable data in recent years and has appeared in many Bioinformatics applications [6]. This works also focused on the parallel process using OpenMP. The effort is still being studied by researchers in finding the best way to align sequences by maximize the matches while reducing the gaps, mismatches, length sizes of the optimal alignment and execution time for large data.

2. Literature Review

There are many techniques proposed to solve sequence alignment problems such as Dynamic Programming, Heuristic, Linear Programming, Hidden Markov models and T-Coffee. Optimal alignment is affected by scoring parameters (gaps, mismatch and match), as given a specific scoring system, so, Dynamic programming methods ensure the optimal global alignment by exploring all possible alignments and choose the best [4]. An alternative approach developed by Nordin et al. [7] in five phases include query initialization, patterns generating, pattern's scanning, ranking, and optimal local alignment. The model considers sequences with highest exact matching scores are the most similar to the query sequence. However, many stages of the proposed algorithm require a huge amount of spaces to store results in every stage, which is a source of wasting time.

Zhang, et al. [8], Luis de la Torre and Jaime Seguel, [9] in a characterized work compare most of Dynamic Programming and Heuristic Method sequences alignment algorithms including Needleman-Wunsch, Smith-Waterman, FASTA, and BLAST. The comparisons are aimed to highlight computations and space complexity in term of performance parameters for optimal alignment. These parameters include speed, running times, and affine gap penalties. As a result, there is a tradeoff between speed and sensitivity, BLAST and FASTA consider subsequences k-tuples and words to achieve higher speeds, while Needleman-Wunsch, Smith-Waterman produce optimal and accurate results, but scarifying with the speed in alignment. Parallelization using OpenMP are also used to reduce the computation time. B. Deepa and V. Nagaveni.[10] proposed Needleman-Wunsch algorithm and Smith-Waterman algorithm for identifying the enhanced local sequencing alignments to reduce the running time and increases accuracy of the sequence matching within two sequences. There are also many alignment tools can be obtained from the internet in solving sequence alignment problems for searching similarity such as BLAST, FASTA, CLUSTAL W, LALIGN, GeneWise and EMBOSS.

3. Methodology

The aim of this work is to apply the global and local alignment besides implementing the parallel computing to speed up the process of comparing sequences without compromise with incomplete results, like missing some optimal results. All problems are analysed and studied its characteristic as well as to choose the significant parameters that affect the result. The research methodology as depicted in Figure 2 is designed to produce an effective optimal alignment by reducing the gaps, mismatch and length of sequences optimal alignment without sacrificing the result of accuracy, efficient in time and space complexity, and practical for sequence alignments of large genomic regions. This work designs two proposed algorithms highlighted as, Needle (Needleman-Wunsch algorithm) and Smith (Smith-Waterman algorithm) programs. The programs aligning sequence 1, seqA with *lenA* length (subject sequence) against sequence 2, seqB with *lenB* length (query sequence). To implement the algorithms, a size of matrix (*lenA x lenB*) is calculated.

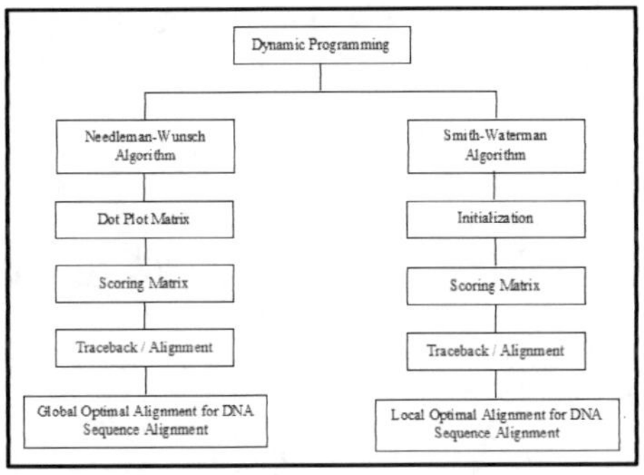

Figure 2. Techniques of Dynamic Programming for Needle program (Parallelization of Needleman-Wunsch algorithm) and Smith program (Parallelization of Smith-Waterman algorithm).

3.1. *Needleman_Wunsch Algorithm for Needle Program*

All the matrices in parallel version of Needle program for Needleman-Wunsch algorithm are places in global memory space so that all available processors can access them at the same time to perform initialization and other calculations.

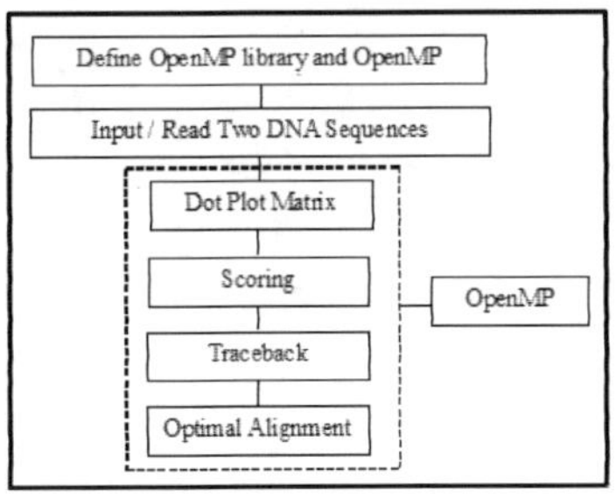

Figure 3. Parallelization of Needle program that run based on the Needleman-Wunsch algorithm.

As illustration in Figure 3, shown the implementation of the parallel programming into the Needle program based on the Needleman-Wunsch algorithm. Needleman-Wunsch can be made to run in linear space, while finding max score is easy but finding path is not.

3.1.2 Dot Plot Matrix of Needle Program. Needle program initialize the *NeedleArray[i][j]* matrix with a Dot Plot matrix. Completed dot plot matrix is using one point (1) for match and zero points (0) for mismatch. Dot plot matrix used parallel pragma in the dot plot process marks the object of *tablefill* and *seq* with parallel shared.

3.1.3 Scoring Matrix of Needle Program. For this work, in cell *NeedleArray[0][0]* is not set to zero, but it is continued with the scoring calculation and all other entries are computed with the scoring formula :

$$NeedleArray[i][j] = max \{ \begin{array}{l} NeedleArray[i-1][j-1] + sub(seqA[i], seqB[j]), \\ NeedleArray[i-1][j] + del(seqA[i]), \\ NeedleArray[i][j-1] + ins(seqB[j]) \end{array} \}$$

This work it is start in the bottom row moving up, where $i >= 0$ and $j = lenB$ until the process of identifying the cell position is ended. This technique is to simplify in calculating the score by referred to the dot plot matrix. A general rule to calculate the score *NeedleArray[i][j]* is illustrated in Figure 4.

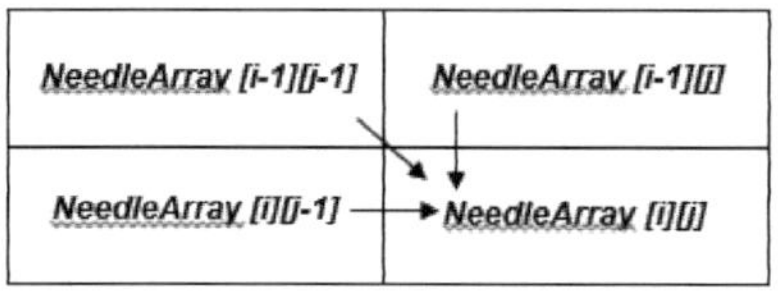

Figure 4. A general rule to calculate the score for Needleman-Wunsch algorithm.

While, parallel processes for scoring are inserting pragma at the outer loop. The scores for *NeedleArray[i][j]* matrix of the alignment of the two sequences by penalty with 0 for a mismatch and a gap and score a match with +1.

3.1.4 Traceback of Needle Program. This work perform a "traceback" by starting at upper-left cell of *NeedleArray[i][j]*, where $i = lenA$ and $j = lenB$, at the highest score. The parallelization of the traceback process is begin with loop that read the scores in the *NeedleArray* matrix until the bottom or far right of the matrix is reached for the size of sequences *lenA* and *lenB*. The traceback process has to identify gap in the alignments, either in Sequence A (*seqA*) or Sequence B (*seqB*).

3.1.5 Optimal Alignment of Needle Program. Needle program produces an optimal sequence alignment based on the Needleman-Wunsch algorithm by exploring all possible alignments and finding the best fit between the two DNA sequences.

3.2. Smith-Waterman Algorithm for Smith Program

There is no constrained to aligning the entire sequences in Smith program. The fact that two zero-length strings is a local alignment with score of 0.

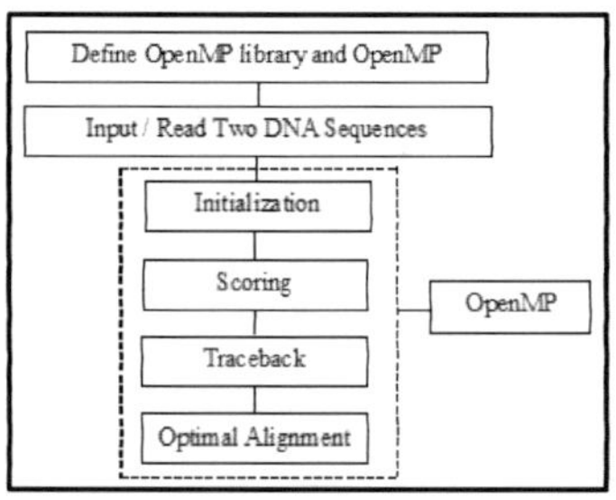

Figure 5. Parallelization of Smith program that run based on the Smith-Waterman algorithm.

As illustration in Figure 5, the implementation of parallel programming to Smith program is based on Smith-Waterman algorithm that uses Dynamic Programming to find the best local alignment between any two given DNA sequences.

3.1.1. Initialization of Smith Program. The first row and column of *SmithArray[i][j]* matrix must be filled with 0. The purpose of the matrix must be filled with 0 is to facilitate the traceback process, where it start with the cell that has the highest score and work back until reached a cell with a score of 0. Besides, the edges of the matrix are initialized to 0 instead of increasing gap penalties. The *SmithArray[i][j]* matrix is first initialized with :

$$SmithArray[i][0] = 0 \text{ and } SmithArray[0][j] = 0, \text{ for all } i \text{ and } j$$

There are parallel double for loop in initialization, so it parallelizes with the outer loop parallel by inserting the private directive into the pragma. Here, can be seen that the execution time did indeed get reduced by half in plotting dot plot in the *SmithArray[i][j]* matrix.

3.1.2. Scoring Matrix of Smith Program. To fill the scores in SmithArray[i][j] matrix, it needs to define the scoring scheme for Match = 1, Mismatch = -1, GapPenalty = 1 and GapExt = 1. The formula for scoring is :

$$SmithArray[i][j] = \max \begin{cases} 0, \\ SmithArray[i-1][j-1] + sub(seqA[i], seqB[j]), \\ SmithArray[i-1][j] + del(seqA[i]), \\ SmithArray[i][j-1] + ins(seqB[j]) \end{cases}$$

A general rule to calculate the score in *SmithArray[i][j]* is illustrated in Figure 6.

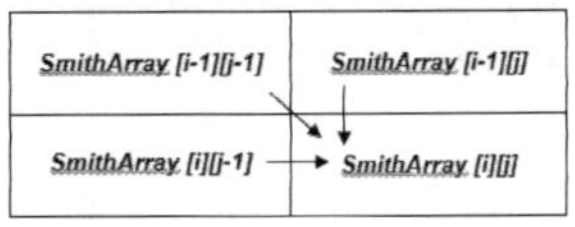

Figure 6. A general rule to calculate the score for Smith-Waterman algorithm.

Parallel insert the pragma at the outer loop, *#pragma* omp parallel shared, and declare shared variables for *tablefill* that shared among the threads. When filling score in the matrix, it do not let any of the matrix values become negative, and thus consider 0 as potentially being the maximum value of the three other cases (where seqA[i] = seqB[j], or there is a gap in seqA or a gap in seqB).

3.1.3. Traceback of Smith Program. To find the optimal alignment it started the traceback with the highest scoring (optimal score) cell in the matrix and it occurs by following the path through the maximum scores back until a 0 value is reached. It finding maximum score in sub row and column of matrix, then trace match in the matrix. When the sequences become aligned their similarity scores align along a diagonal, if there was a gap, the values line up vertically or horizontally.

3.1.4. Optimal Alignment of Smith Program. From the traceback process, it can be seen that when an arrow skips a row, it is a gap in seqA, and when it skips a column, it is a gap in seqB.

3.2. Hardware and Software Development

By using C Programming, Needle and Smith programs are developed based on the algorithms (respectively). For implemented parallel technique on Needle and Smith programs, OpenMP is chosen due to suitability and practically in parallelization. The programs have been tested on HP PC with Intel® Core™ i5-3470 CPU @3.2o GHz for 4 Quad-Core, Memory with 3.2 GiB and LinuxMint 17.1 (rebecca) as the Operating System. The data test is compiled using gcc and run on LinuxMint 17.1 compiler supported from MATE Desktop Environment 1.8.1.

4. Result

Experiments for Needle program (Needleman-Wunsch algorithm) and Smith program (Smith-Waterman algorithm) are conducted on 1000 different pairs of sequences from the Genbank sequence database, National Center for Biotechnology Information (NCBI) (NCBI, 2015). The available data was encoded in the FASTA format (K.D Aumann and R. Huehne, 2008). Dataset in this experiment are made up of DNA sequences from Gorilla gorilla (Western Gorilla) as seqA and Pan Troglodytes Verus (West African Chimpanzee) as seqB. During the sequence alignment process only a DNA sequence from Western Gorilla (seqA) is used and it is held constant. While, seqB consist of 1000 different data from chimpanzees that are run paired with a data from gorilla.

After the data set is analysed, the results parameters of measurement are tabulated. The parameters measured their means, standard deviations and correlation. Spreading of results population within DNA dataset can measured the standard deviation. In determining the effectiveness of the accuracy, space and time it is important to reduce the gap, mismatch and execution time.

The main purpose of the experimental is to evaluate the performance of Needleman-Wunsch and Smith-Waterman algorithms by compared with a publicly available implementation for protein and DNA database searchers, EMBOSS (global) and EMBOSS (local). EMBOSS is an establish system for sequence analysis that has been used quite widely. It is compared to verify the accuracy and consistency of the measurements in space and time complexity.

4.1. Analysis on Similarity

Analysis on similarity generated for Needle against EMBOSS Global and Smith against EMBOSS Local as depicted in Figure 7. According to the result, Needle and Smith programs outperform EMBOSS Global and EMBOSS Local in similarity result. From the analysis, Needle and Smith programs show consistency in similarity result. Needle produced the highest mean with 50.52% where mean is the average of all data plots in the series. The performance of Needle in similarity is better than other methods according to the overall trend of all the 20 datasets. But, in order to determine whether Needle and Smith results are more accurate than others are by comparing both gap and mismatch results.

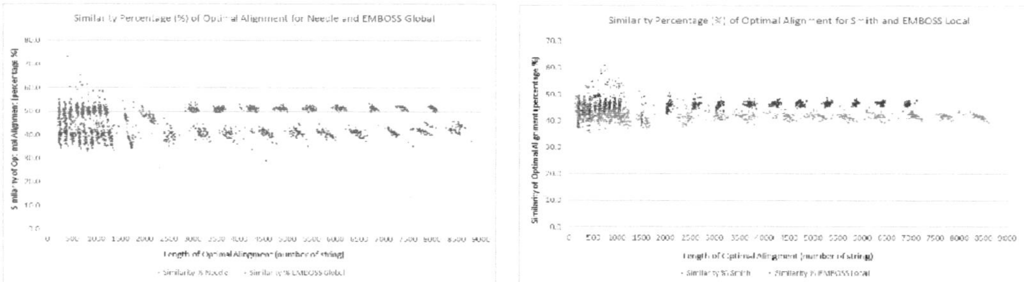

Figure 7. Performance comparison for Similarity (%) between Needle and EMBOSS Global, Smith and EMBOSS Local, respectively.

4.2. Analysis on Gap

The global and local alignment is an alignment such that the matches are maximised and the gaps are minimized [11]. To measure optimal alignment accuracy of Needle and Smith programs, this work compared the output results of gap against EMBOSS Global and EMBOSS Local as shown in Figure 8. According to the result, Needle and Smith programs outperform EMBOSS Global and Local because gap results of Needle and Smith are consistently below from the gap results of EMBOSS Global and Local. The performance of Smith in gap is better than other according to the overall plot trend for the 20 datasets because Smith has the lowest mean of gap, 35.48%.

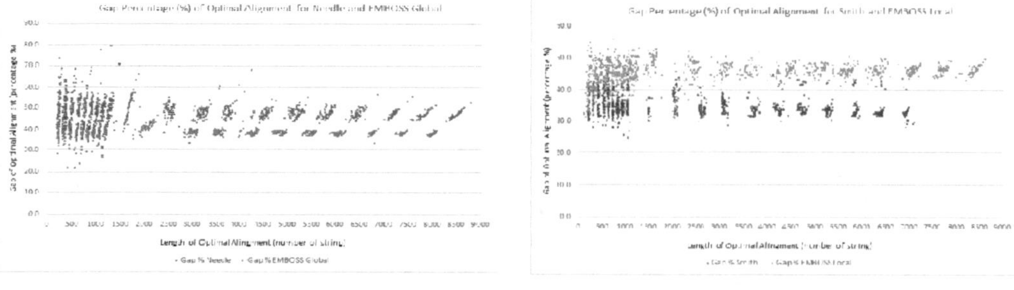

Figure 8. Performance comparison for Gap (%) between Needle and EMBOSS Global, Smith and EMBOSS Local, respectively.

4.3. Analysis on Mismatch

To measure mismatch of Needle and Smith programs, this work compared the output results of mismatch against EMBOSS Global and EMBOSS Local as depicted in Figure 9.

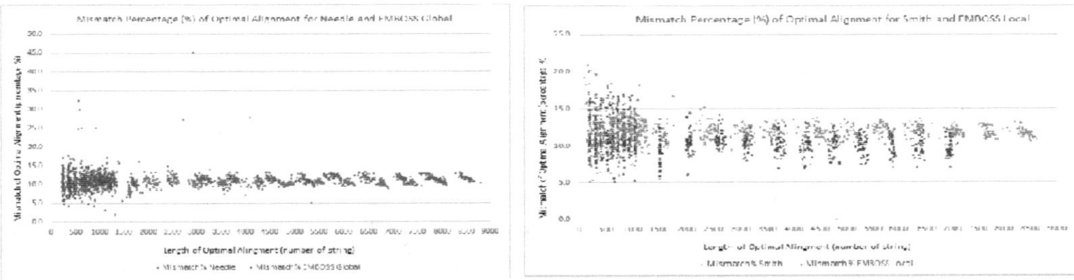

Figure 9. Performance comparison for Mismatch (%) between Needle and EMBOSS Global, Smith and EMBOSS Local, respectively.

According to the result, Needle and Smith programs outperform EMBOSS Global and Local because mismatch result for Needle and Smith are consistently below from the gap result for EMBOSS Global and Local. While Needle has the lowest mean of 10.38%.

4.4. Analysis on Length of Optimal Alignment

To measure length of optimal alignment for Needle and Smith programs, this work compared the output results of optimal length against EMBOSS Global and EMBOSS Local as shown in Figure 10. From the result, Needle and Smith outperform EMBOSS Global and Local in finding the best way to align sequences by searching similarity, reducing gap and mismatch without compromising the sizes of optimal length.

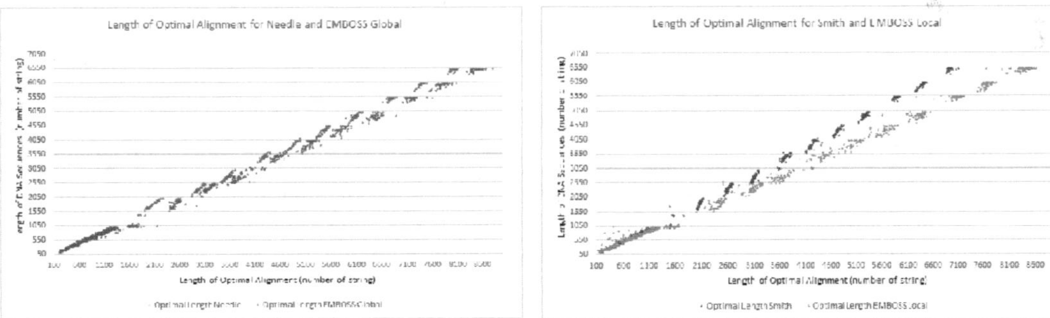

Figure 10. Performance comparison for Lengh of Optimal Alignment between Needle and EMBOSS Global, Smith and EMBOSS Local, respectively.

4.3. Analysis on Execution Time

Figure 11 shows the analysis on execution time is compare the performance of parallel implementation on Needle and Smith using OpenMP. The analysis shows the performance of the larger sequences for Needle and Smith. It is noteworthy to mention that parallelization is very worthwhile for larger sequence, the usage for four and three cores need less time. So, the overhead time is lower than the work that wanted to be parallelized and it is suitable using parallel working under multiple cores.

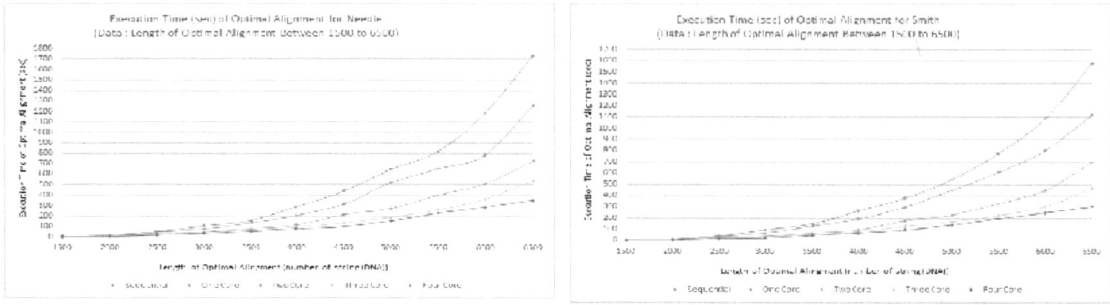

Figure 11. Performance comparison for Execution Time of Optimal Alignment for Needle and Smith from 1500 to 6500 length.

5. Conclusion

With regard to results that gathered from experiment and analysis it shows that Needle and Smith programs are better than EMBOSS Global and EMBOSS Local due to all parameters (gaps, mismatch and match, size of optimal alignment) correlate highly each other in determining accuracy from optimal alignment produced. For larger sequence data, OpenMP directives able to parallelize the code and execute it faster, with four cores it can get an execution time of around 60%. But, for smaller sequences it spent more time to execute the parallel by OpenMP directives.

6. References

[1] S.K. Moore. (2000). Understanding the human genome (vol. 11: pp, 34 – 35). IEEE Press.

[2] Shalini R., Vijay S. and Naveen H. (2011). Bioological computer Model to Solve NP-Complete Problem. International Journal of Information Technology and Knowledge Management. Vol:4, No:1, pp:191-194.

[3] F. Sanchez, E. Salami, A. Ramirez and M.Valero. (2005).Parallel processing in biological sequence comparison usinggeneral purpose processors. Workload Characterization Symposium, 2005. Proceedings of the IEEE International, (pp.99-108).Francis Crick and James Watson. (1953). Molecular Structure Of Nucleic Acids. NATURE, April 25, Volume 171.

[4] S.B. Needleman and Christian D. Wunsch. (1970). A general method applicable to the search for similarities in the amino acid sequence of two sequences. Journal of Molecular Biology, 48(3):443-453.

[5] Temple F. Smith and Michael S. Waterman. (1981). Identification of common molecular subsequences. Journal of Molecular Biology, 147:195-197.

[6] Joshua,L. et al. (2014). BitPAI: A Bit Parallel, General Integer Scoring Sequence Alignment Algorithm. Bioinformatics, 30, 3166–3173

[7] Nordin A., M. Yazid, A. Aziz, and M. Osman. (2009). A guided dynamic programming approach for searching a set of similar DNA sequences. Applications of Digital Information and Web Technologies, 2009. ICADIWT'09. Second International Conference on the, 2009, pp. 512-517.

[8] F. Sanchez, E. Salami, A. Ramirez and M.Valero. (2005).Parallel processing in biological sequence comparison usinggeneral purpose processors. Workload Characterization Symposium, 2005. Proceedings of the IEEE International, (pp.99-108).Francis Crick and James Watson. (1953). Molecular Structure Of Nucleic Acids. NATURE, April 25, Volume 171.

[9] Luis de la Torre and Jaime Seguel, (2015). A Parallel Needleman-Wunsch and Hirschberg Bio-sequence Alignment Algorithm. MIT Press

[10] B. C. Deepa and V. Nagaveni. (2015). Parallel Smith-Waterman Algorithm for Gene Sequencing. International Journal on Recent and Innovation Trends in Computing and Communication. Volume: 3 Issue: 5.

[11] M. Apel, C. Bockermann, and M. Meier. (2009).Measuring similarity of malware behavior. International Proceeding of 34th LCN 2009. IEEE Computer Society.

Ant Colony Optimization Algorithm Parameter Tuning for T-way IOR Testing

N Ramli[1], R R Othman[2,*], S S M Fauzi[3]

[1,2]School of Computer and Communication Engineering, Universiti Malaysia Perlis Pauh Putra Campus, 02600 Arau, Perlis, Malaysia
[3]Faculty of Computer and Mathematical Sciences, Universiti Teknologi MARA, 02600 Arau, Perlis, Malaysia

*rozmie@unimap.edu.my

Abstract. In this study, Ant Colony Optimization (ACO) algorithm's parameters for t-way IOR testing were examined. ACO and its variant have been applied to t-way testing but never to t-way IOR interaction support. Tuning ACO parameters were executed to ensure that ACO could perform for IOR as good as other t-way interaction support. Parameter α, β, τ_0, q_0, ρ value and number of ant were tuned to uniform and non-uniform configuration. Each parameter was executed for 10 independent run. Average best test suite and best test suite were recorded and compared among other parameter values to find which value will produce the best result. The optimum test suite size and average test suite size were generated when the value for parameter $\alpha = 0.5$, $\beta = 3$, $\tau_0 = 0.4$ (uniform configuration) and, 0.2 and 1 (non-uniform configuration), $q_0 = 0.5$, $\rho = 0.5$ and number of ant = 20.

Keywords : t-way testing, input-output based relation, ant colony optimization

1. Introduction

Software applications become complex and crucial in our daily life. To assure the quality of software application, software testing plays an important role. Combinatorial software testing or also known as t-way (where t is the strength) testing is stated to be very effective test design technique in software testing [1]. T-way testing is different from other test design technique (e.g. Boundary Value Analysis and Cause and Effects) because it involves interaction of input parameters and its values. Its purpose is to overcome exhaustive testing problem systematically [2]. Exhaustive testing is impossible to be implemented because it generates enormous number of test cases [3] as it tests all possible values of input parameters. This situation lead to more time [4] and resources allocated for testing activities.

Input-output based Relation (IOR) is categorized as one of the t-way testing support interactions [5], [6]. It involves interactions of multi input parameters that affect an output. This type of support interaction closely related with system under test's specifications. Software tester, who performs the testing activities, must have the knowledge of the system under test to interact related inputs and its respective output [7]. T-way IOR has proven to reduce number of test cases [5], [8], [9] because it ignores any unrelated interaction and eliminates any redundant test cases. Although the number of test cases is being reduced, the strategy is maintaining its capability in finding faults.

Consider an application of Colour Scheme Generator as in Figure 1 to facilitate the understanding on t-way IOR testing. The application involves three inputs, red, blue and yellow. Each input colour consists of two values, either dark or light. It produces two (2) outputs, purple and green. However the outputs are restricted to certain combination of input. For example, combination of input red and blue will produce output purple. This application is an example of IOR t-way testing and can be represented as in (1).

$$P(purple) = f\{red, blue\}$$
$$P(green) = f\{blue, yellow\} \ldots\ldots \quad (1)$$

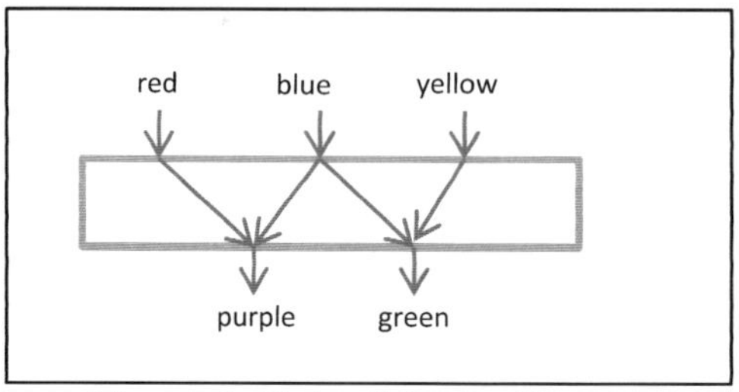

Figure 1. Color Scheme Generator

Test cases generated for Color Scheme Generator is illustrated in Figure 2. Two types of testing design techniques are presented in the figure (i.e. exhaustive testing and t-way IOR testing). Obviously, t-way IOR testing reduces half of number of test cases generated by exhaustive testing. The t-way IOR strategy only generates related input parameters that produce an output and removes any redundant test cases.

Red	Blue	Yellow
Dark red	Dark blue	Dark yellow
Dark red	Dark blue	Light yellow
Dark red	Light Blue	Dark yellow
Dark red	Light Blue	Light yellow
Light red	Dark blue	Dark yellow
Light red	Dark blue	Light yellow
Light red	Light Blue	Dark yellow
Light red	Light Blue	Light yellow

Exhaustive testing

Red	Blue	Yellow
Dark red	Dark blue	Dark yellow
Dark red	Light Blue	Dark yellow
Light red	Dark blue	Light yellow
Light red	Light Blue	Light yellow

T-way IOR

Figure 2. Exhaustive testing and t-way IOR testing

Mathematically, t-way IOR can be represented as covering array. The notation is as in Equation (2).

$$P = IOR(N, C, Rel) \quad (2)$$

From the equation (2), N is the final test suite size, C is value configuration a be represented as $V_0^{p_0}$, $V_1^{p_1}$,……., $V_n^{p_n}$ (p_0 is parameters with V_0 values, p_1 parameters with V_1 values and so on) and *Rel* is the input output relationships involve in generating the test suite.

T-way IOR capability in generating optimum test suite size encourages researchers to further study in this area. There are few strategies that support IOR exists in the market (i.e. Union [9], Greedy [10], Density[11], Test Vector Generator [12], Integrated T-Way Test Data Generator (ITTDG) [13], Aura [14], Para Order and ReqOrder [15]). Those existing strategies apply computational search technique but none of them found to utilize metaheuristic search technique that is proven to produce reasonable test suite size for other type of interaction supports [16].

A metaheuristic algorithm, Ant Colony Optimization (ACO) is known to solve combinatorial problem because the algorithm produces good results in a reasonable time [17]. T-way combinatorial testing also has no exception in using ACO to generate the optimum test suite size. Ant Colony Algorithm [1] and Ant Colony System (ACS) [18] has been introduced for other type of t-way support interactions. Whereas a proposed strategy for IOR using ACO is found in the literature, without stating any suitable parameter settings in the algorithm [19]. There are six parameters (i.e. α, β, τ_0, q_0, ρ and number of ant) to be tuned to get the best settings for the t-way IOR strategy.

This paper presents tuned parameters of Ant Colony Optimization for t-way IOR. The best parameter settings involve in ACO algorithm that can boost up the effectiveness of IOR in generating the best test suite size is examined. Even though ACO has been applied by many researchers, it has not been explored in t-way IOR [16].

The rest of this paper is structured as follows: In section 2, overview of Ant Colony Optimization algorithm is explained in detail. ACO tuning parameter for t-way IOR testing is presented in section 3. Results and discussion is further described in section 4. Finally, section 5 delivers the conclusion.

2. Overview of Ant Colony Optimization Algorithm

Ant Colony Algorithm is an optimization algorithm applied to many combinatorial optimization problem [17] including t-way testing [1], [18], [20]. This algorithm mimics a colony of ant travelling from their nest to find food source. Each ant communicates with other ants by their chemical pheromone deposited at every path travelled. The pheromone deposited by previous ant become guidance to the others in choosing which path to travel. Their aim is to choose the shortest path to the food source. Later on, the colony will build the shortest path from their nest to food source.

The capability of ACO and its family in t-way testing has been proven as it generated very promising results. It has been explored in uniform [1] and variable strength [18] (i.e. Ant Colony System). As a result of good performance by both studies, ACO is introduced to another type of t-way support interaction, IOR. The application of ACO family (i.e. Ant Colony System) towards IOR has been introduced in [19]. To further strengthen the ability of ACO in IOR, we adapted and modified the ACO strategy.

Figure 3 presents the algorithm of our ACO. The algorithm's structure is based on [19] except that some modification has been made in calculating heuristic value as in [1] and the pheromone local update has been removed to suit with IOR strategy. The algorithm begins with initializing heuristic and a constant pheromone value to each edge. Next, each ant starts to construct test case by travelling from one node to another until it reaches food node (i.e. end of the node). After all ants finish travelling to the food node, the best test case constructed is chosen based on fitness function of each test case. Fitness function is calculated based on IOR interaction and function output and it is adopted as in [19]. The process continues until the IOR interaction set is empty.

Input : IOR interaction set, S
Output : Test Suite, TS

Begin
    Set ants, $Iteration_{max}$, bestTest, q
    for interaction set, S is not empty {
        produce single test :

            Set pheromone , τ to a constant value
            Initialize heuristic value for every edge,

$$\text{heuristicValue } (i,j) = \frac{E_{i,max} - E_{i,j} + 1}{E_{i,max} - E_{i,min} + 1}$$

$E_{i,j}$ = number of test cases in S
$E_{i,max} = \max_{1 \leq j \leq \text{edge } i} \{E_{i,j}\}$
$E_{i,min} = \min_{1 \leq j \leq \text{edge } i} \{E_{i,j}\}$

            Repeat {
              Put ants in node N_1
              Repeat{
              Construct test case
                Set q_0 to $0 \leq q \leq 1$
                Set random value, q in [0,1]

If $q \leq q_0$
Exploit path, $argmax_{1 \leq h \leq li}\{[\tau_{i,j}(t)^{\alpha}[\text{heuristicValue}_{i,j}(t)]^{\beta}\}$

If $q > q_0$
Explore new path, $p_{i,j}(t) = \frac{[\tau_{i,j}(t)]^{\alpha}[\text{heuristicValue}_{i,j}(t)]^{\beta}}{\sum_{h=1}^{li}[\tau_{i,h}(t)^{\alpha}[\text{heuristicValue}_{i,h}(t)]^{\beta}}$

Move to N_{i+1} by choosing path that has the highest value of $p_{i,j}$
} Until $N_i == N_{max}$
Calculate fitness function, ft by each ant
Get overall best test
if bestFitnessFunction > bestTest
bestTest = bestFitnessFunction
Update τ ,

$$\tau_{i,j}(t+1) = \begin{cases} (1-\rho)\tau_{i,j}(t) + \rho(ft) \; if \; e_{i,j} \in bestTest \\ \tau_{i,j}(t) \qquad otherwise \end{cases}$$

if bestTest remains unchanged == 5 times
reset τ to τ_0
} Until Iteration > $Iteration_{max}$

if there is interaction covered by single bestTest
remove interactions covered by single test from S
add single bestTest into TS
}
End

Figure 3. Test Suite Generation Algorithm

3. ACO Parameter tuning for t-way IOR

In order to adopt ACO into t-way testing strategy, few parameters tuning has to be made. Five parameters are required to be tuned to find the best value that can produce the optimum or near optimum test suite size. Table 1 shows parameters settings as suggested by [1], [18], [20], [21] and other additional values.

Table 1. Parameters settings for ACO

Parameter	Value
Ants	5,10,20,30
α	0.2,0.4,0.5,0.6,0.8, 1.0,1.5, 1.6, 2.0
β	0.5 , 0.2, 2 – 5 , 2.0
ρ	0.1, 0.2,0.4,0.5, 0.6, 0.8
τ_0	0.2, 0.4, 0.5, 0.8,1.0, 1.5
q_0	0.2, 0.4, 0.5, 0.6, 0.8, 0.9, 1.0

To tune the parameters as in Table 1, two configurations from benchmarked experiments for IOR [16] have been executed. The configurations are as follows:

- a uniform configuration IOR (N, 3^{10} , R) - consists of 10 3-valued parameters and 10 input parameters where the relation, R = [{1,2,7,8}, {0,1,2,9}, {4,5,7,8}, {0,1,3,9}, {0,3,8}, {6,7,8}, {4,9}, {1,3,4}, {0,2,6,7}, {4,6}].
- a non-uniform configuration IOR (N, 2^3 $3^3 4^3 5^1$, R) – consists of 3 2-valued, 3 3-valued, 3 4-valued and 1 5-valued parameters and 20 input parameters where the relation, R = [{1,2,7,8}, {0,1,2,9}, {4,5,7,8}, {0,1,3,9}, {0,3,8}, {6,7,8}, {4,9}, {1,3,4}, {0,2,6,7}, {4,6}, {2, 3, 4, 8}, {2, 3, 5}, {5, 6}, {0, 6, 8}, {8, 9}, {0, 5}, {1, 3, 5, 9}, {1, 6, 7, 9}, {0, 4}, {0, 2, 3}].

Each parameter has to perform 10 independent executions to get the best test suite size, the average of test suite size and average of execution time. This is due to our strategy utilizes a random q value to determine probability value that results different test suite size per execution (i.e. run).

4. Results and Discussion

The experiments' goal is to find the best value for each parameter. Value which generates the smallest average test suite size and best test suite size is the best value. Even though the aim of any strategy developed for t-way testing is to find the best test suite size, this experiments are concern with both average test suite size and best size. The smallest gap between average test suite size and best size shows that the value of parameter is consistent in generating the best size. This section will presents a graph for each parameter. Each graph will present test suite size generated based on value of parameter. The right axis presents an average test suite size and best test suite size for configuration 2 (i.e. non-uniform), while left axis presents an average test suite size and best test suite size for configuration 1 (i.e. uniform). In order to find the best value for each parameter, the differences between average test suite sizes and best sizes are calculated. To ease the graph presentation, configuration 1 represents uniform configuration, while configuration 2 represents non-uniform configuration.

Figure 4 shows the best test suite size, average test suite size and best test suite size based on α values for configuration IOR (N, 3^{10} , R) and IOR (N, 2^3 $3^3 4^3 5^1$, R). As can be seen in the graphs, both lines for average test suite size fluctuate with the smallest size at 0.5. The best test suite size for α parameter is unchanged except for value 1.5, it climbs dramatically at 1.5. Value 2 shows the smallest gap between average test suite size and best size for configuration 1 (i.e. IOR (N, 3^{10}, R)) while value 0.5 is the best value for configuration 2 (i.e. IOR (N, 2^3 $3^3 4^3 5^1$, R).). However, some researchers found that alpha value greater than 1 will cause the stagnation in the algorithm [21]. Although the study was about a general ant colony algorithm, the suggestion should be considered. In that case, 0.5 is most suitable value for α parameter.

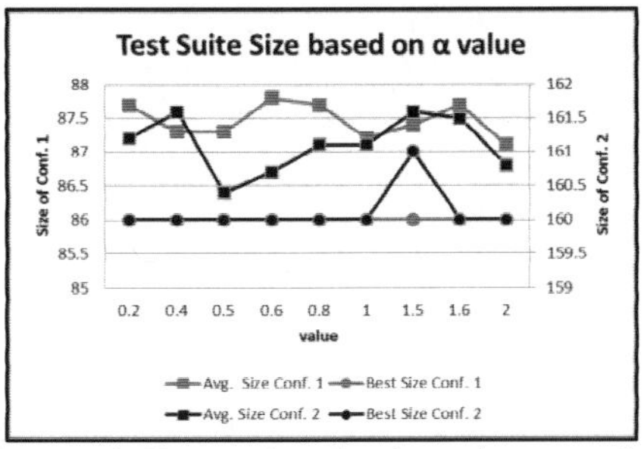

Figure 4. Test suite size based on α value

The second parameter, β parameter is given in Figure 5. A slight fluctuate graph is shown for both configurations except for a linear graph for configuration 2's best size. As discussed earlier, the smallest test suite size the better the value parameter, which implied that value 3 is the most suitable β parameter. The value also has been suggested by [21]. Parameter β, similar to α, involves in calculation of probability.

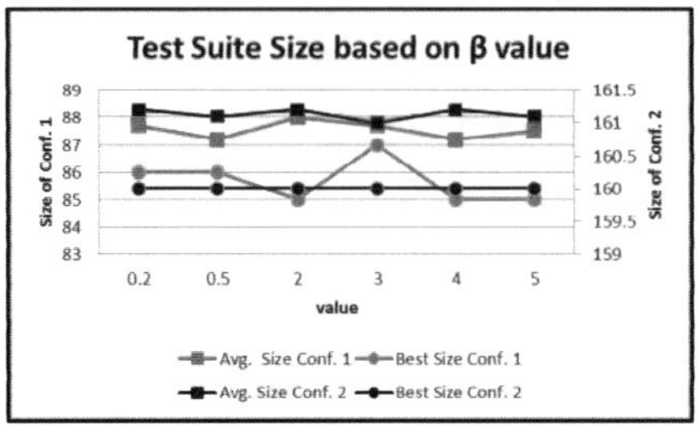

Figure 5. Parameter β values

Next parameter, τ_0 can be seen in Figure 6. It shows a similar trend as in Figure 5. There are three values recorded the smallest gap between average test suite size and best size. Value 0.4 shows the smallest gap for configuration 1, while value 0.2 and 1 hit the smallest gap for configuration 2. Value 0.4 has been suggested by [1]. This parameter is important and required during initialization. It contributes to the first move of the ant.

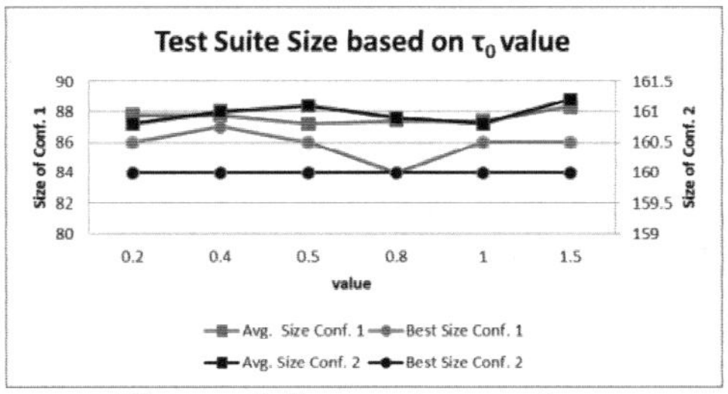

Figure 6. Parameter τ_0 values

Parameter q_0 in Figure 7 illustrates inconsistent results for the graph except the best size for configuration 2. The smallest gap between average test suite size and best size is at value 0.8. The value also reaches the smallest best size for configuration 1. However, the value illustrates the worst gap for configuration 1. Value of 0.5 shows a consistent gap between average test suite size and best size for both configurations. The value is suggested by [18]. Parameter q_0 affects ants to explore new path or exploit similar path chosen by previous ant.

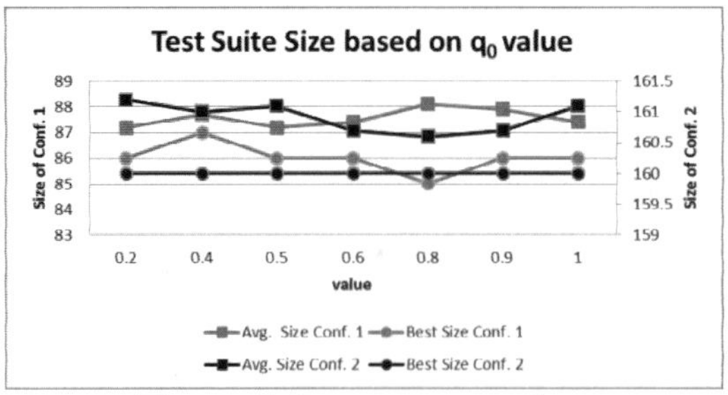

Figure 7. Parameter q_0 values

As compared to other parameters, parameter ρ projected a steady line for both average test suite size and best size from 0.1 until 0.8 for both configurations as in Figure 8. However, there has been a steep rise for ρ value is 1. Value of 0.5 shows the smallest gap between the two sizes. The value also is suggested by [1]. Parameter ρ is a decay value and it affects global pheromone value in the algorithm.

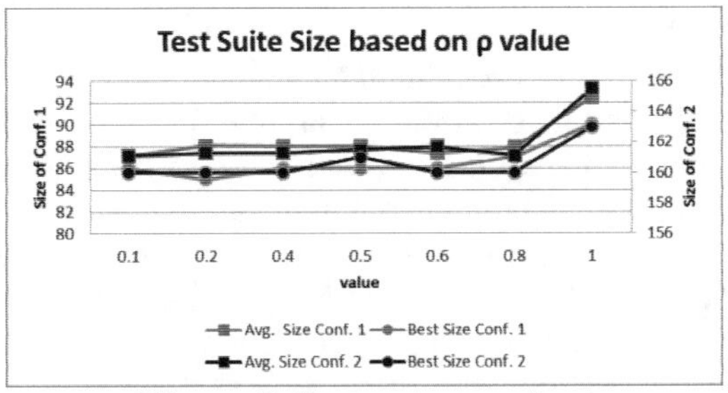

Figure 8. Parameter ρ values

The last parameter is number of ant as presented in Figure 9. As can be seen, the more ants used in the strategy, the smaller test suite size is generated by the strategy. Obviously, value of 40 shows the smallest gap between average test suite size and best size. However, time generation is increasing when the number is ant increased. Because of this, value of 20 is most suitable value as it shows a consistent gap for both configurations. Besides that, more time can be saved rather than value of 40.

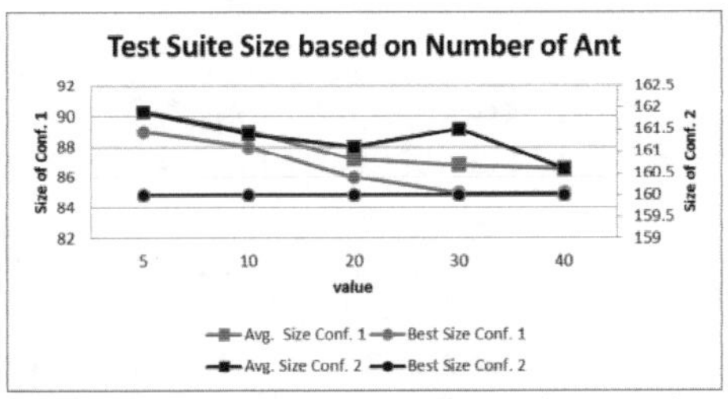

Figure 9. Parameter number of ant values

From 10 independent run of each configuration, the average test suite size and time generated has been calculated. Out of 10 results, the best size was recorded. This is the reason some of the results of the best test suite size are differ from the best average test suite size. Another factor of this situation is due to randomness in the strategy. Even though the value of parameter could generate the best test suite size, it cannot be concluded that the value also will generate the best average test suite size.

5. Conclusion

In this work, six parameters of ACO were tuned to suit t-way IOR. This was performed by experimenting 10 independent runs for each parameters to two configurations, uniform configuration IOR (N, 3^{10}, R) and non- uniform configuration IOR (N, 2^3 $3^3 4^3 5^1$, R). From the experiments, average test suite size and best test were recorded. It was found that the best value for α parameter = 0.5, β = 3, τ_0 = 0.4 (for uniform configuration) and, 0.2 and 1 (for non-uniform configuration), q_0 = 0.5, ρ = 0.5 and number of ant = 20. Those values were recorded to be the best value for both configurations except τ_0 parameter that has two different values for both configurations. Even though value for each parameter has been suggested by some researchers, but none of them has suggested for t-way IOR. Because of that, this work introduced ACO parameters for t-way IOR. Therefore, for further works, experiments based on benchmark experiments will be executed based on the values.

IOP Conf. Series: Journal of Physics: Conf. Series **1019** (2018) 012086 doi:10.1088/1742-6596/1019/1/012086

6. References

[1] T. Shiba, T. Tsuchiya, and T. Kikuno, "Using artificial life techniques to generate test cases for combinatorial testing," in *Proc. of the 28th Annual Int. Computer Software and Applications Conference, 2004. COMPSAC 2004.*, 2004.

[2] R. R. Othman, N. Khamis, and K. Z. Zamli, "Variable Strength t-way Test Suite Generator with Constraints Support," *Malaysian J. Comput. Sci.*, vol. **27**, no. 3, pp. 204–217, 2014.

[3] M. B. Cohen, "A Study in Prioritization for Higher Strength Combinatorial Testing," in *2013 IEEE 6th Int. Conf. on Software Testing, Verification and Validation Workshops*, 2013.

[4] M. Rahman, R. R. Othman, R. B. Ahmad, and M. Rahman, "A Meta Heuristic Search based T-way Event Driven Input Sequence Test Case Generator," *Int. J. Simul. Syst. Sci. Technol.*, vol. **15**, no. 3, pp. 65–71, 2014.

[5] R. Othman and K. Zamli, "T-Way Strategies and Its Applications for Combinatorial Testing," *Int. J. New Comput. Archit. Their Appl.*, vol. **1**, no. 2, pp. 459–473, 2011.

[6] R. R. Othman, K. Z. Zamli, S. Mashita, and S. Mohamad, "T-Way Testing Strategies : A Critical Survey and Analysis," *Int. J. Digit. Content Technol. its Appl.*, vol. **7**, May, pp. 222–235, 2013.

[7] K. Z. Zamli, R. R. Othman, M. I. Younis, and M. H. Mohamed Zabil, "Practical Adoptions of T-Way Strategies for Interaction Testing," in *Int. Conf. on Software Engineering and Computer Systems*, 2011, pp. 1–14.

[8] A. a Al-sewari and K. Z. Zamli, "An Orchestrated Survey on T-Way Test Case Generation Strategies Based on Optimization Algorithms," in *The 8th Int. Conf. on Robotic, Vision, Signal Processing & Power Applications*, 2014, vol. **291**, pp. 255–263.

[9] P. J. Schroeder, "Black-box Test Reduction using Input-Output Analysis," Illinois Institute of Technology, 2001.

[10] C. Cheng, A. Dumitrescu, and P. Schroeder, "Generating small combinatorial test suites to cover input-output relationships," in *Third Int. Conf. on Quality Software*, 2003, pp. 76–82.

[11] W. Ziyuan, X. Baowen, and N. Changhai, "Greedy Heuristic Algorithms to Generate Variable Strength Combinatorial Test Suite," in *The 8th Int. Conf on Quality Soft.*, 2008, pp. 155–160.

[12] P. J. Schroeder, E. Kim, J. Arshem, P. Bolaki, and W. Milwaukee, "Combining Behavior and Data Modeling in Automated Test Case Generation," in *3rd Int. Conf. On Quality Software (QSIC'03)*, 2003, pp. 1–8.

[13] R. R. Othman and K. Z. Zamli, "ITTDG : Integrated T-way test data generation strategy for interaction testing," *Sci. Res. Essays*, vol. **6**, no. 17, pp. 3638–3648, 2011.

[14] H. Y. Ong and K. Z. Zamli, "Development of interaction test suite generation strategy with input-output mapping supports," *Sci. Res. Essays*, vol. **6**, no. 16, pp. 3418–3430, 2011.

[15] W. Ziyuan, C. Nie, and X. Baowen, "Generating Combinatorial Test Suite for Interaction Relationship," in *4th Int. workshop on Soft. Quality Assurance: in conjunction with the 6th ESEC/FSE joint meeting*, 2007, pp. 55–61.

[16] A. A. Alsewari, N. M. Tairan, and K. Z. Zamli, "Survey on Input Output Relation Based Combination Test Data Generation Strategies," *ARPN J. Eng. Appl. Sci.*, vol. **10**, no. 18, pp. 8427–8430, 2015.

[17] M. Dorigo and C. Blum, "Ant colony optimization theory: A survey," *Theor. Comput. Sci.*, vol. **344**, no. 2–3, pp. 243–278, 2005.

[18] X. Chen, Q. Gu, A. Li, and D. Chen, "Variable strength interaction testing with an ant colony system approach," in *Asia-Pacific Soft. Eng. Conf., APSEC*, 2009, pp. 160–167.

[19] N. Ramli, R. R. Othman, and M. S. A. Rashid Ali, "Optimizing Combinatorial Input-Output Based Relations Testing using Ant Colony Algorithm," in *3rd Int. Conf. on Electronic Design (ICED), August 11-12, 2016, Phuket, Thailand*, 2016, pp. 586–590.

[20] X. Chen, Q. Gu, X. Zhang, and D. Chen, "Building prioritized pairwise interaction test suites with ant colony optimization," in *Int. Conf. Qual. Softw.*, pp. 347–352, 2009.

[21] M. Dorigo and T. Stützle, *Ant Colony Optimization*. The MIT Press.

IOP Conf. Series: Journal of Physics: Conf. Series **1019** (2018) 012087 doi:10.1088/1742-6596/1019/1/012087

Developing a Profile of Small-Medium Companies in Leveraging IT for Sustainable Competitiveness

Raja Haslinda Raja Mohd Ali*[1], Rosli Mohamad[1], Ku Maisurah Ku Bahador[1]

[1] Tunku Puteri Intan Safinaz School of Accountancy, Universiti Utara Malaysia, Sintok, 06010, Kedah, Malaysia

E-mail: rj.linda@uum.edu.my, roslim@uum.edu.my, kumaisurah@uum.edu.my

Abstract. As outlined in Malaysia Economic Transformation Program (ETP), business sector is expected to play crucial role to harness the potential of globalization and to escalate national economic growth. Considering its domination over 98% of total established businesses, SME sector become essential to drive the government's agenda. Apart from human resource, financing, innovativeness, and legal environment, availability of timely and relevant information in support of business's decision making is highly crucial for ensuring its competitiveness. Accessibility to the relevant information helps firms to lower their operating cost, to effectively manage available resources and ultimately to improve overall productivity. Information Technology/Information Systems (IT/IS) has been recognized as one of the potential means to supply information for timely business decision making. Despite growing importance of IT/IS for firm's competitiveness, SMEs embrace constant challenges to deploy IT/IS effectively and thus hardly to optimize its strategic values. Consequently, IS planning is getting paramount to promote effective management of IT/IS within SME sector. Thus, this study proposes a study that investigates the current state of IS planning practices and the factors affecting the IS Planning success among Malaysian SMEs. Adapting IS planning framework of large organizations, this study proposes IS planning framework that is applicable for SMEs environment. The developed framework could facilitate SMEs to effectively manage IS planning as to ensure greater strategic benefits of IT/IS deployment.

1. Introduction

Economic Transformation Program has been proposed (ETP) as one of the national agenda to elevate Malaysian economy into developed-nation status as well as to realize high-income nation. To realize the high-income nation aspiration, ability of business sector to compete at global platform emerges as one of the key success factors. Consequently, Malaysian businesses play substantial roles to enhance their productivity and innovativeness. As such, efficient use of available resources should not be side-lined in promoting more productive business environment.

In view of its domination of total registered businesses Small Medium Enterprises (SME) sector has been taken into account in formulating various national strategic plans. The sector becomes one of the key enablers in promoting economic growth as formulated in 11th Malaysia Plan (11MP) and New Economic Model 2010 (NEM 2010). Specifically, 11MP aims to nurture the SMEs as engine of growth and innovation via several strategies; (i) regulatory cost reduction, (ii) capability building,

(iii) nurturing entrepreneurial culture, (iv) extending SMEs support systems, and (v) enhancing access to financing (EPU, 2010). SMEs are also expected to play more active roles in promoting Malaysian business presence at global platform. This corresponds well to the SME Master Plan that aims to mold globally competitive SMEs across all sectors. Moreover, expanding operation into global market is vital considering a stiff competition among businesses in securing domestic market opportunities. Thus, SMEs are expected to strengthen their capabilities and capacities to embrace globalization and liberalization challenges. The SMEs need to be highly competitive and innovative to be able to become part of the global supply chain and to collaborate with giant business entities worldwide.

Nevertheless, SMEs penetration into global competition has been relatively low as indicated by its export contribution. Despite its substantial contribution to the national GDP (36.3%) as well as employment opportunities (57%), more effort needed to promote greater roles of SMEs with respect to their productivity and of becoming high-value added exporters. As reported, Malaysian businesses are currently stuck in the middle and hardly move up to 'high value-added exporters' category as they relatively not ready to embrace global competition.

At present, only 49 SMEs companies are considered as successful exporters in Malaysia. This number is still relatively low as compared to the total numbers of SMEs in Malaysia. Furthermore, a total of 550 SMEs companies were awarded as top firms according to E50. These firms were among the four and five-star ranked firms as indexed by SME Competitiveness Rating for Enhancement (SCORE). SCORE is a diagnostic tool used to rate and enhance competitiveness of SMEs based on seven assessment criteria such as financial strength, business performance, human resource, technology acquisition and adoption, certification and market presence. One of the purposes is to assist the SME Corp Malaysia on improving the capacity building of 3-Star companies and below.

Facing globalization challenges require firms to be equipped with various capabilities. As outlined in SME Master Plan, six major factors deserve special attention to enhance SME competitiveness at global arena. These include human capital development, access to financing, market access, legal and regulatory requirement, infrastructure as well as innovation and technology adoption [1]. To remain highly competitive and survive in their respective industry, SME is under great pressures to respond quickly to the changes surrounding their business. In other words, availability of right information at the right time is vital as to ensure firms to make accurate and timely decision. Thus, deployment of appropriate IT/IS would facilitate firms with right information at the right time. In response, SMEs now depends more on IS to enhance their business effectiveness [2, 3].

Deployment of IT/IS enables provision of both financial and non-financial information for decision support. Consequently, availability of relevant and accurate information on timely manner would enable firm to reduce operating cost, to effectively utilize available resources, to execute strategic plans and ultimately to improve its overall productivity. Thus, promoting extensive IT/IS deployment among firms is in line with government effort in molding globally competitive SME across industries as outlined in SME Development Plan [1]. Despite progress on IS deployment among SMEs, the firms use of IS are relatively non-strategic in nature. The focus is mainly to support operational business instead to support firm's strategic objective. In turn, firms gain limited values out of the IT/IS-related investment. A global study conducted by KPMG and ACCA involving 2,600 finance professionals from 129 countries including Malaysia found that companies are not leveraging technology to drive profitability. The study also highlighted the challenges of four main areas that these companies are facing: organisation structure, operating model, quality of data, and use of technology.

Thus, to ensure successful deployment of IT/IS so as to correspond to the firm's strategic direction and ultimately preparing more competitive SME, proper IS planning deserves further attention. Information systems planning (IS planning) is a process of identifying an organization's portfolio of computer-based applications and technologies that could help the organization's in executing their business plans and achieving their business objectives [4]. Many studies have highlighted the importance of IS planning [5]. IS planning assists the organization in selecting the most suitable databases and systems that could support their operations and identifying the current and future applications. The dependency of business towards the IS practices in the organisation's daily operations requires them to focus more on the IS planning that aims to achieve closer alignment between IS (and the business processes that rely on them) and the organisations strategic goals, resulting in gaining competitive advantage from information technology [6].

Specifically, IT/IS deployment is more widespread for finance, accounting, and inventory management [7]. On other respect, the lack of competent IT experts has make firm to outsource their IS related projects [2]. With limited resources available, strategic IS planning becomes even more crucial for the firms. Ashurst, Cragg and Herring [8] also suggest that high level of internal competencies such as IT leadership by the owner or the top management of the firm are particularly important for progressive IT development to take place in SMEs. Taking into account dominant roles of top management in most of the SMEs decision making, therefore their active involvement in IT/IS related project promotes better planning process which ultimately ensures successful use of IT/IS. In similar vein, taking into consideration the need to ensure IT/IS corresponds well to users requirement, their involvement in any IS planning activities do have considerable impact towards successful IS planning processes [9]. Considering the emerging importance of IS in one hand, and firm challenges to embrace IS effectively on the other, it is critically important to understand the IS leveraging in the SME firms. Generally, this study aims to develop a profile of the IS leveraging among firms with different level of competitiveness.

2. Literature Review

2.1. IS practices amongst Small and Medium-sized Enterprises (SMEs)

SMEs contribute substantially to both the economic and social growth of most countries. This sector plays substantial roles in promoting higher Gross Domestic Product (GDP), greater export activities, and employment opportunities [10]. As they are generally domestic-oriented business, thus their trading activities would have direct impacts to a nation's economy [11]. Considering its vital roles in both developed and developing economies, many international agencies such as the United Nation (UN), World Intellectual Property Organisation (WIPO), and Organisation for Economic Cooperation and Development (OECD) come into play to facilitate firms to have a meaningful role in the economic system worldwide.

The SMEs also play greater roles in ensuring competitiveness of most developing and transitional economies. In the Asia Pacific region, more than 95% of businesses are SMEs with more than 70% in micro firm category [12]. The report further shows that out of 49 million SMEs in this region, about half of them are currently operating in China and Indonesia. Thus, their activities could be an important indicator of entrepreneurial health and competitiveness of a particular country.

Owing to the salient role of SMEs for most nations' economy, the government has initiated various strategies and incentives to extend their efficiency and productivity. Deployment of appropriate IT/IS is one of the possible strategies for such purpose [11]. Nowadays, IT/IS has made it presence in all kinds of businesses of all sizes and has a significant bearing in extending business productivity. Ideally, a firm would receive considerable benefits of deploying IT/IS. IT/IS facilitates firms in managing administrative tasks, enhancing production activities, and expanding product/service market [10]. It further enables rapid innovation activities and levels the playing field with their larger counterpart [13]. However, in Malaysia, the ICT usage in business operation among SMEs is only 27%, and 67% of internet usage.

Due to the overwhelming claims of IT/IS potentials, SMEs has been the subject of many studies over the years. Previous works concentrate on several aspects of IS practices. Earlier works are primarily concerned on drivers/inhibitors of IS adoption [14], intensity of IS deployment [7,15]. Several other works have also examined IS sophistication [16], IT/IS success factors [17. 18]. However, quite a handful of these studies have investigated on IS strategic-related issues. For example, Levy & Powell [19] explored current state of IS alignment in SMEs sector and reported that lack of alignment leads to firms failure to benefit from IS investment. Similarly, Cragg, King & Hussin [20] suggested that appropriate IS alignment with SMEs business strategy ensures greater IS success and firm performance. Meanwhile, Ismail & King [21] reaffirms the essence of IS alignment within a specific business function (accounting information system) to successful deployment of IS.

2.2. *Overview of Malaysian SMEs*

In Malaysia, SMEs is a seedbed to the Malaysian economy, as this sector constitutes virtually the entire business community. They are an important generator of the national economy and social harmony as they account for more than 99% of the established businesses [1]. Micro firms dominate more than 79% of the Malaysian registered business.

Malaysian firms have the greatest presence in the service sector, accounting for 87% of the total established businesses. Meanwhile, the remaining firms involve either in manufacturing or in agriculture business. The service-based firms make most presence in wholesale and retailing businesses. Pertaining to business nature, manufacturing-based firms involve predominantly in textile/apparel and food/beverage subsectors. Overall, the sector contributes about 40% of the nation's GDP and employment opportunities [1]. Consistent with initiatives by the SMEs' institutions worldwide, the Malaysian government is constantly finding ways to extend productivity of the SMEs. As such, the government has initiated various programs and allocated substantial resources to expand SMEs capabilities. The efforts pay particular attention on several core aspects like technology deployment, skills development, global presence, and entrepreneurship talent [22]. Agencies such as Multimedia Development Corporation (MDeC), SME Corporation (SME Corp), National SME Development Council, Malaysia External Trade Development Corporation (MATRADE), and Malaysian Ministry of International Trade and Industry (MITI) then spearhead the implementation of all government approved policies and strategies involving the SMEs.

Constant government's efforts progressively shaped Malaysian firms to embrace IT/IS. Malaysia has shown positive initiatives in promoting IS with respect to regulatory assurance, infrastructure, business climate, and business usage [23]. The Internet penetration has risen slightly from about 48% in 2005 to 53% of the total population in 2010 while broadband is getting more commonplace [24]. The domain name registrations have also surged substantially about 28% from 91,000 in 2009 to 127,000 in 2011 [25]. These are indeed strong signals for the growing acceptance among firms in term of IS in general and Internet-based applications, in particular. Despite the extensive government initiatives, findings from several studies reveal that IS usage is not well received by the SMEs. Most of Malaysian SMEs keep their traditional ways of operating their businesses while IS deployment is somewhat limited [26]. The IS use is relatively more commonplace for supporting informational-related tasks while online transactions are relatively rare in most firms [27]. Various barriers have been attributed for the slow progress of IS acceptance among SMEs.

Factors such as top management support and limited knowledge are among highly cited reasons [28]. SMEs also have inadequate IT infrastructures and technical support to ensure successful deployment of IS [29]. Other barriers such as perceived benefits [27], security, and trust issues have certain roles in promoting IS use amongst SMEs [30]. Ismail and King [21] noted that lack of alignment between IS capabilities and firm's information requirement do also contribute to effective use of IS among firms. More recently, Hussin and Suhaimi [31] found that major proportions of Malaysian SMEs are still struggling to align IS with their strategic objectives and thus limiting the benefits from such investment.

In view of the importance of IS on firm's operation especially for SMEs, adequate IS planning is necessary to ensure its effectiveness in supporting the business strategic objectives. However, limited studies have been conducted on assessing strategic IS issues amongst SMEs. Therefore, the SISP study focusing on SMEs context is greatly needed to establish further understanding of SISP success from industry level perspective. Industry level strategies such as SME sector constantly change over time which therefore demands for investigation at industrial level perspective [34].

3. Methodology

In line with the proposed study objective, data for this research will be collected in two stages, namely questionnaire survey and face to face interview. The questionnaire survey collects data that corresponds the first objective of the study. This study defines SME based on the National SME Development Councils guideline. In view of the study objective that attempts to examine how firms leverage IT for sustainable future, this study focuses on the firms which have been assessed under SME Competitiveness Rating for Enhancement (SCORE) index. SCORE index is an instrument developed to assess firm level of competitiveness. A self-administered questionnaire will be mailed to selected firms as to be identified from the list of firms registered under the SCORE index. The targeted respondent is firm's owner or any related manager whom is expected to have sufficient knowledge about the firm's strategy and IT deployment.

The questionnaire consists of four major parts. The first part gathers descriptive information of the firm. The second part assesses current use of IT applications as adapted from IT indicators [35]. The indicators include ICT indicators for individual as well as business level of IT use. This study will adopt ICT indicators for business checklist, which are more appropriate for the study. The study includes both basic core and extended core components in this present study. The third part examines IS planning tasks and phases as currently practiced by the firms based on Newkirk, Lederer, & Johnson [36] and Cassidy [37]. The final part seeks demographic data on participating respondents.

Stage 2 employs the qualitative field study. This approach is considered to provide additional support for the second objective. The interview will be carried out with selected firms after the survey is successfully completed. The respondents of the interview sessions are the top management of the chosen firms. The top management will be identified based on the survey. At the end of the interview sessions, the top management will be asked to provide several names(at least two) from their employees list that can be contacted. The follow up interview aims to explore in depth on the execution of IS planning tasks/activities in support of its business. A standardized open ended interview method will be utilized as it does not predetermine the answers and allow room for the participants to respond in their own terms [38]. Open ended interview allows the researcher to collect detailed data systematically [39]. To establish better insight of best practice of IT leverage amongst SMEs, this study will identify several firms obtained rating of 4 or 5 stars in their SCORE assessment. Interview with these firms will reveal how highly competitive firms deal with IT planning and use.

Having identified firm's best practices in leveraging IT would enable researchers to propose the most effective IS deployment and planning approach in building IT capability of the firms. Data obtained both from quantitative and qualitative approaches will be used in proposing the most effective IS deployment and planning approaches in preparing for more competitive firms.

4. Proposed data analysis

This study employs cluster analysis technique to explore possible profiles of firms that potentially emerged from the data collected. Cluster analysis is a multivariate approach to 'classify objects such as respondents, products, or other entities, on a set of user selected characteristics' [40]. It is primarily conducted to classify the objects into several distinct groups with high internal homogeneity (withincluster) and high external heterogeneity (intercluster). This technique groups the objects with similar characteristics, while objects with relatively distinct characteristics would stay apart. Another relevant point is that cluster analysis is an interdependent multivariate technique which allows analysis to be executed without the presence of dependent variable [41].

Cluster analysis has been widely applied across various research domains especially for developing and testing taxonomies [42]. Due to its classification ability, it is relatively popular among information systems research [43]. It was also considered for investigating IT/IS practices among SMEs. For example, Vlachos [44] employed cluster analysis to propose SMEs profiles based on selected demographic factors. Meanwhile, Magal, Koslage, & Levenburg [45] employed the cluster analysis to establish meaningful patterns of ebusiness diffusion among SMEs based on its use across value chain activities. The clustering procedures will be carried out based on (i) ICT indicators, (ii) IS planning tasks/activities and (iii) selected demographic variables as the clustering variables. The following diagram (Figure 1) illustrates overall research model.

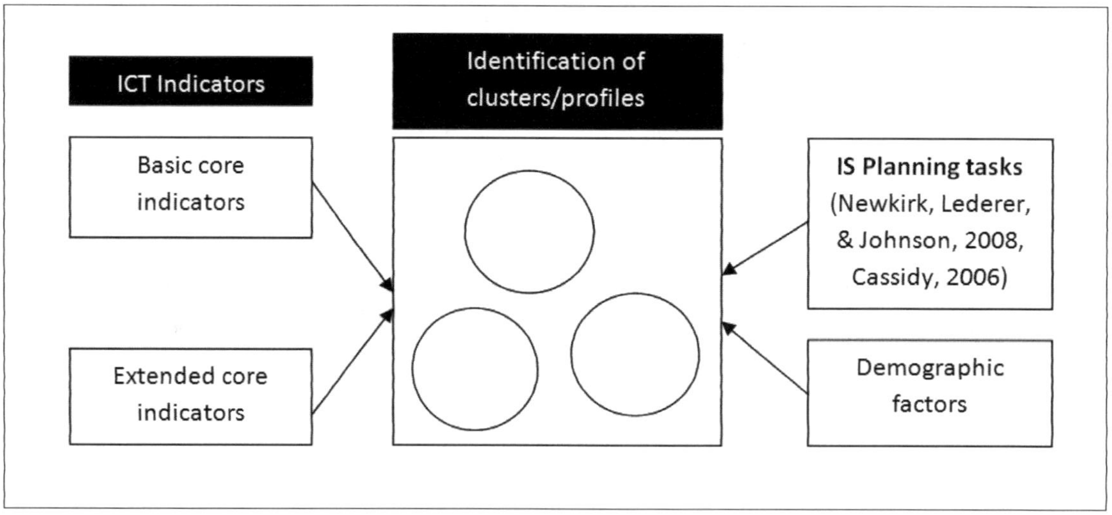

Figure 1 : Research Model

5. Conclusion

SMEs are generally family owned businesses. Therefore, deployment of IT/IS in a more effective manner would extend firm's productivity. More productivity will be reflected to better financial return which in turn promises greater income to the business ingeneral and to the families in particular. This corresponds well to the government aims to promote the high income society. Inanother respect, SMEs presently provides around 40% of employment opportunities to the society. More competitive business ensures a more active business environment and promotes business expansion and finally creates more demands on employment opportunities.

By recognizing the factors affecting IS leverage activities, necessary actions can be initiated to ensure effective use of IS among the SMEs. Effective use of IS as potential tools to support the business operation would then lead to greater productivity. In other words, the SME would provide a higher contribution to the national economic growth with respect to the Gross Domestic Product (GDP). More productive operation also promises greater firm's competitiveness. Highly competitive and innovative firms are more capable to survive in the long term and to compete in global platform. Ultimately, this sector would contribute further to the national income from increasing export activities. This is in line with the continuous government effort to promote the SME sector as an economic backbone of national growth.

References

[1] SME CORP. (2010, October 12). SME Annual Report 2009/2010: Transformation to the new economic model. Retrieved from SME Corporation: http://www.smecorp.gov.

[2] Premkumar, G. (2003). A metaanalysis of research on information technology implementation in small business. Journal of Organizational Computing and Electronic Commerce, 13(2), 91 121. doi:10.1207/S15327744JOCE13022

[3] Schubert, P., & Leimstoll, U. (2007). Importance and use of information technology in small and mediumsized companies. Electronic Markets, 17(1), 38 55. doi:10.1080/10196780601136799

[4] Lederer, A. L., & Salmela, H. (1996). Toward a theory of strategic information systems planning. Journal of Strategic Information Systems, 237253.

[5] Bechor, T., Neumann, S., Zviran, M., &Glezer, C., "A contingency model for estimating success of strategic information systems planning", Information & Management, vol 47, no. 1, pp 17-29, 2010

[6] Basu, V., Hartono, E., Lederer, A. L., & Sethi, V. (2002). The impact of organisational commitment, senior management involvement, and team involvement on strategic information systems planning. Information & Management, 39, 513524.

[7] Ashrafi, R., & Murtaza, M. (2008). Usages and effects of information and communication technologies on small and medium sized enterprises in Oman. Retrieved from http://aisel.aisnet.org/confirm2008/60.

[8] Ashurst, C., Cragg, P. & Herring, P. (2011). The role of IT competencies in gaining value from ebusiness: An SME case study. International Small Business Journal, 30 (6), 640 – 658.

[9] Raymond, L., Croteau, A. M. & Bergeron, F. (2011). The strategic role of IT: An empirical study of its impact on IT performance in manufacturing SMEs. Paper presented at ICIW 2011 : The Sixth International Conference on Internet and Web Applications and Services, Netherland, 20 – 25 March 2011.

[10] World Association for SMEs [WASME] (2006). Annual report for 2004 – 2006. WASME: Uttar Pradesh, India. Retrieved from http://www.wasmeinfo.org

[11] Kotelnikov, V. (2007). Small and medium enterprises and ICT. Bangkok, Thailand: UNDP Asia-Pacific Development Information Programme. Retrieved from http://www.apdip.net/news/sme

[12] Asia Pacific Economic Cooperation [APEC]. (2011). Study on SMEs internationalisation best practices across selected APEC economies. Spring Singapore: Singapore. Retrieved from http://www.apec.org

[13] Dibrell, C., Davis, P. S. & Craig, J. (2008), Fueling Innovation through Information Technology in SMEs. Journal of Small Business Management, 46: 203–218. doi: 10.1111/j.1540-627X.2008.00240.x

[14] Premkumar, G. (2003). A meta-analysis of research on information technology implementation

in small business. Journal of Organizational Computing and Electronic Commerce, 13(2), 91 - 121. doi:10.1207/S15327744JOCE13022

[15] Thong, J. Y. (1999). An integrated model of information systems adoption in small business. Journal of Management Information Systems, 15(4), 187-214.

[16] Kutlu, B., & Ozturan, M. (2008). The usage and adoption of it among SMEs in turkey: An exploratory and longitudinal study. Journal of Information Technology Management, XIX(1), 12 - 24. Retrieved from http://jitm.ubalt.edu.

[17] DeLone, W. H. (1988). Determinants of success for computer usage in small business. MIS Quarterly, 12(1), 51 - 61. Doi: 10.2307/248803

[18] Thong, J. Y. (2001). Resource constraints and information systems implementation in Singaporean small businesses. Omega, 29(2), 143-156. doi:10.1016/S0305-0483(00)00035-9

[19] Levy, M., & Powell, P. (2003). Exploring SME Internet adoption: Towards a contingent model. Electronic Markets, 13(2), 173-181. doi:10.1080/ 1019678032000067163

[20] Cragg, P. B., King, M., & Hussin, H. (2002). IT alignment and firm performance in small manufacturing firms. Journal of Strategic Information Systems, 11, 109–132. doi:10.1016/S0963-8687(02)00007-0

[21] Ismail, N. A., & King, M. (2007). Factors influencing the alignment of accounting information systems in small and medium sized Malaysian manufacturing firms. Journal of Information Systems and Small Business, 1(1 - 2), 1 - 20.

[22] Economic Planning Unit. (2006). Ninth Malaysia plan: 2006 - 2010. Putrajaya, Malaysia: Economic Planning Unit.

[23] Economist Intelligence Unit. (2008). E-readiness rankings 2008: Maintaining momentum - A white paper from the Economist Intelligence Unit. London, England: Economist Intelligence Unit and The IBM Institute.

[24] United Nations Conference on Trade and Development [UNCTAD]. (2011). Information economy report 2011: ICT as an enabler for private sector development. United Nations: New York

[25] Domain Registry. (2011). Domain Registry Statistics. Retrieved from Domain http://www.domainregistry.my/en/statistics.php

[26] Hussin, H., Noor, R. M., & Suhaimi, M. A. (2008). Perceived attributes of e-commerce and the adoption decision: The case of Malaysian SMEs. Jurnal Teknologi Maklumat & Multimedia, 5, 107-125.

[27] Tan, K. S., & Eze, U. C. (2008). An Empirical Study of Internet-Based ICT Adoption among Malaysian SMEs. Communications of the IBIMA, 1, 1 - 12.

[28] Lip-Sam, T., & Chooi-Leng, A. (2008). An integrated model of B2B e-commerce adoption amongst Malaysian SMEs. Journal of Logistics and Supply Chain Management, 1(3), 134 - 144.

[29] Kurnia, S., Alzougool, B., Ali, M., & Alhashmi, S. M. (2009). Adoption of electronic commerce technologies by SMEs in Malaysia. The 42nd Hawaii International Conference on Systems Science. Hawaii, USA

[30] Narayanasamy, K., Santhapparaj, A., & Eze, U. C. (2008). An empirical study of website adoption among Small and Medium Enterprises (SMEs) in Malaysia. Communications of the IBIMA, 2, 50 - 62. Retrieved from http://www. ibimapublishing. com

[31] Hussin, H. & Suhaimi, M. A. (2011). Information technology and business alignment in Malaysian SMEs. IIUM Press: Selangor, Malaysia.

[32] Levy, M., Powell, P. & Galliers, R. (1999). Assessing information systems strategy development frameworks in SMEs. Information & Management, 36, 247 – 261.

[33] Garg, A. & Goyal, D. P. (2012). Striving towards strategic alignment in SMEs: An empirical analysis. Journal of Advances in Management Research, 9 (1), 77 – 95.

[34] Gable, G. (2010) Strategic information systems research: an archival analysis. The Journal of Strategic Information Systems, 19(1). pp. 3-16.

[35] United Nations. (2005). Core ICT indicators: partnership on measuring ICT for development. Printed in UN-ESCWA, Beirut.
[36] Newkirk, H. E., Lederer, A. L., & Johnson, A. M. (2008). Rapid business and IT change: drivers for strategic information systems planning, European Journal of Information Systems, volume 17, pp 198-218.
[37] Cassidy, A. (2006). A practical guide to information systems strategic planning (2nd ed.). Boston, MA: Auerbach Publications
[38] Berry, R. S. Y. (1999). Collecting data by in-depth interviewing. Paper presented at the British Educational Research Association Annual Conference, University of Sussex at Brighton, September 2-5. Retrieved July 1, 2013 from http://www.leeds.ac.uk/educol/documents/000001172.htm.
[39] Gubrium, J. F., & Holstein, J. A. (Eds). (2002). Handbook of interview research. Thousand Oaks, CA: Sage.
[40] Hair, J. F., Black, W. C., Babin, B. J., & Anderson, R. E. (2010). Multivariate data analysis. New Jersey, USA: Pearson Prentice Hall.
[41] Tabachnick, B. G., & Fidell, L. S. (2007). Using multivariate statistics. Boston, MA: Pearson.
[42] Cooper, D., & Schindler, P. (2008). Business research methods (10th ed.). New York, McGraw-Hill/Irwin
[43] Premkumar, G., Ramamurthy, K., & Saunders, C. S. (2005). Information processing view of organizations: An exploratory examination of fit in the context of interorganizational relationships. Journal of Management Information Systems, 22(1), 257 - 294.
[44] Vlachos, I. P. (2011). SMEs e-business behaviour: A demographics and strategic analysis. Journal of Enterprise Resource Planning Studies, 2011, 1 - 21. Retrieved from http://www.ibimapublishing.com
[45] Magal, S. R., Koslage, P., & Levenburg, N. M. (2008). Towards a stage model for e-business adoption among SMEs: Preliminary results for manufacturing and service firms. Fourteenth America's Conference on Information Systems. Toronto, Canada: 14 - 17 August.

IOP Conf. Series: Journal of Physics: Conf. Series **1019** (2018) 012088 doi:10.1088/1742-6596/1019/1/012088

Prediction of crude palm oil solubility in hot compressed water using thermodynamic model prediction

Mohd Sharizan Md Sarip[1]*, Noor Azian Morad[2], Syahrul Affandi Saidi[1], Nazmizan Muhammad[1]

[1]Faculty of Engineering Technology, Universiti Malaysia Perlis, Kampus UNICITI Alam, Sungai Chuchuh, Padang Besar, 02100, Perlis, Malaysia
[2]Shizen Conversion and Separation Technology (Shizen ikohza), Malaysia – Japan International Institute of Technology (MJIIT), Universiti Teknologi Malaysia, Jalan SultanYahya Petra, 54100 Kuala Lumpur, Malaysia

*sharizan@unimap.edu.my

Abstract. The solubility of crude palm oil in hot compressed water was predicted using thermodynamic model prediction. The solubility data was used to understand the phase equilibria in the separation process. In this study, the modified UNIFAC method was used. The solubility prediction using the modified UNIFAC involves several parts that are; pseudo TAG structure development, GC parameter determination and solubility calculation. Pseudo TAG structure was develop based on the crude palm oil composition analyzed using GC-FID. No solubility of CPO in HCW was obtained though this method. Meanwhile, the solubility of HCW in CPO increases from 8.76 to 13.38 mole (%) from the temperatures of 120 to 180°C respectively. The solubility prediction using this method is beneficial due to the less experimental data and inexpensive.

1. Introduction
Phase equilibria is an important perspective that needs to be explored in order to understand the separation process. From this phase equilbria, the distribution of the component can be explained in term of solubility. The efficiency of the process also can be evaluated by comparing the solubility obtained from the experimental data. Phase equilibria is developed from the compounds solubility in the solvent which can be obtained from binary experimental route or thermodynamic model prediction. Generally, there are many experimental routes to obtain phase equilibria data in order to understand the phase behaviour for liquids [1]. The experimental routes includes the static, recirculation and continuous flow methods. Among that, the static method offers the most accurate data compared to the others. However, this method is difficult, expensive, time consuming and is unattainable for the thermo labile compound [2,3]. Therefore, the predictive molecular thermodynamic model is a more preferable method for this purpose. It is a computational method for the phase equilibria prediction which helps researcher to develop the system without or less experimental input. This is beneficial especially for the high pressure process [3] and high molecular weight solute such as triglyceride (TAG). Phase equilibria prediction for systems containing hot compressed water (HCW) is very limited with less than 10 publications found in the literature [4,5,6]. This method includes the empirical model, Hansen solubility parameter, Universal Quasi Chemical Functional Group Activity Coefficient (UNIFAC) and COSMO-RS methods [5,6]. Syed Jaapar et al. (2015) had applied COSMO-RS method to predict the solubility of 6-gingerol and 6-shogaol in HCW with or without the addition of ethanol as entrainer [6]. In their study, the predictive capability of the method had been

validated based on selected hydrocarbons 72 which have similar functional groups with the solute of interest due to the limited experimental solubility data of the compound and obtained AAD not more than 0.2648 [6]. Fornari et al. (2008) applied various versions of UNIFAC model and proposed a modified A-UNIFAC model with temperature dependent associating energy to predict the solubility of polycyclic aromatic hydrocarbon (PAH) in HCW [4]. An average reduction of absolute average standard deviation (AASD) is obtained from 27.7 to 6.1% by the introduction of temperature dependent associating energy in the A-UNIFAC model [4]. However, the study also found that the optimized modified UNIFAC (Dortmund) model had provided the best representation of PAH solubility in HCW compared with UNIFAC, A-UNIFAC and modified A-UNIFAC [4]. Carr et al. (2010) also utilized the modified UNIFAC proposed by Fornari et al. (2008) with some modification on carboxylic acid -water and aromatic hydrocarbon-water interaction parameters to predict the naproxen solubility in HCW between 130 to 170 °C and obtained an average deviation of 6% [5]. Therefore, in this study the modified UNIFAC model are used to predict the solubility of crude palm oil (CPO) in hot compressed water (HCW).

2. Methodology

In order to predict the solubility of crude palm oil (CPO) in hot compressed water (HCW), the modified UNIFAC model was used [8]. The characterization of the component to represent the CPO is one of the important criteria in the solubility prediction since that CPO consist up to 14 different type of triglyceride (TAG). In this study, pseudo-TAG structure was used as proposed by Espinosa et al (2002). The methods show a good performance history and easier method in comparison with the multicomponent system approach. The solubility prediction using the modified UNIFAC involves several parts that are; pseudo TAG structure development to represent CPO, GC parameter determination and solubility calculation as summarized in Figure 1.

2.1. Analysis of CPO using Gas Chromatography -Flame Ionization Detector (GC-FID)

The overall CPO analysis including FFA, MAG, DAG and TAG was conducted using GC-FID (Agilent technologies, USA) based on the AOCS method Cd 11b-91. This analysis was conducted at Sime Darby Research Sdn Bhd, Carey Island, Selangor, Malaysia. In this method, an overall composition of CPO was screened based on the carbon number and composition of fatty acid consist in CPO was identified and used to develop the pseudo-TAG structure for CPO.

2.2 *Calculation of solubility*

From the analysis of TAG composition using GC-FID, the pseudo TAG structure was developed using the equations 1, 2 and 3 as shown in Figure 1 to represent the CPO. This pseudo structure was splitted based on its structural group to predetermine the GC parameter [8]. The GC parameter would be uses to calculate the ln γ as shown in Figure 1. The γ would be used to calculate the mole fraction of the CPO in HCW phases or vice versa using equation 4.

$$(x_i\gamma_i)' = (x_i\gamma_i)'' \tag{4}$$

where x_i is mole fraction of component i, γ_i is activity coefficient and superscripts $'$ and $''$ is mean phase 1 and phase 2 or in this study ; HCW phase and CPO phase. Then, calculated $x_i$was used to constructed the phase diagram for the system containing HCW and CPO. The ICAS 13.1 software (CAPEC, Dermark) was uses to calculate this solubility.

IOP Conf. Series: Journal of Physics: Conf. Series **1019** (2018) 012088 doi:10.1088/1742-6596/1019/1/012088

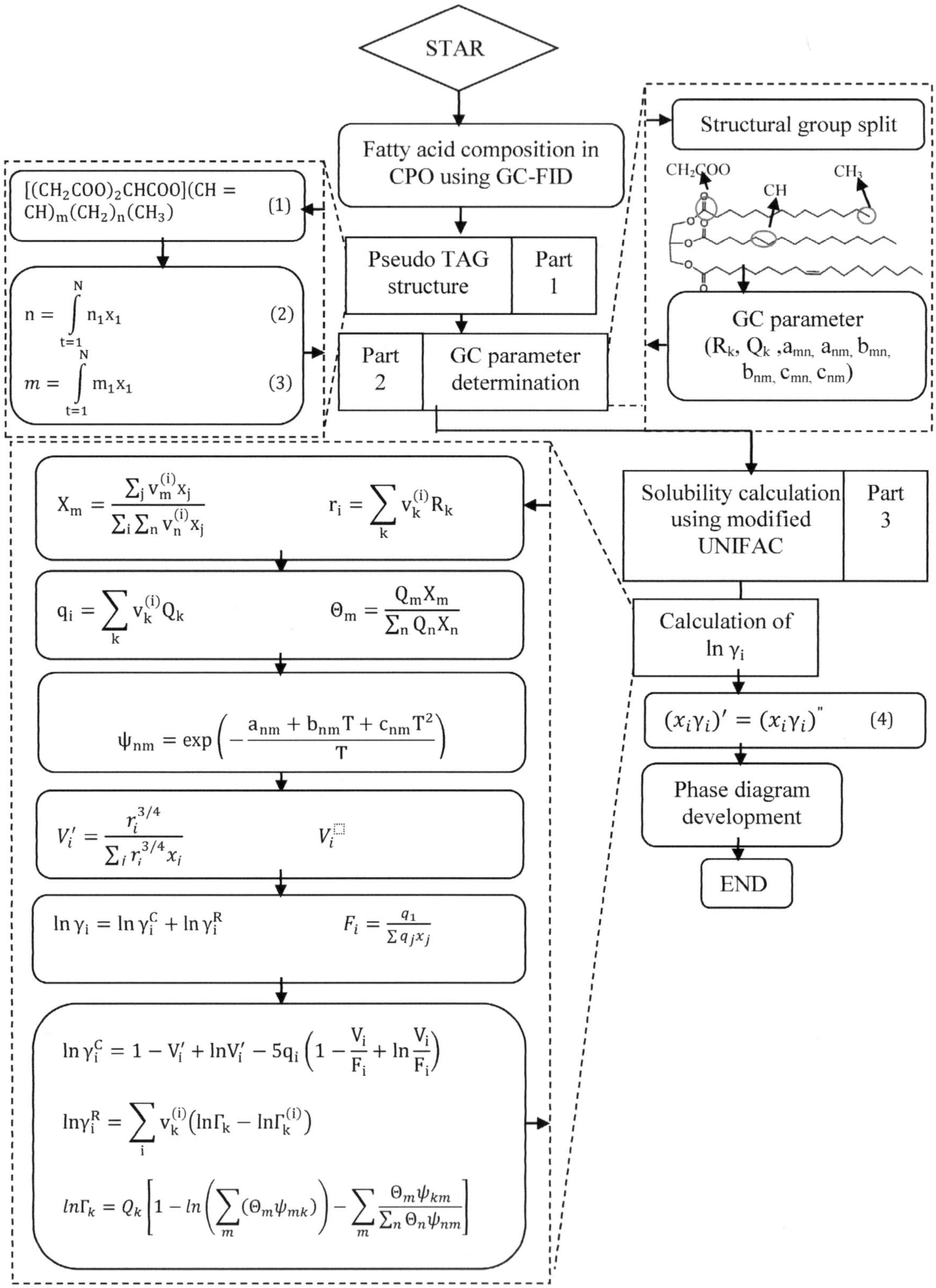

Figure 1: Calculation of TAG solubility in HCW using modified UNIFAC

3 Result and discussion

3.1 Analysis of fatty acid using GC-FID

The fatty acid composition of CPO obtained from HCWE process is tabulated in Table 1. As shown in Table 1, the HCWE CPO consist a slightly higher saturated fatty acids at 50.630.15% in comparison with commercial CPO at 49.91% [9]. However, the compositions are still in the range of the Malaysian CPO (Corley and Tinker, 2003b).

3.2 Calculation and pseudo TAG structure development for crude palm oil (CPO)

Table 1 is shows the calculation of m for CH=CH functional group and n for CH_2 functional group using fatty acid composition in CPO corresponding to the simple TAGs. For example, P(C16:0) represent Tripalmitin (PPP) TAGs.

Table 1: Calculation for pseudo-TAG structure

Fatty acid	x_i	n_i	m_i
La (C12:0)	0.03	30	0
M(C14:0)	1.59	36	0
P (C16:0)	44.89	42	0
S (C18:0)	4.12	48	0
Ar (C20:0)	0.15	54	0
Pi (C16:1)	0.87	36	3
O (C18:1)	37.16	42	3
L (C18:2)	11.55	36	6
Li (C18:3)	0.35	30	9

From Table 1, the values of 35.5~36 and 1.9~2 were obtained for n and m respectively. Therefore, the Pseudo-TAG molecule to represent CPO is $[(CH2COO)_2CHCOO](CH = CH)_2(CH_2)_{36}(CH_3)_3$. The van der Waals sphere representation for this molecule is shown in figure 2.

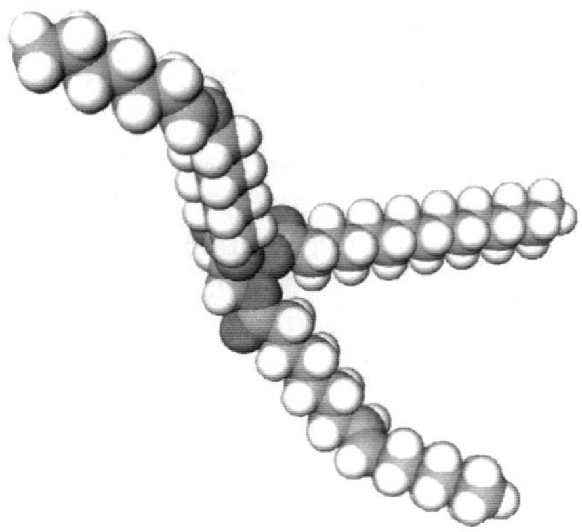

Figure 2: *Pseudo*-TAG structure to represent the crude palm oil

3.3 Solubility of crude palm oil (CPO) in hot compressed water (HCW)

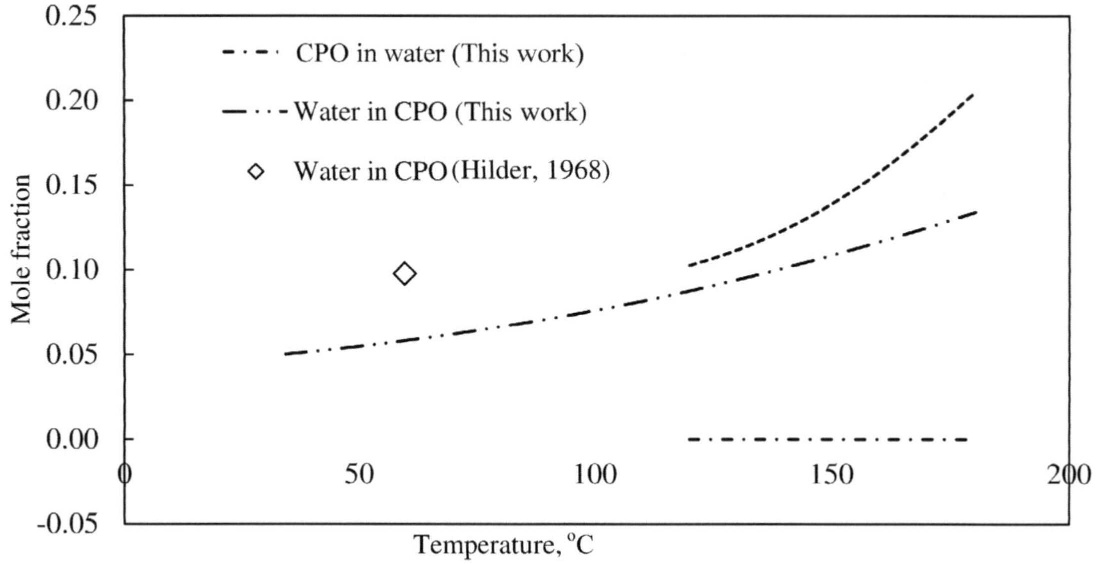

Figure 3: Predicted solubility of CPO in HCW and HCW in CPO.

The predicted solubility of CPO in HCW and HCW in CPO is shown in Figure 3. This solubility was calculated between the temperatures of 60 to 180°C. Water was set as liquid phase at all conditions. No solubility of CPO in HCW was obtained though the this method represent as one solubility line in Figure 3. This indicated that CPO was completely insoluble in the HCW from the temperature of 120 to 180°C. CPO which mainly consist of TAGs is known insoluble compounds with water due to higher hydrophobic chain length of the acid [10]. Meanwhile, the solubility of HCW in CPO increases from 8.76 to 13.38 mole (%) from the temperatures of 120 to 180°C respectively. This data was slightly lower by about 1.7 folds in comparison with binary experimental data as tabulated in Figure 3 [11]. This showing that the solubility is irrelevant to effect the quality of the CPO due to its lower concentration for less than 0.15% (w/w) and the hydrolysis during the process can be ignored.

4 Conclusion

This paper shows that the thermodynamic model prediction can be used to predict the solubility of CPO in hot compressed water with less experimental data. The experimental measurement of solubility was very difficult and expensive for hot compressed water system. This data is important the understanding of the mass transfer mechanism in separation process.

5 Acknowledgments

The help and permission from Prof. Ir. Dr. Sharifah Rafidah Datu Wan Alwi from PROSPECT, UTM for the use of ICAS 3.1 is greatly acknowledged. The help and assistant from Pn Norliza the Plantation Manager from Felda Mempaga Satu, Bentong, Pahang for fresh fruit bunch supply is greatly acknowledged.

References

[1] Bogatu, C., Vilcu, R. and Duta, A. (2005). Experimental Methods for Study HighPressure Phase Behaviour. Part 1: Static Method. AnaleleUniversitii din BucurestiChimie, Anul XIV. 1-2, 193–203.

[2] Cunico,L. P., Hukkerikar, A.S., Ceriani, R.,Sarup, B. and Gani, R. (2013). Molecular Structure-Based Methods of Property Prediction in Application to Lipids: A Reviewand Refinement. Fluid Phase Equilibria. 357, 2–18.

[3] Skjold-Joergensen, S. (1988). Group Contribution Equation of State (GC-EOS):a Predictive Method for Phase Equilibrium Computations Over Wide Ranges ofTemperature and Pressures up to 30 MPa. Industrial & Engineering ChemistryResearch. 27(1), 110–118.

[4] Fornari, T., Stateva, R. P., Seorans, F. J., Reglero, G. and Ibaez, E. (2008). Applying UNIFAC-based Models to Predict the Solubility of Solids in Subcritical Water. *TheJournal of Supercritical Fluids*. 46(3), 245–251.

[5] Carr, A. G., Mammucari, R. and Foster, N. R. (2011). A Review of Subcritical Water asa Solvent and its Utilisation for the Processing of Hydrophobic Organic Compounds.*Chemical Engineering Journal*. 172(1), 1–17.

[6] Syed Jaapar, S. Z., Morad, N. A., Iwai, Y., Md. Sarip, M. S., Yunus, M. A. C. andYusuff, Y. A. M. (2015). Estimation of Ginger Bioactive Compound Solubilities in Hot Compressed Water and Enhancement with Entrainer. Fluid Phase Equilibria.393(0), 40–47.

[7] Carr, A. G., Mammucari, R. and Foster, N. R. (2010). Solubility, Solubility Modeling,and Precipitation of Naproxen from Subcritical Water Solutions. Industrial &Engineering Chemistry Research. 49(19), 9385–9393.

[8] Weidlich, U. and Gmehling, J. (1987). A Modified UNIFAC Model. 1. Prediction ofVLE, hE, and 1. Industrial&Engineering Chemistry Research. 26(7), 1372–1381.

[9] Che Man, Y. B., Haryati, T., Ghazali, H. M. and Asbi, B. A. (1999). Composition andThermal Profile of Crude Palm Oil and its Products. Journal of the American Oil Chemists' Society. 76(2), 237–242.

[10] Bockisch, M. (1998). *Fats and Oils Handbook*. Champaign, Illinois: AOCS Press.

[11] Hilder, M. H. (1968). The solubility of water in edible oils and fats. Journal of theAmerican Oil Chemists Society. 45(10), 703–707

Innovation of flipped learning encouraging better communication and critical thinking skills among accounting students

Marhaiza Ibrahim[1], Norhaiza Khairudin[2] and Danilah Salleh[3]

[123] Tunku Puteri Intan Safinaz School of Accountancy,
Universiti Utara Malaysia,
Malaysia
E-mail:marhaiza@uum.edu.my

***Abstract*:** Currently the trend among the lecturer to adopt innovation and technology as a tool in the lecture hall in fostering and increasing the positive impact among the students. Innovation and technology may encourage better impact to develop soft skills among the students such as communicate effectively, critical thinking skills, demonstrate appropriate interpersonal skills and teamwork, demonstrate leadership qualities and entrepreneurial skill. Many researchers have attempted to study the efficiency of e-learning and blended learning concepts on teaching and learning processes. One of the most recent developments in technology enhanced method in teaching and learning is a method called flipped learning. It is an approach where the traditional one-way lecture is flipped to make room for active learning opportunities where educators shift learning into the individual learning space facilitated by technology. This study implemented flipped learning approach in the effort to enhance communication and critical thinking skills among accounting students in Tunku Puteri Intan Safinaz School of Accounting (TISSA) in Universiti Utara Malaysia, UUM. The objective of this study is to examine how the use of flipped learning can help to improve communication and critical thinking skills in problem solving among students taking the accounting subject. From this study, the results showed that students' communication and critical thinking skill has developed with the use of flipped learning. Sufficient time however, is needed for students to understand the content from the lectures given in the form of a video used as the media in this project used as an e-learning platform in this project, as well as the 'catchy' contents should be created to enhance interest from the students to watch the video.

Keywords: Flipped learning, technology in education, action research

1. Introduction

The use of technology in delivering university's courses has become ubiquitous in this era of knowledge enhancement. Since the 1980's the use of technology to deliver accounting courses were strongly recommended (Friedman, 1981). Nevertheless, the use of technology in university courses has been reportedly to be largely focused in delivering contents rather than as a medium to interact among students. In the traditional teaching method which is also known as the teacher-centred learning approach, the teacher or instructor delivers knowledge in a classroom in a one-way direction (Keller, Hassell, Webber and Johnson, 2009). This method urges students to be passive learners where they solely rely on learning by listening, memorising and knowledge repetition. Flipped learning is an approach where the traditional one-way lecture is flipped to make room for active learning opportunities where educator shift learning into the individual learning space facilitated by technology (Jone and Chen,2008; Kember and Leung,2005)). This paper attempted to implement flipped learning approach in a classroom to enhance active learning opportunities among business accounting students in the Tunku Intan Safinaz School of Accounting (TISSA) in Universiti Utara Malaysia.

This study employed qualitative action research methodology (McNiff, 2010). Action research is utilized in many professional learning contexts where action inquiries begin with the query, 'How do I improve my study? (McNiff, 2010). In this study, the researcher carried out a systematic investigation into her own experience of teaching Business Accounting subject. Business accounting students comprised of 49 participants were under studied. The researcher's reflection, observation checklist and interviews were utilized to collect data throughout three phases; Pre-Implementation, During-Implementation and Post-Implementation. After the activities for each phase are completed, reflections to the students and lecturers are obtained. Qualitative data analyses which include thematic analysis are used to analyse the qualitative data.

Pre – Implementation

During the pre-implementation phase, we produced two sets of videos on the lecture of two chapters from the syllabus. On the first day of class, the explanation on the project and the use of flipped learning were explained to students. There was a questions and answers session which allowed students to really understand the process and that everything are clarified. During this phase we also made all the preparations needed for the class to use flipped learning (Michael, 2006). The videos were developed and uploaded to UUM Online Learning prior to the class sessions.

During Implementation

During the implementation phase, we conducted two phases of activities in each of the classes involved in this project. Two classes were involved. The first one was the class on Budgeting topic and the next one was the class on Variance Analysis topic.

Phase 1 – Students were first asked to watch the video as their homework before the topic is scheduled in class. The students were also given the link to the exercise questions on the topic to be discussed in class at the end of the video. This will prompt the students to finish watching the video and get the link to download the questions to be discussed in class. In the class, the students were first given a set of open ended questions for them to give some reflections on the video of the topic given and how they think flipped learning approach will help to understand more on the topic.

Reflection questions before class activities started- Phase 1

1. What do you think about the lessons given in the video?
2. Do you think the video on the lecture is better than the normal way of conveying lecture (face to face)? Why and why not?
3. Do you think the videos can assist you in answering the exercise questions?

During class, presentations and discussions (pairing, grouping, mapping etc.) were conducted. Evaluation and analysis on the presentations and discussions were conducted based on the rubrics. During the 30 minutes before the class ended, we performed some activities to obtain reflections on the class activities from the students. Students were asked to answer another set of open ended questions which ask the students' views on the class session of the first chapter after their experience using flipped learning approach. The enhancement in critical thinking and problem solving skills and communication skills can be obtained by the open ended questions given as well as the reflections by the lecturer that was gained through the observation process. An overall reflection by the lecturer was obtained based on the overall achievement of the sessions.

Reflection questions after class activities being conducted –Phase 1

1. Do you think that the video has assisted you in answering the exercise questions?

2. Did you discuss and communicate with your friends during the discussion on the questions?

3. Overall, do you like this new teaching and learning method compared to the normal teaching method? Why or Why not?

The data gathered from the data collection activities in Phase 1 were analysed (Du, 2011). The reflections from the students and the lecturer are considered important in order to enhance the process of conducting flipped learning approach in class (Grandzol,2004). All the suggestions and reflections provided from the student were undertaken and applied in the process of producing the next phase of video content. The second video on the preceding topic were then developed. In this video some modification have been made on the preparation of the content including shorter and more precise content, more animation on the slides to make it more interesting and shorter and clearer explanations (Freeman and Hancock,2013).

Phase 2 – A video on the next topic was developed based on the suggestions and observations input gathered and analysed in Phase 1. Students were asked to watch the video before class as their homework. Before the class started, the students were first given a set of open ended questions for them to give some reflections on the video on the topic given and how and how they think the flipped learning approach will help to understand more on the topic (Dowling,Godfrey and Gyles, 2003).

Reflection questions before class activities started – Phase 3

1. Do you think that the content of the video is better delivered that the previous one? How?

2. Do you think that the video will help you better in understanding the lessons and exercises which is going to be done in class?

During the 30 minutes before the class ended, we performed some activities to obtain reflections from the students (Bachhel and Thaman, 2014). Students were asked to answer another set of open ended questions which ask the students' views on the class session of the first chapter after their experience using flipped learning approach after modifications were made on delivery of content in the video. The enhancement in critical thinking and problem solving skills and communication skills can be obtained by the open ended questions given as well as the reflections by the lecturer that was gained through the observation process (Chiu and Cheng, 2017; Cheng and Jone, 2007). An overall reflection by the lecturer was obtained based on the overall achievement of the sessions.

Reflection questions after class activities being conducted – Phase 3

1. Do you think that the video had helped you in understanding the lesson better?

2. Do you think that this method of learning and teaching improves your communication skills with your groupmates? How?

3. Do you think by watching the class video before class can give you more time on the exercise and think more critically in class/exercise session?

Post – Implementation

After the open ended questions data were collected, interviews were conducted to obtain further feedback on the flipped learning approach used in teaching the course. The interview is conducted to obtain the participants' feedback on their experience using flipped learning approach and to understand their perceptions on flipped learning and how they gained the benefits from the sessions. All the data were analysed and reported.

Analysis of Data

There are two types of data were collected and analysed for this paper such as qualitative data from the open ended questions given to the participants before and after the class activities conducted and qualitative data from the interviews providing overall perceptions of the flipped learning method used in the teaching and learning process. The qualitative data collected from these two activities were coded and analysed using thematic analysis (Patton, 2002), which involves assigning each unit of data its own unique code and identifying repetitive patterns of actions and consistencies. Thematic analysis is among the collective forms of analysis in qualitative research.Data coding was performed to identify the themes and categories of qualitative data (Anderson and Moore,2005) . Data coding is a cyclical process, where usually several cycles of coding processes are needed for a good coding result, and the cycles of coding processes are performed to manage, filter, highlight, and focus the salient features of the qualitative data with the aim to generate categories, themes, concepts and build theories (Creswell, 2013; Saldana, 2009).

Encouraging better communication and critical thinking skills among accounting students during post – implementation phase

During the post-implementation phase, interviews were conducted to obtain further feedback on the flipped learning approach used in teaching the course. Six participants were interviewed. Appointments to interview them were made after class and the students were interviewed in the lecturer office.

From the results, it has been found that students have positive thoughts about flipped learning concept (Detlor, Booker, Serenko and Julien, 2012). However they are concerned about the video contents that are too lengthy, boring and uninteresting. Some of their comments are as follows:

"It is good. But the video which is too long and no creativity will make students fell bored and fall sleep. But the discussion is good because we can ask questions, rather than just lecture in class".

"To be honest, when I watched the video, it was actually at night, I got sleepy from watching the video and it was too long. But it gives a picture on what the lecturer is going to teach in class so it helps me a bit on the preparation even though I couldn't understand the whole content".

"Flipped learning concept has its pros and cons. First I think it's ok but when too long and too many details make it boring".

"For me the idea is very good. When I get the video, I did not watch it, I listen to it while having my breakfast. So it saves my time".

The result shows that students able to demonstrate passion for the topic presented. The classroom became lively where the students can actively communicate with each other on the exercise given. This is consistent with previous studies which shows that through flipped learning, the students actually are able to identify issue/ problem in a complex situation and able to assess and justify the situation (Kashefi, Ismail and Mohammad Yusof, 2012). When asked whether they can communicate better with their friends and lecturers through flipped learning activities, the participants give the following answers:

Yes. One day before class we communicate with WhatsApp".

"In class we can communicate".

"Yes, we have the opportunity to discuss in class. Some of my friends can understand more than me so they can teach us during the discussion".

"Yes, whenever we can't hear what the lecturer says in front because I sit at the back, so when we can discuss and communicate, I can ask my friends anything that I don't understand".

"Yes I communicate a lot during discussions with my friends. Moreover I am not good in English so this improves my English because I am force to speak and discuss with my friend at my table who are foreign students".

When asked whether they can communicate better with the lecturer through flipped learning approach, the participants said that:

"Yes, because we can ask the lecturer that time".

"Of course, because most of the times lecturers come in class and teach and we can only do minimal exercises in class, the rest we have to do it ourselves and we don't even have the time to ask the lecturer more detail. So when the class is fully utilised with exercises

activity, it is good, when we don't understand we can straight ask the lecturer questions. That is one advantage for us".

"I like that I can ask the lecturer during class and not like usual where I need to wait until the next class to ask the question on that topic".

"Yes I feel that my time spent in class is so valuable because this is the time I can speak to my lecturer".

However, from the results, it has been found that that not all students are able to watch the video. This is due to limited time given for the students to watch the video. They also stressed that they need more time to watch the video repeatedly in order to understand. Reflectively, not all students can understand the content of the video since their level of understanding are at different level.

The participants were also asked in to give their opinion on how to improve the method used to conduct flipped learning in the future. The participants suggested that the video to convey the topics should be less lengthy and should be more interesting. One participant also added that students can do presentations based on the lessons given in the video. The followings are their suggestions:

"Lessen the length".

"The tone of voice in the video plays an important part. It must be interesting. The content or PowerPoint could be more interesting by adding more animations colours and creativity. We will feel more alive when watching the video and not get sleepy. And the video must not be too long".

2. Conclusion

Researchers have demonstrated positive impacts resulted from their studies done to examine the effect of flipped learning approach on students' learning processes (Ferreira-Meyers, 2015; Musallam, 2010; Warter-Perez & Dong, 2012). From the results of their survey, Warter-Perez and Dong (2012) demonstrated that students felt that the flipped class learning environment was more interactive and that the new learning environment allowed them to gain better hands-on skills and they also agreed that the flipped class helped them to learn the content better. In addition, Ferreira-Meyers (2015) argued that the flipped learning model can enable educators to pull in the transformation from instructor-driven education to student-centered learning where the excellence in students' communication and critical thinking skills are the essence of concern.

References

[1] Anderson, B., & Moore, A. (2005). Active learning for hidden markov models: Objective functions and algorithms. *In Proceedings of the 22nd international conference on Machine learning* (pp. 9-16). ACM.

[2] Bachhel, R., & Thaman, R. G. (2014). Effective use of pause procedure to enhance student engagement and learning. *Journal of clinical and diagnostic research: JCDR*, *8*(8), XM01.

[3] Chiu, P. H. P., & Cheng, S. H. (2017). Effects of active learning classrooms on student learning: a two-year empirical investigation on student perceptions and academic performance. *Higher Education Research & Development*, *36*(2), 269-279.

[4] Chen, C. C., & Jones, K. T. (2007). Blended Learning vs. Traditional Classroom Settings: Assessing Effectiveness and Student Perceptions in an MBA Accounting Course. *Journal of educators online*, 4(1), n1.

[5] Creswell, J. W. (2013). *Research design: Qualitative, quantitative, and mixed methods approaches.* Thousand Oaks CA Sage.

[6] Detlor, B., Booker, L., Serenko, A., & Julien, H. (2012). Student perceptions of information literacy instruction: The importance of active learning. *Education for Information*, *29*(2), 147-161.

[7] Dowling, C., Godfrey §, J. M., & Gyles, N. (2003). Do hybrid flexible delivery teaching methods improve accounting students' learning outcomes? *Accounting Education*, 12(4), 373-391.

[8] Du, C. (2011). A comparison of traditional and blended learning in introductory principles of accounting course. *American Journal of Business Education (AJBE)*, 4(9), 1-10.

[9] Ferreira-Meyers, K. (2015). What emerging technology can do for learning environments: the case of moocs. *Journal of English Literacy Education*, 1(2), 12-24.

[10] Freeman, M., & Hancock, P. (2013). Milking MOOCs: Towards the right blend in accounting education. The Virtual University: Impact on Australian Accounting and Business Education, part B, 86-100.

[11] Friedman, M. E. (1981). The effect on achievement of using the computer as a problem-solving tool in the intermediate accounting course. *Accounting Review*, 137-143.

[12] Grandzol, J. R. (2004). Teaching MBA statistics online: A pedagogically sound process approach. *Journal of Education for Business*, 79(4), 237-244.

[13] Jones, K. T., & Chen, C. C. (2008). Blended-learning in a graduate accounting course: Student satisfaction and course design issues. *The Accounting Educators' Journal*, 18.

[14] Kashefi, H., Ismail, Z., & Yusof, Y. M. (2012). The Impact of Blended Learning on Communication Skills and Teamwork of Engineering Students in Multivariable Calculus. *Procedia-Social and Behavioral Sciences*, 56, 341-347.

[15] Keller, J. H., Hassell, J. M., Webber, S. A., & Johnson, J. N. (2009). A comparison of academic performance in traditional and hybrid sections of introductory managerial accounting. *Journal of Accounting Education*, 27(3), 147-154.

[16] Kember, D., & Leung, D. Y. (2005). The influence of active learning experiences on the development of graduate capabilities. *Studies in Higher Education*, *30*(2), 155-170.

[17] McNiff, J. (2010). *Action research for professional development: Concise advice for new action researchers*. Dorset: September books.

[18] Michael, J. (2006). Where's the evidence that active learning works? *Advances in physiology education*, 30(4), 159-167.

[19] Musallam, R. (2010). The effects of screen casting as a multimedia pre-training tool to manage the intrinsic load of chemical equilibrium instruction for advanced high school chemistry students. University of San Francisco, San Francisco.

[20] Patton, M. Q. (2002). *Qualitative research and evaluation methods*: Sage Publications, Inc.

[21] Saldana, J. (2009). *The coding manual for qualitative researchers*: Sage Thousand Oaks, CA.

[22] Warter-Perez, N., & Dong, J. (2012). Flipping the classroom: How to embed inquiry and design projects into a digital engineering lecture. Paper presented at the,*Proceedings of the 2012 ASEE PSW Section Conference*.

IOP Conf. Series: Journal of Physics: Conf. Series **1019** (2018) 012090 doi:10.1088/1742-6596/1019/1/012090

Sawdust Flour Reinforced Epoxidized Natural Rubber: Effect Of Trans-Polyoctylene Rubber on Curing Characteristics and Tensile Properties

Omar S Dahham[1], N Z Noriman[1], R Hamzah[1], S Z Syed Idrus[2], Z Shayfull[3,4], T Adam[1,5], H Jaya[1]

[1] Center of Excellence Geopolymer and Green Technology (CEGeoGTech), Faculty of Engineering Technology (FETech), Universiti Malaysia Perlis (UniMAP), Level 1 Block S2, UniCITI Alam Campus, Sungai Chucuh, Padang Besar, 02100, Perlis, Malaysia.

[2] School of Human Development and Techno-communication (iKOM), Green Advanced Computing and Technology (GREAT) Research Group – CEGeoGTech, Universiti Malaysia Perlis (UniMAP)

[3] School of Manufacturing Engineering, Universiti Malaysia Perlis, Arau, 02600, Malaysia

[4] Green Design and Manufacture Research Group, Center of Excellence Geopolymer and Green Technology (CEGeoGTech),Universiti Malaysia Perlis, Kangar, 0100, Malaysia

[5] Institute of Nano Electronic Engineering (INEE), Universiti Malaysia Perlis (UniMAP), 01000 Kangar, Perlis, Malaysia

E-mail: niknoriman@unimap.edu.my

Abstract. In this work, the compatibilization effect of trans-polyoctylene rubber (TOR) on the cure characteristics, tensile properties and morphology of epoxidized natural rubber/ sawdust flour (ENR-25/SD) compounds was evaluated. Five different TOR content (2, 4, 6, 8 and 10 phr) was used. Results showed that the increasing of TOR content caused decreasing on scorch and cure time. Moreover, the maximum torque has improved. The minimum torque showed decreasing as TOR content increased indicating that the rubber compound became less viscos and easy to process. The addition of TOR, particularly high TOR content (10 phr) into rubber compounds has increased the interfacial adhesion between SD and ENR-25, which in turn improved the tensile strength and modulus of the rubber compounds. The elongation at break has decreased as TOR content increased indicating that the samples became less flexible. SEM micrographs proved that the high TOR content presented stronger SD/ENR-25 than low TOR content.

1. Introduction

Nowadays, wood polymer Composites is used extensively as a raw material in building such as fences, railings, windows, doorframes, park benches and other. This type of composite gives numerous advantages such as low cost, dimensional stability, good mechanical properties, long service life and less deterioration compared to natural wood it self [1,2]. The sawdust of wood is often used as filler in the production of wood polymer Composites due to the environmental friendly, bio-degradability characteristics and low cost. Furthermore, sawdust fibers are less abrasive to processing machine and abundantly available in the nature [3]. In the past few years, several studies were conducted on the natural fibers using as competently as possible to produce high quality fiber-reinforced polymer (thermoplastic, thermoset and rubber) composites for a wide range of applications [4-8]. Natural fibers are currently used as an alternative material to conventional fibers, such as carbon, glass, aramid and other due to advantages of natural fiber compared to conventional fibers as mention above. Nonetheless, the declining of compatibility of polymer composites as natural fiber added is the major constraint in the polymer industry.

Compatibilization is defined as a process that uses to reduce the high interfacial tension between the thermoplastic and elastomeric phases of the polymer components [9]. Based on researches that found in the literature, the compatibilization effects can increase the interfacial adhesion, which in turn lead to improve the mechanical properties of the polymer components. This process could be achieved either by adding a new component, called a compatibilizer, or by improving the polymer components interaction mechanically or chemically. One of the most effective techniques that used widely for compatibilization is adding a third polymer into the polymer components to be partially miscible in early stage [10].

Trans-polyoctylene rubber (TOR) was used widely as a compatibilizer in incompatible polymer blends, particularly the heterogeneous rubber blends. TOR introduces a good processability in the rubber processing with a temperature range (100-150 ^{0}C). Furthermore, TOR can present a good collapse resistance below the melting temperature because of the recrystallization [11]. Several researchers studied the compatibilization effect of TOR on the properties of polymeric materials such as rubber blends. Bizi et al. have studied the TOR effects on the properties of NR/SBR blends. Results showed that the physical properties of the rubber blends have clearly improved [10]. Our previous researches scrutinized the TOR influence on the Properties different rubber compounds. The results of these researches proved that most of the rubber blends properties were improved as TOR added [8].

The aim of the current research is to evaluate the impact of different TOR content as a compatibilizer on the cure characteristics, tensile properties and morphology of the epoxidized natural rubber / sawdust flour compounds

2. Experimental

2.1. Materials

Epoxidized natural rubber with 25mol% epoxidation (ENR-25) was obtained from Rubber Research Institute of Malaysia (RRIM). Whereas other materials such as zinc oxide, stearic acid, Trans-Polyoctylene Rubber sulfur and N-cyclohexyl-2-benzothiazole sulfonamide (CBS) were obtained from Anchor Chemical Co. (M) Ltd. Sawdust (SD) was collected from Perlis Sawmill Sdn. Bhd. The formulation of the rubber compounds of this study is shown in Table 1.

Table 1. The formulation of ENR-25/SD/TOR compounds at different TOR content.

Ingredients	Phr					
	R0	R02	R04	R06	R08	R10
ENR-25	100	100	100	100	100	100
Sulfur	2	2	2	2	2	2
CBS	1	1	1	1	1	1
Stearic Acid	2	2	2	2	2	2
ZnO	5	5	5	5	5	5
SD	5	5	5	5	5	5
TOR	0	2	4	6	8	10

IOP Conf. Series: Journal of Physics: Conf. Series **1019** (2018) 012090 doi:10.1088/1742-6596/1019/1/012090

2.2. Sample Preparation and analysis

After Sawdust cleaning and drying Crusher model RT34 (Chyun Industrial Co. Ltd.) was utilized for sawdust grinding. Lab sieves were used to obtain SD flour at 100 - 200 μm size range. ENR-25 was mixed together with the other ingredients using two-roll mill X (S K - 160 X 320) based on D 3184-89 and then rubber compounds were put inside freezer for 48 hours to avoid further crosslink. After freezing, hot press machine was used to cure and mold the rubber compounds with 160^0C temperature and under 30 tonne pressure. The cure characteristics test was carried out using Rheometer (MDR 2000) according to ASTM D 2240-93. A dumbbell-shaped mold was used to obtain tensile test samples. According to ASTM D412, tensile test was conducted using universal testing machine (Instron 5582). The tensile fracture surface of samples that obtained from tensile test was analyzed using scanning electron microscopy (SEM) model JEOL JFC6460LA. Before SEM analysis, palladium layer (1.5 - 3 nm) was coated samples using sputter coater machine to increase the image resolution and also to avoid the electrostatic charging during samples analyzing.

3. Results and discussion

3.1. Cure characteristics

The results of scorch time t_2, cure time t_{90}, minimum torque M_L and maximum torque M_H of the rubber compounds with different TOR content are shown in Table 2 respectively. It can be seen that the value of t_2 and t_{90} become shorter as TOR content increased in the rubber compounds. This could be attributed to the nature of TOR that acts as a compatibilizer in the rubber compounds. Additionally, TOR also could behave as an unsaturated rubber, which in turn lead to increasing of bonding between TOR itself and elastomer in short time. The increasing of TOR content decreased the viscosity and improved the processability of the rubber compounds due to the plasticizer effect of TOR [12,13]. Therefore, M_L decreased. By contrast, the value of M_H presented an opposite direction, which increased as TOR content increased in the rubber compounds. The increasing of TOR content has increased the crosslinks between the elastomer and TOR. Thus, M_H value improved.

Table 2. The cure characteristics of ENR-25/SD/TOR compounds.

TOR	t_2 (min)	t_{90} (min)	M_L (dNm)	M_H (dNm)
R0	1.81	9.22	8.30	22.5
R02	1.23	7.71	6.10	26.15
R04	1.15	6.89	5.90	28.00
R06	0.99	6.02	5.12	28.95
R08	0.84	5.54	4.55	31.40
R10	0.57	4.21	3.88	35.10

3.2. Tensile properties

The results of tensile properties, such as tensile strength (Ts), modulus (M100) and elongation at break (Eb) of the rubber compounds at different TOR content are listed in Table 3 respectively. It's obviously seen that the Ts values have increased linearly with the increasing of TOR content in the rubber compounds. The addition of TOR, particularly high TOR content has increased the dispersion of SD in the rubber matrix, which in turn improved the interfacial adhesion of the rubber compounds [14]. M100 values shown same direction of Ts value. The addition of TOR, particularly high TOR content into the rubber compounds has increased the incorporation between SD and ENR-25 matrix, which in turn gave a positive impact on the modulus value. By contrast, Eb shows decline as TOR content increased in the rubber compounds. The addition of TOR, particularly high TOR content hindered the flow and mobility, which means the flexibility of the rubber compounds has reduced.

Table 3. The variation of tensile properties of ENR-25/SD/TOR compounds.

TOR	Ts (MPa)	Eb (%)	M100 (MPa)
R0	12.4	0.55	1245
R02	14.12	0.75	1012
R04	14.90	0.85	989
R06	15.25	0.90	901
R08	16.30	1.00	814
R10	18.95	1.25	694

3.3. Morphology

The SEM micrographs of tensile fracture surface of ENR-25/SD/TOR at 2 phr and 10 phr TOR content are shown in Figure 1 (a) and (b) respectively. It's clearly seen both two samples, low (2 phr) and high (10 phr) TOR content shows several tearing line on the tensile fracture surface of the samples indicating that the samples needed more energy to break the matrix [15,16]. The high TOR content sample shows strong SD bonding in SD ENR-25 due to the TOR effect that increased compatibility between SD in rubber matrix. However, the low TOR content shows less SD bonding in ENR-25 matrix due to the relatively poor compatibility between SD and ENR-25 compared to high TOR content, which caused breakages and pull out of the SD and few holes occurred.

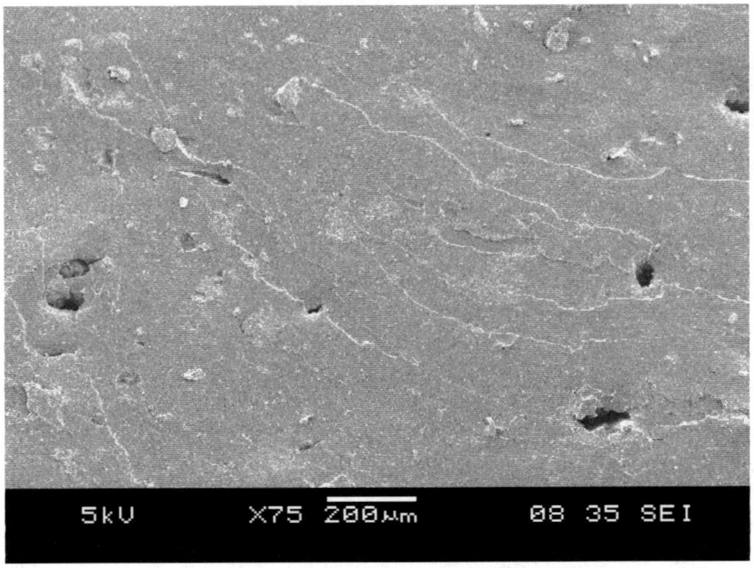

(a)

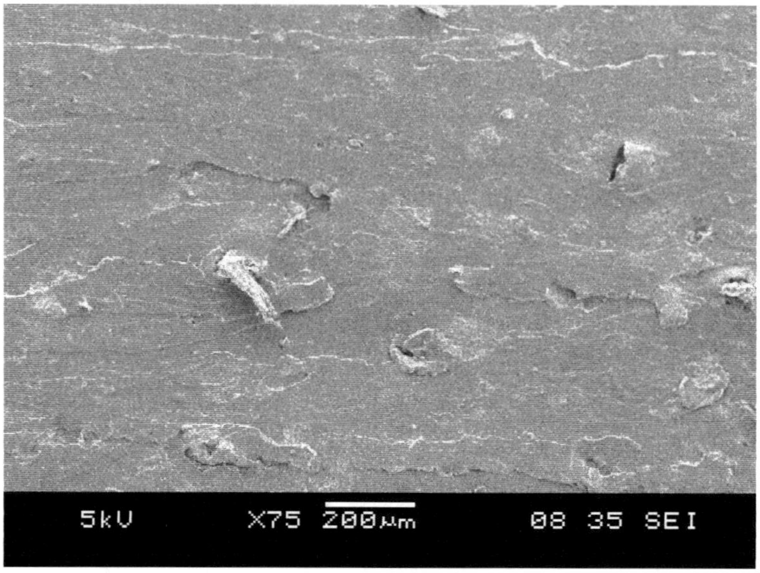

(b)

Figure 1. The SEM of tensile fracture surfaces of ENR-25/ SD/TOR compounds at (a) 2 phr TOR and (b) 10 phr TOR

4. Conclusion

Due to the nature of TOR that can acts as an unsaturated rubber and also as a compatibilizer in the rubber compounds, the curing characteristics have improved as TOR added into rubber compounds. Furthermore, the addition of TOR, particularly high TOR content has increased the dispersion of SD in the rubber matrix, which in turn improved the interfacial adhesion of the rubber compounds. Therefore, The tensile properties have improved. This matter was proven in SEM section.

Acknowledgments

The authors would like to acknowledge the Fundamental Research Grant Scheme (FRGS) Phase 1/2015 (9003-00523) under Ministry of Higher Education, Malaysia. We also would appreciate Universiti Malaysia Perlis (UniMAP) for providing us all the facilities to make this research.

References

1. T. Fakhrul, R. Mahbub and M. A. Islam, *Journal of Modern Science and Technology* **1**, 135 (2013).
2. H. Ismail and M. N. Salmah, *Polymer testing* **20**, 819 (2001).
3. N. Sombatsompop, K. Chaochanchaikul, C. Phromchirasuk and S. Thongsang, *Polymer international* **52**, 1847 (2003).
4. D. N. Saheb and P. P. Jog, *Advances in polymer technology*, **18**, 351 (1999).
5. K. Balaed, N. Z. Noriman, O. S. Dahham, S. T. Sam, R. Hamzah and M. F. Omar, *International Journal of Polymer Analysis and Characterization*, 1-9 (2016).
6. V. K. Thakur, M. K. Thakur and R. K. Gupta, *International Journal of Polymer Analysis and Characterization*, **19**, 256 (2014).
7. S. Taj, M. A. Munawar and S. Khan, Proceedings-Pakistan Academy of Sciences, **44**, 129 (2007).
8. O. S. Dahham, N. Z. Noriman, S. T. Sam, H. Ismail, S. Ragunathan, H. Rosniza and M. N. Al-Samarrai, *Journal of Polymer Materials*, **33**, 647 (2016).
9. S. T. Sam, O. S. Dahham, P. G. Gan, N. Z. Noimam, J. Y. Kuan, and A. M. Alakrach, (2017). *Solid State Phenomena* **264**, 120 (2017).
10. Y. W. Chang, Y. S. Shin, H. Chun and C. Nah, *Journal of applied polymer science*, **73**, 749 (1999).
11. C. M. P. Bizi and N. R. Demarquette, *Journal of Applied Polymer Science*, **109**, 445 (2008).

12. C. Nah, Polymer international, **51**, 245-252 (2002).
13. N. Z. Noriman, H. Ismail, and A. A. Rashid, *Journal of Applied Polymer Science*, **126** 56 (2012).
14. C. Nah, S. C. Han, B. W. Jo, W. D. Kim and Y. W. Chang, Journal of applied polymer science, **86**, 125-134 (2002)
15. J. Zhang, H. Chen, Y. Zhou, C. Ke and H. Lu, Polymer bulletin, **70**, 2829-2841 (2013).
16. M. G. Oliveira, A. C. Gomes, M. S. Almeida and B. G. Soares, Macromolecular Chemistry and Physics, **205**, 465-475 (2004).

The Effects of Different Rice Husk Loading and Size on The Properties of Standard Malaysian Rubber/ Rice Husk Composites

Omar S Dahham[1], N Z Noriman[1], R Hamzah[1], S Z Syed Idrus[2], Z Shayfull[3,4], T Adam[1,5], N R Munirah[1]

[1] Center of Excellence Geopolymer and Green Technology (CEGeoGTech), Faculty of Engineering Technology (FETech), Universiti Malaysia Perlis (UniMAP), Level 1 Block S2, UniCITI Alam Campus, Sungai Chucuh, Padang Besar, 02100, Perlis, Malaysia.
[2] School of Human Development and Techno-communication (iKOM), Green Advanced Computing and Technology (GREAT) Research Group – CEGeoGTech, Universiti Malaysia Perlis (UniMAP)
[3] School of Manufacturing Engineering, Universiti Malaysia Perlis, Arau, 02600, Malaysia
[4] Green Design and Manufacture Research Group, Center of Excellence Geopolymer and Green Technology (CEGeoGTech),Universiti Malaysia Perlis, Kangar, 0100, Malaysia
[5] Institute of Nano Electronic Engineering (INEE), Universiti Malaysia Perlis (UniMAP), 01000 Kangar, Perlis, Malaysia

E-mail: niknoriman@unimap.edu.my

Abstract. This research is focused on the rice husk (RH) that acts as natural filler in Standard Malaysian Rubber (SMR L). Two ranges of RH size were prepared, coarser size at 10-20 mm and fine size at 150- 300 μm. While the RH loading was (10,20,30,40 and 50 Phr). The effects of RH on the tensile properties, physical properties and cure characteristics of the compounds were examined. The results shown that the increasing of rice husk loading in SMR L/RH composites led to decrease in the cure (t_{90}) and scorch time (t_2) and increase in the minimum torque (M_L) and maximum torque (M_H). Despite the values of tensile strength decreased as RH content increase, the physical properties, such as hardness and crosslink density of the composites, particularly the composite with fine size filler were improved due to good filler-rubber interaction than coarser size filler.

1. Introduction

In general, natural fillers are produced from agriculture technology as it's claimed as natural resources. It is subdivided from several factors such as origins; plants, animals or mineral. Stem, leaf, seed, wood and grass are examples of fillers from plant. These kinds of fillers contain cellulose and lignin. The filler structure contents are as they are composed of cellulose, hemicelluloses, waxes, lignin and water-soluble compounds [1]. The natural fillers do has differences in terms of properties as its properties are based on their own species. The exhibit properties are within a species that highly depending on area of growth, climate and age of the plant. Besides, it also varies depending on their processing method used to breakdown to the fiber level. For other resource, which is mineral filler, it is naturally occurring fiber or slightly modified fiber procured from mineral.

There are some of natural filler's advantages benefited compare to synthetic filler, such as high stiffness, biodegradable, low density, recyclable, renewable raw materials and their relatively low cost [2]. Moreover, natural fillers give less health problems for the people who work with the fillers. Natural filler also do no cause skin irritations and not suspected of causing lung cancer. Nevertheless, instead of advantages, it also has some disadvantages, such as high moisture sensitivity and variability of diameter and length. We need to dry it up to release the moisture as it will not definitely affect our natural fillers properties [3]. Rice husk is the largest waste produced from agricultural processing of grains. It has known as significant source of silica and to reduce the quantity of this squander material. RH can be burned in the open air. As the researcher begun to identify the benefits of using rice husk, it always been used in rubber properties research as a natural filler [4]. Santiagoo, R. et al. studied the properties of polypropylene/recycled acrylonitrile butadiene rubber/rice husk powder [5]. They found that the processing torque and tensile modulus increased with increasing rice husk content for all composites, which were attributed to the brittle nature of rice husk filler.

Our preliminary research investigated the effects of different size and content of rice husk on the tensile properties, physical properties and cure characteristics of Epoxidized Natural Rubber/ rice husk composites [6]. It is found that the addition of rice husk into rubber has improved the cure characteristics of the compounds by reducing the cure and scorch time. Other advantages, the physical properties (hardness and crosslink density) were improved as rice husk added. However, the value of elongation at break and tensile strength showed decline as rice husk added.

In our current work, we have investigated the influence of RH on the tensile properties, physical properties and cure characteristics of SMR L/ RH compounds at different RH loading and size.

2. Materials and methods

2.1. Materials and preparation of samples

Rice husks (RH) have been obtained from Bernas Malaysia Sdn. Bhd., Simpang Empat, Perlis. It was dried under sunlight exposure for 72 hours. Dried rice husk were ground and sieved by using Crusher model RT34 (Chyun Industrial Co. Ltd.) and lab sieves to obtain rice husk at 10-20 mm and 150- 300 μm respectively. Standard Malaysian Rubber (SMR L) was used as a raw material in this work, which is obtained from Rubber Research Institute of Malaysia (RRIM). While other chemicals, such as sulfur stearic acid, zinc oxide and N-cyclohexyl-2-benzothiazole sulfonamide (CBS) were purchased from Anchor-Chemical Co. (M) Ltd. All materials were prepared according to formulation as shown in Table 1. Based on ASTM D 3184-89, two-roll mill X-(S)-K/160-X-320 was used for ingredients mixing. Rheometer, model MDR-2000 was utilized for cure characteristics test based on ASTM D 2240-93. Hot-press machine was used to compression accordingly to respective cure time at 150 °C and under 30 tonne pressure.

2.2. Measurement of properties

Tensile test was done based on ASTM D412 using Universal testing machine, model Instron-5582. Hardness test was carried out using Durometer device, model Shore A. Toluene solvent has been used for swelling test purpose based on ASTM D3616. The data of swelling test were then used for crosslinking density calculation of the composites using Flory Rehner equations [7].

Table 1. The ingredients of SMRL/RH composites.

Components	Phr					
	R0	R10	R20	R30	R40	R50
SMRL	100	100	100	100	100	100
Sulfur	2	2	2	2	2	2
Stearic Acid	2	2	2	2	2	2

CBS	1	1	1	1	1	1
ZnO	5	5	5	5	5	5
*RH (FC &CS)	0	10	20	30	40	50

3. Characterization

Figure 1 displays the morphology of raw RH with two different magnifications. Figure 1 (a) proved that the external surface of raw RH was rough and showing several aligned lumps. Meanwhile, Figure 1 (b) displayed the protuberance outer epidermis and silica. It also found that particles of RH were highly agglomerate, which could effect on some properties of the rubber compounds negatively [8].

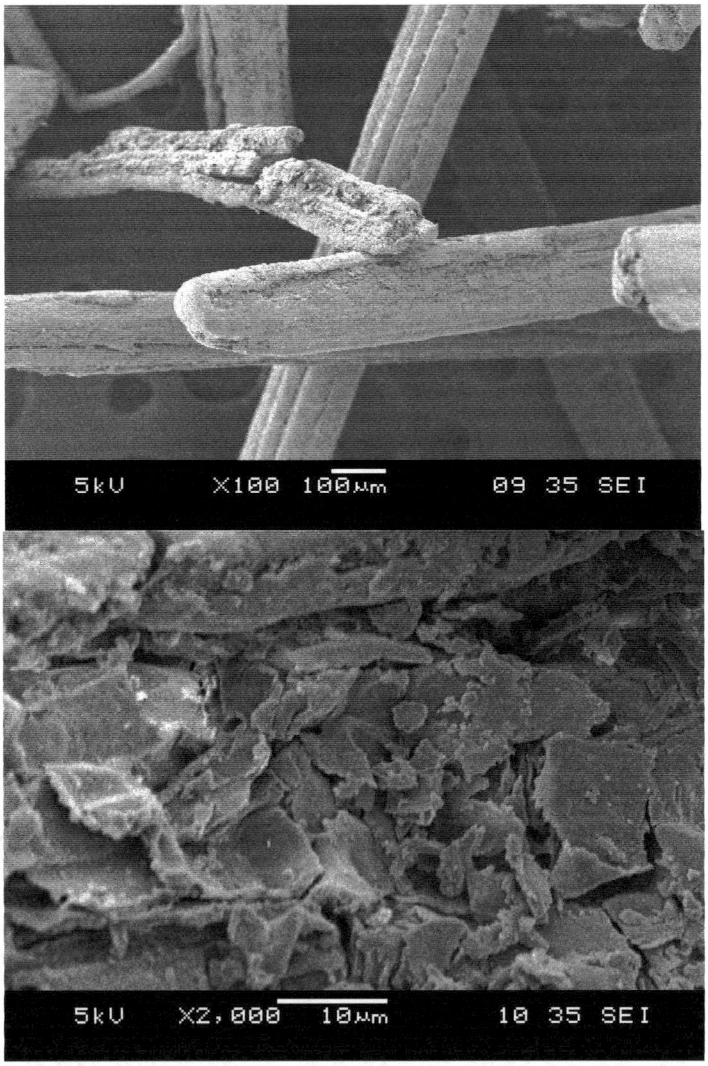

Figure 1. SEM image of raw RH at low (X100) and high (X2000) magnification.

4. Results and discussion

4.1. Cure characteristics

Table 2 shows scorch time t_2, cure time t_{90}, minimum torque M_L and maximum torque M_H respectively. It clearly seen that t_2 and t_{90} (Figure 2) decreased as RH content increase. This was accredited with the presence of more crosslinking network in SMRL which was dominant matrix in compound. As for fine RH loading incorporated, the large surface area of RH filler helps to interact more with rubber matrix and consequently reduce t_2 and t_{90} [9-11]. The increasing of rice husk loading in rubber compound reduced the macromolecular mobility in rubber compound, which in turn caused increase in the viscosity. Furthermore, the processability became more difficult. Thus, M_L increased. Fine RH filler has better processability since the smaller particle contributes to larger surface area and better interaction with matrix. M_H shows same trend of M_L. The increasing of M_H as RH content increase contributed to crosslinking increasing. Fine RH filler was higher M_H than coarser size due to the bigger surface area of fine RH filler that contact with rubber matrix [12].

Table 2. The cure characteristic of SMR L/RH composites.

	RH	t_2 (min)	t_{90} (min)	M_L (dNm)	M_H (dNm)
R0	Control	3.49	6.34	3.20	12.34
R10	FS	2.28	6.25	5.56	22.28
	CS	2. 41	6.75	6.00	36.16
R20	FS	2.15	6.10	7.99	23.00
	CS	2. 31	6.35	11.29	36.94
R30	FS	0.59	5.24	11.04	25.62
	CS	0. 67	5.79	12.41	37.41
R40	FS	0.39	5.15	12.82	26.00
	CS	0. 43	5.65	13.50	38.44
R50	FS	0.28	5.06	13.44	40.21
	CS	0. 32	5.45	13.99	28.8

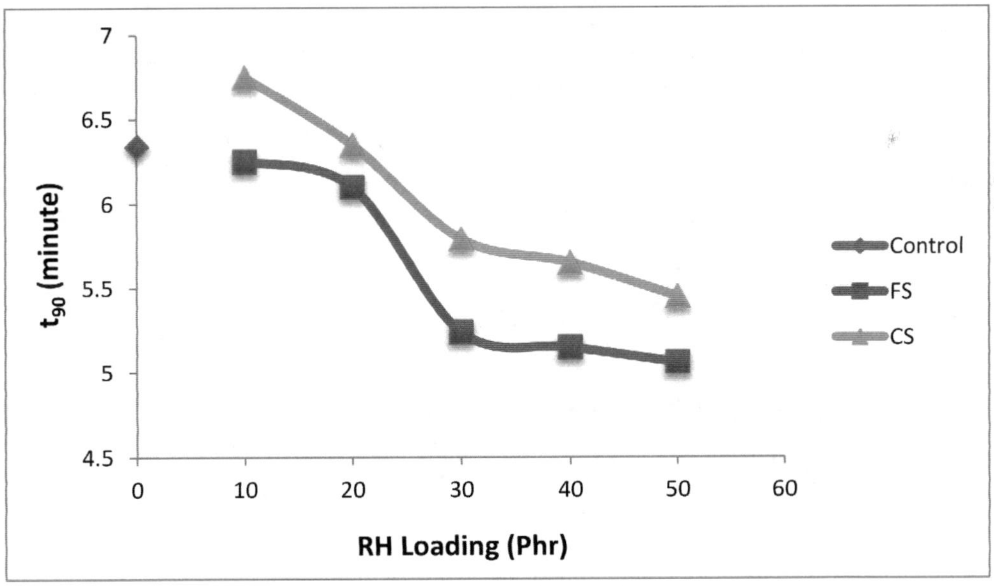

Figure 2. The cure time (t_{90}) of SMR L/RH compounds

4.2. Physical and and tensile properties

The values of physical properties (hardness and crosslink density) and tensile properties (tensile strength Ts, elongation at break Eb, and modulus at 100% elongation M100) are listed in Table 3 respectively. It can bee seen clearly Ts values decreased when RH increase due to the increasing of agglomeration of RH inside the matrix. Moreover, the miss-orientation of RH increased the difficulty of stress transmission from SMR L to RH. The inability of the coarse RH to support stress transferred from SMR L caused highly decreasing in Ts as compared with RH fine size [13,14].

Eb also decreases with increasing RH loading as expected. According to Premalal et al., as RH filler content is increases, a restriction to molecular motion of the macromolecules increase [15].

M100 is shown increasingly in trend and coarser RH has higher M100 than fine RH at same loading. Theoretically, the rigid filler in the rubber compound will significantly enhance the elastic stiffness and increase the modulus. As expected correlate with crosslinking density and hardness values, fine RH shows higher value than coarse RH at same loading. This was attributed to the better fine RH incorporation into SMR L matrix compared to coarse RH which in turn led to enhance the stiffness and rigidity of the rubber compound [16,17].

Table 3. The physical and tensile properties of SMR L/RH composites.

RH		Ts (MPa)	Eb (%)	M100 (MPa)	Crosslink density X 10^{-4} (Mole/cm^3)	Hardness (Shore A)
R0	Control	18.24	1450	0.45	0.75	35.64
R10	FS	14.98	1305	0.50	0.82	41.75
	CS	11.90	1179	0.60	0.76	39.5
R20	FS	12.65	1249	0.80	0.93	46.68
	CS	8.12	1118	0.10	0.86	42.11
R30	FS	10.89	1098	1.00	0.97	50.93
	CS	7.11	993	1.20	0.90	49.21
R40	FS	8.00	955	1.50	1.10	52.05
	CS	5.20	931	1.90	0.94	51.49
R50	FS	6.10	853	1.90	1.36	53.58
	CS	4.14	845	2.40	1.03	53.19

5. Summary

The increasing of rice husk loading led to reduce the cure and scorch time and increased the torque of the rubber compounds. The addition of RH improved the modulus, hardness and crosslink density of the rubber compounds. Elongation at break and tensile strength tend to decrease as rice husk content increased. Nevertheless, fine rice husk showed better properties as compared with coarse rice husk at all loading.

Acknowledgments

The authors would like to acknowledge the Fundamental Research Grant Scheme (FRGS) Phase 1/2015 (9003-00523) under Ministry of Higher Education, Malaysia. We also would appreciate Universiti Malaysia Perlis (UniMAP) for providing us all the facilities to make this research.

References

1. Y. Zhou, M. Fan, L. Chen and J. Zhuang, Composites Part B: Engineering **76,** 180 (2015).
2. K. Balaed, N. Z. Noriman, O. S. Dahham, S. T. Sam, R. Hamzah and M. F. Omar, *International Journal of Polymer Analysis and Characterization* **21**, 590 (2016).
3. Y. Xie, C. A. Hill, Z. Xiao, H. Militz and C. Mai, Composites Part A: Applied Science and Manufacturing **41**, 806 (2010).
4. S. S. N. Khalil Ahmed and Nudrat Zahid Raza., Journal of Advanced Research **5**, 165 (2014).
5. R. Santiagoo, S. S. Ting, H. Ismail and M. Jaafar, Applied Mechanics and Materials **754,** 210 (2015).
6. O. S. Dahham, N. Z. Noriman, S. T. Sam, N. M. Al-Samarrai, Z. Shayfull, A. M. Alakrach and S. A. Elnaid. *MATEC Web of Conferences* **78**, 01075 (2016).
7. P. J. Flory and J. Rehner, Journal Chemistry Physics **11**, 512 (1943).
8. S. L. Fávaro, M. S. Lopes, A. G. V. de Carvalho Neto, R. R. de Santana and E. Radovanovic, *Composites Part A: Applied Science and Manufacturing* **41**, 154 (2010).
9. S. S. Choi, Journal of applied polymer science, **75**, 1378-1384 (2000).
10. H. Ismail, S. Shuhelmy and M. R. Edyham, European Polymer Journal, **38**, 39-47 (2002).
11. H. Ismail, N. Rosnah and H. D. Rozman, Polymer, **38**, 4059-4064 (1997).
12. S. S. N. Khalil Ahmed and Nudrat Zahid Raza, Journal of Industrial and Engineering Chemistry, **19**, 1169 (2013).

13. H. S. Yang, H. J. Kim, J. Son, H. J. Park, B. J. Lee and T. S. Hwang, Composite Structures, **63**, 305-312 (2004).
14. N. Z. Noriman, H. Ismail, and A. A. Rashid, Journal of Applied Polymer Science, 126 56 (2012).
15. H. G. Premalal, H. Ismail and A. Baharin, *Polymer Testing*, **21**, 833 (2002).
16. M. Jacob, S. Thomas and K. T. Varughese, *Composites Science and Technology*, **64**, 955-965 (2004).
17. O. S. Dahham, N. Z. Noriman, S. T. Sam, H. Ismail, S. Ragunathan, H. Rosniza and M. N. Al-Samarrai, *Journal of Polymer Materials* **33**, 647 (2016).

Institute of Physics
Dirac House, Temple Back
Bristol BS1 6BE UK

ISSN: 1742-6588
ISBN 978-1-5108-6587-7

CURRAN ASSOCIATES INC.
proceedings.com